Metal Nanoparticles in Pharma

Mahendra Rai • Ranjita Shegokar
Editors

Metal Nanoparticles in Pharma

Springer

Editors
Mahendra Rai, Ph.D
Nanobiotechnology laboratory
Department of Biotechnology
SGB Amravati University
Amravati, Maharashtra, India

Ranjita Shegokar, Ph.D
Institute of Pharmacy, Department
Pharmaceutics, Biopharmaceutics
& NutriCosmetics
Freie Universität Berlin
Berlin, Germany

ISBN 978-3-319-63789-1 ISBN 978-3-319-63790-7 (eBook)
DOI 10.1007/978-3-319-63790-7

Library of Congress Control Number: 2017958441

Printed on acid-free paper

This Springer imprint is published by Springer Nature
The registered company is Springer International Publishing AG
The registered company address is: Gewerbestrasse 11, 6330 Cham, Switzerland

Preface

Metallic nanoparticles have fascinated scientists for over many centuries and are now widely utilized in biomedical field. These nanoparticles are a focus of interest because of their multipotential aspects and uses in nanotechnology. Today, these materials can be synthesized and modified with various chemical functional groups, which allow them to be conjugated with antibodies, ligands, and drugs of interest, thus opening a wide range of possible applications in biotechnology, magnetic separation and preconcentration of target analytes, targeted drug delivery, and more importantly diagnostic imaging.

Over the years, nanoparticles such as magnetic nanoparticles (iron oxide), gold/silver nanoparticles, nanoshells, and nanocages have been investigated and used for variety of biomedical applications. These nanoparticles are further modified to enable their use as a diagnostic and therapeutic agent. The nanosize of these particles allows enhanced communications with biomolecules on the cell surfaces and within the living cells in a way that can be decoded and designated to various biochemical and physiochemical properties of these cells. Many metal and metalloid elements are able to form nanoscale structures. Some of the better known nanoparticles currently being investigated include those based on silver, which are known for their antimicrobial and anti-inflammatory properties. Silver has long been known for its ability to combat infection, being able to inhibit various aspects of respiration, while gold nanoparticles are utilized in imaging like in transmission electron microscopy (TEM) and magnetic resonance imaging (MRI) scans. In fact, metal nanoparticles have huge potential in biomedical field, and therefore, these are selected as a main theme of this book.

The present book has been divided into two sections: Section 1 deals with the role of different metal nanoparticles in medicine. This includes nanoparticles of metals and their oxides such as silver, gold, iron, and titanium. In addition to this, the role of "Bhasmas" (nanoparticles of different metals) in "Ayurveda" – an Indian traditional system of medicine – has also been discussed. Section 2 discusses about different issues concerning the toxicity of metal nanoparticles, for example, biodistribution of different metal nanoparticles in the human body, and their toxicity issues particularly the effect of cobalt-chromium nanoparticles have been discussed.

The book would be very useful for pharmacologists, microbiologists, biotechnologists, nanotechnologists, and medical experts. Students and researchers would be highly enriched by the book.

Amravati, Maharashtra, India — Dr. Mahendra Rai
Berlin, Germany — Dr. Ranjita Shegokar

Contents

Contributors

Sofia Afonso Alves Brazilian Branch of Institute of Biomaterials, Tribocorrosion and Nanomedicine (IBTN), Chicago, IL, USA

Center for MicroElectroMechanical Systems, Universidade do Minho, Guimarães, Portugal

Vipul Bansal Ian Potter NanoBioSensing Facility, NanoBiotechnology Research Laboratory School of Science, RMIT University, Melbourne, VIC, Australia

Pedro V. Baptista UCIBIO, Departamento de Ciências da Vida, Faculdade de Ciências e Tecnologia, Universidade Nova de Lisboa, Caparica, Portugal

Adriano Brandelli Laboratório de Bioquímica e Microbiologia Aplicada, Instituto de Ciência e Tecnologia de Alimentos, Universidade Federal do Rio Grande do Sul, Porto Alegre, Brazil

Gary Bryant Centre for Molecular and Nanoscale Physics, School of Science, RMIT University, Melbourne, VIC, Australia

Alice Sandra Buteica Department of Pharmacy, Faculty of Pharmacy, University of Medicine and Pharmacy of Craiova, Craiova, Romania

Costica Caizer Department of Physics, Faculty of Physics, West University of Timisoara, Timisoara, Romania

Fábio F. Carlos UCIBIO, Departamento de Ciências da Vida, Faculdade de Ciências e Tecnologia, Universidade Nova de Lisboa, Caparica, Portugal

Catherine Carnovale Ian Potter NanoBioSensing Facility, NanoBiotechnology Research Laboratory, School of Science, RMIT University, Melbourne, VIC, Australia

Centre for Molecular and Nanoscale Physics, School of Science, RMIT University, Melbourne, VIC, Australia

Guillermo R. Castro Nanobiomaterials Laboratory, Applied Biotechnology Institute (CINDEFI, UNLP-CONICET CCT La Plata) Department of Chemistry School of Sciences, Universidad Nacional de La Plata, La Plata, Argentina

Silvio Silvério da Silva Escola de Engenharia de Lorena, University of Sao Paulo, Sao Paulo, Brazil

Kateřina Dědková Nanotechnology Centre, VŠB-Technical University of Ostrava, Ostrava, Poruba, Czech Republic

Regional Materials Science and Technology Centre, VŠB-Technical University of Ostrava, Ostrava, Poruba, Czech Republic

James Drummond Department of Orthopaedics, Western Health, Melbourne, VIC, Australia

Anisha D'Souza Department of Biosciences and Bioengineering, Indian Institute of Technology, Mumbai, India

Marcela Durán Institute of Biology, Urogenital Carcinogenesis and Immunotherapy Laboratory, University of Campinas, Campinas, SP, Brazil

NanoBioss, Institute of Chemistry, University of Campinas, Campinas, SP, Brazil

Nelson Durán NanoBioss, Institute of Chemistry, University of Campinas, Campinas, SP, Brazil

Instituto of Chemistry, Biological Chemistry Laboratory, University of Campinas, Campinas, SP, Brazil

Brazilian Nanotechnology National Laboratory (LNNano-CNPEM), Campinas, SP, Brazil

Institute of Chemistry, University of Campinas, Campinas, SP, Brazil

Kelly J. Dussán Department of Biochemistry and Chemical Technology, Institute of Chemistry, São Paulo State University-UNESP, Araraquara, São Paulo, Brazil

Camdon Fary Department of Orthopaedics, Western Health, Melbourne, VIC, Australia

Australian Institute for Musculoskeletal Science, Melbourne, VIC, Australia

Epworth Musculoskeletal Institute, Melbourne, VIC, Australia

Wagner J. Fávaro Institute of Biology, Urogenital Carcinogenesis and Immunotherapy Laboratory, University of Campinas, Campinas, SP, Brazil

NanoBioss, Institute of Chemistry, University of Campinas, Campinas, SP, Brazil

Alexandra R. Fernandes UCIBIO, Departamento de Ciências da Vida, Faculdade de Ciências e Tecnologia, Universidade Nova de Lisboa, Caparica, Portugal

Lian-Hua Fu Engineering Research Center of Forestry Biomass Materials and Bioenergy, Beijing Key Laboratory of Lignocellulosic Chemistry, College of Materials Science and Technology, Beijing Forestry University, Beijing, People's Republic of China

Sara Gemini-Piperni Biomaterials Laboratory Applied Physics Department, Brazilian Center for Physics Research- CBPF, Rio de Janeiro, Brazil

Brazilian Branch of Institute of Biomaterials, Tribocorrosion and Nanomedicine (IBTN), Chicago, IL, USA

Ellen C. Giese Coordination for Metallurgical and Environmental Process, Centre for Mineral Technology, CETEM, Rio de Janeiro, RJ, Brazil

José Mauro Granjeiro Directory of Life Sciences Applied Metrology, National Institute of Metrology, Quality and Technology, Rio de Janeiro, Brazil

Brazilian Branch of Institute of Biomaterials, Tribocorrosion and Nanomedicine (IBTN), Chicago, IL, USA

Dental School, Fluminense Federal University, Niterói, Brazil

Anna M. Grudniak Department of Bacterial Genetics, Institute of Microbiology, Faculty of Biology, University of Warsaw, Warsaw, Poland

Indarchand Gupta Department of Biotechnology, Government Institute of Science, Aurangabad, Maharashtra, India

Nanobiotechnology Laboratory, Department of Biotechnology, SGB Amravati University, Amravati, Maharashtra, India

Vinod Kumar Gurjar Institute of Pharmaceutical Sciences, Guru Ghasidas Vishwavidyalaya (A Central University), Koni, Bilaspur, Chhattisgarh, India

Veronika Holišová Nanotechnology Centre, VŠB - Technical University of Ostrava, Ostrava, Poruba, Czech Republic

Marianna Hundáková Nanotechnology Centre, VŠB-Technical University of Ostrava, Ostrava, Poruba, Czech Republic

Regional Materials Science and Technology Centre, VŠB-Technical University of Ostrava, Ostrava, Poruba, Czech Republic

Avinash Ingle Nanobiotechnology Laboratory, Department of Biotechnology, SGB Amravati University, Amravati, Maharashtra, India

German A. Islan Nanobiomaterials Laboratory, Applied Biotechnology Institute (CINDEFI, UNLP-CONICET CCT La Plata) Department of Chemistry School of Sciences, Universidad Nacional de La Plata, La Plata, Argentina

Zuzana Konvičková Nanotechnology Centre, VŠB – Technical University of Ostrava, Ostrava, Poruba, Czech Republic

Gabriela Kratošová Nanotechnology Centre, VŠB – Technical University of Ostrava, Ostrava, Poruba, Czech Republic

Ondrej Laššák 4 Medical Innovations CBTD, Ostrava, Poruba, Czech Republic

Dong Gun Lee School of Life Sciences, College of Natural Sciences, Kyungpook National University, Daegu, Republic of Korea

Heejeong Lee School of Life Sciences, College of Natural Sciences, Kyungpook National University, Daegu, Republic of Korea

Lionete N. Lima Department of Chemical Engineering, Federal University of São Carlos – UFSCar, São Carlos, SP, Brazil

Ming-Guo Ma Engineering Research Center of Forestry Biomass Materials and Bioenergy, Beijing Key Laboratory of Lignocellulosic Chemistry, College of Materials Science and Technology, Beijing Forestry University, Beijing, People's Republic of China

Irena Maliszewska Division of Medicinal Chemistry and Microbiology, Faculty of Chemistry, Wroclaw University of Science and Technology, Wroclaw, Poland

Katarzyna Markowska Department of Bacterial Genetics, Institute of Microbiology, Faculty of Biology, University of Warsaw, Warsaw, Poland

Gražyna Simha Martynková Nanotechnology Centre, VŠB-Technical University of Ostrava, Ostrava, Poruba, Czech Republic

Ana S. Matias UCIBIO, Departamento de Ciências da Vida, Faculdade de Ciências e Tecnologia, Universidade Nova de Lisboa, Caparica, Portugal

Ion Mindrila Department of Morphological Sciences, Faculty of Medicine, University of Medicine and Pharmacy of Craiova, Craiova, Romania

Dilipkumar Pal Institute of Pharmaceutical Sciences, Guru Ghasidas Vishwavidyalaya (A Central University), Koni, Bilaspur, Chhattisgarh, India

Raksha Pandit Nanobiotechnology Laboratory, Department of Biotechnology, SGB Amravati University, Amravati, Maharashtra, India

Priti Paralikar Nanobiotechnology Laboratory, Department of Biotechnology, SGB Amravati University, Amravati, Maharashtra, India

P. Pedrosa UCIBIO, Departamento de Ciências da Vida, Faculdade de Ciências e Tecnologia, Universidade Nova de Lisboa, Caparica, Portugal

Katarzyna Popko Division of Medicinal Chemistry and Microbiology, Faculty of Chemistry, Wroclaw University of Science and Technology, Wroclaw, Poland

Mahendra Rai Nanobiotechnology Laboratory, Department of Biotechnology, SGB Amravati University, Amravati, Maharashtra, India

Ana Rosa Ribeiro Directory of Life Sciences Applied Metrology, National Institute of Metrology, Quality and Technology, Rio de Janeiro, Brazil

Physics Department, University Estadual Paulista, Bauru, São Paulo, Brazil

Postgraduate Program in Clinical and Experimental Dentistry (Odontoclinex), University of Grande Rio, Duque de Caxias, Brazil

Ana Carolina Ritter Laboratório de Bioquímica e Microbiologia Aplicada, Instituto de Ciência e Tecnologia de Alimentos, Universidade Federal do Rio Grande do Sul, Porto Alegre, Brazil

Luís Augusto Rocha Brazilian Branch of Institute of Biomaterials, Tribocorrosion and Nanomedicine (IBTN), Chicago, IL, USA

Center MicroElectroMechanical Systems, Universidade do Minho, Guimarães, Portugal

Physics Department, University Estadual Paulista, Faculty of Sciences, São Paulo, Brazil

Amedea B. Seabra Human and Natural Sciences Center, Federal University of ABC, Santo André, SP, Brazil

Ranjita Shegokar Institute of Pharmacy, Department Pharmaceutics, Biopharmaceutics & NutriCosmetics, Freie Universität Berlin, Berlin, Germany

Ravi Shukla Ian Potter NanoBioSensing Facility, NanoBiotechnology Research Laboratory, School of Science, RMIT University, Melbourne, VIC, Australia

Michael G. Shurygin Irkutsk Scientific Center of Surgery and Traumatology, Irkutsk, Russia

Irina A. Shurygina Irkutsk Scientific Center of Surgery and Traumatology, Irkutsk, Russia

Debora D.V. Silva Department of Biochemistry and Chemical Technology, Institute of Chemistry, São Paulo State University-UNESP, Araraquara, São Paulo, Brazil

Kateřina Škrlová Nanotechnology Centre, VŠB – Technical University of Ostrava, Ostrava, Poruba, Czech Republic

Adriana M. Torres Pharmacology, Faculty of Biochemical and Pharmaceutical Sciences, Rosario National University- National Council of Scientific and Technical Research (CONICET), Rosario, Argentina

Phong Tran Department of Orthopaedics, Western Health, Melbourne, VIC, Australia

Australian Institute for Musculoskeletal Science, Melbourne, VIC, Australia

Raúl A. Trbojevich Division of Biochemical Toxicology, National Center for Toxicological Research. U.S. Food and Drug Administration, Jefferson, AR, USA

Flávio Fonseca Veras Laboratório de Bioquímica e Microbiologia Aplicada, Instituto de Ciência e Tecnologia de Alimentos, Universidade Federal do Rio Grande do Sul, Porto Alegre, Brazil

Gustavo N.A. Vieira Department of Biochemistry and Chemical Technology, Institute of Chemistry, São Paulo State University-UNESP, Araraquara, São Paulo, Brazil

Krystyna I. Wolska Department of Bacterial Genetics, Institute of Microbiology, Faculty of Biology, University of Warsaw, Warsaw, Poland

Jun Yang Engineering Research Center of Forestry Biomass Materials and Bioenergy, Beijing Key Laboratory of Lignocellulosic Chemistry, College of Materials Science and Technology, Beijing Forestry University, Beijing, People's Republic of China

Jie-Fang Zhu Department of Chemistry – Ångström Laboratory, Uppsala University, Uppsala, Sweden

Author Biography

Mahendra Rai is a Professor and Head at the Department of Biotechnology, Sant Gadge Baba Amravati University at Amravati, Maharashtra, India. He was a Visiting Scientist at the Department of Bioenergetics, University of Geneva, Switzerland, in 2004 and at the Department of Plant Protection of Debrecen University, Debrecen, Hungary, in 2005 and 2008. He visited Department of Chemical Biology, University of Campinas, Brazil under Indo-Brazil Research Programme (DST-CNPq collaboration) in 2009, 2011, and 2012. In 2012, he was a Visiting Professor in Nicolaus Copernicus University, Torun, Poland. In 2013, he was invited by State University of Campinas, Brazil. He was also a Visiting Professor in 2015 in Ostrava, Czech Republic. His area of expertise includes microbial biotechnology and nanobiotechnology. His present research interest includes application of nanobiotechnology in medicine with particular reference to the development of nanoantimicrobials. He has published more than 350 research papers in national and international journals. In addition, he has edited/authored more than 40 books and 6 patents.

Ranjita Shegokar Dr. Shegokar holds a Ph.D. in Pharmaceutical Technology from the SNDT University, India, and has been postdoctoral researcher at Free University of Berlin, Department of Pharmaceutical Technology, Biopharmaceutics and NutriCosmetics, Berlin, Germany. For the last 6 years, she has been working for pharmaceutical industry in various technical/R&D roles. Currently, she serves as Head of Formulation Development Department at Symrise AG, Germany. Dr. Shegokar has authored more than 45 research articles, 20 book chapters, and 150 research abstracts. She has filed two patent applications on nanoparticles for targeting malaria and HIV/AIDS reservoirs. She is a recipient of several national and international awards for her research. Her areas of interest are oral solid dosage forms, dermal formulations, nanocrystals, lipid nanoparticles (SLNs/NLCs), nanoemulsions, and galenic formulations.

Part I
Metal Nanoparticles in Medicine

Chapter 1
Iron Oxide Magnetic Nanoparticles in Photodynamic Therapy: A Promising Approach Against Tumor Cells

Amedea B. Seabra

Abstract Iron oxide magnetic nanoparticles have been extensively employed in biomedical applications due to their biocompatibility, small size, ability to surface functionalization, superparamagnetism behavior, and targeting properties. Upon the application of external magnetic field, iron oxide nanoparticles can be guided to the target site of application, minimizing possible side effects to nontarget tissues. Recently, several papers describe the combination of iron oxide magnetic nanoparticles with photosensitizer (PS) molecules for photodynamic therapy (PDT). PDT is a clinical treatment based on the administration of a photosensitizer to the tumor site, which under the irradiation with visible-near-infrared light generates reactive oxygen species (ROS) able to cause deleterious effects to the treated tumor site. The main disadvantage of PDT is the lack of selectivity; therefore, the combination of magnetic iron oxide nanoparticles with photosensitizer is a new and promising approach in PDT. In this direction, this chapter discusses the recent advantages in the design and applications of magnetic iron oxide nanoparticles in conjugation with photosensitizer in PDT to combat cancer.

Keywords Iron oxide nanoparticles • Photodynamic therapy • Cytotoxicity • Magnetic nanoparticles • Singlet oxygen • Fe_3O_4 nanoparticles

A.B. Seabra (✉)
Human and Natural Sciences Center, Federal University of ABC,
UFABC, Av dos Estados, 5001. Bairro Bangú, Santo André, SP 09210-580, Brazil
e-mail: amedea.seabra@ufabc.edu.br; amedea.seabra@gmail.com

M. Rai, R. Shegokar (eds.), *Metal Nanoparticles in Pharma*,
DOI 10.1007/978-3-319-63790-7_1

Nomenclature

$^{1}O_2$	Singlet oxygen
$^{3}O_2$	Triplet oxygen
AHP	Hyaluronic acid
AIPcS4	Tetrasulfonic phthalocyanine aluminum
Ce6	Chlorin e6
CSQ	Chitosan quaternary ammonium
DOX	Doxorubicin
Eca-109	Esophageal squamous carcinoma
EPR	Enhanced permeability and retention
ESIONs	Extremely small iron oxide nanoparticles
FDA	Food and Drug Administration
Fe_3O_4	Magnetite
HeLa	Human cervical cancer cell line
HMNSs	Hollow magnetic nanospheres
HP	Hematoporphyrin
IC_{50}	Maximal inhibitory concentration
IONCs	Iron oxide nanoclusters
IR820	Indocyanine green
Jurkat	Human T-cell leukemia cells
LED	Light-emitting diode
MCF-7	Human breast cancer cells
MDA-MB-231	Breast cancer
MLs	Magnetoliposomes
MNCs	Magnetic nanoclusters
$mSiO_2$	Mesoporous silica
NIH3T3	Mouse embryonic fibroblast
NRI	Near infrared
OA	Oleic acid
PC-3	Human prostate cancer
PDT	Photodynamic therapy
PEG	Polyethylene glycol
pheoA	Pheophorbide A
PHPP	2,7,12,18-Tetramethyl-3,8-di(1-propoxyethyl)-12,17-bis-(3-hydroxypropyl) porphyrin
PMNs	Magnetic nanogrenades
PpIX	Protoporphyrin IX
PS	Photosensitizer
RGD	Integrin-binding cell adhesive peptide
rGO	Reduced graphene oxide
ROS	Reactive oxygen species
S-180	Sarcoma cells
SK-OV-3	Ovarian cancer cells
SPIO	Superparamagnetic iron oxide

1.1 Introduction

Photodynamic therapy (PDT) is a clinical treatment that involves the administration of a photosensitizer (e.g., photofrin) in the tumor site (Bechet et al. 2008; Banerjee et al. 2017). Once at the tumor tissue, appropriated light irradiation (visible or near-infrared wavelengths) yields excited photosensitizer. Upon light activation, excited triplet photosensitizer in the presence of molecular oxygen (naturally available triplet-oxygen ($^3O_2$) produces highly reactive singlet oxygen ($^1O_2$), a well-known mediator of cell death (Allison et al. 2008; Chatterjee et al. 2008; Bhattacharyya et al. 2011; Smith et al. 2015). The efficiency of PDT depends on singlet oxygen formation, which is highly reactive and able to readily diffuse through the biological membranes (Bechet et al. 2008). Besides the generation of singlet oxygen, various free radicals are generated. All those reactive species might induce cell death via apoptosis, necrosis, and cell membrane damages with release of contents (Allison et al. 2008; Kim et al. 2016a; Lee et al. 2017). Indeed, singlet oxygen reacts with nucleic acids, lipids, amino acids, and proteins in the tumor tissue preventing angiogenesis in solid tumors (Smith et al. 2015). The generated oxygen-based cytotoxicity is expected to occur upon light irradiation, since in dark conditions, the photosensitizers are harmless. Therefore, in PDT light is used as a tool to trigger and modulate the selective destruction of malignant tissues/organs (Bechet et al. 2008). It should be noted that in PDT, the wavelength of light irradiation is a very important parameter. Irradiation in the near-infrared (NIR) region (700–1,000 nm) is the most appropriated wavelength to be employed. Hemoglobin is the major protein in the blood, which absorbs light up to 600 nm; therefore, tissues/organs must be irradiated at higher wavelengths to allow a suitable light penetration (Bechet et al. 2008; Huang et al. 2012; Chen et al. 2015a).

The clinical uses of PDT started in 1933, although it has been studied for over a century (Smith et al. 2015). Nowadays, there are several photosensitizers approved by the Food and Drug Administration (FDA) (Smith et al. 2015). The major clinical use of PDT is in cancer treatment (Bhattacharyya et al. 2011), although noncancerous applications have also been applied. Against cancer, PDT has been employed in the treatment of superficial tumors, such as melanoma, and tumors in the esophagus and bladder (Bechet et al. 2008). The noncancerous applications of PDT are mainly in the skin to treat diseases such as psoriasis and actinic keratosis, in the eyes in particular to treat macular degeneration, as well as at localized sites of inflammations including rheumatoid arthritis and dental infections (periodontitis) (Chatterjee et al. 2008). For example, a methylene blue (photosensitizer)-containing gel is applied to the infected gum pocket, and upon light irradiation via an optical fiber, singlet oxygen is locally generated in the infected pocket leading to cytotoxic effects (Chatterjee et al. 2008).

The advantages of PDT over traditional therapies are (i) relatively high cure rates, (ii) cost-effectiveness, (iii) localized treatment, (iv) no cumulative toxicity upon repetition of the therapy, and (v) induction of immunity (Chatterjee et al. 2008). In contrast, the disadvantages of PDT are the long in vivo half-life of the photosensitizers, consequently, their nonspecific distribution in vivo, and the limitation to reach deeper tumors in the body of the patient (Smith et al. 2015).

1.2 Why Use Superparamagnetic Iron Oxide Nanoparticles in PDT?

Although there are promising clinical uses of PDT, some drawbacks need to be overcome. For instance, the selective targeting and accumulation of the photosensitizers in the disease organ/tissue are necessary to minimize toxicity to normal cells (Bechet et al. 2008). Moreover, photosensitizers have poor water solubility and tend to aggregate in the biological medium, decreasing their photo-efficiency (Paszko et al. 2011).

The combination of photosensitizer and nanomaterials may overcome these issues. In this context, the combination of nanoparticles as carrier of photosensitizers is a promising approach in PDT (Bechet et al. 2008). The advantages of nanoparticles as nanocarriers of photosensitizers in PDT are:

(i) Ability to carry and to deliver therapeutic amounts of the photosensitizers in deep-seated tumors

(ii) Due to the high surface area/volume ratio, the nanoparticle surface can be used as multifunctional vehicle upon its functionalization with other therapeutic molecules, such as chemotherapeutical agents and/or tumor-seeking molecules

(iii) Enhanced solubility of hydrophobic photosensitizers with proper size to accumulate in the tumor tissues via enhanced permeability and retention effect (EPR effect) (Wang et al. 2004; Paszko et al. 2011; Gangopadhyay et al. 2015)

One of the major advantages of using nanoparticles in PDT is the EPR effect, which is the passive accumulation of nanoparticles in tumor sites (Kim et al. 2016a). This effect can be explained by considering the altered anatomy of tumor tissues characterized by an excessive angiogenesis and leaky blood vessels with impaired lymphatic drainage system. Therefore, nanoparticles can extravasate from blood pool and spontaneously accumulate into tumor sites. Once accumulated, the nanoparticles are retained in tumor tissues due to the compromised lymphatic drainage (Kelkar and Reineke 2011; Kim et al. 2016a, b). Interestingly, therapeutic amounts of photosensitizer-containing nanoparticles are accumulated in solid tumors due to EPR effect (Mbakidi et al. 2013), making nanoparticles promising vehicles for singlet oxygen generators in PDT.

Recently, iron oxide magnetic nanoparticles have been successfully employed as drug carriers for photosensitizers and also for chemotherapeutic agents. Magnetic nanoparticles such as magnetite (Fe_3O_4) have interesting properties that make them suitable nanocarriers in several biomedical applications. These properties can be summarized as: ability to reach the desired site of treatment upon application of an external magnetic field (magnetic targeting), high surface area, small size, biocompatibility, chemical stability, relatively low-cost production, and several different synthetic routes including chemical, physical, and biogenic synthesis (Seabra et al. 2013, 2014; Seabra and Duran 2015, Cheng et al. 2016). Therefore, iron oxide magnetic nanoparticles have been widely employed in several biomedical applications of targeted drug delivery system, magnetic resonance imaging, biosensing, immunoassay, and magnetic hyperthermia (Cheng et al. 2016). Magnetic hyperthermia can be allied to PDT to significantly enhance the cytotoxicity toward cancer tissues. Upon the appropriate targeting of superparamagnetic iron oxide nanoparticles using a magnetic targeting, an alternative magnetic field is applied for a certain period of time, leading to a local increase in the tissue temperature (42–45 °C), which targets cancer cell death by apoptosis (Kandasamy and Maity 2015). Therefore, hyperthermia therapy can be employed to enhance the efficacy of various cancer treatments.

Magnetite is the mostly used for synthesis of iron oxide magnetic nanoparticles due to its superparamagnetism, which allows rapid nanoparticle magnetization upon the influence of a magnetic field; however, once this field is turned off, there is no permanent magnetization. In particular, superparamagnetic iron oxide nanoparticles have been employed in cancer therapy and diagnosis by magnetic targeting (Kandasamy and Maity 2015).

In this scenario, the use of magnetic iron oxide nanoparticles with PDT is based on the combination of a chromophore that is a photosensitizer and a phototrigger, and the ability to target the photosensitizer to the desired site of application, with minimum side effects to healthy cells/tissues (Kandasamy and Maity 2015). In this direction, the next section presents and discusses the recent advantages on the design and uses of superparamagnetic iron oxide nanoparticles as nanocarrier of photosensitizers in PDT.

1.3 Iron Oxide Magnetic Nanoparticles in PDT Applications

Recently, the photosensitizer protoporphyrin IX (PpIX), a metal-free porphyrin, was loaded in magnetoliposomes (MLs), which are comprised by liposomes and Fe_3O_4 magnetic nanoparticles (Basoglu et al. 2016). PpIX-containing MLs have a hydrodynamic size of 221 nm and displayed phototoxicity against human breast cancer cells (MCF-7). Cell irradiation was performed by using a LED (light-emitting diode) downlight white light (visible light irradiation in the range of 400–700 nm, at 20 W). MCF-7 viability was reduced to 65% upon cell incubation with 350 nM of PpIX-containing MLs under dark condition; however, upon 5 min of light irradiation, a significant cell death was observed at nanoparticle concentration of 250 nM. This result demonstrated the effective phototoxicity of PpIX-loaded magnetoliposomes (MLs) at nanomolar concentrations (Basoglu et al. 2016).

As stated before, near infrared (NIR) is a suitable wavelength range for PDT applications. In this direction, gold nanopopcorns containing self-assembled iron oxide cluster were synthesized for magnetic field targeting in PDT applications (Bhana et al. 2015). The nanoparticle surface was coated with the photosensitizer silicon 2,3-naphthalocyanine dihydroxide, followed by stabilization with polyethylene glycol (PEG) linked with 11-mercaptoundecanoic acid. The hybrid nanomaterial demonstrated superparamagnetic behavior and 61% of photothermal conversion efficiency under NIR irradiation. Upon the application of a gradient magnetic field, the hybrid nanomaterial demonstrated superior release of the photosensitizer and cellular uptake. The nanomaterial showed enhanced ability to carry therapeutic molecules deep into tumor tissue through magnetic manipulation and combat tumors locally by PDT (NIR light irradiation of 0.55 W/cm^2) and by photothermal effect (Bhana et al. 2015). NIR and photothermal effect are expected to efficiently eradicate tumors, due to the delivery of high and therapeutic concentrations of photosensitizer-containing magnetic nanocarriers in deep tumors via applying a magnetic field on tumor surface.

The design of hybrid nanocarriers in PDT is a promising strategy. In this context, multifunctional silica-based magnetic nanoparticles were prepared (Chen et al. 2009). The nanoparticles are composed by Fe_3O_4 magnetic nanoparticles coated with 2,7,12,18-tetramethyl-3,8-di(1-propoxyethyl)-12,17-bis-(3-hydroxypropyl) porphyrin, shorted as PHPP, employed as photosensitizer (encapsulation efficiency of PHPP, 20.8%). Magnetic Fe_3O_4 nanoparticles and PHPP were incorporated into silica nanoparticles by sol-gel and microemulsion techniques; the final nanoparticles have a spherical shape and a diameter of 20–30 nm. The magnetic nanoparticles demonstrated good compatibility in dark conditions and ability to generate toxic singlet oxygen upon irradiation with $\lambda = 488$ nm (argon-ion laser) with energy density of 4.35 J/cm^2 during 10 min toward SW480 carcinoma cells (Chen et al. 2009). Therefore, the nanoparticles displayed remarkable antitumor effects under irradiation.

A hybrid nanomaterial composed of reduced graphene oxide (rGO), magnetic iron oxide, and gold nanoparticles was designed. rGO-Fe_2O_3@Au core-shell nanoparticle-graphene nanocomposite was prepared for magnetic targeting in cancer treatment via PDT and photothermal synergist effect and bioimaging agent (Chen et al. 2015b). The presence of Fe_2O_3 is responsible for magnetic field-assisted singlet oxygen generation and chemophotothermal effect direct to the cancer tissue. The hybrid nanomaterial was found to be superparamagnetic at room temperature, demonstrated high capacity to load the chemotherapeutic doxorubicin (DOX, 1.0 mg mg^{-1}), and had considerable photothermal conversion efficiency under 808 nm (NIR) laser irradiation (2 W cm^{-2}). The authors demonstrated that NIR irradiation triggers doxorubicin (DOX) release with an initial burst of the release of the therapeutic molecule. The in vitro cytotoxicity of DOX-rGO-Fe_2O_3@Au NPs (DOX-loaded hybrid nanomaterial) was evaluated against human cervical cancer cell line (HeLa) upon 5 min irradiation with NIR laser. Cancer cells significantly uptake DOX-containing nanoparticles within 4 h of incubation. Moreover, magnetic field-guided chemophotothermal synergistic therapy might find important applications in cancer treatment (Chen et al. 2015b).

In a similar approach, magnetofluorescence nanoparticles composed of Fe_3O_4@SiO_2@APTES@glutaryl-PPa (MF NPs) were synthesized (Cheng et al. 2016), where APTES is (3-aminopropyl)triethoxysilane. The obtained nanoparticles have a mean diameter of 50 nm and can be used for PDT and fluorescence imaging in medical applications. The surface of core-shell Fe_3O_4@SiO_2@APTES NPs was functionalized with the photosensitizer chlorine pyropheophorbide-a (PPa) for PDT application. The cytotoxicity of the nanoparticles was evaluated toward HeLa cells, and the results indicated that the nanoparticles were able to quickly permeate the tumor cells due to their lipo-hydro partition coefficient, leading to cell apoptosis and death. Indeed, due to their small size, the nanoparticles were absorbed on the cancer cell membrane, followed by cell internalization via endocytosis vesicle formation by deformation of cell membrane, which led to the dispersion of the nanoparticles into the cytoplasm of HeLa cells. In addition, the nanoparticles demonstrated light-dependent cytotoxicity to the cancer cells (irradiation with visible light for 10 min at dosage of 25 Jcm^{-2}). As expected, reactive oxygen species (ROS) generated in HeLa cells (Cheng et al. 2016). Hence, these results highlight the promising therapeutic uses of these nanoparticles in PDT treatment with good magnetic targeting to site-specific toxicity.

In an interesting paper, iron oxide magnetic nanoparticles (Fe_3O_4) were synthesized and coated with carboxyl-terminated poly(ethylene glycol)-phospholipid to increase nanoparticle dispersion in water (Chu et al. 2013). The phototoxicity of the nanoparticles was evaluated toward esophageal cancer cells upon irradiation with NIR (λ = 808 nm, power density of 0.25 W/cm^2 for 20 min) light exposition. Figure 1.1 shows the schematic representation of the light-trigged toxicity of Fe_3O_4-based nanoparticles toward cancer cell. The application of magnetic nanoparticle and NIR laser irradiation is responsible to decrease esophageal cancer cell viability and to inhibit mouse esophageal tumor growth, leading to tumor reduction (Chu et al. 2013).

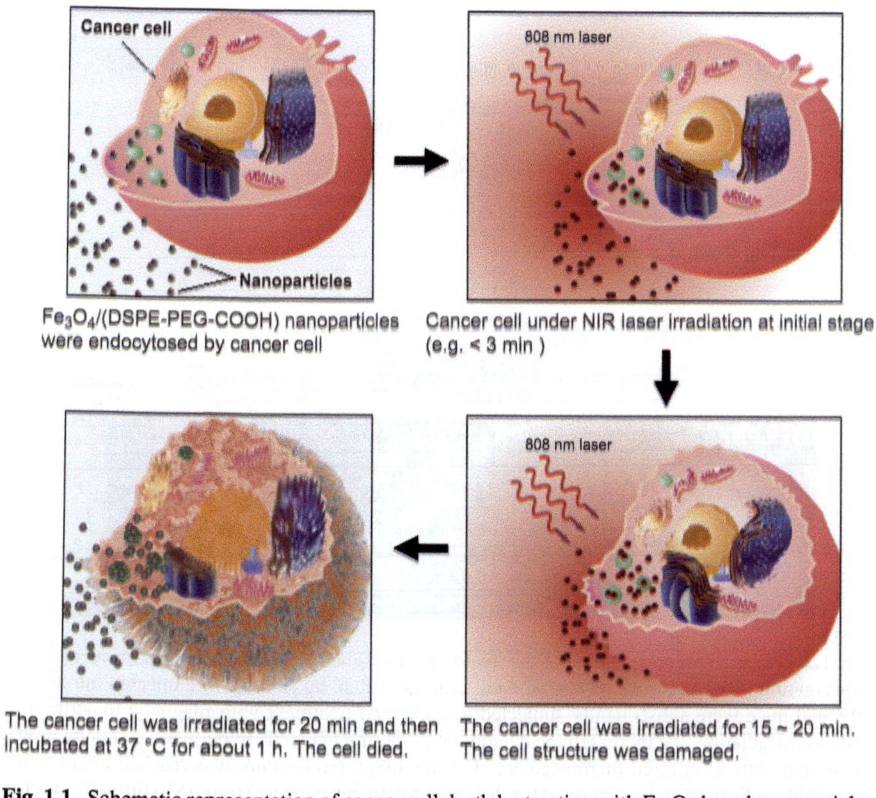

Fig. 1.1 Schematic representation of cancer cell death by treating with Fe_3O_4-based nanoparticles upon NIR laser light irradiation (Reproduced from reference Chu et al. 2013 with permission from Elsevier)

In a promising strategy, magnetic photosensitive liposomes were synthesized by loading aqueous core iron oxide nanoparticles with lipid bilayer containing therapeutic amounts of photosensitizer payload (Corato et al. 2015). Under laser excitation, the obtained nanoparticles were able to generate toxic singlet oxygen and heat upon alternating magnetic field stimulation, for hyperthermia application. Figure 1.2 shows the schematic representation of the biomedical application of magnetic photosensitive liposome nanoparticles (Corato et al. 2015).

In a further example, magnetic nanoclusters (MNCs) are composed of Fe_3O_4 nanoparticles and poly(acrylic acid-co-propargyl acrylate) on the surface for multimodal imaging probes (Daniele et al. 2013). The NRI photosensitizer azide-modified indocyanine green was inserted on the surface of the MNCs. In biological medium, the photosensitizer complexed with bovine serum albumin leading to an extended coating of serum on the MNCs, which induce turn-on of NRI emission for PDT applications (Daniele et al. 2013). Recently, SPIO@SiO_2-Re@PEG magnetic nanoparticles for optical probes and PDT sensitizers were obtained and characterized (Galli et al. 2016). Firstly, superparamagnetic iron oxide (SPIO) nanoparticles with an average size distribution of 10 nm were synthesized by thermal decomposition of iron oleate, followed by nanoparticle coating with silica shell by reverse microemulsion technique. Upon pyridine ligand via triethoxysilane moiety, the surface

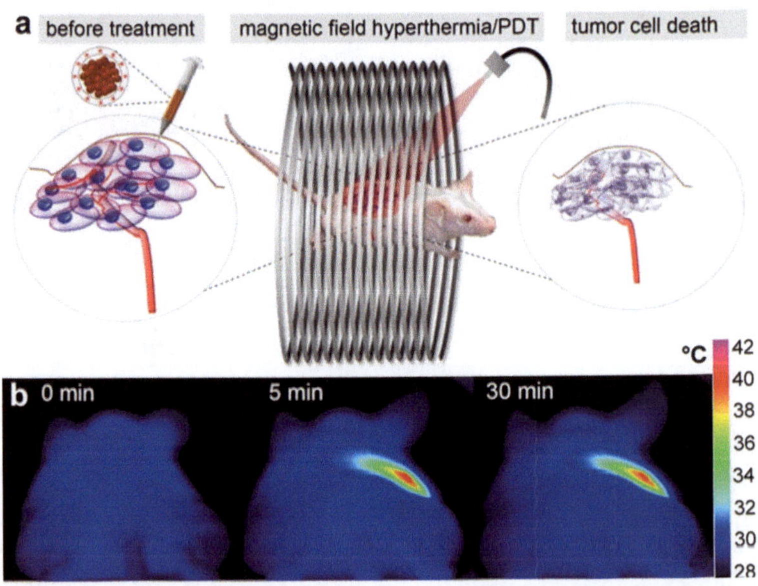

Fig. 1.2 Schematic representation of the treatment of magnetic photosensitive liposome nanoparticles on tumor-bearing mice: (**a**) therapeutic strategy sketch, liposomes were injected intratumorally, and mice were subsequently subjected to combined treatment with magnetic hyperthermia and laser irradiation. (**b**) Increased local temperature during magnetic hyperthermia treatment was monitored with an infrared thermocamera. The maximum temperature was reached about 5 min after field application and maintained for the entire treatment cycle (30 min) (Modified from reference Corato et al. 2015 with permission of American Chemical Society)

of the nanoparticles was functionalized with luminescent [Re(phen)$(CO)_3$(py)] CF_3SO_3 complexes anchored to the silica layer. To increase the stability and dispersion of the nanoparticles in physiological conditions, the nanoparticle surface was further covered with PEG layer leading to the formation of SPIO@SiO_2-Re@PEG nanoparticles. PEG also increases the nanoparticle circulation in vivo by avoiding nanoparticle uptake and clearance by macrophage. The final nanocomposite had an average size of 40 nm and demonstrated satisfactory efficiency to generate singlet oxygen. The toxicity of the nanoparticles was evaluated toward human lung adenocarcinoma A549 cells. The confocal microscopy demonstrated that after 4 h of cell incubation with the nanoparticles, the latter were efficiently internalized and accumulated in the perinuclear region of the cancer cells. Moreover, increased cytotoxicity was revealed upon irradiation with visible light (by using a 150 W/NDL lamp, 390 nm cut off filters, 142 mW/cm^{-2}). No significant cytotoxicity was observed in the dark condition, indicating the promising uses of this nanoparticle in PDT applications (Galli et al. 2016). Guo et al. (2016) described the preparation of monodisperse Fe_3O_4 nanoparticles with sizes in the range of 60–310 nm. The authors observed that smaller Fe_3O_4 nanoparticles displayed enhanced cell internalization and, hence, deeper penetration into multicellular spheroids, which make them suitable for PDT applications. Larger Fe_3O_4 nanoparticles tended to accumulate in tumors increasing tumor growth inhibition (Guo et al. 2016).

Similarly, multifunctional magnetic nanoparticles with average size of 108 nm were synthesized for tumor treatment and diagnosis (Kim et al. 2016b). The magnetic nanoparticle core (MNP) is composed of Fe_3O_4, and the nanoparticle surface was conjugated with hyaluronic acid (AHP), leading to the formation of AHP@ MNPs. The cytotoxicity of AHP@MNPs was evaluated toward mouse embryonic fibroblast (NIH3T3) cells, via heat generation properties of the nanoparticles and generation of singlet oxygen, upon magnetic and laser irradiation conditions. The nanoparticle concentration was fixed at 12 g/mL, and NIH3T3 cells were exposed to PDT (4 Jcm^{-2}) and magnetic heat treatment alone or in combination. Cell viability decreased to 30% with PDT treatment. Live/dead staining experiments revealed that the nanoparticles displayed controllable cell death via targeting CD44 receptor-mediated endocytosis (Kim et al. 2016b). Lee et al. (2016) reported the synthesis of caffeic acid-polyethylene glycol-folic acid; FA-PEG-CA, caffeic acid-polyethylene glycol-pheophorbide-a; and pheoA-PEG-CA-coated Fe_3O_4 nanoparticles for PDT targeting. The presence of caffeic acid on the surface of iron oxide nanoparticles increased the nanoparticle dispersion in aqueous environment, while folate is responsible for mediation of tumor targeting. The cytotoxicity of the nanoparticles was evaluated toward human breast carcinoma cell line (MDA-MB-231). The cell was incubated with the nanoparticles, and light irradiation was performed by using 670 nm laser source (5 mW/cm^2) for 200 s. The irradiation increased twofold the cytotoxicity. Furthermore, due to the presence of folate on the surface of the nanoparticle, the nanoparticles were internalized into tumor cells, which over express folic acid receptors (Lee et al. 2016).

In another study, a widely employed photosensitizer in PDT applications, i.e., chlorin e6 (Ce6), was loaded on PEG-functionalized iron oxide nanoclusters

(IONCs), leading to the formation of IONC-PEG-Ce6 nanoparticles (Li et al. 2013). The absorbance and excitation peaks of the IONC-PEG-Ce6 were found to be located in the NIR (~700 nm), making the nanoparticles suitable for PDT due to improved tissue penetration. The in vivo efficacy of IONC-PEG-Ce6 was performed based on magnetic tumor targeting by using the 4T1 tumor modal on Balb/c mice. The results revealed a significant delay in tumor growth upon a single nanoparticle injection and the application of a magnetic field-enhanced PDT treatment. The light source for PDT application was 704 nm light at 5 mW/cm^2 for 1.5 h (27 J/cm^2). Under dark conditions, tumors of mice that received IONC-PEG-Ce6 injection showed the same growth trend compared to the untreated groups. This result suggests that IONC-PEG-Ce6 is not toxic under dark condition. However, significant inhibition of tumor growth was observed upon light irradiation after IONC-PEG-Ce6 injection. Indeed, littler tumor growth was observed over a course of 16 days post-PDT application, revealing the post-effect of the treatment in the combat to cancer. These nanoparticles have desirable properties for cancer treatment, such as no cytotoxicity under dark condition, high generation of singlet oxygen upon NIR illumination, tumor targeting upon application of externa magnetic field, and high cancer cell uptake (Li et al. 2013).

The same photosensitizer, Ce6, was combined with magnetic nanoparticles for PDT (Ling et al. 2014). The authors reported an elegantly pH-dependent PDT to selectively kill cancer cells. Tumor tissues are acidic with pH range of 6.8 which can also reach lower values (5.0–5.5) in endo-/lysosomes. Therefore, the changes in the pH environment from healthy tissues to tumor sites can be used to trigger the release of therapeutic agents on the surface of nanocarriers. In this context, self-assembled extremely small iron oxide nanoparticles (ESIONs), average size of 3 nm, were prepared and coated with polymeric ligands, leading to the formation of pH-sensitive magnetic nanogrenades (PMNs). PMNs can be activated under low pH values found in tumors for combating resistant heterogeneous tumors in vivo and also as bimodal imaging agent, as represented in Fig. 1.3. According to Fig. 1.3a, the ESIONs were functionalized via catechol-anchored ligand A. Figure 1.3b shows the schematic representation of self-assembly ESIONs into colloidal magnetic core-shell nanomaterial. In this structure, the ESIONs are the hydrophobic core, which is coated with hydrophobic blocks of the tumor-sensing polymeric ligands. The obtained nanoparticles had hydrodynamic diameter of 70 nm, which is suitable for tumor treatments due to the EPR effect. Moreover, the nanoparticle surface allows a pH activation resulting in surface charge reversal in tumor periphery that increases cell adsorption and permeation and endo-/lysosomal pH-dependent theranostic effect of the nanoparticles. Interestingly, the PMN photoactivity (generation of oxygen singlet) was quenched at physiological pH (7.4) due to fluorescence resonance energy transfer. In contrast, at lower pH value found in tumor environment (pH <6), the photoactivity of the PMNs significantly increased due to the disassembly, in a similar manner observed for other self-assembled polymeric ligands. Therefore, pH controls the fluorescence intensity and NIR generation of toxic singlet oxygen. The cytotoxicity of PMNs was demonstrated toward human colorectal carcinoma (HCT116) cells under irradiation and compared to dark conditions. As expected,

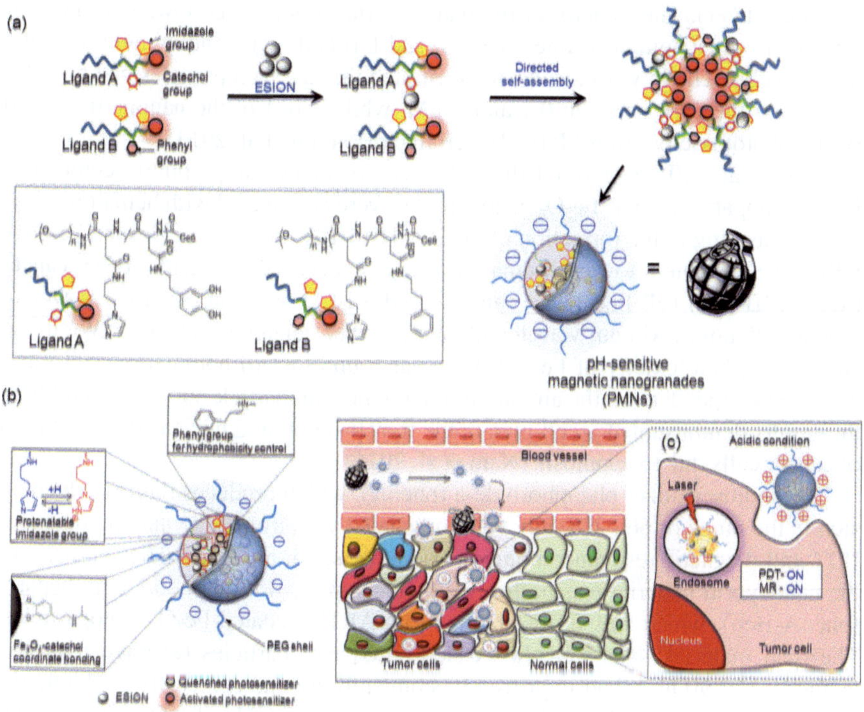

Fig. 1.3 Schematic representation of the design and mechanism of pH-sensitive magnetic nanogrenades (PMNs) for tumor pH activation: (**a**) Schematic representation of pH-responsive, ligand-assisted self-assembly of extremely small iron oxide nanoparticles (*ESIONs*), (**b**) schematic representation of tumor pH-recognizable treatment strategy using PMNs (Reproduced from reference Ling et al. 2014 with permission from American Chemical Society)

PMNs had no toxicity in the dark and significant toxicity under irradiation. In vivo studies were also performed by using HCT116 tumor-bearing mice injected with PMNs, followed by irradiation. A significant tumor regression of treated groups was observed, in comparison with control groups. The nanomaterial was designed to be toxic only at the tumor site, since the photosensitizer on the nanoparticle can be self-quenched in the normal tissues, until it reaches the target tumor site. In the tumor, the low pH value suppresses the self-quenched effect, particularly, the intracellular pH stimulus, resulting to a high specific toxic effect to the target region. One week post-PMN injection, the tumors were completely destroyed, and only a remaining scar tissue was observed in the animals (Ling et al. 2014). Taken together, these results demonstrated the efficacy of tumor-targeting iron oxide nanoparticles for PDT and tumor diagnosis.

In a similar interesting strategy, hydrophilic chlorin-conjugated magnetic nanoparticles were designed for PDT in the combat of melanoma (Mbakidi et al. 2013).The nanomaterial is comprised of an iron oxide magnetic core covered with water-stable and biocompatible dextran shell bearing polyaminated chlorin p6 (the photosensitizer). The toxicity of the nanomaterial was assayed toward variants of B16 mouse melanoma cell line (B16F10 and B16G4F, with and without melanin, respectively). The in vitro results demonstrated that the nanoparticles had no cytotoxicity in the dark between 0.1 and 1 μM, while 1 μM of the nanoparticles had strong phototoxicity toward B16F10 cell line (Mbakidi et al. 2013).

Nam et al. (2016) reported the PDT activity of hematoporphyrin-conjugated Fe_3O_4 nanoparticles. The Fe_3O_4 nanoparticles were conjugated with hematoporphyrin (HP) leading to the formation of Fe_3O_4@HPs. The phototoxicity of the Fe_3O_4@HPs was evaluated toward human prostate cancer (PC-3) and breast cancer (MDA-MB-231) cell lines. The authors used a LED, with an electrical power of 3 W at 505 nm maximum wavelength and a power density of up to 7 mW/cm^2. A significant phototoxicity of Fe_3O_4@HPs was found toward both cancer cell lines, which was dependent on the amount of HP on the nanoparticle surface. In addition, the results demonstrated a cell membrane translocation and nuclear fragmentation of cancer cells due to apoptosis (Nam et al. 2016).

The photosensitizer pheophorbide-a (pheoA), which produces fluorescent emission in the range of 660–670 nm, was conjugated with iron oxide nanoparticles for PDT and magnetic resonance imaging (Nafiujjaman et al. 2015). The synthesis of the nanohybrid material was performed by the conjugation of pheoA with heparin (pheoA-Hep), which was further bound with APTES-coated Fe_3O_4 nanoparticles. The final material, spherical pheoA-Hep-Fe_3O_4 nanoparticles (average hydrodynamic size of 90 nm), had improved biocompatibility and can easily penetrate the cell surface via ERP effect. The phototoxicity of pheoA-Hep-Fe_3O_4 nanoparticles was demonstrated toward epithelial cancer (KB) cells, at nanoparticle concentration of 50 μg mL^{-1} and 24 h of incubation. The irradiation was performed by using a laser source 670 nm (4 mW/cm^{-2}) for 10 min. The results showed that pheoA-Hep-Fe_3O_4 nanoparticles caused 80% cell death, under irradiation, while in the dark, no toxicity was observed. Moreover, a high degree of nanoparticle internalization was reported (Nafiujjaman et al. 2015).

Oleic acid (OA)-coated magnetic iron oxide nanoparticles were synthesized by coprecipitation technique, followed by addition of Mn^{2+}-doped NaYF4:Yb/Er luminescent shell (Qin et al. 2016). The obtained monodispersed Fe_3O_4@Mn^{2+}-doped $NaYF_4$Yb/Er core-shell nanoparticles were able to emit luminescence under NIR excitation indicating their potential uses for PDT (Qin et al. 2016). Similarly, core-shell nanoparticles (size of 99 nm) composed by magnetite/silica/titania were obtained by sol-gel technique (Sakr et al. 2016). The photosensitizer ruthenium polypyridyl dye was added to the surface of the synthesized nanoparticles. The cytotoxicity of the nanoparticles was demonstrated toward lung cancer cells (A549) under a laser source at 532 nm (5 mW/cm^2). Indeed, upon irradiation, the ROS generation increased fivefold compared to bare NPs, which caused cell death (the half maximal inhibitory concentration, IC_{50}, was found to be 3 $\mu g/mL^1$). The mechanisms of cell death are under investigation (Sakr et al. 2016).

The photosensitizer porphyrin was added on the surface of superparamagnetic iron oxide nanoparticles (SPION-TPP) through click chemistry (Thandu et al.

2014). Phototoxicity of the obtained nanomaterial was demonstrated toward murine amelanotic melanoma B78-H1 cells, upon cell irradiation at NIR region (14 Jcm^{-2}, 30 min). Cell internalization of the nanoparticles was improved by the conjugation of the nanoparticles with TAT peptide (a cell-penetrating peptide) (Thandu et al. 2014). Unterweger et al. (2015) reported the synthesis of hypericin-linked-containing dextran coating on the surface of superparamagnetic iron oxide nanoparticles with 55–85 nm of hydrodynamic size. Dextran was used as stabilizing agent for the nanoparticles, and hypericin is a PS. The phototoxicity of the obtained nanoparticles was demonstrated toward human T-cell leukemia cells (Jurkat) upon white irradiation (40 Wm^{-2}) up to 15 min. The results showed a time-dependent and concentration-dependent cytotoxicity due to ROS generation (Unterweger et al. 2015).

A versatile nanomaterial was synthesized for multifunctional anticancer treatment and diagnosis (Wan et al. 2015). The nanomaterial, Fe_3O_4-NS-C_3N_4@mSiO_2-PEG-RGD, is composed of C_3N_4 (PS, a graphitic carbon nitride), mSiO_2 (mesoporous silica to enhance the stability of the nanomaterial and the ability of drug loading), and RGD (integrin-binding cell adhesive peptide, which is a tumor biomarker). The versatile nanomaterial allied PDT and low pH-triggered anticancer properties. The phototoxicity of the nanomaterial was evaluated toward A549 and HeLa cancer cells, upon visible light irradiation (450 nm, 40 mW/cm^{-2} for 10 min). Figure 1.4 shows images, obtained by confocal laser scanning microscopy, of the cells after incubation with the nanomaterial and PDT treatment, compared with control groups. As can be observed, a significant green fluorescence was emitted from the cells that were irradiated, in comparison with a less intensive green fluorescence light emission from the cells in the dark (no irradiated). This result illustrates the ability of the nanoparticles to generate ROS under irradiation with visible light. Interestingly, the confined beam of light source employed in this experiment caused a desired site-targeted cytotoxic effect on the treated cells, which might be able to decrease potential side effects toward noncancer cells (Fig. 1.4) (Wan et al. 2015).

Core-shell nanomaterial was obtained by coating hollow magnetic nanospheres (HMNSs) with silica (SiO_2) shells followed by conjugation with carboxylated graphene quantum dots (GQDs) leading to HMNS/SiO_2/GQDs (Wo et al. 2016). The anticancer drug DOX was added to the nanomaterial, which was further stabilized with liposomes. The phototoxicity of the multimodal nanoparticle was proved toward human esophageal squamous carcinoma (Eca-109) cells upon NIR laser irradiation (671 nm or 808 nm for 20 min). The authors demonstrated that liposome-stabilized nanomaterial was adsorbed by the tested cancer cells, which led to membrane and nuclei damages. As expected, by DOX loading on the nanoparticle, the cytotoxicity toward cancer cells was significantly increased (Wo et al. 2016).

Other multifunctional iron oxide-based nanoparticles were synthesized and applied in PDT. Firstly, Fe_3O_4@$NaYF_4$:Yb/Er core-shell nanomaterial was synthesized, followed by addition of tetrasulfonic phthalocyanine aluminum (AlPcS4), as the photosensitizer and PEG coating (Zeng et al. 2013). The phototoxicity of the

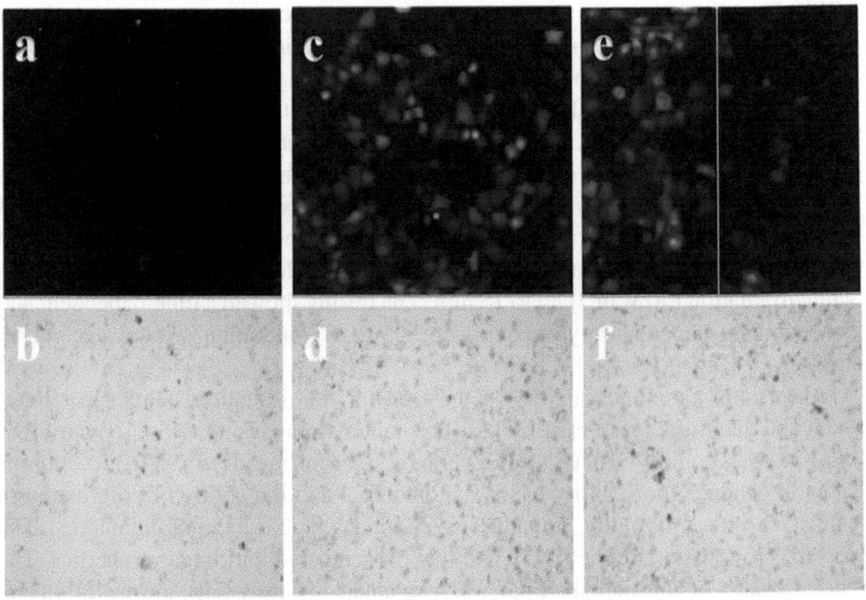

Fig. 1.4 Detection of intracellular ROS in A549 cancer cells treated with Fe_3O_4-NS-C_3N_4@ $mSiO_2$-PEG-RGD either (**c**) under visible light irradiation or (**a**) not; (**e**) site-targeted PDT. Parts **b**, **d**, and **f** correspond to the optical images of **a**, **c**, and **e** (Reproduced from reference Wan et al. 2015 with permission from American Chemical Society)

nanomaterial was shown toward MCF-7 cells, upon irradiation with 980 nm laser source (20 mW/cm^2) for 3 min. The authors reported a decrease of 70% of cancer cell viability upon this treatment. Moreover, the nanoparticles were found to accumulate on the MCF-7 surfaces. The toxicity to cancer cells was attributed to the ability of AIPcS4 to generate singlet oxygen (Zeng et al. 2013).

The nanocomposite Fe_3O_4@NaYF4:Yb-E conjugated with folic acid (FA) was synthesized and applied as PDT agent (under NIR irradiation) toward in vitro and in vivo (MCF-7 tumor-bearing nude mice) evaluations (Zeng et al. 2015). In vitro studies demonstrated that the viabilities of HeLa and MCF-7 cells decreased to 18% and 31%, upon incubation with the nanocomposite and 980 nm laser irradiation. Moreover, in vivo studies revealed 95% of MCF-7 tumors in bearing nude mice, indicating the promising application of this nanomaterial to combat cancer (Zeng et al. 2015). A DNA damage caused by Fe_3O_4 nanoparticles in PDT application was reported (Zhang and Chai 2012). Magnetite-silica core-shell nanomaterial was synthesized by sol-gel technique to promote DNA damage by a photoactive platinum-diimine complex under red light (Zhang and Chai 2012).

Zhao et al. (2014) reported the incorporation of the photosensitizer methylene blue (MB) into folate-conjugated Fe_3O_4/mesoporous silica core-shell nanoparticles. The toxicity of the nanoparticles was demonstrated toward different cell lines: HeLa cells, mouse fibroblast cells (NIH3T3 cells), mouse sarcoma cells (S-180 cells), and human ovarian cancer cells (SK-OV-3 cells). Folic acid (FA) was used to target the nanoparticles to the cancer cells, since cancer cells are known to overexpress folate receptors. The cell irradiation was performed by using a 650 nm laser beam (70 mW/ cm^2 for 4 min). The results demonstrated a significant decrease in cell viability (reduction of 80% of viable cells compared to control group) by incubating SK-OV-3 cells with 200 μg/mL[1] of the nanoparticle. Moreover, the nanoparticles were found to internalize to cancer cells. In vivo experiments were performed with S-180 tumor-bearing mice treated with the nanoparticles and laser irradiation at 650 nm. The results demonstrated the potential anticancer activity of the nanoparticles by suppressing tumor growth (Zhao et al. 2014).

Finally, the photosensitizer indocyanine green (IR820) was conjugated with chitosan quaternary ammonium (CSQ) magnetic iron oxide nanoparticles, leading to the formation of IR820-CSQ-Fe nanoparticles (Zhou et al. 2016). The phototoxicity of the nanoparticles (808 laser source, 8 Wcm^{-2}, 7 min) was demonstrated with breast cancer cell line (MDA-MB-231 cells) (Zhou et al. 2016). Taking together, several important and recent publications described successfully the synthesis of magnetic iron oxide-based nanoparticles as targeting nanocarriers in PDT to combat cancer.

1.4 Conclusions

The recent advantages in the design of iron oxide magnetic nanoparticles as nanocarriers of photosensitizers are able to generate therapeutic amounts of singlet oxygen directly in the target site of application. The targeting therapy can be achieved upon the application of an external magnetic field due to the superparamagnetic grapheme of iron oxide nanoparticles. It is clear from the reviewed literature that there is a great interest on the preparation of hybrid nanomaterials with multifunctions (e.g., cancer treatment and cancer diagnosis).

It should be noted that although tremendous progress has been made in this promising field, still more studies are required to translate the research to the clinical settings. In this direction, the ideal nanocarrier needs to meet the following requirements: (i) biocompatibility, (ii) biodegradability, (iii) small size and high loading capacity of the photosensitizer, (iv) minimum side effects, (v) prolonged body circulation with minimum tendency to aggregation, (vi) enhanced ERP effect, and (vii) ability to generate therapeutic amounts of singlet oxygen directed to the desired site and no dark toxicity.

Although PDT has been used in oncology for several years, some challenges need to be overcome, for instance, (i) minimization of systemic toxicity, (ii) adequate generation and permeation of singlet oxygen, and (iii) no dark toxicity.

Finally, it can be concluded that the combination of magnetic iron oxide nanoparticles as nanocarriers of photosensitizers in PDT is a promising field in cancer treatment. However, more studies are necessary in this area of nanomedicine.

References

Allison RR, Mota HC, Bagnato VS, Sibata CH. Bio-nanotechnology and photodynamic therapy—State of the art review. Photodiagn Photodyn Ther. 2008;5:19–28.

Banerjee SM, MacRobert AJ, Mosse CA, Periera B, Bown SG, Keshtgar MRS. Photodynamic therapy: inception to application in breast cancer. Breast. 2017;31:105–13.

Basoglu H, Bilgin MD, Demir MM. Protoporphyrin IX-loaded magnetoliposomes as a potential drug delivery system for photodynamic therapy: fabrication, characterization and in vitro study. Photodiagn Photodyn Ther. 2016;13:81–90.

Bechet D, Couleaud P, Frochot C, Viriot ML, Guillemin F, Barberi-Heyob M. Nanoparticles as vehicles for delivery of photodynamic therapy agents. Trends Biotechnol. 2008;26:612–21.

Bhana S, Lin G, Wang L, Starring H, Mishra SR, Liu G, Huang X. Near-infrared-absorbing gold nanopopcorns with iron oxide cluster core for magnetically amplified photothermal and photodynamic cancer therapy. ACS Appl Mater Interfaces. 2015;7:11637–47.

Bhattacharyya S, Kudgus RA, Bhattacharya R, Mukherjee P. Inorganic nanoparticles in cancer therapy. Pharm Res. 2011;28:237–59.

Chatterjee DV, Fong LS, Zhang Y. Nanoparticles in photodynamic therapy: an emerging paradigm. Adv Drug Deliv Rev. 2008;60:1627–3.

Chen ZL, Sun Y, Huang P, Yang XX, Zhou XP. Studies on preparation of photosensitizer loaded magnetic silica nanoparticles and their anti-tumor effects for targeting photodynamic therapy. Nanoscale Res Lett. 2009;4:400–8.

Chen G, Agren H, Ohulchanskyy TY, Prasad PN. Light upconverting core–shell nanostructures: nanophotonic control for emerging applications. Chem Soc Rev. 2015a;44:1680–713.

Chen H, Liu F, Lei Z, Ma L, Wang Z. Fe_2O_3@Au core@shell nanoparticle–graphene nanocomposites as theranostic agents for bioimaging and chemo-photothermal synergistic therapy. RSC Adv. 2015b;5:84980–7.

Cheng J, Tan G, Li W, Li J, Wang Z, Jin Y. Preparation, characterization and in vitro photodynamic therapy of a pyropheophorbide-a-conjugated Fe_3O_4 multifunctional magnetofluorescence photosensitizer. RSC Adv. 2016;6:37610–20.

Chu M, Shao Y, Peng J, Dai X, Li H, Wu Q, Shi D. Near-infrared laser light mediated cancer therapy by photothermal effect of Fe_3O_4 magnetic nanoparticles. Biomaterials. 2013;34:4078–88.

Corato DR, Béalle G, Kolosnjaj-Tabi J, Espinosa A, Clément O, Silva AK, Ménager C, Wilhelm C. Combining magnetic hyperthermia and photodynamic therapy for tumor ablation with photoresponsive magnetic liposomes. ACS Nano. 2015;9:2904–16.

Daniele MA, Shaughnessy ML, Roeder R, Childress A, Bandera YP, Foulger S. Magnetic nanoclusters exhibiting protein-activated near-infrared fluorescence. ACS Nano. 2013;7:203–13.

Galli M, Moschini E, Dozzi MV, Arosio P, Panigati M, D'Alfonso L, D'Alfonso L, Mantecca P, Lascialfari A, D'Alfonso G, Maggioni D. SPIO@SiO_2–Re@PEG nanoparticles as magneto-optical dual probes and sensitizers for photodynamic therapy. RSC Adv. 2016;6:38521–32.

Gangopadhyay M, Mukhopadhyay SK, Karthik S, Barmana S, Pradeep Singh ND. Targeted photoresponsive TiO_2–coumarin nanoconjugate for efficient combination therapy in MDA-MB-231 breast cancer cells: synergic effect of photodynamic therapy (PDT) and anticancer drug chlorambucil. Med Chem Commun. 2015;6:769–77.

Guo X, Wu Z, Li W, Wang Z, Li Q, Kong F, Zhang H, Zhu X, Du YP, Jin Y, Du Y, You J. Appropriate size of magnetic nanoparticles for various bioapplications in cancer diagnostics and therapy. ACS Appl Mater Interfaces. 2016;8:3092–106.

Huang Y, He S, Cao W, Cai K, Liang X-J. Biomedical nanomaterials for imaging-guided cancer therapy. Nanoscale. 2012;4:6135–49.

Kandasamy G, Maity D. Recent advances in superparamagnetic iron oxide nanoparticles (SPIONs) for in vitro and in vivo cancer nanotheranostics. Int J Pharm. 2015;496:191–218.

Kelkar SS, Reineke TM. Theranostics: combining imaging and therapy. Bioconjug Chem. 2011;22:1879–903.

Kim H, Chung K, Lee S, Kim DH, Lee H. Near-infrared light-responsive nanomaterials for cancer theranostics. WIREs Nanomedicine Nanobiotechnol. 2016a;8:23–45.

Kim KS, Kim J, Lee JY, Matsuda S, Hideshima S, Mori Y, Osaka T, Na K. Stimuli-responsive magnetic nanoparticles for tumor-targeted bimodal imaging and photodynamic/hyperthermia combination therapy. Nanoscale. 2016b;8:11625–34.

Lee J, Kim KS, Na K. Caffeic acid-coated multifunctional magnetic nanoparticles for the treatment and bimodal imaging of tumours. J Photochem Photobiol B. 2016;160:210–6.

Lee HH, Choi MG, Hasan T. Application of photodynamic therapy in gastrointestinal disorders: an outdated or re-emerging technique? Korean J Intern Med. 2017;32:1–10.

Li Z, Wang C, Cheng L, Gong H, Yin S, Gong Q, Yonggang L, Zhuang L. PEG-functionalized iron oxide nanoclusters loaded with chlorin e6 for targeted, NIR light induced, photodynamic therapy. Biomaterials. 2013;34:9160–70.

Ling D, Park W, Park SJ, Lu Y, Kim KS, Hackett MJ, Kim BH, Yim H, Jeon YS, Na K, Hyeon T. Multifunctional tumor pH-sensitive self-assembled nanoparticles for bimodal imaging and treatment of resistant heterogeneous tumors. J Am Chem Soc. 2014;136:5647–5.

Mbakidi JP, Drogat N, Granet R, Ouk TS, Ratinaud MH, Rivière E, Verdier M, Sol V. Hydrophilic chlorin-conjugated magnetic nanoparticles-potential anticancer agent for the treatment of melanoma by PDT. Bioorg Med Chem Lett. 2013;23:2486–90.

Nafiujjaman M, Revuri V, Nurunnabi M, Cho KJ, Lee YK. Photosensitizer conjugated iron oxide nanoparticles for simultaneous in vitro magneto-fluorescent imaging guided photodynamic therapy. Chem Commun. 2015;51:5687–90.

Nam KC, Choi KH, Lee KD, Kim JH, Jung JS, Park BJ. Particle size dependent photodynamic anticancer activity of hematoporphyrin-conjugated Fe3O4 particles. J Nanomater. 2016;2016:1278393.

Paszko E, Ehrhardt C, Senge MO, Kelleher DP, Reynolds JV. Nanodrug applications in photodynamic therapy. Photodiagn Photodyn Ther. 2011;8:14–29.

Qin Z, Du S, Luo Y, Liao Z, Zuo F, Luo J. Hydrothermal synthesis of superparamagnetic and red luminescent bifunctional Fe_3O_4@Mn^{2+}-doped $NaYF_4$:Yb/Er core@shell monodisperse nanoparticles and their subsequent ligand exchange in water. Appl Surf Sci. 2016;378:174–80.

Sakr MH, Halabi MN, Kalash LN, Al-Ghadban SI, Rammah MK, Sabban MEE, Bouhadir KH, Ghaddar TH. Synthesis and in vitro cytotoxicity evaluation of ruthenium polypyridyl-sensitized paramagnetic titania nanoparticles for photodynamic therapy. RSC Adv. 2016;6:47520–9.

Seabra AB, Duran N. Nanotoxicology of metal oxide nanoparticles. Metals. 2015;5:934–75.

Seabra AB, Haddad PS, Duran N. Biogenic synthesis of nanostructured iron compounds: applications and perspectives. IET Nanobiotechnol. 2013;7:90–9.

Seabra AB, Pasquôto T, Ferrarini AC, Santos Mda C, Haddad PS, de Lima R. Preparation, characterization, cytotoxicity, and genotoxicity evaluations of thiolated- and S-nitrosated superparamagnetic iron oxide nanoparticles: implications for cancer treatment. Chem Res Toxicol. 2014;27:1207–18.

Smith BE, Roder PB, Hanson JL, Manandhar S, Devaraj A, Perea DE. Singlet-oxygen generation from individual semiconducting and metallic nanostructures during near-infrared laser trapping. ACS Photon. 2015;2(4):559–64.

Thandu M, Rapozzi V, Xodo L, Albericio F, Comuzzi C, Cavalli S. "Clicking" porphyrins to magnetic nanoparticles for photodynamic therapy. Chem Plus Chem. 2014;79:90–8.

Unterweger H, Subatzus D, Tietze R, Janko C, Poettler M, Stiegelschmitt A, Schuster M, Maake C, Boccaccini AR, Alexiou C. Hypericin-bearing magnetic iron oxide nanoparticles for selective drug delivery in photodynamic therapy. Int J Nanomedicine. 2015;10:6985–96.

Wan H, Zhang Y, Zhang W, Zou H. Robust two-photon visualized nanocarrier with dual targeting ability for controlled chemo-photodynamic synergistic treatment of cancer. ACS Appl Mater Interfaces. 2015;7:9608–18.

Wang S, Gao R, Zhou F, Selke M. Nanomaterials and singlet oxygen photosensitizers: potential applications in photodynamic therapy. J Mater Chem. 2004;14:487–93.

Wo F, Xu R, Shao Y, Zhang Z, Chu M, Shi D, Liu S. A multimodal system with synergistic effects of magneto-mechanical, photothermal, photodynamic and chemo therapies of cancer in Grapheme-quantum dot-coated hollow magnetic nanospheres. Theranostics. 2016;6:485–500.

Zeng L, Xiang L, Ren W, Zheng J, Li T, Chen B. Multifunctional photosensitizer-conjugated core–shell Fe_3O_4@$NaYF_4$:Yb/Er nanocomplexes and their applications in T_2-weighted magnetic resonance/upconversion luminescence imaging and photodynamic therapy of cancer cells. RSC Adv. 2013;3:13915–25.

Zeng L, Luo L, Pan Y, Luo S, Lu G, Wu A. In vivo targeted magnetic resonance imaging and visualized photodynamic therapy in deep-tissue cancers using folic acid-functionalized superparamagnetic-upconversion nanocomposites. Nanoscale. 2015;7:8946–54.

Zhang Z, Chai A. Core-shell magnetite-silica composite nanoparticles enhancing DNA damage induced by a photoactive platinum-diimine complex in red light. J Inorg Biochem. 2012;117:71–6.

Zhao X, Chen Z, Zhao H, Zhang D, Tao L, Lan M. Multifunctional magnetic nanoparticles for simultaneous cancer near-infrared imaging and targeting photodynamic therapy. RSC Adv. 2014;4:62153–9.

Zhou H, Hou X, Liu Y, Zhao T, Shang Q, Tang J, Liu J, Wang Y, Wu Q, Luo Z, Wang H, Chen C. Superstable magnetic nanoparticles in conjugation with near-infrared dye as a multimodal theranostic platform. ACS Appl Mater Interfaces. 2016;8:4424–33.

Chapter 2
Nanoparticles in Wound Healing and Regeneration

Irina A. Shurygina and Michael G. Shurygin

Abstract Silver, gold, platinum, selenium, and copper nanoparticles, as well as zinc oxide, tantalum oxide, iron oxide, and titanium dioxide nanoparticles, have shown potential therapeutic effects. A wide range of studies are devoted to the use of nanosilver for wound healing. The possibility of using noble metal nanoparticles for regenerative medicine is considered. Noble metal nanoparticles such as aurum (Au), palladium (Pd), and platinum (Pt) nanoparticles are considered to function as antioxidants due to their strong catalytic activity. There are some works devoted to the use of zinc oxide (ZnO) for regeneration and wound healing. The use of selenium in nanoform is promising for regenerative medicine.

Materials with metal nanoparticles are widely used in regenerative medicine now. Pharmaceutical nanotechnologies are going to be widely used for regenerative medicine in the near future not only using direct effects of metal nanocomposites but also adding the ability to deliver drugs to damaged area. We can see it in the increasing number of publications dedicated to the issue.

Keywords Metal nanoparticles • Silver nanoparticles • Gold nanoparticles • Selenium nanoparticles • Wound healing • Regenerative medicine

2.1 Introduction

Today, nanotechnology is widely used in various spheres of life – electronic, environment, agriculture, etc.; medicine is not an exception. In recent years, these technologies are practiced on a wider scale for regenerative medicine and for the development of the new ways to stimulate regeneration (Kalashnikova et al. 2015).

In the past few years, nanotechnology has been constantly revolutionizing the treatment and management of wound care, by offering novel solutions that include

I.A. Shurygina (✉) • M.G. Shurygin
Irkutsk Scientific Center of Surgery and Traumatology,
1, Bortsov Revolutsii St., Irkutsk, Russia, 664003
e-mail: irinashurygina@gmail.com; mshurygin@gmail.com

M. Rai, R. Shegokar (eds.), *Metal Nanoparticles in Pharma*,
DOI 10.1007/978-3-319-63790-7_2

but are not limited to state-of-the-art materials, so-called "smart" biomaterials and theranostic nanoparticles (Kalashnikova et al. 2015). In this area, nanoparticles have emerged as important tools for regeneration stimulation and treatment of wounds. Silver (Ag), Au, Pt, selenium (Se), and copper (Cu) nanoparticles, as well as ZnO, tantalum oxide, iron oxide, and titanium dioxide nanoparticles, have shown potential therapeutic effects. Wound healing and regeneration are the fastest growing areas in regenerative medicine (Rahimnejad et al. 2017). A wide range of studies are devoted to the use of nanosilver for wound healing. The ability of the silver nanoparticles to have a significant antibacterial effect is well known in medicine (Shurygina et al. 2011a, b; Fadeeva et al. 2015; Kon and Rai 2016).

This effect is extremely helpful and widely used while including nanocomposites in the composition of ointments and bandages.

The use of nanocomposites containing metals and metal oxides allows the therapeutic use of both unique properties of nanoparticles and polymer matrix properties. In this case, often the use of nanoparticles in nanocomposite structure allows not only to increase the stability of nanoparticles but also to reduce their potential cytotoxicity (Li et al. 2015). All this opens new horizons in the use of nanocomposite materials for accelerating wound healing and damage reparation.

The aim of the chapter is to summarize the data on the application of nanocomposites of metal and metal oxides for the purposes of regenerative medicine.

2.2 Nanosilver and Regenerative Medicine

The use of nanosilver as a component of ointments and bandages (Guthrie et al. 2012; Jenwitheesuk et al. 2013; Fries et al. 2014; Pei et al. 2015) specifically for preventing wound contamination (Keen et al. 2012) or for curing infected wounds (Asz et al. 2006) is described in a large amount of research works. The high efficiency of silver nanoparticles is well known (Dhapte et al. 2014, Hebeish et al. 2014), the nanocomposites consisting of nanosized silver (Im et al. 2013; Singh et al. 2015) and nanofibers (Wu et al. 2014; Ghavami Nejad et al. 2015) as antibacterial agents are widely used.

However, apart from antibacterial effect, Ag NPs also have a cytotoxic effect. Correlation between composite concentration and antibacterial and cytotoxic properties is typical for silver nanocomposite. As shown for composite poly(L-lactic acid)-co-poly(ϵ-caprolactone) nanofibers containing silver nanoparticles, the concentration of 0.25 wt% Ag NPs was not toxic for the cell culture of human skin fibroblasts and the cells preserve its composition and divide very good (Jin et al. 2012). The same concentration was required for realization of antimicrobial activity against *S. aureus* and *S. enterica*. Pauksch et al. (2014) noted dose-dependent antibacterial and cytotoxic effect of silver nanoparticles toward mesenchymal stem cells and osteoblasts. After 21 days of cell incubation with silver nanoparticles, impairment of cell viability of mesenchymal stem cells and osteoblasts occurred at a concentration of 10 μg/g of Ag NPs (Pauksch et al. 2014).

Our studies have shown that the bacteriostatic concentration of zerovalent metallic silver nanoparticles stabilized by sulfated arabinogalactan toward *E. coli* ATCC 25922, *E. coli* ESBL1224, *S. aureus* MRSA34R, *S. aureus* ATCC 29213, and *P. aeruginosa* ATCC 27853 and hospital strains *E. coli*, *P. aeruginosa*, and *P. mirabilis* ranges from 3 to 50 μg mL^{-1}. The bactericidal activity of these nanocomposites varied from 5 to 100 μg mL^{-1} (Shurygina et al. 2015b).

The nanocomposite was obtained using an earlier described procedure (Fadeeva et al. 2015). This drug represents the metallic silver nanoparticles encapsulated into macromolecules of arabinogalactan sulfate with average sizes of 10–25 nm according to the X-ray phase diffraction and electronic transmission microscopy data.

The potential cytotoxicity of this composite was tested by the incubation method with the mononuclear fraction of the peripheral human blood (which consists of lymphocytes (95%) and monocytes (5%)) in the cell culture medium for 1 day at the final concentrations of the nanocomposite in the studied samples equal to 0.5, 0.05, and 0.005 mg mL^{-1}. The incubation of cells without introducing the nanocomposite into the medium served as a control (Fig. 2.1a). It is found that after 1 day of incubation with the nanocomposite concentration of 0.5 mg mL^{-1}, no viable lymphocytes were observed. Single retained lymphocytes exhibit pronounced morphological changes, and the nucleus loses compactness, increases sharply in size, and has signs of destruction (Fig. 2.1b). When using a concentration of 0.05 mg mL^{-1}, after 24 h of incubation, the number of cells in the sample decreases to 13.37% compared to the control. For the most part, mononuclears were mostly damaged, cell nuclei increased in sizes and lose compactness, and traces of decomposed cells appeared (Fig. 2.1c). For the incubation of mononuclears with the nanocomposite in a concentration of 5 μg mL^{-1}, in 1 day of incubation of the cell, concentration was 38.59% compared to the control; for the most part, lymphocytes are retained and morphologically undamaged and monocytes had properties of insignificant damage (Fig. 2.1d).

Thus, the concentrations of the silver nanocomposite with sulfated arabinogalactan are overlapped for toxic effect toward microorganisms and isolated cells of the peripheral human blood. This indicates that the mechanism of cytotoxicity of the studied silver nanocomposite toward both microbial cells and human lymphocytes is approximately the same. The most important universal shock factor of cells can be a plasmon polariton electric field of silver nanoparticles that acts distantly upon cell membranes disturbing the physiologically transmembrane potential up to the electric breakdown of the cell membrane. The observed overlap of the nanocomposite concentrations toxic for both microbes and lymphocytes predetermines the use of the considered composite as an antiseptic for external administration (Shurygina et al. 2015b).

To decrease Ag NP cytotoxicity, Chu et al. (2012) have developed a new original silver nanocomposite. The nanocomposite consists of 80 × 80 × 1 nm silicate platelets with 5 nm Ag NP on the surface. The weight ratio of Ag NP to silicate platelets is 7:93. The authors demonstrated superior antibacterial properties of developed nanocomposite in acute burn and excision wound healing models.

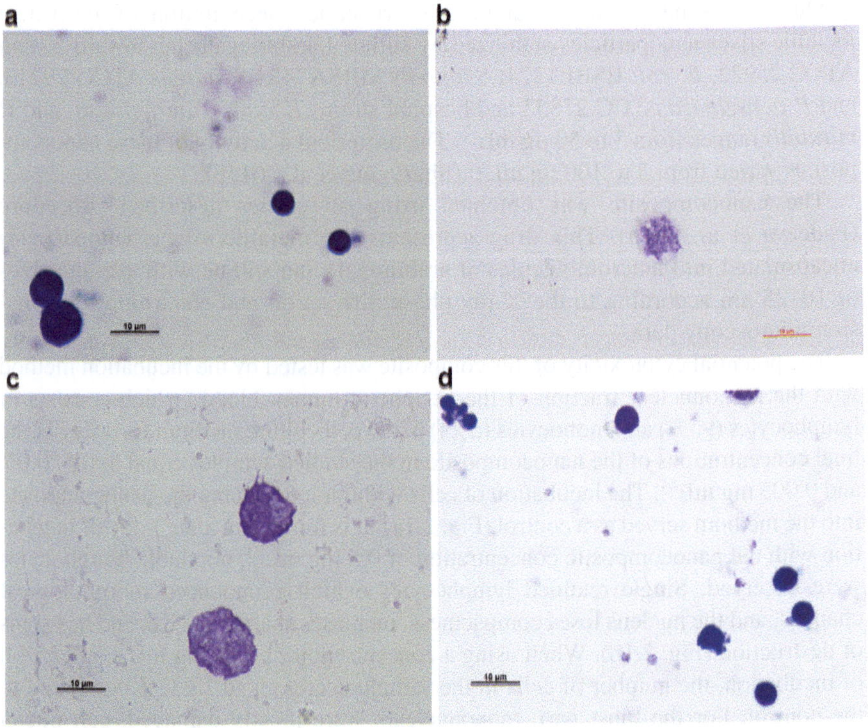

Fig. 2.1 Effect of the nanocomposite of zero valent metallic silver stabilized by sulfated arabinogalactan on the mononuclear fraction of the peripheral blood (color according to May–Grünwald): (a) control group and (b–d) nanocomposite concentrations 0.5 (b), 0.05 (c), and 0.005 mg/mL^{-1} (d)

However, some researchers succeeded in combining antibacterial and proliferative effects of silver nanocomposites (Table 2.1). By doing so, various polymers were explored as nanostabilized matrix. The polymers can supplement antibacterial properties of silver nanoparticles with the ability to enhance reparative effect and promote wound epithelialization.

Particularly, it is illustrated that silver nanoparticles, stabilized by Mexidol and polyvinylpyrrolidone, significantly reduce microbial contamination and promote wound healing process. It was tested in white rats using injection of 10% solution of calcium chloride in submandibular region with further opening of necrotic foci and open management of the wound (Lyakhovskyi et al. 2016).

It is illustrated that high-valence silver-pyridoxine nanoparticles are not only active toward microbial agents, which are mostly effused from burn wound, but also can induce the proliferation and migration of keratinocyte and fibroblast cells and promote wound healing in diabetic mice (Rangasamy et al. 2016).

Abdel-Mohsen et al. (2017) have demonstrated a similar effect for hyaluronan/silver nanoparticles in wound healing process models in nondiabetic and diabetic rats.

Table 2.1 The use of silver nanocomposites for wound healing

Matrix	Model	Antibacterial effect	Wound healing	Source
2-Ethyl-6-methyl-3-hydroxypyridine succinate (Mexidol) and polyvinylpyrrolidone	Injection of 10% solution of calcium chloride	Reduce during 10 days microbial contamination of exudate in 24 times	Wound square – in three times in comparison with original indices	Lyakhovskyi et al. (2016)
Pyridoxine	Diabetic mice in vitro assays	Antibacterial activities in 8 different pathogenic bacteria	Proliferation and migration of keratinocyte and fibroblast cells	Rangasamy et al. (2016)
Chitosan oligosaccharide/poly(vinyl alcohol)		Inhibited bacteria growth	Wound healing at an early stage	Li et al. (2013)
Hyaluronan	Nondiabetics/diabetic rat model		Accelerated the healing process	Abdel-Mohsen et al. (2017)
Dendrimer	RAW264.7 and J774.1 cells, burn wound model in mice	NA	Anti-inflammatory efficacy	Liu et al. (2014)
Guar gum alkylamine	Punch wound models in rodents	NA	Faster healing and improved cosmetic appearance	Ghosh et al. (2013)
Nanoscale silicate platelets	Acute burn and excision wound healing models	Inhibited bacteria growth	Superior for wound appearance	Chu et al. (2012)
Collagen	Mouse femoral facture model	NA	Promoted the formation of fracture callus, induced early closure of the fracture gap	Zhang et al. (2015)

It is noted that silver nanoparticle/chitosan oligosaccharide/poly(vinyl alcohol) nanofibers not only inhibit bacterial growth but also have the ability to promote wound reparation (Li et al. 2013). The authors connect this phenomenon with additional properties which are given by chitosan and polyvinyl alcohol contained in nanocomposite. The authors further revealed the supposed mechanism of action silver nanoparticle/chitosan oligosaccharide/poly(vinyl alcohol) nanofibers promoted wound healing and upregulated the expression levels of cytokines associated with the TGFβ1/Smad signaling pathway such as TGFβ1, TGFβRI, TGFβRII, collagen I, collagen III, pSmad2, and pSmad3 (Li et al. 2016).

In addition, it was shown that silver nanoparticles stabilized by dendrimer demonstrate significant anti-inflammatory properties. This effect was discovered not only in cell line models RAW264.7 and J774.1 stimulated by the introduction of lipopolysaccharide but also in burn wound model in mice (Liu et al. 2014). Silver nanocomposites based on the original matrix – new cationic biopolymer guar gum alkylamine – have demonstrated faster healing and improved cosmetic appearance in the wound healing process model. The authors consider that new material induces proliferation and migration of the keratinocytes at the wound site (Ghosh Auddy et al. 2013).

Our another research was dedicated to the possibility of using hydrophilic gel for curing wounds and burns where silver-containing systems were used as pharmacologically active substances. The systems are presented as zerovalent metallic silver nanoclusters with the size of the particles in the range of 10–25 nm. They were stabilized by arabinogalactan (Ag-AG) or its sulfated derivative (Ag-sAG), i.e., silver nanocomposites.

In the model of uninfected linear skin wound in female Wistar rats at the age of 9 months (weight 220–250 g), the anti-inflammatory activity of the gel containing 1% of silver nanocomposite with arabinogalactan was evaluated. It was found that its use reduced the intensity of neutrophil infiltration in the wound area compared to the control group (Figs. 2.2 and 2.3). In addition, the test product had anti-edematous properties, which were demonstrated by the reduction of the intensity of stromal edema and decrease in cell swelling compared to the control (Figs. 2.4 and 2.5).

The dynamics of the epithelialization process of the wound is an important regeneration characteristic in the wound area. It was found that the use of 1% silver nanocomposite with arabinogalactan caused early epithelialization on the day 3 of the wound healing process. Meanwhile, the epithelization time in the control group started on the day 7 of the wound healing process (Figs. 2.6 and 2.7). Thus, gel containing 1% silver nanocomposite on the arabinogalactan matrix has demonstrated the ability to stimulate epithelialization of the wound healing activity in the noninfected wound (Kostyro et al. 2012).

In a burn wound model in female Wistar rats at the age of 9 months (weight 220–250 g), the rate of epithelialization is evaluated. Scoring system for evaluation of the wound healing process was used. So, 0 point means the absence of epithelialization, 1 point means the beginning of the epithelialization on the wound, 2 points means the epithelialization of more than 80% of the wound surface, and 3 points means complete epithelialization of the wound (Kostyro et al. 2012).

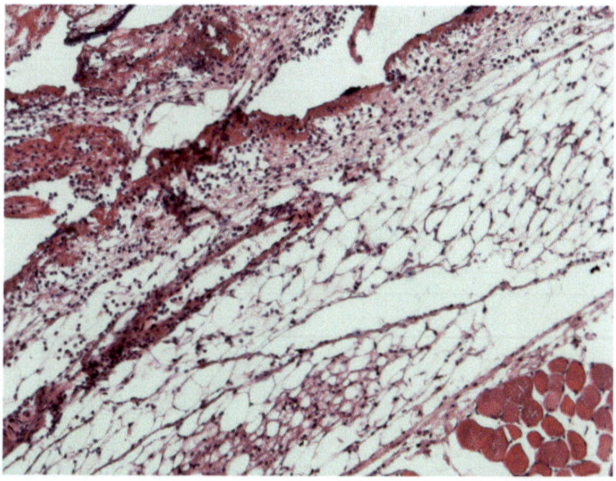

Fig. 2.2 Significant neutrophilic infiltration in the skin wound area, day 1, the control group, hematoxylin and eosin staining

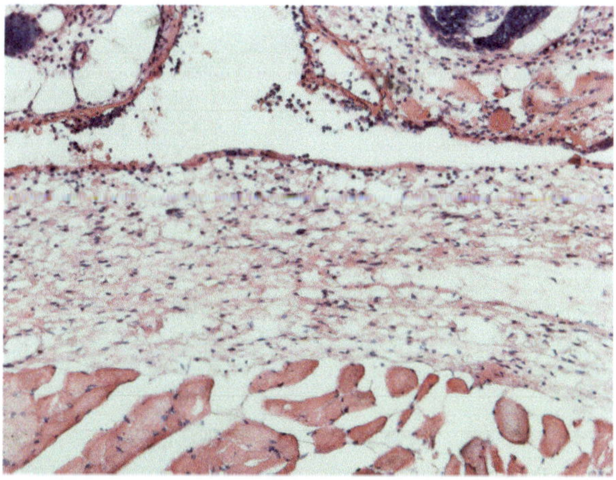

Fig. 2.3 Moderate neutrophilic infiltration in the skin wound area, day 1, the use of 1% silver nanocomposite with arabinogalactan, hematoxylin and eosin staining

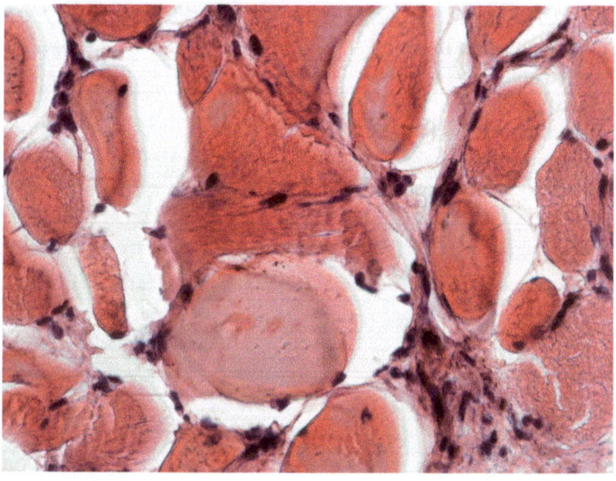

Fig. 2.4 Significant edema in the wound area, day 1, the control group, hematoxylin and eosin staining

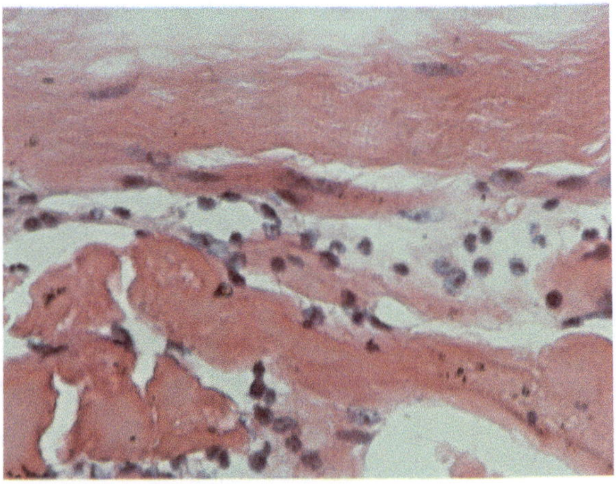

Fig. 2.5 Low-grade edema in the wound area, day 1, the use of 1% silver nanocomposite with arabinogalactan, hematoxylin and eosin staining

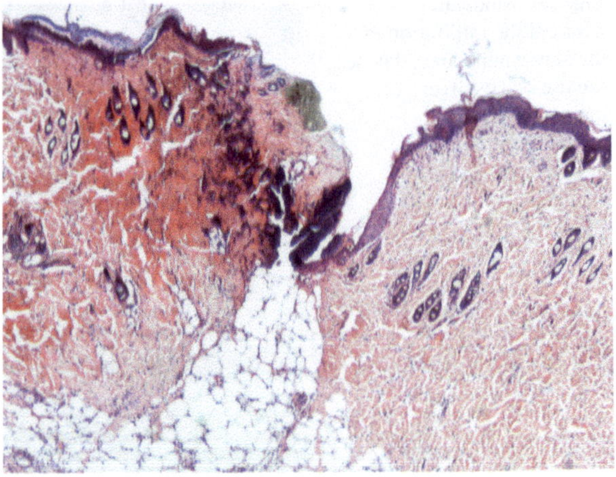

Fig. 2.6 Absence of the epithelialization in the wound area, day 3, the control group, hematoxylin and eosin staining

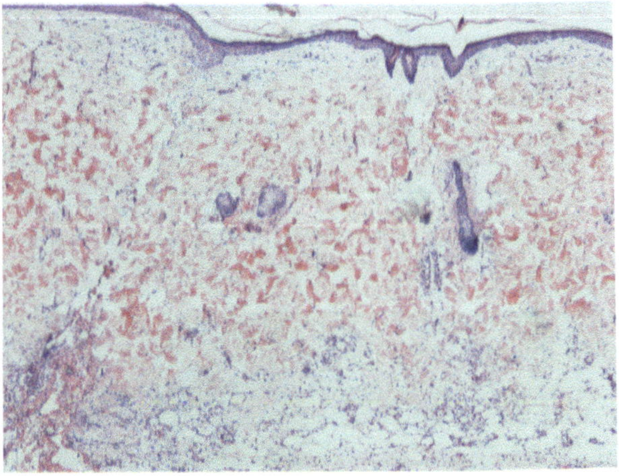

Fig. 2.7 Early epithelialization in the wound area, day 3, the use of 1% silver nanocomposite with arabinogalactan, hematoxylin and eosin staining

It was found that the use of gels containing 1% silver nanocomposite with arabinogalactan or 1% silver nanocomposite with sulfated arabinogalactan accelerates the epithelialization process of a burn wound compared to control group. The statistical analysis revealed that the hydrophilic gels based on silver nanocomposites and arabinogalactan or its sulfated derivative have significant wound healing activity compared to control (comparison: "control" – "Ag-AG" $z = -2.774$, $p = 0.0055$; "control" – "Ag-sAG" $z = -2.05$, $p = 0.04$).

Therefore, original compositions in the form of hydrophilic gels based on silver nanocomposites and arabinogalactan or its sulfated derivative have significant antiinflammatory, anti-exudative, and wound healing properties, which make the product promising for surgical and combustiological practice to treat wounds and burns. It can also be used in veterinary medicine (Kostyro et al. 2012).

In another study, it was found that the silver nanoparticles get localized in the mitochondria in HepG2 cell line and have an IC_{50} value of 251 μg/mL. Even though they elicit an oxidative stress, cellular antioxidant systems (reduced glutathione content, superoxide dismutase and catalase) are triggered and prevent oxidative damage. Moreover, silver nanoparticles induce apoptosis at concentrations of up to 250 μg/mL, which could favor wound healing. Acute dermal toxicity studies on silver nanoparticle gel formulation in Sprague-Dawley rats showed complete safety for topical application (Jain et al. 2009).

The original study by Zhang et al. (2015) showed that Ag NPs promoted mesenchymal stem cell proliferation and osteogenic differentiation in vitro. Ag NPs encapsulated in collagen promoted the formation of fracture callus and induced early closure of the fracture gap at the mouse femoral fracture model. The authors believe that Ag NPs may promote the formation of the callus via multiple routes: chemoattraction of mesenchymal stem cells and fibroblasts to migrate to the fracture site, induction of the proliferation of mesenchymal stem cells, and induction of osteogenic differentiation of mesenchymal stem cells via induction/activation of TGF-β/BMP signaling in mesenchymal stem cells (Zhang et al. 2015).

2.3 Noble Metal Nanoparticles and Wound Healing

The possibility of using noble metal nanoparticles for regenerative medicine is considered. Noble metal nanoparticles such as Au, Pd, and Pt nanoparticles are considered to function as antioxidants due to their strong catalytic activity. Particularly, gold nanoparticles are considered a promising solution. However, potential toxicity of the given nanoparticles necessitates the development of the materials containing gold nanoparticles with controlled nanoparticle realization from composites and reduced toxicity (Table 2.2).

It is shown that silica-gold core-shell materials could promote the proliferation of mouse embryonic fibroblast cells (NIH/3T3). Moreover, nanoparticles are

Table 2.2 The use of gold nanocomposites for wound healing

Matrix	Model	Wound healing	Source
SiO_2	Mouse embryonic fibroblast cells (NIH/3T3)	Promote the proliferation of mouse embryonic fibroblast cells	Li et al. (2015)
SiO_2	Cutaneous full-thickness excisional wound rat model	Promote wound healing	Li et al. (2015)
Epigallocatechin gallate and α-lipoic acid	Diabetic cutaneous mouse wound	Accelerated wound healing	Chen et al. (2012)

restrained by the matrix of the nanocomposite and do not penetrate the fibroblasts. In vivo, a cutaneous full-thickness excisional wound rat model SiO_2@Au NPs could promote wound healing, which was potentially related to the anti-inflammatory and antioxidation of Au NPs (Li et al. 2015).

Gold nanoparticles acting as the antioxidant agents are successfully tested using cutaneous wound model in diabetic mice. It is shown that combination of Au nanoparticles, epigallocatechin gallate, and α-lipoic acid accelerated wound healing on diabetic mouse skin and decreased the advanced glycation end product expression. During 7 days of the treatment, the level of vascular endothelial growth factor increased significantly. Angiopoietin-2 significantly decreased at day 7. CD68 (cluster of differentiation 68) expression significantly decreased from day 3 to day 7. The results suggest that combination of Au nanoparticles, epigallocatechin gallate, and α-lipoic acid significantly accelerated cutaneous wound healing in diabetes through angiogenesis regulation and anti-inflammatory effects (Chen et al. 2012).

Pd and Pt nanoparticles also are promising for regenerative medicine. Dr. Ishizuka made the solution containing palladium and platinum nanoparticles in Japan in 1936. The given nanoparticle solution was called PAPLAL® (Toyokose Pharmaceuticals, Japan) (Shibuya et al. 2014). PAPLAL is composed of a mixture of 0.3 mg/mL (2.82 mM) of Pd nanoparticles and 0.2 mg/mL (1.03 mM) of Pt nanoparticles. Shibuya et al. (2014) investigated the protective effects of PAPLAL® against aging-related skin pathologies in mice at transdermal application. It is demonstrated that the given solution reversed skin thinning associated with increased lipid peroxidation and normalized the gene expression levels of Col1a1, Mmp2, Has2, TNF-α, IL-6, and p53 in the skin.

2.4 Metal Oxide Nanoparticles and Wound Healing

There are some unique studies devoted to the use of ZnO for regeneration and wound healing. For instance, it is demonstrated that cefazolin nanofiber mats loaded with zinc oxide show significant activity toward *Staphylococcus aureus* compared to separate use of zinc oxide nanoparticles and cefazolin. The use of the same nanofibers promotes wound healing in Wistar rats. Macroscopical and histological evaluations demonstrated that ZnO NP hybrid cefazolin nanofiber showed enhanced cell adhesion and epithelial migration, leading to faster and more efficient collagen synthesis (Rath et al. 2016).

Kumar et al. (2012) used chitosan hydrogel/nano zinc oxide composite bandages. Good antibacterial properties of the bandages were shown with no toxicity for normal human dermal fibroblast cells. In some studies, in vivo evaluations in Sprague-Dawley rats revealed that these nanocomposite bandages enhanced the wound healing and helped for faster reepithelialization and collagen deposition (Kumar et al. 2012).

ZnO NP-loaded sodium alginate-gum acacia hydrogels have shown significant antibacterial effect on *Pseudomonas aeruginosa* and *Bacillus cereus* and biocompatibility on peripheral blood mononuclear/fibroblast cells (Raguvaran et al. 2017).

The use of titanium dioxide nanoparticles is quite promising. The green synthesized titanium dioxide NP wound healing activity was examined in the excision wound model by measuring wound closure, histopathology, and protein profiling and revealed significant wound healing activity in albino rats (Sankar et al. 2014).

2.5 Nanoselenium and Regenerative Medicine

In fact, selenium is a nonmetal that has some metal properties. The use of selenium in nanoform is promising for regenerative medicine. It is well known that nanoselenium is a highly effective long-acting antioxidant. Its local introduction in the area of injury can lead to violation of redox signaling. Selenium showed significant antiproliferative activity against HeLa and HepG2 cell lines. The wound healing activity of Se nanoparticles reveals that 5% selenium ointment heals the excision wound of Wistar rats up to 85% within 18 days compared to the standard ointment (Ramya et al. 2015).

In our study, an original nanocomposite of specially synthesized selenium and arabinogalactan was used (Shurygina et al. 2015a; Rodionova et al. 2016). The selenium content determined by titrimetry in a nanocomposite sample was 0.54% (Rodionova et al. 2015). Experimental studies were conducted on model of buttonhole fracture of the shin bone (Chinchilla male rabbits). Markers of the rate of the regeneration of the bone tissues were an oxytetracycline hydrochloride solution, injected intramuscularly into the animals on the 7th and 21st days of the experiment in doses of 250 mg/kg body weight, and a solution of Alizarin Red S, administered intraperitoneally on the 14th and 28th days in doses of 100 mg/kg body weight.

The rabbits were eliminated from the experiment on the 35th day via the intrapleural introduction of sodium thiopental solution in doses of 167 mg/kg. The material was fixed in a FineFix solution (Milestone, Italy). The decalcification and subsequent preparation of bone tissues were carried out for a histological study according to the method we developed (Shurygina and Shurygin 2013). The material was prepared and embedded into paraffin blocks; multiple sections 7 μm wide were prepared, and the sections were deparaffinized and embedded in Fluoromount (Diagnostic Biosystems, REF K024). The specific fluorescence of the labels was visualized using a Zeiss LSM710 laser confocal microscope and a Nikon Eclipse 80 microscope with an DIH-M epifluorescence attachment. Nikon B2A (excitation at 450–490 nm, dichroic mirror 505 LP, emission at >515 nm) and TRITC (excitation at 528–553 nm, dichroic mirror 565 LP, emission at 590–650 nm) filters were used to select the required ranges of fluorescence. Recording was performed using a Nikon DSFi1c camera connected to a Nikon DSU2 controller using the Nikon Elements program software.

Preparations of the slices for visualization of oxytetracycline hydrochloride, Alizarin Red S, and the nanocomposite of elemental selenium on arabinogalactan were performed in the same manner. Pure oxytetracycline hydrochloride has needle-shaped crystals and has 490–565 nm emissions with a peak at 553 nm when excited by a 405 nm laser. Upon laser excitation at 561 nm, the solution of Alizarin Red S produced emissions in the range of 570–675 nm with a maximum at 605 nm. Heavy calcium salt deposits formed in the bone calluses of the rabbits control group were observed with the help of oxytetracycline and Alizarin Red S that formed calcium complex when introduced in vivo.

It was found that the studied nanocomposite of elemental selenium had the widest spectrum of fluorescence. We founded that Se NPs had spectroscopy properties of quantum dots by LSM710 confocal microscopy. The nanocomposite particles had emission with maxima at 480 nm when exited at 405 nm, 538 nm with 458 nm excitation, and 555 nm at 514 nm laser radiation. Larger particles absorbed only shortwave radiation and upon excitation by the laser at 405 nm produced an emission peak at 446 nm.

The animals injected with the selenium nanocomposite containing arabinogalactan had poorly regenerated bones at the fracture site. The trabeculae of the bones were very thin. Barely visible deposits of oxytetracycline and Alizarin Red S at the fracture site were observed with the oxytetracycline deposits predominating. A large amount of fluorescent amorphous masses with intense fluorescence was noted outside the fracture site and in the Haversian spaces. Fluorescence of these masses had the emissions in regions of 515–560 nm (covering the emission maximum of oxytetracycline) and 590–650 nm (covering almost half of the Alizarin Red S spectrum). It was difficult to make any conclusions of their origin. These masses could have been either calcium complexes of oxytetracycline and Alizarin Red S introduced in vivo or deposits of the nanocomposite itself. The use of epifluorescence thus did not allow us to obtain clear answers to the questions raised. Studying these preparations by confocal microscopy in the mode of spectral detection allowed us to see clearly that the fluorescent labels in the regenerated bone were almost entirely oxytetracycline.

The fluorescent amorphous masses outside the fracture sites and in the Haversian spaces regarding the radiation spectrum were closer to the spectra of the nanocomposite of selenium and arabinogalactan (from 430 to 630 nm, with a maximum in the region of 490 nm).

Our study allowed us to establish that the local application of nanoselenium at the fracture sites significantly impairs the reparative processes, slowing down bone regeneration and impairing mineralization in formed bone calluses (Shurygina et al. 2015a). No such deviations in the natural course of the reparative process were observed in earlier studies of the arabinogalactan matrix biological effects in altering tissue. This allows us to exclude the toxic effect of the matrix substance on an organism's cells. The local use of selenium nanoparticles associated with macromolecules of arabinogalactan on a fracture site seems to lead above all to significant diffusion impediments to the evacuation of the nanocomposite from this area and thus to pronounced prolonged local effects of nanoselenium on the trauma site.

2.6 Other Application of Nanoparticles

Copper nanoparticles modulate cells, cytokines, and growth factors involved in wound healing in a better way than copper ions (Gopal et al. 2014). Chitosan-based copper nanocomposite in open excision wound model in adult Wistar rats significantly promoted the wound healing process. Also, there was upregulation of vascular endothelial growth factor (VEGF) and transforming growth factor beta 1 (TGF-β1). The tumor necrosis factor-α (TNF-α) was significantly decreased and interleukin-10 (IL-10) was significantly increased. Histological evaluation showed more fibroblast proliferation, collagen deposition, and intact reepithelialization after chitosan-based copper nanocomposite treatment. Thus, the chitosan-based copper nanocomposite efficiently enhanced cutaneous wound healing by modulation of various cells, cytokines, and growth factors during different phases of healing process (Gopal et al. 2014).

Sankar et al. (2015) have reported the antibacterial activity of green synthesized copper oxide nanoparticles against *Klebsiella pneumoniae*, *Shigella dysenteriae*, *Staphylococcus aureus*, *Salmonella typhimurium*, and *Escherichia coli* as well as the ability to promote the wound healing process in Wistar Albino rats (Sankar et al. 2015).

Due to their specific characteristics, nanoparticles such as nanocapsules, polymersomes, solid lipid nanoparticles, and polymeric nanocomplexes are ideal vehicles to improve the effect of drugs (antibiotics, growth factors, etc.) aimed at wound healing (Oyarzun-Ampuero et al. 2015). A very promising prospect is the use of the magnetic nanoparticle properties to deliver biologically active substances and regenerative medicine drugs (Sensenig et al. 2012). For example, some magnetic nanoparticles were used to deliver antioxidant enzymes (catalase and superoxide dismutase) to endothelial cells (Chorny et al. 2010).

The application of nanotechnology has considerable potential in cell-based therapies for regenerative medicine, e.g., in localizing, recruiting, and labeling stem cells to begin the regeneration process (Atala et al. 2011).

2.7 Conclusion

The use of metal nanoparticles is common in regenerative medicine. The most promising area currently is the use of nanomaterials containing Ag nanoparticles because they combine antibacterial and proliferative effect that is widely used for a contaminated wound.

Moreover, nanomaterials that include nanoparticles of noble metals and selenium are also very important because of antioxidant and proliferative effect.

Metal oxide nanoparticles like zinc and titanium that combine antibacterial and proliferative activity are now widely used for regenerative medicine purposes.

Thus, nanotechnologies see heavier use in regenerative medicine. Pharmaceutical nanotechnologies are going to be widely used for regenerative medicine in the near future using not only direct effects of nanocomposites but also its ability to deliver drugs to damaged area.

References

Abdel-Mohsen AM, Jancar J, Abdel-Rahman RM, Vojtek L, Hyršl P, Dušková M, Nejezchlebová H. A novel in situ silver/hyaluronan bio-nanocomposite fabrics for wound and chronic ulcer dressing: in vitro and in vivo evaluations. Int J Pharm. 2017;520(1–2):241–53.

Asz J, Asz D, Moushey R, Seigel J, Mallory SB, Foglia RP. Treatment of toxic epidermal necrolysis in a pediatric patient with a nanocrystalline silver dressing. J Pediatr Surg. 2006;41(12):e9–12.

Atala A, Lanza R, Nerem R, Thomson JA. Principles of regenerative medicine. 2nd ed. London: Academic Press; 2011. p. 733–1105.

Chen SA, Chen HM, Yao YD, Hung CF, Tu CS, Liang YJ. Topical treatment with anti-oxidants and Au nanoparticles promote healing of diabetic wound through receptor for advance glycation end-products. Eur J Pharm Sci. 2012;47(5):875–83.

Chorny M, Hood E, Levy RJ, Muzykantov VR. Endothelial delivery of antioxidant enzymes loaded into non-polymeric magnetic nanoparticles. J Control Release. 2010;146(1):144–51.

Chu CY, Peng FC, Chiu YF, Lee HC, Chen CW, Wei JC, Lin JJ. Nanohybrids of silver particles immobilized on silicate platelet for infected wound healing. PLoS One. 2012;7(6):e38360.

Dhapte V, Kadam S, Moghe A, Pokharkar V. Probing the wound healing potential of biogenic silver nanoparticles. J Wound Care. 2014;23(9):431–6.

Fadeeva TV, Shurygina IA, Sukhov BG, Rai MK, Shurygin MG, Umanets VA, Lesnichaya MV, Konkova TV, Shurygin DM. Relationship between the structures and antimicrobial activities of argentic nanocomposites. Bull Russ Acad Sci Phys. 2015;79(2):273–5.

Fries CA, Ayalew Y, Penn-Barwell JG, Porter K, Jeffery SL, Midwinter MJ. Prospective randomised controlled trial of nanocrystalline silver dressing versus plain gauze as the initial post-debridement management of military wounds on wound microbiology and healing. Injury. 2014;45(7):1111–6.

Ghavami Nejad A, Rajan Unnithan A, Ramachandra Kurup Sasikala A, Samarikhalaj M, Thomas RG, Jeong YY, Nasseri S, Murugesan P, Wu D, Hee Park C, Kim CS. Mussel-inspired electrospun nanofibers functionalized with size-controlled silver nanoparticles for wound dressing application. ACS Appl Mater Interfaces. 2015;7(22):12176–83.

Ghosh Auddy R, Abdullah MF, Das S, Roy P, Datta S, Mukherjee A. New guar biopolymer silver nanocomposites for wound healing applications. Biomed Res Int. 2013;2013:912458.

Gopal A, Kant V, Gopalakrishnan A, Tandan SK, Kumar D. Chitosan-based copper nanocomposite accelerates healing in excision wound model in rats. Eur J Pharmacol. 2014;731:8–19.

Guthrie KM, Agarwal A, Tackes DS, Johnson KW, Abbott NL, Murphy CJ, Czuprynski CJ, Kierski PR, Schurr MJ, McAnulty JF. Antibacterial efficacy of silver-impregnated polyelectrolyte multilayers immobilized on a biological dressing in a murine wound infection model. Ann Surg. 2012;256(2):371–7.

Hebeish A, El-Rafie MH, El-Sheikh MA, Seleem AA, El-Naggar ME. Antimicrobial wound dressing and anti-inflammatory efficacy of silver nanoparticles. Int J BiolMacromol. 2014;65:509–15.

Im AR, Kim JY, Kim HS, Cho S, Park Y, Kim YS. Wound healing and antibacterial activities of chondroitin sulfate- and acharan sulfate-reduced silver nanoparticles. Nanotechnology. 2013;24(39):395102.

Jain J, Arora S, Rajwade JM, Omray P, Khandelwal S, Paknikar KM. Silver nanoparticles in therapeutics: development of an antimicrobial gel formulation for topical use. Mol Pharm. 2009;6(5):1388–401.

Jenwitheesuk K, Surakunprapha P, Chowchuen B. The use of nanocrystalline silver for the treatment of massive soft tissue defects with exposed bone. J Med Assoc Thail. 2013;96(Suppl 4):S177–84.

Jin G, Prabhakaran MP, Nadappuram BP, Singh G, Kai D, Ramakrishna S. Electrospun Poly(L-Lactic Acid)-co-Poly(ε-Caprolactone) nanofibres containing silver nanoparticles for skin-tissue engineering. J Biomater Sci Polym Ed. 2012;23(18):2337–52.

Kalashnikova I, Das S, Seal S. Nanomaterials for wound healing: scope and advancement. Nanomedicine (London). 2015;10(16):2593–612.

Keen JS, Desai PP, Smith CS, Suk M. Efficacy of hydrosurgical debridement and nanocrystalline silver dressings for infection prevention in type II and III open injuries. Int Wound J. 2012;9(1):7–13.

Kon K, Rai M. Antibiotic resistance. Mechanisms and new antimicrobial approaches. 1st ed. London: Academic Press; 2016.

Kostyro YaA, Alekseev KV, Petrova EN, Gumennikova EN, Romanko TV, Romanko VG, Lepekhova SA, Shurygina IA, Korjakina LB, Fadeeva TV, Vereshchagina SA, Koval EV, Shurygin MG, Babkin VA, Ganenko TV, Grishchenko LA, Ostroukhova LA, Sukhov BG, Trofimov BA. Agent for burn and wound healing. RU Patent 2513186; 2012.

Kumar PT, Lakshmanan VK, Anilkumar TV, Ramya C, Reshmi P, Unnikrishnan AG, Nair SV, Jayakumar R. Flexible and microporous chitosan hydrogel/nanoZnO composite bandages for wound dressing: in vitro and in vivo evaluation. ACS Appl Mater Interfaces. 2012;4(5):2618–29.

Li C, Fu R, Yu C, Li Z, Guan H, Hu D, Zhao D, Lu L. Silver nanoparticle/chitosan oligosaccharide/poly(vinyl alcohol) nanofibers as wound dressings: a preclinical study. Int J Nanomedicine. 2013;8:4131–45.

Li X, Wang H, Rong H, Li W, Luo Y, Tian K, Quan D, Wang Y, Jiang L. Effect of composite SiO_2@AuNPs on wound healing: in vitro and vivo studies. J Colloid Interface Sci. 2015;445:312–9.

Li CW, Wang Q, Li J, Hu M, Shi SJ, Li ZW, Wu GL, Cui HH, Li YY, Zhang Q, Yu XH, Lu LC. Silver nanoparticles/chitosan oligosaccharide/poly(vinyl alcohol) nanofiber promotes wound healing by activating TGFβ1/Smad signaling pathway. Int J Nanomedicine. 2016;11:373–86.

Liu X, Hao W, Lok CN, Wang YC, Zhang R, Wong KK. Dendrimer encapsulation enhances anti-inflammatory efficacy of silver nanoparticles. J Pediatr Surg. 2014;49(12):1846–51.

Lyakhovskyi VI, Lobahn GA, Gancho OV, Vazhnycha OM, Kolomiyets SV, Jaber VK. Dynamics of bacteriological and planimetric indices of the wound under the action of the silver nanoparticles, stabilized by mexidol and polyvinylpyrrolidone. KlinKhir. 2016;4:67–9. Ukrainian

Oyarzun-Ampuero F, Vidal A, Concha M, Morales J, Orellana S, Moreno-Villoslada I. Nanoparticles for the treatment of wounds. Curr Pharm Des. 2015;21(29):4329–41.

Pauksch L, Hartmann S, Rohnke M, Szalay G, Alt V, Schnettler R, Lips KS. Biocompatibility of silver nanoparticles and silver ions in primary human mesenchymal stem cells and osteoblasts. Acta Biomater. 2014;10(1):439–49.

Pei Z, Sun Q, Sun X, Wang Y, Zhao P. Preparation and characterization of silver nanoparticles on silk fibroin/carboxymethylchitosan composite sponge as anti-bacterial wound dressing. Biomed Mater Eng. 2015;26(Suppl 1):S111–8.

Raguvaran R, Manuja BK, Chopra M, Thakur R, Anand T, Kalia A, Manuja A. Sodium alginate and gum acacia hydrogels of ZnO nanoparticles show wound healing effect on fibroblast cells. Int J Biol Macromol. 2017;96:185–91.

Rahimnejad M, Derakhshanfar S, Zhong W. Biomaterials and tissue engineering for scar management in wound care. Burns Trauma. 2017;5:4.

Ramya S, Shanmugasundaram T, Balagurunathan R. Biomedical potential of actinobacterially synthesized selenium nanoparticles with special reference to anti-biofilm, anti-oxidant, wound healing, cytotoxic and anti-viral activities. J Trace Elem Med Biol. 2015;32:30–9.

Rangasamy S, Tak YK, Kim S, Paul A, Song JM. Bifunctional therapeutic high-valence silver-pyridoxine nanoparticles with proliferative and antibacterial wound-healing activities. J Biomed Nanotechnol. 2016;12(1):182–96.

Rath G, Hussain T, Chauhan G, Garg T, Goyal K. Development and characterization of cefazolin loaded zinc oxide nanoparticles composite gelatin nanofiber mats for postoperative surgical wounds. Mater Sci Eng C Mater Biol Appl. 2016;58:242–53.

Rodionova LV, Shurygina IA, Sukhov BG, Popova LG, Shurygin MG, Artemev AV, Pogodaeva NN, Kuznetsov SV, Gusarova NK, Trofimov BA. Nanobiocomposite based on selenium and arabinogalactan: synthesis, structure, and application. Russ J Gen Chem. 2015;85(2):485–7.

Rodionova LV, Shurygina IA, Samoilova LG, Sukhov BG, Shurygin MG. Effect of intraosseous introduction of selenium/arabinogalactan nanoglycoconjugate on the main indicators of primary metabolism in consolidation of bone fracture. Bul East Sib Sci Cent RAMS. 2016;110(4):104–8. Russian

Sankar R, Dhivya R, Shivashangari KS, Ravikumar V. Wound healing activity of Origanum vulgare engineered titanium dioxide nanoparticles in Wistar Albino rats. J Mater Sci Mater Med. 2014;25(7):1701–8.

Sankar R, Baskaran A, Shivashangari KS, Ravikumar V. Inhibition of pathogenic bacterial growth on excision wound by green synthesized copper oxide nanoparticles leads to accelerated wound healing activity in Wistar Albino rats. J Mater Sci Mater Med. 2015;26(7):214.

Sensenig R, Sapir Y, MacDonald C, Cohen S, Polyak B. Magnetic nanoparticle-based approaches to locally target therapy and enhance tissue regeneration in vivo. Nanomedicine (London). 2012;7(9):1425–42.

Shibuya S, Ozawa Y, Watanabe K, Izuo N, Toda T, Yokote K, Shimizu T. Palladium and platinum nanoparticles attenuate aging-like skin atrophy via antioxidant activity in mice. PLoS One. 2014;9(10):e109288.

Shurygina IA, Shurygin MG. Method of preparing bone tissue preparation and set for its realization. RU Patent 2013; 2500104.

Shurygina IA, Sukhov BG, Fadeeva TV, Umanets VA, Shurygin MG, Verecshagina SA, Ganenko TV, Kostyro YA, Trofimov BA. Mechanism of bactericidal action Ag (0)-nanobiocomposite: the evolution of the original composite and live microbial cells in the new nanocomposite. Russ Phys J. 2011a;54(2):285–8. Russian

Shurygina IA, Sukhov BG, Fadeeva TV, Umanets VA, Shurygin MG, Ganenko TV, Kostyro YA, Grigoriev EG, Trofimov BA. Bactericidal action of Ag(0)-antithrombotic sulfated arabinogalactan nanocomposite: coevolution of initial nanocomposite and living microbial cell to a novel nonliving nanocomposite. Nanomedicine. 2011b;7(6):827–33.

Shurygina IA, Rodionova LV, Shurygin MG, Popova LG, Dremina NN, Sukhov BG, Kuznetsov SV. Using confocal microscopy to study the effect of an original pro-enzyme Se/arabinogalactan nanocomposite on tissue regeneration in a skeletal system. Bull Russ Acad Sci Phys. 2015a;79(2):256–8.

Shurygina IA, Shurygin MG, Dmitrieva LA, Fadeeva TV, Ganenko TV, Tantsyrev AP, Sapozhnikov AN, Sukhov BG, Trofimov BA. Bacterio- and lymphocytotoxicity of silver nanocomposite with sulfated arabinogalactan. Russ Chem Bull. 2015b;64(7):1629–32.

Singh D, Singh A, Singh R. Polyvinyl pyrrolidone/carrageenan blend hydrogels with nanosilver prepared by gamma radiation for use as an antimicrobial wound dressing. J Biomater Sci Polym Ed. 2015;26(17):1269–85.
Wu J, Zheng Y, Wen X, Lin Q, Chen X, Wu Z. Silver nanoparticle/bacterial cellulose gel membranes for antibacterial wound dressing: investigation in vitro and in vivo. Biomed Mater. 2014;9(3):035005.
Zhang R, Lee P, Lui VC, Chen Y, Liu X, Lok CN, To M, Yeung KW, Wong KK. Silver nanoparticles promote osteogenesis of mesenchymal stem cells and improve bone fracture healing in osteogenesis mechanism mouse model. Nanomedicine. 2015;11(8):1949–59.

Chapter 3
Silver Nanoparticles for Treatment of Neglected Diseases

Marcela Durán, Wagner J. Fávaro, German A. Islan, Guillermo R. Castro, and Nelson Durán

Abstract The study of neglected diseases is an important topic and deeply discussed in the newspapers, publications, and research foundations in the world. However, unfortunately no public or private attention has been paid on this issue. Still old drugs are being used, and very few are new for these diseases. Nanobiotechnology has appeared as a new strategy for the treatment of neglected diseases. The new developments in nanostructured carrier systems appear as promising in the treatment of many diseases with low toxicity, better efficacy and bioavailability, prolonged release of drugs, and reduction in the dosage of administration. This chapter is related to the use of nanobiotechnology in the treatment of neglected diseases by application of metallic nanoparticles on dengue virus, leishmaniosis, malaria, schistosomiasis, and trypanosomiasis.

Keywords Neglected diseases • Silver • Dengue virus • Leishmaniosis • Malaria • Schistosomiasis • Trypanosomiasis

---

M. Durán • W.J. Fávaro
Institute of Biology, Urogenital Carcinogenesis and Immunotherapy Laboratory, University of Campinas, Campinas, SP, Brazil

NanoBioss, Institute of Chemistry, University of Campinas, Campinas, SP, Brazil

G.A. Islan • G.R. Castro
Nanobiomaterials Laboratory, Applied Biotechnology Institute (CINDEFI, UNLP-CONICET CCT La Plata) Department of Chemistry School of Sciences, Universidad Nacional de La Plata, La Plata, Argentina

N. Durán (✉)
NanoBioss, Institute of Chemistry, University of Campinas, Campinas, SP, Brazil

Instituto of Chemistry, Biological Chemistry Laboratory, University of Campinas, Campinas, SP, Brazil

Brazilian Nanotechnology National Laboratory (LNNano-CNPEM), Campinas, SP, Brazil

Institute of Chemistry, University of Campinas, CP. 6152, CEP 13083-970, Campinas, SP, Brazil
e-mail: duran@iqm.unicamp.br

M. Rai, R. Shegokar (eds.), *Metal Nanoparticles in Pharma*,
DOI 10.1007/978-3-319-63790-7_3

3.1 Introduction

In general, the neglected diseases are getting slow but constant attention, than few years ago. Several reviews on this subject were published recently showing the importance of nanobiotechnology as a new approach to solve old problems in developing countries (Durán et.al. 2009, 2016a, b; Rai et al. 2014b; Rai and Kon 2015).

Metal nanoparticles play a pivotal role since they exhibit unique optoelectronic and physicochemical properties (Rai et al. 2014b). These properties depend on morphological aspects (e.g., shape, size, structure, crystallinity) (Duran et al. 2016b) and thus have led to a large range of applications in various areas such as electronics, molecular diagnostics, drug release, catalysis, and nanosensing (Rai et al. 2014a). Preparation methods for metal nanoparticles are diverse (e.g., physical and chemical methods). The use of biogenic synthesis of nanoparticles has drawn much attention, since they are green, efficient synthetic processes, and cost-effective procedure. There are many organisms able to synthesize nanoparticles, such as yeasts, bacteria, actinomycetes, fungi, and plants. In biomedicine, most importantly, it will play a crucial role in diagnostics, drug delivery, bandages, related cosmetics, etc. Although they are important in remediation through the absorption of pollutants, filtration, sterilization, etc., the most relevant example is the use of these nanoparticles as antimicrobial (Rai and Durán 2011). Silver nanoparticles are used as new generation of antimicrobials, with significant activity against many types of pathogens including multidrug-resistant organisms. Although there is interest in extensive applications, their possible toxicities must be studied (El-Nour et al. 2010; Durán et al. 2010, 2011a, b; Rai and Durán 2011; Rai et al. 2014a, b; Castro et al. 2014).

Nanobiotechnology is an important tool in order to develop new strategies for neglected diseases, which are of great importance in many countries in the world. This chapter will deal with role of silver nanoparticles in treatment of neglected diseases.

3.2 Neglected Diseases

3.2.1 Dengue

Dengue virus infection (DVI) exhibits a spectrum of illnesses from asymptomatic although in apparent infection, or a flu-like mild fever, to classic dengue fever (DF) or worst to DF with hemorrhagic consequences. Many other diseases or nonspecific viral syndrome can mimic DVI (Mungrue 2014). DVI is the wildest mosquito-borne infection which appeared on 2.5 billion people in many regions, including tropical and subtropical areas in the world. It is transmitted by female *Aedes aegypti* or *Aedes albopictus* to humans (Beatty et al. 2010) (Fig. 3.1). Unfortunately, the only treatment against dengue is the prevention and a supportive care; although some attempts were made, still now they are not proved to be efficient (Idrees and Ashfag 2013; Durán et al. 2016a).

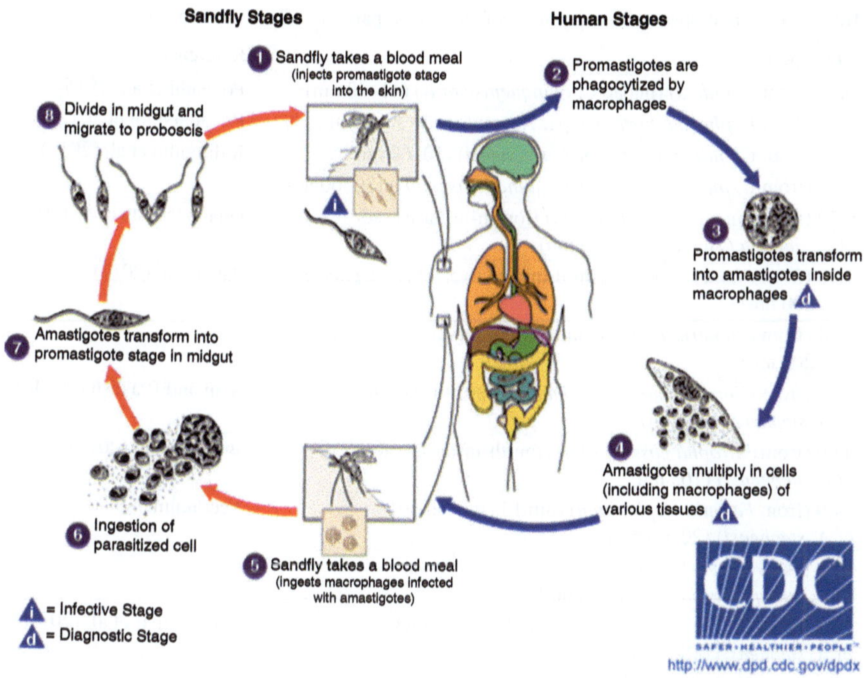

Fig. 3.1 Graphic panel of dengue virus infection (From http://factsanddetails.com/world/cat52/sub334/item1195.html)

Metallic nanoparticles have demonstrated efficacy against mosquito larvae (Salukhe et al. 2011; Soni and Prakash 2012a, b, c, d, 2013, 2014a). The leaf extracts from plants were used for silver nanoparticle (AgNP) production as an eco-friendly alternative for adulticidal activity against filarial, dengue, and malaria vector mosquitoes (Suganya et al. 2013; Veerakumar et al. 2014a, b).

Reviews on biogenic AgNPs against mosquitoes were recently published (Hajra and Mondal 2015; Rai and Kon 2015; Durán et al. 2016a, b), together with important results on biogenic AgNPs on biological systems (Durán et al. 2010; Gaikwad et al. 2013; Mashwani et al. 2015).

The efficacy of biogenic silver nanoparticles on *Aedes aegypti* and *Culex quinquefasciatus* demonstrated that the median lethal concentrations (LC_{50}) of AgNPs that killed fourth instars of *Aedes aegypti* and *Culex quinquefasciatus* were 0.30 and 0.41 μg mL^{-1}, respectively. Adult longevity (days) was reduced by 30% in male and female mosquitoes exposed as larvae to 0.1 μg mL^{-1}AgNPs, whereas the number of eggs laid by female larvae decreased in 36% when exposed to this concentration (Arjunan et al. 2012).

Related to mosquito larvae mortality with metallic nanoparticles, it has been found LC_{50} at the range of 0.56–606.5 μg mL^{-1} and also lower than those values cited on Table 3.1 (Hajra and Mondal 2015; Durán et al. 2016a). These large differences are probably due to synthesis of AgNPs with different kinds of capped pro-

Table 3.1 Lethal dose (LD_{50}) (μgmL^{-1}) of silver nanoparticles effective on mosquito

LD_{50} (μgmL^{-1})	Reference
0.05 (from *A. indica*) (pupae *C. quinquefasciatus*) (41–60 nm)	Poopathi et al. (2015)
34.5 (from *Euphorbia hirta*) (pupae *A. stephensi*) (30–60 nm)	Priyadarshini et al. (2012)
25.3 (from *C. indica*) (pupae of *A. stephensi*) (30–60 nm)	Kalimuthu et al. (2013)
23.8 (from *C. indica*) (pupae of *C. quinquefasciatus*) (30–60 nm)	
0.59 (from *Rhizophora mucronata*) (fourth-instar larvae of *C. quinquefasciatus*) (60–95 nm)	Gnanadesigan et al. (2011)
1.10 (from *Plumeria rubra*) (third-instar larvae of *A. stephensi*) (32–200 nm)	Patil et al. (2012)
1.74 (from *Plumeria rubra*) (fourth-instar larvae of *A. stephensi*) (32–200 nm)	
10.0 (from *Cinnamomum zeylanicum*) (fourth-instar larvae of *A. stephensi*) (12 nm)	Soni and Prakash (2014b)
4.90 (from *Jatropha gossypifolia*) (fourth-instar larvae of *A. stephensi*) (163 nm)	Borase et al. (2013)
54.9 (from *Feronia elephantum*) (third-instar larvae of *A. stephensi*) (20–60 nm)	Veerakumar et al. (2014b, c)
67.1 (from *Feronia elephantum*) (third-instar larvae of *C. quinquefasciatus*) (20–60 nm)	
26.7 (from *Heliotropium indicum*)(adult mosquitoes of *A. stephensi*) (18–45 nm)	Veerakumar et al. (2014a)
32.1 (from *Heliotropium indicum*) (adult mosquitoes of *C. quinquefasciatus*) (18–45 nm)	
21.9 (from *Sida acuta*) (late third-instar larvae of *A. stephensi*) (18–35 nm)	Veerakumar et al. (2013)
26.1 (from *Sida acuta*) (late third-instar larvae of *C. quinquefasciatus*) (18–45 nm)	
32.1 (from *Murraya koenigii*) (pupae of *A. stephensi*) (20–35 nm)	Suganya et al. (2013)
1.0 (from *Hibiscus rosasinensis*) (fourth-instar larvae of *Aedes albopictus*) (35 nm)	Sareen et al. (2012)

teins on them or due to the presence of different silver nanostructures (e.g., silver chloride or/and silver oxides nanoparticles) (Durán et al. 2016b).

Silver nanoparticles (AgNPs) were prepared from the latex of the plant *Euphorbia milii.* Latex-synthesized AgNPs were evaluated against the second- and fourth-instar larvae of *Aedes aegypti* and *Anopheles stephensi. E. milii* AgNPs showed a LC_{50} of 8.76±0.46 and 8.67±0.47 μgmL^{-1}, for second instars of *Ae. aegypti* and *An. stephensi*, respectively, showing similar activities to different mosquitoes (Borase et al. 2014) (Table 3.2).

3.2.2 *Leishmaniasis*

Leishmaniases are vector-borne zoonotic diseases caused by various species of the genus *Leishmania* (protozoa). These pathogens are transmitted by sandflies (e.g., phlebotomine) and infect humans where the vectors and reservoirs coexist (Fig. 3.2).

Table 3.2 Mortality of *Aedes aegypti* with biogenic silver nanoparticles

Plants used for synthesis	Life stages	Size (nm)	LD_{50} (μg mL^{-1})	Ref.
Murraya koenigii	Instar I	20–35	10.8	Suganyaet al. (2013)
	Instar II		14,7	
	Instar III		53.7	
	Instar IV		63.6	
	Pupa		75.3	
Feronia elephantum	Adult (3–4 days)	18–45	20.4	Veerakumaret al. (2014a, b, c)
				Veerakumar and Govindarajan (2014)
Azadirachta indica	Instar III	41–60	0.006	Poopathi et al. (2015)
Bacillus thuringiensis (*Bacteria*)	Instar III	44–143	0.14	Banu et al. (2014)
Rhizophora mucronata	Instar IV	60–95	0.89	Gnanadesigan et al. (2011)
Plumeria rubra	Instar II	32–220	181,7	Patil et al. (2012)
	Instar IV		287.5	
Pedilanthus tithymaloides	Instar I	15–30	0.029	Sundaravadivelan et al. (2013)
	Instar II		0.027	
	Instar III		0.047	
	Instar IV		0.086	
	Pupa		0.018	
Jatropha gossypifolia	Instar II	30–60	5.9	Borase et al. (2013)
	Instar IV		4.44	
Euphorbia milii (*Latex*)	Instar II	208	8.76	Borase et al. (2014)
Sida acuta	Instar IV	18–35	23.9	Veerakumar et al. (2013)

Modified from Durán et al. (2016a)

Anthroponotic cycles have been documented for some species of *Leishmania* (e.g., *Leishmania donovani* in India, *Leishmania major* in Afghanistan). Visceral leishmaniasis (VL) is caused by *L. donovani* (Indian and East Africa) and by *L. infantum* or *L. chagasi* (e.g., Asia, Europe, Africa, and the New World). Tegumentary leishmaniasis (TL) is caused by many species of parasites in Europe (e.g., *L. major*, *L. tropica*, *L. aethiopica*, and sometimes *L. infantum*), in America (e.g., *L. (Viannia) braziliensis*, *L. amazonensis*, *L. (V.) guyanensis*, *L. (V.) panamensis*, *L. mexicana*, *L. pifanoi*, *L. venezuelensis*, *L. (V.) peruviana*, *L. (V.) shawi*, and *L. (V.) lainsoni*), in Mexico, Argentina, and Brazil (e.g., subgenus *Viannia* and *L. amazonensis*), and in Mexico and Central American countries (e.g., *L. mexicana*) (Lindoso et al. 2012). The recent treatment for VL involved miltefosine and paromomycin. These compounds were evaluated only few times and should be evaluated in different epidemiological scenarios (Marinho et al. 2015).

AgNPs as an alternative therapy for leishmaniasis are effective, specifically by subcutaneous intralesional administration for cutaneous leishmaniasis (CL). AgNPs, as discussed in many publications, can be prepared by chemical, physical, or

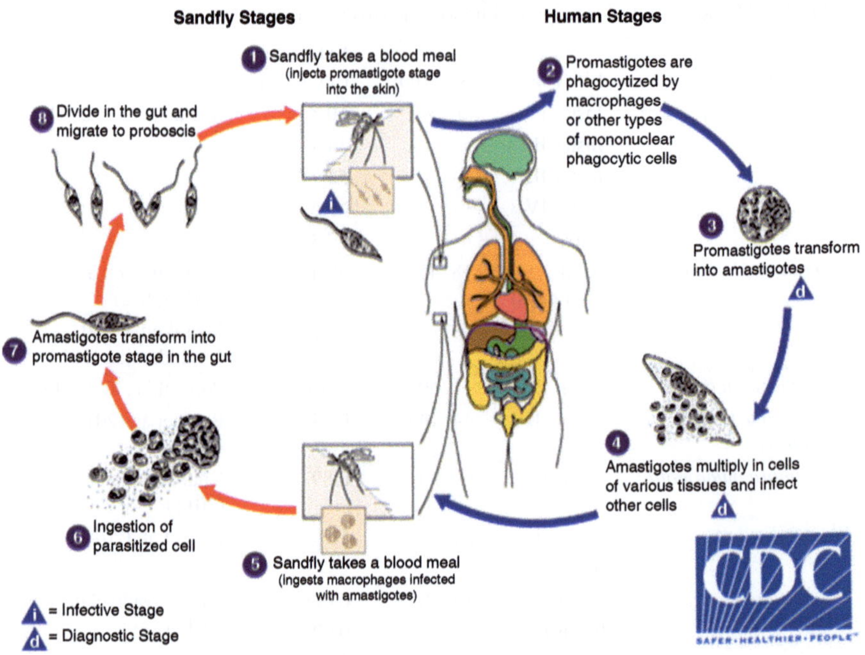

Fig. 3.2 Graphic panel of leishmaniasis (From https://www.cdc.gov/dpdx/leishmaniasis/)

biological procedures (Durán et al. 2011a, b). Besides these, biosynthetic methods generate more effective NPs in medical applications because of their protein-coated surface (Prasad et al. 2011). In addition, both chemically and biologically synthesized NPs were studied first by in vitro experiments against *Leishmania amazonensis* promastigotes. Biologically generated AgNPs (Bio-AgNPs) were shown to be threefold more effective than chemically generated ones (Chem-AgNPs). Later, in vivo studies in infected mice demonstrated that Bio-AgNPs dose showed the same effectiveness than 300-fold higher doses of amphotericin B and also threefold higher doses of Chem-AgNP. Important results such as no hepato- and nephrotoxicity were found in comparison with amphotericin B and Chem-AgNPs (Rossi-Bergmann et al. 2012).

The viability of *L. tropica* promastigotes after 72 h recorded maximum cytotoxic effect of AgNPs (no size was described) at a concentration of 2.1 μg/mL with an IC50 of 1.749 μg/mL, and from *L. tropica* amastigote phase the IC50 was 1.148 μg/mL (Gharby et al. 2017).

3.2.3 Malaria

Plasmodium species *P. malariae*, *P. knowlesi*, *P. ovale*, *P. falciparum*, and *P. vivax* infect human with malaria (Fig. 3.3). A decrease of 42% in malaria death was achieved due to many efforts to control and eradicate malaria through insecticides

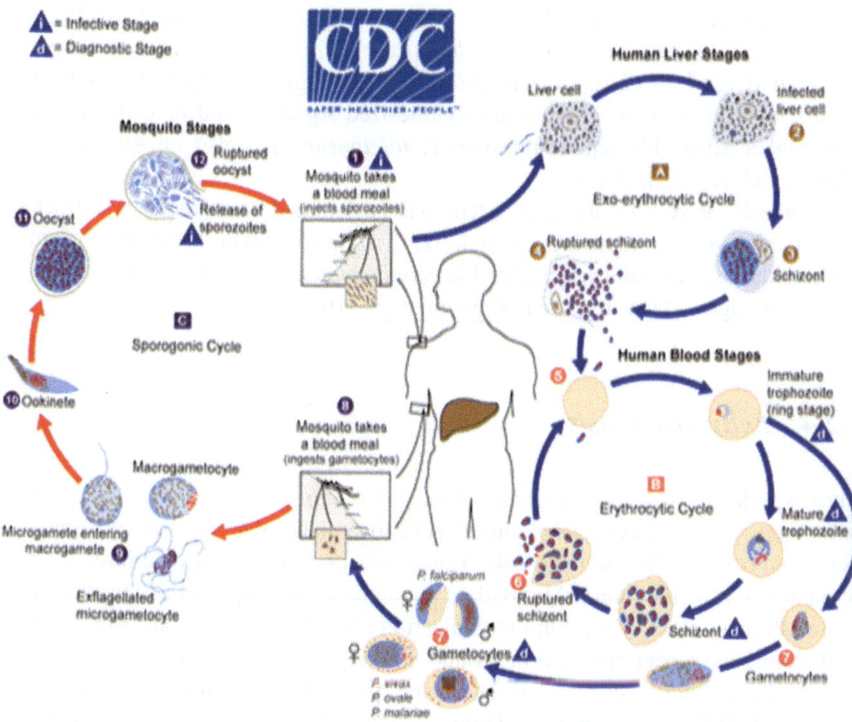

Fig. 3.3 Graphic panel for malaria (From https://www.cdc.gov/malaria/about/biology)

and antimalarial treatments (e.g., artemisinin-combined therapies). However, one of the challenges is the increasing drug resistance, and no effective malaria vaccine exists today. One malaria vaccine in phase III is under testing by GSK (Glaxo Smith KlineRTS,S/AS01 vaccine), but its vaccine efficacy is around 30% (Siu and Ploss 2015).

AgNPs produced from aqueous extracts of leaves and bark of *Azadirachta indica* (neem) (10.5 nm in leaf and 19.2 nm in bark) showed activities against mosquito (e.g., larvicides, pupicides, and adulticides) and against the malaria vector *Anopheles stephensi* (*An. stephensi*) and filariasis vector *Culex quinquefasciatus* at different doses. The larvae, pupae, and adults of *C. quinquefasciatus* were found to be more susceptible to AgNPs than *An. stephensi*. The I and the II instar larvae of *C. quinquefasciatus* show a mortality rate of 100% after 30 min of exposure. The results against the pupa of *C. quinquefasciatus* were recorded as LC_{50} 4 μg mL^{-1} (3 h exposure). In the case of adult mosquitoes, LC_{50} 1.06 μL/cm^2 was obtained (4 h exposure). The authors suggested that biogenic AgNPs were environment friendly for controlling malarial and filarial vectors (Soni and Prakash 2014a).

AgNP synthesis using plant extract of *Pteridium aquilinum*, acting as a reducing and capping agent, showed that AgNP (10 × LC_{50}) led to 100% larval reduction after 72 h and reduced longevity and fecundity of *An. stephensi* adults. Furthermore, the antiplasmodial activity of AgNPs was evaluated against CQ-resistant (CQ-r) and CQ-sensitive (CQ-s) strains of *P. falciparum*. *P. aquilinum*-synthesized AgNPs achieved IC_{50} of 78.12 μgmL^{-1} (CQ-s) and 88.34 μgmL^{-1} (CQ-r). Overall, their results highlighted that *P. aquilinum*-synthesized AgNPs could be candidate as a new tool against chloroquine-resistant *P. falciparum* and also on *An. stephensi* (Panneerselvam et al. 2016).

Synthesis of AgNPs using β-caryophyllene isolated from the leaf extract of *Murraya koenigii*, as reducing and stabilizing agent (5–100 nm), exhibited promising activity on chloroquine-sensitive *Plasmodium falciparum* (3D7) with an IC50 of 2.34 ± 0.07 μg/mL) was reported (Kamaraj et al. 2017).

3.2.4 *Schistosomiasis*

Three species of parasitic flatworms of the genus *Schistosoma* (*S. mansoni*, *S. haematobium*, and *S. japonicum*) caused schistosomiasis that is also considered a neglected tropical disease (Fig. 3.4). These parasites cause a chronic infection and often debilitating the infected individual that impairs development and productivity. In an estimate, the World Health Organization (WHO) indicated that over 250 million people have been infected in around 80 endemic countries (e.g., sub-Saharan Africa, the Middle East, the Caribbean, and South America) resulting in approximately 200,000 deaths annually. Unfortunately, praziquantel (PZQ) is the actual product used due to the absence of an effective vaccine. PZQ offers oral administration, high efficacy, tolerability, low transient side effects, and a low cost. However, PZQ resistance is actually detected (Neves et al. 2015).

Another strategy for controlling schistosomiasis is combating the vector *Biomphalaria glabrata* (mollusk) through the use of AgNPs as a molluscicidal with low toxicity to other aquatic organisms (Yang et al. 2011; Guang et al. 2013).

3.2.5 *Trypanosomiasis*

Human African trypanosomiasis (HAT) caused by infection with the parasite *Trypanosoma brucei gambiense* or *T. b. rhodesiense* and its vector is tsetse fly (Fig. 3.5). Around 70 million people worldwide were at risk of infection, and probably over 20,000 people in Africa are infected with HAT (Nagle et al. 2014; Sutherland et al. 2015).

American trypanosomiasis or Chagas disease is caused by the protozoan parasite *Trypanosoma cruzi*. This disease is endemic in 21 Latin American countries, with a strong economic impact because it affects economically active people. Over ten million people are infected, and over 25 million people are within the endemic countries. After many years of infection, 10–30% of infected people develop symptoms of chronic phase. In general, the effect on the heart is the most common organ

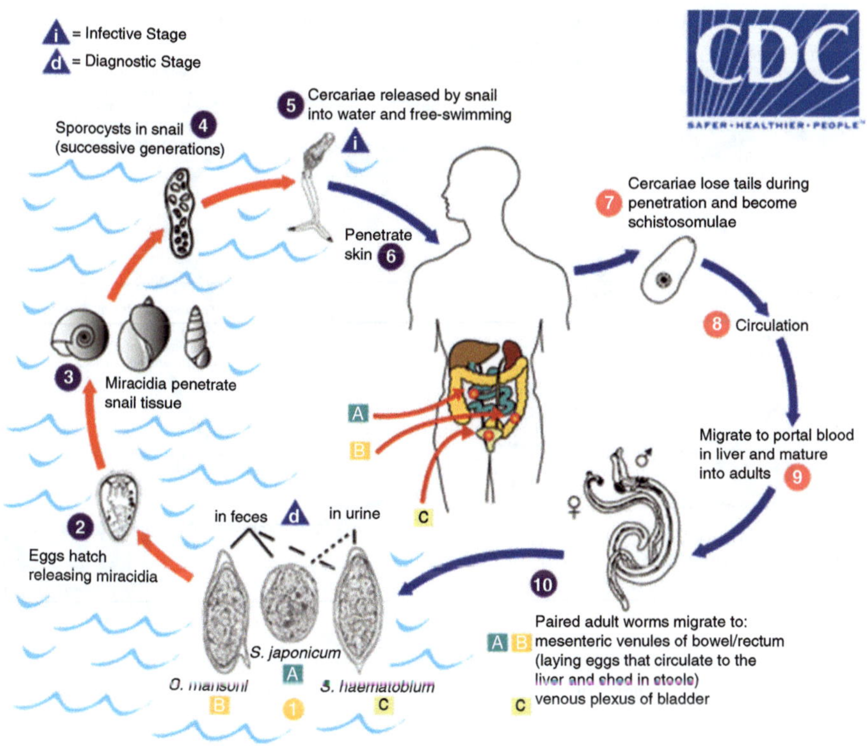

Fig. 3.4 Graphic panel of schistosomiasis (From https://www.cdc.gov/dpdx/schistosomiasis/)

problem; symptoms include cardiomyopathy and thromboembolism. The patient's death usually occurs from heart failure (Rodrigues-Morales et al. 2015).

Unfortunately, very few compounds are in development, and drug discovery efforts are limited. Nifurtimox and SCYX-7158 are the only two compounds in clinical trials for HAT showing the need for novel chemical entities or new strategies (Nagle et al. 2014).

Against Chagas disease, in human trials, are nifurtimox and benznidazole. Benznidazole is still being used in Brazil (Pereira and Navarro 2013). Unfortunately, limited human data and better supported by the findings in animal models suggest that *T. cruzi* strains may vary in their drug susceptibility (Bern 2015).

The enzyme arginine kinase (AK) is absent in humans, and important for the Trypanosoma development, fact that makes it an attractive target choice for a trypanocide development. Adeyemi and Whiteley (2014) performed a thermodynamic and spectrofluorimetric study on the interaction of metal nanoparticles (i.e., AuNPs and AgNPs) with AK. AgNPs and AuNPs bound tightly to the arginine substrate through a sulfur atom of a cysteine residue (Cys^{271}). This interaction controls the electrophilic and nucleophilic profile of the substrate arginine-guanidinium group, absolutely important for enzyme phosphoryl transfer from ATP to *Trypanosoma*.

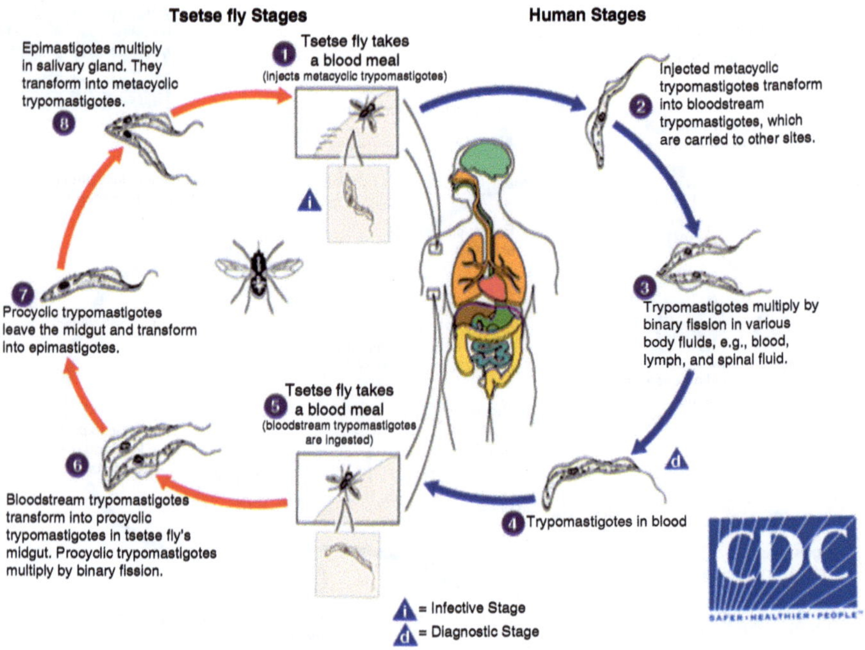

Fig. 3.5 Graphic panel of trypanosomiasis (From https://www.cdc.gov/dpdx/trypanosomiasisafrican/)

3.3 Conclusion

Actually, there are a few recently published reviews focusing on dengue virus, leishmaniasis, malaria, schistosomiasis, and trypanosomiasis with alternative strategies. The present review pointed out the most important advances in the metallic nanoparticles action on these diseases. Silver nanoparticles appeared as a possible alternative in the therapy of many of these diseases and also on vectors with low toxicity and with enhanced efficacy. This revision dealt with the current status of nanobiotechnology through silver nanoparticles acting on neglected diseases. Therefore, it is clear that it is possible to use nanotechnology to manage safety to these humans' diseases.

Acknowledgments Support by FAPESP, INOMAT (MCTI/CNPq), NanoBioss (MCTI), Brazilian Network on Nanotoxicology (MCTI/CNPq) is acknowledged.

References

Adeyemi OS, Whiteley CG. Interaction of metal nanoparticles with recombinant arginine kinase from *Trypanosoma brucei*: thermodynamic and spectrofluorimetric evaluation. Biochim Biophys Acta. 2014;1840:701–6.

Arjunan NK, Murugan K, Rejeeth C, Madhiyazhagan P, Barnard DR. Green synthesis of silver nanoparticles for the control of mosquito vectors of malaria, filariasis, and dengue. Vector Borne Zoonotic Dis. 2012;12:262–8.

Banu AN, Balasubramanian C, Moorthi PV. Biosynthesis of silver nanoparticles using *Bacillus thuringiensis* against dengue vector, *Aedes aegypti* (Diptera: Culicidae). Parasitol Res. 2014;113:311–6.

Beatty ME, Stone A, Fitzsimons DW, Hanna JN, Lam SK, Vong S, Guzman MG, Mendez-Galvan JF, Halstead SB, Letson GW, Kuritsky J, Mahoney R, Margolis HS. Best practices in dengue surveillance: a report from the Asia-Pacific and Americas Dengue prevention boards. PloS Neglected Trop Dis. 2010;4:e890.

Bern, G.. http://www.uptodate.com/contents/chagas-disease-antitrypanosomal-drug-therapy. Accessed on 12 July 2015.

Borase HP, Patil D, Salunkhe RB, Narkhede CP, Salunke BK, Patil SV. Phyto-synthesized silver nanoparticles: a potent mosquito biolarvicidal agent. J Nanomed Biother Disc. 2013;3:1.

Borase HP, Patil CD, Salunkhe RB, Narkhede CP, Suryawanshi RK, Salunke BK, Patil SV. Mosquito larvicidal and silver nanoparticles synthesis potential of plant latex. J Entomol Acaralological Res. 2014;46:59–65.

Castro L, Belázquez ML, Muñoz JA, González FG, Ballester A. Mechanism and applications of metal nanoparticles prepared by bio-mediated process. Rev Adv Sci Eng. 2014;3:1–18.

Durán N, Marcato PD, Teixeira Z, Durán M, Costa FTM, Brocchi M. State of art of nanobiotechnology applications in neglected diseases. Curr Nanosci. 2009;5:396–408.

Durán N, Marcato PD, De Conti R, Alves OL, Costa FTM, Brocchi M. Potential use of silver nanoparticles on pathogenic bacteria, their toxicity and possible mechanisms of action. J Braz Chem Soc. 2010;21:949–59.

Durán N, Marcato PD, De Conti R, Alves OL, Costa FTM, Brocchi M. Potential use of silver nanoparticles on pathogenic bacteria, their toxicity and possible mechanisms of action. J Braz Chem Soc. 2011a;21:949–59.

Durán N, Marcato PD, Durán M, Yadav A, Gade A, Rai M. Mechanistic aspects in the biogenic synthesis of extracellular metal nanoparticles by peptides, bacteria, fungi and plants. Appl Microbiol Biotechnol. 2011b;90:1609–24.

Durán N, Islan GA, Durán M, Castro GR. Nanotechnology solutions against *Aedes aegypti*. J Braz Chem Soc. 2016a;27:1139–49.

Durán N, Durán M, de Jesus MB, Fávaro WJ, Nakazato G, Seabra AB. Silver nanoparticles: a new view on mechanistic aspects on antimicrobial activity. Nanomedicine: NBM. 2016b;12:789–99.

El-Nour KMMA, Eftaiha A, Al-Warthan A, Ammar RAA. Synthesis and applications of silver nanoparticles. Arab J Chem. 2010;3:135–40.

Gaikwad SC, Birla SS, Ingle AP, Gade AK, Marcato PD, Rai M, Durán N. Screening of different *Fusarium* species to select potential species for the synthesis of silver nanoparticles. J Braz Chem Soc. 2013;24:1974–82.

Gharby MA, AL-Qadhi BN, Jaafar SM. Evaluation of silver nanoparticles (Ag NPs) activity against the viability of *Leishmania tropica* promastigotes and amastigotes in vitro. Iraqi J Sci. 2017;58:13–21.

Gnanadesigan M, Anand M, Ravikumar S, Maruthupandy M, Vijayakumar V, Selvam S, Dhineshkumar M, Kumaraguru AK. Biosynthesis of silver nanoparticles by using mangrove plant extract and their potential mosquito larvicidal property. Asian Pac J Trop Med. 2011;4:799–803.

Guang XY, Wang JJ, He ZG, Chen GX, Ding L, Dai JJ, Yang XH. Molluscicidal effects of nanosilver biological molluscicide and niclosamide. Chin J Schistosomiasis Control. 2013;25:503–5.

Hajra A, Mondal NK. Silver nanoparticles: an eco-friendly approach for mosquito control. Int J Sci Res Environ Sci. 2015;3:47–61.

Idrees S, Ashfaq UA. RNAi: antiviral therapy against dengue vírus. Asian Pac J Trop Biomed. 2013;3:232–6.

Kalimuthu K, Panneerselvam C, Murugan K, Hwang J-S. Green synthesis of silver nanoparticles using Cadaba indica lam leaf extract and its larvicidal and pupicidal activity against Anopheles stephensi and Culex quinquefasciatus. J Entomol Acaralological Res. 2013;45:e11 57–64.

Kamaraj C, Balasubramani G, Siva C, Raja M, Balasubramanian V, Raja RK, Tamilselvan S, Benelli G, Perumal P. Ag nanoparticles synthesized using β-caryophyllene isolated from

Murraya koenigii: antimalarial (Plasmodium falciparum 3D7) and anticancer activity (A549 and HeLa cell lines). J Clust Sci. 2017, 2017; doi:10.1007/s10876-017-1180-6.
Lindoso JAL, Costa JML, Queiroz IT, Goto H. Review of the current treatments for leishmaniases. Res Rep Trop Med. 2012;3:69–77.
Marinho DS, Casas CNPR, Pereira CCA, Leite IC. Health economic evaluations of visceral leishmaniasis treatments: a systematic review. PLoS Negl Trop Dis. 2015;9:e0003527.
Mashwani Z, Khan T, Khan MA, Nadhman A. Synthesis in plants and plant extracts of silver nanoparticles with potent antimicrobial properties: current status and future prospects. Appl Microbiol Biotechnol. 2015;99:9923–34.
Mungrue K. The laboratory diagnosis of dengue virus infection, a review. Adv Lab Med Int. 2014;4:1–8.
Nagle AS, Khare S, Kumar AB, Supek F, Buchynskyy A, Mathison CJN, Chennamaneni NK, Pendem N, Buckner FS, Gelb MH, Molten V. Recent developments in drug discovery for leishmaniasis and human African Trypanosomiasis. Chem Rev. 2014;114:11305–47.
Neves BJ, Andrade CH, Cravo PVL. Natural products as leads in schistosome drug discovery. Molecules. 2015;20:1872–903.
Panneerselvam C, Murugan K, Roni M, Aziz AT, Suresh U, Rajaganesh R, Madhiyazhagan P, Subramaniam J, Dinesh D, Nicoletti M, Higuchi A, Alarfaj AA, Munusamy MA, Kumar S, Desneux N, Benelli G. Fern-synthesized nanoparticles in the fight against malaria: LC/MS analysis of *Pteridium aquilinum* leaf extract and biosynthesis of silver nanoparticles with high mosquitocidal and antiplasmodial activity. Parasitol Res. 2016;115:997–1013.
Patil CD, Patil SV, Borase HP, Salunke BK, Salunkhe RB. Larvicidal activity of silver nanoparticles synthesized using Plumeria rubra plant latex against *Aedes aegypti* and *Anopheles stephensi*. Parasitol Res. 2012;110:1815–22.
Pereira PCM, Navarro EC. Challenges and perspectives of Chagas disease: a review. J Venom Anim Toxins Trop Dis. 2013;19:34–5.
Poopathi S, De Britto LJ, Praba VL, Mani C, Praveen M. Synthesis of silver nanoparticles from *Azadirachta indica*-A most effective method for mosquito control. Environ Sci Pollut Res. 2015;22:2956–63.
Prasad TNVKV, Kambala VSR, Naidu R. A critical review on biogenic silver nanoparticles and their antimicrobial activity. Curr Nanosci. 2011;4:531–44.
Priyadarshini KA, Murugan K, Panneerselvam C, Ponarulselvam S, Hwang JS, Nicoletti M. Biolarvicidal and pupicidal potential of silver nanoparticles synthesized using *Euphorbia hirta* against *Anopheles stephensi* Liston (Diptera: Culicidae). Parasitol Res. 2012;111:997–1006.
Rai M, Durán N, editors. Metal nanoparticles in microbiology. Germany: Springer; 2011. p. 330.
Rai M, Kon K. Silver nanoparticles for the control of vector-borne infections. In: Rai M, Kon K, editors. Nanotechnology in diagnosis, treatment and prophylaxis of infectious diseases. London: Elsevier; 2015. ch. 3.
Rai M, Kon K, Ingle A, Durán N, Galdiero S, Galdiero M. Broad-spectrum bioactivities of silver nanoparticles: the emerging trends and future prospects. Appl Microbiol Biotechnol. 2014a;98:1951–61.
Rai M, Birla S, Ingle A, Gupta I, Gade A, Abd-Elsalam K, Marcato PD, Durán N. Nanosilver: an inorganic nanoparticle with myriad potential applications. Nanotechnol Rev. 2014b;3:281–309.
Rodríguez-Morales OR, Monteón Padilla VN, Carrillo-Sánchez SC, Rios-Castro M, Martínez-Cruz M, Carabarin-Lima A, Arce-Fonseca M. Experimental vaccines against Chagas disease: a journey through history. J Immunol Res. 2015;2015:489758.
Rossi-Bergmann B, Pacienza-Lima W, Marcato PD, De Conti R, Durán N. Therapeutic potential of biogenic silver nanoparticles in murine cutaneous leishmaniasis. J Nano Res. 2012;20:89–97.
Salunkhe RB, Patil SV, Patil CD, Salunke BK. Larvicidal potential of silver nanoparticles synthesized using fungus *Cochliobolus lunatus* against *Aedes aegypti* (Linnaeus, 1762) and *Anopheles stephensi* Liston (Diptera; Culicidae). Parasitol Res. 2011;109:823–31.
Sareen SJ, Pillai RK, Chandramohanakumar N, Balagopalan M. Larvicidal potential of biologically synthesized silver nanoparticles against *Aedes albopictus*. Res J Recent Sci. 2012;1:52–6.

Siu E, Ploss A. Modeling malaria in humanized mice. Ann N Y Acad Sci. 2015;1342:29–36.
Soni N, Prakash S. Efficacy of fungus mediated silver and gold nanoparticles against *Aedes aegypti* larvae. Parasitol Res. 2012a;110:175–84.
Soni N, Prakash S. Synthesis of gold nanoparticles by the fungus *Aspergillus niger* and its efficacy against mosquito larvae. Rep Parasitol. 2012b;2:1–7.
Soni N, Prakash S. Fungal-mediated nano silver: an effective adulticide against mosquito. Parasitol Res. 2012c;111:2091–8.
Soni N, Prakash S. Entomopathogenic fungus generated nanoparticles for enhancement of efficacy in *Culex quinquefasciatus* and *Anopheles stephensi*. Asian Pac J Trop Dis. 2012d;2:S356–61.
Soni N, Prakash S. Possible mosquito control by silver nanoparticles synthesized by soil fungus (*Aspergillus niger* 2587). Adv Nanopart. 2013;2:125–32.
Soni N, Prakash S. Silver nanoparticles: a possibility for malarial and filarial vector control technology. Parasitol Res. 2014a;113:4015–22.
Soni N, Prakash S. Green nanoparticles for mosquito control. Sci World J. 2014b;2014:Article ID 496362.
Suganya A, Murugan K, Kovendan K, Kumar PM, Hwang JS. Green synthesis of silver nanoparticles using *Murraya koenigii* leaf extract against *Anopheles stephensi* and *Aedes aegypti*. Parasitol Res. 2013;112:1385–97.
Sundaravadivelan C, Padmanabhan MN, Sivaprasath P, Kishmu L. Biosynthesized silver nanoparticles from *Pedilanthusti thymaloides* leaf extract with anti-developmental activity against larval instars of *Aedes aegypti* L. (Diptera; Culicidae). Parasitol Res. 2013;12:303–11.
Sutherland CS, Yukich J, Goeree R, Tediosi FA. Literature review of economic evaluations for a neglected tropical disease: human African trypanosomiasis ("Sleeping Sickness"). PLoS Negl Trop Dis. 2015;9:e0003397.
Veerakumar K, Govindarajan M. Adulticidal properties of synthesized silver nanoparticles using leaf extracts of *Feronia elephantum* (Rutaceae) against filariasis, malaria, and dengue vector mosquitoes. Parasitol Res. 2014;13:4085–96.
Veerakumar K, Govindarajan M, Rajeswary M. Green synthesis of silver nanoparticles using Sida acuta (Malvaceae) leaf extract against *Culex quinquefasciatus*, *Anopheles stephensi*, and *Aedes aegypti* (Diptera: Culicidae). Parasitol Res. 2013;112:4073–85.
Veerakumar K, Govindarajan M, Hoti SL. Evaluation of plant-mediated synthesized silver nanoparticles against vector mosquitoes. Parasitol Res. 2014a;113:4567–677.
Veerakumar K, Govindarajan M, Rajeswary M, Muthukumaran U. Low-cost and eco-friendly green synthesis of silver nanoparticles using *Feronia elephantum* (Rutaceae) against *Culex quinquefasciatus*, *Anopheles stephensi*, and *Aedes aegypti* (Diptera: Culicidae). Parasitol Res. 2014b;113:1775–85.
Veerakumar K, Govindarajan M, Rajeswary M, Muthukumaram U. Mosquito larvicidal properties of silver nanoparticles synthesized using *Heliotropium indicum* (Boraginaceae) against *Aedes aegypti*, *Anopheles stephensi* and *Culex quinquefasciatus* (Diptera:Culicidae). Parasitol Res. 2014c;113:2663–73.
Yang XH, Wang JJ, Ge JY, Wang JS, Dong FQ. Controlled synthesis of water soluble silver nanoparticles and the molluscicidal effect. Adv Mater Res. 2011;399–401:527–31.

Chapter 4
Encapsulation and Application of Metal Nanoparticles in Pharma

Anisha D'Souza and Ranjita Shegokar

Abstract Recent years have witnessed an extraordinary increase in number of research studies on understanding fundamental and applied uses of metal nanoparticles. In all these studies, various aspects of metal nanoparticles like type of metal, nanoparticle size, and shape of nanoparticle and shape-dependent properties are studied. Metal nanoparticles are being applied in a variety of medical and nonmedical areas. To confer multifunctionality and enhance safety, these metal nanoparticles are often encapsulated using polymers or lipids. This chapter overviews metal nanoparticles, their encapsulation and applications in pharmaceuticals.

Keywords Toxicity of metal nanoparticles • Encapsulation • Coating iron nanoparticles • Coating gold nanoparticles • Coating silver nanoparticles • Dendrimer-encapsulated • Metal nanoparticles, surface modification

4.1 Introduction

Although the word "nano" is buzzed lately its existence was well known centuries ago. Colloidal gold had been used as a traditional alchemy preparation (Weissig et al. 2014). The buzz started during 1970s with the first approval of Doxil by USFDA in 1995. Till then, colloidal systems were not as famous as "nano." Nano indicates any dimensions between approximately 1 and 100 nm exhibiting unique properties which bulk material or single atoms or molecules cannot (NNI frequently asked questions). Gold nanoparticles exhibit different colors in different sizes, which bulk cannot. The change in surface plasmon resonance caused by change in size and shape of the metal is responsible for it.

A. D'Souza
Department of Biosciences and Bioengineering, Indian Institute of Technology, Mumbai, India

R. Shegokar (✉)
Institute of Pharmacy, Department Pharmaceutics, Biopharmaceutics & NutriCosmetics, Freie Universität Berlin, Kelchstraße 31, 12169 Berlin, Germany
e-mail: ranjita@arcslive.com

M. Rai, R. Shegokar (eds.), *Metal Nanoparticles in Pharma*,
DOI 10.1007/978-3-319-63790-7_4

Nanoparticles thus exhibit properties different from the atoms/molecules or bulk of the material. Nanopharmaceuticals are produced by nanoengineering process with additional and unique therapeutic activity than the active drug. Nanotechnology does not end with nanoengineering but begins with understanding the unique physicochemical properties, optical properties, and mechanisms at this scale (Marty et al. 1978). The concept of liposomes, PEGylated proteins, aptamers, polymeric nanoparticles, and other organic particles has been widely explored; however, the importance of nonphysiological inorganic particles and metal and metal oxide nanoparticles cannot be neglected. These particles also play a pivotal role in therapy, biomarkers, devices, and biosensors, and the interest continues to expand.

Metallic nanoparticles are nanosized metal containing inorganic particles. Their nanoscale size confer unique quantum effects on their physical, chemical, and biological properties, resulting in different electronic and optical characteristics. They can be functionalized with various molecules like nucleic acids, genetic material, enzymes, ligands, and antibodies with wide applicability in drug delivery, diagnosis, and imaging as magnetic resonance imaging, fluorescent, radioisotope, or photoluminescence tracking with multimodal imaging (Cole 2011). Additionally, they have been studied for treatment in cancer. Their ability to form reactive oxygen species, exhibit hyperthermia, and photothermal effect increases their cytotoxicity against cancer cells. Metals can be easily excreted from the body (Sharma et al. 2015). They could range from gold nanoparticles to silver or iron nanoparticles. These noble metals have been in contact of human body for a very long time. Silver had been used as an antimicrobial in various critical in vivo devices like catheters, prostheses, etc. Gold has been reported effective against arthritis and an inflammation (Brown et al. 2007; Leonaviciene et al. 2012). Moreover, iron is well known for its magnetic resonance imaging as well as for thermal ablation during cancer (Pranatharthiharan et al. 2013). These metals have also been reported for synergistic activity along with existing antimicrobials, antibiotics, and many other therapies (Kumar et al. 2002).

Metals though have a magnificent bioactivity; however, they meet with limitations of toxicity and incompatibility. Masking with biocompatible polymers or lipids decreases the bio-adversities and toxicities associated with unmasked nanoparticles.

The current chapter discusses in detail the different encapsulating material commonly reported for encapsulating popularly known metal nanoparticles, namely, iron nanoparticles, gold nanoparticles, and silver nanoparticles. The authors also discussed the realm of commercialization of such encapsulated metal nanoparticles.

4.2 Need for Encapsulation of Metal Nanoparticles

Excessive stress caused by the reactive oxygen species is harmful and generates oxidative stress in enzymatic and defense mechanism of body. Once the stress increases beyond a level, the cellular redox equilibrium is disturbed resulting in irreparable tissue damage, necrosis, and apoptosis. Pro-inflammatory responses mainly tumor necrosis factor and interleukin (IL-2) are activated. If uncontrolled, it may lead to pulmonary diseases, atherosclerosis, and cancers. Disturbance in intracellular calcium ion concentration disturbs and inhibits the calcium influx and cellular signaling. Designing site-specific targeting mechanisms or the use of biodegradable coats on the surface of the nanoparticles can decrease their toxicity (Sharma et al. 2015).

Silver nanoparticles have been studied for their toxicity in capped and naked forms. Bare silver nanoparticles and polyvinylpyrrolidone (PVP)-decorated silver nanoparticles sized 8 and 38 nm on nematode *Caenorhabditis elegans* were studied. Silver nitrate and silver nanoparticles exhibited similar toxicities, while PVP-coated silver nanoparticles and silver nanoparticles sized 8 nm induced deoxyribonucleic acid (DNA) damage and did not form polymerase which inhibits DNA lesions in mitochondria and nucleus (Ahn et al. 2014). Bioavailability of silver nanoparticles is depend on hardness of water in aquatic animals. But PVP-loaded silver nanoparticles were ineffective of hardness of water when studied on *Lymnaea stagnalis*, fresh water snail (Oliver et al. 2014). Coating of silver nanoparticles and their size is thus highly influential in deciding the toxicity of the nanoparticles.

Gold nanoparticles of average size 18 nm did not show any toxicity in leukemia cell line (Connor et al. 2005). It did not exhibit toxicity upon crossing the blood-brain barrier and human prostate cancer cell line (Arnida et al. 2010). Gold nanoparticles with negative charge like polyallylamine hydrochloride-capped gold nanoparticles, citrate-stabilized gold nanoparticles, and mercaptopropionic acid-stabilized gold nanoparticles on *Daphnia magna* are less toxic than positively charged gold nanoparticles (Bozich et al. 2014). Surface capping also influences the toxicity of gold nanoparticles (Rai et al. 2015).

4.3 Encapsulation of Metal Nanoparticles

Functionalization of metal nanoparticles on the surface is a common technique to improve biocompatibility and could vary from ionic ligands like citrate or phosphates (Nguyen and Luke 2010). However, being charged molecules, they tend to aggregate. Hemocompatibility increases and cytotoxicity can be decreased upon functionalization with polymers like polyethylene glycol (PEG) (D'souza and Shegokar 2016). For instance, iron oxide (Fe_2O_3) nanoparticles deposited on carbon nanotubes. Some other coating examples on iron/magnetic nanoparticles are depicted in Fig. 4.1. Cisplatin was conjugated to block copolymers like poly(citric acid)-PEG-poly(citric acid). The prodrug could easily penetrate through cell membrane with high drug loading, improved water solubility, and biocompatibility. The iron nanoparticles could facilitate targeting to the tumor cells. Metal nanoparticles can be hybridized with dendrimers, polymers, lipids, etc.

Metal nanoparticles can be tagged with biopolymers by (a) covalent, i.e., either chemical bonding or by grafting, and (b) noncovalent, i.e., physically linked by van der Waals, hydrogen bonding, and ionic interaction. Grafting indicates building up

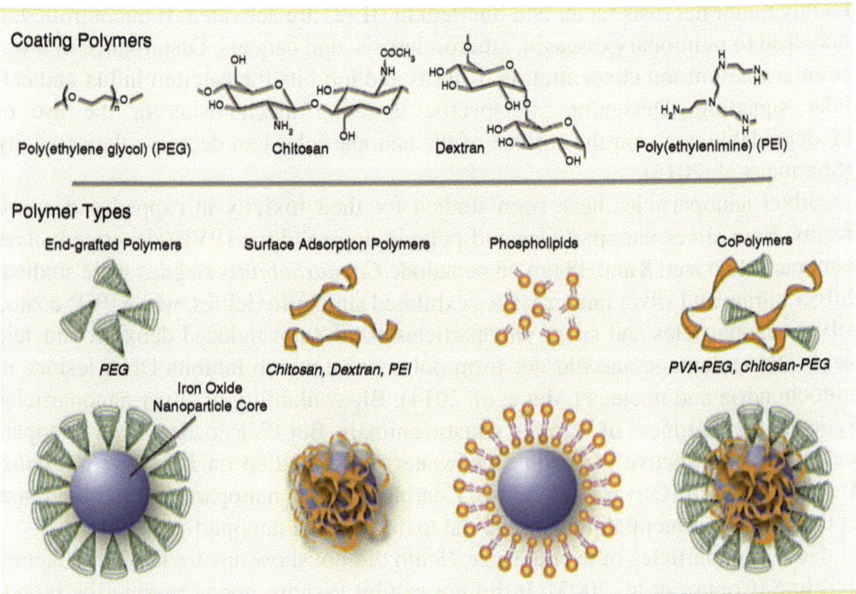

Fig. 4.1 Arrangement of polymers over metal nanoparticles like iron (Reproduced with permission from Veiseh et al. (2010))

from monomers by polymerization, while chemical interaction means covalent bonding between functional groups and biopolymer. This coating can be either post-synthesis of adsorption or post-synthesis of grafting. Post-synthesis end-grafting provides a control over density of biopolymer and its density with uniform coating. Lipids, liposomes, or micelles can coat upon the surface of nanoparticles. Methods of coating polymers upon noble metals synthesized by citrate reduction in the presence of salt have been reviewed in Zhang et al. (2016).

4.3.1 Coating and Carrier Materials for Encapsulation

Polymers are excellent physical barriers around the core nanoparticles as they protect the core nanoparticles from direct contact with environment. Encapsulation increases the particle size (Thanh 2010). This plays an important role for particles needed for systemic circulation albeit control on particle size for site-specific biodistribution. Water solubility can be increased by dextran, pullulan, chitosan, starch, as well as PEG. Too thin coatings of polymers are poor barriers leading to aggregation due to ionic interaction (Thanh 2010) but at the same time increased thickness have their properties affected specially in semiconducting materials, e.g., magnetism.

Hydrocarbon chains as in lipids are also well known as hydrophobic barriers (Sahoo et al. 2001). However, the complex interactions and forces rearrange the structures to form bilayers and micelles over the nanoparticles.

Dendrimers are also well suited for encapsulation of metal in the three-dimensional void space structure of dendrimers and thus can stabilize the metals nanoparticles. Metals are encapsulated via electrostatic interactions, van der Waals, hydrogen bonding, complexation reactions, etc., and hence, they do not agglomerate (Crooks et al. 2001). Dendrimers easily encapsulate silver or gold through electrostatic forces (Esumi et al. 2003; He et al. 1999).

4.4 Iron Nanoparticles

Iron nanoparticles are exceptional nanoparticles as the properties exhibited by the bulk iron oxide are the same in nanosized iron oxide nanoparticles. They are biodegradable in nature, biocompatible and exhibit superior contrast agent during imaging via magnetic resonance imaging (Pranatharthiharan et al. 2013). Iron oxides could be called as very small super paramagnetic iron oxide nanoparticles (VSPIONs) if their diameter is less than 10 nm, ultra-small superparamagnetic iron oxide nanoparticles (USPIONs) if they of size 10–50 nm, and superparamagnetic iron oxide nanoparticles (SPIONs) if the size if bigger than 50 nm. Iron nanoparticles are commonly used for imaging, thermal ablation and magnetically targetable system. SPIONs exhibit zero magnetism only in the presence of a magnetic field (Singh and Sahoo 2014). This results in decreased agglomeration of SPIONs and thus reduced chance of thromboembolism. Iron oxide nanoparticles could be of two types: Fe_3O_4 (magnetite) and Fe_2O_3 (maghemite). Magnetite is the most promising among both due to its better biocompatibility (Parveen et al. 2012).

The superparamagnetic iron oxide decreases the spin-spin relaxation time. Shorter relaxation time is depicted as dark images. However, iron oxide is rapidly sequestered by the mononuclear phagocytic system (MPS) (Weinstein et al. 2010). The rate of phagocytosis increases with increase in the particle size (Weinstein et al. 2010). Hence, they are used for tagging the activated macrophages and microglia detectable by magnetic resonance imaging. Inflammation of MPS in brain tumors, stroke, and plaques in carotid artery thus can be labeled with iron oxide nanoparticles due to phagocytosis (Wang 2011). Once they are internalized within lysosomes of reticuloendothelial system (RES), they breakdown into hemosiderin and/or ferritin, the different antiferromagnetic iron oxide forms (Briley-Saebo et al. 2004). Besides, unmodified SPIONs are found to be stable at extreme pHs.

For in vivo administration, SPIONs need to be coated with either organic molecules or polymers like PEG, chitosan, dextran (carboxydextran and carboxymethyldextran), phospholipids, or polyethyleneimine (PEI). This would decrease the agglomeration of nanoparticles. Also, it provides a starting point for chemical conjugation with different ligands or labels. Nonspecific interactions are thus minimized. Coating alters the thickness and hence alters the R2 relaxivities. Normally coating decreases the R2 relaxivities (Duan et al. 2008; LaConte et al. 2007). PEGylated iron nanoparticles also known as stealth prevent the uptake by RES. This can thus increase their systemic circulation (Harris and Chess 2003). PEG below

100,000 Da is amphiphilic and are assembled via various chemistries. Some have precipitated PEG onto iron nanoparticles (Lutz et al. 2006), or some have grafted via silane group (Kohler et al. 2004), while some have grafted them to surface and another end to targeting ligands. Dextran interacts with iron nanoparticles via hydrogen bonding and chelation of polar side groups. Hydrogen bonding however causes detachment which can be prevented by cross-linking with SPIONs using epichlorohydrin and ammonia (Josephson et al. 1999). Epichlorohydrin tends to have a negligible chance of clinical use as epichlorohydrin cannot degrade or can be cleared from the body (McCarthy and Weissleder 2008). Chitosan has also been used for SPION coating, which is difficult due to its poor solubility at pH values required for precipitation of SPIONs (Kumar et al. 2004). Cationic chitosan can complex with genetic materials incorporated as coating on SPIONs (Kim et al. 2005). Amino and hydroxyl functional groups of chitosan are useful for functionalization of SPIONs. Likewise, PEI, a cationic water-soluble polymer, has been coated onto SPIONs (McBain et al. 2007; Corti et al. 2008). The intrinsic toxicity of PEI and poor colloidal stability is a matter of concern (Kircheis et al. 2001; Park et al. 2008; Petri-Fink et al. 2008).

Amphiphilic molecules as in liposome and micelles can also coat either by post-synthesis (SPIONs) i.e. aqueous core of liposome or hydrophobic SPIONs core surrounded by micelles (Martina et al. 2005; Yang et al. 2007) or synthesizing with SPIONs directly with core (Decuyper and Joniau 1988). The former one though commonly studied has the disadvantage of having agglomerates of coating without any iron nanoparticles entrapped (Dagata et al. 2008).

After administration of magnetic nanoparticles intravenously, a high magnetic field is generated externally which attracts the nanoparticles toward the site of magnetic field (Nishijima et al. 2007). Creating a hydrophilic coating over the nanoparticles enables aqueous dispersions of the nanoparticles, e.g., PEG, dextran, and Pluronic F-127 (Jain et al. 2008).

US Food and Drug Administration (USFDA) approved iron dextran injection USP and iron sucrose injection USP for intravenous administration, which is a complex of ferric hydroxide and polysaccharides like dextran and sucrose, respectively, for iron-deficient subjects not responding to oral administration of iron. Complexation with polysaccharides decreases the excretion of iron (USFDA 2017).

4.4.1 Polymer Encapsulated

USFDA approved USPIONs – ferumoxytol (Feraheme) manufactured by AMAG Pharmaceuticals. They are magnetite cores coated by a carbohydrate mainly polyglucose sorbitol carboxymethylether to decrease immunological response (Provenzano et al. 2009). Clinical testing in patients with anemic chronic kidney disease was safe at 510 mg administered intravenously with a dose of 2 and 6 mg per kg (Spinowitz et al. 2005). Likewise, Combidex, USPIONs coated with dextran, has been studied for enhancing contrast during magnetic resonance imaging (MRI). Toxicity of dextran-coated iron oxide nanoparticles showed dose-dependent decrease in cell viability and loss of membrane integrity in Jurkat cells up to 72 h (Balas et al. 2017).

Likewise, in ferumoxtran-10, the carbohydrate selected is dextran T-10. Ferumoxtran-10 is proposed for differentiation between cancerous and noncancerous lymph nodes. It specifically accumulates in lymph nodes 24 and 36 h after intravenous administration of 2.6 and 3.4 mg of iron/kg studied in healthy humans for extra cranial neck MRI (Hudgins et al. 2002). On administration, it accumulates in non-cancerous organs causing dark areas in MRI images. Cancer cells fail to show the nanoparticles uptake (Groman et al. 1991). The hydrodynamic coating of carbohydrates on iron oxide though incomplete was 8–12 nm in thickness (Jung 1995; Jung and Jacobs 1995). Dextran-coated iron oxide encapsulated up to 95% of drug and is used in cancer treatment (Jain et al. 2008). SPIONs coated with polyvinyl benzyl-O-b-D-galactopyranosyl-D-gluconamide (PVLA) bearing galactose moieties have been studied for liver targeting (Yoo et al. 2007). Thermal stability of dextran-coated iron nanoparticles is also higher than dextran (Can et al. 2017).

PEG-g-chitosan-g-PEI encapsulating SPIONs complexed with DNA was effective in gene transfection (Kievit et al. 2009). Moreover, copolymer of PEG-poly(methacrylic acid) and poly(glycerol monomethacrylate) used for coating SPIONs layer by layer is pH-sensitive polymer releasing the drug in acidic endosomes of the cell (Guo et al. 2008). Biocompatible nanometric metal-organic frameworks (MOFs) of mesoporous iron(III) trimesate nanoparticles and bioadhesive polysaccharide chitosan showed promising results in oral uptake with improved intestinal barrier bypass (Hidalgo et al. 2017). Similarly other MOFs of iron(III) trimesate NPs functionalized with heparin (Bellido et al. 2015) or biocompatible cyclodextrins improved biocompatibility (Agostoni et al. 2015).

Cross-linked carboxymethylated chitosan and iron nanoparticles covalently linked to genistein increased inhibition effect of genistein on SGC-7901 cancer cells than free genistein solution. The presence of iron nanoparticles could further increase the efficacy by hyperthermia effect (Si et al. 2010). Iron nanoparticles modified with polyethylenimine as carrier for gene increased tumor entrapment up to 30-fold compared to simple intravenous administration (Chertok et al. 2010). Functionalized dendrimers with iron nanoparticles assisted in tracking the transvascular transport and uptake by glioma cells. Particles in the range of 11.7–11.9 nm traversed the blood-brain tumor barrier easily, while the larger-sized dendrimers failed (Sarin et al. 2008).

Magnetite nanoparticles radiolabeled with ^{188}Re sized 200 nm and coated with human serum albumin have been studied for targeted delivery in cancer. Hyperthermia arising from magnetite nanoparticles causes death of tumor cells locally (Ito et al. 2005). Magnetic nanoparticles conjugated to near-infrared fluorescence followed by coating the lipid with Gal-P123, a hepatocyte targeting polymer, achieved high and specific targeting to liver cancer cells. The magnetic resonance signal enhanced 5.4-fold than only dye-conjugated iron nanoparticles (Liang et al. 2017). Peptide glycine coated upon iron nanoparticles encapsulating amphotericin B demonstrated high efficacy against visceral leishmaniasis and was almost twice effective than plain amphotericin B (Kumar et al. 2017).

4.4.2 Lipid Encapsulated

Lately, iron nanoparticles widely explored as ferrofluids have also been studied for stimulus response providing targeted drug delivery. Temperature-responsive magneto-micelles comprising of an iron oxide core functionalized with undecylenic acid are surrounded by an amphiphilic temperature-responsive polymer coat of NIPAAm (N-isopropyl acrylamide). The magneto-micelles exhibited a size of 8 nm in water with a high potential for trigger-controlled release (Kim et al. 2008). PEGylated magnetite functionalized with specific monoclonal antibodies could be used for hyperthermia treatment in cancer (Ito et al. 2005).

Iron nanoparticles consisting of hyaluronic acid and peptides sized 160 nm delivered peptides up to 100% level in HEK293 and A549 cell lines (Kumar et al. 2007). Water-dispersible oleic acid-Pluronic-coated iron oxide nanoparticles sustained release of doxorubicin intracellularly and in vitro over a period of 2 weeks compared to drug solution. Dose-dependent antiproliferative effect was also observed (Jain et al. 2005). Iron oxide nanoparticles coated with double layer of oleic acid also enhanced bioavailability besides being a biocompatible theranostic agent in skin pathology. Both in vitro toxicity studies on human keratinocytes (HaCat cells) and acute dermal toxicity test on female and male SKH-1 hairless did not compromise skin function (Coricovac et al. 2017). NanoMOFs, as mentioned in Sect. 4.5, and lipid coating of MOFs have also been studied (Wuttke et al. 2015). A schematic diagram of the same is given in Fig. 4.2.

4.4.3 Dendrimer Encapsulated

Dendrimers of iron oxide prepared by layer-by-layer film technique, coated with poly(lysine)/poly(glutamic acid) followed by cross-linking with carbodiimide. It was further functionalized with folic acid-tagged poly(amidoamine) (PAMAM) (Shi et al. 2008). Poly(lysine)/poly(glutamic acid) stabilizes the iron oxide nanoparticles, while folic acid targets the dendrimers to cancer cells overexpressed with folate receptors (Wang et al. 2007).

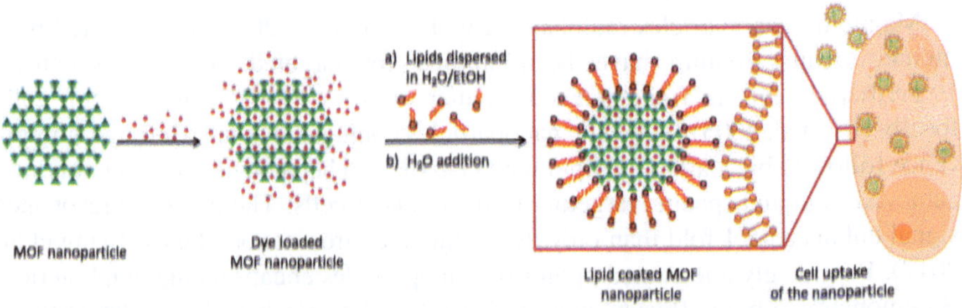

Fig. 4.2 Schematic representation of dye-encapsulated bilayer-coated NanoMOF and their uptake (Reproduced with permission from Wuttke et al. (2015))

4.5 Gold Nanoparticles

The unique optical property of gold nanoparticles and photothermal therapy is the basis for their extensive use in cancer. Gold belongs to the category "plasma" which contains fixed and equal number of positive ions and free and highly mobile conducting electrons. In irradiation with an electromagnetic wave, the electrons interact with oscillating electric fields and result in oscillation of electrons driven by electrical field. Such oscillatory electrons are called as plasmons. The plasmons present on the surface of gold interact with visible light to result in surface plasmon resonance (SPR). SPR causes enhanced absorption and scattering radiation and non-radiative properties of gold. The ratio of scattering to absorption increases with increase in size of nanoparticles (Jain 2006). The properties of SPR are highly sensitive to composition, shape, and size of the nanoparticles and thus are optical transducers for sensing.

4.5.1 Polymer Encapsulated

Gold nanoparticles covered with cyclodextrin have been reported as a carrier for anticancer drug through noncovalent interactions. Cyclodextrin-harbored β-lapachone was also functionalized with anti-epidermal growth factor and PEG as a targeting moiety and antifouling shell respectively. Release was dependent upon the glutathione concentration as demonstrated in MCF-7 and A549 cells (Xiaohua et al. 2006). Gold nanoparticles coated with chitosan easily adsorb which can be useful in delivery of insulin through transmucosal route (Bhumkar et al. 2007). Chitosan-coated gold nanoparticles have also been studied for delivery of cisplatin (Guo et al. 2010). A brief example of polymer-encapsulated gold nanoparticles has been listed in Table 4.1.

Gold nanoparticles immobilized with doxorubicin and carboxymethyl chitosan increased uptake by cancer cells at acidic pH. At acidic pH, carboxylic groups ionize and release drug. The technology is useful in cervical cancer uptake and wherever pH-triggered drug release is needed (Madhusudhan et al. 2014). Enzyme triggered a platinum (IV) prodrug release was possible iron nanoparticles coated with gelatin permitting MRI imaging too (Cheng et al. 2014).

Poly(lactic-co-glycolic acid) (PLGA) nanoparticles encapsulated with perfluorocarbon and silica-coated gold nanoparticles demonstrated increased photoacoustic signal intensity. The particles were also taken up by MDA-MB-231 breast cancer cells visualized by fluorescence microscopy followed by vaporization forming bubbles and cell destruction. The particles thus provided imaging as well as optically triggered delivery system (Wang et al. 2014). Nanoemulsion templating using class III organic solvent, ethyl acetate, and low-energy phase inversion composition emulsification method for coating gold nanoparticles within PLGA nanoparticles decreased the cytotoxicity and hemolysis of gold nanoparticles. Increasing the concentration of gold nanoparticles greater than 100pM decreased the stability of nanoparticles (Fornaguera et al. 2017).

Table 4.1 Applications of encapsulated gold nanoparticles in cancer

Encapsulated gold nanoparticles	Observation	Reference
Doxorubicin-grafted PEGylated gold nanoparticles	Overcomes uptake in doxorubicin resistant cell lines	Gu et al. (2012)
Oxaliplatin onto PEGylated gold nanoparticles	Increased efficacy and distributed up to nucleus	Brown et al. (2010)
mPEG-SH 5000 conjugated to gold nanorods	Tumor could be identified due to NIR absorption by nanorods. Exposure to NIR diode laser at 808 nm for 10 min at 1–2 W/cm^2 intensity, growth of tumor was significantly inhibited. Thermal transient heat increased by over 20-fold to aid tissue destruction	Dickerson et al. (2008) Liao et al. (2005)
PEGylated gold nanorods	Half-life time over 17 h in mice. Blood circulation time for rod-shaped nanoparticles was increased compared to spheres. Also increased cellular affinity. Eradicated all irradiated tumors with no regrowth over 50 days in mice	Park et al. (2008b) and von Maltzahn et al. (2009)
mPEG-SH5000 conjugated to gold nanoparticles and plasmid DNA	Gene expression detected in major organs of mouse. Naked DNA showed tenfold lower level of detection due to destruction in blood	Kawano et al. (2006)
PEGylated gold colloids with adsorbed tumor necrosis factor (TNF)	Intravenous administration accumulated the nanoparticles in MC-38 colon carcinoma tumors rather than other organs. TNF provided specificity and cytotoxicity of targeted cells	Paciotti et al. (2004)
Thiol-functionalized amphiphilic copolymer PEG-poly(n-butyl acrylate) coated on gold nanoparticles	Good stability seen in aqueous medium. High drug loading capacity observed with these nanoparticles rather than plain gold nanoparticles	Wang et al. (2013)
Gold nanoparticles encapsulated in polylactic acid-co-ethyl cellulose containing fluorouracil	Controlled release of fluorouracil was observed. Release was controlled with the presence of gold in nanocapsules rather than those without it	Sathishkumar (2012)
Gold nanoparticles surrounded by oligo (ethylene glycol) and proteins	Oligo (ethylene glycol) stabilized the nanoparticles in aqueous medium of broad pH. Protein also decreased the attraction. Addition of protein beyond a certain point caused flocculation which could be depleted with aid of salt	Zhang et al. (2007a)
Gold nanoparticles decorated with neoglycoconjugates of mannose-6-phosphate	Anti-angiogenic activities observed by chorioallantoic membrane (CAM) assay. Multivalent ligands were more effective in angiogenic activity than azide-mannose-6-phosphate monomer	Combemale et al. (2014)

(continued)

Table 4.1 (continued)

Encapsulated gold nanoparticles	Observation	Reference
Gold nanoparticles decorated with hyaluronic acid	The nanoparticles identified metastatic tumors for diagnosis and treat them. Multifunctional particles with hyaluronic acid-curcumin and folic acid-polyethylene glycol bioconjugate increased targeting and improved efficacy compared to free curcumin	Manju et al. (2012)
Gold nanoparticles bearing adamantane moieties and cyclodextrin-grafted hyaluronic acid	High cellular uptake by hyaluronic acid-mediated endocytosis. Effectively inhibited the growth of MCF-7 cells. pH-responsive drug release observed and no toxicity to normal cells	Li et al. (2014a, b)
Gold nanocomposites with bacterial cellulose	Nanofibers of bacterial cellulose are robust	Zhang et al. (2010)

Gold nanoparticles coated with methoxy-PEG-graft-poly(l-lysine) imparted stealth properties. Radiolabeling of this gold nanoparticle with ^{99m}Tc showed longer systemic circulation times (Bogdanov 2015). Doxorubicin-loaded fucoidan gold nanoparticles have dual mode of drug delivery and photoacoustic imaging. Doxorubicin was released higher in pH 4.5 than that of neutral pH. Doxorubicin-fucoidan-gold nanoparticles inhibited proliferation of human breast cancer cells at 24 h than any plain gold nanoparticles or doxorubicin solution. It also induced apoptosis (early as well as late) (Manivasagan et al. 2016).

Intravenous injection of PEGylated gold nanoshells in PC3 tumor bearing athymic (nu/nu) mice followed by a laser irradiation for 3 min at 4 W/cm^2 elevated the temperature up to 65.4 °C and showed regression of tumor up to 93% after 3 weeks. The same was also observed in orthotopic canine model with selective temperature increase at tumor site due to high accumulation of nanoparticles in tumor attributed to enhanced permeability and retention effect (Schwartz et al. 2009).

While stability of gold nanorods increased with PEGylation, cytotoxicity also reduced. Almost 54% of injected PEGylated nanorods retained up to 30 min in mice. Intratumor administration of PEGylated gold nanorods decreased tumor growth up to 96% upon laser irradiation at tumor site of nude (nu/nu) xenograft mice induced with subcutaneous squamous cell carcinoma (Niidome et al. 2006). Moreover, intravenous administration resulted in only 74% decrease in tumor growth. Biodistribution of PEGylated gold nanorods reflected that molar ratio of PEG to gold of 1:5 was optimum for maintaining prolonged systemic circulation and enhanced permeation and retention effect (Akiyama et al. 2009; Niidome et al. 2006). Likewise, nanorods coated with 5 kDa and 10 kDa of PEG were more effective in maintaining a prolonged systemic circulation than PEG with molecular weight of 2 kDa and 20 kDa (Zhang et al. 2009). PLGA conjugated to gold nanocrystals, PLGA lecithin/DSPE(1,2-Distearoyl-sn-glycero-3-phosphoethanolamine), and PEG gold nanocrystals have also been successful in medical computed tomography imaging (Huang et al. 2011).

4.5.2 *Dendrimer Encapsulated*

Gold nanoparticles have also been studied as enzymatic sensors (Hernandez-Santos et al. 2002). Gold nanoparticles conjugated to dendrimers are also proved to be a better nanobiosensor too (Hasanzadeh et al. 2014). Gold nanoparticles and cadmium sulfide nanoparticles deposited on dendrimers immobilized with anti-chloramphenicol acetyl transferase antibody deposited on layer of poly-TTCA (poly(terthiophene- 3-carboxylic acid)), responsible for conductivity and enhancing sensitivity, are fabricated on glassy-carbon electrode. It could easily detect chloramphenicol as low as 45 ± 5.0 pg/mL in meat samples of chicken, pork, and beef (Kim et al. 2010). Gold nanoparticles decorated on amine-terminated poly(amidoamine) dendrimers immobilized with brevetoxin B-bovine serum albumin conjugate can screen brevetoxin in food up to levels of 0.01 ng/mL by differential pulse voltammetry (Tang et al. 2011). PAMAM dendrimers and glucose oxidase are constructed on gold by layer-by-layer film assemblies via Schiff's base bonds (Yoon and Kim 2000). Increasing the number of layers up to 5 linearly increases the output current. Glucose easily permeates through the porous dendrimers. A glucose biosensor based on gold nanoparticles with FcSH (6-(Ferrocenyl) hexanethiol) + Cyst (cysteamine)/PAMAM/Glucose oxidase system detects glucose in beverages from 1.0 to 5.0 mM (Karadag et al. 2013). PAMAM dendrimers tethered with ferrocene and gold nanoparticles have also been studied for amperometric biosensors (Suk et al. 2004). Poly(propyleneimine) dendrimers encapsulate gold nanoparticles and myoglobin on graphite electrodes. Response with incorporation of gold nanoparticles increased than those without them (Zhang and Hu 2007). Likewise, a hydrogen peroxide biosensor containing gold nanoparticles encapsulated in PAMAM dendrimers on cysteamine-modified gold electrode is sensitive in detection of peroxides in clinical laboratories and environment (Ferapontova and Gorton 2002). Nanocomposites of PAMAM and gold formed three-dimensional immobilized matrix for horseradish peroxidase-based peroxide biosensor on multiwalled carbon nanotube-modified glassy-carbon electrode for quantifying hydrogen peroxide from 6.72 μM up to 20.80 mM by amperometry (Luo et al. 2011). Poly(N-isopropylacrylamide) polymer has also been studied for reducing toxicity and improving stability (Wei et al. 2008). Higher grafting of PEG on gold nanorods bypassed the uptake by RES and provided longer circulation. The nanorods accumulated in the spleen, liver, and tumor tissues (Akiyama et al. 2009). The same PEGylated gold nanorods did not show perfusion through the human placenta within 6 h (Myllynen et al. 2008); PEGylated gold nanoparticles conjugated to tamoxifen via thiol-PEG are currently in phase I clinical trial (Libutti et al. 2010). The PEGylation monitored accumulation of nanoparticles by surface-enhanced Raman scattering (SERS) (Qian et al. 2008).

4.6 Silver Nanoparticles

Silver has been known since time immemorial for its antibacterial activity. Decreasing the size increases the antimicrobial activity (Ranghar et al. 2014). The increased resistance of bacteria to antibiotics has gained popularity of silver nanoparticles and prevents its reoccurrence. Silver nanoparticles are safe in vivo up to a dose of 1,000 mg/kg bodyweight/day administered for 28 days. Moderate hepatotoxicity was, however, observed. Decrease in weight was seen after 90 days dosing with dose-dependent altered alkaline phosphatase and cholesterolemia (Kim et al. 2011; Lee et al. 2013). A lowest-observed-adverse-effect level (LOAEL) of 125 mg/kg has thus been suggested for silver nanoparticles, beyond which liver damage can be seen (Gaillet and Rouanet 2015). They exhibit minimal skin penetration. Long-term occupational exposure showed relatively nontoxic nature of silver nanoparticles (Trop 2009). There has been widespread use of silver nanoparticles in toothpaste, shampoos, washing machines, kitchen utensils, toys, etc. (Gaillet and Rouanet 2015). Like gold nanoparticles, silver nanoparticles are also plasmonic entities. They can also be used for thermal killing, while scattered light is used for imaging. Silver exhibits the highest plasmon excitation among gold, silver, and copper and is the only metal whose SPR can be tuned to visible wavelength (Evanoff and Chumanov 2005). Silver nanoparticles neutralize the reactive oxygen species mainly glutathione and thioredoxin. It initiates the inflammation process and destroys mitochondria. Apoptogenic factors are released after mitochondria are destroyed and result in programmed cell death (Mohammadzadeh 2012). They have been used in development of wound dressings, water purification systems, medical devices, etc.

4.6.1 *Polymer Encapsulated*

Complexes of azithromycin-silver and ofloxacin-silver nanoparticles were stabilized with chitosan and demonstrated no cytotoxicity against human peripheral blood cells with no lysis or morphological changes of red blood cell (Namasivayam and Samrat (2016).

Nanofibrous antibacterial mats of chitosan-polyethylene oxide containing 0.25% and 0.50% of silver nanoparticles sized 70 nm revealed 100% bactericidal activities against *Escherichia coli* and *Staphylococcus aureus*. Silver nanoparticles released increasingly up to the first 8 h followed by a sustained released. Burst release could be attributed to the hydrophilic nature of silver nanoparticles present in hydrophilic gel (Kohsaria et al. 2016). Scaffolds of silver nanoparticles and chitin have been studied for wound healing and blood clotting against *Staphylococcus aureus* and *Escherichia coli* (Madhumathi et al. 2010).

Silver nanoparticles synthesized in situ by reduction of silver nitrate were impregnated into cellulose nanocrystals obtained from bamboos for silver-loaded film and ointments. Cellulose obtained from plant sources is known to possess higher water absorption capacity with synergistic wound healing and wound closure in mice. Lesser inflammation and increase in fibroblasts and collagen from day 3 to day 8 increased the epithelialization by day 14 (Singla et al. 2016).

Poly(N-vinyl-2-pyrrolidone) and bovine serum albumin conjugated to silver nanoparticles were low in inhibition of human immunodeficiency virus than that foamy carbon silver nanoparticles (Elechiguerra et al. 2005). PVP, bovine serum albumin, and RF 412 (recombinant F protein) were effective against respiratory syncytial virus in Hep-2 cell lines. PVP specifically interacted with the G proteins present on the viral surface with regular spatial arrangement unlike the nonspecific association observed with bovine serum albumin and PF412 (Sun et al. 2008).

Tannic acid-modified silver nanoparticles have been studied in stationary cell lines and spheroids obtained from human-derived keratinocyte VK2-E6/E7 and HaCaT. Tannic acid-modified nanoparticles exhibited higher cytotoxicity with higher levels of reactive oxygen species and increased JNK stress kinase in VK2-E6/E7 cells, while opposite effect was seen in HaCaT cells with ERK kinase. Both types of nanoparticles showed cellular uptake. However, particles above 30 nm did not lyse DNA. Tannic acid-modified silver nanoparticles could downregulate IL-8 and (tumor necrosis factor). TNF-α triggered by LPS in VK2-E6/E7 but failed to activate the same in HaCaT cells. Thus tannic acid-modified nanoparticles could demonstrate immunomodulatory properties and can be used for dermal applications (Orlowski et al. 2016).

Mercaptoethane sulfonate-coated silver nanoparticles strongly inhibited HSV-1 infections. It prevented the entry of virus into host cell by preventing its binding to host cells (Baram-Pinto et al. 2010). While polysaccharide-coated silver nanoparticles were effective against monkeypox virus, different particle sizes of 10 nm, 25 nm, and 80 nm were evaluated for inhibition on monkeypox virus plaque formation. Particle size of 25 nm exhibited dose-dependent inhibition. They could additionally disrupt the intracellular replication of virus along with blocking the host cell uptake of virus (Rogers et al. 2008). Polysaccharide-coated silver nanoparticles were also checked for Vero cells infected against Tacaribe virus. Uncoated nanoparticles showed 50% inhibition in the virus progeny while polysaccharide coated showed little infectivity reduction. Nanoparticles bound to the membrane glycoproteins of virus (Speshock et al. 2010). Folic acid-coated silver nanoparticles have also been studied for cancer treatment (Wang et al. 2012a; Wang et al. 2012b).

Chitosan stabilized upon silver nanoparticles and conjugated to azathioprine has been studied for treatment of rheumatoid arthritis. Studies on 3T3 NIH fibroblast showed dual role of nanoparticles i.e. synergism for anti-inflammation, targeting to specific site and controlled release of drug (Prasad et al. 2013). Chitosan-coated silver nanoparticles exhibited better compatibility and lower systemic absorption compared to silver nanoparticles alone and polyvinylpyrrolidone-coated silver nanoparticles in methicillin-resistant *Staphylococcus aureus* wound infection mouse model. Aspartate aminotransferase and alanine aminotransferase levels, an indicator of liver dysfunction, were also significantly reduced (Peng et al. 2017).

Starch-coated silver nanoparticles have also been developed for selective interaction with Hg^{+2}, useful in monitoring toxic metals in environment. The intensity of surface plasmonic band alters with change in concentration of Hg^{+2} irrespective of the presence of other metal ions like sodium, potassium, calcium, copper, iron, nickel, cadmium, etc. (Vasileva et al. 2017). Silver nanoparticles coated with beta-cyclodextrin and modified with para-aminothiophenol and folic acid have specific affinity for folate receptors overexpressed in cancerous cells. These nanoparticles could be used as both invaders due to uptake via folate receptor-mediated endocytosis (Fig. 4.3) as well as a probing agent via SERS (Zhai et al. 2017).

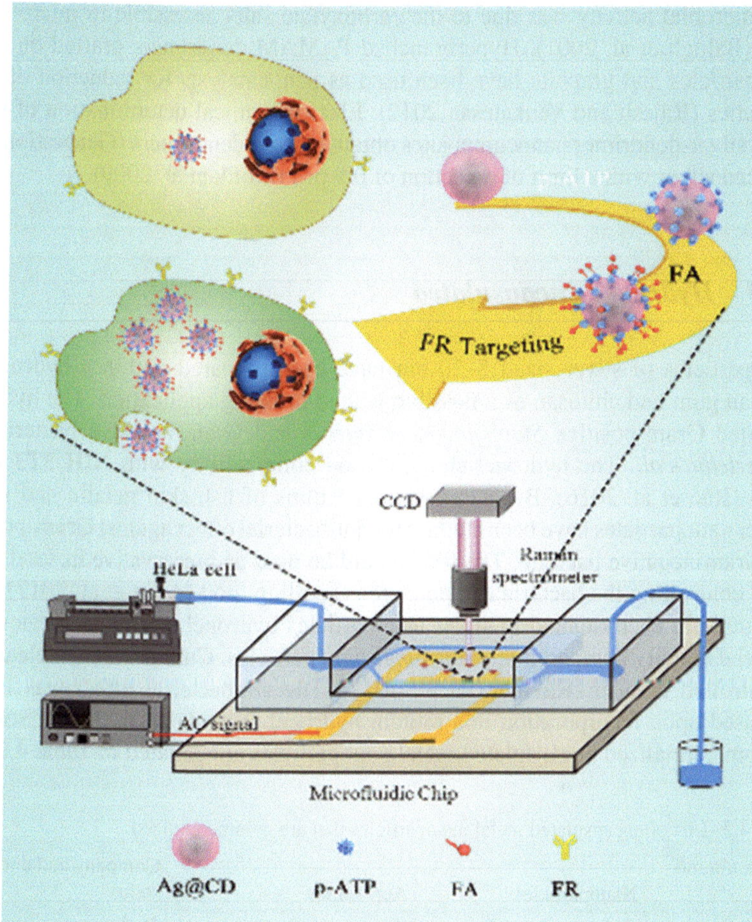

Fig. 4.3 Schematic representation of beta-cyclodextrin modified with para-aminothiophenol (*p-ATP*) and folic acid (*FA*) coated upon silver nanoparticles (*Ag@CD*) targeting folate receptors (*FR*) used in imaging (Reproduced with permission from Zhai et al. (2017))

4.6.2 Dendrimer Encapsulated

Dendrimers of silver have been studied as an electrochemical label for metallo-immunnoassays. Nanocomposites of silver encapsulated in generation 5–7 of PAMAM dendrimers could detect silver up to 0.9 pM after silver nanoparticles were dissolved in diluted nitric acid (Stofik et al. 2009). The dendrimers also exhibited slow release of solver and exhibited antimicrobial activity against Gram-positive bacilli such as *Pseudomonas aeruginosa*, *Staphylococcus aureus*, and *Escherichia coli*. Silver is encapsulated within the dendrimers due to the formation of silver carboxylate and complex formed between the nitrogens of dendrimers and silver ions. The complex is also photolyzed slowly forming dark-brown solution.

Antimicrobial activity was due to the carboxylate salts accessible to microorganisms (Balogh et al. 2001). Hyperbranched PAMAM dendrimers grafted on silver nanoparticles and graphite have been used as nanocatalysts for reduction of nitro aromatics (Rajesh and Venkatesan 2012). Electrochemical determination of silver using silver-dendrimer nanocomposites obtained from dendrimers (Generation 5–7) detected silver with a limit of detection of 0.9 pM (Stofik et al. 2009).

4.6.3 Hydrogel Encapsulated

Nanoparticles of silver sized 5–20 nm have been prepared within a hydrogel of xanthan gum and chitosan as a possible wound dressing application. The hydrogel inhibited Gram-positive *Staphylococcus aureus* and Gram negative bacteria like *Escherichia coli*. The hydrogel showed good compatibility with NIH 3T3 fibroblasts (Rao et al. 2016). Bio-nanocomposite films of fish skin gelatin and silver-copper nanoparticles have been studied for antibacterial effect against Gram-positive and Gram-negative bacteria. The films could be used as preservative in food packaging controlling the bacteria and pathogens spoiling food (Arfat et al. 2017).

Curcumin encapsulated in silver nanoparticles hydrogel exhibited greater antibacterial activity compared to plain silver nanoparticles. Curcumin was released in a controlled manner (Ravindra et al. 2012). The antibacterial effect was further increased upon incorporation in a protein hydrogel (Vimala et al. 2014). Some of the commercialized encapsulated metal nanoparticles are enlisted in Table 4.2.

Table 4.2 List of encapsulated metal nanoparticles that are commercialized

Product/brand name	Nanoparticles	Application	Company and current status
Aurimmune	Gold nanoparticles coupled to TNF-α, PEG-thiol sized ~27 nm	Intravenous for solid tumor	CytImmune sciences in phase – II
Resovist	SPIONs coated with carboxydextran	Imaging	Commercialized
Feridex	SPIONs coated with dextran sized 80–150 nm	Liver/spleen lesion MRI. Targeting MPS with 80% taken up by liver and 10% by spleen. Tumors show no uptake	Commercialized since 1996 but discontinued in 2008
Feraheme	SPIONs coated with dextran	Iron deficiency with chronic kidney disease	Commercialized since FDA approval in 2009
NanoTherm	SPIONs coated with aminosilane	Thermal ablation in prostate, pancreatic cancer after intratumoral administration	Approved in Europe since 2013

4.7 Conclusion

Metals are new active for next-generation anticancer therapy. They offer several benefits over the existing chemotherapeutic agents with the advantage of being a diagnostic and imaging agent along being with therapeutic agent. Some of them are already commercialized. Biopolymer can also provide synergism to metal nanoparticles when conjugated or encapsulated. The shortcomings of metal nanoparticles mainly the toxicology effect can be overcome by its combination with polymer or lipid coating. Biopolymers could increase colloidal stability or target specific targets or increase biocompatibility. Encapsulation can effectively solve issue of biocompatibility and toxicity. In the future, encapsulated metal nanoparticles will have a wide implication in therapeutic and imaging arena. However, extensive research and collaborations among research institutions are also needed in this direction to explore efficacious use of noble metals upon encapsulation.

References

Agostoni V, Horcajada P, Noiray M, Malanga M, Aykac A, Jicsinszky L, Vargas-Berenguel A, Semiramoth N, Daoud-Mahammed S, Nicolas V, Martineau C, Taulelle F, Vigneron J, Etcheberry A, Serre C, Gref R. A "green" strategy to construct non-covalent, stable and bioactive coatings on porous MOF nanoparticles. Sci Rep. 2015;5:7925.

Ahn J, Eom H, Yang X, Meyer J, Choi J. Comparative toxicity of silver nanoparticles on oxidative stress and DNA damage in the nematode, *Caenorhabditis elegans*. Chemosphere. 2014;108:343–52.

Akiyama Y, Mori T, Katayama Y, Niidome T. The effects of PEG grafting level and injection dose on gold nanorod biodistribution in the tumor-bearing mice. J Control Release. 2009;139:81–4.

Arfat Y, Ahmed J, Hiremath N, Auras R, Joseph A. Thermo-mechanical, rheological, structural and antimicrobial properties of bionanocomposite films based on fish skin gelatin and silver-copper nanoparticles. Food Hydrocoll. 2017;62:191–202.

Arnida A, Malugin A, Ghandehari H. Cellular uptake and toxicity of gold nanoparticles in prostate cancer cells: a comparative study of rods and spheres. J Appl Toxicol. 2010;30:212–7.

Balas M, Ciobanu C, Burtea C, Stan M, Bezirtzoglou E, Predoi D, Dinischiotu A. Synthesis, characterization, and toxicity evaluation of dextran-coated iron oxide nanoparticles. Metals. 2017;7(63):1–17.

Balogh L, Swanson D, Tomalia D, Hagnauer G, McManus A. Dendrimer–silver complexes and nanocomposites as antimicrobial agents. Nano Lett. 2001;1:3.

Baram-Pinto D, Shukla S, Gedanken A, Sarid R. Inhibition of HSV-1 attachment, entry, and cell-to-cell spread by functionalized multivalent gold nanoparticles. Small. 2010;6:1044–50.

Bellido E, Hidalgo T, Lozano MV, Guillevic M, Simon-Vazquez R, Santander-Ortega MJ, Gonzalez-Fernandez A, Serre C, Alonso MJ, Horcajada P. Heparin-engineered mesoporous iron metal-organic framework nanoparticles: toward stealth drug nanocarriers. Adv Healthc Mater. 2015;4(8):1246–57.

Bhumkar D, Joshi H, Sastry M, Pokharkar V. Chitosan reduced gold nanoparticles as novel carriers for transmucosal delivery of insulin. Pharm Res. 2007;24:11.

Bogdanov A. Gold nanoparticles stabilized with MPEG-grafted poly(l-lysine): in vitro and in vivo evaluation of a potential theranostic agent. Bioconjug Chem. 2015;26:39–50.

Bozich J, Lohse S, Torelli M, Murphy C, Hamers R, Klaper R. Surface chemistry, charge and ligand type impact the toxicity of gold nanoparticles to *Daphnia magna*. Environ Sci Nano. 2014;1:260–70.

Briley-Saebo K, Bjornerud A, Grant D, Ahlstrom H, Berg T, Kindberg G. Hepatic cellular distribution and degradation of iron oxide nanoparticles following single intravenous injection in rats: implications for magnetic resonance imaging. Cell Tissue Res. 2004;316:8.

Brown CL, Bushell G, Whitehouse MW, Agrawal DS, Tupe SG, Paknikar KM, Tiekink ERT. Nanogold pharmaceutics. Gold Bull. 2007;40(3):5.

Brown S, Nativo P, Smith J, Stirling D, Edwards P, Venugopal B. Gold nanoparticles for the improved anticancer drug delivery of the active component of oxaliplatin. J Am Chem Soc. 2010; 132: 4678–4684.

Can H, Kavlak S, ParviziKhosroshahi S, Güner A. Preparation, characterization and dynamical mechanical properties of dextran-coated iron oxide nanoparticles (DIONPs). Artif Cells Nanomedicine Biotechnol. 2017:1–11.

Cheng Z, Dai Y, Kang X, Li C, Huang S, Lian H, Hou Z, Ma P, Lin J. Gelatin-encapsulated iron oxide nanoparticles for platinum (IV) prodrug delivery, enzyme-stimulated release and MRI. Biomaterials. 2014;35:6359–60.

Chertok B, David A, Yang V. Polyethyleneimine-modified iron oxide nanoparticles for brain tumor drug delivery using magnetic targeting and intra-carotid administration. Biomaterials. 2010;31:6317–24.

Cole A. Cancer theranostics: the rise of targeted magnetic nanoparticles. Trends Biotechnol. 2011;29:323–32.

Combemale S, Assam-Evoung J, Houaidji S, Bibi R, Barragan-Montero V. Gold nanoparticles decorated with mannose-6-phosphate analogues. Molecules. 2014;19:1120–1149.

Connor E, Mwamuka J, Gole A, Murphy C, Wyatt M. Gold nanoparticles are taken up by human cells but do not cause acute cytotoxicity. Small. 2005;1:325–7.

Coricovac D, Moaca E, Pinzaru I, Cîtu C, Soica C, Mihali C, Pacurariu C, Tutelyan V, Tsatsakis A, Dehelean C. Biocompatible colloidal suspensions based on magnetic iron oxide nanoparticles: synthesis, characterization and toxicological profile. Front Pharmacol. 2017;8(154):1–18.

Corti M, Lascialfari A, Marinone M, Masotti A, Micotti E, Orsini F, Ortaggi G, Poletti G, Innocenti C, Sangregorio C. Magnetic and relaxometric properties of polyethylenimine-coated superparamagnetic MRI contrast agents. J Magn Magn Mater. 2008;320:E316–9.

Crooks R, Zhao M, Sun L, Chechik V, Yeung. Dendrimer-encapsulated metal nanoparticles: synthesis, characterization, and applications to catalysis. Acc Chem Res. 2001;34(3):10.

D'souza A, Shegokar R. Polyethylene glycol (PEG): a versatile polymer for pharmaceutical applications. Expert Opin Drug Deliv. 2016;13:1257–75.

Dagata J, Farkas N, Dennis C, Shull R, Hackley V, Yang C, Pirollo K, Chang E. Physical characterization methods for iron oxide contrast agents encapsulated within a targeted liposome-based delivery system. Nanotechnology. 2008;19:305101.

Decuyper M, Joniau M. Magnetoliposomes – formation and structural characterization. Eur Biophys J Biophys Lett. 1988;15:311–9.

Dickerson E, Dreaden E, Huang X, El-Sayed I, Chu H, Pushpanketh S. Gold nanorod assisted near-infrared plasmonic photothermal therapy (PPTT) of squamous cell carcinoma in mice. Cancer Lett. 2008; 269(1): 57–66.

Duan H, Kuang M, Wang X, Wang Y, Mao H, Nie S. Reexamining the effects of particle size and surface chemistry on the magnetic properties of iron oxide nanocrystals: new insights into spin disorder and proton relaxivity. J Phys Chem C. 2008;112:8127–31.

Elechiguerra J, Burt J, Morones J, Camacho-Bragado A, Gao X, Lara H, Yacaman M. Interaction of silver nanoparticles with HIV-1. J Nanobiotechnol. 2005;3(6):10.

Esumi K, Akiyama S, Yoshimura T. Multilayer formation using oppositely charged gold- and silver- dendrimer nanocomposites. Langmuir. 2003;19:76779–7681.

Evanoff D, Chumanov G. Synthesis and optical properties of silver nanoparticles and arrays. ChemPhysChem. 2005;6:1221–31.

Ferapontova E, Gorton L. Effect of pH on direct electron transfer in the system gold electrode-recombinant horseradish peroxidase. Bioelectrochemistry. 2002;55(4):83.

Fornaguera C, Feiner-Gracia N, Dols-Perez A, Garcia-Celma MJ, Solans C. Versatile methodology to encapsulate gold nanoparticles in PLGA nanoparticles obtained by Nano-emulsion templating. Pharm Res. 2017;34(5):1093–103.

Gaillet S, Rouanet J. Silver nanoparticles: their potential toxic effects after oral exposure and underlying mechanisms – a review. Food Chem Toxicol. 2015;77:58–63.

Groman E, Josephson L, Lewis J. Biologically degradable superparamagnetic materials for use in clinical applications. Magnetic Resonance Imaging. 1991;9(2):II.

Gu Y, Cheng J, Man C, Wong W, Cheng S. Gold-doxorubicin nanoconjugates for overcoming multidrug resistance. Nanomedicine. 2012;8: 204–211.

Guo M, Yan Y, Zhang H, Yan H, Cao Y, Liu K, Wan S, Huang J, Yue W. Magnetic and pH-responsive nanocarriers with multilayer core-shell architecture for anticancer drug delivery. J Mater Chem. 2008;18:5104–12.

Guo R, Zhang L, Qian H, Li R, Jiang X, Liu B. Multifunctional nanocarriers for cell imaging, drug delivery, and near-IR photothermal therapy. Langmuir. 2010;26:5428–34.

Harris J, Chess R. Effect of pegylation on pharmaceuticals. Nat Rev Drug Discov. 2003;2:214–21.

Hasanzadeh M, Shadjou N, Eskandani M, Soleymani J, Jafari F, de la Guardia M. Dendrimer-encapsulated and cored metal nanoparticles for electrochemical nanobiosensing. Trends Anal Chem. 2014;53:12.

He J, Valluzzi R, Yang K, Dolukhanyan T, Sung CKJ, Tripathy S. Electrostatic multilayer deposition of a gold-dendrimer nanocomposite. Chem Mater. 1999;11:3268–74.

Hernandez-Santos D, Gonzalez-Garcia M, Garcia A. Metal nanoparticles based electroanalysis. Electroanalysis. 2002;14:10.

Hidalgo T, Giménez-Marqués M, Bellido E, Avila J, Asensio MC, Salles F, Lozano MV, Guillevic M, Simón-Vázquez R, González-Fernández A, Serre C, Alonso MJ, Horcajada P. Chitosan-coated mesoporous MIL-100(Fe) nanoparticles as improved bio-compatible oral nanocarriers. Sci Rep. 2017;7:43099.

https://www.nature.com/articles/srep43099#supplementary-information

Huang H, Barua S, Sharma G, Dey S, Rege K. Inorganic nanoparticles for cancer imaging and therapy. J Control Release. 2011;155:344–57.

Hudgins P, Anzai Y, Morris M, Lucas M. Ferumoxtran-10, a superparamagnetic iron oxide as a magnetic resonance enhancement agent for imaging lymph nodes: a phase 2 dose study. AJNR Am J Neuroradiol. 2002;23(4):7.

Ito A, Shinkai M, Honda H, Kobayashi T. Medical application of functionalized magnetic nanoparticles. J Biosci Bioeng. 2005;100:11.

Jain P. Calculated absorption and scattering properties of gold nanoparticles of different size, shape, and composition: applications in biological imaging and biomedicine. J Phys Chem B. 2006;110:7238–48.

Jain T, Morales M, Sahoo S, Leslie-Pelecky D, Labhasetwar V. Iron-oxide nanoparticles for sustained delivery of anticancer agents. Mol Pharm. 2005;2:194–205.

Jain T, Richey J, Strand M, Leslie-Pelecky D, Flask C, Labhasetwar V. Magnetic nanoparticles with dual functional properties: drug delivery and magnetic resonance imaging. Biomaterials. 2008;29:9.

Josephson L, Tung C, Moore A, Weissleder R. High-efficiency intracellular magnetic labeling with novel superparamagnetic-tat peptide conjugates. Bioconjug Chem. 1999;10:186–91.

Jung C. Surface properties of superparamagnetic iron oxide MR contrast agents: ferumoxides, ferumoxtran, ferumoxsil. Magn Reson Imaging. 1995;13(5):6.

Jung C, Jacobs P. Physical and chemical properties of superparamagnetic iron oxide MR contrast agents: ferumoxides, ferumoxtran, ferumoxsil. Magn Reson Imaging. 1995;13(5):13.

Karadag M, Geyik C, Demirkol D, Nil F, Ertas, Timur S. Modified gold surfaces by 6-(ferrocenyl) hexanethiol/dendrimer/gold nanoparticles as a platform for the mediated biosensing applications. Mater Sci Eng C. 2013;33:6.

Kievit F, Veiseh O, Bhattarai N, Fang C, Gunn J, Lee D, Ellenbogen R, Olson J, Zhang M. PEI-PEG-Chitosan-Copolymer-Coated iron oxide nanoparticles for safe gene delivery: synthesis, complexation, and transfection. Adv Funct Mater. 2009;19:2244–51.

Kim E, Lee H, Kwak B, Kim B. Synthesis of ferrofluid with magnetic nanoparticles by sonochemical method for MRI contrast agent. J Magn Magn Mater. 2005;289:328–30.

Kim G, Li Y, Chu Y, Cheng S, Zhuo R, Zhang X. Nanosized temperature-responsive Fe3O4-UA-g-P(UA-co-NIPAAm) magnetomicelles for controlled drug release. Eur Polym J. 2008;44:7.

Kim D, Aminur M, Rahman DM, Ban C, Shim Y. An amperometric chloramphenicol immunosensor based on cadmium sulfide nanoparticles modified-dendrimer bonded conducting polymer. Biosens Bioelectron. 2010;25:8.

Kim J, Sung J, Ji J, Song K, Lee J, Kang C, Yu I. In vivo genotoxicity of silver nanoparticles after 90-day silver nanoparticle inhalation exposure. Safety Health Work. 2011;2(1):34–8.

Kircheis R, Wightman L, Wagner E. Design and gene delivery activity of modified polyethylenimines. Adv Drug Deliv Rev. 2001;53:341–58.

Kohler N, Fryxell G, Zhang M. A bifunctional poly(ethylene glycol) silane immobilized on metallic oxide-based nanoparticles for conjugation with cell targeting agents. J Am Chem Soc. 2004;126:7206–11.

Kohsaria I, Shariatinia Z, Pourmortazavi S. Antibacterial electrospun chitosan–polyethylene oxidenanocomposite mats containing bioactive silver nanoparticles. Carbohydr Polym. 2016;140:287–98.

Kawano T, Yamagata M, Takahashi H, Niidome Y, Yamada S, Katayama Y, Niidome T. Stabilizing of plasmid DNA in vivo by PEG-modified cationic gold nanoparticles and the gene expression assisted with electrical pulses. J Controlled Release. 2006;111:382–389.

Kumar M, Behera A, Lockey R, Zhang J, Bhullar G, De La Cruz C, Chen L, Leong K, Huang S, Mohapatra S. Intranasal gene transfer by chitosan-DNA nanospheres protects BALB/c mice against acute respiratory syncytial virus infection. Hum Gene Ther. 2002;13:10.

Kumar M, Muzzarelli R, Muzzarelli C, Sashiwa H, Domb A. Chitosan chemistry and pharmaceutical perspectives. Chem Rev. 2004;104:6017–84.

Kumar A, Sahoo B, Montpetit A, Behera S, Lockey R, Mohapatra S. Development of hyaluronic acid-Fe2O3 hybrid magnetic nanoparticles for targeted delivery of peptides. Nanomedicine. 2007;3:132–7.

Kumar R, Pandey K, Sahoo GC, Das S, Das V, Topno RK, Das P. Development of high efficacy peptide coated iron oxide nanoparticles encapsulated amphotericin B drug delivery system against visceral leishmaniasis. Mater Sci Eng C Mater Biol Appl. 2017;75:1465–71.

LaConte L, Nitin N, Zurkiya O, Caruntu D, O'Connor C, Hu X, Bao G. Coating thickness of magnetic iron oxide nanoparticles affects R-2 relaxivity. J Magn Reson Imaging. 2007;26:1634–41.

Lee J, Kim Y, Song K, Ryu H, Sung J, Park J, Park H, Song N, Shin B, Marshak D, Ahn K, Lee J, Yu I. Biopersistence of silver nanoparticles in tissues from Sprague–Dawley rats. Part Fibre Toxicol. 2013;10:36.

Leonaviciene L, Kirdaite G, Bradunaite R, Vaitkiene D, Vasiliauskas A, Zabulyte D, Ramanaviciene A, Ramanavicius A, Asmenavicius T, Mackiewicz Z. Effect of gold nanoparticles in the treatment of established collagen arthritis in rats. Medicina (Kaunas). 2012;48(2):91–101.

Li N, Chen Y, Zhang Y, Yang Y, Su Y, Chen JT, Liu Y. Polysaccharide-gold nanocluster supramolecular conjugates as a versatile platform for the targeted delivery of anticancer drugs. Sci Rep. 2014a;4:4164.

Li Y, Yang Y, Zhang W, Chen X. Nanotechnology: Advanced materials and nanotechnology for drug delivery. Adv Mater. 2014;26(31):5576.

Liang J, Zhang X, Miao Y, Li J, Gan Y. Lipid-coated iron oxide nanoparticles for dual-modal imaging of hepatocellular carcinoma. Int J Nanomedicine. 2017;12:2033–44.

Libutti S, Paciotti G, Byrnes A, Alexander H, Gannon W, Walker M. Phase I and pharmacokinetic studies of CYT-6091, a novel PEGylated colloidal gold-rhTNF nanomedicine. Clin Cancer Res. 2010;16:6139–49.

Liao H, Hafner J. Synthesis and applications of gold nanorod bioconjugates. Abstr Pap Am Chem Soc. 2005; 230: U1099.

Luo J, Dong M, Lin F, Liu M, Tang H, Li H, Zhang Y, Yao S. Three dimensional network polyamidoamine dendrimer-Au nanocomposite for the construction of a mediator-free horseradish peroxidase biosensor. Analyst. 2011;136:6.

Lutz J, Stiller S, Hoth A, Kaufner L, Pison U, Cartier R. One-pot synthesis of PEGylated ultrasmall iron-oxide nanoparticles and their invivo evaluation as magnetic resonance imaging contrast agents. Biomacromolecules. 2006;7:3132–8.

Madhumathi K, Sudheesh-Kumar P, Abhilash S, Sreeja V, Tamura H, Manzoor K, Nair S, Jayakumarm R. Development of novel chitin/nanosilver composite scaffolds for wound dressing applications. J Mater Sci Mater Med. 2010;21:807–13.

Madhusudhan A, Reddy G, Venkatesham M, Veerabhadram G, Kumar D, Natarajan S, Yang M, Hu A, Singh S. Efficient pH dependent drug delivery to target cancer cells by gold nanoparticles capped with carboxymethyl chitosan. Int J Mol Sci. 2014;15:8216–34.

Manju S, Sreenivasan K. Gold nanoparticles generated and stabilized by water soluble curcumin-polymer conjugate: Blood compatibility evaluation and targeted drug delivery onto cancer cells. Colloid Interface Sci. 2012;368:144–151.

Manivasagan P, Bharathiraja S, Bui N, Jang B, Oh Y, Lim I, Oh J. Doxorubicin-loaded fucoidan capped gold nanoparticles for drugdelivery and photoacoustic imaging. Int J Biol Macromol. 2016;91:578–88.

Martina M, Fortin J, Menager C, Clement O, Barratt G, Grabielle-Madelmont C, Gazeau F, Cabuil V, Lesieur S. Generation of superparamagnetic liposomes revealed as highly efficient MRI contrast agents for in vivo imaging. J Am Chem Soc. 2005;127:10676–85.

Marty J, Oppenheim R, Speiser P. Nanoparticles – a new colloidal drug delivery system. Pharm Acta Helv. 1978;53(1):5.

McBain S, Yiu H, El Haj A, Dobson J. Polyethyleneimine functionalized iron oxide nanoparticles as agents for DNA delivery and transfection. J Mater Chem. 2007;17:2561–5.

McCarthy J, Weissleder R. Multifunctional magnetic nanoparticles for targeted imaging and therapy. Adv Drug Deliv Rev. 2008;60:1241–51.

Mohammadzadeh R. Hypothesis: silver nanoparticles as an adjuvant for cancer therapy. Adv Pharm Bull. 2012;2:133.

Myllynen P, Loughran M, Howard C, Sormunen R, Walsh A, Vahakangas K. Kinetics of gold nanoparticles in the human placenta. Reprod Toxicol. 2008;26:130–7.

Namasivayam S, Samrat K. Cytotoxicity of chitosan stabilized silver and gold nanoparticles loaded azithromycin and ofloxacin nanodrug conjugate AgNp-AZ, OF and AuNp-AZ, OF against blood cells. Pharm Lett. 2016;8(2):4.

Nguyen T, Luke A. Functionalisation of nanoparticles for biomedical applications. Nano Today. 2010;5(3):213–30.

Niidome T, Yamagata M, Okamoto Y, Akiyama Y, Takahashi H, Kawano T, Katayama Y, Niidome Y. PEG-modified gold nanorods with a stealth character for in vivo applications. J Control Release. 2006;114(3):343–7.

Nishijima S, Mishima F, Terada T, Takeda S. A study on magnetically targeted drug delivery system using superconducting magnet. Physica C. 2007;463:4.

NNI Frequently asked questions – National Nanotechnology Initiative. In: http://www.nanogov/nanotech-101/nanotechnology-facts.

Oliver A, Croteau M, Stoiber T, Tejamaya M, Romer I, Lead J, Luoma S. Does water chemistry affect the dietary uptake and toxicity of silver nanoparticles by the freshwater snail *Lymnaea stagnalis*? Environ Pollut. 2014;189:87–91.

Orlowski P, Soliwoda K, Tomaszewska E, Bien K, Fruba A, Gniadek M, Labedz O, Nowak Z, Celichowski G, Grobelny J, Krzyzowska M. Toxicity of tannic acid-modified silver nanoparticles in keratinocytes: potential for immunomodulatory applications. Toxicol Vitro. 2016;35:43–54.

Park I, Ng C, Wang J, Chu B, Yuan C, Zhang S, Pun S. Determination of nanoparticle vehicle unpackaging by MR imaging of a T-2 magnetic relaxation switch. Biomaterials. 2008;29:724–32.

Park J, Von Maltzahn G, Zhang L, Schwartz M, Ruoslahti E, Bhatia S. Magnetic iron oxide nanoworms for tumor targeting and imaging. Adv Mater. 2008b;20(9):1630–1635.

Parveen S, Misra R, Sahoo S. Nanoparticles: a boon to drug delivery, therapeutics, diagnostics and imaging. Nanomedicine. 2012;8:147–66.
Peng Y, Song C, Yang C, Guo Q, Yao M. Low molecular weight chitosan-coated silver nanoparticles are effective for the treatment of MRSA-infected wounds. Int J Nanomedicine. 2017;12:295–304.
Petri-Fink A, Steitz B, Finka A, Salaklang J, Hofmann H. Effect of cell media on polymer coated superparamagnetic iron oxide nanoparticle (SPIONs): colloidal stability, cytotoxicity, and cellular uptake studies. Eur J Pharm Biopharm. 2008;68:129–37.
Paciotti G, Myer L, Weinreich D, Goia D, Pavel N, McLaughlin R, Tamarkin L. Colloidal gold: a novel nanoparticle vector for tumor directed drug delivery. Drug Deliv. 2004;11:169–183.
Pranatharthiharan S, Patel M, D'Souza A, Devarajan P. Inorganic nanovectors for nucleic acid delivery. Drug Deliv Transl Res. 2013;3:446–70.
Prasad S, Elango K, Damayanthi D, Saranya J. Formulation and evaluation of azathioprine loaded silver nanoparticles for the treatment of rheumatoid arthritis. Asian J Biomed Pharm Sci. 2013;3(23):28–32.
Provenzano R, Schiller B, Rao M, Coyne D, Brenner L, Pereira B. Ferumoxytol as an intravenous iron replacement therapy in hemodialysis patients. Clin J Am Soc Nephrol. 2009;4(2):7.
Qian X, Peng X, Ansari D, Yin-Goen Q, Chen G, Shin D. In vivo tumor targeting and spectroscopic detection with surface-enhanced Raman nanoparticle tags. Nat Biotechnol. 2008;26:83–90.
Rai M, Ingle A, Gupta I, Brandelli A. Bioactivity of noble metal nanoparticles decorated with biopolymers and their application in drug delivery. Int J Pharm. 2015;496:159–72.
Rajesh R, Venkatesan R. Encapsulation of silver nanoparticles into graphite grafted with hyperbranched poly(amidoamine) dendrimer and their catalytic activity towards reduction of nitro aromatics. J Mol Catal A Chem. 2012;359:88–96.
Ranghar S, Sirohi P, Verma P, Agarwal V. Nanoparticle-based drug delivery systems: promising approaches against infections. Braz Arch Biol Technol. 2014;57(2):209–22.
Rao K, Kumar A, Haider A, Han S. Polysaccharides based antibacterial polyelectrolyte hydrogels with silver nanoparticles. Mater Lett. 2016;184:189–92.
Ravindra S, Mulaba-Bafubiandi A, Rajinikanth V, Varaprasad K, Reddy N, Raju K. Development and characterization of curcumin loaded silver nanoparticle hydrogels for antibacterial and drug delivery applications. J Inorg Organomet Polym. 2012;22:1254–62.
Rogers J, Parkinson C, Choi Y, Speshock J, Hussain S. A preliminary assessment of silver nanoparticle inhibition of monkeypox virus plaque formation. Nanoscale Res Lett. 2008;3:129–33.
Sahoo Y, Pizem H, Fried T, Golodnitsky D, Burstein L, Sukenik C. Alkyl phosphonate/phosphate coating on magnetite nanoparticles: a comparison with fatty acids. Langmuir. 2001;17(25):7907–11.
Sarin H, Kanevsky A, Wu H, Brimacombe K, Fung S, Sousa A, Auh S, Wilson C, Sharma K, Aronova M. Effective transvascular delivery of nanoparticles across the blood-brain tumor barrier into malignant glioma cells. J Transl Med. 2008;6:80.
Sathishkumar K. Gold nanoparticles decorated polylactic acid-co-ethyl cellulose nanocapsules for 5-fluorouracil drug release. Int J Nano Biomater. 2012;4:12–20.
Schwartz J, Shetty A, Price RSR, Wang J, Uthamanthil R, Pham K, McNichols R, Coleman C, Payne J. Feasibility study of particleassisted laser ablation of brain tumors in orthotopic canine model. Cancer Res. 2009;69(4):1659–67.
Sharma H, Mishra P, Talegaonkar S, Vaidya B. Metal nanoparticles: a theranostic nanotool against cancer. Drug Discov Today. 2015;20(9):1143–51.
Shi X, Wang S, Swanson S, Ge S, Cao Z, Van Antwerp M, Landmark K, Baker J. Dendrimer-functionalized shell-crosslinked iron oxide nanoparticles for in-vivo magnetic resonance imaging of tumors. Adv Mater. 2008;20:1671–8.
Si H, Li D, Wang T, Zhang H, Ren F, Xu Z, Zhao Y. Improving the anti-tumor effect of genistein with a biocompatible superparamagnetic drug delivery system. J Nanosci Nanotechnol. 2010;10:2325–31.

Singh A, Sahoo S. Magnetic nanoparticles: a novel platform for cancer theranostics. Drug Discov Today. 2014;19:474–81.

Singla R, Soni S, Markand K, Kumari A, Mahesh S, Patial V, Padwad Y, Yadav S. In situ functionalized nanobiocomposites dressings of bamboo cellulose nanocrystals and silver nanoparticles for accelerated wound healing. Carbohydr Polym. 2016. doi: http://dx.doi.org/doi:10.1016/j.carbpol.2016.08.065.

Speshock J, Murdock R, Braydich-Stolle L, Schrand A, Hussain S. Interaction of silver nanoparticles with Tacaribe virus. J Nanobiotechnol. 2010;8:–19.

Spinowitz B, Schwenk M, Jacobs P. The safety and efficacy of ferumoxytol therapy in anemic chronic kidney disease patients. Kidney Int. 2005;68(4):7.

Stofik M, Stryhal Z, Maly J. Dendrimer-encapsulated silver nanoparticles as a novel electrochemical label for sensitive immunosensors. Biosens Bioelectron. 2009;24:5.

Suk J, Lee J, Kwak J. Electrochemistry on alternate structures of gold nanoparticles and ferrocene-tethered polyamidoamine dendrimers. Bull Kor Chem Soc. 2004;25:1681–6.

Sun L, Singh A, Vig K, Pillai S, Singh S. Silver nanoparticles inhibit replication of respiratory syncytial virus. J Biomed Biotechnol. 2008;4:149–58.

Tang D, Tang J, Su B, Chen G. Gold nanoparticles-decorated amineterminated poly (amidoamine) dendrimer for sensitive electrochemical immunoassay of brevetoxins in food samples. Biosens Bioelectron. 2011;26:6.

Thanh NGL. Functionalisation of nanoparticles for biomedical applications. Nano Today. 2010;5:213–30.

Trop M. The safety of nanocrystalline silver dressing on burns: a study of systemic silver absorption. Burns. 2009;35:3068.

USFDA FDA approved drug products. In: Drugs@FDA, vol. U.S. Food and Drug Administration, US 2017.

Vasileva P, Alexandrova T, Karadjova I. Application of starch-stabilized silver nanoparticles as a colorimetric sensor for mercury(II) in 0.005 mol/L nitric acid. J Chem. 2017;2017:9.

Veiseh O, Gunn J, Zhang M. Design and fabrication of magnetic nanoparticles for targeted drug delivery and imaging. Adv Drug Deliv Rev. 2010;62(3):284–304.

Vimala K, Varaprasad K, Sadiku R, Ramam K, Kanny K. Development of novel protein-Ag nanocomposite for drug delivery and inactivation of bacterial applications. Int J Biol Macromol. 2014;63:75–82.

Von Maltzahn G, Park J, Agrawal A, Bandaru N, Das S, Sailor M. Computationally guided photothermal tumor therapy using long-circulating gold nanorod antennas. Cancer Res. 2009;69(9):3892–3900.

Wang Y. Superparamagnetic iron oxide based MRI contrast agents: current status of clinical application. Quant Imaging Med Surg. 2011;1(1):5.

Wang S, Shi X, Van Antwerp M, Cao Z, Swanson S, Bi X, Baker J. Dendrimer-functionalized iron oxide nanoparticles for specific targeting and imaging of cancer cells. Adv Funct Mater. 2007;17:3043–50.

Wang Y, Newell B, Irudayaraj J. Folic acid protected silver nanocarriers for targeted drug delivery. J Biomed Nanotechnol. 2012a;8(5):751–9.

Wang H, Zheng L, Guo R, Peng C, Shen M, Shi X, Zhang G. Dendrimer-entrapped gold nanoparticles as potential CT contrast agents for blood pool imaging. Nanoscale Res Lett. 2012b;7:190.

Wang Z, Jia L, Li M. Gold nanoparticles decorated by amphiphilic block copolymer as efficient system for drug delivery. J Biomed Nanotechnol. 2013;9:61–68.

Wang Y, Strohm E, Sun Y, Niu C, Zheng Y, Wang Z, Kolios M. PLGA/PFC particles loaded with gold nanoparticles as dual contrast agents for photoacoustic and ultrasound imaging. Proc of SPIE. 2014;8943:89433M.

Wei Q, Ji J, Shen J. Synthesis of near-infrared responsive gold nanorod/PNIPAAm core/shell nanohybrids via surface initiated ATRP for smart drug delivery. Macromol Rapid Commun. 2008;29:645–50.

Weinstein J, Varallyay C, Dosa E. Superparamagnetic iron oxide nanoparticles: diagnostic magnetic resonance imaging and potential therapeutic applications in neurooncology and central nervous system inflammatory pathologies, a review. J Cereb Blood Flow Metab. 2010;30(1):20.
Weissig V, Pettinger T, Murdock N. Nanopharmaceuticals (part 1): products on the market. Int J Nanomedicine. 2014;9:6.
Wuttke S, Braig S, Prei ZA, Sicklinger J, Bellomo C, Radler J, Vollmar A, Bein T. MOF nanoparticles coated by lipid bilayers and their uptake by cancer cells. Chem Commun. 2015;51(87):15752–5.
Xiaohua H, Prashant K, Ivan H, Mostafa A. Determination of the minimum temperature required for selective photothermal destruction of cancer cells with the use of immunotargeted gold nanoparticles. Photochem Photobiol. 2006;82:5.
Yang J, Lee T, Lee J, Lim E, Hyung W, Lee C, Song Y, Suh J, Yoon H, Huh Y, Haam S. Synthesis of ultrasensitive magnetic resonance contrast agents for cancer imaging using PEG-fatty acid. Chem Mater. 2007;19:3870–6.
Yoo M, Kim I, Kim E, Jeong H, Lee C, Jeong Y, Akaike T, Cho C. Superparamagnetic iron oxide nanoparticles coated with galactose-carrying polymer for hepatocyte targeting. J Biomed Biotechnol. 2007;10:94740.
Yoon H, Kim H. Multilayered assembly of dendrimers with enzymes on gold: thickness-controlled biosensing interface. Anal Chem. 2000;72:922–6.
Zhai Z, Zhang F, Chen X, Zhong J, Liu G, Tian Y, Huang Q. Uptake of silver nanoparticles by DHA-treated cancer cells examined by surface-enhanced Raman spectroscopy in a microfluidic chip. Lab Chip. 2017;17(7):1306–13.
Zhang H, Hu N. Assembly of myoglobin layer-by-layer films with poly(propyleneimine) dendrimer-stabilized gold nanoparticles and its application in electrochemical biosensing. Biosens Bioelectron. 2007;23:393–9.
Zhang F, Skoda M, Jacobs R, Zorn S, Martin R, Martin C, Clark G, Goerigk G, Schreiber F. Gold nanoparticles decorated with oligo(ethylene glycol) thiols: Protein resistance and colloidal stability. J Phys Chem B. 2007a;111:12229–12237.
Zhang G, Yang Z, Lu W, Zhang R, Huang Q, Tian M, Li L, Liang D, Li C. Influence of anchoring ligands and particle size on the colloidal stability and in vivo biodistribution of polyethylene glycol-coated gold nanoparticles in tumor xenografted mice. Biomaterials. 2009;30(10):1928–36.
Zhang T, Wang W, Zhang D, Zhang X, Ma Y, Zhou Y, Qi L. Biotemplated synthesis of gold nanoparticle-bacteria cellulose nanofiber nanocomposites and their application in biosensing. Adv Function Mater. 2010;20:1152–1160.
Zhang Z, Maji S, da Fonseca AA, De Rycke R, Hoogenboom R, De Geest B. Salt-driven deposition of thermoresponsive polymer-coated metal nanoparticles on solid substrates. Angew Chem Int Ed. 2016;55(25):7086–90.

Chapter 5
Pharmaceutical and Biomedical Applications of Magnetic Iron-Oxide Nanoparticles

Kelly J. Dussán, Ellen C. Giese, Gustavo N.A. Vieira, Lionete N. Lima, and Debora D.V. Silva

Abstract In the past few years, due to the rapid development of the advances in the pharmaceutical and biomedical field, the magnetic iron-oxide nanoparticles have received considerable attention for their attractive properties. Magnetic nanoparticles are perfect candidates for use in diagnosis and disease treatment because they have properties as superparamagnetic behavior, a high superficial area that allows functionalizing, biocompatibility, nanometric size (10–100 nm), low toxicity, possibility of in vivo manipulation by a low external magnetic field, and placement in a specific place. The key point of magnetic iron-oxide nanomaterials is to develop effective synthesis techniques that allow particles to have with a uniform size, high magnetic saturation, and stability, preventing aggregation and oxidation with air since these result in the loss of its magnetic properties. This chapter presents a review of various strategies to synthesis of magnetic nanoparticles and their use in pharmaceutical and biomedical field.

Keywords Magnetic iron-oxide nanoparticles • Synthesis techniques • Pharmacological activity • Biomedical application

K.J. Dussán (✉) • G.N.A. Vieira • D.D.V. Silva
Department of Biochemistry and Chemical Technology, Institute of Chemistry, São Paulo State University-UNESP, Av. Prof. Francisco Degni, 55 – Jardim Quitandinha, CEP 14800-900 Araraquara, São Paulo, Brazil
e-mail: kelly.medina@iq.unesp.br

E.C. Giese
Coordination for Metallurgical and Environmental Process, Centre for Mineral Technology, CETEM, Pedro Calmon 900, CEP 21941-908 Rio de Janeiro, RJ, Brazil

L.N. Lima
Department of Chemical Engineering, Federal University of São Carlos – UFSCar, Rodovia Washington Luís KM 235, CEP 13565-905 São Carlos, SP, Brazil

M. Rai, R. Shegokar (eds.), *Metal Nanoparticles in Pharma*,
DOI 10.1007/978-3-319-63790-7_5

5.1 Introduction

At present, magnetism plays an important role in the study and manufacture of nanostructured and magnetic materials (Abolfazl et al. 2012). In order to make nanostructured materials, it is necessary to manipulate objects with a similar size to groupings of molecules and even molecules and atoms; the diameter of these materials is smaller than 1,000 nanometers (nm). The purpose of manipulating these materials in nanometric scale, must to the possibility of creating devices and systems with new properties allowing specific functions that represent nature behaviors (Martín et al. 2003; Abolfazl et al. 2012).

The use of magnetic iron-oxide nanoparticles (IONPs) offers advantages due to the chemical, physical, and pharmacological properties. These properties include magnetic behavior, chemical composition, the structure of the crystal, granulometric uniformity, properties related to adsorption processes, surface structure, and solubility (Raz et al. 2012). Among the magnetic nanostructured materials, magnetite (Fe_3O_4) and maghemite (Fe_2O_3) are commonly used due to their strong magnetic properties and low toxicity, which motivate their application in the field of biotechnology and medicine (Reddy et al. 2012; Wu et al. 2015).

In the past few years, due to the rapid development of the nanotechnology and magnetic nanostructured materials in biotechnology and medicine areas, magnetic particles of nanometric sizes have received considerable attention (Curtis and Wilkinson 2001; Pankhurst et al. 2003; Tartaj et al. 2003). These particles may have sizes that are comparable to virus (20–500 nm), proteins (5–50 nm), or genes (2 nm to wide it and 10–100 nm along). Moreover, it is possible to control using an external magnetic field and has a great surface area that can be modified to attach biological species via chemical interactions (Tartaj et al. 2005).

For example, in the medicine field, drugs' guided transport to a specific site has been studied by the application of an external magnetic field, and this has been achieved with a minimum amount of magnetic particles obtaining that magnetic drugs are safe and effective (Reddy et al. 2012). The transport guided of biologically active substances to a specific organ allows creating an optimal therapeutic concentration of the drug in the desired part of the organism while maintaining the injection total dose at low levels (Kuznetsov et al. 1999; Koneracká et al. 2002). The use of biocompatible magnetic particle as carrying drugs seems to be a promising technique (Lübbe et al. 1999). The superparamagnetic properties of fine magnetic particles have great importance from a practical point of view because it means that these magnetic particles can be located in a convenient position, transported to specific places, and controlled in desirable parts of organs or blood vessels with the assistance of an external magnetic field (Koneracká et al. 2002).

Some of the advantages of these magnetic materials in this area are (1) the size of the particle because of tolerated small size in order to improve the tissular diffusion, allow lower sedimentation rates, and obtain high effective surface areas. (2) The superficial characteristics allow the particles to be easily functionalized and encapsulated with various compounds (biomolecules, inorganic and organic materials, polymers, surfactants). Consequently, the particles are more resistant to degradation, and their biocompatibility and stability are increased. (3) The good magnetic response makes possible that the concentration of the nanomagnets decreases in the blood and therefore decreases the collateral effects (Reddy et al. 2012; Wu et al. 2015).

The magnetic separation is a technological advancement recently developed and used in the bioseparation area (Marszałł 2011). For example, magnetic particles can be used for immobilization of molecules via binding to form a magnetic biocatalyst that can be separated from the solution by an external magnetic field gradient. Recently, this magnetic property of the particles allows using in the immobilization of proteins, peptides, and enzymes (Bagheri et al. 2016; Gao et al. 2016), microorganism detection (Shi et al. 2014; Du et al. 2016; Yin et al. 2016), bioseparation (Paulus et al. 2015; Gu et al. 2016), immunoassays (Ahn et al. 2016; Vidal et al. 2016), pathogen detection (Brandão et al. 2015; Chen et al. 2015), controlled release drugs (Hyun 2015; Müller et al. 2017), biosensors (Xu and Wang 2012; Jamshaid et al. 2016), cellular classification (Thornhill et al. 1994; Carinelli et al. 2015), adsorption and purification of proteins (Mirahmadi-Zare et al. 2016), diagnostic magnetic resonance imaging (MRI) (Valdora et al. 2016), magnetic fluid hyperthermia therapy (Yang et al. 2015; Farzin et al. 2017), and separation of nucleic acid (Sun et al. 2014; Ali et al. 2016), among others.

On the other hand, in separation processes (in vitro), magnetic particles must be stable units composed of a high concentration of superparamagnetic nanoparticles (Medeiros et al. 2011). The main difficulty in the synthesis of these ultrafine particles is to control the size of the particles on a nanometric scale. This difficulty is the result of the high superficial energy of these systems. The interface tension acts as force guides for the spontaneous reduction of the superficial area by growth during the initial precipitation step and during aging (Sugimoto 1987). Therefore, the search for easy and flexible routes to synthesize and to produce magnetic iron-oxide nanoparticles with the acceptable size distribution, with desired size, and with high dispersibility, without having particle aggregate and low reactivity with the air, is of extreme importance to understand the potentials that are these materials in biomedicine and biotechnology (Tartaj et al. 2005).

5.2 Synthesis of Magnetic Iron-Oxide Nanoparticles (IONPs)

The synthetic methods used to synthesize magnetic IONPs are based on aqueous and nonaqueous routes on coprecipitation (Wu et al. 2008), thermal decomposition (Sun and Zeng 2002), hydrothermal and solvothermal syntheses (Wu et al. 2008), sol-gel synthesis (Dong and Zhu 2002), microemulsion (López Pérez et al. 1997), ultrasound irradiation (Ali et al. 2016), and biological synthesis (Zhu and Chen 2014) approaches using microorganisms.

The actual challenge in the area is synthesized to lead and improve its magnetic properties while keeping their small size, which is complex due to the colloidal nature of IONPs. A high biocompatibility and stability are also required to extend

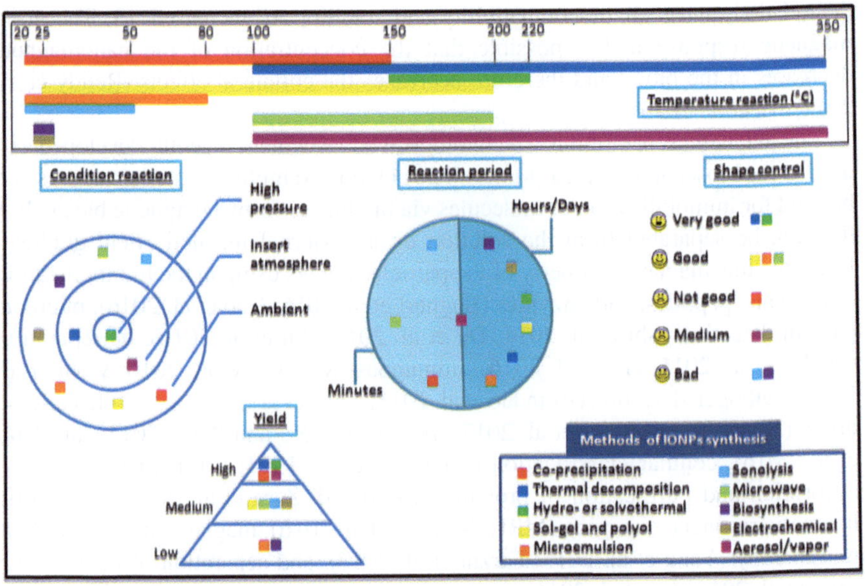

Fig. 5.1 Summary comparison of the synthetic methods for the production of IONPs

the possibility of its applications. Figure 5.1 summarizes a comparison of different reaction characteristics between the main methodologies used in the synthesis of magnetic IONPs according to Wu et al. (2015).

5.2.1 Coprecipitation

The coprecipitation is the simplest and most conventional method and consists of mixing ferric and ferrous ions in a 2:1 molar ratio (Fe^{3+}/Fe^{2+}) resulting in magnetite (Fe_3O_4) nucleus growth at very basic solutions (pH >11) at room or elevated temperature, as the reaction mechanism below:

$$Fe^{2+} + 2Fe^{3+} + 8OH \leftrightarrow Fe(OH)_2 + 2Fe(OH)_3 \rightarrow Fe_3O_4 \downarrow + 4H_2O$$

Magnetite (Fe_3O_4) may also be transformed into maghemite (γ-Fe_2O_3) depending on reaction conditions. Massart (1981) described for the first time the method for peptizing magnetite both in alkaline and acidic media. Coprecipitation method has been used to nanoparticle (NP) production, generally associated with different techniques to making magnetic Fe_3O_4 NPs, and, as these particles generated by this method present a wide size distribution, a secondary size selection sometimes is required (Wu et al. 2008).

5.2.2 *High-Temperature Thermal Decomposition*

The nonaqueous thermal decomposition has been adopted to try to control the particle size and distribution of IONPs formed. The IONPs obtained from high-temperature thermal decomposition display superior properties since they present narrow size distribution and high crystallinity in comparison to coprecipitation, where the particles obtained exhibit low crystallinity once the reactions are carried out at room temperature (Sun and Zeng 2002).

Thermal decomposition approaches include (a) hot-injection method (Tian et al. 2011), where the precursors of IONPs are injected into a hot reaction mixture, and (b) conventional method (Wu et al. 2015), where a reaction mixture is prepared at room temperature and then heated in a closed or open reaction vessel. Both methods are based on the decomposition of different ferric sources as described in Table 5.1.

Organic molecules are also used as reaction stabilizers to obtain monodisperse IONPs. Compounds as oleic acid or ether derivatives can affect the nucleation process decreasing the time spent to growing nanocrystals growth time, favoring the formation of small spherical IONPs (≤30 nm) (Demortiere et al. 2011). The use of specific solvents could contribute for the preparation of strongly faceted iron-oxide nanocrystals with nanocube or octahedron structures (Shavel and Liz-Marzan 2009).

5.2.3 *Hydrothermal and Solvothermal Synthesis*

Many techniques of crystallization using high-temperature solution (130–250 °C) under high vapor pressure (0.3–4.0 MPa) have been described. The hydrothermal method allows a crystal growth through different crystalline phases resulting in IONPs with controlled size and shape (Sun et al. 2009; Gao et al. 2010; Xu and Zhu 2011).

Table 5.1 Common ferric salts used in high-temperature thermal decomposition reactions

Ferric salts	Temperature requirement (K)	Reference
$Fe(CO)_5$	300	Woo et al. (2004)
$Fe_3(CO)_{12}$	100	Maity et al. (2009)
$Fe(acetylacetonate)_3$	200	Wang et al. (2012)
$Fe(N\text{-}nitrosophenylhydroxylamine)_3$	300	Rockenberger et al. (1999)
$Fe_4[Fe(CN)_6 14H_2O]$ (Prussian blue)	300–600	Hu et al. (2012)
$[Fe(CON_2H_4)_6](NO_3)_3$ (Fe-urea)	470	Asuha et al. (2009)
$Fe(C_5H_5)_2$ (Ferrocene)	770–970	Amara and Margel (2011)
$C_{54}H_{99}FeO_6$ (iron oleate)	650	Bronstein et al. (2007)
$FeCl_3$	650	Park et al. (2004)

The synthesis for γ-Fe_2O_3 involves a controlled oxidation of Fe_3O_4 and direct mineralization of Fe^{3+} ions and ensures a better crystallinity of single crystal particles of IONPs (α-Fe_2O_3, γ-Fe_2O_3, and Fe_3O_4) (Daou et al. 2006). In the hydrothermal (aqueous) and solvothermal (nonaqueous) processes, Fe^{3+} is used as the iron source and acetate, and urea and sodium citrate are mixed in ethylene glycol, resulting in a homogeneous dispersion which is transferred to a Teflon-lined stainless steel autoclave and sealed to heat at about 200 °C for 8–24 h (Hu et al. 2009; Lin et al. 2012).

The solvothermal method (Walton 2002) is used to synthesize IONPs containing ionic conductivity, magnetism, giant magnetoresistance, low thermal expansion, and ferroelectricity, properties that make it advantageous in comparison to the traditional ceramic synthetic routes.

5.2.4 *Sol-Gel Method*

The sol-gel method uses a colloidal solution as a stable dispersion precursor for an agglomeration of colloidal or sub-colloidal particles to form the IONPs through at least a two-step phase: $Fe(OH)_3 \rightarrow \beta$-$FeOOH \rightarrow \gamma$-$Fe_2O_3$. In sol systems, these particles interact by van der Waals forces or hydrogen bonds forming linking polymer chains. In gel systems, interactions are of a covalent nature collaborates to an irreversible process (Qi et al. 2011; Lemine et al. 2012).

The precursors of hydrolysis and polycondensation reactions include iron alkoxides and iron salts, and the reaction is performed at room temperature following by heating to obtain the IONPs in a final crystalline state (Dong and Zhu 2002; Pandey and Mishra 2011).

5.2.5 *Polyol Method*

The polyol method consists of an inverse sol-gel method, since the sol-gel uses an oxidation and polyol uses a reduction reaction. In this method, polyols act as solvents and reducing agents, as well as stabilizers, controlling particle growth and preventing interparticle aggregation (Lemine et al. 2012). In polyol synthesis, the iron precursor compound is stirred in suspension and heated to a given temperature of the boiling point of the polyol, generating different IONP sizes according to polyol nature (Caruntu et al. 2007; Shen et al. 2009).

As the surface of IONPs produced by both methods contains many hydrophilic ligands, the nanoparticles can be easily dispersed in aqueous solution and other polar solvents (Fig. 5.2). These IONPs present also a higher crystallinity and saturation magnetization obtained under high reaction temperature (Wu et al. 2015).

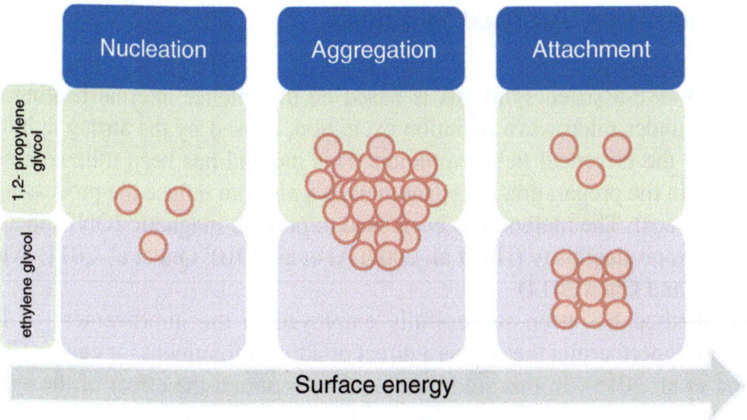

Fig. 5.2 Synthesis of iron-oxide NPs using two different polyols (Adapted from Dhand et al. 2015)

5.3 Microemulsion

In the microemulsion method, a monolayer of surfactant molecules forms an interface between the oil and water, with the hydrophobic tails from surfactant molecules dissolved in the oil phase and the hydrophilic head groups in the aqueous phase, or vice versa, at room temperature. The aggregates formed are called as reverse micelles, which can be formed in the presence or absence of water (López Pérez et al. 1997).

The aqueous phase may contain metal salts and/or other ingredients, and the hydrophobic phase can be a complex mixture of different hydrocarbons and olefins. The proportion of these components and the hydrophilic-lipophilic balance value of the surfactant used may result in the formation of IONPs in different systems as oil-in-water (O/W), water-in-oil (W/O), bicontinuous, and supercritical CO_2 microemulsions (Malik et al. 2012).

5.4 Sonolysis

The sonolysis (sonochemical or ultrasound irradiation) method uses high-intensity ultrasound for the production of IONP structures based on bubble/cavity formation and oscillation, growing to a certain size. In this method, the IONPs are synthesized by the sonication of an aqueous Fe^{3+} or Fe^{2+} salt solution combined with different polymers under room temperature. Another advantage of this method is the possibility of the use of volatile organometallic precursors (Hee Kim et al. 2005; Laurent et al. 2008). Sonolysis has been used also in the synthesis of superparamagnetic iron-oxide nanoparticles (SPIONs) having high magnetization and crystallinity (Mahmoudi et al. 2011; Yoffe et al. 2013).

5.5 Microwave-Assisted Synthesis

The microwave-assisted synthesis is based on the intense internal heating of the molecules under microwave radiation excitation, caused by the strong agitation in phase with the electrical field excitation. This method has been utilized since the late 1980s in the preparation of organometallics and can reduce the processing time and energy cost. The method has been used to prepare magnetic IONPs increasing yields and reproducibility (Hu et al. 2007; Ai et al. 2010; Qiu et al. 2011; Wu et al. 2011; Zhu and Chen 2014).

This method has been successfully employed in the improvement of IONPs applied in hyperthermia therapy as a direct or adjunct treatment for cancer (Blanco-Andujar et al. 2015). In this study, the authors reported the effect of the synthesis conditions on the properties of IONPs obtained by a coprecipitation method in a microwave reactor and also the beneficial effects of the microwave route with regard to the citric acid coatings that were produced.

Preparation of ferrite magnetic nanoparticles of different particle sizes by controlling the reaction temperature using microwave-assisted synthesis was reported with a temperature variation between 45 and 85 °C. These IONPs exhibited superparamagnetic behavior at room temperature, and their size could be varied by controlling the reaction temperature inside a microwave reactor (Kalyani et al. 2015).

5.6 Biosynthesis

Biosynthesis of IONPs is based on reduction and oxidation reactions catalyzed by microbial enzymes or plant phytochemicals. In this method, the special ability of Gram-negative magnetotactic bacteria in biomineralize magnetosomes is exploited in the obtainment of IONPs with uncontrolled shapes (Yan et al. 2012). The magnetosomes consist of intracellular crystals with high levels of purity and crystallinity of magnetic iron mineral, such as magnetite (Fe_3O_4), which can oxidize into maghemite (γ-Fe_2O_3), or greigite (Fe_3S_4) (Alphandéry 2014).

The magnetosomes present a great advantage in biomedical applications once the biomembrane surrounding the crystals is composed of lipids and proteins, and these functional groups make them suitable for use in living organisms (Prabhau and Kowshik 2016).

It is known that the key functions of magnetosome biogenesis are encoded by about 30 genes and, recently, a non-magnetotactic bacterium was capable of expressing these genes and encoding the magnetosome biogenesis pathway (Uebe and Schuler 2016).

Due to its ferrimagnetism propriety, bacterial magnetite magnetosomes have been used in the immobilization of bioactive substances such as glucose oxidase and uricase enzymes (Matsunaga and Kamiya 1987), antibodies for fluoroimmunoassays (Bazylinski and Schübbe 2007), and immunoassays (Tang et al. 2012).

5.7 Electrochemical

The electrochemical method generates IONPs with high purity and controlled particle size by adjusting the current or the potential applied to the reaction system. In a galvanostatic synthesis, the cell potential deviates by decreasing the reactant activity, and the reaction is suitable to supply a pure single-phase product by chosen potential. On the other hand, in a potentiostatic synthesis, a linear increasing of the cell potential from an initial to a final value if the reaction occurs in an intermediate value first carries out a linear voltammetry (Ramimoghadam et al. 2014).

In this method, IONPs are synthesized by an anodic polarization of iron in the transpassive range of the potential generating mainly maghemite (γ-Fe_2O_3) with a certain addition of magnetite (Fe_3O_4) in neutral pH (~7) (Starowicz et al. 2011; Ramimoghadam et al. 2014). In comparison to other abiotic methods as thermal decomposition, the electrochemical method produced the largest iron oxide in terms of mean particle size (Jung et al. 2007). IONPs have been synthesized within the pores of mesoporous silica (MS) microspheres by an electrochemical method to produce particles with a diameter of 20 nm inside the pore of MS spheres (Liberman et al. 2014).

5.8 Flow Injection Synthesis

The flow injection synthesis (FIS) is a modified coprecipitation method based on flow injection where different precursors can be added by pumping with a controllable flow rate in a capillary reactor under laminar flow. The advantage is the high reproducibility and high mixing homogeneity in a continuous synthesis reaction (Mohapatra and Anand 2010; Ramimoghadam et al. 2014).

A novel technique based on a flow injection was developed using continuous or segmented mixing of reagents under laminar flow regime in a capillary reactor. The obtained IONPs had a narrow size distribution in the range 2–7 nm. It was observed that the variation of reagent concentrations and the flow rates allowed the manipulation of the particle size and narrow down the particle size distribution without decreasing the quality of the particles (Salazar-Alvarez et al. 2006).

5.9 Aerosol/Vapor

The aerosol method consists of using of spray and laser pyrolysis techniques for the continuously controlled production of IONPs. In spray pyrolysis, ultrafine particles are aggregated into larger particles and are obtained from evaporation of ferric salts, drying, and pyrolysis reaction of liquid drops (a reducing agent in organic solvent) inside a high-temperature atmosphere. In laser pyrolysis, ultrafine particles are less aggregated and are obtained from a heating a flowing mixture of gasses with a continuous wave CO_2 laser, which initiates and sustains a chemical reaction (Wu et al. 2008, 2015).

5.10 Pharmaceutical Application of Iron-Oxide Nanoparticles (IONPs)

The main disadvantage of most chemotherapeutic agents is its side effects since these are non-specific compounds. For example, most of the anticancer drugs have characteristics such as hydrophobicity, low water solubility, high clearance, short residence time, and some systemic side effects (Akash and Rehman 2015).

As a way to bypass these negative effects, mainly avoid side effects, the use of magnetic iron-oxide nanoparticles (IONPs) as carriers to target-specific drug delivery has been studied. Exploring the attraction of IONP carriers to an external magnetic field could increase delivery of a drug in a specific site. Such particles can be also used to form complexes with other materials. Besides, those systems formed by magnetic nanoparticles have also been evaluated to be used as theranostic, that is, materials that combine therapeutic and diagnostic functions in a nanostructured complex (Zhou et al. 2016).

In general, the use of target-specific drug delivery involves binding of a drug to biocompatible IONP carrier, injection of magnetic target carries (MTC) as a colloidal suspension, application of magnetic field gradient to MTC be direct to the specific site, and release of drug from MTC (Sun et al. 2008).

Magnetic delivery of drugs or magnetic drug targeting (MDT) is a technique that adds drugs in nano-/micro-magnetic particles and then applies an external magnetic field to direct and concentrate these particles in disease sites such as solid tumors, infection regions, or blood clots. Magnetic drug targeting allows its dosage to be increased without side effects in healthy tissues (Do et al. 2015).

Molecular transport employing nanoparticles has been used as a strategy to improve the drug delivery and reduce its toxicity in different areas including cardiology, hemostasis, ophthalmology, and oncology. Magnetic nanoparticles have been developed because of their ability to respond to magnetic fields, including magnetic hyperthermia, their controllable movements, and their utilization as contrast agents in magnetic resonance imaging (Bhandari et al. 2016; Do et al. 2016; Zhou et al. 2016).

As reported by Bixner and Reimhult (2016), combining magnetic nanoparticles with the liposome drug delivery technology could be a reasonable pharmaceutical formulation technology alternative. Among the main advantages, one might point out hydrolytic degradation in nontoxic ions, high compatibility with in vivo applications, and the low susceptibility of tissue to magnetic fields, favoring their use to direct the drug delivery and in diagnostic bioimaging techniques (Liu et al. 2013; Amjad et al. 2015).

There are many types of target-specific drug delivery employing IONPs, most of them with patents disclosed, as reported by Daniel-da-Silva et al. (2013). Here, some examples of these systems are presented. Tehrani et al. (2014) related the use of an electromagnetic system, composed by six electromagnets powered by currents, for directing magnetic nanoparticles in blood vessels, which presented reduction of cost and lower power consumption as advantages.

The use of IONPs also has been studied to deliver diclofenac, a drug used for the treatment of inflammatory diseases. According to Agotegaray et al. (2014), chitosan-linked magnetic nanocarriers have been studied to target diclofenac delivery, with a satisfactory efficiency of drug loading in vitro assays. The authors affirmed that system would be suitable for in vivo assays. Posteriorly, Agotegaray et al. (2016) evaluated the effects of these magnetic IONPs on rat aortic endothelial cells. Results demonstrated that even after different doses (1, 10, and 100 μg/ml), endothelial cell metabolism was not affected by IONPs and these nanoparticles neither induced cytotoxicity in the cells nor accumulated in the organs. According to the authors, chitosan-linked magnetic IONPs could be a successful alternative to personalized treatments with site-specific drug delivery.

Georgiadou et al. (2016) prepared $CoFe_2O_4$ IONPs as carriers for the naproxen (NAP), a nonsteroidal anti-inflammatory drug, and its biological behavior was evaluated in vitro in rat serum and in vivo in mice. The authors observed that the use of magnetic IONPs-NAP carriers avoided the undesirable drug release and their accumulation at the inflammation site, possibly because of the increased vascular permeability of the inflamed muscle.

Hybrid beads composed of magnetic nanoparticles and alginate (Alg-IONPs) were synthesized and evaluated as a carrier for dopamine release, in the absence and the presence of an external magnetic field. The results suggested that Alg-IONP beads presented potential utility for loading of dopamine and its controlled release in the presence of external magnetic field (Kondaveeti et al. 2016).

Stocke et al. (2015) reported the use of a spray drying formulated with magnetic nanocomposite microparticles (MnMs) composed by IONPs and D-mannitol. According to the authors, these materials presented moderate cytotoxicity in vitro studies on a human lung cell line and have potential applications for thermal treatment of the lungs through targeted pulmonary inhalation aerosol delivery of IONPs.

The combination of magnetic IONPs (Fe_3O_4) with amino acids (L-lysine and L-arginine) to the construction of fluorescent magnetic nanoparticles was also reported that could be biocompatible and nontoxic to be used in biological systems (Ebrahiminezhad et al. 2013).

There are many works related to the use of magnetic IONPs for the treatment of diseases of central nervous system (CNS) and cancer. Do et al. (2016) reported that magnetic nanoparticles (MNPs) could be directed by external magnetic forces to cross the blood-brain barrier and delivery drugs to a disease region.

Akash and Rehman (2015) reported the use of polymeric-based targeted particulate carrier system, which has been showed efficient in delivering anticancer encapsulated drug directly at the desired action site avoiding interaction of encapsulated drug to normal cells. According to the authors, the use of pluronic F127 (PF127) conjugated with MNPs could result in increased stability of incorporated hydrophobic drugs with higher cytotoxicity in vitro, improvement of cell assimilation of anticancer drugs, and higher specific distribution with minimum toxicity.

Montha et al. (2016) described that poly-lactide-co-glycolic acid-coated chitosan stabilized (Mn, Zn) ferrite nanoparticles could be used as an efficient carrier of the doxorubicin (anticancer drug) because the carrier presented biological activity and pH-responsive controlled release, i.e., low pH around the tumor (pH 4.0) favored drug release.

Chen et al. (2016) studied a cis-diamminedichloridoplatinum(II) (CDDP), also known as cisplatin, loaded magnetic nanoparticle system as an intelligent drug delivery system that target malignant tumors of the head and neck, particularly nasopharyngeal cancer. This system showed stable and exhibited magnetic responsiveness, which released CDDP in a low pH environment.

Farjadian et al. (2016) evaluated the production of hydroxyl-modified magnetite as a nanocarrier for methotrexate conjugation (an anticancer drug). The in vitro cell assays in the presence of free methotrexate and conjugated form showed an excellent anticancer effect of magnetic IONPs when compared with the soluble drug.

Another anticancer chemotherapy medicine, docetaxel, had its targeted delivery evaluated employing magnetic IONPs prepared with poly-N-5-acrylamido isophthalic acid grafted on to Fe_3O_4 magnetic nanoparticles and conjugated with β-cyclodextrin and tumor-targeting folic acid to increase the site-specific intracellular delivery (Tarasi et al. 2016). The effect of magnetic IONPs on the cell viability was evaluated for the human embryonic kidney normal cell line and in different cancerous cell lines. According to authors, the new magnetically nano-drug delivery system did not show any apparent cytotoxic effect and besides reduced the growth of cancerous cell lines.

Tariq et al. (2016) studied the effect of surface decoration on pharmacokinetic and pharmacodynamic profile of epirubicin (EPI), an anthracycline drug used for chemotherapy, which elicits poor oral bioavailability. EPI-loaded poly-lactide-co-glycolic acid nanoparticles (PLGA-NPs) were prepared, and their superficies were decorated with polyethylene glycol (EPI-PNPs) and mannosamine (EPI-MNPs). Cytotoxicity studies were performed against human breast adenocarcinoma cell lines. The results showed that epirubicin linked to the nanocarriers had superior in vitro and in vivo activities than free epirubicin solution. Besides, EPI-MNPs showed better pharmacokinetic and pharmacodynamic profile when compared with EPI-PNPs.

Kaushik et al. (2016) explored a noninvasive magnetically guided central nervous system (CNS) delivery of magnetoelectric nanocarriers (MENCs) for on-demand controlled release of anti-HIV drugs as a potential therapy against NeuroAIDS (a neurodegenerative disorder), using in vitro model and in vivo assays (adult mice). The results demonstrated that delivered MENCs were uniformly distributed inside the brain and were nontoxic to the brain and other major organs, such as the kidney, lung, liver, and spleen, and did not affect hepatic, kidney, and neurobehavioral functioning. The authors considered blood-brain barrier (BBB) delivery method as noninvasive and completely safe for in vivo application and the nanocarriers as potential to deliver therapeutic agents across the BBB to treat CNS diseases such as Alzheimer's, brain tumors, and NeuroAIDS.

5.11 Biomedical Application

The development of magnetic nanoparticles (MNPs) has opened many opportunities for potential biomedical applications. The main reason MNPs are increasingly being tested for such applications is their size itself: MNPs are smaller than cells, but their size is comparable to those of viruses, proteins, or genes. Consequently, many potential applications are related to considerably localized action in biological systems (e.g., within cells). Additionally, their magnetic properties provide the possibility of manipulation by an external field, improving the property of locally interacting with biological systems and reducing side effects, caused by unintentional interaction with other biological entities (Nikiforov and Filinova 2009).

Reported biomedical applications of MNPs include its usage as alternative contrast agents in magnetic resonance imaging (MRI), due to their low toxicity, better colloidal stability, and magnetic properties. Other properties of MNPs, such as the possibility of functionalization, surface modification, and the potential use of heat sources, provide other applications.

For biomedical applications, additional procedures for the preparation of MNPs are required to increase their biocompatibility. Due to the importance of such procedures, this section begins with general considerations about biocompatibility of MNPs. Afterward, each discussion of the biomedical and pharmaceutical application of MNPs is started by a brief description of the basic principles of each application, followed by a discussion of recently published papers on the subject.

5.12 Biocompatibility of MNPs and Its Relation to Surface Modification

The concept of biocompatibility is more commonly related to medical devices, which are intended to remain inside the body for a long time (e.g., implants, pacemakers, etc.). Even though MNPs are not intended for such applications a priori, their application in any treatment or diagnosis procedure must minimize or eliminate possible side effects while performing the intended function; thus they must present biocompatibility. MNP biocompatibility is related to the different paths they can be eliminated from the body, which is in turn related to their physical and chemical properties. In order to interact with its target, MNPs should undergo a coating process, yielding a core-shell structure for MNPs. Different coating layers result in different MNPs for different applications. As expected, MNP biocompatibility is related to their physical and chemical properties, which are related to their synthetic route.

In addition to the general requirements for an effective performance of MNPs on applications (as aforementioned in this chapter), MNPs should present additional properties for biomedical applications, which can be attained by the choice of a suitable coating. Different coating materials for MNPs are compared and contrasted by Gupta et al. (2007) and by Gun'ko and Brougham (2009). Common considerations are (a) preventing the degradation of MNPs or particle agglomeration in physiological conditions, such as intravenous media; (b) reducing MNP toxicity, even though non-coated iron-oxide MNPs present IC50 (19.1 mm (red), 4.8 mm (yellow), 3 mm (green) CdTe quantum dots) suitable for biomedical applications (Lewinski et al. 2008); (c) the necessity (or not) of opsonization of MNPs, i.e., adsorption of serum proteins onto the nanoparticles, which is the first step required by the major defense system (reticuloendothelial system, RES) to perform phagocytosis; and (d) increasing specificity to certain cells by attaching, e.g., antibodies, hormones, etc., to its surface (Gupta and Gupta 2005; Hola et al. 2015).

5.13 Magnetic Resonance Imaging (MRI)

Magnetic resonance imaging (MRI) is an in vivo imaging procedure, which provides detailed images of soft tissues. One of its main advantages over other imaging diagnostic methods (such as X-ray or ultrasound techniques) is that the contrast between different tissues is a consequence of the physicochemical environment of water and also due to local biochemical and metabolic activity in each tissue, being thus much more sensitive to tissue composition. Consequently, MRI allows a more detailed tissue analysis. Another important advantage of MRI is that both spatial and temporal analyses might be carried out, hence being capable of providing dynamic imaging (Gun'ko and Brougham 2009; Qiao et al. 2009).

The physical basis for MRI is the reduction of magnetic relaxation times by MRI contrast agents: hydrogen protons (mainly those from water) tend to align themselves with an external magnetic field. When another radio-frequency pulse sequence is applied to the external magnetic field, a magnetic moment is created onto particles. When the pulse sequence ends, the protons tend to align themselves again to the previous magnetic field. This process of protons returning to the original position is known as magnetic relaxation. Relaxation occurs by either longitudinal (T1-recovery time) or transversal (T2-decay time) relaxation. Reduction in T1 relaxation times is observed for positive contrasts, whereas reduction in T2 relaxation times is observed for negative contrasts. More details of fundamental aspects of magnetic relaxation are found elsewhere (Aharoni 1992).

The most commonly used MRI contrast agents are based on gadolinium chelates, such as Gd-DTPA (gadolinium-diethylenetriaminepentaacetic acid). However, MNPs are more efficient in promoting relaxation (Kim 2009), by reducing relaxation times in water even at nanomolar concentrations. Additionally, the possibility of physical and chemical modification of MNPs enables the customization of their properties, according to the applications (Gun'ko and Brougham 2009).

Iron oxide-based MNPs for MRI applications are classified according to their hydrodynamic size. Particles with hydrodynamic size larger than 40 nm are quickly opsonized and uptaken by the reticuloendothelial system (as aforementioned in Sect. 4.1) and are then processed, e.g., in the liver or spleen, in order to eliminate MNPs from the body. Such materials are classified as small particles of iron oxide (SPIOs) and are suitable for liver MRI applications. On the other hand, particles smaller than 40 nm (typically smaller than 10 nm, known as ultrasmall particles of iron oxide – USPIOs) are more difficult to opsonize; hence USPIOs have longer blood circulation times and have the tendency of accumulation within lymph nodes. Such property renders suitability for the detection of lymph nodes metastases by MRI (Qiao et al. 2009).

Novel applications of MNPs for tumor imaging commonly involve the addition of antibodies to MNP surfaces, increasing their specificity to the affected region. For instance, Tse et al. (2015) evaluated the potential MNPs conjugated with an antibody to an extracellular epitope of prostate-specific membrane antigen. In comparison to MNPs without the attached antibody, preclinical imaging of prostate cancer was improved with MNPs coated with antibodies. The prepared MNPs were proven to be nontoxic to prostate cells.

An MRI contrast agent for unstable atherosclerotic plaques was developed by Meier et al. (2015). The researchers developed magnetoliposomes linked to an antibody, which targets activated receptors of platelets. The specific binding to such receptors was confirmed ex vivo and presents an opportunity for further development of a contrast agent for early detection of unstable atherosclerotic plaques.

There has been a tendency of including both antibodies and specific drugs onto MNP coatings, in the context of the theranostics, a new research area which deals with simultaneous therapy and diagnosis of tumors. Consequently, both drug delivery and MR imaging might be carried out simultaneously, allowing an early detection and treatment of degenerative diseases.

5.14 Magnetic Fluid Hyperthermia

In the context of this chapter, hyperthermia or thermotherapy is a treatment of malignant diseases based on the increase of the temperature. Hyperthermia is a medical procedure recognized by the US National Cancer Institute for the treatment of tumors, despite being still under study in clinical trials. In general, such clinical studies are carried out in combination with other forms of cancer therapy, as a complementary treatment to traditional chemotherapy and radiotherapy (Wust et al. 2002).

Magnetic fluid hyperthermia (MFH) consists of the use of MNPs suspended in a biocompatible fluid for local temperature increase. Such temperature increase is based on applying external alternating current magnetic field onto magnetic particles (Nikiforov and Filinova 2009). The external magnetic field causes alignment of magnetic spins, thus reducing magnetic entropy. If the magnetic field is adiabatically removed, magnetic spins tend to their previous random orientation (magnetic relaxation). Hence entropy is generated and temperature is increased (Shull et al. 1992). An alternating magnetic field at frequencies of several hundreds of MHz repeats the cycle of orientation and randomness of magnetic spins and is responsible for the temperature increase on which MFH is based.

Heat is generated locally in MFH treatments by, e.g., implanted RF electrodes, causing concentrated thermal damage on a pathological tissue (e.g., tumor cells). Considering the local effect of MFH, side effects are reduced in comparison to other common treatments such as chemotherapy or radiotherapy (Laurent et al. 2011).

Since the first applications of MFH, many papers on the subject have been published. In fact, the International Journal of Hyperthermia is the official journal of the Society of Thermal Medicine, the European Society for Hyperthermic Oncology, and the Japanese Society for Thermal Medicine. Many articles have been published in the Journal since its first issue in 1985, coping with clinical and biological studies (either in vivo, ex vivo, or in vitro studies), including procedures for production of hyperthermia, modeling analysis, and calibration and control of hyperthermia equipment.

Di Corato et al. (2015) combined MFH and photodynamic therapy for the treatment of tumors (human adenocarcinoma cells grown in vitro and epidermoid carcinoma cells grown in vitro or inoculated in mice) by using photoresponsive magnetoliposomes. The photoresponsive magnetoliposomes were synthesized by alkaline coprecipitation of iron salts (magnetic core), followed by a reverse-phase evaporation method, i.e., an addition of the magnetic suspension (MNPs suspended in buffer) into an emulsion, sonication, and evaporation of organic solvents. Each treatment – either photodynamic therapy or MFH – yielded similar tumor cell death rates in vitro and in vivo when performed separately, whereas the combined MFH-photodynamic therapy was able to completely eradicate the tumor.

A slightly different procedure for the synthesis of MNPs was carried out by Munoz de Escalona et al. (2016). Solid lipid nanoparticles were formulated by the authors by synthesizing magnetic nanoparticles embedded within a lipidic (glyceryl trimyristate) solid matrix. Human HT29 colon adenocarcinoma cells were submitted to MFH in vitro, and the particles presented promising hyperthermia characteristics.

5.15 Conclusion

There are numerous applications for magnetics iron-oxide nanoparticles (IONPs) in pharmaceutical and biomedical areas. The development of new methodologies for the improvement of the synthesis techniques of these IONPs will allow progressing in their utilization. Magnetic nanoparticles allow functionalization with various biomolecules without losing their magnetic properties or change in the original structure, and it is possible to prevent the formation of aggregates and biodegradation. These features are useful for their application, for example, in magnetic resonance contrast media and therapeutic agents in cancer and neurodegenerative disorder treatments, magnetic fluid hyperthermia, and drug and gene delivery, without side effects. The field of iron-oxide nanoparticles has grown rapidly. Many types of researchers have realized, and the results demonstrated that monitoring of the medication could be possible by magnetic resonance imaging and finally, drug delivery in the specific area without affecting other tissue and/or organ, reducing dosages and increasing rapid action.

References

Abolfazl A, Mohamad S, Soodabeh D. Magnetic nanoparticles: preparation, physical properties, and applications in biomedicine. Nanoscale Res Lett. 2012;7:144–1.

Agotegaray M, Palma S, Lassalle V. Novel chitosan coated magnetic nanocarriers for the targeted Diclofenac delivery. J Nanosci Nanotechnol. 2014;14:3343–7.

Agotegaray M, Campelo A, Zysler R, Gumilar F, Bras C, Minetti A, Massheimer V, Lassalle V. Influence of chitosan coating on magnetic nanoparticles in endothelial cells and acute tissue biodistribution. J Biomater Sci Polym Ed. 2016;27:1069–85.

Aharoni A. Magnetic properties of fine particles, Chap. 1. Relaxation processes. Amsterdam: Elsevier; 1992.

Ahn S, Cho J, Kim SI, Yim J, Lee S-G, Kim J-H. Characterization of circulating antibodies with affinity to an epitope used in antibody-conjugated magnetic immunoassays from a case of falsely elevated cyclosporine A. Clin Chim Acta. 2016;458:35–9.

Ai Z, Deng K, Wan Q, Zhang L, Lee S. Facile microwave-assisted synthesis and magnetic and gas sensing properties of Fe3O4 nanoroses. J Phys Chem C. 2010;114:6237–42.

Akash MS, Rehman K. Recent progress in biomedical applications of Pluronic (PF127): pharmaceutical perspectives. J Control Release. 2015;209:120–38.

Ali ME, Rahman MM, Dhahi TS, Kashif M, Sarkar MS, Basirun WJ, Hamid SBA, Bhargava SK. Reference module in materials science and materials engineering, Chap. 10. Nanostructured materials. Amsterdam: Elsevier; 2016.

Alphandéry E. Applications of magnetosomes synthesized by magnetotactic bacteria in medicine. Front Bioeng Biotechnol. 2014;2:5.

Amara D, Margel S. Solventless thermal decomposition of ferrocene as a new approach for the synthesis of porous superparamagnetic and ferromagnetic composite microspheres of narrow size distribution. J Mater Chem. 2011;21:15764–72.

Amjad MS, Sadiq N, Qureshi H, Fareed G, Sabir S. Nano particles: an emerging tool in biomedicine. Asian Pac J Trop Dis. 2015;5:767–71.

Asuha S, Zhao S, Wu HY, Song L, Tegus O. One step synthesis of maghemite nanoparticles by direct thermal decomposition of Fe–urea complex and their properties. J Alloys Compd. 2009;472:L23–5.

Bagheri H, Ranjbari E, Amiri-Aref M, Hajian A, Ardakani YH, Amidi S. Modified fractal iron oxide magnetic nanostructure: a novel and high performance platform for redox protein immobilization, direct electrochemistry and bioelectrocatalysis application. Biosens Bioelectron. 2016;85:814–21.

Bazylinski DA, Schübbe S. Controlled biomineralization by and applications of magnetotactic bacteria. Adv Appl Microbiol. 2007;62:21–62.

Bhandari R, Gupta P, Dziubla T, Hilt JZ. Single step synthesis, characterization and applications of curcumin functionalized iron oxide magnetic nanoparticles. Mater Sci Eng C Mater Biol Appl. 2016;67:59–64.

Bixner O, Reimhult E. Controlled magnetosomes: embedding of magnetic nanoparticles into membranes of monodisperse lipid vesicles. J Colloid Interface Sci. 2016;466:62–71.

Blanco-Andujar C, Ortega D, Southern P, Pankhurst QA, Thanh NT. High performance multi-core iron oxide nanoparticles for magnetic hyperthermia: microwave synthesis, and the role of core-to-core interactions. Nanoscale. 2015;7:1768–75.
Brandão D, Liébana S, Pividori MI. Multiplexed detection of foodborne pathogens based on magnetic particles. New Biotechnol. 2015;32:511–20.
Bronstein LM, Huang X, Retrum J, Schmucker A, Pink M, Stein BD, Dragnea B. Influence of iron Oleate complex structure on iron oxide nanoparticle formation. Chem Mater. 2007;19:3624–32.
Carinelli S, Martí M, Alegret S, Pividori MI. Biomarker detection of global infectious diseases based on magnetic particles. New Biotechnol. 2015;32:521–32.
Caruntu D, Caruntu G, O'Connor CJ. Magnetic properties of variable-sized Fe 3 O 4 nanoparticles synthesized from non-aqueous homogeneous solutions of polyols. J Phys D Appl Phys. 2007;40:5801.
Chen J, Lin Y, Wang Y, Jia L. Cationic polyelectrolyte functionalized magnetic particles assisted highly sensitive pathogens detection in combination with polymerase chain reaction and capillary electrophoresis. J Chromatogr B. 2015;991:59–67.
Chen SJ, Zhang HZ, Wan LC, Jiang SS, Xu YM, Liu F, Zhang T, Ma D, Xie MQ. Preparation and performance of a pH-sensitive cisplatin-loaded magnetic nanomedicine that targets tumor cells via folate receptor mediation. Mol Med Rep. 2016;13:5059–67.
Curtis A, Wilkinson C. Nantotechniques and approaches in biotechnology. Trends Biotechnol. 2001;19:97–101.
Daniel-da-Silva AL, Carvalho RS, Trindade T. Magnetic hydrogel nanocomposites and composite nanoparticles – a review of recent patented works. Recent Pat Nanotechnol. 2013;7:153–66.
Daou TJ, Pourroy G, Bégin-Colin S, Grenèche JM, Ulhaq-Bouillet C, Legaré P, Bernhardt P, Leuvrey C, Rogez G. Hydrothermal synthesis of monodisperse magnetite nanoparticles. Chem Mater. 2006;18:4399–404.
Demortiere A, Panissod P, Pichon BP, Pourroy G, Guillon D, Donnio B, Begin-Colin S. Size-dependent properties of magnetic iron oxide nanocrystals. Nanoscale. 2011;3:225–32.
Dhand C, Dwivedi N, Loh XJ, Jie Ying AN, Verma NK, Beuerman RW, Lakshminarayanan R, Ramakrishna S. Methods and strategies for the synthesis of diverse nanoparticles and their applications: a comprehensive overview. RSC Adv. 2015;5:105003–37.
Di Corato R, Béalle G, Kolosnjaj-Tabi J, Espinosa A, Clément O, Silva AKA, Ménager C, Wilhelm C. Combining magnetic hyperthermia and photodynamic therapy for tumor ablation with photoresponsive magnetic liposomes. ACS Nano. 2015;9:2904–16.
Do TD, Noh Y, Kim MO, Yoon J. An optimized field function scheme for nanoparticle guidance in magnetic drug targeting systems, IEEE/RSJ International Conference on Intelligent Robots and Systems: Hamburg; 2015.
Do TD, Noh Y, Kim MO, Yoon J. In silico magnetic nanocontainers navigation in blood vessels: a feedback control approach. J Nanosci Nanotechnol. 2016;16:6368–73.
Dong W, Zhu C. Use of ethylene oxide in the sol-gel synthesis of [small alpha]-Fe2O3 nanoparticles from Fe(iii) salts. J Mater Chem. 2002;12:1676–83.
Du J, Zhao Y, Yang Z, Xu C, Lu Y, Pan Y, Shi D, Wang Y. Influence of controlled surface functionalization of magnetic nanocomposites on the detection performance of immunochromatographic test. Sensors Actuators B Chem. 2016;237:817–25.
Ebrahiminezhad A, Ghasemi Y, Rasoul-Amini S, Barar J, Davaran S. Preparation of novel magnetic fluorescent nanoparticles using amino acids. Colloid Surf B. 2013;102:534–9.
Farjadian F, Ghasemi S, Mohammadi-Samani S. Hydroxyl-modified magnetite nanoparticles as novel carrier for delivery of methotrexate. Int J Pharm. 2016;504:110–6.
Farzin A, Fathi M, Emadi R. Multifunctional magnetic nanostructured hardystonite scaffold for hyperthermia, drug delivery and tissue engineering applications. Mater Sci Eng C. 2017;70(1):21–31.
Gao G, Liu X, Shi R, Zhou K, Shi Y, Ma R, Takayama-Muromachi E, Qiu G. Shape-controlled synthesis and magnetic properties of monodisperse Fe3O4 nanocubes. Cryst Growth Des. 2010;10:2888–94.

Gao R, Hao Y, Zhang L, Cui X, Liu D, Zhang M, Tang Y, Zheng Y. A facile method for protein imprinting on directly carboxyl-functionalized magnetic nanoparticles using non-covalent template immobilization strategy. Chem Eng J. 2016;284:139–48.

Georgiadou V, Makris G, Papagiannopoulou D, Vourlias G, Dendrinou-Samara C. Octadecylamine-mediated versatile coating of CoFe2O4 NPs for the sustained release of anti-inflammatory drug naproxen and in vivo target selectivity. ACS Appl Mater Interfaces. 2016;8:9345–60.

Gu J-L, Tong H-F, Lin D-Q. Evaluation of magnetic particles modified with a hydrophobic charge-induction ligand for antibody capture. J Chromatogr A. 2016;1460:61–7.

Gun'ko YK, Brougham DF. Nanotechnologies for the life sciences, Chap. Magnetic nanomaterials. Wiley-VCH Verlag GmbH & Co. KGaA, Weinheim, Germany. 2009.

Gupta AK, Gupta M. Synthesis and surface engineering of iron oxide nanoparticles for biomedical applications. Biomaterials. 2005;26:3995–4021.

Gupta AK, Naregalkar RR, Vaidya VD, Gupta M. Recent advances on surface engineering of magnetic iron oxide nanoparticles and their biomedical applications. Nanomedicine-UK. 2007;2:23–39.

Hee Kim E, Sook Lee H, Kook Kwak B, Kim B-K. Synthesis of ferrofluid with magnetic nanoparticles by sonochemical method for MRI contrast agent. J Magn Magn Mater. 2005;289:328–30.

Hola K, Markova Z, Zoppellaro G, Tucek J, Zboril R. Tailored functionalization of iron oxide nanoparticles for MRI, drug delivery, magnetic separation and immobilization of biosubstances. Biotechnol Adv. 2015;33:1162–76.

Hu X, Yu JC, Gong J, Li Q, Li G. α-Fe2O3 nanorings prepared by a microwave-assisted hydrothermal process and their sensing properties. Adv Mater. 2007;19:2324–9.

Hu P, Yu L, Zuo A, Guo C, Yuan F. Fabrication of monodisperse magnetite hollow spheres. J Phys Chem C. 2009;113:900–6.

Hu M, Jiang J-S, Bu F-X, Cheng X-L, Lin C-C, Zeng Y. Hierarchical magnetic iron (iii) oxides prepared by solid state thermal decomposition of coordination polymers. RSC Adv. 2012;2:4782–6.

Hyun DC. Magnetically-controlled, pulsatile drug release from poly(ε-caprolactone) (PCL) particles with hollow interiors. Polymer. 2015;74:159–65.

Jamshaid T, Neto ETT, Eissa MM, Zine N, Kunita MH, El-Salhi AE, Elaissari A. Magnetic particles: from preparation to lab-on-a-chip, biosensors, microsystems and microfluidics applications. Trends Anal Chem. 2016;79:344–62.

Jung H, Park H, Kim J, Lee J-H, Hur H-G, Myung NV, Choi H. Preparation of biotic and abiotic iron oxide nanoparticles (IOnPs) and their properties and applications in heterogeneous catalytic oxidation. Environ Sci Technol. 2007;41:4741–7.

Kalyani S, Sangeetha J, Philip J. Microwave assisted synthesis of ferrite nanoparticles: effect of reaction temperature on particle size and magnetic properties. J Nanosci Nanotechnol. 2015;15:5768–74.

Kaushik A, Jayant RD, Nikkhah-Moshaie R, Bhardwaj V, Roy U, Huang Z, Ruiz A, Yndart A, Atluri V, El-Hage N, Khalili K, Nair M. Magnetically guided central nervous system delivery and toxicity evaluation of magneto-electric nanocarriers. Sci Rep. 2016;6:25309.

Kim KY. Nanotechnologies for the life sciences, Chap. Magnetic nanomaterials. Wiley-VCH Verlag GmbH & Co. KGaA, Weinheim, Germany. 2009.

Kondaveeti S, Cornejo DR, Petri DF. Alginate/magnetite hybrid beads for magnetically stimulated release of dopamine. Colloid Surface B. 2016;138:94–101.

Koneracká M, Kopčanský P, Timko M, Ramchand CN. Direct binding procedure of proteins and enzymes to fine magnetic particles. J Magn Magn Mater. 2002;252:409–11.

Kuznetsov AA, Filippov VI, Kuznetsov OA, Gerlivanov VG, Dobrinsky EK, Malashin SI. New ferro-carbon adsorbents for magnetically guided transport of anti-cancer drugs. J Magn Magn Mater. 1999;194:22–30.

Laurent S, Forge D, Port M, Roch A, Robic C, Vander Elst L, Muller RN. Magnetic iron oxide nanoparticles: synthesis, stabilization, vectorization, physicochemical characterizations, and biological applications. Chem Rev. 2008;108:2064–110.

Laurent S, Dutz S, Häfeli UO, Mahmoudi M. Magnetic fluid hyperthermia: focus on superparamagnetic iron oxide nanoparticles. Adv Colloid Interf. 2011;166:8–23.

Lemine OM, Omri K, Zhang B, El Mir L, Sajieddine M, Alyamani A, Bououdina M. Sol–gel synthesis of 8 nm magnetite (Fe3O4) nanoparticles and their magnetic properties. Superlattice Microst. 2012;52:793–9.

Lewinski N, Colvin V, Drezek R. Cytotoxicity of nanoparticles. Small. 2008;4:26–49.

Liberman A, Mendez N, Trogler WC, Kummel AC. Synthesis and surface functionalization of silica nanoparticles for nanomedicine. Surf Sci Rep. 2014;69:132–58.

Lin X, Ji G, Liu Y, Huang Q, Yang Z, Du Y. Formation mechanism and magnetic properties of hollow Fe3O4 nanospheres synthesized without any surfactant. Cryst Eng Comm. 2012;14:8658–63.

Liu G, Gao J, Ai H, Chen X. Applications and potential toxicity of magnetic iron oxide nanoparticles. Small. 2013;9:1533–45.

López Pérez JA, López Quintela MA, Mira J, Rivas J, Charles SW. Advances in the preparation of magnetic nanoparticles by the microemulsion method. J Phys Chem B. 1997;101:8045–7.

Lübbe AS, Bergemann C, Brock J, McClure DG. Physiological aspects in magnetic drug-targeting. J Magn Magn Mater. 1999;194:149–55.

Mahmoudi M, Sant S, Wang B, Laurent S, Sen T. Superparamagnetic iron oxide nanoparticles (SPIONs): development, surface modification and applications in chemotherapy. Adv Drug Deliv Rev. 2011;63:24–46.

Maity D, Choo S-G, Yi J, Ding J, Xue JM. Synthesis of magnetite nanoparticles via a solvent-free thermal decomposition route. J Magn Magn Mater. 2009;321:1256–9.

Malik MA, Wani MY, Hashim MA. Microemulsion method: a novel route to synthesize organic and inorganic nanomaterials: 1st Nano update. Arab J Chem. 2012;5:397–417.

Marszałł MP. Application of magnetic nanoparticles in pharmaceutical sciences. Pharm Res. 2011;28:480–3.

Martín JI, Nogués J, Liu K, Vicent JL, Schuller IK. Ordered magnetic nanostructures: fabrication and properties. J Magn Magn Mater. 2003;256:449–501.

Massart R. Preparation of aqueous magnetic liquids in alkaline and acidic media. IEEE Trans Magn. 1981;17:1247–8.

Matsunaga T, Kamiya S. Use of magnetic particles isolated from magnetotactic bacteria for enzyme immobilization. Appl Microbiol Biotechnol. 1987;26:328–32.

Medeiros SF, Santos AM, Fessi H, Elaissari A. Stimuli-responsive magnetic particles for biomedical applications. Int J Pharm. 2011;403:139–61.

Meier S, Putz G, Massing U, Hagemeyer CE, von Elverfeldt D, Meissner M, Ardipradja K, Barnert S, Peter K, Bode C, Schubert R, von zur Muhlen C. Immuno-magnetoliposomes targeting activated platelets as a potentially human-compatible MRI contrast agent for targeting atherothrombosis. Biomaterials. 2015;53:137–48.

Mirahmadi-Zare SZ, Allafchian A, Aboutalebi F, Shojaei P, Khazaie Y, Dormiani K, Lachinani L, Nasr-Esfahani M-H. Super magnetic nanoparticles NiFe2O4, coated with aluminum–nickel oxide sol-gel lattices to safe, sensitive and selective purification of his-tagged proteins. Protein Expr Purif. 2016;121:52–60.

Mohapatra M, Anand S. Synthesis and applications of nano-structured iron oxides/hydroxides – a review. Int J Eng Sci Technol. 2010;2:127–46.

Montha W, Maneeprakorn W, Buatong N, Tang IM, Pon-On W. Synthesis of doxorubicin-PLGA loaded chitosan stabilized (Mn, Zn)Fe2O4 nanoparticles: biological activity and pH-responsive drug release. Mater Sci Eng C Mater Biol Appl. 2016;59:235–40.

Müller R, Zhou M, Dellith A, Liebert T, Heinze T. Meltable magnetic biocomposites for controlled release. J Magn Magn Mater. 2017; 431(1): 289–93.

Munoz de Escalona M, Saez-Fernandez E, Prados JC, Melguizo C, Arias JL. Magnetic solid lipid nanoparticles in hyperthermia against colon cancer. Int J Pharm. 2016;504:11–9.

Nikiforov VN, Filinova EY. Magnetic nanoparticles, Chap. 10. Wiley-VCH Verlag GmbH & Co. KGaA; 2009.

Pandey S, Mishra SB. Sol–gel derived organic–inorganic hybrid materials: synthesis, characterizations and applications. J Sol-Gel Sci Technol. 2011;59:73–94.

Pankhurst QA, Connolly J, Jones SK, Dobson J. Applications of magnetic nanoparticles in biomedicine. J Phys D Appl Phys. 2003;36:R167.

Park J, An K, Hwang Y, Park J-G, Noh H-J, Kim J-Y, Park J-H, Hwang N-M, Hyeon T. Ultra-large-scale syntheses of monodisperse nanocrystals. Nat Mater. 2004;3:891–5.

Paulus A, Till N, Franzreb M. Controlling the partitioning behavior of magnetic micro-particles via hydrophobization with alkylamines: tailored adsorbents for continuous bioseparation. Appl Surf Sci. 2015;332:631–9.

Prabhau NN, Kowshik M. Magnetosomes: the bionanomagnets and its potential use in biomedical applications. J Nanomed Res. 2016;3:00057.

Qi H, Yan B, Lu W, Li C, Yang Y. A non-alkoxide sol-gel method for the preparation of magnetite (Fe3O4) nanoparticles. Curr Nanosci. 2011;7:381–8.

Qiao R, Yang C, Gao M. Superparamagnetic iron oxide nanoparticles: from preparations to in vivo MRI applications. J Mater Chem. 2009;19:6274–93.

Qiu G, Huang H, Genuino H, Opembe N, Stafford L, Dharmarathna S, Suib SL. Microwave-assisted hydrothermal synthesis of nanosized α-Fe2O3 for catalysts and adsorbents. J Phys Chem C. 2011;115:19626–31.

Ramimoghadam D, Bagheri S, Hamid SBA. Progress in electrochemical synthesis of magnetic iron oxide nanoparticles. J Magn Magn Mater. 2014;368:207–29.

Raz M, Moztarzadeh F, Ansari Hamedani A, Ashuri M, Tahriri M. Controlled synthesis, characterization and magnetic properties of magnetite (Fe3O4) nanoparticles without surfactant under N2 gas at room temperature. Key Eng Mater. 2012;493-494:746–51.

Reddy LH, Arias JL, Nicolas J, Couvreur P. Magnetic nanoparticles: design and characterization, toxicity and biocompatibility, pharmaceutical and biomedical applications. Chem Rev. 2012;112:5818–78.

Rockenberger J, Scher EC, Alivisatos AP. A new nonhydrolytic single-precursor approach to surfactant-capped nanocrystals of transition metal oxides. J Am Chem Soc. 1999;121:11595–6.

Salazar-Alvarez G, Muhammed M, Zagorodni AA. Novel flow injection synthesis of iron oxide nanoparticles with narrow size distribution. Chem Eng Sci. 2006;61:4625–33.

Shavel A, Liz-Marzan LM. Shape control of iron oxide nanoparticles. Phys Chem Chem Phys. 2009;11:3762–6.

Shen YF, Tang J, Nie ZH, Wang YD, Ren Y, Zuo L. Preparation and application of magnetic Fe3O4 nanoparticles for wastewater purification. Sep Purif Technol. 2009;68:312–9.

Shi D, Huang J, Chuai Z, Chen D, Zhu X, Wang H, Peng J, Wu H, Huang Q, Fu W. Isothermal and rapid detection of pathogenic microorganisms using a nano-rolling circle amplification-surface plasmon resonance biosensor. Biosens Bioelectron. 2014;62:280–7.

Shull RD, McMichael RD, Swartzendruber LJ, Bennett LH. Magnetic properties of fine particles., Chap. 19. Amsterdam: Elsevier; 1992.

Starowicz M, Starowicz P, Żukrowski J, Przewoźnik J, Lemański A, Kapusta C, Banaś J. Electrochemical synthesis of magnetic iron oxide nanoparticles with controlled size. J Nanopart Res. 2011;13:7167–76.

Stocke NA, Meenach SA, Arnold SM, Mansour HM, Hilt JZ. Formulation and characterization of inhalable magnetic nanocomposite microparticles (MnMs) for targeted pulmonary delivery via spray drying. Int J Pharm. 2015;479:320–8.

Sugimoto T. Preparation of monodispersed colloidal particles. Adv Colloid Interface. 1987;28:65–108.

Sun S, Zeng H. Size-controlled synthesis of magnetite nanoparticles. J Am Chem Soc. 2002;124:8204–5.

Sun C, Lee JSH, Zhang M. Magnetic nanoparticles in MR imaging and drug delivery. Adv Drug Deliv Rev. 2008;60:1252–65.

Sun X, Zheng C, Zhang F, Yang Y, Wu G, Yu A, Guan N. Size-controlled synthesis of magnetite (Fe3O4) nanoparticles coated with glucose and gluconic acid from a single Fe(III) precursor by a sucrose bifunctional hydrothermal method. J Phys Chem C. 2009;113:16002–8.

Sun N, Deng C, Liu Y, Zhao X, Tang Y, Liu R, Xia Q, Yan W, Ge G. Optimization of influencing factors of nucleic acid adsorption onto silica-coated magnetic particles: application to viral nucleic acid extraction from serum. J Chromatogr A. 2014;1325:31–9.
Tang YS, Wang D, Zhou C, Ma W, Zhang YQ, Liu B, Zhang S. Bacterial magnetic particles as a novel and efficient gene vaccine delivery system. Gene Ther. 2012;19:1187–95.
Tarasi R, Khoobi M, Niknejad H, Ramazani A, Ma'mani L, Bahadorikhalili S, Shafiee A. β-cyclodextrin functionalized poly (5-amidoisophthalicacid) grafted Fe3O4 magnetic nanoparticles: a novel biocompatible nanocomposite for targeted docetaxel delivery. J Magn Magn Mater. 2016;417:451–9.
Tariq M, Alam MA, Singh AT, Panda AK, Talegaonkar S. Surface decorated nanoparticles as surrogate carriers for improved transport and absorption of epirubicin across the gastrointestinal tract: pharmacokinetic and pharmacodynamic investigations. Int J Pharm. 2016;501:18–31.
Tartaj P, Morales MDP, Veintemillas-Verdaguer S, González-Carreño T, Serna CJ. The preparation of magnetic nanoparticles for applications in biomedicine. J Phys D Appl Phys. 2003;36:R182.
Tartaj P, Morales MP, González-Carreño T, Veintemillas-Verdaguer S, Serna CJ. Advances in magnetic nanoparticles for biotechnology applications. J Magn Magn Mater. 2005:290–291, Part 1: 28–34.
Tehrani MD, Kim MO, Yoon J. A novel electromagnetic actuation system for magnetic nanoparticle guidance in blood vessels. IEEE Trans Magn. 2014;50:1–12.
Thornhill RH, Grant Burgess J, Sakaguchi T, Matsunaga T. A morphological classification of bacteria containing bullet-shaped magnetic particles. FEMS Microbiol Lett. 1994;115:169–76.
Tian Q, Tao K, Li W, Sun K. Hot-injection approach for two-stage formed hexagonal NaYF4:Yb,Er nanocrystals. J Phys Chem C. 2011;115:22886–92.
Tse BW, Cowin GJ, Soekmadji C, Jovanovic L, Vasireddy RS, Ling MT, Khatri A, Liu T, Thierry B, Russell PJ. PSMA-targeting iron oxide magnetic nanoparticles enhance MRI of preclinical prostate cancer. Nanomedicine-UK. 2015;10:375–86.
Uebe R, Schuler D. Magnetosome biogenesis in magnetotactic bacteria. Nat Rev Microbiol. 2016;14:621–37.
Valdora F, Cutrona G, Matis S, Morabito F, Massucco C, Emionite L, Boccardo S, Basso L, Recchia AG, Salvi S, Rosa F, Gentile M, Ravina M, Pace D, Castronovo A, Cilli M, Truini M, Calabrese M, Neri A, Neumaier CE, Fais F, Baio G, Ferrarini M. A non-invasive approach to monitor chronic lymphocytic leukemia engraftment in a xenograft mouse model using ultra-small superparamagnetic iron oxide-magnetic resonance imaging (USPIO-MRI). Clin Immunol. 2016;172:52–60.
Vidal JC, Bertolín JR, Bonel L, Asturias L, Arcos-Martínez MJ, Castillo JR. Rapid determination of recent cocaine use with magnetic particles-based enzyme immunoassays in serum, saliva, and urine fluids. J Pharm Biomed. 2016;125:54–61.
Walton RI. Subcritical solvothermal synthesis of condensed inorganic materials. Chem Soc Rev. 2002;31:230–8.
Wang Y, Zhu Z, Xu F, Wei X. One-pot reaction to synthesize superparamagnetic iron oxide nanoparticles by adding phenol as reducing agent and stabilizer. J Nanopart Res. 2012;14:1–7.
Woo K, Hong J, Choi S, Lee H-W, Ahn J-P, Kim CS, Lee SW. Easy synthesis and magnetic properties of iron oxide nanoparticles. Chem Mater. 2004;16:2814–8.
Wu W, He Q, Jiang C. Magnetic iron oxide nanoparticles: synthesis and surface functionalization strategies. Nanoscale Res Lett. 2008;3:397–415.
Wu L, Yao H, Hu B, Yu S-H. Unique lamellar sodium/potassium iron oxide nanosheets: facile microwave-assisted synthesis and magnetic and electrochemical properties. Chem Mater. 2011;23:3946–52.
Wu W, Wu Z, Yu T, Jiang C, Kim W-S. Recent progress on magnetic iron oxide nanoparticles: synthesis, surface functional strategies and biomedical applications. Sci Technol Adv Mater. 2015;16:023501.
Wust P, Hildebrandt B, Sreenivasa G, Rau B, Gellermann J, Riess H, Felix R, Schlag PM. Hyperthermia in combined treatment of cancer. Lancet Oncol. 2002;3:487–97.

Xu Y, Wang E. Electrochemical biosensors based on magnetic micro/nano particles. Electrochim Acta. 2012;84:62–73.

Xu J-S, Zhu Y-J. [small alpha]-Fe2O3 hierarchically hollow microspheres self-assembled with nanosheets: surfactant-free solvothermal synthesis, magnetic and photocatalytic properties. Cryst Eng Comm. 2011;13:5162–9.

Yan L, Zhang S, Chen P, Liu H, Yin H, Li H. Magnetotactic bacteria, magnetosomes and their application. Microbiol Res. 2012;167:507–19.

Yang C, Bian X, Qin J, Guo T, Zhao S. Fabrication and hyperthermia effect of magnetic functional fluids based on amorphous particles. Appl Surf Sci. 2015;330:216–20.

Yin B, Wang Y, Dong M, Wu J, Ran B, Xie M, Joo SW, Chen Y. One-step multiplexed detection of foodborne pathogens: combining a quantum dot-mediated reverse assaying strategy and magnetic separation. Biosens Bioelectron. 2016;86:996–1002.

Yoffe S, Leshuk T, Everett P, Gu F. Superparamagnetic iron oxide nanoparticles (SPIONs): synthesis and surface modification techniques for use with MRI and other biomedical applications. Curr Pharm Des. 2013;19:493–509.

Zhou H, Hou X, Liu Y, Zhao T, Shang Q, Tang J, Liu J, Wang Y, Wu Q, Luo Z, Wang H, Chen C. Superstable magnetic nanoparticles in conjugation with near-infrared dye as a multimodal Theranostic platform. ACS Appl Mater Interfaces. 2016;8:4424–33.

Zhu Y-J, Chen F. Microwave-assisted preparation of inorganic nanostructures in liquid phase. Chem Rev. 2014;114:6462–555.

Xu Y, Wang E. Electrochemical biosensors based on magnetic micro/nano particles. Electrochim Acta. 2012;84:62–73.
Li S, Zhu Y, [illegible] alpha-Fe2O3 hierarchically hollow microspheres self-assembled with nanosheets: [illegible] magnetic and photocatalytic properties. CrystEngComm. 2011;13:5602–8.
Yan L, Zhang S, Chen P, Liu H, Yin H, Li H. Magnetotactic bacteria, magnetosomes and their application. Microbiol Res. 2012;167:507–19.
Yang C, [illegible] X, Guo L, Zhao S. Fabrication and hyperthermia effect of magnetic functional fluids based on amorphous particles. Appl Surf Sci. 2015;330:[illegible].
Yin H, Wang Y, Dong M, Wu J, Ran B, Xie M, Joo SW, Chen Y. One-step multiplexed detection of foodborne pathogens: combining a quantum dot-mediated reverse assaying strategy and magnetic separation. Biosens Bioelectron. 2016;86:996–1002.
Yoffe S, Leshuk T, Everett P, Gu F. Superparamagnetic iron oxide nanoparticles (SPIONs): synthesis and surface modification techniques for use with MRI and other biomedical applications. Curr Pharm Des. 2013;19:493–509.
Zhou H, Hou X, Liu Y, Zhao T, Shang Q, Tang J, Liu J, Wang Y, Wu Q, Luo Z, Wang H, Chen C. Superstable magnetic nanoparticles in conjugation with near-infrared dye as a multimodal theranostic platform. ACS Appl Mater Interfaces. 2016;8:4424–33.
Zhu YJ, Chen F. Microwave-assisted preparation of inorganic nanostructures in liquid phase. Chem Rev. 2014;114:6462–555.

Chapter 6
Titanium Dioxide Nanoparticles and Nanotubular Surfaces: Potential Applications in Nanomedicine

Ana Rosa Ribeiro, Sara Gemini-Piperni, Sofia Afonso Alves, José Mauro Granjeiro, and Luís Augusto Rocha

Abstract Titanium dioxide nanotubes and nanoparticles are believed to be stable, possess antibacterial properties, and biocompatible and less toxic than other nanostructures, making them excellent candidates for biomedical applications. Among others, they have been widely used as drug-delivery systems, components for articulating orthopaedic implants or cosmetics for dermatological and skin lesion treatments. However, when exposed to the biological environment, selective proteins

A.R. Ribeiro (✉)
Directory of Life Sciences Applied Metrology, National Institute of Metrology, Quality and Technology, Rio de Janeiro, Brazil

Physics Department, University Estadual Paulista, Bauru, São Paulo, Brazil

Postgraduate Program in Clinical and Experimental Dentistry (Odontoclinex), University of Grande Rio, Duque de Caxias, Brazil
e-mail: arribeiro@inmetro.gov.br

S. Gemini-Piperni
Biomaterials Laboratory Applied Physics Department, Brazilian Center for Physics Research-CBPF, Rua Dr. Xavier da Sigaud 150 CEP: 22271110, Rio de Janeiro, Brazil

Brazilian Branch of Institute of Biomaterials, Tribocorrosion and Nanomedicine (IBTN), Chicago, IL, USA

S.A. Alves
Brazilian Branch of Institute of Biomaterials, Tribocorrosion and Nanomedicine (IBTN), Chicago, IL, USA

Center for MicroElectroMechanical Systems, Universidade do Minho, Guimarães, Portugal

J.M. Granjeiro
Directory of Life Sciences Applied Metrology, National Institute of Metrology, Quality and Technology, Rio de Janeiro, Brazil

Brazilian Branch of Institute of Biomaterials, Tribocorrosion and Nanomedicine (IBTN), Chicago, IL, USA

Dental School, Fluminense Federal University, Niterói, Brazil

M. Rai, R. Shegokar (eds.), *Metal Nanoparticles in Pharma*,
DOI 10.1007/978-3-319-63790-7_6

and ions may adsorb to the nanostructures creating a dynamic nano-bio interface that mediate a cellular response. This complex nano-bio interface depends on the physical-chemical characteristics of the nanostructures as well as the specific biological environment. In this chapter, the formation of these biocomplexes (protein and ions) is discussed together with its impact on cellular behaviour. Finally, the potential application of TiO_2 nanoparticles and nanotubes in nanomedicine will be addressed.

Keywords Titanium dioxide • Nanoparticles • Nanotubes • Nano-bio interface • Nanotechnology • Nanomedicine • Nanotoxicology • Regulation

6.1 Introduction

Nanomedicine is the knowledge of preventing, diagnosing, and treating diseases in order to improve the health of patients by means of nanosized materials (Buzea et al. 2007; Moghimi and Farhangrazi 2013). Nanomaterials may have dimensions comparable to some biological molecules, such as proteins, nucleic acids and viruses, and can be classified based on parameters such as origin, chemical composition, size, shape and also application. Compared to bulk materials, nanomaterials have much smaller dimensions leading to a very high surface/volume ratio and surface reactivity. Together with chemical reactivity, mechanical, optical, electrical and magnetic properties are also altered and in some cases improved (Buzea et al. 2007). The photocatalytic activity of titanium dioxide (TiO_2) is a result of the absorption of high energy photons (near UV radiation), resulting in the excitation of electrons from the valence band to the conduction band. The produced electron-hole pairs migrate and react with the organic compounds adsorbed on the surface (Diebold et al. 2003). This photodecomposition involves one or more radicals and superoxide species that are highly reactive leading to rupture of cell and bacteria cell wall, oxidative stress, mutagenesis, DNA damage and respiratory disturbance (Diebold et al. 2003; Buzea et al. 2007).

The goals of the use of nanostructures such as nanoparticles (NPs) in nanomedicine include the improvement of imaging and diagnostic tools, or drug distribution at targeted sites, with an increase of the bioavailability and stability as well as its uptake by tissues and organs (Moghimi and Farhangrazi 2013). The small size of NPs may give them the ability to penetrate biological barriers and translocate to distal parts of

L.A. Rocha
Brazilian Branch of Institute of Biomaterials, Tribocorrosion and Nanomedicine (IBTN), Chicago, IL, USA

Center MicroElectroMechanical Systems, Universidade do Minho, Guimarães, Portugal

Physics Department, University Estadual Paulista, Faculty of Sciences, Bauru, São Paulo, Brazil the human system. Titanium dioxide is the naturally occurring oxide of titanium (Ti) that can exhibit three distinct crystalline forms: anatase, rutile, brookite or a mixture of them (Diebold et al. 2003). Rutile and anatase are the most stable phases being investigated in detail, to be applied in several biomedical applications. Anatase and rutile NPs may be synthesized by different techniques. Solid-state processing routes include, for instances, mechanical alloying and milling or RF thermal plasma. Liquid-state processing routes involve mainly sol–gel and hydrothermal treatments (Diebold et al. 2003). Titanium nanoparticles may also be synthesized by microorganisms (Jha et al. 2009). The processing route has a strong influence on the size distribution of the NPs, shape and crystallographic form. As frequently happens, TiO_2 NPs have distinct characteristics when compared to fine particles of the same composition. Due to their high specific surface area, TiO_2 NPs present a higher reactivity being considered more bioactive. Comparing both crystal forms, anatase is considered more reactive than rutile since charge carriers are excited deeper in the bulk, contributing to high photocatalytic activity and giving rise to a superior toxic capability, namely, against microorganisms (Diebold et al. 2003; Shi et al. 2013). The excellent biocompatibility, high strength-to-weight ratio, wear/corrosion resistance, antimicrobial/drug release capacities, and potential tissue regeneration abilities are some of the key characteristics of TiO_2 NPs that make them attractive for biomedical applications. Applications range from nanotherapeutics for some dermatological diseases, photodynamic therapy for specific cancer diseases or novel sophisticated imaging tools (Buzea et al. 2007; Moghimi and Farhangrazi 2013).

Today the most clinically available osseointegrated Ti implants try to mimic bone's hierarchical structure by incorporating some type of surface modification combining micro- and nanoscale features (Gittens et al. 2014). In particular, the decoration of Ti-based metallic implants with TiO_2 nanotubes (NTs) has attracted broad interest due to their enhanced bio-functionality over conventional Ti implants, which may be ascribed to their morphological and physical-chemical properties. Beyond their promising application in biomedical implants, TiO_2 NTs are also potential candidates to be used as specific drug-delivery systems and biosensors and in photodynamic therapy (Flak et al. 2015; Wang et al. 2016a, b). TiO_2 NTs can be produced by different methods including sol-gel, hydrothermal or anodization methods (Roy et al. 2011). Among them, electrochemical anodization of Ti or Ti alloys is generally preferred, as it enables a good control over the geometry of the NTs, being an easy, low cost and versatile technique (Roy et al. 2011). This technique allows to tune nanotube characteristics regarding their morphology and physical-chemical properties. By adjusting the electrochemical parameters, namely, the anodizing voltage and treatment duration, the diameter and length of the tubes may be easily controlled: while anodization time set the tube length, the diameter is controlled linearly by the applied voltage. The nanotube length may vary from few nanometres to hundred micrometres and the diameter from a few to hundred nanometres (Roy et al. 2011). The nature of the electrolyte (i.e. organic or water-based electrolyte) and its composition are also parameters of upmost importance that beyond nanotube geometry will strongly dictate its chemistry (Wang et al. 2016a, b). TiO_2 NTs produced by anodization are generally characterized by the absence of a crystalline structure, i.e.

an amorphous state, since these are formed at room temperature. The efficiency of many chemical and physical reactions may be optimized by the development of TiO_2 crystalline phases, namely, anatase or rutile (Regonini et al. 2010). Once in contact with the complex biological environment, the physico-chemical properties of TiO_2 nanotubular surfaces may have a strong effect on the biological responses by influencing the biological or cellular interactions. Interestingly, multiple functionalities can be added to NTs by interfacing them with other chemical elements in accordance with a specific application. For dental and orthopaedic applications, the doping of TiO_2 NTs with bioactive elements natively present in bone, such as calcium (Ca) and phosphorus (P), is an interesting strategy to improve the bioactivity of Ti-based metallic implants and so to improve the osteointegration process.

In this chapter, titanium dioxide NPs and NTs will be discussed as the point of view of physical-chemical features, biomedical applications as well as in vitro and in vivo toxicology.

6.2 TiO_2 Nano-bio Interactions and Its Possible Biological Consequences

As soon as nanostructures (NPs or NTs) are into the human body, a complex series of biological events occur through the interactions established between their surface and the surrounding medium. Within a few nanoseconds, water molecules bound to the nanostructure forming a water layer surrounding it (Chug et al. 2013; Sptarshi et al. 2013). At the same time, the adhesion of proteins and other macromolecules (firstly derived from blood, interstitial or cerebrospinal fluid or even saliva) takes place onto the surface, and it is initiated within milliseconds after insertion in the human body. The adsorption of proteins is governed by hydrodynamic, electrodynamic and electrostatic forces, and they may be considered as the biological identity of nanostructures (Sptarshi et al. 2013). This protein corona can be divided in a "soft" layer, in which a fast and dynamic exchange of biomolecules predominates, and a "hard" corona, whose biological molecules have a long lifetime and high affinity to the nanostructure. The hard corona is not found in all kinds of nanostructures, and it is now hypothesized that peptides of the soft corona bind to the hard corona (Sptarshi et al. 2013). After protein adsorption, the subsequent formation of bonds between cell-surface receptors (through integrins) and the functional groups of proteins (ligands) occurs. During adhesion, cells alternately interact with nanostructures through integrins, which transduce external chemical signals or mechanical stimulation, modulating the expression of specific genes and transduction of proteins. This process is widely known as a bidirectional signal transduction, in which cellular functions and material surface properties correlate intimately (Tian et al. 2015). Besides protein adsorption, the high specific surface area of nanostructures results in a strong adhesion of biological ions leading to alteration of surface characteristics, as well as the biological fate. For instances, Ribeiro et al. (2016) observed that TiO_2 NPs with different crystal structures (anatase and rutile) have the

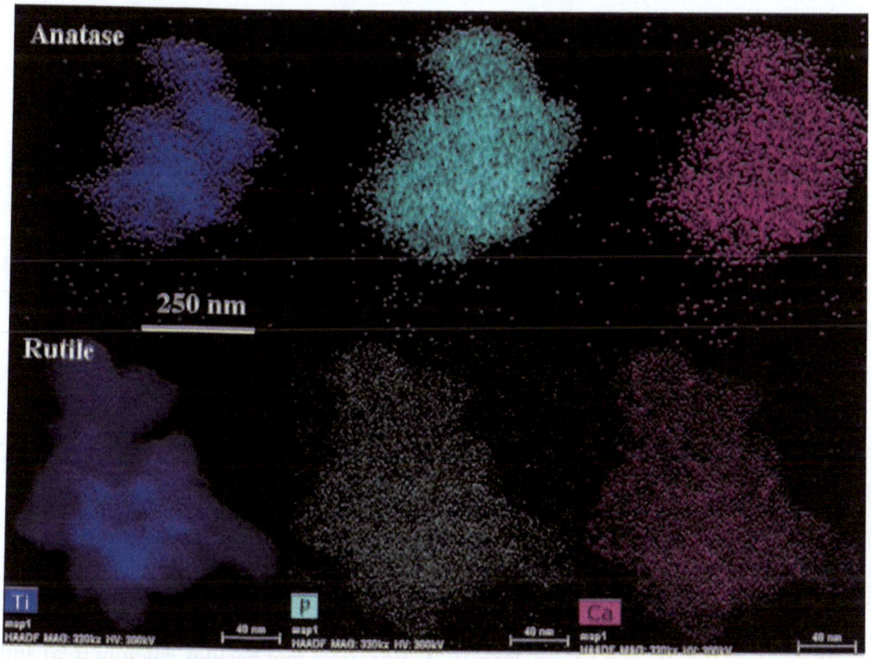

Fig. 6.1 Biocomplex formation in relevant biological milieu: STEM/EDS Ti-K map; STEM/EDS P-K map, STEM/EDS Ca-K map

ability to bound calcium (Ca) and phosphorous (P) ions from conventional medium culture to form an ionic corona, as shown in Fig. 6.1. This phenomenon appears to be independent of the presence of proteins, and it is not exclusive of NPs, as it was already observed for bulk titanium (Pattanayak et al. 2012). Whatever the nature of the protein and ion corona (also called biocomplex), it will become the biological identity seen by living organisms, as the original surface will be hidden by the bio-complex. In other words, a bio-nano interface is developed, promoting several cellular/tissue responses such as cellular adhesion, uptake, migration, differentiation, signalling, accumulation, transport or toxicity.

6.2.1 Nano-bio Interactions at Nanoparticles Surfaces

One of the first events that can occur together with the process of protein and ion adsorption is aggregation and/or agglomeration of NPs, which is related with their specific physico-chemical characteristics, being strongly dependent of temperature, pH, ionic strength and surfactants of the biological medium (Taurozzi et al. 2013). Therefore, for in vitro research purposes, the optimization of a specific protocol of dispersion is required, which result will influence the surface area available for biomolecules adsorption. Nanoparticles bound to biological molecules in biological fluids experience dynamic changes as the particles subsequently move or interact with cells. However, it is already known that for a given nanoparticle, a small change in size, shape and/or surface properties is able to completely alter the nature of the biologically active proteins in the corona and thereby possibly the biological impacts (Spatarshi et al. 2013). Different plasma proteins were adsorbed on TiO_2 nanorods and NTs (Ruh et al. 2012). Both immunoglobulins (IgM and IgG) were the major bound proteins to nanorods, while fibrinogen bound to NTs (Deng et al. 2009). The identification of serum proteins bound to nanomaterials was described by Ruh et al. (2012). They found that fibrinogen showed the strongest affinity for TiO_2 and that the recognized proteins from human serum were diverse comparing the different NPs. Among all the blood proteins, only apolipoprotein A1 was found to be adsorbed to all NPs. Borgognoni et al. *(*2015*)* revealed that the concentration of the corona covering TiO_2 NPs induces IL-6 secretion, which is responsible for the development of autoimmune diseases. The nature of protein corona, its relative abundance and conformation at the NP surface are supposed to disturb bio-distribution, cellular toxicity, uptake as well as intracellular localization of NPs (Caracciolo et al. 2017). The recognition of specific peptides determines particular biological processes, such as receptor-mediated cellular association; particle internalization in cells, tissues and organs; and ability to cross biological barriers. Ribeiro et al. (2016) used a proteomic approach to understand ion and protein corona surrounding TiO_2 in Dulbecco's cell culture medium. Results demonstrate that besides calcium (Ca) and phosphorus (P) adsorption, serum albumin and other glycoproteins such as Alpha 2HS dominate particle surface. Both glycoproteins have a good binding affinity to Ca and are involved in endocytic processes. Moreover, immunoglobulins, complement, fibrinogen and apolipoproteins also adsorb in a lower amount, and it is also known that they are involved at receptor-mediated phagocytosis.

6.2.2 Nano-bio Interactions at Nanotubes Surfaces

Considering implant materials covered with TiO_2 NTs, the exact constitution of adsorbed proteins and their conformational state(s) are strongly controlled by their surface features such as surface energy, hydrophilicity, morphology, topography and chemistry, and these proteins mediate the subsequent cellular adhesion process (Spatarshi et al. 2013). It is known that the surface energy of TiO_2 NTs leads to an increase of initial protein adsorption compared to the amount adsorbed on a compact film (Kullkarni et al. 2016), and therefore, it might play a critical role on the biological responses. Various studies found that hydrophilic surfaces tend to enhance the early stages of cell adhesion, proliferation, differentiation and bone mineralization, influencing the integration of hard and soft tissues with implant, the healing and osseointegration processes (Gittens et al. 2014). It has been shown that the nanoscale dimensions of TiO_2 NTs (e.g. diameter, length and geometry) influence cellular functions such as adhesion, proliferation and differentiation. This behaviour might be related with the initial interactions established between proteins and TiO_2

NTs. Kulkarni et al. (2016) examined the influence of anodic NTs with specific morphology, by controlling simultaneously their diameter (15, 50 or 100 nm) and length (250 nm up to 10 μm), on the amount of bound protein. The authors found out that the NT surface area has a direct influence on the quantity of bound protein. Nonetheless, it was also found that protein physical properties, such as the electric charge and size (in relation to nanotopography and biomaterial's electric charge), are crucial too. Positively charged proteins such as histone, showed an improved adhesion on TiO_2 NT surface because of its negative charge and also due to the existence of more adhesion sites without hindering steric/charge effects. Tian et al. (2015) also investigated the effect of TiO_2 NTs with different morphologies and structures on the biological behaviour of glioblastoma and osteosarcoma cells. The results demonstrated that the nanotube diameter, rather than the crystalline structure of the films, was a major factor influencing cell functions, namely, adhesion, migration and proliferation. The authors proposed that the nanotube diameter, geometry and size could modulate cell behaviour by regulating the degree of transduction of specific proteins. A special focus has been granted to the investigation of nanostructured surfaces, since topographical cues on the nanometre scale have been linked to promote cellular functions. Nevertheless, surface chemistry is also a parameter of upmost importance that regulates biological responses. Different combinations of surface features such as morphology and chemistry might lead to synergistic effects and influence cell behaviour in a unique way due to distinctive physical-chemical properties (Frandsen et al. 2014). Frandsen et al. (2014) studied the effect of TiO_2 NT surface chemistry by coating it with a thin film of tantalum (Ta), which is known to possess excellent osseointegrative properties. The authors concluded that the Ta-coated NTs promoted a faster rate of matrix mineralization and bone-nodule formation, indicating that a surface chemistry modification superimposed on nanotopography provides additional benefit to the bone growth behaviour. Besides, the coating of TiO_2 NTs with gold nanoparticles was achieved by Wang et al. (2016a, b), showing superior osteoblastic cell attachment and proliferation. Electrochemical deposition was implemented by Chen et al. (2014) to introduce nanoscale calcium phosphate (CaP) into well-ordered TiO_2 NTs. The authors observed that a greater amount of BSA was also found on CaP NTs and pre-osteoblasts exhibited enhanced adhesion, proliferation and differentiation on these surfaces. The introduction of nanoscale CaP into self-organized TiO_2 NTs enhanced the bioactivity and osteogenic response, owing to the combined effects of the nanostructured surface topography, chemical composition and hydrophilicity. One key advantage of using TiO_2 NTs is that their features may be adjusted to give a specific feedback in accordance with a specific application. For example, by adjusting the nanotube diameter to 15 nm, it is known that the behaviour of mesenchymal stem cells (MSCs) is significantly improved regarding adhesion, spreading, proliferation and differentiation, and so, these dimensions would be interesting to be used in osseointegrated implants (Kulkarni et al. 2015). Earlier and more stable osseointegration can promote the long-term health of peri-implant tissues, possibly extending the lifetime of the implant (Gittens et al. 2014). On the other hand, it is known that on nanotubes with 100 nm diameter, the cell functions are hindered, leading to cell death and apoptosis (Kulkarni et al. 2015), and therefore, these could be useful for cancer treatment applications.

The mechanism that regulates the interaction between protein corona with nanostructures and biological systems is still largely unknown. A better knowledge of the mechanisms behind those interactions would allow the design of specific nanostructures that could build a specific protein corona when in contact with blood, which could eventually be recognized by specific cells. In fact, the interface between nanostructures and biological systems may involve a dynamic cascade of events which are quite important to explore, since they can regulate whether the nanomaterial is bioavailable and may participate in biocompatible or bio adverse interactions. Mass spectrometry-based techniques are being used to understand the composition of protein corona and to predict biological reactivity of nanostructures, i.e. their cytotoxicity, inflammatory potential and other key properties of these novel nanostructures. However, a major concern is that most of the times, in vitro predictions do not match with observed in vivo behaviour, probably due to the dynamic nature of body fluids, which introduces shear stresses and provides a continual source of new biomolecules (Caracciolo et al. 2017). The use of 'organs-on-a-chip' or multichannel 3D microfluidic cell culture chips, aimed to simulate the physiological response of organs, may contribute for the understanding of protein corona formation and evolution characteristics in more realistic conditions (Caracciolo et al. 2017).

6.3 Toxicology of Titanium Dioxide Nanostructures

6.3.1 Nanoparticles

As it was referred before, the main routes of entrance of titanium dioxide NPs in the body are through oral, dermal and inhalation (Buzea et al. 2007; Shi et al. 2013). As TiO_2 NPs are used in cosmetic formulations due to their photoactivity, skin is considered as the principal tissue in contact with TiO_2. Comparing the different crystallographic structures of NPs, rutile seems to be less phototoxic when compared to anatase. Studies with human keratinocytes reveal that TiO_2 NPs were able to induce mitochondrial damage. Also under UV irradiation, TiO_2 exhibits a dose-dependent phototoxicity in human keratinocytes due to the formation of reactive oxygen species (ROS) (Jaeger et al. 2012; Shi et al. 2013). However, the potential toxicity of TiO_2 NPs depends on their ability to penetrate skin. Regarding this issue, literature shows controversial results. Some in vivo studies carried out with porcine skin exposed TiO_2 NPs of different sizes and showed no penetration in the stratum corneum. TiO_2 NPs (20 nm) present in oil-in-water emulsion formulations were able to penetrate in hairy skin, possibly passing through hair follicles or pores (Zhao et al. 2013). Moreover, a study conducted on damaged pigs skin by ultraviolet B (UVB) showed that treatment with sunscreen formulations containing rutile 14–16 nm TiO_2 enhanced NP's penetration compared to normal skin (Senzui et al. 2010; Shi et al. 2013). Thus, the effect of NPs seems to be dependent on the type of skin and size of NPs.

A major concern in the field of dermatology is that in most of the articles, there is an incipient physiochemical characterization of the nanostructures in the biological relevant matrices. Although in sunscreen formulations NPs are fully agglomerated, there are some NPs that are isolated; they can easily penetrate the skin and enter possibly in blood circulation giving rise to some cytotoxicity events in other organs of the human body. Several studies demonstrate TiO_2 NP's ability to reach blood circulation, accumulating in organs such as liver, kidney and spleen (Shi et al. 2013). The effect of nanoparticles on the organ function is, of course, important.

In vitro studies conducted on hepatoma cell line (HepG2) showed that, even at low concentration (1 μg/ml), TiO_2 NPs induced oxidative stress with DNA damage and apoptosis leading to an increase in the level of several proteins involved in caspase-dependent pathway (Shukla et al. 2012). Moreover, HepG2 showed increased nuclear factor Kappa light-chain enhancer of activated B cell levels after 24 h exposition to TiO_2, suggesting inflammatory activity induction (Prasad et al. 2014). In vivo studies with intraperitoneal injection administrated every 2 days during 20 days of TiO_2 NPs demonstrated that NPs seem to accumulate in rat liver after inducing liver cells apoptosis and hepatic tissue fibrosis around the central vein (Younes et al. 2015). Also, TiO_2 NP accumulation in the liver leads to changes in total protein, glucose, aspartate transaminase, alkaline transaminase and alkaline phosphatase levels, indicating liver damage.

TiO_2 NP toxicity was also demonstrated to affect the kidney, both by in vitro and in vivo studies. Human embryonic kidney cell proliferation was inhibited in a time and dose-dependent manner after exposure to TiO_2 NPs (Meena et al. 2012). Oxidative stress and DNA damage are induced by p53 activation and overexpression of BAX2 and caspase 3, both associated to apoptosis induction. Another study showed TiO_2 NP exposition effects using IP15 glomerular mesangial cells and LLCPK epithelial proximal tubular cells. Results showed cytotoxicity in both cell lines, with abnormal cell size, morphological changes, increased number of vesicle inside the cell and cell detach. Moreover, exposition to high concentrations of TiO_2 NPs showed higher ROS levels (L'Azou et al. 2008). In vivo, TiO_2 NPs increased nephrotoxicity markers, inflammation and fibrosis through Wnt pathway activation (Shi et al. 2013). Furthermore, TiO_2 NP intragastric administration induced increased inflammation resulting in dysfunction, inflammation and cell necrosis by NFKB activation, inflammatory cytokines induction and heat shock protein inhibition (Shi et al. 2013).

TiO_2 NP accumulation in spleen was reported in in vivo studies. Aziz and Awaad (2014) demonstrate a dose-dependent vacuolar degeneration, splenocyte apoptosis leading to decrease in splenic follicle size and organ injury, probably due to caspase 3 increase after a 2-week TiO_2 NP intraperitoneal injection. Moreover, intragastric administration of TiO_2 NPs during 1 month resulted in a lymph node proliferation caused by oxidative stress, lipid peroxidation and ROS accumulation (Wang et al. 2014). In the same way, more extended administration (15–90 days) also results in time-dependent splenic inflammation and necrosis, probably due to activation of mitogen-activated protein kinase and phosphoinositide 3-kinase signalling pathways and to increase cyclooxygenase 2 levels (Sang et al. 2013).

Brain functions may also be affected by TiO_2 NPs (Shi et al. 2013). In vitro studies using a cell line from rat adrenal gland phaeochromocytoma, used as a neurosecretion model, showed time and dose-dependent cytotoxicity after anatase TiO_2 NP exposition. Both rat and human glial cell lines present ability to internalize TiO_2 NPs after 96 h exposition, causing cell proliferation inhibition, morphological changes and apoptosis (Liu et al. 2010). Human corneal epithelial cell also showed cytotoxicity associated with ROS production and DNA breaks correlating with autophagy. Rat primary hippocampal neurons 24 h exposed to anatase TiO_2 showed a dose-dependent reduced viability, lactate dehydrogenase release and apoptosis, associated with mitochondria and endoplasmic reticulum signalling pathway induction (Shi et al. 2013). Furthermore, several in vivo studies demonstrate TiO_2 NP's ability to cross the blood-brain barrier and accumulate in the brain. Essentially, intranasal TiO_2 NP inhalation results in accumulation in the hippocampus.

Titanium is a material commonly used in the orthopaedic and dental field (Simchi et al. 2015). However, its low wear resistance together with corrosion can lead to the degradation of the material resulting in the formation of metallic ions but also micron and nanosized debris (Mathew et al. 2009). This is a concern from the biological point of view since nanosized wear debris can affect different cell types near the vicinity of the osseointegrated implant and, at the same time, enter in blood circulation giving rise to the problems already referred in this chapter. Some of the in vitro studies in this field are with bone cells, where studies with 6 h of TiO_2 NP exposition to human osteoblasts show an improvement in cell survival in a size-dependent manner (Gutwein and Webster 2004). Ribeiro et al. (2016) demonstrated that anatase nanoparticles exposed to primary human osteoblast were very easily internalized by the cells (see Fig. 6.2) and did not interfere with cell toxicity. However, at higher concentration it was possible to observe DNA damage by cell cycle analysis. Controversially, another study indicates a decrease in alkaline phosphatase levels, but at the same time increase in alkaline phosphatase activity and collagen type I production, resulting in an increase in mineralization.

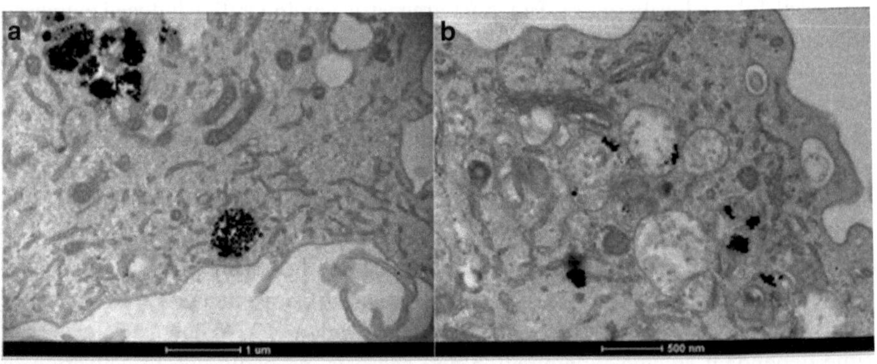

Fig. 6.2 Internalization of NPs: Electron micrographs of osteoblast cells exposed to 100 μg/mL (**a, b**) 5 μg/mL of anatase nanoparticles where they can be seen inside different vesicles

However, these phenomena appear to be dependent on the cell type. A study with MSCs exposed to TiO_2 NPs shows a decrease in osteogenic differentiation with osteopontin, osteocalcin and alkaline phosphatase reduction resulting in a compromise mineralization capability (Hou et al. 2013). Moreover, nanometre-sized Ti showed a higher internalization in periodontal ligament cell line and a decrease in adhesion, migration, viability, proliferation and differentiation of MSCs (Hou et al. 2013). Furthermore, studies demonstrate that TiO_2 exposition to osteoclast precursors induces an increase in cytokines (Shi et al. 2013).

As seen, there are several reports demonstrating toxicity of TiO_2 NPs at higher concentration and time exposure rates. However, there is an absent regulation to testing nanomaterials (Heringa et al. 2016). Although at the moment, there are no reports about human death upon exposure to TiO_2 NPs, a better understanding of the molecular mechanism responsible to TiO_2 NP toxicity is mandatory, leading to the possibility to protect the body against those adverse effects. The functionalization of TiO_2 NPs with specific molecules is an alternative strategy that can be used to improve the use of these nanoparticles (Faure et al. 2013).

6.3.2 Nanotubes

The promising effects of TiO_2 NT surfaces on bone regeneration and repair were reported by Brammer et al. (2012), in line with the studies conducted by Wang et al. (2016a, b), who concluded that TiO_2 NTs can modulate bone formation events at the bone-implant interface as to reach favourable molecular response and osseointegration. The communication of cells with a biomaterial surface is supported by interactions with nanoscale dimensioned extracellular matrix (ECM) components, which stimulate and interact with cell physiology. Nano-featured artificial substrates can impact cell behaviour likewise as the natural ECM, since they dictate cell adhesion, spreading, biochemical functioning and differentiation (Brammer et al. 2012). Several studies have reported the benefits of TiO_2 NTs on cellular functions, which are strongly dependent on their features. Oh et al. (2009) studied the effect of nanotube diameter on cellular functions, and they found out that MSC fate is dictated exclusively by TiO_2 NT dimensions. In accordance with the authors, NTs with shorter diameter promoted cell adhesion without noticeable differentiation, whereas NTs with larger diameter elicited a dramatic stem cell elongation, inducing cytoskeletal stress and selective differentiation into osteoblast-like cells. Lv et al. (2015) also explored the influence of TiO_2 NT diameter on the osteogenic behaviour of human adipose-derived stem cells (hASCs). Both in vitro and in vivo studies demonstrated that the nanoscale geometry influenced cell differentiation. The authors observed that TiO_2 NTs promoted the osteogenic differentiation of hASCs, by upregulating methylation level of histone in the promoter regions of osteogenic gene runt-related transcription factor 2 (Runx2) and osteocalcin. In addition, in vitro and in vivo studies performed by Bjursten et al. (2010) showed that TiO_2 NTs improved osteoblast adhesion and proliferation. TiO_2 NTs implanted in rabbit tibias also improved the bone bonding strength, and histological analysis confirmed greater bone-implant contact, new bone formation and higher calcium and phosphorus levels on the nanotube surfaces. Also, Wilmowsky et al. (2009) explored the effects of TiO_2 NTs implanted in pigs on peri-implant bone formation and concluded that the nanostructured implants influenced bone formation and development, by enhancing osteoblast functions.

Orally administered NTs are suggested to be absorbed from gastrointestinal tract to the blood circulation in mice and rats. However, the toxicity of TiO_2 NTs on cells, tissues and organs has not been investigated yet, and this is an aspect of main importance. To further study TiO_2 NT toxicity and risk assessment, studies should be conducted following standard protocols (Buzea et al. 2007; Roy et al. 2011).

6.4 Applications of Titanium Dioxide Nanostructures in Nanomedicine

6.4.1 Photocatalytic Activity for Cancer, Theranostics and Antibacterial Applications

Cancer incidences have been increasing over time. Although the enormous evolution was observed in the last decades regarding therapies to treat cancer, they still present poor specificity, high cost and sometimes a high toxicity towards healthy cells or organs (Buzea et al. 2007; Moghimi and Farhangrazi 2013). Alternative therapies based on nanomaterials emerged in the last years. Nanoparticles have been widely investigated for cancer therapy. Engineered nanoparticles have been tested to deliver therapeutic compounds directly to specific cells and tissues, or targeted to eliminate cancer cells throughout photothermal, oxidative therapy or autophagic mechanisms (Seo et al. 2007; Rozhkova et al. 2009; Radad et al. 2012). TiO_2 NPs of different sizes and morphologies are in between the NPs investigated for cancer treatment in applications such as skin, bladder, lung, colon brain, oral, osteosarcoma and chondrosarcoma (Seo et al. 2007; Rozhkova et al. 2009; Lucky et al.2016), among others. Cancer cell viability seems to be dependent of NP concentration, surface chemistry, time exposure to UV as well as cell model (Rozhkova et al. 2009; Lucky et al. 2016). However, there are already some reports indicating that TiO_2 NPs may be toxic to cancer cells, in the absence of UV irradiation (Vinardell and Mitjans 2015). In this case, it is suggested that NPs induce oxidative stress due to pro-oxidative functional groups at the NP surface. From the literature, it is known that oxygen species oxidize DNA as well as cellular organelles and interfere with cellular signal pathways leading to necrosis or apoptosis of the cells in target tissue, e.g. tumour, inflammation or infections. In most studies UV radiation is used. Some limitations appear in this strategy, since UV light cannot reach all internal tissues and organs (e.g. gastrointestinal system) and that reactive oxygen species (ROS) generation is brief and insufficient to keep a prolonged cancer-killing effect (Vinardell and Mitjans 2015). To overcome this limitation, new strategies, namely, surface functionalization and the use of Ti core shell nanocomposites, are being explored by near-infrared light (NIR) activation that penetrates deeper in tissues (Zhang et al. 2015; Zeng et al. 2015). Ti-based NPs have been explored also as radiosensitizers to enhance the effect of x-ray in cancer treatments, where Ti peroxide NPs seem to be promising agents to enhance the in vitro and in vivo effects of radiation against a human pancreatic cancer model (Nakayama et al. 2016). The potential use of TiO_2 NTs for localized cancer therapy has been also explored. TiO_2 NTs have revealed potential to be used as customizable and targeted drug-delivery vehicles, capable of carrying large doses of chemotherapeutic agents or therapeutic genes into malignant cells while sparing healthy tissues (Lv et al. 2015,). Beyond that, a wide range of therapeutic agents have been investigated to be conjugated with these nanostructures such as antibiotics, anti-inflammatory drugs, anticancer drugs, bone proteins, peptides, enzymes, vitamins, hormones, genes, antibodies, neurotransmitters, drug nanocarriers and NPs (Gulati et al. 2012; Indira et al. 2014).

Recently, NPs evolved to combine both diagnostic and therapeutic properties, the so-called multifunctional nanoparticles or theranostics (Xie et al. 2010; Zhang et al. 2015; Zeng et al. 2015). This term is referred to as personalized therapy that can be applied to several diseases, where it combines diagnostic and therapeutic capabilities into a single agent. Comparable to other photosensitizers, Ti can be injected in the tumour or infection site. It has previously been suggested that TiO_2 could be utilized as a photodynamic therapeutic agent in cancer but also for other microbial infections (Jha et al. 2009). Roshkova et al. (2009) used 5 nm TiO_2 particles bounded to an antibody that exhibited distinct anti-glioblastoma cell phototoxicity. Some research has developed multifunctional theranostic TiO_2-based nanocomposites which can be used for magnetic resonance imaging (Zhang et al. 2015) or incorporating chemotherapeutic agents for combined chemotherapy and photodynamic therapy (Zeng et al. 2015). Ribeiro et al. (2016) recently studied the interaction of anatase NPs with primary human bone cells. The extensive characterization of NPs in culture medium demonstrates the formation of a biocomplex (ions and protein adsorption) that works as a Trojan-like system for NP internalization in osteoblasts. For higher concentrations of NPs, this system could be used as a photodynamic therapeutic agent for bone cancer, since bone cell cycle analysis reveals some DNA damage possibly due to the presence of hydroxyl radical activity triggering reactive oxygen species formation. Anatase NPs could be injected into the tumour or infection site and excited by light (visible range).

Nowadays, nanostructures are being widely used in nanomedical devices to prevent bacterial infection, since antibiotics already reveal increasing prevalence of antimicrobial resistance and biofilm formation. Therefore, there is an urgent clinical need to develop novel antibacterial therapies to destroy biofilms and reduce healthcare infections. TiO_2 NPs demonstrate antibacterial properties that are reactive with microorganisms having effects such as killing bacteria and virus, sterilization and deodorization (Zhang and Webster 2009; Jesline et al. 2014; Kumeria et al. 2015). TiO_2 NPs display substantial antimicrobial activity against *E. coli, S. aureus, S. mutans, P. gingivalis, P. aeruginosa, E. faecium, B. subtilis*, and *Klebsiella pneumoniae*

(Jesline et al. 2015; Mazare et al. 2016). The cytotoxicity appears to increase in the presence of UV irradiation, where photoactivation of TiO_2 already demonstrates activity against *Bacteroides fragilis, E. coli, E. hire, P. aeruginosa, S. typhimurium* and *S. aureus, A. actinomycetemcomitans* and *F. nucleatum* (Campoccia et al. 2013). Also, TiO_2 NPs were efficient against methicillin-resistant *S. aureus* that is a pathogen responsible for infections which includes skin and soft tissue infections, pneumonia, bacteraemia, surgical site infections and catheter-related infections (Jesline et al. 2014). In skin care products, TiO_2 NPs were proposed as a new therapeutic treatment for acne vulgaris, atopic dermatitis as well as hyperpigmented skin lesions. The use of metallic NPs, namely, TiO_2 on biomedical surfaces such as implants, seems to be an alternative approach to decrease the microorganism colonization and device-associated infection.

In dentistry and orthopaedic fields, bacterial infection is one of the main reasons for Ti-based implant failure. Mazare et al. (2016) recently studied the antibacterial activity of annealed TiO_2 NTs, showing their potential to inhibit *S. aureus* and *P. aeruginosa* growth. Beyond that, the nanostructured surfaces demonstrated enhanced haemocompatibility, highlighting their potential to be used in medical devices that come in contact with blood, as dental and orthopaedic implants. The drug-eluting features of TiO_2 NT decorated implants have gained increased attention to act as localized delivery systems of different therapeutics. The NTs may be tuned with antimicrobial properties by loading or doping them with different antibacterial agents such as antibiotics (e.g. gentamicin and vancomycin), peptides or nanoparticles (e.g. silver, zinc and gold) (Jesline et al.2014; Roguska et al. 2015; Kumeria et al. 2015; Mazare et al. 2016). Several strategies have been found to control drug release from NTs up to several days, by coating the surface with biocompatible, biodegradable and antibacterial polymers, and so to overcome one of the main limitations of the existing drug-delivery systems, which is the drug release in an unpredictable way (Bjursten et al. 2010). Roguska et al. (2015) loaded TiO_2 NTs with metallic (silver) and metallic oxide (ZnO-zinc oxide) NPs. The authors observed that an increase of loaded ZnO amount unexpectedly altered the structure of nanoparticle-nanolayer, caused partial closure of nanotube interior and significantly reduced the solubility of ZnO and antibacterial efficacy. Zhang and Webster (2009) loaded TiO_2 NTs with vancomycin, and both in vitro and in vivo studies showed good biocompatibility and antibacterial efficacy. The excellent in vitro bioactivity and enhanced corrosion resistance of TiO_2 NTs were also shown by Indira et al. (2014), after incorporation of strontium ions into NTs. Gulati et al. (2012) produced biocompatible polymer-coated TiO_2 NTs, previously loaded with water-insoluble anti-inflammatory drug indomethacin. The results showed the capability of this system for delivering a drug to a bone site over an extended period of time and with predictable kinetics and, simultaneously, induced to an excellent osteoblast adhesion and proliferation. Another example is the work recently carried out by Kumeria et al. (2015), who have reported the development of biopolymer-coated drug-releasing NTs able to promote osteoblast adhesion and proliferation, while effectively preventing *S. epidermidis* colonization by hindering their proliferation and biofilm formation.

6.4.2 *Biocompatibility and Bio-reactivity of TiO_2 Nanostructures for Tissue Engineering and Regenerative Medicine*

Nowadays, the diseases with high significant clinical problems are associated with bone ruptures, osteoarthritis, osteoporosis and also bone cancers (Salgado et al. 2004). In the field of bone regenerative medicine, proteins of interest adsorbed into biodegradable but also non-biodegradable NPs are being used to be injected directly on the site of the injury. For example, bone morphogenetic proteins (BMPs) and transforming growth factor β (TGF-β) are being used for the enhancement of osteogenic differentiation (Carreira et al. 2015). A potential role of TiO_2 NPs is the release control of BMPs, since one important issue in BMP therapy is the risk of tumour induction due the use of high doses.

Nevertheless, it is essential that the therapeutic combined with NPs does not compromise the biological activity of all the cells that constitute bone. Taking this approach in consideration, and the reactivity of TiO_2 surface to adsorb proteins, it is believed that the combination of anatase NPs with a pharmacologic drug could be explored as a drug-delivery system for bone tissue regeneration. However, TiO_2 NPs could be also used combined with osteoporosis-antagonizing drugs to control and decrease the osteoclast activity that leads to bone resorption. Gene therapy is also an interesting concept, where specific plasmids (encoding transcription factors, growth factors, and even hormones) can be bound to NPs to be delivered into bone cells. Most of the studies are related with the use of transfection to enhance bone regeneration in the case of fractures but also malignant bone tumours. Ribeiro et al. (2016) also found that calcium ion adsorption is not exclusive from bulk Ti. Anatase NPs can adsorb calcium and phosphate ions from the biological milieu to its surface, and more interesting is the nucleation of hydroxyapatite crystals that occur that can be applied for bone tissue therapies. Gupta and Reviakine (2012) studied the effect of platelet activation in Ti surfaces and the role of calcium in its activation; they could show that in the presence of calcium, TiO_2 activate adhering platelets. These results are very interesting for osseointegrated implants where it is beneficial blood coagulation and platelet activation for implant integration.

By rational design, this bio-camouflage may create protein and ion-enriched environment that can, alone or in combination with drugs/molecules, promote regeneration and improve implant-tissue interfaces. For instance, this bio-camouflage that also occurs in human plasma could be used as a neuroprotective system to capture the massive calcium influx that occurs in ischemic stroke that is a major cause of death or disability of humans. Nowadays, the new tendency in regenerative medicine is to use ions released from metallic biomedical devices (e.g. hip prostheses, dental implants) to immunomodulate and heal tissues and organs (Vasconcelos et al. 2016). This approach can create ionic, debris and protein environment that in combination with specific molecules (e.g. drugs) can promote tissue regeneration but also improve the implant-tissue interface and the quality of life of the patient.

Frequently, NPs can be also combined with different scaffolds to regenerate tissues and organs (Feng et al. 2016). For example, TiO_2 NPs were introduced in PLLA scaffolds to improve the physicochemical characteristics of the scaffolds. Results demonstrate that NPs improve the polymeric degradation, highlighting the potential application of this system in tissue engineering.

The selection of the nanotube features is of crucial importance, and it must be accomplished based on its final application. These nanostructures may be easily produced with different morphology, topography, length and chemistry. The functionalization of NTs may be achieved by several techniques, and it may strongly dictate the cellular responses. A very interesting functionalization study was carried out by Shen et al. (2015) in which TiO_2 NT surface was covered by ECM secreted from human umbilical artery muscle cells, which significantly improved the proliferation of human umbilical vein endothelial cells. An additional motivating work was recently conducted by Khoshroo et al. (2017), who have incorporated TiO_2 NTs into a polycaprolactone (PCL) scaffold for improving mechanical, biological and physicochemical properties of a composite for bone tissue engineering. The results showed that PCL-TiO_2 NT scaffold presented an increased cell viability and evidenced to be a good substrate for cell differentiation leading to ECM mineralization. Kendall and Lynch (2016) have created a Ti implantable device by growing TiO_2 NTs on Ti for loading bone morphogenetic protein 2 (BMP2). The nano-reservoirs were covered with biomimetic multilayered coatings of gelatin/chitosan for controlled release of BMP2, which has promoted osteoblastic differentiation of MSCs.

6.5 Regulatory Challenges Applied to Nanostructures

Nanotechnology advances have led to the use of nanomaterials and nanodevices for diagnostic and therapeutic purposes. At the moment, there is already several FDA-approved nanotherapeutic and nanodiagnostic structures for clinical use; however, some of them in clinical trials had denied authorization due to poor specificity but also high number of false negatives. In some cases, the unique physical and chemical properties that make nanomaterials so appealing may have a dual effect that is associated with their potentially toxic effect on cells, tissues and organs, causing side effects. The knowledge of their metabolism and distribution in the human body is mandatory. In the regulatory background, NPs are mostly treated as chemicals where there is no specific guideline for their regulation, although some scientific opinions were already reported. Currently, the EU guidelines for testing medicines do not require any nanospecific protocols; however, reviewing published literature of nanoparticle for therapeutic applications relates some cases of toxicity (Kendall and Lynch 2016). Therefore, the need to include information about size, exposure, dose, the population involved, and the life cycle evaluation of their impact in human health is urgent and mandatory. Till the moment, the risk assessment and management of TiO_2 NPs and its implications in human health effects are still incipient although TiO_2 NPs have been studied significantly in recent years. Till the moment it was not defined occupational or environmental limits to TiO_2 NPs by any other regulatory agency (Shi et al. 2013). The International Agency for Research on Cancer indicates that TiO_2 NPs have possible human carcinogen (group 2B), where there is a vast literature demonstrating that TiO_2 nanoparticle exposure leads to cell death (Shi et al. 2013). Most of the toxicological studies suggest that anatase NPs are cytotoxic and genotoxic; however, in vitro and in vivo results are conflicting possibly due to the high doses that were administrated as well as by the incipient physicochemical characterization of the NPs in the biological milieu. Whether human exposure to TiO_2 NPs triggers reproductive and developmental toxicities is unclear. There is an emerging concern about the potential effects of TiO_2 NPs on health; however, specific mechanisms and pathways related to its carcinogenicity remain unknown (Shi et al. 2013).

There is a world consensus that nanomaterials require a more arduous assessment of their potential effect in human health due to their surface area. The specific mechanisms and pathways through which NPs can exert their toxicity are multiple and not so well clarified. NPs can induce intracellular (DNA damage, mitochondrial dysfunction, ROS formation) but also extracellular (release of cytokines) processes that can go beyond cellular to tissue and organ level compromising patient's health. Basic issues concerning the exposure of nanomaterials in the human body have not been explored. Till the moment, scientists are not able to understand how NPs penetrate and how they are distributed inside the body; what are the mechanisms of action, the toxico-kinetics and transport across biological barriers; as well as how NPs are degraded and excreted. The gap in knowledge is elevated, where there is the need to develop precise methods both analytical and biological, as well as validated in vitro and in vivo models for the safety assessment of NPs. The tools for in situ evaluation and detection of therapeutic NPs must be established. Profound investigations of the use of TiO_2 NPs in human health are necessary to fully elucidate whether these NPs should be used in biomedical applications. It is known that cytotoxicity, as in any pharmacological substance, depends on the dose and time of the contact. Most of the in vitro and in vivo data with low doses of TiO_2 NP exposure demonstrate absent toxicity. Besides in vitro studies, before animal studies, the use of in silico nanotoxicology systems such as 'organ-on-a- chip' devices may be useful. These approaches will follow the 3Rs approach improving cell culture models and giving rise to accurate, high-throughput and cost-effective semi-in vivo stages for toxicology applications. In fact, Organization for Economic Co-operation and Development (OCDE) guides are accepted as a starting point to evaluate the hazard of NPs, although some adaptation in the methods could be required. Although the methods for safety assessment of therapeutic NPs for human health were not established and defined, long exposure analysis (chronic exposure) in human should be assessed as well as NP biotransformation, degradation, excretion, adsorption and distribution in the human body. Likewise, it is necessary that patients understand the risk, if any, related with the use of therapeutic NPs and decide based on the information available to them.

6.6 Conclusion

Nanotubular structures of TiO_2 have demonstrated unique morphological and physicochemical features to stimulate implant bone integration, both in vitro and in vivo, and therefore become very promissory candidates to improve the clinical outcomes of osseointegrated implants. Additionally, both nanostructures (nanotubes and nanoparticles) have potential to behave as a platform for drug-eluting and local delivery and may be tailored with multiple functionalities due to their ability to incorporate several bioactive elements into their structure. Beyond that, TiO_2 NTs and NPs are also promising candidates to be used for localized cancer thermotherapy. However, the potential local and systemic toxicity effects of TiO_2 nanostructures remain unclear, and therefore, the study of the mechanisms and pathways involved on the interactions between TiO_2 NTs and the complex biological structures existing in the human body must be urgently conducted following standard protocols.

References

Aziz H, Awaad A. Titanium dioxide (TiO_2) nanoparticles induced apoptosis of splenocytes in adult male albino rat and the protective role of milk thistle seeds extract. Int J Adv Res. 2014;2:732–46.

Bjursten LM, Rasmusson L, Oh S, Smith GC, Brammer KS, Jin S. Titanium dioxide nanotubes enhance bone bonding in vivo. J Biomed Mater Res. 2010;92:1218–24.

Borgognoni CF, Mormann M, Qu Y, Schäfer M, Langer K, Öztürk C, Wagner S, Chen C, Zhao Y, Fuchs H, Riehemann K. Reaction of human macrophages on protein corona covered Tio_2 nanoparticles. Nanomedicine. 2015;2:275–82.

Brammer KS, Frandsen CJ, Jin S. TiO_2 nanotubes for bone regeneration. Trends Biotechnol. 2012;30:315–22.

Buzea C, Pacheco II, Robbie K. Nanomaterials and nanoparticles: sources and toxicity. Biointerphases. 2007;2(4):MR17.

Campoccia D, Montanaro L, Arciola CR. A review of the biomaterials technologies for infection-resistant surfaces. Biomaterials. 2013;34:8533–54.

Caracciolo G, Farokhzad OC, Mahmoudi M. Biological identity of nanoparticles in vivo: clinical implications of the protein corona. Trends Biotechnol. 2017;35:257–64.

Carreira ACO, Zambuzzi WF, Rossi MC, Filho RA, Sogayar MC, Granjeiro JM. Bone morphogenetic proteins: promising molecules for bone healing, bioengineering, and regenerative medicine. Vitam Horm. 2015;99:293–322.

Chen J, Zhang Z, Ouyang J, Chen X, Xu Z. Bioactivity and osteogenic cell response of TiO_2 nanotubes coupled with nanoscale calcium phosphate via ultrasonification-assisted electrochemical deposition. Appl Surf Sci. 2014;305:24 32.

Chug A, Shukla S, Mahesh L, Jadwani S. Osseointegration—molecular events at the bone–implant interface: a review. J Oral Maxillofac Surg Med Pathol. 2013;25:1–4.

Deng ZJ, Mortimer G, Schiller T, Musumeci A, Martin D, Minchin RF. Differential plasma protein binding to metal oxide nanoparticles. Nanotechnology. 2009;20:455101.

Diebold U. The surface science of titanium dioxide. Surf Sci Rep. 2003;48:53–229.

Faure B, Salazar-Alvarez G, Ahniyaz A, Villaluenga I, Berriozabal G, De Miguel YR, Bergström L. Dispersion and surface functionalization of oxide nanoparticles for transparent photocatalytic and UV-protecting coatings and sunscreens. Sci Technol Adv Mater. 2013;14:023001–24.

Feng T, Xufeng N, Xiaoming L, Qingling F, Yubo F. Porous poly(L-lactic acid) scaffold reinforced by titanium dioxide nanoparticles. J Biomat Tissue Eng. 2016;6(6):478–483.

Flak D, Coy E, Nowaczyk G, Yate L, Jurga S. Tuning the photodynamic efficiency of TiO_2 nanotubes against HeLa cancer cells by Fe-doping. RSC Adv. 2015;5:85139–52.

Frandsen CJ, Brammer KS, Noh K, Johnston G, Sungho J. Tantalum coating on TiO_2 nanotubes induces superior rate of matrix mineralization and osteofunctionality in human osteoblasts. Mater Sci Eng C Mate Biol Appl. 2014;37:332–41.

Gittens RA, Scheideler L, Rupp F, Hyzy SL, Geis-Gerstorfer J, Schwartz Z, Boyan B. A review on the wettability of dental implant surfaces II: biological and clinical aspects. Acta Biomater. 2014;10:2907–18.

Gulati K, Ramakrishnan S, Aw MS, Atkins GJ, Findlay DM, Losic D. Biocompatible polymer coating of titania nanotube arrays for improved drug elution and osteoblast adhesion. Acta Biomater. 2012;8:449–56.

Gupta S, Reviakine I. Platelet activation profiles on TiO_2: effect of Ca^{2+} binding to the surface. Biointerphases. 2012;7:1–12.

Gutwein LG, Webster TJ. Increased viable osteoblast density in the presence of nanophase compared to conventional alumina and titania particles. Biomaterials. 2004;25:4175–83.

Heringa MB, Geraets L, van Eijkeren JCH, Vandebriel RJ, de Jong WH, Oomen AG. Risk assessment of titanium dioxide nanoparticles via oral exposure, including toxicokinetic considerations. Nanotoxicology. 2016;11:1–11.

Hou Y, Cai K, Li J, Chen X, Lai M, Hu Y, Uo Z, Ding X, Xu D. Effects of titanium nanoparticles on adhesion, migration, proliferation, and differentiation of mesenchymal stem cells. Int J Nanomedicine. 2013;8:3619–30.

Indira K, Mudali UK, Rajendran N. In-vitro biocompatibility and corrosion resistance of strontium incorporated TiO_2 nanotube arrays for orthopaedic applications. J Biomater Appl. 2014;1:113–29.

Jaeger A, Weiss DG, Jonas L, Kriehuber R. Oxidative stress-induced cytotoxic and genotoxic effects of nano-sized titanium dioxide particles in human HaCaT keratinocytes. Toxicology. 2012;296:27–36.

Jesline A, John NP, Narayanan PM, Vani C, Murugan S. Antimicrobial activity of zinc and titanium dioxide nanoparticles against biofilm-producing methicillin-resistant Staphylococcus aureus. Appl Nanosci. 2014;5:157–62.

Jesline et al 2015: A. JeslineNeetu P. JohnP. M. NarayananC. VaniSevanan Muruga, Antimicrobial activity of zinc and titanium dioxide nanoparticles against biofilm-producing methicillin-resistant Staphylococcus aureus,Appl Nanosci (2015) 5:157–162.

Jin Xie, Seulki Lee, and Xiaoyuan Chen, Nanoparticle-based theranostic agents, Adv Drug Deliv Rev. 2010 August 30; 62(11): 1064–1079. doi:10.1016/j.addr.2010.07.009.

Jha AK, Prasad K, Kulkarni AR. Synthesis of TiO_2 nanoparticles using microorganisms. Colloids Surf B: Biointerfaces. 2009;71:226–9.

Kendall M, Lynch I. Long-term monitoring for nano- medicine implants and drugs. Nat Nanotecnol. 2016;1:206–10.

Khoshroo K, Jafarzadeh Kashi TS, Moztarzadeh F, Tahriri M, Jazayeri HE, Tayebi L. Development of 3D PCL microsphere/TiO_2 nanotube composite scaffolds for bone tissue engineering. Mater Sci Eng C Mater Biol Appl. 2017;1:586–98.

Kulkarni M, Mazare A, Gongadze E, Perutkova Š, Kralj-Iglič V, Milošev I, Schmuki P, Iglic A, Mozetic M. Titanium nanostructures for biomedical applications. Nanotechnology. 2015;22:1–18.

Kulkarni M, Mazare A, Park J, Gongadze E, Killian MS, Kralj S, Von der Mark K, Iglič A, Schmuki P. Protein interactions with layers of TiO_2 nanotube and nanopore arrays: morphology and surface charge influence. Acta Biomater. 2016;45:357–66.

Kumeria T, Mon H, Aw MS, Gulati K, Santos A, Griesser HJ, Losic D. Advanced biopolymer-coated drug-releasing titania nanotubes (TNTs) implants with simultaneously enhanced osteoblast adhesion and antibacterial properties. Colloids Surf B: Biointerfaces. 2015;30:255–63.

L'Azou B, Jorly J, On D, Sellier E, Moisan F, Fleury-Feith J, Cambar J, Brochard P, Ohayon-Courtès C. In vitro effects of nanoparticles on renal cells. Part Fibre Toxicol. 2008;5:22–14.
Liu S, Xu L, Zhang T, Ren G, Yang Z. Oxidative stress and apoptosis induced by nanosized titanium dioxide in PC12 cells. Toxicology. 2010;12:172–7.
Lucky SS, Idris NM, Huang K, Kim J, Li Z, Thong PSP, Xu R, Soo KC, Zhang Y. In vivoBiocompatibility, biodistribution and therapeutic efficiency of Titania coated Upconversion nanoparticles for photodynamic therapy of solid oral cancers. Theranostics. 2016;6:1844–65.
Lv L, Liu Y, Zhang P, Zhang X, Liu J, Chen T, Su P, Li H, Zhou Y. The nanoscale geometry of TiO_2 nanotubes influences the osteogenic differentiation of human adipose-derived stem cells by modulating H3K4 trimethylation. Biomaterials. 2015;39:193–205.
Mathew MT, Srinivasa Pai P, Pourzal R, Fischer A, Wimmer MA. Significance of Tribocorrosion in biomedical applications: overview and current status. Adv Tribol. 2009;9:1–12.
Mazare A, Totea G, Burnei C, Schmuki P, Demetrescu I, Ionita D. Corrosion, antibacterial activity and haemocompatibility of TiO_2 nanotubes as a function of their annealing temperature. Corros Sci. 2016;103:215–22.
Meena R, Rani M, Pal R, Rajamani P. Nano-TiO_2-induced apoptosis by oxidative stress-mediated DNA damage and activation of p53 in human embryonic kidney cells. Appl Biochem Biotechnol. 2012;67:791–808.
Moghimi SM, Farhangrazi ZS. Nanomedicine and the complement paradigm. Nanomedicine: nanotechnology, biology and medicine. Elsevier. 2013;9:458–60.
Nakayama M, Sasaki R, Ogino C, Tanaka T, Morita K, Umetsu M. Titanium peroxide nanoparticles enhanced cytotoxic effects of X-ray irradiation against pancreatic cancer model through reactive oxygen species generation in vitro and in vivo. Radiat Oncol. 2016;11:91–1.
Oh S, Brammer KS, Li YSJ, Teng D, Engler AJ, Chien S, Sungho J. Stem cell fate dictated solely by altered nanotube dimension. Proc Natl Acad Sci. 2009;106:2130–5.
Pattanayak DK, Yamaguchi S, Matsushita T, Nakamura T, Kokubo T. Apatite-forming ability of titanium in terms of pH of the exposed solution. J R Soc Interface. 2012;9:2145–55.
Prasad RY, Simmons SO, Killius MG, Zucker RM, Kligerman AD, Blackman CF, Fry RC, Demarini DM. Cellular interactions and biological responses to titanium dioxide nanoparticles in HepG2 and BEAS-2B cells: role of cell culture media. Environ Mol Mutagen. 2014;55:336–42.
Radad K, Al-Shraim M, Moldzio R, Rausch W-D. Recent advances in benefits and hazards of engineered nanoparticles. Environ Toxicol Pharmacol. 2012;34:661–72.
Regonini D, Jaroenworaluck A, Stevens R, Bowen CR. Effect of heat treatment on the properties and structure of TiO_2 nanotubes: phase composition and chemical composition. Surf Interface Anal. 2010;42:139–44.
Ribeiro AR, Gemini-Piperni S, Travassos R, Lemgruber L, Silva RC, Rossi AL, Farina M, Anselme K, Shokuhfar T, Shahbazian-Yassar R, Borojevic R, Rocha LA, Werckmann J, Granjeiro JM. Trojan-like internalization of Anatase titanium dioxide nanoparticles by human osteoblast cells. Sci Rep. 2016;6:23615.
Roguska A, Belcarz A, Pisarek M, Ginalska G, Lewandowska M. TiO_2 nanotube composite layers as delivery system for ZnO and Ag nanoparticles – an unexpected overdose effect decreasing their antibacterial efficacy. Mat Sci Eng C. 2015;51:158–66.
Roy P, Berger S, Schmuki P. TiO_2 nanotubes: synthesis and applications. Angew Chem Int Ed Engl. 2011;50:2904–39.
Rozhkova EA, Ulasov I, Lai B, Dimitrijevic NM, Lesniak MS, Rajh T. A high-performance nanobio photocatalyst for targeted brain cancer therapy. Nano Lett. 2009;9:3337–42.
Ruh H, Kühl B, Brenner-Weiss G, Hopf C, Diabaté S, Weiss C. Identification of serum proteins bound to industrial nanomaterials. Toxicol Lett. 2012;208:41–50.
Salgado AJ, Coutinho OP, Reis RL. Bone tissue engineering: state of the art and future trends. Macromol Biosci. 2004;4:743–65.
Sang X, Li B, Ze Y, Hong J, Ze X, Gui S, Liu H, Zhao X, Sheng L, Liu D, Yu X, Wang L, Hong F. Toxicological mechanisms of nanosized titanium dioxide-induced spleen injury in mice after repeated peroral application. J Agric Food Chem. 2013;61:5590–9.

Saptarshi SR, Duschl A, Lopata AL. Interaction of nanoparticles with proteins: relation to bioreactivity of the nanoparticle. J Nanobiotechnol BioMed. 2013;11:26.
Senzui M, Tamura T, Miura K, Ikarashi Y, Watanabe Y, Fujii M. Study on penetration of titanium dioxide (TiO_2) nanoparticles into intact and damaged skin in vitro. J Toxicol Sci. 2010;35:107–13.
Seo J-W, Chung H, Kim M-Y, Lee J, Choi I-H, Cheon J. Development of water-soluble single-crystalline TiO_2 nanoparticles for photocatalytic cancer-cell treatment. Small. 2007;3:850–3.
Shen F, Zhu Y, Li X, Luo R, Tu Q, Wang J. Vascular cell responses to ECM produced by smooth muscle cells on TiO_2 nanotubes. Appl Surf. 2015;349:589–98.
Shi H, Magaye R, Castranova V, Zhao J. Titanium dioxide nanoparticles: a review of current toxicological data. Part Fibre Toxicol. 2013;l10:15.
Shukla RK, Kumar A, Gurbani D, Pandey AK, Singh S, Dhawan A. TiO_2 nanoparticles induce oxidative DNA damage and apoptosis in human liver cells. Nanotoxicology. 2012;7:48–60.
Simchi A, Mazaheri M, Eslahi N, Ordikhani F, Tamjid E. Nanomedicine applications in orthopedic medicine: state of the art. Int J Nanomedicine. 2015;10:6039–53.
Taurozzi JS, Hackley VA, Wiesner MR. A standardised approach for the dispersion of titanium dioxide nanoparticles in biological media. Nanotoxicology. 2013;7:389–401.
Tian A, Qin X, Wu A, Zhang H, Xu Q, Xing D, He Y, Qiu B, Xue X, Zhang D, Dong C. Nanoscale TiO_2 nanotubes govern the biological behavior of human glioma and osteosarcoma cells. Int J Nanomed. 2015;10:2423–39.
Vasconcelos DM, Santos SG, Lamghari M, Barbosa MA. The two faces of metal ions: from implants rejection to tissue repair/regeneration. Biomaterials. 2016;22:1–47.
Vinardell M, Mitjans M. Antitumor activities of metal oxide nanoparticles. Nano. 2015;5:1004–21.
Wang Y, Cui H, Zhou J, Li F, Wang J, Chen M. CytotoxicityDNA damage, and apoptosis induced by titanium dioxide nanoparticles in human non-small cell lung cancer A549 cells. Environ Sci Pollut Res. 2014;22:5519–30.
Wang C, Bai Y, Bai Y, Gao J, Ma W. Enhancement of corrosion resistance and bioactivity of titanium by Au nanoparticle-loaded TiO_2 nanotube layer. Surf Coatings Technol. 2016a;286:327–34.
Wang J, Li H, Sun Y, Bai B, Zhang Y. Anodization of highly ordered TiO_2 nanotube arrays using orthogonal design and its wettability. Int J Electrochem. 2016b;11:710–23.
Wilmowsky von C, Bauer S, Lutz R, Meisel M, Neukam FW, Toyoshima T, Nkenke E, Schlegel KA. In vivo evaluation of anodic TiO_2 nanotubes: an experimental study in the pig. J Biome Mater Res Part B Appl Biomater. 2009;89:165–71.
Younes NRB, Amara S, Mrad I, Ben-Slama I, Jeljeli M, Omri K, Ghoul E, Rhouma KB, Abdelmelek H, Sakly M. Subacute toxicity of titanium dioxide (TiO_2) nanoparticles in male rats: emotional behavior and pathophysiological examination. Environ Sci Pollut Res. 2015;22:8728–37.
Zeng L, Pan Y, Tian Y, Wang X, Ren W, Wang S, Lu G, Wu A. Doxorubicin-loaded NaYF4:Yb/Tm-TiO_2 inorganic photosensitizers for NIR-triggered photodynamic therapy and enhanced chemotherapy in drug-resistant breast cancers. Biomaterials. 2015;57(C):93–106.
Zhang L, Webster TJ. Nanotechnology and nanomaterials: promises for improved tissue regeneration. Nano Today. 2009;4:66–80.
Zhang L, Zeng L, Pan Y, Luo S, Ren W, Gong A, Ma X, Liang H, Lu G, Wu A. Inorganic photosensitizer coupled Gd-based up conversion luminescent nanocomposites for in vivo magnetic resonance imaging and near-infrared-responsive photodynamic therapy in cancers. Biomaterials. 2015;44:82–90.
Zhao Y, Howe JLC, Yu Z, Leong DT, Chu JJH, Loo JSC, Ng KW. Exposure to titanium dioxide nanoparticles induces autophagy in primary human keratinocytes. Small. 2013;9:387–92.

Chapter 7
Inhibition of Bacterial Quorum Sensing Systems by Metal Nanoparticles

Krystyna I. Wolska, Anna M. Grudniak, and Katarzyna Markowska

Abstract Quorum sensing (QS) is a commonly used way for intercellular communication utilizing small, self-generated signal molecules called autoinducers. QS-controlled genes can constitute as much as 10% of bacterial genome. This system controls a variety of bacterial physiological traits, including biofilm formation and pathogenesis. QS signaling pathways are well described in several species including *Pseudomonas aeruginosa* and *Staphylococcus aureus*. The inhibition of QS, called quorum quenching (QQ), can be a potential and promising strategy to combat bacterial infections. Nanoparticles (NPs) are considered to be potential QS inhibitors which was proved mainly by in vitro experiments. The QQ potential was proved for silver nanoparticles (AgNPs) which was demonstrated by their ability to inhibit bacterial biofilms. Antibiofilm activity of gold nanoparticles (AuNPs), zinc oxide nanoparticles (ZnONPs), copper nanoparticles (CuNPs), and platinum nanoparticles (PtNPs) was also shown. As an antibacterial potential of metal NPs is well documented, the ability to inhibit QS additionally argues toward their future usage as an alternative of antibiotics. Moreover, the development of bacterial resistance to NPs was not yet well documented.

Keywords Biofilm • Nanoparticles • *Pseudomonas aeruginosa* • Quorum quenching • Quorum sensing • *Staphylococcus aureus* • Virulence factors

Abbreviations

AgNPs	Silver nanoparticles
Agr	Accessory gene regulator
AIPs	Autoinducing peptides
AuNPs	Gold nanoparticles
CF	Cystic fibrosis

K.I. Wolska (✉) • A.M. Grudniak • K. Markowska
Department of Bacterial Genetics, Institute of Microbiology, Faculty of Biology, University of Warsaw, Warsaw, Poland
e-mail: izabelaw@biol.uw.edu.pl

M. Rai, R. Shegokar (eds.), *Metal Nanoparticles in Pharma*,
DOI 10.1007/978-3-319-63790-7_7

CuNPs	Copper nanoparticles
HSLs	Homoserine lactones
MRSA	Methicillin-resistant *S. aureus*
PIA	Polysaccharide intercellular antigen
PQS	*Pseudomonas* quorum sensing
PtNPs	Platinum nanoparticles
QQ	Quorum quenching
QS	Quorum sensing
RIP	RNAIII inhibiting peptide
ROS	Reactive oxygen species
ZnONPs	Zinc oxide nanoparticles

7.1 Introduction

Quorum sensing (QS) is a common communication mechanism in bacteria. This mechanism allows intra- and interspecies cell communication. QS based on chemical signaling is considered as the main communication mechanism and along with relatively recently described electrical signaling, allows bacterial populations to communicate in coordinated manner. It was originally described in the genus *Vibrio fischeri* (Nealson et al. 1970), and since then this system has been intensively studied. The literature of this subject is really huge and includes a variety of review papers (e.g., de Kevit and Iglewski 2000; Whitehead et al. 2001; Ng and Bassler 2009; Monnet et al. 2014; Majumdar and Pal 2017). Single bacterium synthesizes and releases the special, chemical compounds named autoinducers which can be transported through cell membranes in the passive or active manners. When certain cell density is reached and the concentration of autoinducers exceeds the threshold level, the gene expression profile is altered and the number of genes is switched on or off. The details of QS systems differ between Gram-negative and Gram-positive bacteria, mainly by the chemical nature of autoinducers. Gram-negative bacteria utilize L-**h**omoserne **l**actones (HSLs) which usually, but not always, diffuse passively in and out the cell, while Gram-positive species preferentially utilize actively transported short peptides (Miller and Bassler 2001). The schematic diagram of QS systems in both types of bacteria is shown in Fig. 7.1, and their functioning is described in the legend to the figure. Gram-negative bacteria utilize yet another QS system mediated by alkyl quinolones (Pesci et al. 1999).

QS-controlled genes constitute around 10% of bacterial genome (Wagner et al. 2003) and comprise the information encoding adapting, collective traits which allow bacteria to change their metabolism in the way beneficial for the group (Goo et al. 2015). The list of functions regulated by QS is long and includes sporulation, enzyme production, resistance to antimicrobials, and siderophore synthesis (for review, see Jimenez et al. 2012; Monnet et al. 2014). Recently, it was shown that the production of bacteriocins and the entry into competent state which allows natural transformation in *Streptococcus pneumoniae* are also regulated by QS (Shanker and Federle 2017). Regulation of bacterial virulence determinants by QS with the focus put on biofilm formation will be described in the following chapters. It should be

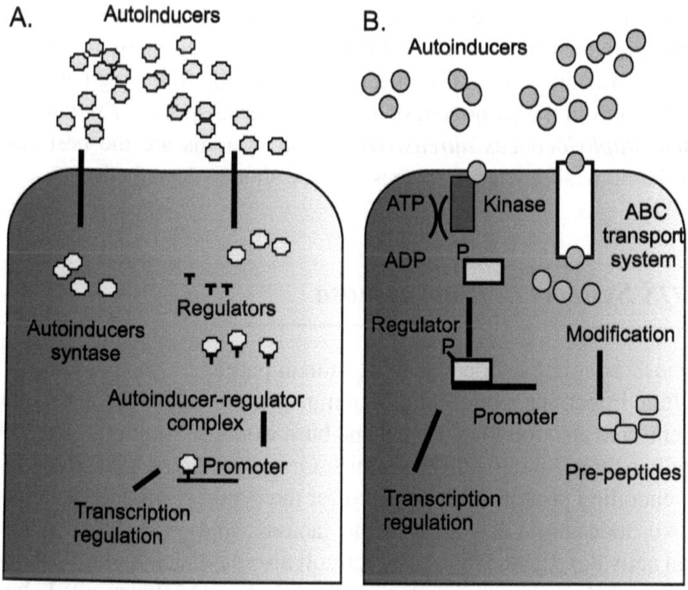

Fig. 7.1 QS signal transduction in (**a**) Gram-negative (*Pseudomonas aeruginosa*) and (**b**) Gram-positive (*Staphylococcus aureus*) bacteria. (**a**) L-homoserine lactones serve as the system inducers. These particles freely diffused in and out the cell. When their threshold concentration is reached, inducer molecule forms the complex with transcription regulator protein. Complex, after binding to promoter sequence in DNA, activates transcription of QS-regulated operons (**b**) Short peptides serve as system inducers. They are actively translocated trough the membranes which is concomitant with peptide modification. When the threshold concentration is achieved, modified peptides activate typical two-component system of signal transduction composed of protein kinase and transcription regulator. After phosphorylation, regulator protein binds to promoter sequence in DNA and activates transcription of QS-regulated operons (Adapted from Wolska K.I. et al. 2016b)

noted that QS communication systems are also utilized by eukaryotic microorganisms and human cells for the development of several biological functions (Hornby et al. 2001; Chen and Fink 2006; Hickson et al. 2009).

Search for the compounds able to inhibit bacterial QS systems is carried out in many laboratories. The list of QQ factors includes, among others, metal nanoparticles with the special emphasis put on AgNPs. In this review, we describe the ability of metal nanoparticles to inhibit QS, especially the inhibition of biofilm formation that creates the perspective to interfere with bacterial pathogenicity.

7.2 Role of Quorum Sensing Systems in Bacterial Pathogenesis

QS systems are utilized by several bacterial pathogens to coordinate the expression of multiple virulence factors and the traits influencing their pathogenesis, such as biofilm formation and swarming motility (Castillo-Juárez et al. 2015). It was demonstrated both in vitro and in vivo, in animal models and in human infections, that

QS enhances bacterial virulence and coordinates bacterial attack against the host, thus maximizing the probability to establish infection, especially for pathogens that relay on high infection doses (Gama et al. 2013). In the following part of this chapter the role of QS in the pathogenesis of two bacterial species, *Pseudomonas aeruginosa* and *Staphylococcus aureus*, whose QS systems are the best studied and understood among all bacterial species, will be shortly described.

7.2.1 QS Systems in *P. aeruginosa*

P. aeruginosa is a major, opportunistic, human pathogen causing chronic nosocomial and lung infections, preferentially in transplantation, cancer, and cystic fibrosis (CF) patients and also found in dermal and burn wounds (Ammons et al. 2009). This bacterium has three functional QS circuits. One is mediated by HSLs and contains gene *lasI* encoding protein LasI, required for the synthesis of an autoinducer which is N-(3-oxo-dodecanopyl)-L-homoserine lactone, and gene *lasR* encoding transcriptional activator, LasR. The second circuit also mediated by homoserine lactone contains gene *rhlI*, an enzyme synthesizing autoinducer N-(butanoyl)-L-homoserine lactone and gene *rhlR* coding for transcriptional activator RhlR (Pearson et al. 1994, 1995). The third QS system, called *Pseudomonas* QS (PQS), is mediated by alkyl quinolones, especially 2-heptyl-3-hydroxy-4-quinolone synthesized by the products of *pqsABCDEH* genes and controlled by PqsR regulator (Pesci et al. 1999). All three systems are regulated hierarchically: LasR positively regulates RhlR, and RhlR negatively regulates PQS (Wade et al. 2005).

It was shown by in vitro studies that Las circuits positively regulate the synthesis of several *P. aeruginosa* virulence factors, such as elastases LasA and LasB, exotoxin A, and alkaline protease (Gambello and Iglewski 1991; Passador et al. 1993). In turn Rhl circuits control the expression of pyocyanin – green/blue pigment – enhancing the synthesis of **r**eactive **o**xygen **s**pecies (ROS) and thus causing the oxidative damage of host eukaryotic cell (Gloyne et al. 2011). PQS system increases the expression of elastase and pyocyanin (Déziel et al. 2005). The QS-dependent regulation of all *P. aeruginosa* virulence factors can be influenced by environmental conditions, such as the presence of calcium, iron, and several bacterial metabolites like 2,3-butanediol (Jensen et al. 2006; Sarkisova et al. 2014). It should be also mentioned that QS systems regulate several physiological traits leading to the enhanced stress tolerance which in turn allows to avoid host immune system attack and therefore indirectly enhances bacterial virulence (Telford et al. 1998; Garcia-Contraras et al. 2015).

P. aeruginosa is an example of bacterium able to form biofilm. Biofilm is a structured bacterial community characterized by an increased tolerance to the exogenous stress, including treatment with antibiotics and other antibacterial factors and strategies, and severely influences bacterial pathogenesis (Hall-Stoodley and Stoodley 2009). The involvement of QS signaling in biofilm development is well established. It was shown that QS is important in the late stage of biofilm development however also participates in the former stages. It is executed by its involvement in rhamnolipid synthesis and allows to form the channels in mushroom-shaped structures which results in proper distribution of nutrients and oxygen and the removal of waste products (Davey and O'Toole 2000). The overproduction of rhamnolipids in the late stage of biofilm development causes the biofilm detachment from the surface and its subsequent dispersal, which is documented for several bacterial species (Boles and Horswill 2005). Biofilm dispersal (in contrary to its eradication) is very important to the colonization of new environmental niches. QS, precisely PQS system, is involved in cell autolysis of bacterial subpopulation and the release of a large amount of extracellular DNAs (eDNA) being one of the main compounds of biofilm matrix (Allesen-Holm et al. 2006). The regulation by QS of previously mentioned *P. aeruginosa* virulence factors, such as lectins and siderophores, pyoverdine and pyochelin, has also an impact on biofilm formation. Siderophores exert their action through the participation in iron metabolism, and it was demonstrated that not optimal concentration of iron results in the inhibition of biofilm formation (Singh et al. 2002). Yet another factor regulated by QS and involved in the first stage of biofilm formation is the ability of swarming-type movement. QS-controlled production of extracellular polysaccharides and biosurfactants by the "swarmers" allows bacteria to colonize the new environmental surfaces (Rashid and Kornberg 2000).

QS controls the pathogenesis of *P. aeruginosa* also in vivo which was proved in various animal models, such as nematode *Caenorhabditis elegans*, fruit fly (*Drosophila melanogaster*), zebra fish (*Danio rerio*), and mouse (*Mus musculus*) (Bjarnsholt et al. 2010; Papaioannou et al. 2013). The in vivo experiments were conducted using two main strategies: the utilization of mutant strains with QS genes disrupted and the quenching of QS systems. The results of these studies demonstrated that the disruption of QS diminishes the production of phenazines thus decreasing the concentration of active oxygen species and subsequently decreasing the killing rate of *C. elegans* (Tan et al. 1999). In turn various *P. aeruginosa* mutants defective in QS circuits showed a substantially lower *D. melanogaster* killing rate than wild-type strain (Lutter et al. 2012). It was also demonstrated, applying murine model, that the inhibition of quinolone signals restricts the systematic spread of PA14 strain and decreases the animal death by 30–50% (Lesic et al. 2007). The importance of QS in human infections caused by *P. aeruginosa* is also well documented. Many experiments showed that about 90% of *P. aeruginosa* isolates able to cause infections have active HSL systems. For example, it was demonstrated that the main *P. aeruginosa* autoinducer, N-(3-oxododecanoyl)-HSL is often detected in the patients with cystic fibrosis (Sing et al. 2000). The QS-controlled virulence factor, the polysaccharide alginate, which is an important compound of matrix produced by mucoid *P. aeruginosa* strains protects biofilm from macrophage killing (Leid et al. 2005).

7.2.2 QS Systems in *S. aureus*

S. aureus is a Gram-positive, nonmotile coccus, forming cell clusters and producing yellow pigment. This bacterium is responsible for the infections of skin and other tissues, bacteremia, endocarditis, sepsis, and many other diseases (Lowy 1998). It is also characterized by the fluent evolvement of multidrug-resistant strains, known as **m**ethicillin-**r**esistant *S. aureus* (MRSA) (Lowy 2003). The infections caused by *S. aureus* are facilitated by the production of broad repertoire of various virulence factors including enzymes, exotoxins, and adhesions (Gordon and Lowy 2008). Many of them are controlled by QS-dependent **a**ccessory **g**ene **r**egulator – Agr system (Painter et al. 2014). Agr locus contains *agrA*, *agrC*, *agrD*, and *agrB* genes constituting the so-called RNAII transcript and *hld* gene. The pheromone AgrD is exported and modified by AgrB along with SpsB peptidase leading to the pro-peptide AgrD conversion into **a**uto**i**nducing **p**eptides – AIPs (Qui et al. 2005). The length of these peptides varies between seven and nine amino acids, but all contain thiolactone ring at C-terminus (Malone et al. 2007). AIPs activate a typical two-component system AgrC/AgrA, where AgrC has the activity of histidine kinase and AgrA serves as transcriptional activator and, when phosphorylated, regulates *hld* and *agr* operons providing an autofeedback loop (Ji et al. 1995).

The *hld* gene encodes RNAIII effector molecule that posttranscriptionally regulates several virulence factors (Fechter et al. 2014). In total, RNAIII (and AgrA) regulates the transcription of around 200 genes encoding virulence factors. Among the upregulated genes are these coding for α-hemolysin which destroys the membrane structures and can cause pneumonia (Bubeck Wardenburg and Schneewind 2008), proteases, toxins, and components of bacterial capsule. Downregulated genes are responsible for the synthesis of adhesins and protein A – a factor allowing *S. aureus* to evade opsonization (Novick and Geisinger 2008). RNAIII is also involved in the regulation of biofilm formation by *S. aureus*. This bacterium, due to its lack of motility, forms flatter biofilm than the one formed by motile genera (Mann et al. 2009). *S. aureus* biofilm is embedded with glycocalyx or slime layer composed of teichoic acid (Husain et al. 2013). Besides, the significant biofilm constituents are **p**olysaccharide **i**ntercellular **a**ntigen (PIA) and eDNA released in consequence of massive cell lysis by holin homolog, CidA (Wolska et al. 2016a). The appreciable strain-dependent variations in biofilm composition are noticed (Kiedrowski and Horswill 2011).

The involvement of the ability to form biofilm on *S. aureus* pathogenesis is not unambiguous. The formation of biofilm is closely related to the character of infection; therefore, the determinants of acute and chronic infections are regulated by QS in an opposite fashion. QS is important in acute virulence characterized by the formation of differentiated biofilm with the capacity to disseminate, which is facilitated by the QS-dependent synthesis of surfactants (Otto 2014), peptidases and nucleases (Lauredale et al. 2009). In chronic infections, biofilm is downregulated with concomitant presence of mutations in the genes coding for QS components (Shopsin et al. 2010).

The participation of QS in *S. aureus* pathogenesis was proved using numerous animal models and the experimental approaches described above for *P. aeruginosa*. It was demonstrated in the mice model that *agr* mutants showed strongly reduced ability to induce septic arthritis in comparison to the wild-type strain (Abdelmour et al. 1993). Also *C. elegans* killing in consequence of feeding with *S. aureus arg* mutant was much lower with respect to wild-type strain (Sifri et al. 2003). At the other site, the inhibitors of QS, such as autoinducer analog AIP-II peptide, decreased formation of abscesses (Wright et al. 2005), and yet another RNAIII-inhibiting peptide, RIP, significantly reduced the bacterial load and mortality in mouse sepsis model when administrated with antibiotic, e.g., vancomycin (Giacometti et al. 2005). Until now the importance of QS for the *S. aureus* ability to cause human infections has not been experimentally proved; however, the link between QS and biofilm formation suggests that QS is very important for the establishment of chronic infections (Yardwood and Schlievert 2003).

7.3 Inhibition of Bacterial QS Systems

It is well proved that bacterial QS systems can constitute the target for antibacterial treatment, and their inhibition can be exploited against infections caused by antibiotic-resistant bacteria (Castillo-Juarez et al. 2015). Targeting QS in order to combat bacterial infections has an advantage over the usage of antibiotics because the disruption of communication mechanisms without killing should generate a lower selective pressure and therefore reduce the rate of resistance developing (Clatworty et al. 2007). Various chemical compounds which attack the different steps of QS transduction signal are described. The mechanisms of action of QS inhibitors are usually repression of signal generation, blockade of signal receptors, and disruption of QS signals (Reuter et al. 2016). The efficiency of quorum quenchers is genus dependent. The largest number of literature positions described anti-QS potential of organic inhibitors inducing those of plant, animal, and bacterial origin (for review, see Reuter et al. 2016; Wolska et al. 2016a, b). In the following chapters, antibacterial potential of metal nanoparticles with the emphasis put on their quorum quenching activity will be described.

7.3.1 *Antibacterial Activity of Metal Nanoparticles*

Nanoparticles are the clusters of atoms of size ranging from 1 to 100 nm (Lemire et al. 2013). The antimicrobial potential of NPs is considered to be stronger than respective metal ions, and their effect depends on size and shape. The global surface area of NPs is inversely correlated with their size, so small particles have a greater interaction with the surrounding environment; the triangular-shape NPs are more bactericidal than rods or spherical forms (Pal et al. 2007). Another characteristic affecting the activity of NPs is zeta potential which plays a significant role in the ability of nanoparticles to penetrate into the cell (Seil and Webster 2012). NPs can be obtained from the natural sources, but more often they are synthesized using physical, chemical, and biological methods; the last are considered to be the most friendly and safe for the environment (Iravani et al. 2014; Patil and Kim 2017). Among all metal nanoparticles, the best studied are silver nanoparticles (Durán et al. 2016; Zhang et al. 2016).

AgNPs exert a pleiotropic effect on bacterial cell and have multiple mechanisms of antibacterial activity (Dakal et al. 2016). The multiplicity of AgNPs' targets and mechanisms of activity contributes to the fact that they can overcome existing mechanisms of microbial drug resistance (Pelgrift and Friedman 2013). Two mechanisms are believed to be a primary cause of bacterial killing. The first is based on AgNPs' effect on bacterial envelope permeability which is correlated with their ability to modify the potential of cell membranes (Morones et al. 2005). AgNPs are able to interact with membrane proteins but also with cytoplasmic proteins, preferentially those containing thiol groups which are found, e.g., in respiratory enzymes and also with phosphate residues in DNA (Sondi and Salopek-Sondi 2004). The second mechanism of antibacterial activity is based on the induction of ROS production (Hwang et al. 2008). ROS, as highly oxidative agents, inactivate many bacterial components which invariably leads to bacterial death (Khan 2012). Schematic representation of AgNPs' effect on bacterial cell is presented in Fig. 7.2. Generally, Gram-negative bacteria are more susceptible to NPs than Gram-positive species due to the differences in their cell wall. The cell wall of Gram-negative bacteria consists of thin peptidoglycan layer and outer membrane containing lipopolysaccharide complex. Gram-positive bacterial cell wall primarily consists peptidoglycan and is much thicker. Nanoparticles are very effective not only against planktonic cells but also bacterial biofilms which will be described in the next chapter (Markowska et al. 2013). Due to the pleiotropic mechanism of action, the evolution of nanosilver resistance is slow and of less concern than the resistance to conventional antibiotics.

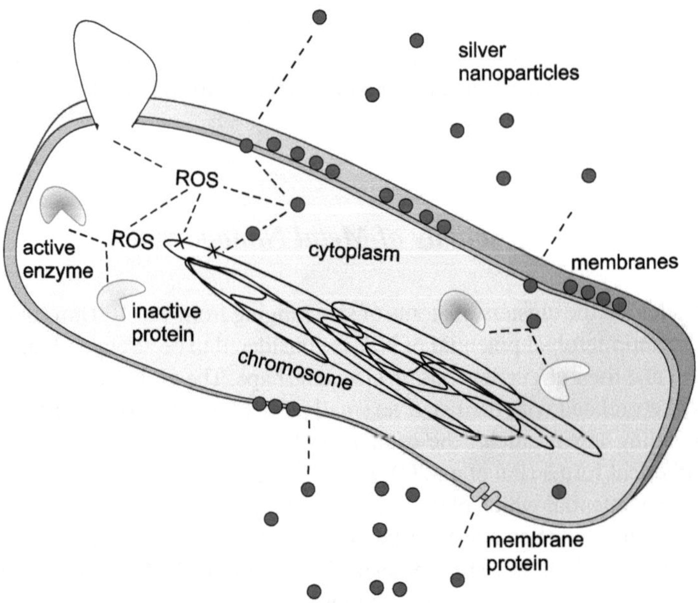

Fig. 7.2 Schematic representation of silver nanoparticles mechanism of action and their cellular targets (Adapted from Markowska et al. 2013)

However silver resistance determinants located on mobile genetic elements were found in *Pseudomonas stutzeri*, *Klebsiella pneumoniae*, and *Salmonella enterica* (Durán et al. 2016). It is believed that the main mechanism of resistance is based on the efflux of silver ions arising in result of AgNPs' oxidation (Percival et al. 2005).

Apart from AgNPs, the antibacterial potential was proved also for gold, copper, zinc oxide, titanium oxide, and organic nanoparticles among which polycations, chitosan, triclosan, poly-ε-lysine, and 5-chloro-8-hydroxyqinolines should be mentioned (Wolska et al. 2016b).

7.3.2 Quorum Quenching Potential of Metal NPs

The literature treating directly on QQ ability of NPs is scarce, and the molecular basis of this ability is not known until now. However, quite recognized knowledge about antibiofilm potential of NPs together with the QS involvement on the various steps of biofilm development allows to argue that QS signal transduction can be at least one of the several targets of NPs' effect on bacterial cells. The literature of this subject deals mainly, but not only, with AgNPs and their inhibitory potential against *P. aeruginosa* and *S. aureus* biofilms.

The early studies of Fabrega and coworkers (2009) demonstrated that AgNPs significantly inhibited *Pseudomonas putida* 24-h biofilm. The average diameter of AgNPs used in this study, 65 ± 30 nm, was quite high considering that the most active nanoparticles are within the range 1–10 nm. The study of Kalishwaralal et al. (2010) showed that AgNPs of mean diameter of 50 nm almost completely prevented biofilm formation by *P. aeruginosa* and *Staphylococcus epidermidis* impeding the initial step – bacterial adhesion to the colonized surface. Nanosilver of smaller particle size (average diameter: 25.2 ± 4 nm) was found to effectively prevent the formation of *P. aeruginosa* biofilms and also effectively kill bacteria in the established biofilm structure; 4-log reduction in the number of colony-forming units was observed (Martinez-Gutierrez et al. 2013). Another study showed that AgNPs are also effective against *Mycobacterium smegmatis* biofilm, preventing its formation on the membranes coated with nanosilver. AgNPs decreased the biofilm formation by 98% and reduced bacterial survival to 0.03% (Islam et al. 2013). In another study, it was shown that the spherical AgNPs with diameter ranging from 15 to 34 nm substantially inhibited biofilms formed by marine bacteria (Inbakandan et al. 2013). The results of our group demonstrated that AgNPs of size ranging from 1 to 10 nm applied in concentration of 4 μg mL^{-1} efficiently reduced the early stages of biofilm formation by *P. aeruginosa* inhibiting bacterial adhesion to the polystyrene surfaces. Reduction in the production of biofilm matrix components along with decreasing cell viability in biofilm and changes in cell morphology, e.g., condensation of cytoplasm, were also observed (Markowska 2016). The antibacterial effect of nanosilver is usually studied in static conditions, but it was proved that AgNPs can also kill bacteria under turbulent fluid condition in bioreactors (Martinez-Gutierrez et al. 2013).

Many studies on antibacterial potential of nanosilver were carried out using medical devices such as catheters and other biomaterials. Roe et al. (2008), when studying the efficacy of coating the catheters surface with nanosilver, demonstrated the positive effect of this procedure against Gram-positive and Gram-negative bacteria and *Candida albicans*. In addition, these authors showed that no significant accumulation of silver was detected in the major organs of mice fitted with the treated catheters which suggested the lack of nanosilver toxicity. Subsequently, Cheng et al. (2012) showed that dental nanocomposites containing AgNPs reduced the metabolic activity and lactic acid production by *Streptococcus mutans* biofilm. The successful usage of nanosilver has also been confirmed for coating surgical masks (Li et al. 2006). Recently the stabilized or modified preparations of nanosilver are under study; the amendments impede AgNPs' aggregation thus extending the period of their successful usage. It was shown that starch-stabilized AgNPs disrupted *P. aeruginosa* and *S. aureus* biofilms at a very low concentration, 1–2 mM (Mohanty et al. 2012). Park et al. (2013) showed that citrate-capped AgNPs were efficient against *P. aeruginosa* biofilms, and Hartman and coworkers (2013) demonstrated that AgNPs anchored on the agarose beads inhibited biofilm formed by *P. putida*. Recently it was shown that the combination of AgNPs and curcumin nanoparticles has an enhanced antibiofilm activities when comparing to the compounds applied separately (Loo et al. 2016). In yet another study, it was demonstrated that AgNPs synthesized using fresh leaf extract of *Cymbopogon citratus* (lemongrass) enhanced QQ activity against *S. aureus* biofilm and prevented biofilm formation (Masurkar et al. 2012).

Several papers reported antibiofilm activity of gold nanoparticles, AuNPs. The significant reduction of *S. aureus* and *P. aeruginosa* biofilms by AuNPs applied in high concentration, exceeding 50 μg mL^{-1}, was demonstrated (Sathyanarayanan et al. 2013). In opposite, the changes in morphology of *Legionella pneumophila* biofilm were observed even after the exposure to very low AuNPs' concentration (Raftery et al. 2013, 2014). Recently it was demonstrated that AuNPs inhibited biofilm formation and invasion of dental pulp stem cells by *Candida albicans* and *P. aeruginosa* (Ramasamy et al. 2016). In another study it was shown that nanogold, when applied in concentration 100 μg mL^{-1}, was efficient in the inhibition of biofilm formed by the multiple antibiotic-resistant *Enterococcus faecalis* (Vijayakumar et al. 2017). The comparison of AgNP and AuNP antibiofilm activity revealed that the latter was substantially less effective in biofilm inhibition (Vijayan et al. 2014).

Strong antibiofilm properties were reported also for ZnONPs (Appelrot et al. 2012). Seil and Webster (2012) suggested that ZnONPs are more effective against Gram-positive than Gram-negative bacteria. This suggestion can be questioned in light of the results obtained by Lee et al. (2014). These authors in their excellent study demonstrated that ZnONPs efficiently inhibited *P. aeruginosa* biofilm formation without inhibiting planktonic growth. ZnONPs hampered also several virulence determinants – pyocyanin, pyochelin, and hemolytic activity. Moreover, the authors showed that ZnONPs were active in inhibiting one of *P. aeruginosa* QS circuits, PQS, and suggested that the functions controlled by RhL circuits were also disrupted. The ability of ZnONPs to inhibit quorum-sensing-dependent virulence factors and biofilm formation by clinical and environmental *P. aeruginosa* strains was also confirmed by Garcia-Lara et al. (2015). The broad-spectrum inhibition of QS-regulated functions was recently demonstrated in *Chromobacterium violaceum*, *P. aeruginosa*, *Listeria monocytogenes*, and *Escherichia coli* (Al-Shabib et al. 2016).

The number of publications describing antibiofilm activity of other metal nanoparticles is scarce. For example, the efficient inhibition of oral biofilm formation was proved for CuNPs (Allaker 2010), and the eradication of *L. pneumophila* biofilm by citrate-coated PtNPs was also demonstrated (Raftery et al. 2014).

7.4 Conclusions

The ability of nanoparticles to inhibit QS indicates their good potential to treat the infections caused by bacterial biofilms. Until now the molecular mechanism of QQ activity just started to be resolved, and the details of the multistep QS signal transduction impairment by nanoparticles are still under study. The antibacterial activity of metal nanoparticles including their antibiofilm potential is well documented. Their practical applications are however still limited, and NPs are used mainly to cure skin infections and as components of medical devices. It should be noted that several disappointing clinical trials with medical devices coated with nanosilver were reported. The problem was probably due to the possible covering of their surfaces with dead cells or blood which highly impede antimicrobial potential of nanoparticles (Rai et al. 2009). The restricted usage of NPs is undoubtedly due to their poor specificity as compared to the conventional antibiotics (Pauksch et al. 2014). The full knowledge about the mechanism of QQ activity and thus the NPs interference with bacterial virulence will be very valuable for their clinical applications. To understand the beneficial and adverse effects of NPs, further detailed studies are required before the approval of broad clinical use of nanoparticles.

References

Abdelmour A, Arvidson S, Bremell T, Ryden C, Tarkowski A. The accessory gene regulator (*agr*) controls *Staphylococcus aureus* virulence in murine arthritis model. Infect Immun. 1993;61:3879–85.

Allaker RP. The use of nanoparticles to control oral biofilm formation. J Dent Res. 2010;89:1175–86.

Allesen-Holm M, Barken KB, Yang L, Klausen M, Webb JS, Kjelleberg S, Molin S, Givskov M, Tolker-Nielsen T. A characterization of DNA release in *Pseudomonas aeruginosa* cultures and biofilms. Mol Microbiol. 2006;59:1114–28.

Al-Shabib NA, Husain FM, Ahmed F, Khan RA, Ahmad I, Aksharaed E, Khan MS, Hussain A, Rehman MT, Yusuf M, Hassan I, Khan JM, Ashraf GM, Alsamle AM, Al-Aimi MF, Tarasov VV, Aliev G. Biogenic synthesis of zinc oxide nanostructures from *Nigella sativa* seed: prospective role as food packaging material inhibiting broad-spectrum quorum sensing and biofilm. Sci Rep. 2016;6:36761.

Ammons MC, Ward LS, Fisher ST, Wolcott RD, James GA. *In vitro* susceptibility of established biofilms composed of a clinical wound isolate of *Pseudomonas aeruginosa* treated with lactoferrin and xylitol. Int J Antimicrob Agents. 2009;33:230–6.

Appelrot G, Lellouche J, Perkas N, Nitzan Y, Gedanken A, Banin F. ZnO nanoparticle-coated surfaces inhibit biofilm formation and increase antibiotic susceptibility. RSC Adv. 2012;2:2314.

Bjarnsholt T, van Gennip M, Jakobsen TH, Christensen LD, Jensen PO, Givskov M. In vitro screens for quorum sensing inhibitors and in vivo confirmation of their effect. Nat Protoc. 2010;5:282–93.

Boles BR, Horswill AR. Agr-mediated dispersal of *Staphylococcus aureus* biofilms. PLoS Phatog. 2005;4:e1000052.

Bubeck Wardenburg J, Schneewind O. Vaccine protection against *Staphylococcus aureus* pneumonia. J Exp Med. 2008;205:287–94.

Castillo-Juárez I, Maeda T, Mandujano-Tinoco EA, Tomás M, Pérez-Eretza B, Garcia-Contreras SJ, Wood TK, Garcia-Contreras R. Role of quorum sensing in bacterial infections. World J Clin Cases. 2015;3:575–98.

Chen H, Fink GR. Feedback control of morphogenesis in fungi by aromatic alcohols. Genes Dev. 2006;20:1150–61.

Cheng L, Weir MD, Xu HHK, Antonucci JM, Kraigsley AM, Lin NJ, Lin-Gibson S, Zhou X. Antibacterial amorphous calcium phosphate nanocomposites with a quaternary ammonium dimethacrylate and silver nanoparticles. Dent Mater. 2012;28:561–72.

Clatworty AE, Pierson E, Hung DT. Targeting virulence: a new paradigm for antimicrobial therapy. Nat Chem Biol. 2007;3:541–8.

Dakal TC, Kumar A, Majumdar RS, Yadav V. Mechanistic basis on antimicrobial actions of silver nanoparticles. Front Microbiol. 2016;7:1831.

Davey ME, O'Toole GA. Microbial biofilms: from ecology to molecular genetics. Microbiol Mol Biol Rev. 2000;64:847–67.

Déziel E, Gopalan S, Tampakaki AP, Lépine F, Padfield KE, Saucier M, Xiao G, Rahme LG. The contribution of MvfR to *Pseudomonas aeruginosa* pathogenesis and quorum sensing circuitry regulation: multiple quorum sensing-regulated genes are modulated without affecting *lasRI, rhlRl* or the production of N-acyl-L-homoserine lactones. Mol Microbiol. 2005;55:998–1014.

Durán N, Durán M, Dejesus MB, Seabra AB, Fávaro WJ, Nakazato G. Silver nanoparticles: a new view on mechanistic aspects on antimicrobial activity. Nanomedicine. 2016;12:789–99.

Fabrega J, Renshaw JC, Lead JR. Interaction of silver nanoparticles with *Pseudomonas putida* biofilms. Environ Sci Technol. 2009;43:9004–9.

Fechter P, Caldelari I, Lioliou E, Romby P. Novel aspects of RNA regulation in *Staphylococcus aureus*. FEBS Lett. 2014;588:2523–9.

Gama JA, Abby SS, Vieira-Silva S, Dionisio F, Rocha EP. Immune subversion and quorum-sensing shape in variation in infectious dose among bacterial pathogens. PLoS Pathog. 2013;8:e1002503.

Gambello MJ, Iglewski BH. Cloning and characterization of the *Pseudomonas aeruginosa lasR* gene, a transcriptional activator of elastase expression. J Bacteriol. 1991;173:3000–9.

Garcia-Contraras R, Nuñez-López L, Jasso-Chávez R, Kwan BW, Belmont JA, Rangel-Vega A, Maeda T, Wood TK. Quorum sensing enhancement of the stress response promotes resistance to quorum quenching and prevents social cheating. ISME J. 2015;9:115–25.

Garcia-Lara B, Saucedo-Mora MA, Roldán-Sánchez JA, Pérez-Eretza B, Ramasamy M, Lee J, Coria-Jimenez R, Tapia M, Varela-Guerrero V, Garcia-Contrras R. Inhibition of quorum-sensing-dependent virulence factors and biofilm formation of clinical and environmental *Pseudomonas aeruginosa* strains by ZnO nanoparticles. Lett Appl Microbiol. 2015;61:299–305.

Giacometti A, Cirioni O, Ghiselli R, Dell'Acqua G, Orlando F, D'Amato G, Mocchegiani F, Silvestri C. RNAIII-inhibiting peptide improves efficacy of clinically used antibiotics in a murine model of staphylococcal sepsis. Peptides. 2005;26:169–75.

Gloyne LS, Grant GD, Perkins AV, Powell KL, McDermott CM, Johnson PV, Anderson GJ, Kiefel M, Anoopkumar-Dukie S. Pyocyanin-induced toxicity in A549 respiratory cells is causally linked to oxidative stress. Toxicol in Vitro. 2011;25:1353–8.

Goo E, An JH, Kang Y, Hwang I. Control of bacterial metabolism by quorum sensing. Trends Microbiol. 2015;23:567–76.

Gordon RJ, Lowy FD. Pathogenesis of methicillin-resistant *Staphylococcus aureus* infections. Clin Infect Dis. 2008;46:S350–9.

Hall-Stoodley L, Stoodley P. Evolving concepts of biofilm infections. Cell Microbiol. 2009;11:1034–43.

Hartmann T, Mühling M, Wolf A, Mariana F, Maskow T, Mertens F, Neu TR, Lerchner J. A chip-calorimetric approach to the analysis of Ag nanoparticle caused inhibition and inactivation of beads-grown bacterial biofilms. J Microbiol Methods. 2013;95:129–37.

Hickson J, Diane Yamada S, Berger J, Alverdy JO, Keefe J, Brassler B, Rinker-Schaeffer C. Social interactions in ovarian cancer metastasis: a quorum-sensing hypothesis. Clin Exp Metastasis. 2009;26:67–76.

Hornby JM, Jensen EC, Lisec AD, Tasto JJ, Jahnke B, Shoemaker R, Dussault P, Nickerson KW. Quorum sensing in the dimorphic fungus *Candida albicans* is mediated by farnesol. Appl Environ Microbiol. 2001;67:2982–92.

Husain FM, Ahmad I, Asif M, Tahseen Q. Influence of clove oil on certain quorum-sensing regulated functions and biofilm of *Pseudomonas aeruginosa* and *Aeromonas hydrophila*. J Biosci. 2013;38:835–44.

Hwang ET, Lee JH, Chae YJ, Kim YS, Kim BC, Sang BI. Analysis of the toxic mode of action of silver nanoparticles using stress-specific bioluminescent bacteria. Small. 2008;4:746–50.

Inbakandan D, Kumar C, Abraham LS, Kirubagaran R, Venkatesan R, Khan SA. Silver nanoparticles with anti microfouling effect: a study against marine biofilm forming bacteria. Coll Surf B: Biointerfases. 2013;111C:636–43.

Iravani S, Korbekandi H, Mirmohammadi SV, Zolfaghari B. Syntesis of silver nanoparticles: chemical, physical and biological methods. Res Pharm Sci. 2014;9:385–406.

Islam MS, Larimer C, Ojha A, Nettleship I. Antimycobacterial efficacy of silver nanoparticles as deposited on porous membrane filters. Mater Sci Eng C Mater Biol Appl. 2013;33:4575–81.

Jensen V, Lons D, Zaoui C, Bredenbruch F, Meissner A, Dieterich G, Münch R, Häussler S. RhlR expression in *Pseudomonas aeruginosa* is modulated by the *Pseudomonas* quinolone signal via PhoB-dependent and independent pathways. J Bacteriol. 2006;188:8601–6.

Ji G, Beavis RC, Novick RP. Cell density control of staphylococcal virulence mediated by an octapeptide pheromone. Proc Natl Acad Sci U S A. 1995;92:12055–9.

Jimenez PN, Koch G, Thompson JA, Xavier KB, Cool RH, Quax WJ. The multiple signaling systems regulating virulence in *Pseudomonas aeruginosa*. Microbiol Mol Biol Rev. 2012;76:46–65.

Kalishwaralal K, Barath Mani Kanh S, Pandian SRK, Deepak V, Gurunathan S. Silver nanoparticles impede the biofilm formation by *Pseudomonas aeruginosa* and *Staphylococcus epidermidis*. Coll Surf B: Biointerfases. 2010;79:340–4.

de Kevit TR, Iglewski BH. Bacterial quorum sensing in pathogenic relationship. Infect Immun. 2000;68:4839–49.

Khan AU. Medicine at nanoscale: a new horizon. Int J Nanomedicine. 2012;7:2997–8.

Kiedrowski MR, Horswill AR. New approaches for treating staphylococcal biofilm infections. Ann N Y Acad Sci. 2011;1241:104–21.

Lauredale KJ, Boles BR, Cheung AL, Horswill AR. Interconnections between Sigma B, *agr*, and proteolytic activity in *Staphylococcus aureus* biofilm maturation. Infect Immun. 2009;77:1623–35.

Lee J-H, Kim Y-G, Cho MH, Lee J. ZnO nanoparticles inhibit *Pseudomonas aeruginosa* biofilm formation and virulence factor production. Microbiol Res. 2014;169:888–96.

Leid JG, Wilson CJ, Shirliff ME, Hassett DJ, Parsek MR, Jeffers AK. The exopolysaccharide alginate protects *Pseudomonas aeruginosa* biofilm bacteria from IFN-gamma-mediated macrophage killing. J Immunol. 2005;175:7512–8.

Lemire JA, Harrison JJ, Turner RJ. Antimicrobial activity of metals: mechanisms, molecular targets and applications. Nat Rev Microbiol. 2013;11:371–84.

Lesic B, Lépine F, Déziel E, Zhang J, Zhang Q, Padfield K, Castonguay MH, Milot S, Stachel S, Tzika AA, Tompkins RG, Rahme LG. Inhibitors of pathogen intercellular signals as selective anti-infective compounds. PLoS Pathog. 2007;3:1229–39.

Li Y, Leung P, Yao L, Song QW, Newton E. Antimicrobial effect of surgical masks coated with nanoparticles. J Hosp Infect. 2006;62:58–63.

Loo CY, Rohanizadeh R, Young PM, Trani D, Cavaliere R, Whitchurch CB, Lee WH. Combination of silver nanoparticles and curcumin nanoparticles for enhanced anti-biofilm activities. J Agric Food Chem. 2016;64:2513–22.

Lowy FD. *Staphyloccocus aureus* infections. N Engl J Med. 1998;339:520–32.

Lowy FD. Antimicrobial resistance: the example of *Staphylococcus aureus*. J Clin Invest. 2003;111:1265–73.

Lutter EI, Purighalla S, Duong J, Storey DG. Lethality and cooperation of *Pseudomonas aeruginosa* quorum-sensing mutants in *Drosophila melanogaster* infection models. Microbiology. 2012;158:2125–32.

Majumdar S, Pal S. Cross-species communication in bacterial world. J Cell Commun Signal Doi. 2017; doi:10.1007/s12079-017-0383-09.

Malone CL, Boles BR, Horswill AR. Biosynthesis of *Staphylococcus aureus* autoinducing peptides by using the *Synechocystis* DnaB mini-intein. Appl Environ Microbiol. 2007;73:6036–44.

Mann EE, Rice KC, Boles BR, Endres JL, Ranjit D, Chandramohan L, Tsong LH, Smeltzer MS, Horswill AR, Bagles KW. Modulation of eDNA release and degradation affects *Staphylococcus aureus* biofilm maturation. PLoS One. 2009;4:e5822.

Markowska K. Antibacterial activity of silver nanoparticles – effect on the structure and the functions of bacterial cells. (PhD dissertation, in Polish). Poland: University of Warsaw; 2016.

Markowska K, Grudniak AM, Wolska KI. Siver nanoparticles as an alternative strategy against bacterial biofilms. Acta Biochim Pol. 2013;60:523–30.

Martinez-Gutierrez F, Boegli L, Agostinho A, Sánchez EM, Bach H, Ruiz F, James G. Anti-biofilm activity of silver nanoparticles against different microorganisms. Biofouling. 2013;29:651–60.

Masurkar SA, Chaudhari PR, Shidore VB, Kamble SP. Effect of biologically synthesized silver nanoparticles on *Staphylococcus aureus* biofilm quenching and prevention of biofilm formation. IET Nanobiotechnol. 2012;6:110–4.

Miller MB, Bassler BL. Quorum sensing in bacteria. Annu Rev Microbiol. 2001;55:165–99.

Mohanty S, Mishra S, Jena P, Jacob B, Sarkar B, Sonawane A. An investigation on the antibacterial, cytotoxic, and antibiofilm efficacy of starch-stabilized silver nanoparticles. Nanomedicine. 2012;8:916–24.

Monnet V, Juillard V, Garden R. Peptide conversation in Gram-positive bacteria. Crit Rev Microbiol. 2014;8:1–13.

Morones JR, Elechiguerra JL, Camacho A, Holt K, Kouri JB, Ramirez JT, Yacaman MJ. The bactericidal effect of silver nanoparticles. Nanotechnology. 2005;16:2346.

Nealson KH, Platt T, Hastinds JW. Cellular control of synthesis and activity of bacterial luminescent system. J Bacteriol. 1970;104:313–22.

Ng WL, Bassler BL. Bacterial quorum sensing network architectures. Annu Rev Genet. 2009;43:197–222.

Novick RP, Geisinger E. Quorum sensing in staphylococci. Annu Rev Genet. 2008;42:541–64.

Otto M. Phenol-soluble modulins. Int J Med Microbiol. 2014;304:164–9.

Painter KL, Krishna A, Wigneshweraraj S, Edwards AM. What role does the quorum-sensing accessory gene regulator system plays during *Staphylococcus aureus* bacteremia? Trends Microbiol. 2014;22:676–85.

Pal S, Tak YK, Song JM. Does antibacterial activity of silver nanoparticles depend on the shape of the nanoparticles? A study of the Gram-negative bacterium *Escherichia coli*. Appl Environ Microbiol. 2007;73:1712–20.

Papaioannou E, Utari PD, Quax WJ. Choosing an appropriate infection model to study quorum sensing inhibition in *Pseudomonas* infections. Int J Mol Sci. 2013;14:19309–40.

Park H-J, Park S, Roh J, Kim S, Choi K, Yi J, Kim Y, Yoon J. Biofilm-inactivating activity of silver nanoparticles: a comparison with silver ions. J Ind Eng Chem. 2013;19:614–9.

Passador L, Cook JM, Gambello MJ, Rust L, Iglewski BH. Expression of *Pseudomonas aeruginosa* virulence genes requires cell-to-cell communication. Science. 1993;260:1127–30.

Patil MP, Kim GD. Eco-friendly approach for nanoparticles synthesis and mechanism behind antibacterial activity of silver and anticancer activity of gold nanoparticles. Appl Microbiol Biotechnol. 2017;101:79–92.
Paukscb L, Hartmann S, Rohnke M, Szalay G, Alt V, Schnettler R, Lips KS. Biocompatibility of silver nanoparticles and silver ions in primary human mesenchymal stem cells and osteoblasts. Acta Biomater. 2014;10:439–49.
Pearson JP, Gray KM, Passador L, Tucker KD, Eberhard A, Iglewski BH, Greenberg EP. Structure of the autoinducer required for expression of *Pseudomonas aeruginosa* virulence genes. Proc Natl Acad Sci U S A. 1994;91:197–201.
Pearson JP, Passador L, Iglewski BH, Greenberg EP. A second N-acylhomoseine lactone signal produced by *Pseudomonas aeruginosa*. Proc Natl Acad Sci U S A. 1995;92:1490–4.
Pelgrift RY, Friedman AJ. Nanotechnology as a therapeutic tool to combat microbial resistance. Adv Drug Deliv Rev. 2013;65:1803–15.
Percival SL, Bower PG, Russell D. Bacterial resistance to silver in wound care. J Hosp Infect. 2005;60:1–7.
Pesci EC, Milbank JB, Pearson JP, McKnight S, Kende AS, Greenberg EP, Iglewski BH. Quinolone signaling in the cell-to-cell communication system of *Pseudomonas aeruginosa*. Proc Natl Acad Sci U S A. 1999;96:11229–34.
Qui R, Pei W, Zhang L, Lin J, Ji G. Identification of the putative staphylococcal AgrB catalytic residues involving the proteolytic cleavage of AgrD to generate autoinducing peptide. J Biol Chem. 2005;280:16695–704.
Raftery TD, Lindler H, McNealy TL. Altered host cell – bacteria interaction due to nanoparticles interaction within a bacterial biofilm. Microb Ecol. 2013;65:496–503.
Raftery TD, Kerscher P, Hart AE, Saville SL, Qi B, Kichens CL, Mefford OT, McNeely TL. Discrete nanoparticles induce loss of *Legionella pneumophila* biofilms from surfaces. Nanotoxicology. 2014;8:477–84
Rai M, Yadav A, Gade A. Silver nanoparticles as a new generation of antimicrobials. Biotechnol Adv. 2009;27:76–83.
Ramasamy M, Lee JH, Lee J. Potent antimicrobial and antibiofilm activities of bacteriogenically synthesized gold-silver nanoparticles against pathogenic bacteria and their physiochemical characterizations. J Biomater Appl. 2016;31:366–78.
Rashid MH, Kornberg A. Inorganic phosphate is needed for swimming, swarming, and twitching motilities in *Pseudomonas aeruginosa*. Proc Natl Acad Sci U S A. 2000;97:4885–90.
Reuter K, Steinbach A, Helms V. Interfering with bacterial quorum sensing. Perspect Med Chem. 2016;8:1–15.
Roe D, Karandikar B, Bonn-Savage N, Gibbins B, Roullet JB. Antimicrobial surface functionalization of plastic catheters by silver nanoparticles. J Antimicrob Chemother. 2008;61:869–76.
Sarkisova SA, Lotikar SR, Guragain M, Kubat R, Cloud J, Franklin MJ, Patrauchan MA. A *Pseudomonas aeruginosa* EF-hand protein, EfhP (PA4107), modulates stress responses and virulence at high calcium concentration. PLoS One. 2014;9:e98985.
Sathyanarayanan MB, Balachandranath R, Genji Srinivasulu Y, Kannaiyan SK, Subiadhoss G. The effect of gold and iron-oxide nanoparticles on biofilm forming pathogens. ISRN Microbiol. 2013;2013:e272086.
Seil JT, Webster TJ. Antimicrobial applications of nanotechnology: methods and literature. Int J Nanomedicine. 2012;7:2767–81.
Shanker E, Federle MJ. Quorum sensing regulation of competence and bacteriocins in *Streptococcus pneumoniae* and mutants. Gene. 2017;8(1):pli:E15.
Shopsin B, Eaton C, Wasserman GA, Mathema B, Adhikari RP, Agolory S, Altman DR, Holzman RS, Kreiswirth BN, Novick RP. Mutations in *agr* do not persist in natural populations of methicillin-resistant *Staphylococcus aureus*. J Infect Dis. 2010;202:1593–9.
Sifri CD, Begun J, Ausubel FM, Calderwood SB. *Caenorhabditis elegans* as a model host for *Staphylococcus aureus* pathogenesis. Infect Immun. 2003;71:2208–17.
Singh PK, Schaefter AL, Parsek MR, Moninger TO, Welsh MJ, Greenberg EP. Quorum-sensing signals indicate that cystic fibrosis lungs are infected with bacterial biofilms. Nature. 2000;407:762–4.

Singh PK, Parsek MR, Greenberg EP, Welsh MJ. A component of innate immunity prevents bacterial biofilm development. Nature. 2002;417:552–5.

Sondi I, Salopek-Sondi B. Silver nanoparticles as antimicrobial agent; a case study in *E. coli* as a model for Gram-negative bacteria. J Colloid Interface Sci. 2004;275:177–82.

Tan MW, Mahajan-Mikols S, Ausubel FM. Killing of *Caenorhabditis elegans* by *Pseudomonas aeruginosa* used to model bacterial pathogenesis. Proc Natl Acad Sci U S A. 1999;96:715–20.

Telford G, Wheeler D, Williams P, Tomkins PT, Appleby P, Sewell H, Stewart GS, Bycroft BW, Pritchard DI. The *Pseudomonas aeruginosa* quorum-sensing signal molecule n-(3-oxododecanoyl)-L-homoserine lactone has immunomodulatory activity. Infect Immun. 1998;66:36–42.

Vijayakumar S, Vaseenharan B, Malaikozhundan B, Gopi N, Ekambaram P, Velusamy P, Murugan K, Benelli G, Suresh Kumar R, Suryanarayanamoorthy M. Therapeutic effects of gold nanoparticles synthesized using *Musa paradisiacal* peel extract against multiple antibiotic resistant *Enterococcus faecalis* biofilms and human lung cancer cells (A549). Microb Pathog. 2017;102:173–83.

Vijayan SR, Santhiyagu P, Singamuthu M, Kumari Ahila N, Jayaraman R, Ethiraj K. Synthesis and characterization of silver and gold nanoparticles using aqueous extract of seaweed *Turbinaria conoides*, and their antimicrofouling activity. Sci World J. 2014;2014:e938272.

Wade DS, Calfee MW, Rocha ER, Ling EA, Engstrom E, Coleman JP, Pesci EC. Regulation of *Pseudomonas* quinolone signal synthesis in *Pseudomonas aeruginosa*. J Bacteriol. 2005;187:4372–80.

Wagner VE, Bushnell D, Passador L, Brooks AI, Iglewski BH. Microarray analysis of *Pseudomonas aeruginosa* quorum-sensing regulons: effects of growth phase and environment. J Bacteriol. 2003;185:2080–95.

Whitehead NA, Barnard AM, Slater H, Simpson NJ, Salmond GP. Quorum-sensing in Gram-negative bacteria. FEMS Microbiol Rev. 2001;25:365–404.

Wolska KI, Grudniak AM, Rudnicka Z, Markowska K. Genetic control of bacterial biofilms. J Appl Genet. 2016a;57:225–38.

Wolska KI, Grudniak AM, Markowska K. Inhibitors of bacterial quorum sensing systems and their role as potential therapeutics. In Polish. Postepy Mikrobiol. 2016b;55:300–8.

Wright JS, Jin R, Novick RP. Transient interference with staphylococcal quorum sensing blocks abscesses formation. Proc Natl Acad Sci U S A. 2005;102:1691–6.

Yardwood JM, Schlievert PM. Quorum sensing in *Staphylococcus infections*. J Clin Invest. 2003;112:1620–5.

Zhang XF, Liu ZG, Shen W, Gurunathan S. Silver nanoparticles: synthesis, characterization, properties, applications and thertapeutic approaches. Int J Mol Sci. 2016;17(9):1534. pli:E1534

Chapter 8
A Novel Strategy for Antimicrobial Agents: Silver Nanoparticles

Heejeong Lee and Dong Gun Lee

Abstract Metals, such as copper, gold, and silver, have typically been used for the synthesis of stable dispersions of nanoparticles. Metallic nanoparticles have high biocompatibility owing to their high surface area to lack of charge, volume ratio, and lack of toxicity to humans. In particular, silver nanoparticles (nano-Ag) are widely used in the field of pathogenic microbiology owing to their potent antimicrobial activities. Their antimicrobial efficacy has been evaluated against bacteria, viruses, and other eukaryotic microbes based on interactions with membranes or cell walls, DNA, or proteins. Nano-Ag exerts antimicrobial activity via a variety of mechanisms, e.g., membrane disruption, apoptosis, and synergy. Accordingly, nano-Ag has wide applications ranging from burn treatments to silver-coated medicinal devices, such as coating stainless steel materials, dental materials, nanogels, nanolotions, etc. Nano-Ag is a potential alternative strategy for the development of cosmetic and pharmacological agents for pathogens that are resistant to existing antibiotics.

Keywords Silver nanoparticles • Pathogenic microorganism • Mechanism of action • Cell membrane • Apoptosis

Nomenclature

DHR-123	Dihydrorhodamine
$DiBAC_4(3)$	Bis-(1,3-dibutylbarbituric acid)-trimethine oxonol
$DiOC_6(3)$	3,3′-dihexyloxacarbocyanine iodide
DPH	1,6-diphenyl-1,3,5-hexatriene
HPF	3′-(*p*-hydroxyphenyl) fluorescein
Nano-Ag	Silver nanoparticles
ROS	Reactive oxygen species
TUNEL	Terminal deoxynucleotidyl transferase dUTP nick end labeling

H. Lee • D.G. Lee (✉)
School of Life Sciences, College of Natural Sciences, Kyungpook National University, 80 Daehakro, Bukgu, Daegu 41566, Republic of Korea
e-mail: dglee222@knu.ac.kr

M. Rai, R. Shegokar (eds.), *Metal Nanoparticles in Pharma*,
DOI 10.1007/978-3-319-63790-7_8

8.1 Introduction

Infectious diseases caused by pathogenic bacteria represent one of the greatest health problems worldwide and the main causes of worldwide morbidity and mortality (Li et al. 2015). Life-threatening human diseases caused by the bacteria *Escherichia coli*, *Pseudomonas aeruginosa*, and *Salmonella spp.* and the fungi *Aspergillus fumigatus* and *Candida albicans* with increased virulence and an enhanced ability to induce disease have emerged; therefore, a variety of antibiotics have been developed and used to cure human diseases (Mantareva et al. 2007; Paulo et al. 2010; Li et al. 2015). Unfortunately, their continued usage in clinical applications raises a serious problem, i.e., the development of microbial resistance to conventional drugs, with unique properties (Sharma et al. 2009).

Silver is a rare, basic, and naturally occurring element with high electrical and thermal conductivity; it has been used to produce utensils, monetary currency, jewelry, photography, dental alloy, and explosives since ancient times (Rai et al. 2016). Silver has been known for centuries to be curative and has been used extensively to fight against infections (Rai et al. 2016). Since the discovery of antibiotics, however, silver use as an antimicrobial agent has been limited. Nanotechnology provides a platform to modify and develop the properties of metals by reducing them to metal nanoparticles, which have huge applications in a range of fields, such as diagnostics, biomarkers, drug delivery systems, antimicrobial agents, and other nanomedicines for the treatment of various diseases (Singh and Nalwa 2011). Owing to their high surface area to volume ratio and unique chemical and physical properties, with no charge and lack of toxicity to humans, nanoscale materials are considered novel antimicrobial agents (Retchkiman-Schabes et al. 2006; Aruguete et al. 2013). Nanoparticles are particularly promising as they show good antimicrobial effects against pathogenic microorganisms. There are various types of nanoparticles, including copper, zinc, titanium, gold, and silver, with antimicrobial activity (Oberdörster et al. 2005; Gong et al. 2007; Gurunathan et al. 2009).

The mechanisms by which silver nanoparticles (nano-Ag) affect pathogenic microorganisms have been examined by many researchers. Additionally, nano-Ag is nontoxic to humans, like silver ions that are nontoxic to humans according to centuries of use (Oberdörster et al. 2005; Gong et al. 2007; Gurunathan et al. 2009). It is generally reported that nano-Ag interacts with three main components of microorganisms to exert an antimicrobial effect: membranes or cell walls (Yamanaka et al. 2005; Kim et al. 2009), DNA (Yang et al. 2009), and microbial proteins (Yamanaka et al. 2005). Therefore, in the present chapter, the mechanisms by which nano-Ag functions against microorganisms, including the genetic mechanisms and the roles of cell walls, proteins, and bacterial enzymes, are described. The properties and applications of nano-Ag are also summarized.

8.2 Antifungal Activity and Mechanism

8.2.1 Decrease of Membrane Integrity and Changes in Membrane Depolarization

Nano-Ag has the ability to anchor the microbial cell wall and then penetrate it, thereby affecting the permeability of the cell membrane that causes structural changes, eventually resulting in cell death (Prabhu and Poulose 2012; Ortega et al. 2015). Moreover, proteins and polymeric carbohydrates that accompany to nanoparticles would favor interactions with the cytoplasmic membrane (Fernández et al. 2016). Among membrane functions, maintaining membrane integrity is important for cell survival because the membrane protects the cells from the external environment and regulates cellular responses (Benyagoub et al. 1996; Portet et al. 2009). 1,6-Diphenyl-1,3,5-hexatriene (DPH), an indicator of changes in membrane integrity, interacts with an acyl group of the membrane lipid bilayer and does not disturb the membrane. When cells experience severe damage, such as membrane disruption and pore formation, DPH becomes detached from the membrane (Vincent et al. 2004). Nano-Ag decreases fungal membrane integrity, and this effect tends to increase depending on the concentration.

Fungal cells establish various ion gradients across the membrane to retain their membrane potential. The ion gradients are sustained by restricted membrane permeability to small solutes and ions (Yu 2003). However, when a membrane loses its stability after severe damage, increased membrane permeability leads to the leakage of a variety of ions from the cell. Then, undesired ion leakage causes the depolarization of the fungal membrane (Bolintineanu et al. 2010). Alterations to fungal membranes could be detected by a bis-(1,3-dibutylbarbituric acid)-trimethine oxonol [$DiBAC_4(3)$]. Increased $DiBAC_4(3)$ fluorescence has been observed after treatment with nano-Ag, suggesting that nano-Ag induces membrane depolarization via increased membrane permeability.

8.2.2 Release of Intracellular Glucose and Trehalose and Cell Cycle Arrest

During stress challenge such as toxic agents, dehydration, desiccation, and oxidation, fungal cells exhibit the inactivation and denaturation of proteins and membranes. Trehalose appears to stabilize membranes and native proteins, suppress the aggregation of denatured proteins, and further repair damaged proteins (Alvarez-Peral et al. 2002; Leaper 2006; Landolfo et al. 2008). The effect of nano-Ag on fungal membranes could be detected by measuring trehalose and glucose released in a cell suspension (Alvarez-Peral et al. 2002). After treatment with nano-Ag, the

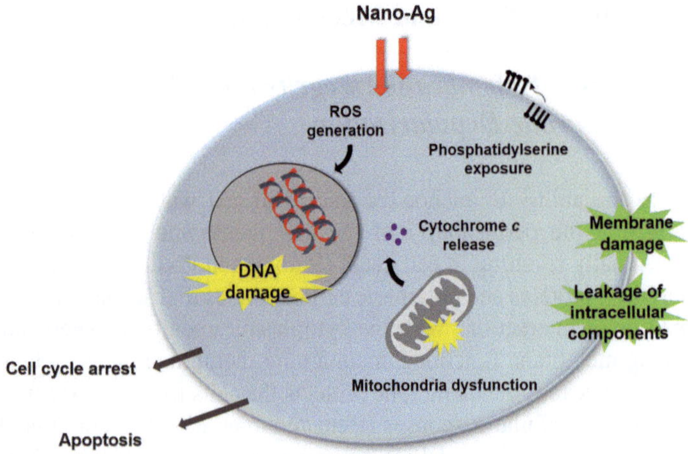

Fig. 8.1 Antifungal mechanisms of silver nanoparticles. Membrane disruption leading to leakage of intracellular components and cell cycle arrest was caused by silver nanoparticles. Furthermore, silver nanoparticles promoted apoptosis through mitochondria dysfunction and DNA damage, according to ROS generation

amounts of both released and intracellular glucose and trehalose from *C. albicans* cells increase. These results confirm that nano-Ag induces cellular stress on membranes, leading to the release of several intracellular components, such as glucose and trehalose.

As cells progress through the cell cycle, they undergo cell cycle arrest to progress to programmed cell death or repair cellular damage when intracellular damage is too extensive to be recovered. The position of arrest within the cell cycle varies depending upon the phase in which the damage is sensed (Elledge 1996). DNA-damaged cells are prevented from entering mitosis at the G2/M or G1/S cell cycle checkpoint (Samuel et al. 2002). The growth inhibition of budding cells is correlated with membrane damage, and cell cycle arrest can occur by the disruption of membrane integrity (Endo et al. 1997). After treatment with nano-Ag, the amount of cells in the G2/M phase increases while that in the G1 phase significantly decreases. This demonstrates that nano-Ag induces physiological damage, leading to cell cycle arrest at the G2/M phase of *C. albicans* (Fig. 8.1).

8.2.3 Reactive Oxygen Species (ROS) Accumulation

ROS, including hydroxyl radicals, hydrogen peroxide, and superoxide anions, are continuously produced during normal aerobic metabolism and act as second messengers in signal transduction pathways in fungal cells (Costa and Moradas-Ferreira 2001). Increases in intracellular ROS bring about reduced mitochondrial enzyme activity and damage to lipids, nucleic acids, and cellular contents (Costa and Moradas-Ferreira 2001; Landolfo et al. 2008). ROS accumulation could be detected by the ROS-sensitive fluorescence probe dihydrorhodamine (DHR-123). A significant increase in DHR-123 fluorescence has been observed after treatment with nano-Ag, indicating excessive ROS levels and the generation of oxidative stress. Hydroxyl radicals are the neutral form of hydroxide ions and are highly reactive, promoting cellular oxidative damage (Haruna et al. 2002). In the presence of hydrogen peroxide and metal ions, hydroxyl radicals are produced by the Fenton reaction and damage intracellular molecules, such as lipids, DNA, and proteins, resulting in cell death (Rollet-Labelle et al. 1998). Increased 3′-(*p*-hydroxyphenyl) fluorescein (HPF) fluorescence, indicating hydroxyl radical formation, has been detected in nano-Ag-treated cells, suggesting that hydroxyl radical formation is induced by nano-Ag in *C. albicans* cells. To investigate the effect of hydroxyl radicals, a potent hydroxyl radical scavenger, thiourea, was used to establish hydroxyl radical scavenging effects in both eukaryotes and prokaryotes (Novogrodsky et al. 1982). Nano-Ag and thiourea co-treated cells significantly reduced hydroxyl radical formation. These results indicate that hydroxyl radical formation plays an essential role in nano-Ag-induced cell death (Fig. 8.1).

8.2.4 *Fungal Apoptosis*

8.2.4.1 Phosphatidylserine Exposure and DNA Fragmentation

In most cells, the plasma membrane consists of asymmetric phospholipids, such as phosphatidylcholine, phosphatidylethanolamine, phosphatidylserine, and sphingomyelin (Fadok et al. 1998). Phosphatidylcholine and sphingomyelin are concentrated on the outer leaflet of the plasma membrane, whereas phosphatidylethanolamine and phosphatidylserine are concentrated on the inner leaflet of the cytoplasmic membrane (Mantareva et al. 2007). Almost 90% of phosphatidylserines are oriented toward the inner leaflet of the cytoplasmic membrane, but by inhibiting flip-flop action, they can be translocated to the outer leaflet in yeast (Cerbon and Calderon 1991; Suzuki et al. 2013). In apoptotic cells, phosphatidylserine exposure is observed in the early stage. Annexin V detects to phosphatidylserine with high affinity in the presence of Ca^{2+} and therefore is used as an early apoptosis marker based on the detection of phosphatidylserine exposure (Smrz et al. 2007). The cell population exhibiting early apoptosis significantly increases after treatment with nano-Ag, whereas no increase in early apoptosis is detected in thiourea co-treated cells. Nano-Ag induces phosphatidylserine exposure in *C. albicans* cells and apoptotic death in the presence of hydroxyl radical formation (Hwang et al. 2012).

Apoptosis involves diverse cellular processes, including phosphatidylserine externalization to the outer leaflet of the plasma membrane, active chromatin condensation, and DNA fragmentation (Kyrylkova et al. 2012; Léger et al. 2015). In particular, DNA damage from oxidative stress triggers cell death, and the fragmentation of

DNA by activated endonucleases is considered a feature of late-stage apoptosis (Wadskog et al. 2004). DNA fragmentation could be revealed by a terminal deoxynucleotidyl transferase dUTP nick end labeling (TUNEL) assay. Nano-Ag treatment results in TUNEL-positive cells, showing an increase in fluorescence or intense fluorescent spots. However, these cells are not observed after co-treatment with thiourea. Nano-Ag promotes the oxidative condition, resulting in phosphatidylserine and DNA fragmentation (Hwang et al. 2012).

8.2.4.2 Attenuation of Mitochondrial Functions

Mitochondria play a central role in yeast apoptosis because they contain many pro-apoptotic factors and are the major organelle of ROS production (Reiter et al. 2005). Especially, enhancement of mitochondrial permeability and the loss of mitochondrial membrane potential are hallmarks of the early stage of apoptosis (Jakubowski et al. 2000). The opening of membrane pores that are located in the mitochondrial membrane induces a decrease in mitochondrial inner membrane potential. Additionally, as a consequence of the decline in mitochondrial membrane potential, mitochondrial structures are altered, including cristae unraveling and matrix condensation, and these structural changes result in the redistribution of cytochrome *c* from the cristae to the intermembrane spaces, facilitating release (Anantharaman et al. 2010). To investigate whether nano-Ag decreases the mitochondrial membrane potential, the fluorescent voltage-dependent probe 3,3′-dihexyloxacarbocyanine iodide [$DiOC_6(3)$] was used. It accumulates in mitochondria and fluoresces green. The dissipation of mitochondrial membrane potential was revealed by nano-Ag-treated cells. In contrast, cells treated with nano-Ag and thiourea did not undergo substantial changes (Hwang et al. 2012). Moreover, changes in mitochondrial inner membrane potential were investigated by a JC-1 assay to confirm the effect of nano-Ag on mitochondrial membrane potential. Nano-Ag induced the breakdown of the mitochondrial inner membrane potential, which is a crucial stage in apoptotic cells, and increased mitochondrial permeability (Yao et al. 2006).

Many studies have demonstrated that mitochondria play a key role in fungal apoptosis via cytochrome *c* release. A decrease in the mitochondrial membrane potential causes the translocation and activation of the apoptotic factor cytochrome *c* (Gottlieb et al. 2003). Cytochrome *c* release from mitochondria to the cytosol is a crucial event in the apoptotic pathway. It binds electrostatically to the mitochondrial outer membrane and is an essential component of the respiratory chain in mitochondria (Dejean et al. 2006). Additionally, cytochrome *c* is leaked to the cytosol during apoptosis and subsequently provokes the caspase cascade, representative of apoptotic proteases. Released cytochrome *c* from mitochondria binds to apoptotic protease-activating factors in cytosol (Pereira et al. 2007). A large amount of cytochrome *c* is found in the cytosol after treatment with nano-Ag, and a small amount of cytochrome *c* is found in thiourea co-treated cells. Nano-Ag causes the release of cytochrome *c* from mitochondria to the cytosol, and hydroxyl radical formation affects cytochrome *c* release (Fig. 8.1).

8.3 Antibacterial Activity and Mechanism of Action

Silver has been used as a therapeutic agent against various diseases; in the past, it has been used as an antimicrobial and antiseptic agent against Gram-positive and Gram-negative bacteria (Costa and Moradas-Ferreira 2001; Mantareva et al. 2007; Landolfo et al. 2008) because it exhibits low cytotoxicity (Rai et al. 2009). Although the highly antibacterial effect of nano-Ag has been widely described, the mechanism of action is yet to be fully elucidated. In fact, the potent antibacterial and broad-spectrum activity against metabolically and morphologically diverse microorganisms seem to be correlated with a multifaceted mechanism by interacting with microbes (Fig. 8.2).

8.3.1 Bacterial Cell Membrane

8.3.1.1 Leakage of Intracellular Contents

Protection from outside is granted by two structures used to classify bacteria as Gram-positive or Gram-negative (Durán et al. 2016). Gram-negative bacteria reveal only a thin peptidoglycan layer between the outer membrane and the cytoplasmic membrane (Morones et al. 2005); in contrast, Gram-positive bacteria lack the outer membrane but have a peptidoglycan layer of approximately 30 nm thick (Morones et al. 2005). The outer membrane is important for protecting bacteria from harmful

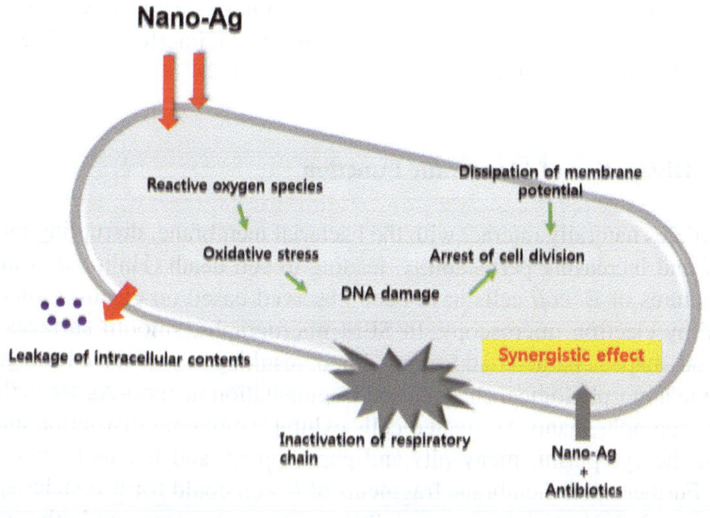

Fig. 8.2 Antibacterial mechanisms of silver nanoparticles. Silver nanoparticles exerted the antibacterial effect via membrane damage and process according to oxidative stress. Also, silver nanoparticles showed the synergistic effect in combination with conventional antibiotics

agents, such as degradative enzymes, toxins, drugs, and detergents. The outer leaflet of the outer membrane is composed of lipopolysaccharide molecules that cover approximately three quarters of the surface, and the remaining quarter is composed of membrane proteins (Amro et al. 2000). Reducing sugars exhibit leakage from nano-Ag-treated cells. Similarly, nano-Ag elevates the leakage of proteins through the membrane of *E. coli*. Furthermore, nano-Ag reduces the levels of intracellular adenosine triphosphate, inducing bacterial cell death (Mukherjee et al. 2014). These results demonstrate that nano-Ag induces bacterial membrane damage and permeabilizes the outer membrane, resulting in the leakage of cellular materials, such as reducing sugars and proteins.

8.3.1.2 Effect on Respiratory Chain Dehydrogenases

Bactericidal mechanisms include binding to thiol group agents, involving the induction of hydroxyl radical formation and the inactivation of respiratory chain and TCA cycle enzymes (Gordon et al. 2010). Since respiratory chain enzymes are bound to the cell membrane, they can be considered strongly, early affected targets of nano-Ag. Additionally, the respiratory chain contains iron, mainly coordinated in iron-sulfur clusters (Panáček et al. 2016). The most widely known bactericidal mechanism of nano-Ag is its interaction with the thiol groups of the L-cysteine residue of proteins and the consequent inactivation of their enzymatic functions (You et al. 2012). Nano-Ag interacts with the thiol group of cysteine by replacing the hydrogen atom, thus inactivating the function of protein and inhibiting the growth of *E. coli* and *S. aureus* (Durán et al. 2016). It is assumed that nano-Ag breaks the barriers of outer membrane permeability, periplasm, and peptidoglycan and enters the inner membrane to destroy respiratory chain dehydrogenases (Li et al. 2010, 2011). Accordingly, respiration is inhibited, interfering with the normal growth and metabolism of bacteria cells (Li et al. 2011).

8.3.1.3 Dissipation of Membrane Function

Nanoparticles naturally interact with the bacterial membrane, disrupting membrane integrity and increasing permeability, leading to cell death (Hajipour et al. 2012). The structures of *E. coli* cells have been observed based on electron micrographs obtained by electron microscopy. In SEM micrographs, smooth surfaces and the typical rod-shaped characteristic are modified, resulting in extensive leakage as well as additional morphological changes and fragmentation in nano-Ag-treated cells. In TEM micrographs, nano-Ag-treated cells exhibit membrane disruption and severe efflux of the cytoplasm; many pits and gaps appear, and the membrane is fragmented. Furthermore, membrane fragments of *E. coli* could form vesicles spontaneously (Li et al. 2010; Grigor'eva et al. 2013). Thus, ribosomes and other structural components of the cytoplasm lose their structure and aggregate in clumps of high electron density located in the electron-lucent cytoplasm (Grigor'eva et al. 2013).

8.3.2 ROS Generation and Oxidative Stress

Nanotoxicity is generally caused by the induction of oxidative stress by free radical formation, i.e., ROS, following the administration of nanoparticles (Hajipour et al. 2012). ROS result from the damage of the respiration chain and lead to protein, DNA, and lipid peroxidation damage (Radzig et al. 2013). Thus, ROS generation and oxidative stress can be used as a paradigm to assess nano-Ag toxicity (You et al. 2012). Nano-Ag induces an increase in the respiration rate, thus accumulating intracellular ROS (You et al. 2012). The generation of ROS by nano-Ag results in free radical formation and powerful bactericidal action. Since antioxidative enzymes rely on thiol groups, they most likely do not detoxify ROS generated from the damaged respiratory chain (Gordon et al. 2010). Glutathione contributes to cellular protection from oxidative stress (Mukherjee et al. 2014). The nano-Ag inside cells may interact with cellular glutathione and oxidize it. This oxidized glutathione results in the formation of extensive ROS, which may promote bacterial cellular toxicity and consequently inhibit bacterial growth (Mukherjee et al. 2014).

8.3.3 Cell Filamentation and Change in Membrane Potential

Filamentation describes anomalous bacterial growth in which cells do not undergo septum formation and continue to elongate with multiple chromosomal copies, resulting in a lack of cell division (Chatterjee et al. 2014). The incubation of *Salmonella typhimurium* with nano-Ag arrested cell division; the number of dividing cells decreased significantly. Nano-Ag increased the size of *E. coli* cells due to the inhibition of cell division progression. Bacterial cell filamentation occurs by the action of an external agent that induces either a dissipated cell membrane potential or an SOS-like stress response in cells. A change in membrane potential could be measured by $DiBAC_4(3)$ staining, which is sensitive to cell membrane potential and enters depolarized cells (Liao et al. 1999). An increase in $DiBAC_4(3)$ fluorescence has been observed in nano-Ag-treated cells, indicating a change in membrane potential. This result indicates that nano-Ag exerts its antibacterial effect via membrane depolarization.

8.3.4 DNA Damage and Damage Repair System

When bacterial cells suffer from severe DNA damage, the SOS response is operated by RecA to repair DNA damage (Li et al. 2011). DNA fragmentation and damage by nano-Ag have been observed (Lee et al. 2014). During DNA damage repair, SulA stops cell division by binding to FtsZ, and this causes filamentation (Lee et al. 2014). The integrity of damaged DNA can be repaired by DNA repair mechanisms.

A primary defense mechanism for cells is the SOS response, which plays a role in the exposure of prokaryotic cells to DNA-damaging or DNA replication-interfering agents by producing about 30 proteins involved in damage tolerance (Chatterjee et al. 2014). RecA may act as a major regulator in the SOS repair system. RecA has a co-protease function in the autocatalytic cleavage of the LexA repressor, which restrains the expression of genes involved in the bacterial repair system (Li et al. 2011). The activation of the SOS response could be confirmed by measuring RecA expression by Western blotting. Blots from nano-Ag-treated cells showed higher RecA expression levels than those of untreated cells. This result indicates that bacterial cell death involves severe DNA damage by nano-Ag and also suggests that the DNA repair system is activated by the RecA protein. Additionally, the influence of mutations in genes in the base excision repair system on bacterial sensitivity to nano-Ag has been explored. BER is the most essential mechanism for cellular DNA protection from oxidative damage caused by nano-Ag-induced ROS in bacterial cells (Radzig et al. 2013). The bases in DNA are susceptible to oxidative damage; the most common lesion is 8-oxoguanine, leading to oxidative mutagenesis (Radzig et al. 2013).

8.4 Synergistic Effect

In order to move toward practical uses in the medical field, the synergistic effects of nano-Ag have been examined in many studies using conventional antibiotics (Hwang et al. 2012; Ghosh et al. 2013; Markowska et al. 2014). The minimum inhibitory concentration and fractional inhibitory concentration have been used to confirm synergistic effects and antibacterial susceptibility (Hwang et al. 2012). These synergistic activities of nano-Ag in the presence of conventional antibiotics may enable reductions in the viability of bacterial strains at lower concentrations (Ghosh et al. 2012; Panáček et al. 2016). Nano-Ag and antibiotics can lead to bacterial cell death via different mechanisms. Therefore, the synergistic effects can act as an influential tool against resistant microorganisms. A bonding reaction between antibiotics and nano-Ag ultimately increases the sensitivity of antimicrobial agents at specific points on the cell membrane. This may be attributed to the selectivity of nano-Ag toward the cell membrane, consisting of phospholipids and glycoprotein. Thus, nano-Ag facilitates the transport of antibiotics to the cell surface. The permeability of the membrane increases, facilitating enhanced infiltration of antibiotics into the cell as nano-Ag binds to sulfur-containing proteins of the bacterial cell membrane (Ghosh et al. 2012). Based on a different approach, it can be demonstrated that the generation of ·OH is a common mechanism by which antibiotics cause bacterial cell death. Indeed, nano-Ag is capable of ·OH generation, and two different classes of bactericidal antibiotics, kanamycin and ampicillin, induce OH formation according to a study of synergistic effects of nano-Ag with conventional, indicating that ·OH is an important cause of synergistic effects (Grigor'eva et al. 2013).

8.5 Applications

Many studies focused on clinical and industrial applications of nano-Ag have been performed over the past few decades. The application of nano-Ag demonstrates the possibility of additional clinical and industrial uses (Chen and Schluesener 2007). Nano-Ag can be used to prepare and develop nanomedicines, new generations of biosensors, antimicrobials, drug delivery systems, silver-coated medicinal devices, silver-based dressings, nanogels, and nanolotions (Sahu et al. 2013). Dressings are important in the management of wounds, and newly designed wound dressings are required to prevent pathogenic infections. Silver nanocrystalline dressings reduce bacterial infections in chronic wounds (Ip et al. 2006; Leaper 2006). Medical applications in the form of nanolotions, nanogels, silver-coated medicinal devices, and silver-based dressings have been incorporated (Oberdörster et al. 2005). Moreover, the advantages of silver-coated medical devices include the protection of both the outer and inner surfaces of devices. The continuous release of silver ions provides antimicrobial activity (Rai et al. 2009).

8.6 Conclusion

Owing to its effective antimicrobial properties, silver has been used for centuries to treat various diseases, most notably infections, and its biological safety has been demonstrated. Although silver is a practical antimicrobial agent, its use has a major drawback, i.e., is easily inactivated by complexation and precipitation. However, nanotechnology has resolved this issue by the development of nano-Ag, which exhibits unique chemical and physical properties that are beneficial for the development of new antimicrobial agents. Antimicrobial mechanisms against pathogenic microorganisms have been reported. Nano-Ag induces membrane disruption in fungal and bacterial cells. Apoptosis hallmarks have been observed in yeast cells. Efficiency with respect to ROS has been indicated, such as DNA damage, cell filamentation, and the induction of DNA repair systems in bacterial cells. Its synergistic capacity against various representative pathogenic bacteria has been confirmed. Considering the significant antimicrobial efficacy of nano-Ag in clinical and industrial applications, it has emerged as a potent and practical antimicrobial agent. These studies clarify the efficient antimicrobial activity and mechanism of nano-Ag. Further investigations of the specific clinical implications are needed.

References

Alvarez-Peral FJ, Zaragoza O, Pedreno Y, Argüelles JC. Protective role of trehalose during severe oxidative stress caused by hydrogen peroxide and the adaptive oxidative stress response in *Candida albicans*. Microbiology. 2002;148:2599–606.

Amro NA, Kotra LP, Wadu-Mesthrige K, Bulychev A, Mobashery S, Liu G. High-resolution atomic force microscopy studies of the *Escherichia coli* outer membrane: structural basis for permeability. Langmuir. 2000;16:2789–96.

Anantharaman A, Rizvi MS, Sahal D. Synergy with rifampin and kanamycin enhances potency, kill kinetics, and selectively of de novo-designed antimicrobial peptides. Antimicrob Agents Chemother. 2010;54:1693e1699.

Aruguete DM, Kim B, Hochella MF Jr, Ma Y, Cheng Y, Hoegh A, Liu J, Pruden A. Antimicrobial nanotechnology: its potential for the effective management of microbial drug resistance and implications for research needs in microbial nanotoxicology. Environ Sci Process Impacts. 2013;5:93–102.

Benyagoub M, Willemot C, Bélanger RR. Influence of a subinhibitory dose of antifungal fatty acids from *Sporothrix flocculosa* on cellular lipid composition in fungi. Lipids. 1996;31:1077–82.

Bolintineanu D, Hazrati E, Davis HT, Lehrer RI, Kaznessis YN. Antimicrobial mechanism of pore-forming protegrin peptides: 100 pores to kill *E. coli*. Peptides. 2010;31:1–8.

Cerbon J, Calderon V. Changes of the compositional asymmetry of phospholipids associated to the increment in the membrane surface potential. Biochim Biophys Acta. 1991;1067:139–44.

Chatterjee AK, Chakraborty R, Basu T. Mechanism of antibacterial activity of copper nanoparticles. Nanotechnology. 2014;25:135101.

Chen X, Schluesener HJ. Nanosilver: a nanoproduct in medical application. Toxicol Lett. 2007;176:1–12.

Costa V, Moradas-Ferreira P. Oxidative stress and signal transduction in *Saccharomyces cerevisiae*: insights in to aging, apoptosis and disease. Mol Asp Med. 2001;22:217–46.

Dejean LM, Martinez-Caballero S, Kinnally KW. Is MAC the knife that cuts cytochrome *c* from mitochondria during apoptosis? Cell Death Differ. 2006;13:1387–95.

Durán N, Durán M, de Jesus MB, Seabra AB, Fávaro WJ, Nakazato G. Silver nanoparticles: a new view on mechanistic aspects on antimicrobial activity. Nanomedicine. 2016;12:789–99.

Elledge SJ. Cell cycle checkpoints: preventing an identity crisis. Science. 1996;274:1664–72.

Endo M, Takesako K, Kato I, Yamaguchi H. Fungicidal action of aureobasidin A, a cyclic depsipeptide antifungal antibiotic, against *Saccharomyces cerevisiae*. Antimicrob Agents Chemother. 1997;41:672–6.

Fadok VA, Bratton DL, Frasch SC, Warner ML, Henson PM. The role of phosphatidylserine in recognition of apoptotic cells by phagocytes. Cell Death Differ. 1998;5:551–62.

Fernández JG, Fernández-Baldo MA, Berni E, Camí G, Durán N, Raba J, Sanz MI. Production of silver nanoparticles using yeasts and evaluation of their antifungal activity against phytopathogenic fungi. Process Biochem. 2016;51:1306–13.

Ghosh S, Patil S, Ahire M, Kitture R, Kale S, Pardesi K, Cameotra SS, Bellare J, Dhavale DD, Jabgunde A, Chopade BA. Synthesis of silver nanoparticles using *Dioscorea bulbifera* Tuber extract and evaluation of its synergistic potential in combination with antimicrobial agents. Int J Nanomedicine. 2012;7:483–96.

Ghosh IN, Patil SD, Sharma TK, Srivastava SK, Pathania R, Navani NK. Synergistic action of cinnamaldehyde with silver nanoparticles against spore-forming bacteria: a case for judicious use of silver nanoparticles for antibacterial applications. Int J Nanomedicine. 2013;8:4721–31.

Gong P, Li H, He X, Wang K, Hu J, Tan W. Preparation and antibacterial activity of Fe3O4@Ag nanoparticles. Nanotechnology. 2007;18:604–11.

Gordon O, Vig Slenters T, Brunetto PS, Villaruz AE, Sturdevant DE, Otto M, Landmann R, Fromm KM. Silver coordination polymers for prevention of implant infection: thiol interaction, impact on respiratory chain enzymes, and hydroxyl radical induction. Antimicrob Agents Chemother. 2010;54:4208–18.

Gottlieb E, Armour SM, Harris MH, Thompson CB. Mitochondrial membrane potential regulates matrix configuration and cytochrome c release during apoptosis. Cell Death Differ. 2003;10:709–17.

Grigor'eva A, Saranina I, Tikunova N, Safonov A, Timoshenko N, Rebrov A, Ryabchikova E. Fine mechanisms of the interaction of silver nanoparticles with the cells of *Salmonella typhimurium* and *Staphylococcus aureus*. Biometals. 2013;26:479–88.

Gurunathan S, Lee KJ, Kalimuthu K, Sheikpranbabu S, Vaidyanathan R, Eom SH. Antiangiogenic properties of silver nanoparticles. Biomaterials. 2009;30:6341–50.

Hajipour MJ, Fromm KM, Ashkarran AA, Jimenez de Aberasturi D, de Larramendi IR, Rojo T, Serpooshan V, Parak WJ, Mahmoudi M. Antibacterial properties of nanoparticles. Trends Biotechnol. 2012;30:499–511.

Haruna S, Kuroi R, Kajiwara K, Hashimoto R, Matsugo S, Tokumaru S, Kojo S. Induction of apoptosis in HL-60 cells by photochemically generated hydroxyl radicals. Bioorg Med Chem Lett. 2002;12:675–6.

Hwang IS, Lee J, Hwang JH, Kim KJ, Lee DG. Silver nanoparticles induce apoptotic cell death in *Candida albicans* through the increase of hydroxyl radicals. FEBS J. 2012;279:1327–38.

Ip M, Lui SL, Poon VKM, Lung I, Burd A. Antimicrobial activities of silver dressings: an in vitro comparison. J Med Microbiol. 2006;55:59–63.

Jakubowski W, Bilinski T, Bartosz G. Oxidative stress during aging of stationary cultures of the yeast *Saccharomyces cerevisiae*. Free Radic Biol Med. 2000;28:659–64.

Kim KJ, Sung WS, Suh BK, Moon SK, Choi JS, Kim JG, Lee DG. Antifungal activity and mode of action of silver nano-particles on *Candida albicans*. Biometals. 2009;22:235–42.

Kyrylkova K, Kyryachenko S, Leid M, Kioussi C. Detection of apoptosis by TUNEL assay. Methods Mol Biol. 2012;887:41–7.

Landolfo S, Politi H, Angelozzi D, Mannazzu I. ROS accumulation and oxidative damage to cell structures in *Saccharomyces cerevisiae* wine strains during fermentation of high-sugar-containing medium. Biochim Biophys Acta. 2008;1780:892–8.

Leaper DL. Silver dressings: their role in wound management. Int Wound J. 2006;3:282–94.

Lee W, Kim KJ, Lee DG. A novel mechanism for the antibacterial effect of silver nanoparticles on *Escherichia coli*. Biometals. 2014;27:1191–201.

Léger T, Garcia C, Ounissi M, Lelandais G, Camadro JM. The metacaspase (Mca1p) has a dual role in farnesol-induced apoptosis in *Candida albicans*. Mol Cell Proteomics. 2015;14:93–108.

Li WR, Xie XB, Shi QS, Zeng HY, Ou-Yang YS, Chen YB. Antibacterial activity and mechanism of silver nanoparticles on *Escherichia coli*. Appl Microbiol Biotechnol. 2010;85:1115–22.

Li WR, Xie XB, Shi QS, Duan SS, Ouyang YS, Chen YB. Antibacterial effect of silver nanoparticles on *Staphylococcus aureus*. Biometals. 2011;24:135–41.

Li H, Chen Q, Zhao J, Urmila K. Enhancing the antimicrobial activity of natural extraction using the synthetic ultrasmall metal nanoparticles. Sci Rep. 2015;5:11033.

Liao RS, Rennie RP, Talbot JA. Assessment of the effect of amphotericin B on the vitality of *Candida albicans*. Antimicrob Agents Chemother. 1999;43:1034–41.

Mantareva V, Kussovski V, Angelov I, Borisova E, Avramov L, Schnurpfeil G, Wöhrle D. Photodynamic activity of water-soluble phthalocyanine zinc(II) complexes against pathogenic microorganisms. Bioorg Med Chem. 2007;15:4829–35.

Markowska K, Grudniak AM, Krawczyk K, Wróbel I, Wolska KI. Modulation of antibiotic resistance and induction of a stress response in *Pseudomonas aeruginosa* by silver nanoparticles. J Med Microbiol. 2014;63:849–54.

Morones JR, Elechiguerra JL, Camacho A, Holt K, Kouri JB, Ramírez JT, Yacaman MJ. The bactericidal effect of silver nanoparticles. Nanotechnology. 2005;16:2346–53.

Mukherjee S, Chowdhury D, Kotcherlakota R, Patra S, BV, Bhadra MP, Sreedhar B, Patra CR. Potential theranostics application of bio-synthesized silver nanoparticles (4-in-1 system). Theranostics. 2014;4:316–35.

Novogrodsky A, Ravid A, Rubin AL, Stenzel KH. Hydroxyl radical scavengers inhibit lymphocyte mitogenesis. Proc Natl Acad Sci U S A. 1982;79:1171–4.

Oberdörster G, Oberdörster E, Oberdörster J. Nanotoxicology: an emerging discipline evolving from studies of ultrafine particles. Environ Health Perspect. 2005;113:823–39.

Ortega F, Fernández-Baldo M, Fernández G, Serrano M, Sanz MI, Diaz-Mochón J, Lorente J, Raba J. Study of antitumor activity in breast cell lines using silver nanoparticles produced by yeast. Int J Nanomedicine. 2015;10:2021–31.

Panáček A, Smékalová M, Večeřová R, Bogdanová K, Röderová M, Kolář M, Kilianová M, Hradilová Š, Froning JP, Havrdová M, Prucek R, Zbořil R, Kvítek L. Silver nanoparticles strongly enhance and restore bactericidal activity of inactive antibiotics against multi resistant *Enterobacteriaceae*. Colloids Surf B: Biointerfaces. 2016;142:392–9.

Paulo L, Ferreira S, Gallardo E, Queiroz JA, Domingues F. Antimicrobial activity and effects of resveratrol on human pathogenic bacteria. World J Microbiol Biotechnol. 2010;26:1533–8.

Pereira C, Camougrand N, Manon S, Sousa MJ, Côrte-Real M. ADP/ATP carrier is required for mitochondrial outer membrane permeabilization and cytochrome *c* release in yeast apoptosis. Mol Microbiol. 2007;66:571–82.

Portet T, Camps I, Febrer F, Escoffre JM, Favard C, Rols MP, Dean DS. Visualization of membrane loss during the shrinkage of giant vesicles under electropulsation. Biophys J. 2009;96:4109–21.

Prabhu S, Poulose E. Silver nanoparticles: mechanism of antimicrobial action, synthesis, medical applications, and toxicity effects. Int Nano Lett. 2012;2:32–5.

Radzig MA, Nadtochenko VA, Koksharova OA, Kiwi J, Lipasova VA, Khmel IA. Antibacterial effects of silver nanoparticles on gram-negative bacteria: influence on the growth and biofilms formation, mechanisms of action. Colloids Surf B: Biointerfaces. 2013;102:300–6.

Rai M, Yadav A, Gade A. Silver nanoparticles as a new generation of antimicrobials. Biotechnol Adv. 2009;27:76–83.

Rai M, Deshmukh SD, Ingle AP, Gupta IR, Galdiero M, Galdiero S. Metal nanoparticles: the protective nanoshield against virus infection. Crit Rev Microbiol. 2016;42:46–56.

Reiter J, Herker E, Madeo F, Schmitt MJ. Viral killer toxins induce caspase mediated apoptosis in yeast. J Cell Biol. 2005;168:353–8.

Retchkiman-Schabes PS, Canizal G, Becerra-Herrera R, Zorrilla C, Liu HB, Ascencio JA. Biosynthesis and characterization of Ti/Ni bimetallic nanoparticles. Opt Mater. 2006;29:95–9.

Rollet-Labelle E, Grange MJ, Elbim C, Marquetty C, Gougerot-Pocidalo MA, Pasquier C. Hydroxyl radical as a potential intracellular mediator of polymorphonuclear neutrophil apoptosis. Free Radic Biol Med. 1998;24:563–72.

Sahu N, Soni D, Chandrashekhar B, Sarangi BK, Satpute D, Pandey RA. Synthesis and characterization of silver nanoparticles using Cynodon dactylon leaves and assessment of their antibacterial activity. Bioprocess Biosyst Eng. 2013;36:999–1004.

Samuel T, Weber HO, Funk JO. Linking DNA damage to cell cycle checkpoints. Cell Cycle. 2002;1:162–8.

Sharma VK, Yngard RA, Lin Y. Silver nanoparticles: green synthesis and their antimicrobial activities. Adv Colloid Interface Sci. 2009;145:83–96.

Singh R, Nalwa HS. Medical applications of nanoparticles in biological imaging, cell labeling, antimicrobial agents, and anticancer nanodrugs. J Biomed Nanotechnol. 2011;7:489–503.

Smrz D, Dráberová L, Dráber P. Non-apoptotic phosphatidylserine externalization induced by engagement of glycosylphosphatidylinositol-anchored proteins. J Biol Chem. 2007;282:10487–97.

Suzuki J, Denning DP, Imanishi E, Horvitz HR, Nagata S. Xk-related protein 8 and CED-8 promote phosphatidylserine exposure in apoptotic cells. Science. 2013;341:403–6.

Vincent M, England LS, Trevors JT. Cytoplasmic membrane polarization in gram-positive and Gram-negative bacteria grown in the absence and presence of tetracycline. Biochim Biophys Acta. 2004;1672:131–4.

Wadskog I, Maldener C, Proksch A, Madeo F, Adler L. Yeast lacking the SRO7/SOP1-encoded tumor suppressor homologue show increased susceptibility to apoptosis-like cell death on exposure to NaCl stress. Mol Biol Cell. 2004;15:1436–44.

Yamanaka M, Hara K, Kudo J. Bactericidal actions of a silver ion solution on *Escherichia coli*, studied by energy-filtering transmission electron microscopy and proteomic analysis. Appl Environ Microbiol. 2005;71:7589–93.
Yang W, Shen C, Ji Q, An H, Wang J, Liu Q, Zhang Z. Food storage material silver nanoparticles interfere with DNA replication fidelity and bind with DNA. Nanotechnology. 2009;20:085102.
Yao G, Yang L, Hu Y, Liang J, Liang J, Hou Y. Nonylphenol-induced thymocyte apoptosis involved caspase-3 activation and mitochondrial depolarization. Mol Immunol. 2006;43:915–26.
You C, Han C, Wang X, Zheng Y, Li Q, Hu X, Sun H. The progress of silver nanoparticles in the antibacterial mechanism, clinical application and cytotoxicity. Mol Biol Rep. 2012;39:9193–201.
Yu SP. Regulation and critical role of potassium homeostasis in apoptosis. Prog Neurobiol. 2003;70:363–86.

Chapter 9
Synthesis of Gold Nanoparticles and Their Applications in Drug Delivery

Lian-Hua Fu, Jun Yang, Jie-Fang Zhu, and Ming-Guo Ma

Abstract Nanotechnology is a rapidly growing area of research because of its wide biomedical applications. Exploring effective drug-delivery systems in nanomedicine is still a challenging task. Gold nanoparticles (AuNPs) have been widely used in drug-delivery systems because of their high stability, high dispersity, tunable monolayers, functional flexibility, non-cytotoxicity, and biocompatibility There are numerous recent reports on the applications of AuNPs in the delivery of antitumor drugs, showing enhanced efficacy of treatment. AuNPs can be functionalized with active targeting ligands to increase their specific binding to the desired drug-delivery systems. This chapter focuses essentially on the synthesis, functionalization strategies, and potential applications of AuNPs in the field of drug delivery. We review current progress with AuNPs in the drug-delivery field, present their unique properties in drug-delivery applications, discuss their toxicity and biological interactions, and suggest future work in this area.

Keywords Nanotechnology • Gold nanoparticles • Drug delivery • Applications

9.1 Introduction

Nanomaterials have been a source of particular interest for scientists over the course of the past few decades. Gold nanoparticles (AuNPs) are one of the most widely investigated nanomaterials, owing to their unique optical and physical properties, as well as their safe use in the biological environment (Austin et al. 2015). Compared with other nanomaterials, AuNPs are chemically stable, non-toxic, and easy to functionalize

L.-H. Fu • J. Yang • M.-G. Ma (✉)
Engineering Research Center of Forestry Biomass Materials and Bioenergy, Beijing Key Laboratory of Lignocellulosic Chemistry, College of Materials Science and Technology, Beijing Forestry University, Beijing 100083, People's Republic of China
e-mail: mg_ma@bjfu.edu.cn

J.-F. Zhu
Department of Chemistry – Ångström Laboratory, Uppsala University, Uppsala 75121, Sweden

M. Rai, R. Shegokar (eds.), *Metal Nanoparticles in Pharma*,
DOI 10.1007/978-3-319-63790-7_9

(Lu et al. 2012). With the rapid development of nanoscience, the synthesis of AuNPs has made great progress, with AuNPs tailored as different morphologies, such as nanospheres, nanorods, nanoshells, nanocages, nanocubes, and nanostars (Alkilany et al. 2012; Khan et al. 2013; Tian et al. 2016). DNA, enzymes, antibodies, and some functional polymers can be easily conjugated with AuNPs without, in most cases, affecting their activities (Lu et al. 2012; Capek 2017), a feature that has encouraged researchers to explore the ultimate potential of AuNPs for biomedical purposes, especially for cancer therapy and drug delivery (Kumar et al. 2013; Abadeer and Murphy 2016). Different functionalized AuNPs can be designed and assembled as required; thus, AuNPs provide a promising platform for various types of applications in bio-nanotechnology. Mainly because of the high surface-to-volume ratio and the functional versatility of their monolayers, AuNPs work efficiently as drug-delivery vehicles, and the drugs can be delivered to specific areas if targeting ligands are utilized (Cheng et al. 2011). AuNPs are 100–10,000 times smaller than human cells and, consequently, they can offer unprecedented interactions with biomolecules, both on the surface of and inside biological cells (Fratoddi et al. 2014). Thus, the application of AuNPs in drug-delivery systems is a rapidly expanding field.

Because this field has already been covered by excellent reviews, herein we narrow the focus to consider only recent studies in which AuNPs have been used to bind specific targeting drugs and recent studies in which AuNPs have been used as new agents for drug delivery. In this chapter, we first discuss AuNPs as a versatile platform for drug-delivery systems, followed by an outline of progress in the synthesis of AuNPs. Current progress in the application of AuNPs with different morphologies (including nanospheres, nanorods, nanoclusters, nanoshells, nanocages, nanocubes, and other shapes) in the drug-delivery field is also presented in this chapter, showing the unique properties of AuNPs in drug-delivery applications, and illustrating how these nanostructures can be exploited as promising agents in drug delivery. Finally, we discuss the toxicity of AuNPs and suggest future work in this area.

9.2 Synthesis of Gold Nanoparticles

The fabrication of AuNPs with different shapes, sizes, and decorations with different functionalities makes them an appealing platform for drug-delivery systems. AuNPs can be produced by the reduction of gold salts in the presence of stabilizing agents in order to prevent agglomeration. Khan et al. (2013) classified the synthesis of AuNPs into four different methods.

(I) Physical methods. These include microwave irradiation, sonochemical methods, ultraviolet (UV) radiation, laser ablation, thermolytic processes, and photochemical processes. (Shang et al. 2013; Wei et al. 2013; Li et al. 2014; Raveendran et al. 2017).

(II) Chemical methods. These methods often involve chemical reactions during the synthesis of AuNPs in the presence of reducing agents and/or stabilizing agents, such as sodium citrate, heparin, hyaluronan, ascorbic acid, and sodium borohydride ($NaBH_4$) (Kemp et al. 2009; Muddineti et al. 2016; Piella et al. 2016; Li et al. 2017a). AuNPs show relative ease of functionalization with various targeting ligands, such as peptides, proteins, and oligosaccharides (Ding et al. 2015; Peng et al. 2016; Manivasagan et al. 2016).

(III) Supercritical fluid technique (Smetana et al. 2008; Machmudah et al. 2011). Supercritical fluid is defined as a solvent at a temperature and pressure above its critical point, at which the fluid remains as a single phase regardless of pressure. Supercritical CO_2 is the most widely used typical supercritical fluid, owing to its mild critical conditions, non-toxicity, non-flammability, and low cost (Khan et al. 2013).

(IV) Biological methods (Das et al. 2012; Vala 2015). Biological methods use fungi or bacteria as a source to produce AuNPs, avoiding the use of organic solvents, compared with chemical and physical processes, thus making biological processes an ecofriendly approach for the production of AuNPs.

Pandey et al. (2013a) synthesized a blend of carbon dots and gold nanorods as a complex (C-dots/AuNRs) for the controlled release of doxorubicin (Dox) under physiological conditions. C-dots were prepared by dissolving gum arabic in pre-chilled double-distilled water. Three distinct bands, labeled as F1, F2, and F3, were observed and they displayed exceptionally sharp contrast under both white and UV light (λ_{ex} = 365 nm) (Fig. 9.1b(i)). Pandey et al. (2013a) obtained AuNRs by a seed-mediated method, using $HAuCl_4$ and $NaBH_4$ in cetyltrimethylammonium bromide (CTAB), and the experiment was performed at 20 °C to initiate rod formation. In order to prepare the C-dots/AuNR complex, $AgNO_3$ and ascorbic acid were added to the $HAuCl_4$ solution with CTAB. Then, the purified C-dots complex from F3 was added to the above solution and the solution was vigorously agitated for 5 min. After that, 4 μL of seed solution was added to this mixture, which was stored for 2 h until the solution turned royal blue.

In the transmission electron microscopy (TEM) image of C-dots/AuNRs (Fig. 9.2a), complete coating on AuNRs and the bridging of C-dots between AuNR surfaces are observed. This observation indicated that selective adsorption of C-dots on the surface of AuNRs showed the presence of irregular C-dot-nanoparticle conjugates. Figure 9.2b shows the C-dots/AuNRs present in the supernatant separated during centrifugation, while Fig. 9.2c, d shows a magnified view of a representative AuNR coated with C-dots and a further enlarged image to highlight the coating of the AuNR surface by C-dots. The lattice constant of the coat confirmed the presence of C-dots on the surface. Figure 9.2e depicts spherical nanoparticles formed during the growth of AuNRs in the solution. Owing to the mild reducing capacity of the C-dots, the spherical AuNRs also became covered with C-dots from darker zones (Fig. 9.2e); a magnified high-resolution (HR) TEM image is shown in Fig. 9.2f.

The prepared C-dots/AuNRs were used for anchoring Dox via covalent and non-covalent pH-sensitive chemical bonds. Under physiological conditions, the drug-loading capacity of the C-dots/AuNRs was 94.08%, which was greater than previously reported (Pandey et al. 2013b). The authors (Pandey et al. 2013a) thought that the high drug-loading capacity shown by C-dots/AuNRs occurred

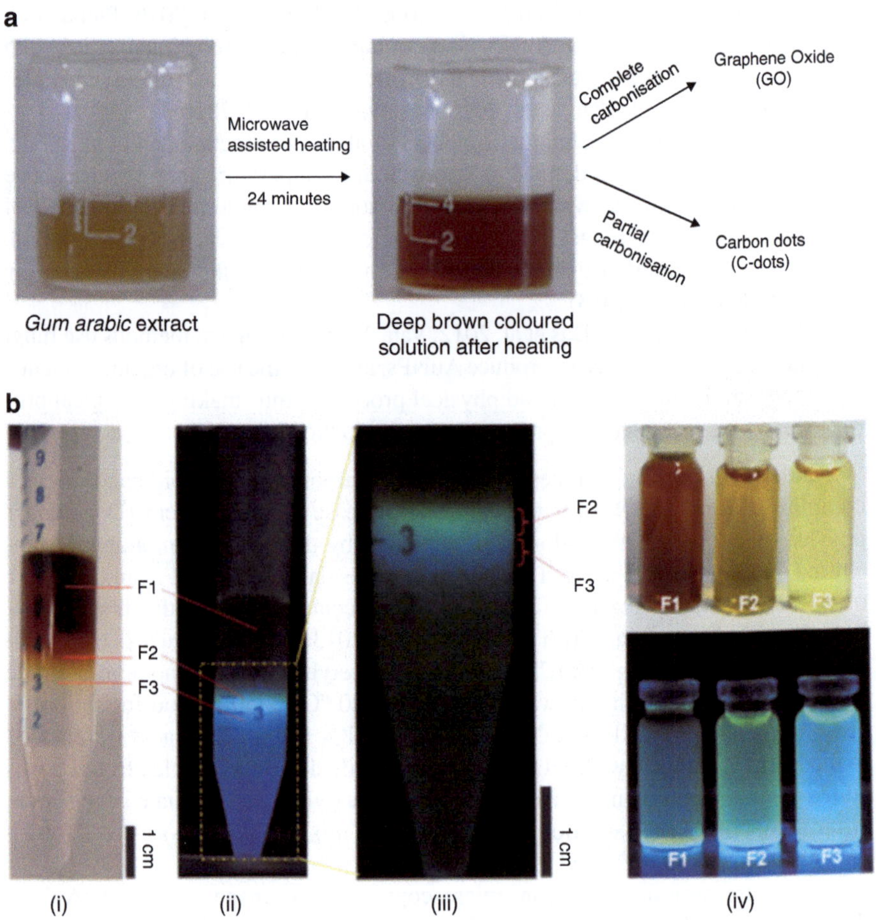

Fig. 9.1 (**a**) Microwave-assisted pyrolysis of gum arabic extract and the formation of *dark brown*-colored solution containing graphene oxide and a blend of carbon dots and gold nanorods as a complex (C-dots). (**b**) Sucrose density gradient with distinct separation of fractions in (i) ambient light and (ii) ultraviolet (UV) light (365 nm); (iii) a closer view of the fluorescent fractions, (iv) separated fractions in ambient light (*top*) and UV light (*bottom*) (From Pandey et al. 2013a. Reprinted with permission from the Royal Society of Chemistry)

owing to the porous nature of the C-dots. Another beneficial effect of the Dox-loaded C-dots/AuNRs was found to be a rapid burst of drug release under near infrared radiation (NIR, 808 nm), thus showing highly biocompatible thermo-chemotherapy for solid tumors. Cytotoxicity studies showed that protection of the AuNRs surface with bio-friendly C-dots reduced the obvious toxicity of the nanorods owing to the presence of the cationic surfactant CTAB. In vitro photothermal therapy and bioimaging were also carried out to evaluate the in-vitro killing capacity of the C-dots/AuNRs-Dox conjugate. All the experimental results indicated that the obtained C-dots/AuNR-Dox complex was a tri-pronged 'molecular weapon' for drug delivery, photothermal therapy, and biological imaging.

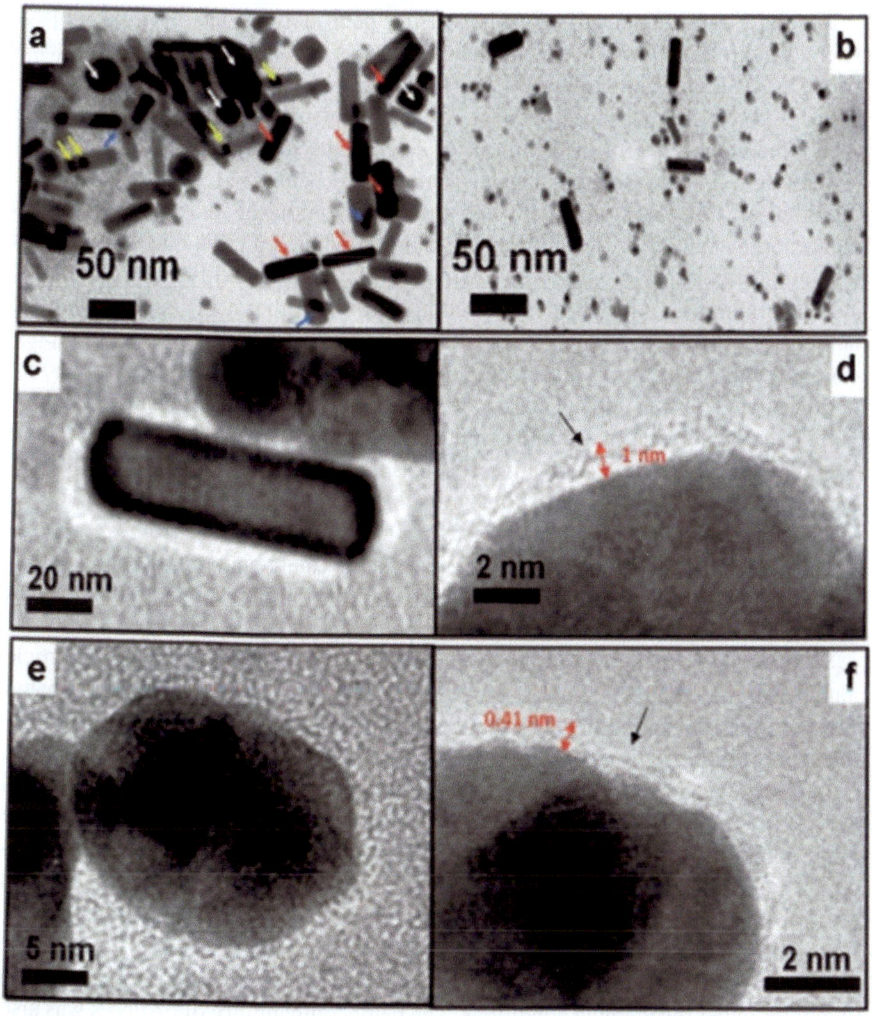

Fig. 9.2 (**a**) Transmission electron microscopy (TEM) images of C-dots/gold nanorods (AuNR) complex with various degrees of coating (*red arrows* show complete coating of C-dots on AuNR, *blue arrows* show bridging of C-dots between two rod surfaces, *yellow arrows* show the selective adsorption of C-dots onto the rod surface, and *white arrows* show the presence of irregular C-dot nanoparticle conjugates). (**b**) TEM of post-centrifugation supernatant displaying the presence of C-dots with a minimal amount of rods. (**c**) Representative high-resolution (HR) TEM of single C-dots/AuNR complex, which is magnified and shown in (**d**) and has a 1-nm-thick layer of C-dots. (**e**) Representative HR TEM of C-dots/AuNR complex; magnified image is shown in (**f**) with a 0.41-nm-thick layer of adsorbed C-dots (From Pandey et al. 2013a. Reprinted with permission from CE: Royal Society of Chemistry (RSC))

Moreover, this research could be extended by introducing synaptic targeting molecules such as folic acid (FA) to further improve the drug efficacy of the complex. This is yet another theranostic application of hybrid nanocomplex molecules combining chemotherapy and diagnosis in nanomedicine.

Recently, green synthesis of AuNPs using plant extracts has been developed, where the extracts themselves simultaneously act as stabilizing and reducing agents for AuNPs (Siddiqi and Husen 2017). The use of plant extracts meets the requirements of green chemistry perspectives. Green chemistry perspectives have emphasized three main steps for the synthesis of AuNPs: the selection of solvent medium for the synthesis, the selection of an environmentally benign reducing agent, and the selection of non-toxic materials for the stabilization of metal nanoparticles (Raveendran et al. 2003). In addition, the plant-based biological synthesis of AuNPs is gaining importance owing to its low cost, high reproducibility, ecofriendliness, and elaborate process of purification, compared with other ecofriendly biological methods (Ganeshkumar et al. 2013). Ganeshkumar et al. (2013) reported the spontaneous synthesis of mono-dispersed punicalagin-stabilized gold nanoparticles) by mixing a gold solution with an extract of *Punica granatum* (pomegranate) fruit peel. Pomegranate peel, one of the most valuable by-products of the food industry, consists of ellagic acid and its derivatives, such as the ellagitannins, gallic acid, quercetin, kaempferol, and naringenin. Pomegranate peel has been used in the preparation of tinctures, cosmetics, therapeutic formulae, and in food recipes. Moreover, pomegranate peel extract has been reported to have antioxidant and antimicrobial activities, and high chemopreventive potential for skin, colon, prostate, and lung cancer, and it is also used as adjuvant therapy for human breast cancer.

With the above method reported by Ganeshkumar et al. (2013), the mean particle diameter of the obtained pAuNPs was 70.90 ± 8.42 nm, and the punicalagin-coated shells on AuNPs resulted in the formation of small and stable colloidal Au nanoparticles. Punicalagin was able to simultaneously reduce Au^{3+} to form PAu^0 and serve as a stabilizing agent to prevent the agglomeration of pAuNPs. In vitro stability was analyzed by incubating the pAuNPs with saline (1, 2, 3, and 4 M) and phosphate buffer solution (PBS; pH 4.5 and 7.4) at 37 ± 1 °C for different times. The experimental results confirmed that the pAuNPs had excellent stability in biological fluids at various physiological pH values. Temperature-dependent stability was also investigated by storing the pAuNPs in a refrigerator. The pAuNPs were found to be stable, confirmed by UV-visible spectroscopy, and they were stable for more than 12 months under refrigerated conditions. The hemocompatibility of pAuNPs was also evaluated in human blood samples, and the prepared pAuNPs exhibited excellent hemocompatibility.

5-Fluorouracil (5-Fu) has been used as a chemotherapeutic agent for several decades; however, the problem in using this drug is its toxicity to bone marrow and the gastrointestinal tract when it is used in conventional dosage form. In a study by Ganeshkumar et al. (2013), casein was employed to couple pAuNPs for FA functionalization . These pAuNPs functionalized by FA (pAuNPs-FA) were used for targeted drug delivery to cancer cells, using 5-Fu as a model drug. The toxicity of free 5-Fu, 5Fu/pAuNPs, and 5Fu/pAuNPs-FA was analyzed in zebrafish embryos, an ideal model in which human diseases, nanoparticle toxicity, and drug screening can be studied. The tumor targeting efficiency of free 5-Fu, 5Fu/pAuNPs, and 5Fu/pAuNPs-FA was also evaluated in the MCF-7 cell line. Regarding the in vitro cytotoxicity of free 5-Fu and 5-Fu/pAuNPs-FA, the amount of 5-Fu required to

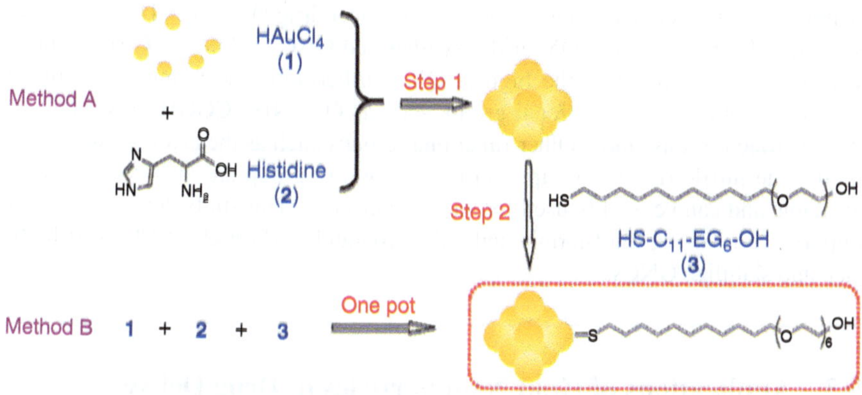

Fig. 9.3 Schematic illustration of the formation of gold nanoclusters (GNCs) (From Zhang et al. 2015. Reprinted with permission from Elsevier)

achieve 50% growth inhibition (IC_{50}) was observed. The in vitro drug release, cytotoxicity, and therapeutic efficacy results showed that 5-Fu/pAuNPs-FA could be a promising alternative carrier for targeting breast cancer.

Gold nanoclusters (GNCs), which consist of several to several hundred gold atoms, have drawn much attention owing to their unique physicochemical properties, catalytic activities, and potential applications in biomedical fields (Zhang et al. 2015; Zhao et al. 2016). The key factor for the practical application of GNCs is their stability. Zhang et al. (2015) developed a synthetic method for preparing ultrastable and multifunctional GNCs. They fabricated GNCs based on thiol-containing short-chain polyethylene glycols (PEGs) "HS-C_{11}-EG_6-X" by using two methods (Fig. 9.3). Each HS-C_{11}-EG_6-X molecule consists of four parts: A mercapto group (−SH), a methylene group (C_{11}), a short-chain ethylene glycol (EG_6), and a terminal functional group (X, where X = OH, COOH, NH_2, or glycine-arginine-glycine-aspartic peptides (GRGD). In method A, histidine and $HAuCl_4$ were blended and incubated for more than 2 h to synthesize GNC precursors. After that, HS-C_{11}-EG_6-OH was added to the above-prepared GNC precursors to form the final GNCs. In method B, $HAuCl_4$, histidine, and HS-C_{11}-EG_6-OH were mixed together in one pot to form the final GNCs. Both the products prepared by methods A and B exhibited ultrahigh stability in aqueous solutions with different pH values, in high salt concentrations, and even in cell medium. Moreover, the GNCs were still stable even when conjugated with the anticancer drug Dox. Confocal imaging experiments were carried out to demonstrate that the drug-carrying GNCs could enter cells, even entering the cell nucleus.

Zhang et al. (2015) noted that these two synthetic methods have their respective pros and cons. For example, the quantum yield (QY) of GNCs prepared by method A was higher than that for method B, while the preparation of GNCs by method B was easier. The GNCs prepared by both methods were ultrastable and had good monodispersity in aqueous media. It was found that the HS-C_{11}-EG_6-X-capped GNCs exhibited superior stability under various solution conditions for at least

6 months, even after conjugation with the anticancer drug Dox. Also, HS-C_{11}-EG_6-X significantly increased the QY of the synthesized GNCs, while the GNCs capped with carboxyl groups had the highest QY. Furthermore, the terminal functional group X could be any functional group, such as OH, NH_2, COOH, or GRGD, as demonstrated in this study. Other functional groups such as the azido group ($-N_3$), biotin, the nitrilotriacetic group, and the acylamino group are also commercially available and can be readily used for various purposes. This study demonstrates the importance of using thiol-terminated, alkyl-containing short-chain PEGs to fabricate and stabilize GNCs.

9.3 Applications of Gold Nanoparticles in Drug Delivery

9.3.1 Gold Nanorods

Gold nanorods (AuNRs) have received much attention as a drug-delivery vehicle owing to their low cytotoxicity and their biocompatibility, excellent stability, and suitable physicochemical parameters (Marangoni et al. 2016). The length and diameter of nanorods have to be determined; the diameter is completely dependent on the pore diameter of the template membrane, whereas the length can be controlled by keeping a check on the amount of gold deposited within the pores of the template membrane (Khan et al. 2013). Hu et al. (2013) fabricated AuNR-covered kanamycin-loaded hollow SiO_2 nanocapsules herpes stromal keratitis gold nanorods (HSKAuNR). They prepared CTAB-stabilized AuNRs by a seed-mediated and silver($_I$)-assisted growth method, with slight modifications. As shown in Fig. 9.4a, the AuNRs had an average aspect ratio (length/diameter) of 4.6 ± 0.7, resulting in an average longitudinal surface plasmon resonance wavelength of 770 nm and a weak transverse plasmon band at 512 nm. The spherical silica nanoparticles were prepared by tetraethyl orthosilicate (TEOS) hydrolysis in ethanol in the presence of ammonia, based on the Stöber process, with some modification. By regulating the volume of ammonia at fixed TEOS and ethanol/deionized (DI) water concentrations, monodispersed SiO_2 nanospheres were obtained with an average hydrodynamic size of *ca.* 205 nm (Fig. 9.4b). Hollow SiO_2 nanocapsules were obtained by etching the nanocapsules with $NaBH_4$ at 51 °C for 6 h. As shown in Fig. 9.4c, the hollow SiO_2 nanocapsules had well-defined hollow nanostructures with smooth SiO_2 shells and discernable pores on the surface, which allow kanamycin to pass through and load onto the SiO_2 shells. In the etching process, polyvinylpyrrolidone (PVP) was added to prevent aggregation of the hollow spheres and to serve as a bridging agent between the hollow SiO_2 nanocapsules and AuNRs. The SiO_2 nanoparticles exhibited an obvious solid-to-hollow structure conversion after etching in $NaBH_4$-PVP solution (Fig. 9.4e). Hollow SiO_2 nanocapsules were used as carriers for drug delivery, with kanamycin as a model drug, and the nanocapsules were covered with AuNRs to avoid drug leakage and realize photothermal treatment; as a consequence, HSKAuNR was obtained. The sterilizing effect was investigated

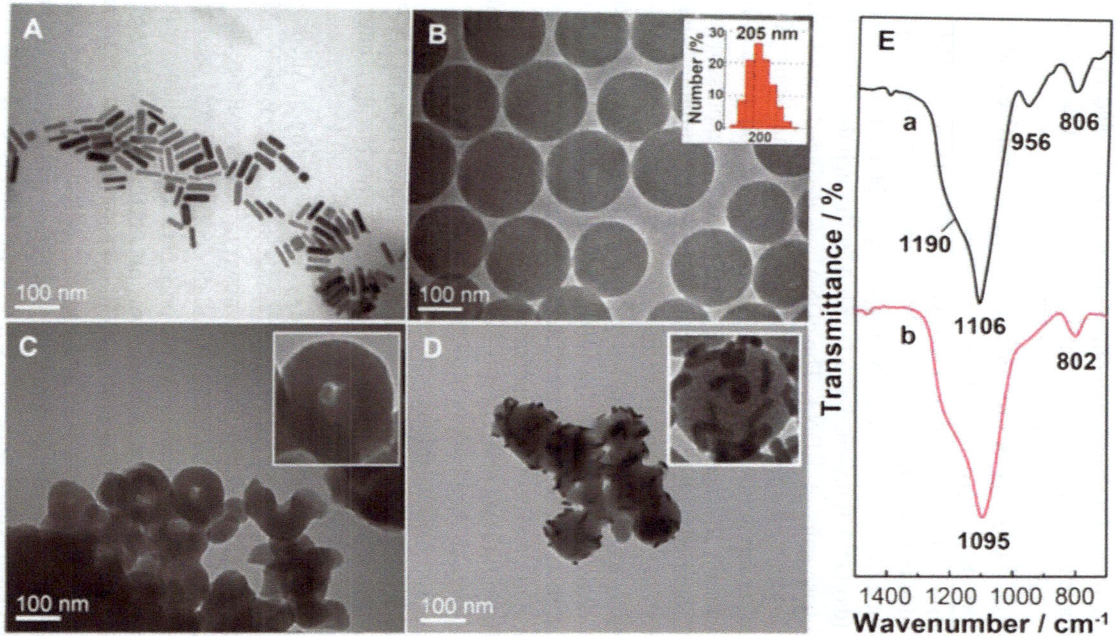

Fig. 9.4 TEM images of (**a**) AuNRs, (**b**) SiO_2 nanospheres, (**c**) hollow SiO_2 nanocapsules, and (**d**) AuNR-covered hollow SiO_2 nanocapsules. (**e**) Fourier transform infrared (FT-IR) spectra of *a* the prepared SiO_2 nanospheres, and *b* the hollow SiO_2 nanocapsules after reaction with 0.06 g mL^{-1} $NaBH_4$ at 51 °C for 6 h (From Hu et al. 2013. Reprinted with permission from Royal Society of Chemistry (RSC))

by incubating *Escherichia coli* BL21 with HSKAuNR under NIR for 20 min. The sterilizing rate was found to be 53.47% under NIR, with respect to a net sum sterilizing rate of 34.49% for the individual components of the HSKAuNR nanocapsules. This result may have occurred owing to the synergistic effect of the chemical drug (kanamycin) and physical sterilization (AuNRs under NIR). This hybrid material can realize chemical drug/physical photothermal sterilization and drug release; thus, it has great potential as an alternative adjuvant therapeutic material for sterilization or even for the control of disease.

Shanmugam et al. (2014) synthesized external stimuli-responsive dual drug carriers with AuNRs as the platform. Functionalized single-stranded (ss) DNA with 5′ thiol (SH-5′-$(CH_2)_6$-T20-ATCGCATGCTAGCGATCGTCGTCGTCGTCG (T20-ssDNA)) ? were added to the AuNR surface by thiol conjugation; thus, complementary DNA (Gold nanorods (AuNRs):cDNA) strands were obtained. This hybridized double-stranded (ds) DNA functioned as a Dox binder through the intercalation of Dox into the CG base pairs of dsDNA. The cDNA was designed with a 5′ amine functional group to tether platinum (IV) [Pt (IV)] prodrugs by establishing an amide bond with the acid group in the axial ligand. The other axial acid group in Pt (IV) was conjugated with FA to target folate receptors that are overexpressed in cancer cells (Fig. 9.5). Compared with Pt (II) compounds, the Pt (IV) prodrug is inert and would cause fewer side effects. This targeting vehicle provided remote-controlled delivery of the highly toxic cargo cocktail at the tumor site, ensuring extra specificity that can avoid acute toxicity, where the release of Dox and Pt (IV) was achieved upon NIR 808-nm diode laser irradiation. The dehybridization set the Dox free to bind to the cell nucleus, and cellular reductants reduced Pt (IV) to yield toxic Pt (II), an active drug. The significant effect of this external stimulus-responsive combination drug delivery was studied both in vitro and in vivo. The drugs were released with the NIR-laser stimulation, through the dehybridization of the dsDNA, which caused significant cell toxicity in vitro. In vivo studies showed that the FA-Pt (IV)/Dox-dsDNA-AuNRs with laser assistance effectively controlled solid tumor growth in a mouse model.

Liu et al. (2015) synthesized a drug-delivery system consisting of AuNRs coated with a mesoporous silica shell (AuNR/mesoporous silica nanoparticle (MSNP)) according to the classical protocol, which is known to produce disordered mesopores in the silica shell over each individual AuNR core. CTAB was used as a soft template to generate the mesopores during the polymerization of silanes around the AuNRs, and a shell of ≈ 35 nm in thickness was obtained. The AuNR/MSNP was partially uploaded with phase-changing molecules (1-tetradecanol [TD]) as gatekeepers to regulate drug (Dox) release, and a remote NIR was used to activate the phase change and subsequent drug release. Almost zero premature release was evidenced at physiological temperature (37 °C), whereas Dox release was efficiently achieved at higher temperatures upon external and internal heating generated by the AuNR core under NIR. Considering the high efficiency of the photothermal conversion of AuNRs, prominent NIR-triggered release is expected to be accomplished

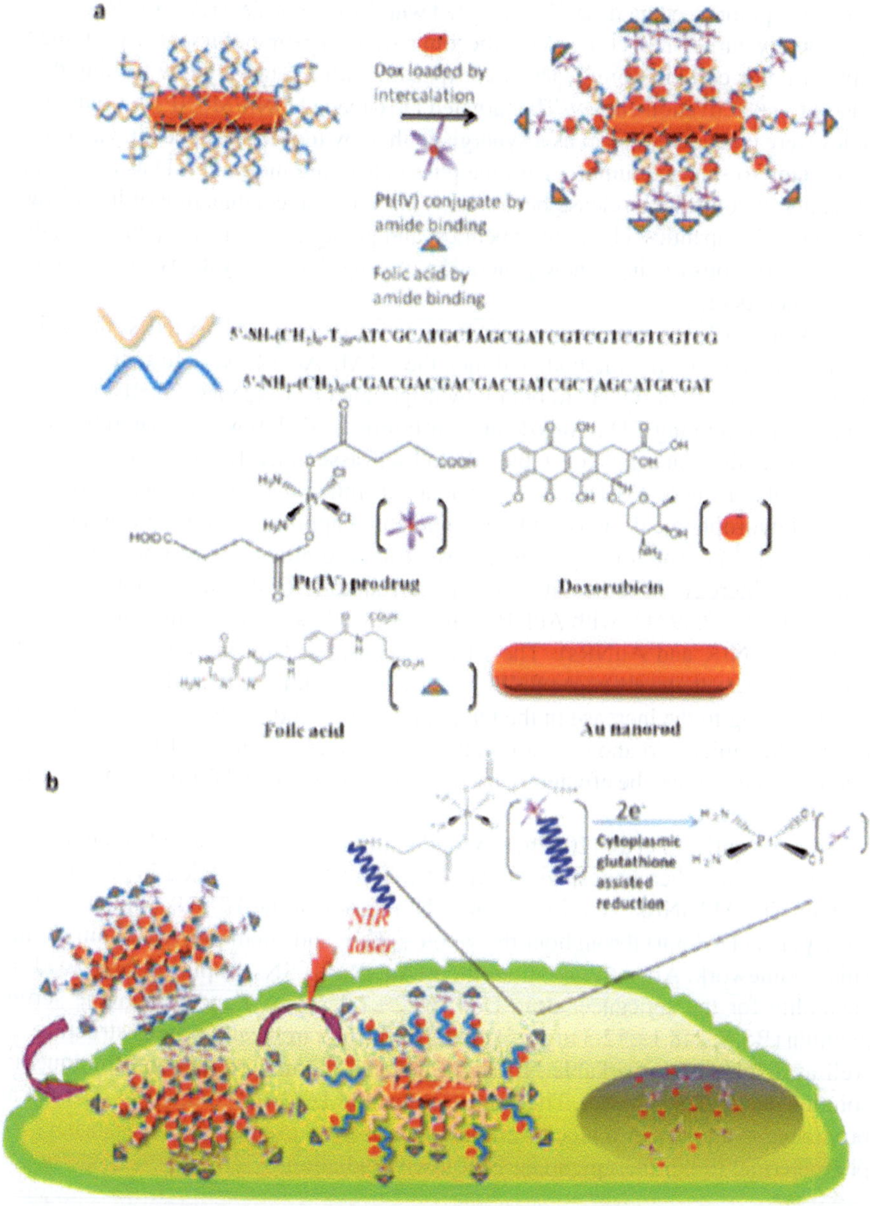

Fig. 9.5 (**a**) Doxorubicin (Dox)/platinum (Pt (IV)) loading and folic acid (FA) conjugation on double-stranded (ds) DNA-AuNRs. (**b**) Targeting drugs, laser-stimulated dehybridization, drug release, and hyperthermia; the released Pt (IV) prodrugs were reduced inside the cancer cells to form cytotoxic Pt (II) complexes (From Shanmugam et al. 2014. Reprinted with permission from the American Chemical Society)

under an optimal feeding dose. When tagged with folate moieties (FA), AuNR/MSNPs specifically target KB cells. With the improved targeting efficiency, prominent cancer/tumor cell killing efficacy was expected with a relatively low feeding dose and optimal activation energy. The targeted Dox-loaded GNR/MSNPs-FA nanoparticles were considered to act as a synergistic therapy tool for local heating and chemotherapy to selected tumors, with the phase-changing molecule TD as the gating modality to control the release behaviors. This versatile combination of local heating, phototherapeutics, chemotherapeutics, and gating components opens new possibilities for constructing a new generation of drug-delivery systems for antitumor chemotherapeutics.

Park et al. (2016) synthesized CTAB-coated AuNRs (CTAB-AuNRs) through a seed-mediated growing method, and then the CTAB-AuNRs were further modified by 1-octadecanethiol (ODT) to obtain hydrophobic ODT-AuNRs. Finally, a micro-organogel containing ODT-AuNRs and flurbiprofen (FBIU) was prepared by ultrasonic emulsification using vegetable oils and self-assembled 12-hydroxystearic acid (HSA) gelator fibers. FBIU, a non-steroidal anti-inflammatory drug, was chosen as a model hydrophobic drug to explore the potential application of micro-organogels in drug delivery. The micro-organogel was found to efficiently respond after NIR caused an increase in temperature in the surrounding GNRs, as the temperatures recorded were 25.9 °C (with AuNRs only), 29.3 °C (with NIR only), and 61.4 °C (with both NIR and AuNRs). The FBIU encapsulated in micro-organogels was released slowly in the absence of NIR, while FBIU rapidly increased in the presence of NIR owing to the increase in the temperature surrounding the AuNRs that transformed the micro-organogel into liquid. Thus, Park et al. (2016) believe that micro-organogels can be efficiently used as a versatile scaffold for on-demand drug-delivery systems.

More recently, Li et al. (2017b) reported multiple AuNRs encapsulated in hierarchically porous silica nanospheres (MAuNRs/HPSNs) synthesized in an oil/water system. The MAuNRs were loaded into the interior of the HPSNs, which contain large pores (13.2 nm) throughout the whole sphere and small pores (2.7 nm) in the silica framework. After PEGylation, the MAuNRs/HPSNs displayed high loading capability for the anticancer drug Dox (69.2 ± 7.2 mg g^{-1}) and for bovine serum albumin (BSA; 248.1 ± 12.3 mg g^{-1}) owing to the hierarchically porous structure, as well as displaying remarkable NIR light-induced thermal effects both in aqueous solution and at the cell level. Thus, these prepared MAuNRs/HPSNs not only act as promising drug/protein nanocarriers, but can also be used as photo absorbers for photothermal tumor therapy under NIR laser irradiation.

9.3.2 *Gold Nanocages*

Gold nanocages (AuNCs) with hollow interiors, porous walls, and tunable localized surface plasmon resonance (LSPR) peaks in the NIR region represent a promising platform for therapeutic applications, and can be used for orthogonally triggered release by choosing the right laser according to the AuNC's LSPR (Shi et al. 2014). AuNCs are hollow porous AuNPs, whose size basically ranges from 10 to 150 nm. They are generally produced by silver nanoparticles with $HAuCl_4$ in boiling water. The nanocages are prepared by a galvanic replacement reaction, which occurs between solutions containing metal precursor salts and silver nanostructures. The size of the nanocages can be easily controlled by adjusting the molar ratio of silver to $HAuCl_4$ (Khan et al. 2013).

Shi et al. (2014) reported NIR light-encoded orthogonally triggered and logically controlled intracellular release based on an AuNC-copolymer (poly(*N*-isopropylacrylamide-*co*-acrylamide)) shell by combining photothermal-sensitive release and a prodrug activation process. AuNCs were prepared with different LSPRs (670 and 808 nm) and were covered with a smart polymer shell. Experimental results showed that laser irradiation in resonance with the LSPR triggered the release of a pre-loaded effector. As a proof of concept, enzyme and substrate (prodrug) were selectively released from two different AuNCs. Enzymatic reactions occurred only after successful opening of both types of AuNC capsules (670- and 808-nm AuNC-copolymers). The system acted as an "AND" logic gate. Furthermore, if the AuNC was loaded with isoenzyme or enzyme inhibitor, it achieved an "OR" or "INHIBIT" logic gate operation. Combining NIR light-encoded orthogonally triggered release with prodrug activation processes achieved different defined logical operations for regulating the dosage of active drug in a specific region. Importantly, this strategy is simple in design, sophisticated to control, and provides new insights into developing NIR light-encoded, logically controlled, intracellular release systems.

The achievement of noninvasive and pinpointed intracellular drug release that responds to multiple stimuli is still a formidable challenge for cancer therapy. In their study in this field, Wang et al. (2014) reported a multi-stimuli-responsive platform based on drug-loaded AuNCs/hyaluronic acid (AuNCs-HA) for pinpointed intracellular drug release and targeted synergistic therapy. The key difference of their delivery system from previously reported multi-stimuli-responsive platforms was that the pH and NIR stimuli could be activated to trigger the release of the encapsulated drug only after the nanoparticles were internalized (Fig. 9.6). The multi-stimuli-responsive system released Dox from Dox-AuNCs-HA only in intracellular environments, such as endosomal and lysosomal vesicles, where HA was degraded by intracellular hyaluronidases. HA has been used as a drug carrier or targeting moiety because of its specific interaction with CD44, and so AuNCs-HA nano-hybrids can specifically recognize cancer cells via HA-CD44 interactions and can be efficiently endocytosed by a receptor-mediated process. Subsequently, the coated HA molecules can be degraded in lysosomes, resulting in the release of the encapsulated drug. In addition, because of their excellent photothermal properties, the AuNCs can accelerate the release of the encapsulated drug and produce higher therapeutic efficacy upon NIR. Both in vitro and in vivo results demonstrated that the combination of photothermal therapy and chemotherapy resulted in high cell toxicity, while chemotherapy or photothermal treatment alone did not achieve this outcome. Another exciting feature of the design of this study is the excellent synergistic effect of combining chemotherapy and photothermal therapy. The exceptional

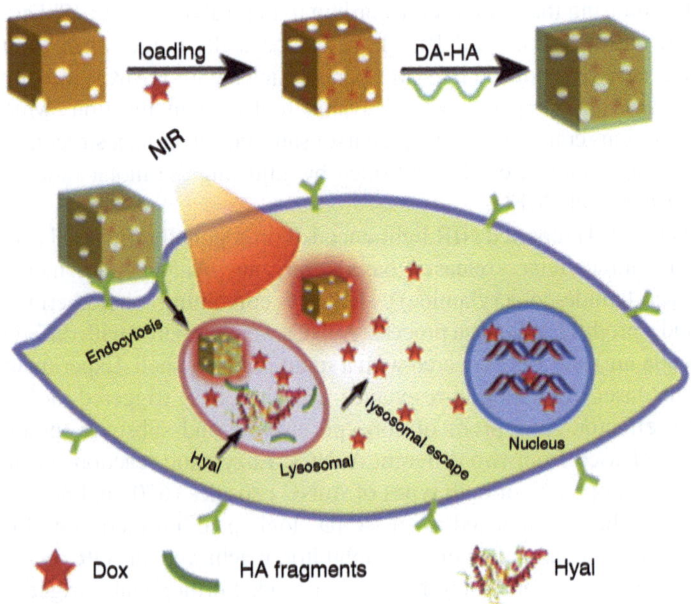

Fig. 9.6 Schematic representation of a multi-stimuli-responsive platform based on Dox-loaded gold nanocages (AuNCs)-hyaluronic acid (HA) nanoparticles for pinpointed intracellular drug release and synergistic therapy (From Wang et al. 2014. Reprinted with permission from Elsevier)

synergistic therapy was evidenced by the ablation of tumor both in vitro and in vivo. The rational integration of these functionalities could markedly enhance therapeutic efficiency while minimizing side effects. Taken together, the results of this study provide new insights into the development of pinpointed, multi-stimuli-responsive intracellular drug release systems for synergistic cancer therapy.

Hu et al. (2015) fabricated double-walled AuNC/SiO_2 nanorattles by combining two hollow-excavated strategies—galvanic replacement and surface-protected etching. As shown in Fig. 9.7, first, Au nanocubes were prepared using a sodium hydrosulfide (NaHS)-mediated polyol method with PVP as a capping agent, and then the nanocubes were transformed into AuNCs using $HAuCl_4$ in boiling water via galvanic replacement. This process obtained AuNCs with an outer edge length of 39 ± 5 nm and wall thickness of ~4 nm. Secondly, the AuNCs were modified with a layer of *p*-aminothiophenol (*p*ATP), followed by the fabrication of a dense SiO_2 shell via the base-catalyzed hydrolysis of TEOS in a water/isopropanol mixture to prepare SiO_2-coated AuNCs. Each AuNC was well encapsulated inside a uniform and dense SiO_2 layer with a thickness of ~45 nm. Then Au/SiO_2 nanorattles, with a uniform size of 120 ± 10 nm, were synthesized with the protection of a PVP coating based upon a surface-protected etching strategy, and an internal cavity was observed between each AuNC and the porous SiO_2 shell. Finally, the Au/SiO_2 nanorattles were functionalized with a Tat peptide to improve cellular uptake efficacy.

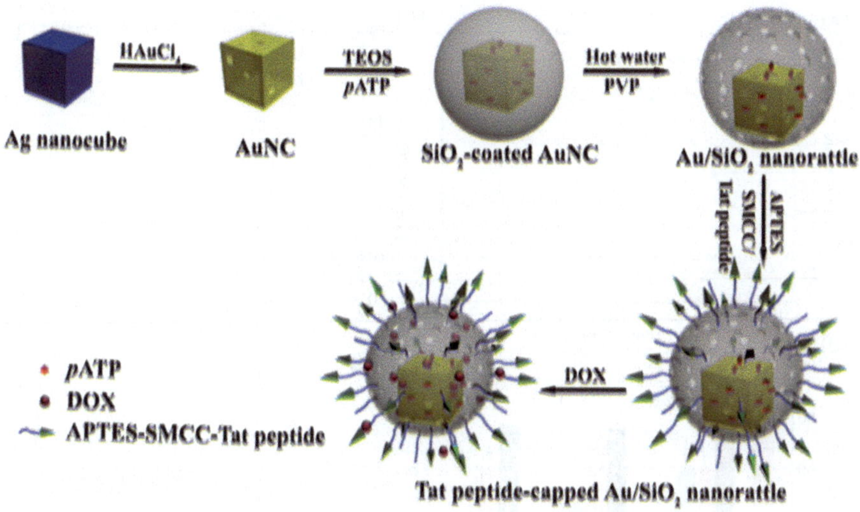

Fig. 9.7 Schematic illustration of the synthesis of double-walled Au/SiO$_2$ nanorattles (From Hu et al. 2015. Reprinted with permission from Wiley)

The obtained Tat peptide-capped Au/SiO$_2$ nanorattles exhibited a Dox loading efficiency of 53%, and showed a pH-dependent drug-release response. A small amount of Dox (8.7%) was initially released from the Dox-loaded Au/SiO$_2$ nanorattles in a slow fashion, while the release of Dox eventually became much faster and was then steady, with a great amount of Dox (46.2%) being released after 22-h incubation (Fig. 9.8b). The therapeutic effect of Dox-loaded Au/SiO$_2$ nanorattles against MCF-7 breast cancer cells was evaluated. As shown in Fig. 9.8c, the Dox-loaded Au/SiO$_2$ nanorattles induced more cell death than free Dox, indicating better uptake of Dox-loaded nanorattles by MCF-7 cells through endocytosis compared with the uptake of free Dox, which enters the cells via a passive diffusion process. The Dox-loaded Au/SiO$_2$ nanorattles, which combined Dox chemotherapy and the nanorattles' photothermal therapy , showed the best treatment effect under NIR compared with the effect of chemotherapy (nanorattles + Dox, without NIR) and the effect of photothermal therapy (nanorattles + laser, without Dox), suggesting the advantages of the multimodality of the reported Au/SiO$_2$ nanorattles. A TEM image further proved that the nanorattles were located around the nuclei of the MCF-7 cells (Fig. 9.8d), indicating that they facilitated the transportation of Dox into the nucleus. This therapeutic effect was consistent with surface-enhanced Raman scattering (Gold nanocages (AuNCs):SERS mapping) mapping and μ-X-ray fluorescence observations (XRF).

Huang et al. (2016) reported a novel strategy for the effective therapy of hepatocellular carcinoma (HCC) by loading a microRNA-181b inhibitor (anti-miR-181b) into polyethyleneimine-modified and folate receptor (FR)-targeted PEGylated AuNCs to obtain a novel tumor-targeting miRNA delivery system, which was designated as anti-miR-181b/PTPAuNCs nanocomplexes. The synergistic effects of gene-photothermal

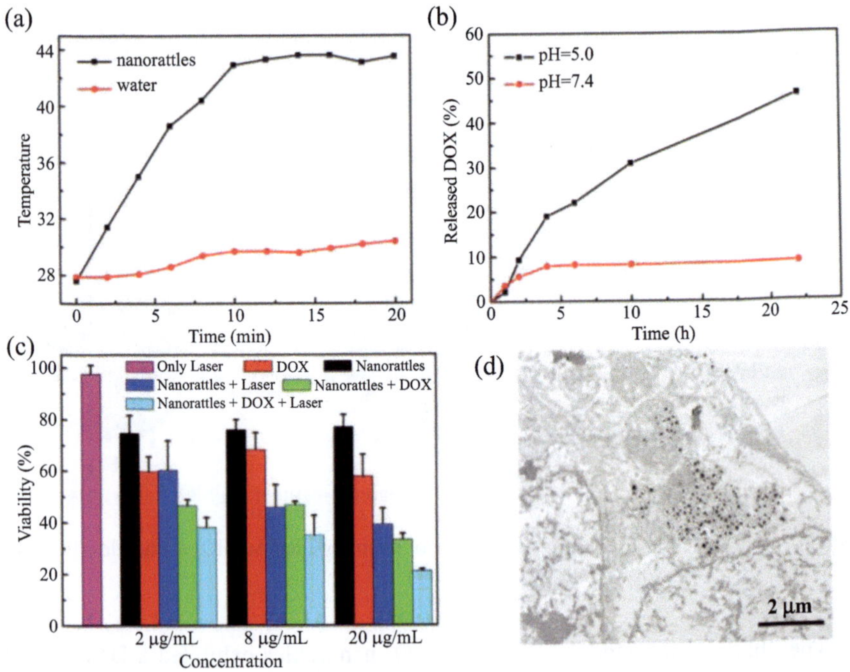

Fig. 9.8 (**a**) Photothermal curves of Au/SiO_2 nanorattles and H_2O under a near infrared radiation (NIR) laser (808 nm). (**b**) Dox release curves from the Au/SiO_2 nanorattles in phosphate-buffer solution (PBS) at pH 7.4 and pH 5.0. (**c**) Viability of MCF-7 cells treated with Dox-loaded Au/SiO_2 nanorattles for 24 h. (**d**) TEM image of MCF-7 cells incubated with the Au/SiO_2nanorattles for 24 h (From Hu et al. 2015. Reprinted with permission from Wiley)

therapy induced by the anti-miR-181b/PTPAuNCs nanocomplexes were investigated in vitro (in a cultured SMMC-7721 cell line) and in vivo (in SMMC-7721 tumor-bearing nude mice). The experimental results showed that this gene delivery system efficiently delivered anti-miR-181b into target sites to suppress tumor growth, and considerably decreased tumor volumes in SMMC-7721 tumor-bearing nude mice under NIR. Moreover, the anti-miR-181b/PTPAuNCs nanocomplexes with NIR exhibited better therapeutic effects against HCC than single gene or sole photothermal therapy, both in vitro and in vivo. Thus, Huang et al. (2016) believe that an anti-miR-181b/PTPAuNCs-mediated synergistic therapeutic strategy could enhance conventional treatments for patients afflicted with HCC.

Cell-specific targeted drug delivery and controlled release of the drug into cancer cells are the main challenges for anti-breast cancer metastasis therapy. Sun et al. (2017) incorporated an anticancer drug (Dox) into AnNCs as an inner core (DAuNCs), and then cancer-cell membranes, derived from 4T1 breast cancer cells (a typical metastatic tumor cell line with high metastatic capability), were coated on the surface of the DAuNCs as outer shells. The obtained biomimetic drug-delivery system consisting of cancer-cell membrane-DAuNCs was designated as CDAuNCs. The CDAuNCs exhibited stimuli-provoked release of Dox under NIR, which indicated hyperthermia-responsive drug-release behavior. Also, the superior targeting

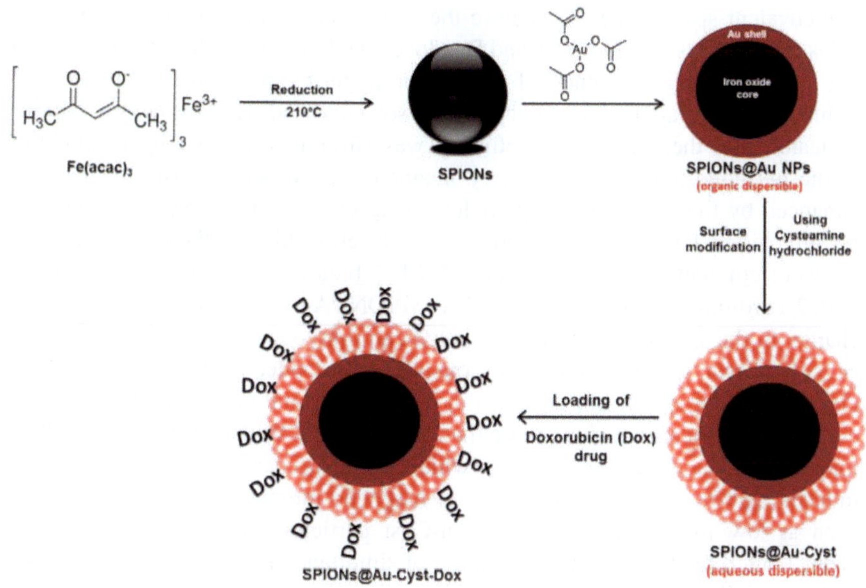

Fig. 9.9 Schematic representation of the multiple steps involved in the formation of superparamagnetic nanoparticles (SPIONs)-Dox (From Mohammad and Yusof 2014. Reprinted with permission from Elsevier)

efficiency of the 4T1 cells and the higher accumulation of CDAuNCs in tumor tissues demonstrated that the obtained CDAuNCs achieved selective targeting of the 4T1 tumor cells in vitro. Moreover, in 4T1 orthotopic mammary tumor models, the excellent combination therapy showed an inhibitory rate of 98.9% on tumor volume and an inhibitory rate of 98.5% in metastatic nodules. All of the experimental results showed that the reported CDAuNCs with tumor-directed and hyperthermia-responsive drug-release capacity could be promising drug-delivery systems for the future therapy of breast cancer.

9.3.3 Gold Nanoshells

Gold nanoshells (AuNSs) are spherical nanoparticles with diameters of 10–200 nm. They can be easily modified to absorb/scatter light at specific wavelengths in the visible and near infrared regions of the spectrum (Khan et al. 2013). Mohammad and Yusof (2014) prepared Dox-loaded gold-coated superparamagnetic Fe_3O_4 nanoparticles (SPIONs/Au) for the combination therapy of cancer by means of both hyperthermia and drug delivery. The magnetic properties of these particles were greatly enhanced (Ms of 84 emu/g at 5 K) on stabilization with a cysteamine (Cyst) biomolecule and the particles were also found to exhibit magnetic susceptibility at higher temperatures (201 K than the SPIONs/Au particles. Dox molecules were conjugated with the SPIONs/Au nanoparticles, with the help of Cyst as a non-covalent space linker, to achieve the non-covalent conjugation of Dox molecules with this system (Fig. 9.9), and Dox loading efficiency was found to be 0.32 mg mg^{-1} (i.e., loading efficiency of 63%). The hyperthermia studies were strongly influenced by the applied frequency and the solvents used. The Dox delivery studies indicated that the drug-release efficacy was strongly improved by maintaining acidic pH conditions and oscillatory magnetic fields. Namely, Dox release was enhanced by the oscillation of particles owing to the applied frequency, and was not affected by heating of the solution. In vitro cell viability and proliferation studies were conducted using cancerous (MCF-7 breast cancer) and non-cancerous (H9c2 cardiomyoblast) cell types. The SPIONs/Au-Cyst did not induce much change in the viability of either the H9c2 cardiomyoblasts or the MCF-7 cancer cells. However, under similar conditions, the drug-loaded SPIONs/Au-Cyst particles exhibited a significant decrease in the viability and proliferation of both the cell types at concentrations of more than 100 μg mL^{-1}. With their non-toxic behavior and high magnetic susceptibility (which helps to generate enough heat and also induces Dox release by the coupling of magnetic moment with the applied field even at low frequencies), SPIONs/Au-Cyst particles could be employed as a sophisticated method for the application of simultaneous hyperthermia and externally controlled drug delivery.

AuNS/silica core nanoparticles have a broad spectrum of applications owing to their unique tunable plasmon resonance (Jain et al. 2008). In recent years, AuNSs have been used for the photothermal ablation of tumors and for imaging, and their use has reached the clinical trial stage. Nguyen and Shen (2016) compared the capacity of bare AuNSs and PEGylated AuNSs (PEG-AuNSs) to stimulate the production of interleukin (IL)-1β in a human macrophage cell line (THP-1). IL-1β is related to several diseases, such as silicosis, gout, asbestosis, and type II diabetes mellitus, and many other disorders. In the in vivo studies, AuNSs formed large aggregates in cell culture medium and induced the production of IL-1β, while PEG-AuNSs did not form aggregates in cell culture medium, nor did they induce the production of IL-1β in macrophage cell lines. Confirming the in vitro results, AuNSs induced a significant level of neutrophil influx in the peritoneal cavity, and PEG-AuNSs reduced the level to one-quarter of that seen with AuNSs. The density of PEG on the particle surface had little effect on either the induction of IL-1β or the neutrophil influx induced by PEG-AuNSs. Nguyen and Shen (2016) explored the capacity of AuNSs and PEG-AuNSs to induce and scavenge reactive oxygen species (ROS); the AuNSs induced and scavenged ROS, while the PEG-AuNSs did not show this capacity. The excess of ROS induced by AuNSs could potentially cause the activation of inflammasomes, and thus the secretion of IL-1β. Aggregations of AuNSs can lead to ROS generation and IL-1β secretion, possibly owing to incomplete phagocytosis. Nguyen and Shen (2016) concluded that surface modification of AuNSs or other nanoparticles by PEGylation could potentially reduce the probability of stimulating ROS and IL-1β production. Finally, they thought their finding on the reduction of IL-1β production by the PEGylation of nanoparticles had implications for the use of other particulates in drug delivery, imaging, and therapy.

9.3.4 Gold Nanoclusters

Metal nanoclusters, which consist of several to hundreds of atoms with a size smaller than 3 nm, have drawn much attention owing to their unique physicochemical properties, catalytic activities, ion detection capacity, and potential applications in biolabeling and sensing (Zhang et al. 2015). Gold nanoclusters (GNCs) have attracted wide attention owing to their outstanding surface and physical properties, such as high fluorescence, unique charging properties, optical chirality, and ferromagnetism (Zhang et al. 2016). Furthermore, GNCs exhibit other fascinating features, including ease of synthesis, good water solubility, low toxicity, surface functionality, biocompatibility, and excellent stability, features which indicate their great promise in biology and medicine (Chen et al. 2012). GNCs have a wide range of applications, such as in single-molecule photonics, sensing, and biolabeling. In contrast to organic dyes and quantum dots, GNCs do not contain toxic heavy metals and have no chemical functions. Their near-infrared range of emission avoids interference from many biological moieties, making GNCs ideal for biological assays and cell imaging; thus they may become a powerful alternative to the usual fluorescence labels (Pan et al. 2007; Wang et al. 2011).

Wang et al. (2011) prepared inorganic phase GNCs by a conventional Brust-Schiffrin procedure and transformed them into aqueous phase in one step. The GNCs readily interacted with reduced graphene oxide (RGO), while their size, narrow size distribution, morphology, and fluorescent properties were maintained. In vitro dose-dependent MTT toxicity tests and microscopic analyses in HepG2 cells showed that GNC-RGO inhibited the HepG2 cells at a high concentration, but more interestingly for oncotherapy, te GNC-RGO nanocomposites carried Dox inside the cells, leading to some synergy in inducing karyopyknosis. Despite their moderate fluorescence intensity, the swift absorption of GNC-RGO nanocomposites by the cells allowed clear imaging of the edges and morphology of the cells, thus showing that these nanocomposites have interesting prospects for cellular imaging, as well as acting as synergistic drug carriers. Moreover, Raman spectroscopic investigations revealed several characteristic features substantiating interactions between GNC-RGO nanocomposites and proteins and DNA, thus affording mechanistic clues about the origin of their inhibition of cancer cells. Generally speaking, GNC-RGO nanocomposites could be employed as multimodal probes and drug carriers for targeting, detection, and oncotherapy.

Chen et al. (2012) reported core-shell structured multifunctional nanocarriers (GNCs-PLA-GPPS-FA) of GNCs with a core and an FA-conjugated amphiphilic hyperbranched block copolymer acting as a shell, based on a poly L-lactide (PLA) inner arm and an FA-conjugated sulfated polysaccharide (GPPS-FA) outer arm (GNCs-PLAGPPS-FA) for targeted anticancer drug delivery. This was an attempt to prepare drug carriers with certain antitumor activities that were achieved with GPPS as a hydrophilic outer shell. In this drug-delivery system, the GNCs were used for fluorescence labeling, and the amphiphilic hyperbranched block copolymer improved the nanocarriers' stability and drug-loading capacity. Camptothecin

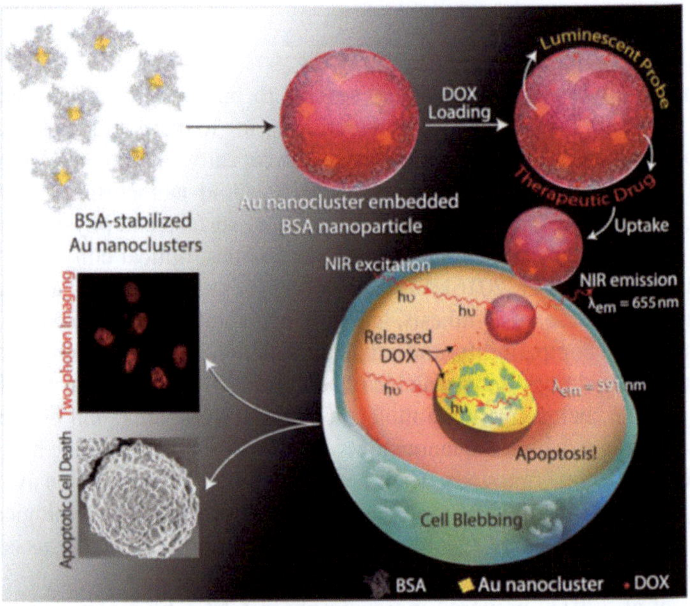

Fig. 9.10 Schematic illustration of the formation of Dox-loaded Au nanocluster-embedded bovine serum albumin (BSA) nanoparticles, followed by the uptake and release of Dox inside HeLa cells, leading to apoptotic cell death; two-photon imaging is also shown (From Khandelia et al. 2015. Reprinted with permission from Wiley)

(CPT) was used as a model hydrophobic anticancer drug, and the drug-loading and in vitro-release behaviors were studied at different pH values. Both types of nanocarriers (i.e., GNCs-PLA-GPPS-FA, CPT-loaded GNCs) showed a relatively rapid release of CPT in the first stage (up to 1 h), followed by a sustained release period (up to 15 h), then reaching a plateau at pH 5.3, 7.4, and 9.6. The slow and steady release of CPT from the nanocarriers may have resulted from the strong interactions of the hydrophobic PLA with the drug molecules. Moreover, the release rate of CPT was significantly accelerated by increasing the pH, and the CPT release from the two types of nanocarriers at pH 9.6 was much greater than that at both pH 5.3 and 7.4. The in vitro cytotoxicity of GNCs-PLA-GPPS-FA copolymers was investigated by employing human cervical cancer Hela cells. The CPT-loaded nanocarriers provided high anticancer activity against the Hela cells and showed specific targeting of some cancer cells owing to the enhanced cell uptake mediated by the FA moiety. These results indicated that not only could the GNCs-PLA-GPPS-FA copolymers act as an excellent tumor-targeted drug-delivery nanocarriers, but that they could also have an adjuvant role in the treatment of cancer.

Khandelia et al. (2015) reported the generation of nanotheranostic BSA) nanoparticles with GNCs as a luminescent probe (henceforth referred to as a composite nanoparticle) and the use of these composite nanoparticles for the in vitro delivery of an anticancer drug (Dox) to HeLa cells. These composite nanoparticles were biocompatible, non-cytotoxic, highly photostable, had a suitable QY, and showed high Stokes-shifted emission. For example, the QY of GNCs embedded in BSA nanoparticles was found to be 1.9% at pH 8.2, while BSA-stabilized GNCs showed a QY of 5.8% at pH 12.0. In addition, excitation (by two-photon) and emission wavelengths of the GNCs were in the NIR 650- to 900-nm window, indicating that the GNCs embedded in BSA nanoparticles could be candidates for in vivo imaging. Further, the drug efficiency of these composite nanoparticles was found to be 83.05% when explored by using Dox as a model drug through a fluorescence spectroscopic technique. This high efficiency may have resulted from electrostatic interactions and the hydrogen bonding of Dox with several BSA amino acids that have hydrophilic and hydrophobic characteristics. An in vitro cell viability assay indicated that the drug-loaded composite nanoparticles released Dox to the HeLa cells, leading to apoptotic cell death.

Figure 9.10 shows a schematic description of the use of Dox-loaded GNC-embedded BSA nanoparticles for in vitro drug delivery and the effect of these nanoparticles on HeLa cells, capturing the essence of the work. The luminescence of GNCs as well as Dox in the BSA nanoparticles was found to be useful. GNCs helped in tracking the uptake of the nanoparticles by the HeLa cells, while the luminescence of Dox helped in probing the intracellular release of the drug. Interestingly, the Dox-loaded composite nanoparticles are known to be suitable for the passive targeting of tumor cells through an enhanced permeation and retention effect and they are stable and retain their luminescence in human blood serum. These nanoparticles have the potential for use in clinical applications, especially for in vivo imaging and combination therapy, where the nanoclusters in conjunction with a conventional drug could be employed for therapy using radiation.

Yahia-Ammar et al. (2016) prepared monodisperse and stable self-assembled particles with sizes around 100–150 nm at room temperature in aqueous medium, using a cationic polymer to cross-link GNCs. The prepared GNCs exhibited pH-dependent swelling and shrinking properties, excellent colloidal stability and photostability in water, buffer, and culture medium, as well as a fourfold fluorescence enhancement, compared with fluorescein, owing to aggregation-induced emission (AIE) occurring with the electrostatic interaction between polyelectrolyte and GNC-stabilizing surface ligands. The AIE effect involving electrostatic interaction inside the self-assembled particles was confirmed by controlling the particle swelling, followed by steady-state and time-resolved fluorescence detection. Multiple imaging techniques (electron and fluorescence intensity and lifetime microscopy) and flow cytometry demonstrated efficient cellular uptake without compromising the integrity of the self-assembly inside the cytoplasm. Peptides and antibodies could be loaded in the GNCs using a single-step reaction, and this led to strongly enhanced biomolecule uptake with a clear co-localization of the nanoparticle carrier and the biomolecules. The versatility of the design of the self-assembled GNCs was demonstrated by selecting GNCs that were stabilized using different molecules, glutathione and zwitterions, and two different cationic polyelectrolytes, polyallylamine hydrochloride and polyethylenimine. These results showed a promising approach for the easy production of self-assembled nanoparticles from metal nanoclusters with different types of polymers as cross-linking agents, with these nanoparticles exhibiting stimuli-responsive properties to temperature and enzymatic or redox reactions. The multimodal imaging properties of the nanoparticles, the strong interaction of GNCs with light or X-ray irradiation, and the ability of the nanoparticle drug carriers to disassemble into non-toxic ultrasmall gold particles (<2 nm) with high renal clearance offer an elegant strategy for the design of new drug-delivery systems for improved theranostic applications.

Functional nanocarriers capable of transporting a high drug content without premature leakage and capable of the controlled delivery of various drugs are needed for better cancer treatments. Croissant et al. (2016) described the co-delivery of the anticancer drugs gemcitabine (GEM) and Dox via an inorganic nanocarrier. This involved mesoporous silica nanoparticles (MSNs) encapsulating a high content of GEM (40 wt%) and electrostatically gated with a GNC-protein shell acting as a reservoir for Dox (32 wt%). The co-delivery of Dox and GEM was achieved via an inorganic nanocarrier, with zero premature leakage, as well as a drug-loading capacity seven times higher than that of polymersome nanoparticles. Unlike the majority of strategies used to cap the pores of MSN, AuNC/BSA nanogates are bio-tools that were applied for targeted red nuclear staining and in vivo tumor imaging. The straightforward non-covalent combined MSN and gold-protein cluster bioconjugates thus led to a multifunctional nanotheranostic for the next generation of cancer treatments. Also, unlike the previous systems with covalently attached MSN-BSA obtained by multi-step methods, negatively charged GNCs/BSA were electrostatically attached to ammonium-functionalized MSN to effectively block the MSN pores, and provided pH-responsiveness for cargo release and delivery.

The interaction of Dox with BSA proteins was found to induce the formation of a highly loaded shell around the GEM-loaded MSN nanoparticles, leading to exceptionally high payloads (Fig. 9.11). The MSN-GNC/BSA drug-delivery system has many advantages, such as a straightforward preparation method; marked robustness in blood serum, with less than 3% drug leakage after 1 week; and higher dual loading of Dox and GEM (32 + 40 wt%) than that reported for liposomes (7 + 3 wt%) and polymersomes (6 + 6 wt%). This drug-delivery system also forms an ideal drug combination for the treatment of ovarian and breast cancers, with the spatial segregation of one drug in the pores of MSN, and a second drug inside the GNC/BSA layer. Furthermore, the dual drug-release experiments and the Dox and GEM multidrug delivery were performed in an inorganic carrier, with almost total cancer cell killing. Unlike previous designs for capping on MSNs, the GNC/BSA nanogates were applied not only to provide a secondary means of drug loading, but also for nuclear staining and in vivo tumor dual imaging with Hoechst-loaded MSN-GNC/BSA, suggesting preferential tumor accumulation.

9.3.5 Gold Nanoparticle Vesicles

In recent years, approaches to the assembly of various kinds of nanoparticles have been developed for electronic and bio-applications. For example, the vesicular assembly of nanoparticles, which assume a hollow structure with a nanoparticle shell, has been reported as a new type of capsule (He et al. 2012; Seo et al. 2015;

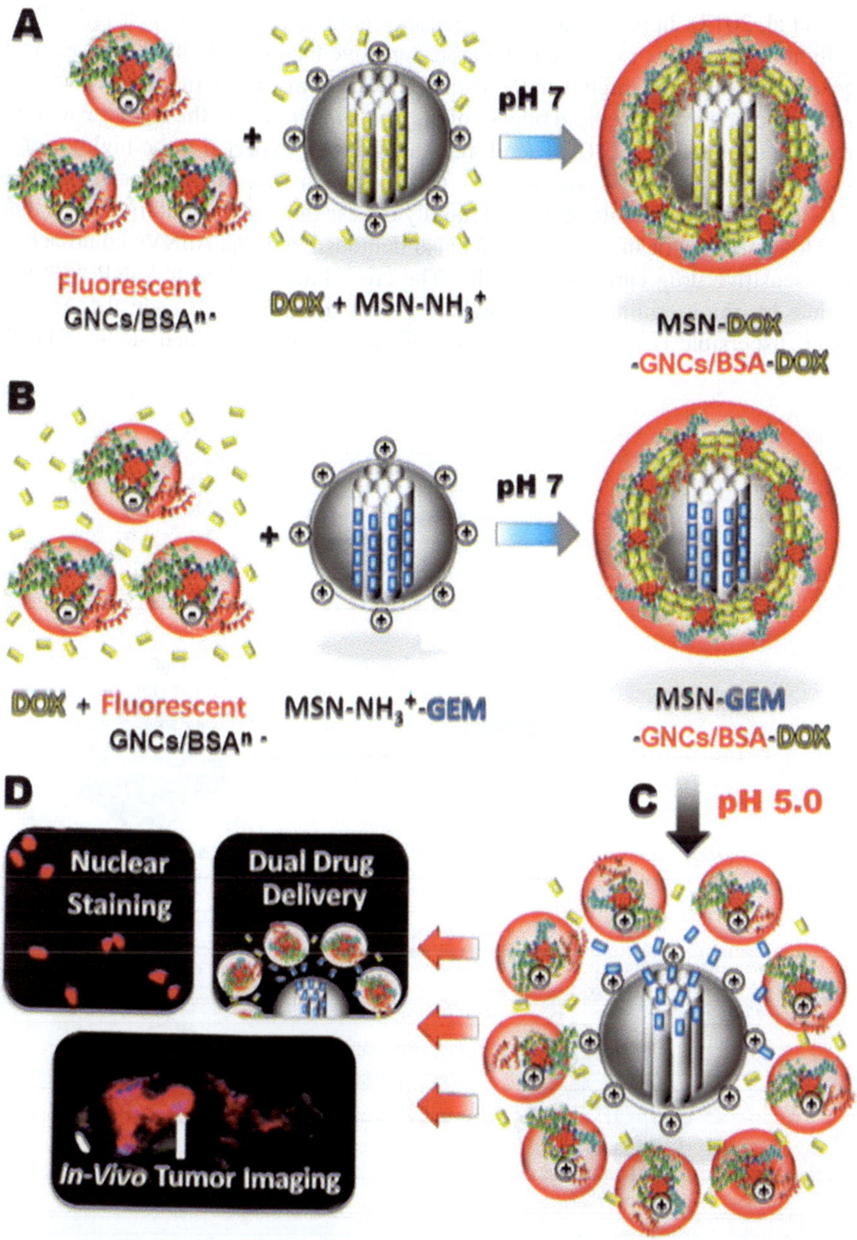

Fig. 9.11 Self-assembly of fluorescent negatively charged gold nanoclusters (GNCs)/BSA with positively charged Dox-loaded mesoporous silica nanoparticles (MSN)-NH_3^+ leading to Dox-loaded MSN-GNCs/BSA carrier (**a**). Self-assembly of fluorescent negatively charged GNCs/BSA and Dox with positively charged gemcitabine (GEM)-loaded MSN-NH_3^+ leading to an MSN-GNCs/BSA carrier loaded with two drugs (**b**), which could be disrupted by an acidic pH trigger (**c**). Applications of multifunctional dual drug-loaded MSN-GNCs/BSA nanocarriers (**d**) (From Croissant et al. 2016. Reprinted with permission from Elsevier)

Hou et al. 2017). In particular, the response of gold nanoparticle vesicles (AuNVs) to light irradiation is expected to be more sensitive than that of single nanoparticle systems, owing to the high AuNP content (Niikura et al. 2013). In fact, the increase in the gold content in the lipid bilayer of the liposome improves the response to light irradiation (An et al. 2010). Therefore, AuNVs were expected to be highly light-responsive drug carriers. Niikura et al. (2013) demonstrated that water-dispersible AuNVs could encapsulate drugs and rapidly release them upon light irradiation that was low enough for in vivo use. This also demonstrated that AuNVs could act as light-sensitive drug carriers into cells. The cross-linking of each AuNP within a vesicle by thiol-terminated polyethylene glycol (dithiol-PEG) made the AuNVs water dispersible (Fig. 9.12) and their hollow structure remained stable even in

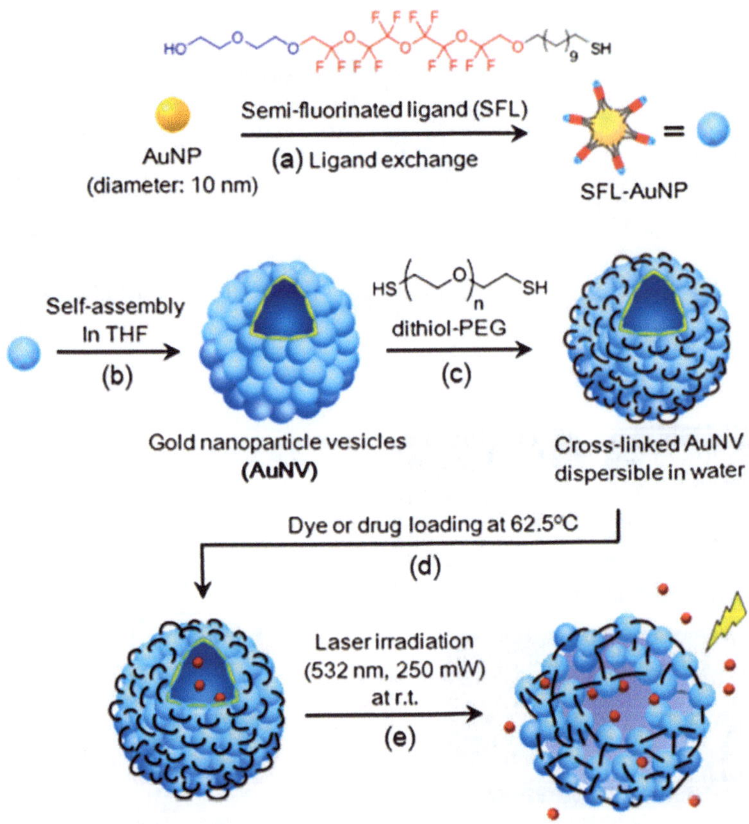

Fig. 9.12 Fabrication of gold nanoparticle vesicles (AuNVs) encapsulating dye or drug molecules and their light-triggered release. (**a**)Surface modification of gold nanoparticles (AuNPs) with semi-fluorinated ligands (SFL). (**b**) Self-assembly of SFL-AuNPs to form AuNVs. (**c**) Cross-linking of AuNPs of the AuNVs with dithiol-polyethyleneglycol (PEG) in tetrahydrofuran (THF) and the cross-linked AuNVs dispersed in water. (**d**) Encapsulation of dye or drug molecules at 62.5 °C, and subsequent cooling down to room temperature. (**e**) Dye release through opened interparticle nanogaps induced by laser irradiation (From Niikura et al. 2013. Reprinted with permission from ACS)

water, so that the cross-linked vesicles could work as drug-delivery carriers enabling light-triggered release. Rhodamine dyes and an anticancer drug (Dox) were encapsulated within the cross-linked vesicles by heating to 62.5 °C. At this temperature, the gaps between nanoparticles open, confirmed by a blue shift in the plasmon peak and more efficient encapsulation than that observed at room temperature. The cross-linked AuNVs released encapsulated drugs upon short-term laser irradiation (5 min, 532 nm) by again opening the nanogaps between each nanoparticle in the vesicle. The vesicles were more efficiently internalized into cells, compared with discrete AuNPs, and they released anticancer drugs upon laser irradiation in the cells. The drug delivery of Dox into cells and its light-triggered release were proven by using cross-linked AuNVs. This system, using a 532-nm laser, could open the possibility of using AuNVs as drug-delivery systems in combination with optical fiber techniques for localized therapy.

Song et al. (2012) reported the development of bio-conjugated plasmonic vesicles assembled from SERS-encoded amphiphilic AuNPs for cancer-targeted drug delivery. Their research demonstrated that amphiphilic AuNPs carrying mixed polymer brush coatings with distinctly different hydrophobicities could self-assemble into vesicles, in which the functional nanoparticles embedded in the shell (formed by the hydrophobic brushes and the hydrophilic brushes) extended into an aqueous environment and stabilized the structures, as illustrated in Fig. 9.13. A key finding was that this type of plasmonic assembly with a hollow cavity could play multifunctional roles as delivery carriers for anticancer drugs and plasmonic imaging probes to specifically label targeted cancer cells, when tagged with cancer-targeting ligands. More interestingly, the pH-responsive disassembly of the plasmonic vesicle (stimulated by the hydrophobic-to-hydrophilic transition of the hydrophobic brushes in

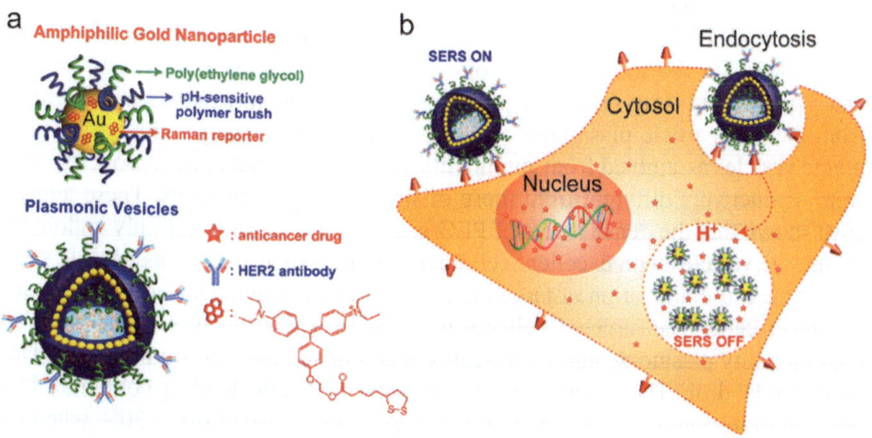

Fig. 9.13 (**a**) Schematic illustration of amphiphilic AuNPs coated with Raman reporter business group for latin america (BGLA) and mixed polymer brushes of hydrophilic PEG and pH-sensitive hydrophobic poly(methyl methacrylate) 4-vinylpyridine (PMMAVP) grafts, and a drug-loaded plasmonic vesicle tagged with HER2 antibody for cancer cell targeting. (**b**) The cellular binding, uptake, and intraorganelle disruption of surface-enhanced Raman scattering (SERS)-encoded pH-sensitive plasmonic vesicles (From Song et al. 2012. Reprinted with permission from ACS)

acidic intracellular compartments) allows for triggered intracellular drug release, which can be monitored in real time by both plasmonic imaging and Raman spectroscopy. Song et al. (2012) prepared Dox-loaded plasmonic vesicles by a modified film-rehydration method with PBS solution (pH 7.4) or sodium bicarbonate buffer solution (pH 10). They found that the use of pH 7.4 PBS for vesicle preparation consistently led to loading efficiencies that were about 10% lower than those in the experiments performed at pH 10, and the highest loading efficiency, 71%, was obtained at a Dox feeding weight ratio of 25% at pH 10. They thought that the higher loading efficiency of Dox at pH 10 may have been a result of the efficient entrapment of insoluble Dox molecules in the hydrophobic shell of the plasmonic vesicles. In contrast, at pH 7.4, water-soluble Dox had a tendency to partition into the aqueous medium, leading to Dox being partially encapsulated in the aqueous cavity, whereas those Dox molecules that leaked into the exterior environment would be lost. The drug-release profile of the plasmonic vesicles also showed strong pH-dependence. In vitro cytotoxicity tests showed that the targeted plasmonic vesicles were biocompatible in nature and had no adverse effect on the proliferation of SKBR-3 cells. When loaded with Dox, the targeted plasmonic vesicles became highly toxic to SKBR-3 cells, with an estimated half maximal inhibitory concentration (IC_{50}) of 0.31 μg mL^{-1}, which was about fivefold lower than that of the non-targeted vesicles. The targeted pH-insensitive AuNP vesicles coated with PEG and polymethyl methacrylate grafts did not lead to obvious cell toxicity, further demonstrating that the effective cancer cell killing was a result of pH-triggered Dox release. Thus, the authors thought that the plasmonic vesicles could have considerable potential for targeted combination therapy, by offering the possibility for cargo loading in both the hydrophobic shell and the aqueous cavity, and by integrating photothermal therapy based on the plasmonic nanostructures and chemotherapy of the therapeutic agents loaded inside the vesicles.

Song et al. (2013a) also reported a type of photo-responsive plasmonic vesicle that allowed for the active delivery of anticancer payloads to specific cancer cells and personalized drug release regulated by external photo-irradiation. Benefiting from the interparticle plasmonic coupling of AuNPs in close proximity, the plasmonic vesicles assembled from amphiphilic AuNPs exhibited optical properties that were distinctively different from those of single nanoparticle units. These authors have shown that the dense layers of PEG grafts on the vesicles not only endow the plasmonic vesicles with excellent colloidal stability, but also serve as flexible spacers for the bio-conjugation of targeting ligands to facilitate the specific recognition of cancer cells. The targeted delivery of an anticancer drug was investigated by dual-modality plasmonic and fluorescence imaging and toxicity studies, using Dox as the model drug. The experimental results showed that the loading content of Dox released in response to increasing Dox concentrations leveled off at 30% when the weight ratio of Dox and the vesicles reached 50%. Cytotoxicity tests in MDA-MB-435 cells showed that the plasmonic vesicles were highly biocompatible. Cell viability was over 80% for the drug-loaded vesicles, as well as for the targeted and non-targeted vesicles. When the vesicles were exposed to light irradiation, they showed greatly enhanced cytotoxicity owing to the drug release, and the targeted vesicles were more efficient than the non-targeted vesicles. The half-maximal inhibitory concentration (IC_{50}) of folate-targeted vesicles against MDA-MB-435 cells was 0.44 μg mL^{-1}, which was fivefold lower than that of the non-targeted vesicles, demonstrating fairly high therapeutic effectiveness. Therefore, the authors concluded that these folate-targeted photolabile plasmonic vesicles provide the opportunity to selectively deliver anticancer drugs to cancer cells by taking advantage of ligand-directed active targeting and triggered release regulated by light irradiation.

9.3.6 Other Gold Nanoparticles

Vivero-Escoto et al. (2009) reported the synthesis of a gold nanoparticle-capped mesoporous silica nanosphere (AuNP-MSN) material for the photo-induced intracellular controlled release of an anticancer drug, paclitaxel, inside human fibroblasts and liver cells, as depicted in Fig. 9.14. In this study, they functionalized the surface of the AuNP with a photoresponsive linker (thioundecyl-tetraethyleneglycolester-*o*-nitrobenzylethyldimethyl ammonium bromide; TUNA). The organically derivatized AuNPs (PR-AuNPs) with an average particle diameter of 5 nm were positively charged (ξ-potential = + 4.2 ± 1.4 mV) in PBS (pH 7.4). Then these authors synthesized spherical MSN material with an average diameter of

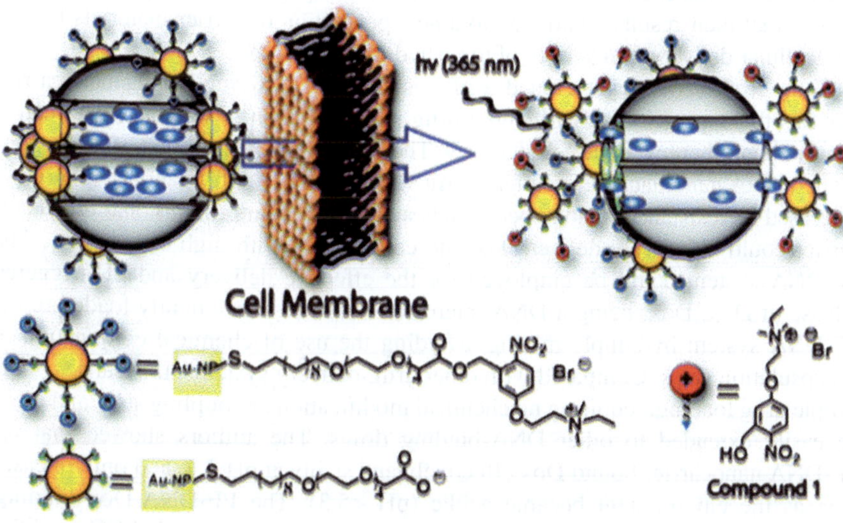

Fig. 9.14 Schematic illustration of the photoinduced intracellular controlled release of photoresponsive (PR)-AuNPs-MSN. Upon UV irradiation, the photolabile linker on the PR-AuNPs was cleaved, changing the surface charge property (-potential) of these gold nanoparticles from positive to negative. The charge repulsion between the AuNPs and MSN would then uncap the mesopores and allow the release of guest molecules (From Vivero-Escoto et al. 2009. Reprinted with permission from ACS)

100 nm, a total surface area of 1083 m^2 g^{-1}, and a narrow Barrett-Joyner-Halenda (BJH) pore size distribution, with an average pore size of 3.0 nm. They proposed that the capping mechanism of this PR-AuNPs-MSN system was based on electrostatic interaction between the positively charged PR-AuNPs and the negatively charged MSN material (ξ-potential = -23.8 ± 1.8 mV) in water. As illustrated in Fig. 9.14, upon photoirradiation, the photolabile linker covalently attached to the surface of the PR-AuNPs was cleaved, resulting in the formation of a cationic compound, as well as the formation of negatively charged thioundecyltetraethyleneglycolcarboxylate (TUEC)-functionalized AuNPs (NC-AuNPs). The charge repulsion between the NC-AuNPs and MSN uncapped the mesopores and allowed the release of guest molecules. In order to explore the feasibility of using the PR-AuNPs-MSN system for intracellular drug delivery in live human cells, an anticancer drug (paclitaxel) was chosen as a model drug for a controlled release study in human liver and fibroblast cells. Vivero-Escoto et al. (2009) found that the paclitaxel-loaded PR-AuNPs-MSN material was rapidly endocytosed by these two cell types. After UV irradiation for 10 min, significant decreases in the cell viability, of 44.2% and 43.5%, respectively, were observed for the liver and fibroblast cells containing paclitaxel-loaded PR-AuNPs-MSN. This result indicated that PR-AuNPs-MSN could indeed transport and release paclitaxel inside these live human cells under photoirradiation control. This "zero premature release" characteristic is of importance for the delivery of toxic drugs in chemotherapy. Furthermore, the authors proved that the cargo-release property of the PR-AuNPs-MSN system could be easily controlled by low-power photoirradiation under biocompatible and physiological conditions. They believe that their results could lead to a new generation of carrier materials for the intracellular delivery of a variety of toxic hydrophobic drugs.

Song et al. (2013b) developed a pH-responsive DNA-AuNP drug nanocarrier that has the capacity of pH-triggered drug release, which is suitable for effective cancer chemotherapy at the cellular level. They found that PEGylation of the proton-fuelled DNA nanomachine-conjugated AuNP (PF-DNA-AuNP) nanocarrier greatly increased its resistance to non-specific adsorption of serum proteins, and this nanocarrier could effectively deliver Dox into cancer cells with high cytotoxicity. The PF-DNA system could be employed for the effective delivery and pH-triggered release of Dox. Dox, being a DNA intercalator, can be conveniently loaded to the PF-DNA system by simple mixing, avoiding the use of chemical conjugation or encapsulation steps. Compared with other drug-delivery systems, this carrier offers simple drug loading, requiring no chemical modification or coupling step, and it can be easily extended to other DNA-binding drugs. The authors showed that the PF-DNA nanocarrier bound Dox effectively and stably at pH 7.4, and only released Dox as the environment became acidic (pH <5.3). The PF-DNA-Dox binding/release process is efficient, rapid, and reversible. The conjugation of a PEG-modified DNA to AuNP produces a multivalent PF-DNA-AuNP nanocarrier that resists non-specific adsorption and can be used for the efficient delivery and pH-responsive release of Dox into model cancer cells, leading to high cytotoxicity. This nanocarrier has numerous features required for a drug nanocarrier, i.e., it has uniform small nanoscale size (~60 nm), which resists non-specific serum protein adsorption; it is non-toxic and biocompatible; it is water-soluble; it is stable with a high drug-loading capacity (~1000 Dox/AuNP); and it has the capacity for controlled release (in intracellular endosomal/lysosomal acidic environments). Song et al. (2013b) developed an active targeting version of the nanocarrier by incorporating cancer cell-specific targeting agents, and extending the drug delivery and toxicity evaluation to preclinical models.

Park et al. (2014) constructed a drug-delivery system by developing gold nanospheres (GNSs) conjugated to four α-helical cell-penetrating peptides (CPPs). They examined the applicability of this cell-selective drug-delivery system by evaluating its cell-penetrating (CP) and cell death activities and comparing them with these activities in the Tat peptide. Using a 25-nm-diameter GNS, they obtained higher cell death induction activity for the anticancer drug Dox than that obtained with a 41-nm GNS. After entering the cell, the peptide-conjugated GNS (25 nm) complexes accumulated around the cell nucleus. High cell selectivity by the α-helical CPP sequences was also demonstrated. Their results indicated that these α-helical peptide and 25 nm GNS conjugates were useful elements in an efficient cell-selective drug-delivery system. They constructed five nanoprobes by conjugating either the Tat peptide or one of four α-helical peptides (RF, RA, EF, and EA) to a 25-nm GNS (P-GNS25). These five nanoprobes were coated with PEG (P-PEG-GNS25) to maintain high stability during long incubations for nuclear targeting. The P-PEG-GNSs were expected to be taken up by the cell via endocytosis, and the P-PEG-GNSs in endosomes or lysosomes would then be expected to migrate to the nucleus. Therefore, a pH-sensitive Dox-PEG was synthesized, using a hydrazone linkage, and conjugated with a 25-nm GNS (P-Dox-PEG-GNS25) to allow the controlled release of Dox from the vesicles at pH 5 (Fig. 9.15). The four α-helical peptide-conjugated nanoprobes showed similar CP and cell death activities compared with the Tat peptide in three cell lines, HeLa, A549, and 3T3-L1, with high cell selectivity shown according to cell type and peptide sequence. Moreover, P-Dox-PEG-GNS25, which contained a pH-sensitive Dox linkage, showed levels of cell death induction by Dox (20–40%) that were approximately two times higher than those (10–20%) reported in a previous study, in which the EF and EA peptides were conjugated with 41-nm GNS (Park et al. 2013). The 25-nm GNS complexes accumulated around the cell nucleus, where Dox was effectively released from the GNS, resulting in a higher induction of cell death than that seen with 41-nm GNS complexes. These data indicate that P-PEG-GNSs25 could be applied to the construction of highly efficient and less toxic cell-selective drug-delivery systems for cancer research, including cancer diagnosis and therapy development.

Peptide-based capping agents for AuNPs are possible alternatives for capping and derivatizing AuNPs, but they have a major disadvantage, sensitivity to non-specific proteases, which may limit their in vivo utility. Using non-natural analogs of natural α-amino acids offers an attractive alternative strategy to circumvent this potential bottleneck in realizing the full potential of peptide-based capping agents for AuNP biological applications. Parween et al. (2013) synthesized and investigated the capping efficiency of AuNPs with a series of pentapeptides incorporating non-natural amino acids (α,β-dehydrophenylalanine [ΔPhe]; and α-aminoisobutyric

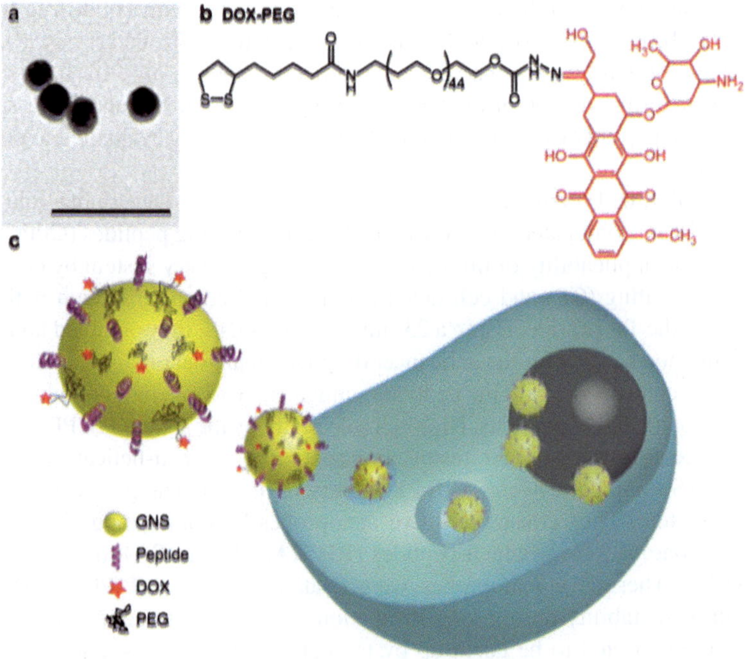

Fig. 9.15 (**a**) TEM image of 25-nm GNS; *scale bar*, 100 nm. (**b**) Dox-PEG with pH-sensitive hydrazone linkage and a lipoic acid moiety for GNS conjugation. (**c**) Illustration of migration of peptides and Dox-PEG-conjugated 25-nm GNS (P-Dox-PEG-GNS25) inside a cell (From Park et al. 2014. Reprinted with permission from Elsevier)

acid [Aib]). The introduction of ΔPhe and Aib in peptide sequences is well-known to provide them with increased resistance to enzymatic degradation. All these peptides were able to efficiently cap AuNPs and, moreover, peptide-induced aggregation was not observed. The peptide-capped AuNPs showed excellent cytocompatibility with mammalian cell lines (HeLa and L929), as well as with mouse splenocytes. Parween et al. (2013) encapsulated small hydrophobic drugs (such as Dox, mitoxantrone, and chloroquine) and found that the cytotoxicity of the drugs entrapped in the peptide-capped AuNPs to HeLa cells was higher than that of the free drugs. These results indicate that these non-natural amino acid-containing peptide-capped AuNPs may be further developed as promising alternative drug-delivery vehicles.

Multifunctional biocompatible nanomaterials with both fluorescent and vehicle functions are highly favored for bioimaging, therapeutic, and drug-delivery applications. Nevertheless, both the rational design and the synthesis of highly biocompatible multifunctional materials remain challenging. Ding et al. (2015) reported the development of a method for the synthesis and assembly of protein-gold hybrid nanocubes (PGHNs) that consisted of GNCs, BSA, and tryptophan, and examined their application in cell imaging and drug delivery (Fig. 9.16). The green-synthesized PGHNs were blue-emitting under UV irradiation and cube-like in structure, with a

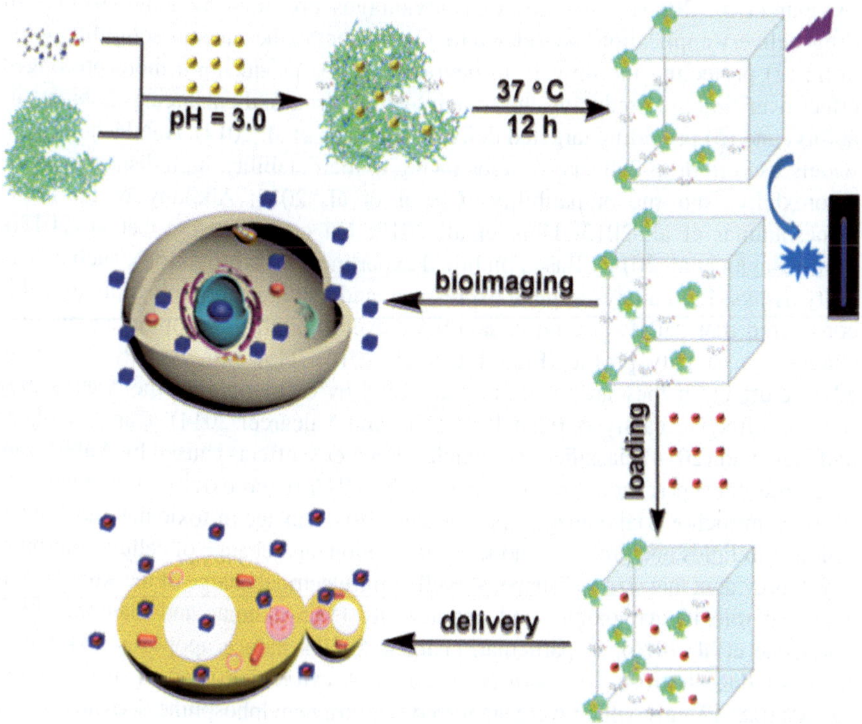

Fig. 9.16 Schematic presentation of protein-gold hybrid nanocube (PGHN) synthesis and its application in cell imaging and drug-delivery studies. The *yellow dot*, the *blue cube*, and the *red circle* indicate AuNCs, PGHNs, and Rh 6G (or Dox), respectively (From Ding et al. 2015. Reprinted with permission from ACS)

size of ~100 nm. The PGHNs were highly biocompatible and the whole synthetic process was ecofriendly and toxicity-free. The biocompatibility of PGHNs ensures that they can be readily internalized by different types of cells, which suggests that this material has high endocytosis efficiency. Moreover, PGHNs can function as nanovehicles that can deliver dyes and drugs into cells. This protein-metal hybrid nanomaterial can serve as a dual-purpose tool: as a blue-emitting cell marker in bioimaging investigations, and as a nanovehicle for drug-delivery studies.

9.4 Toxicity of Gold Nanoparticles

The basis of a targeted drug-delivery system is to concentrate the drug in the tissues of interest while reducing the concentration of the drug in other surrounding tissues. Nanoparticle-based therapeutic approaches have great potential for application in this field, because nanoparticles can be delivered at specific disease sites and then the drug release is triggered for alteration of the physiological microenvironment

(Kumar et al. 2013). The principal advantages provided by nanoparticles in drug-delivery applications are related to: (1) enhancing the aqueous solubility of the drug; (2) protecting the drug from degradation; (3) producing a more prolonged effect over time; (4) reducing the toxicity of the drug and enhancing its bioavailability; and (5) providing targeted delivery (Parveen et al. 2012). AuNPs have been widely used in drug-delivery systems owing to their stability, high dispersity, non-cytotoxicity, and biocompatibility (Cheng et al. 2011; Alkilany et al. 2012; Ganeshkumar et al. 2013; Ding et al. 2015; Hu et al. 2015; Li et al. 2017b; Manivasagan et al. 2016). Based on broad experimental investigations, such as viability assays, ROS analyses, gene expression, and cellular morphology assays, it is considered that AuNPs are fairly non-toxic even though the capping agents may change the toxicity profile (Fratoddi et al. 2014). However, AuNPs can cause adverse effects if they are taken up and stored by cells, because the AuNPs may become effective catalysts (Caballero-Díaz and Valcárcel 2014). Caballero-Díaz and Valcárcel (2014) classified the toxicity or adverse effects caused by AuNPs into nine different types, including interaction with DNA (genotoxicity), the generation of ROS, mitochondrial damage, apoptosis/necrosis, leakage of toxic materials, interaction with lipids and proteins, endocrine disruption, and change of cellular morphology. It is evident that size and shape, as well as the electrical charge of the AuNPs, play a critical role in cytotoxicity and genotoxicity (Vijayakumar and Ganesan 2013; Carnovale et al. 2016). In particular, charged AuNPs induce apoptosis and neutral AuNPs induce necrosis (Fratoddi et al. 2014). A cytotoxicity test of water-soluble AuNPs (0.8–15 nm), which were stabilized with triphenylphosphine derivatives, was carried out in four cell lines that represent major functional cell types (Pan et al. 2007). Sung et al. (2011) studied the toxicity of AuNPs in Sprague Dawley rats by using inhalation; cellular differential counts were investigated and cytotoxicity was measured. The results revealed that there was no statistically significant difference in cellular differential counts; however, histopathological results showed minimal alveoli, an inflammatory infiltrate, and increased macrophages in the rats that were highly dosed.

Although there are numerous publications concerning the toxicological effects of AuNPs, the findings reported are disparate, in that they have been obtained by diverse methods, by testing AuNPs with different sizes or shapes in different cell lines, and under diverse incubation conditions and doses (Caballero-Díaz and Valcárcel 2014). Besides, one must take into account that AuNPs undergo stability changes in the physiological environment as a consequence of their interaction with biomolecules and other components. Regarding toxicological assessment, there are some limitations in the current studies related to the lack of the exhaustive characterization of the AuNPs involved. This characterization should include, at least, the size distribution, shape, specific surface area and chemistry, and stability of the AuNPs. In most cases, the degree of characterization performed is dependent on the facilities available at each laboratory. Differences in the results may also originate in the use of various measurement techniques or even in the use of different operating conditions. Errors associated with different sample preparation methods or with the fact that the samples taken for characterization are not representative of the entire sample can lead to inconsistent results. Furthermore, it is important to mention that in many in vivo studies, organisms are treated with quantities of nanoparticles that are much higher than those that would be expected in the actual scenarios. High concentrations of nanoparticles favor the formation of aggregates and lead to unrealistic scenarios for toxicity assessment (Dhawan and Sharma 2010). Taking into account the potential application of AuNPs in biomedicine, it would be reasonable to perform systematic studies that investigate a broad range of exposure conditions involving nanoparticles with different physicochemical features. This would help us to construct an accurate picture of the variables that influence nanostructure pharmacokinetics (absorption, distribution, metabolism, and excretion) and would help to determine whether toxicity will be observed (Fischer and Chan 2007).

9.5 Conclusion

AuNPs show great potential for the creation of drug-delivery systems, owing to features such as their stability, high dispersity, tunable monolayers, functional flexibility, low toxicity, and ability to control the release of drugs. Currently, there are numerous reports on the synthesis of AuNPs and the development of applications for these entities. Although much investigation remains to be done, at present the physiological destination of AuNPs in vivo is still far from completely understood, although it is considered that nanoparticles definitely have the potential to revolutionize medical therapies, particularly in the case of multimodal cancer treatment. More investigations are still needed to gain a better understanding of how we can extrapolate the potential effects of nanoparticles on human health from the biological behavior that the particles exhibit in animal studies, and how the characteristics and properties of these nanoparticles influence their fate and behavior in vivo. Hence, additional investigations of drug-delivery systems based on AuNPs will be required to fully understand their pharmacokinetics, interactions with the immune system, and the extent of their cytotoxicity owing to their surface and size. Continued research into nanoscale delivery vehicles will expand the understanding of the interactions of these materials with biological systems, promoting the development of more effective drug-delivery systems.

Acknowledgments Financial support from the Fundamental Research Funds for the Central Universities (No. 2015ZCQ-CL-03, JC2013-3) and the Beijing Higher Education Young Elite Teacher Project (No. YETP0763) is gratefully acknowledged.

References

Abadeer NS, Murphy CJ. Recent progress in cancer therapy using gold nanoparticles. J Phys Chem C. 2016;120:4691–716.

Alkilany AM, Thompson LB, Boulos SP, Sisco PN, Murphy CJ. Gold nanorods: their potential for photothermal therapeutics and drug delivery, tempered by the complexity of their biological interactions. Adv Drug Deliv Rev. 2012;64:190–9.

An XQ, Zhang F, Zhu YY, Shen WG. Photoinduced drug release from thermosensitive AuNPs-liposome using a AuNPs-switch. Chem Commun. 2010;46:7202–4.
Austin LA, Kang B, El-Sayed MA. Probing molecular cell event dynamics at the single-cell level with targeted plasmonic gold nanoparticles: a review. Nano Today. 2015;10:542–58.
Caballero-Díaz E, Valcárcel M. Toxicity of gold nanoparticles. In: Gold nanoparticles in analytical chemistry. 2014. p 207–54. http://dx.doi.org/10.1016/B978-0-444-63285-2.00005-5.
Capek I. Polymer decorated gold nanoparticles in nanomedicine conjugates. Adv Colloid Interface. 2017. doi:10.1016/j.cis.2017.01.007.
Carnovale C, Bryant G, Shukla R, Bansal V. Size, shape and surface chemistry of nano-gold dictate its cellular interactions, uptake and toxicity. Prog Mater Sci. 2016;83:152–90.
Chen T, Xu S, Zhao T, Zhu L, Wei DF, Li YY, Zhang HX, Zhao CY. Gold nanocluster-conjugated amphiphilic block copolymer for tumor-targeted drug delivery. ACS Appl Mater Interfaces. 2012;4:5766–74.
Cheng Y, Meyers JD, Agnes RS, Doane TL, Kenney ME, Broome AM, Burda C, Basilion JP. Addressing brain tumors with targeted gold nanoparticles: a new gold standard for hydrophobic drug delivery? Small. 2011;7:2301–6.
Croissant JG, Zhang D, Alsaiari S, Lu J, Deng L, Tamanoi F, AlMalik AM, Zink JI, Khashab NM. Protein-gold clusters-capped mesoporous silica nanoparticles for high drug loading, autonomous gemcitabine/doxorubicin co-delivery, and in-vivo tumor imaging. J Controlled Release. 2016;229:183–91.
Das SK, Liang J, Schmidt M, Laffir F, Marsili E. Biomineralization mechanism of gold by zygomycete fungi *Rhizopous oryzae*. ACS Nano. 2012;6:6165–73.
Dhawan A, Sharma V. Toxicity assessment of nanomaterials: methods and challenges. Anal Bioanal Chem. 2010;398:589–605.
Ding H, Yang DY, Zhao C, Song ZK, Liu PC, Wang Y, Chen ZJ, Shen JC. Protein-gold hybrid nanocubes for cell imaging and drug delivery. ACS Appl Mater Interfaces. 2015;7:4713–9.
Fischer HC, Chan WC. Nanotoxicity: the growing need for in vivo study. Curr Opin Biotech. 2007;18:565–71.
Fratoddi I, Venditti I, Cametti C, Russo MV. Gold nanoparticles and gold nanoparticle-conjugates for delivery of therapeutic molecules. Progress and challenges. J Mater Chem B. 2014;2:4204–20.
Ganeshkumar M, Sathishkumar M, Ponrasu T, Dinesh MG, Suguna L. Spontaneous ultra fast synthesis of gold nanoparticles using *Punica granatum* for cancer targeted drug delivery. Colloids Surf B. 2013;106:208–16.
He J, Liu YJ, Babu T, Wei ZJ, Nie ZH. Self-assembly of inorganic nanoparticle vesicles and tubules driven by tethered linear block copolymers. J Am Chem Soc. 2012;134:11342–5.
Hou WX, Xia FF, Alfranca G, Yan H, Zhi X, Liu YL, Peng C, Zhang CL, de la Fuente JM, Cui DX. Nanoparticles for multi-modality cancer diagnosis: simple protocol for self-assembly of gold nanoclusters mediated by gadolinium ions. Biomaterials. 2017;120:103–14.
Hu B, Zhang LP, Chen XW, Wang JH. Gold nanorod-covered kanamycin-loaded hollow SiO_2 ($HSKAu_{rod}$) nanocapsules for drug delivery and photothermal therapy on bacteria. Nanoscale. 2013;5:246–52.
Hu F, Zhang Y, Chen GC, Li CY, Wang QB. Double-walled au nanocage/SiO_2 nanorattles: integrating SERS imaging, drug delivery and photothermal therapy. Small. 2015;11:985–93.
Huang SN, Duan SF, Wang J, Bao SJ, Qiu XJ, Li CM, Liu Y, Yan LJ, Zhang ZZ, Hu YR. Folic-acid-mediated functionalized gold nanocages for targeted delivery of anti-miR-181b in combination of gene therapy and photothermal therapy against hepatocellular carcinoma. Adv Funct Mater. 2016;26:2532–44.
Jain PK, Huang XH, El-Sayed IH, El-Sayed MA. Noble metals on the nanoscale: optical and photothermal properties and some applications in imaging, sensing, biology, and medicine. Acc Chem Res. 2008;41:1578–86.
Kemp MM, Kumar A, Mousa S, Park TJ, Ajayan P, Kubotera N, Mousa SA, Linhardt RJ. Synthesis of gold and silver nanoparticles stabilized with glycosaminoglycans having distinctive biological activities. Biomacromolecules. 2009;10:589–95.

Khan MS, Vishakante GD, Siddaramaiah H. Gold nanoparticles: a paradigm shift in biomedical applications. Adv Colloid Interface. 2013;199–200:44–58.

Khandelia R, Bhandari S, Pan UN, Ghosh SS, Chattopadhyay A. Gold nanocluster embedded albumin nanoparticles for two-photon imaging of cancer cells accompanying drug delivery. Small. 2015;11:4075–81.

Kumar A, Zhang X, Liang XJ. Gold nanoparticles: emerging paradigm for targeted drug delivery system. Biotechnol Adv. 2013;31:593–606.

Li JY, Han QS, Wang XH, Yu N, Yang L, Yang R, Wang C. Reduced aggregation and cytotoxicity of amyloid peptides by graphene oxide/gold nanocomposites prepared by pulsed laser ablation in water. Small. 2014;10:4386–94.

Li H, Huang H, Wang AJ, Feng H, Feng JJ, Qian ZS. Simple fabrication of eptifibatide stabilized gold nanoclusters with enhanced green fluorescence as biocompatible probe for in vitro cellular imaging. Sensor Actuat B-Chem. 2017a;241:1057–62.

Li N, Niu DC, Jia XB, He JP, Jiang Y, Gu JL, Li Z, Xu SA, Li YS. Multiple nanorods@hierarchically porous silica nanospheres for efficient multi-drug delivery and photothermal therapy. J Mater Chem B. 2017b;5:1642–9.

Liu J, Detrembleur C, De Pauw-Gillet MC, Mornet S, Jerome C, Duguet E. Gold nanorods coated with mesoporous silica shell as drug delivery system for remote near infrared light-activated release and potential phototherapy. Small. 2015;11:2323–32.

Lu F, Doane TL, Zhu JJ, Burda C. Gold nanoparticles for diagnostic sensing and therapy. Inorg Chim Acta. 2012;393:142–53.

Machmudah S, Wahyudiono, Kuwahara Y, Sasaki M, Goto M. Nano-structured particles production using pulsed laser ablation of gold plate in supercritical CO_2. J Supercrit Fluids. 2011;60:63–8.

Manivasagan P, Bharathiraja S, Bui NQ, Lim IG, Oh J. Paclitaxel-loaded chitosan oligosaccharide-stabilized gold nanoparticles as novel agents for drug delivery and photoacoustic imaging of cancer cells. Int J Pharm. 2016;511:367–79.

Marangoni VS, Cancino-Bernardi J, Zucolotto V. Synthesis, physico-chemical properties, and biomedical applications of gold nanorods. a review. J Biomed Nanotechnol. 2016;12:1136–58.

Mohammad F, Yusof NA. Doxorubicin-loaded magnetic gold nanoshells for a combination therapy of hyperthermia and drug delivery. J Colloid Interf Sci. 2014;434:89–97.

Muddineti OS, Kumari P, Ajjarapu S, Lakhani PM, Bahl R, Ghosh B, Biswas S. Xanthan gum stabilized PEGylated gold nanoparticles for improved delivery of curcumin in cancer. Nanotechnology. 2016;27:325101.

Nguyen HT, Shen H. The effect of PEGylation on the stimulation of IL-1β by gold (Au) nanoshell/silica core nanoparticles. J Mater Chem B. 2016;4:1650–9.

Niikura K, Iyo N, Matsuo Y, Mitomo H, Ijiro K. Sub-100 nm gold nanoparticle vesicles as a drug delivery carrier enabling rapid drug release upon light irradiation. ACS Appl Mater Interfaces. 2013;5:3900–7.

Pan Y, Neuss S, Leifert A, Fischler M, Wen F, Simon U, Schmid G, Brandau W, Jahnen-Dechent W. Size-dependent cytotoxicity of gold nanoparticles. Small. 2007;3:1941–9.

Pandey S, Thakur M, Mewada A, Anjarlekar D, Mishra N, Sharon M. Carbon dots functionalized gold nanorod mediated delivery of doxorubicin: tri-functional nano-worms for drug delivery, photothermal therapy and bioimaging. J Mater Chem B. 2013a;1:4972–82.

Pandey S, Shah R, Mewada A, Thakur M, Oza G, Sharon M. Gold nanorods mediated controlled release of doxorubicin: nano-needles for efficient drug delivery. J Mater Sci-Mater M. 2013b;24:1671–81.

Park H, Tsutsumi H, Mihara H. Cell penetration and cell-selective drug delivery using α-helix peptides conjugated with gold nanoparticles. Biomaterials. 2013;34:4872–9.

Park H, Tsutsumi H, Mihara H. Cell-selective intracellular drug delivery using doxorubicin and alpha-helical peptides conjugated to gold nanoparticles. Biomaterials. 2014;35:3480–7.

Park H, Yang S, Kang JY, Park MH. On-demand drug delivery system using micro-organogels with gold nanorods. ACS Med Chem Lett. 2016;7:1087–91.

Parveen S, Misra R, Sahoo SK. Nanoparticles: a boon to drug delivery, therapeutics, diagnostics and imaging. Nanomedicine. 2012;8:147–66.

Parween S, Ali A, Chauhan VS. Non-natural amino acids containing peptide-capped gold nanoparticles for drug delivery application. ACS Appl Mater Interfaces. 2013;5:6484–93.
Peng LH, Huang YF, Zhang CZ, Niu J, Chen Y, Chu Y, Jiang ZH, Gao JQ, Mao ZW. Integration of antimicrobial peptides with gold nanoparticles as unique non-viral vectors for gene delivery to mesenchymal stem cells with antibacterial activity. Biomaterials. 2016;103:137–49.
Piella J, Bastus NG, Puntes V. Size-controlled synthesis of sub-10-nanometer citrate-stabilized gold nanoparticles and related optical properties. Chem Mater. 2016;28:1066–75.
Raveendran P, Fu J, Wallen SL. Completely "green" synthesis and stabilization of metal nanoparticles. J Am Chem Soc. 2003;125:13940–1.
Raveendran S, Sen A, Maekawa T, Kumar DS. Ultra-fast microwave aided synthesis of gold nanocages and structural maneuver studies. Nano Res. 2017;10:1078–91.
Seo JM, Kim EB, Hyun MS, Kim BB, Park TJ. Self-assembly of biogenic gold nanoparticles and their use to enhance drug delivery into cells. Colloid Surface B. 2015;135:27–34.
Shang YZ, Min CZ, Hu J, Wang TM, Liu HL, Hu Y. Synthesis of gold nanoparticles by reduction of $HAuCl_4$ under UV irradiation. Solid State Sci. 2013;15:17–23.
Shanmugam V, Chien YH, Cheng YS, Liu TY, Huang CC, Su CH, Chen YS, Kumar U, Hsu HF, Yeh CS. Oligonucleotides – assembled Au nanorod-assisted cancer photothermal ablation and combination chemotherapy with targeted dual-drug delivery of doxorubicin and cisplatin prodrug. ACS Appl Mater Interfaces. 2014;6:4382–93.
Shi P, Ju E, Ren J, Qu X. Near-infrared light-encoded orthogonally triggered and logical intracellular release using gold nanocage@smart polymer shell. Adv Funct Mater. 2014;24:826–34.
Siddiqi KS, Husen A. Recent advances in plant-mediated engineered gold nanoparticles and their application in biological system. J Trace Elem Med Biol. 2017;40:10–23.
Smetana AB, Wang JS, Boeckl JJ, Brown GJ, Wai CM. Deposition of ordered arrays of gold and platinum nanoparticles with an adjustable particle size and interparticle spacing using supercritical CO_2. J Phys Chem C. 2008;112:2294–7.
Song JB, Zhou JJ, Duan HW. Self-assembled plasmonic vesicles of SERS-encoded amphiphilic gold nanoparticles for cancer cell targeting and traceable intracellular drug delivery. J Am Chem Soc. 2012;134:13458–69.
Song JB, Fang Z, Wang CX, Zhou JJ, Duan B, Pu L, Duan HW. Photolabile plasmonic vesicles assembled from amphiphilic gold nanoparticles for remote-controlled traceable drug delivery. Nanoscale. 2013a;5:5816–24.
Song L, Ho VH, Chen C, Yang ZQ, Liu DS, Chen RJ, Zhou DJ. Efficient, pH-triggered drug delivery using a pH-responsive DNA-conjugated gold nanoparticle. Adv Healthc Mater. 2013b;2:275–80.
Sun HP, Su JH, Meng QS, Yin Q, Chen LL, Gu WW, Zhang ZW, Yu HJ, Zhang PC, Wang SL. Cancer cell membrane-coated gold nanocages with hyperthermia-triggered drug release and homotypic target inhibit growth and metastasis of breast cancer. Adv Funct Mater. 2017;27:1604300.
Sung JH, Ji JH, Park JD, Song MY, Song KS, Ryu HR, Yoon JU, Jeon KS, Jeong J, Han BS, Chung YH, Chang HK, Lee JH, Kim DW, Kelman BJ, Yu JL. Subchronic inhalation toxicity of gold nanoparticles. Part Fibre Toxicol. 2011;8:16.
Tian FR, Conde J, Bao CC, Chen YS, Curtin J, Cui DX. Gold nanostars for efficient in vitro and in vivo real-time SERS detection and drug delivery *via* plasmonic-tunable Raman/FTIR imaging. Biomaterials. 2016;106:87–97.
Vala AK. Exploration on green synthesis of gold nanoparticles by a marine-derived fungus *Aspergillus sydowii*. Environ Prog Sustain. 2015;34:194–7.
Vijayakumar S, Ganesan S. Size-dependent in vitro cytotoxicity assay of gold nanoparticles. Toxicol Environ Chem. 2013;95:277–87.
Vivero-Escoto JL, Slowing II, Wu CW, Lin VSY. Photoinduced intracellular controlled release drug delivery in human cells by gold-capped mesoporous silica nanosphere. J Am Chem Soc. 2009;131:3462–3.
Wang CS, Li JY, Amatore C, Chen Y, Jiang H, Wang XM. Gold nanoclusters and graphene nanocomposites for drug delivery and imaging of cancer cells. Angew Chem Int Edition. 2011;50:11644–8.

Wang Z, Chen Z, Liu Z, Shi P, Dong K, Ju E, Ren J, Qu X. A multi-stimuli responsive gold nanocage-hyaluronic platform for targeted photothermal and chemotherapy. Biomaterials. 2014;35:9678–88.

Wei MY, Famouri L, Carroll L, Lee Y, Famouri P. Rapid and efficient sonochemical formation of gold nanoparticles under ambient conditions using functional alkoxysilane. Ultrason Sonochem. 2013;20:610–7.

Yahia-Ammar A, Sierra D, Merola F, Hildebrandt N, Le Guevel X. Self-assembled gold nanoclusters for bright fluorescence imaging and enhanced drug delivery. ACS Nano. 2016;10:2591–9.

Zhang X, Wu FG, Liu P, Wang HY, Gu N, Chen Z. Synthesis of ultrastable and multifunctional gold nanoclusters with enhanced fluorescence and potential anticancer drug delivery application. J Colloid and Interface Sci. 2015;455:6–15.

Zhang EQ, Xiang ST, Fu AL. Recent progresses of fluorescent gold nanoclusters in biomedical applications. J Nanosci Nanotechnol. 2016;16:6597–610.

Zhao CQ, Du TY, Rehman FU, Lai LM, Liu XL, Jiang XR, Li XQ, Chen Y, Zhang H, Sun Y. Biosynthesized gold nanoclusters and iron complexes as scaffolds for multimodal cancer bioimaging. Small. 2016;12:6255–65.

Chapter 10
Magnetic Hyperthermia-Using Magnetic Metal/Oxide Nanoparticles with Potential in Cancer Therapy

Costica Caizer

Abstract Magnetic hyperthermia (MHT) is heat dissipation in magnetic nanoparticles (MNPs) in an alternating magnetic field (AMF) of hundreds of kHz. This is an alternative method, non-invasive, and appears to be non-toxic compared to the usual methods – chemotherapy and radiotherapy used today in the therapy of malignant tumours – and with great future potential in cancer therapy. For the MHT to be effective in destroying tumour cells the temperature of biocompatible magnetic nanoparticles (Bio-MNPs) in the tumour must reach 42–43 °C, leading to local necrosis. There are many factors that increase the temperature of MNPs, depending on the nanoparticles' size (NPs) and type, metals or oxides with metallic ions. In this chapter, I will discuss application of metal/oxide magnetic nanoparticles (Me/O-MNPs) in MHT as possible candidates for increasing the hyperthermic efficiency in cancer therapy. I will present the basic physical aspects of MHT, the magnetic properties of Me/O-MNPs with potential use in MHT, the results obtained in vitro and in vivo and future development trends of the MHT technique, such as superparamagnetic hyperthemia (SPMHT), proposed as alternative, non-invasive and non-toxic cancer therapy.

Keywords Metal/oxide magnetic nanoparticles • Superparamagnetic hyperthermia • In vitro/in vivo • Cancer thermotherapy

C. Caizer (✉)
Department of Physics, Faculty of Physics, West University of Timisoara, Bv. V. Parvan no. 4, 300223, Timisoara, Romania
e-mail: costica.caizer@e-uvt.ro; ccaizer@physics.uvt.ro

M. Rai, R. Shegokar (eds.), *Metal Nanoparticles in Pharma*,
DOI 10.1007/978-3-319-63790-7_10

Nomenclature

AMF	Alternating magnetic field
BioC	Bioconjugation
BioFS	Biofunctionalised surface
BioFS-NPs	Nanoparticles with biofunctionased surface
Bio-L	Biocompatible layer
Bio-MNPs	Biocompatible magnetic nanoparticles
CDs	Cyclodextrins
FeM	Ferromagnetic
FeMNPs	Ferromagnetic nanoparticles
FiM	Ferrimagnetic
FiMNPs	Ferrimagnetic nanoparticles
FS	Functionalised surface
FS-NPs	Nanoparticles with functionalised surface
LPs	Liposomes
MeIs	Metallic ions
Me-MNPs	Matallic magnetic nanoparticles
Me-FeMNPs	Metallic ferromagnetic nanoparticles
Me-SPMNPs	Metallic superparamagnetic nanoparticles
MFHT	Magnetic fluid hyperthermia
MHT	Magnetic hyperthermia
MNPs	Magnetic nanoparticles
NPs	Nanoparticles
O-FiMNPs	Oxide ferrimagnetic nanoparticles
O-MNPs	Oxide magnetic nanoparticles
SAR	Specific absorption rate
SPM	Superparamagnetic
SPMHT	Superparamagnetic hiperthermia
SPMNPs	Superparmagnetic nanoparticles

10.1 Introduction

Magnetic hyperthermia (MHT) is today one of the most promising alternative methods, that is non-invasive and has low toxicity upon the living organism, and with a great future potential in cancer therapy (Gordon et al. 1979; Pankhurst et al. 2003; Ito et al. 2005a; Kawai et al. 2006; Gazeau et al. 2008; Kobayashi et al. 2014; Datta et al. 2015). The method relies on getting a higher temperature inside the tumor where the magnetic nanoparticles (MNPs) are found, ~43 °C respectively, compared to the normal temperature of the surrounding tissues, by using an external alternating magnetic field (AMF) (Rosensweig 2002; Pankhurst et al. 2003; Hergt et al. 2006). Local heating occurs through the Néel – Brown magnetic relaxation

(Néel 1949; Brown 1963; Rosensweig 2002). The high temperatures destroy the tumour cells through apoptosis or even necrosis (up to 45–46 °C) (Prasad et al. 2007; Gazeau et al. 2008). Since Gordon et al. (1979) revealed for the first time in 1979, the hyperthermic magnetic effect on tumour cells through the use of MNPs coated with dextran exposed to an AMF, it has been considered that MHT with nanoparticles (NPs) can be a viable alternative method in cancer therapy compared to chemotherapy and radiotherapy (Gordon et al. 1979). Chemo- and radiotherapy used today to treat cancer are highly toxic, and in many advanced cases these are even inefficient. Therefore, finding an alternative method, which is non-invasive, with a reduced or no toxicity, and more efficient than the current methods is desirable. Even more, the importance of finding better alternatives for cancer therapy is emphasized by the high rates of cancer diagnosis in many European countries.[1] According to the World Health Organisation findings, cancer is the second leading cause of death globally with an expectancy of 70% increase in deaths over the next two decades.

By using a natural thermal effect instead of chemicals or radiations which are highly destructive for health cells, MHT has the advantage of a very low toxicity rate from the beginning. In addition, under the action of AMF, thermal effects occur only locally, inside only the tumour where are MNPs, and can be controlled by using MNPs without affecting the other surrounding healthy tissue. Moreover, AMF used in MHT has no negative biological effects on healthy tissues, as long as a certain threshold is not passed (Hergt and Dutz 2007). Some MNPs, such as magnetite, are well tolerated by the body (Lacava et al. 1999; Hilger et al. 2003; Naqvi et al. 2010). Furthermore, to eliminate the possible toxicity of the MNPs and to increase their affinity to the tumor cells, they are made biocompatible (Safarik and Safarikova 2002; Kobayashi et al. 2014) by various techniques used in bionanotechnology. Usually, these techniques are used for coating/covering the surface of NPs with a non-toxic organic layer (polymer, non-polymeric or different biological molecules) (Molday and MacKenzie 1982; Portet et al. 2001; Zhang et al. 2002; Berry et al. 2003; Mart et al. 2009; Guandong et al. 2010). Given all this, MHT with MNPs appears to be toxicity free.

However, following hyperthermic effect, MNPs can remain interstitial in the tumour area or in the bloodstream and can possibly migrate towards some organs (liver, spleen, etc.) (Johannsen et al. 2005). Regarding intracellular MNPs in the tumor, those can be eliminated naturally due to their reduced sizes (a few nanometers). These issues and others that can affect the toxicity and efficacy of thermotherapy are taken into account when studying the efficiency of the magnetic method.

Another important aspect of the MHT is finding the most suitable MNPs to get a maximum hyperthermic effect and a reduced toxicity. From this point of view, besides using ferrimagnetic nanoparticles (FiMNPs) of Fe_3O_4 and γ-Fe_2O_3 with

[1] World Cancer Research Fund International, *Data for Cancer Frequency by Country,* accessed at http://www.wcrf.org/int/cancer-facts-figures/data-cancer-frequency-country.

Fe^{+2}/Fe^{+3} metallic ions (MeIs), which seem to be most suitable, metallic ferromagnetic nanoparticles (Me-FeMNPs) of small sizes (only a few nm), such as Fe, Co, etc., are also considered for future use in superparamagnetic hyperthermia (SPMHT). This would greatly increase the hyperthermic effect and ultimately, the efficiency of the method (Zeisbergera et al. 2007; Habib et al. 2008; Mehdaoui et al. 2010a, 2010b; Kappiyoor et al. 2010; McNerny et al. 2010; Wu et al. 2011).

Also, another aspect worthy of attention is the homogenous dispersion and accumulation of MNPs inside tumours. Moreover, if until now the more solid localized tumors were targeted through MHT, being easily accessible by injecting the biocompatible magnetic nanoparticles (Bio-MNPs) suspension directly in the tumour, with the current development of bionanotechnology, also the hidden and harder to reach tumours or extensively metastatic have started to be widely addressed. Through encapsulating NPs in liposomes (LPs) or coating them with a biocompatible layer (Bio-L) and then functionalising the surface (FS) of the NPs (FS-NPs) with molecules/bionanostructures and specific ligands with a high selectivity of tumor cells, and then administer them intravenously, these new techniques become viable alternatives (Shinkai and Ito 2004; Malhi et al. 2013; Le Renard et al. 2010; 2011). As a result of this type of thermotherapy by MHT there are also other advanced effects such as induced anti-tumor immunity (Kobayashi et al. 2014). Bioconjugation/biofunctionalisation the surface of nanoparticles (BioC/BioFS-NPs) with specific antibody (Ito et al. 2005a, b), or by encapsulating the drug/anti-tumor substances in cyclodextrins (CDs) (Fagui et al. 2011; Li et al. 2011; Nigam et al. 2011; Yallapu et al. 2011; Hayashi et al. 2014) with double action (thermal and biochemical) are current approaches in MHT. In vivo results so far have clearly demonstrated tumour regression following MHT therapy (Hergt et al. 1998; Ito et al. 2003a; Johannsen et al. 2005; Tanaka et al. 2005; Hilger et al. 2005) with promising future use in the clinical trial and as alternative method in cancer thermotherapy.

This chapter discusses the MHT with Me-MNPs (FeM) and O-MNPs (FiM) that contain MeIs, and presents some results obtained in vitro and in vivo about the possible use in therapy of malignant tumours. The physical principle of MHT, and in particular, of the *SPMHT*, which is the best method currently used for obtaining the magnetic hyperthermia effect with a maximum efficiency, will be discussed in detail further on. Also, other aspects of the MHT efficiency and real prospects of using this alternative method in the future cancer therapy will be further addressed.

10.2 Principle of Magnetic Hyperthermia

10.2.1 Hyperthermia with Magnetic Nanoparticles

MHT consists in increasing the temperature environment (physical or biological) that contains a magnetic material, metal (FeM) or oxides (generally based on irons) with MeIs (FiM), under the action of an external AMF. In the case of a biological environment, when aiming at destroying tumour cells by using hyperthermic effect

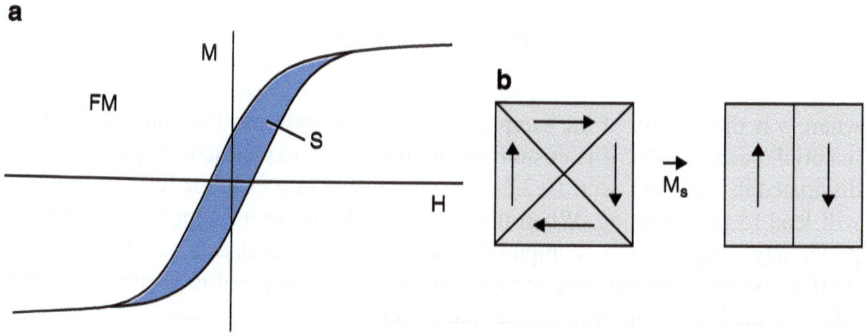

Fig. 10.1 (**a**) The magnetization (M) of multidomain nanoparticles in the external magnetic field (H); (**b**) magnetic structures in multidomain nanoparticles of cubic (*left*) and uniaxial (*right*) anisotropy

(more than the normal value of ~37 °C), the goal is to increase the tissue's temperature at target values of 42–43 °C.

The magnetic effects that can lead to heating the magnetic material (FeM or FiM) located in an external AMF are: (*i*) *eddy currents* that occur through electromagnetic induction in metallic magnetic materials, (*ii*) *magnetic hysteresis*, (*iii*) *magnetic relaxation* and (*iv*) *magnetic resonance* (ferro- or ferrimagnetic). In the case of MNP_S, because they are very small (10^{-7} to 10^{-9} m) and generally isolated from each other in a suspension, turbionari currents (*i*) are negligible, even in conductive MNPs (Fe, Ni, Co, etc.).

In addition, for obtaining MHT, the AMFs are used at a frequency of tens – hundreds of kHz. Having in view that the magnetic resonance in a fero- or ferrimagnetic material (FeM when NPs are metals or FiM when NPs are metal oxides) is achieved at much higher frequencies (hundreds of MHz – tens of GHz), MHT can not be generated by magnetic resonance (*iv*). Only the hysteresis effect (*ii*) (Pankhurst et al. 2003) and the magnetic relaxation (*iii*) (Néel 1949; Brown 1963) may produce heat to the material.

Magnetic hysteresis (*ii*) (Fig. 10.1a is obtained, in general, in the case of large MNPs, that have a structure of magnetic domains (Fig. 10.1b).

Power dissipation per unit of material volume in AMF can be expressed by the formula (Pankhurst et al. 2003)

$$P = \mu_0 f \oint H dM, \left(\text{in W/m}^3\right) \tag{10.1}$$

where f is the frequency of AFM with H magnitude and M is the magnetization of the material. Basically, this power is the surface S of hysteresis loop (Fig. 10.1a) which gives the power dissipated during a hysteresis loop, on a period of time $t(t=1/f)$ of the magnetic field multiplied by frequency ($P=\mu_0 fS$). When the sample is comprised of MNPs the *specific* power dissipation is used (Habib et al. 2008)

$$P_S = \frac{P}{\rho}, \quad (\text{in W/g}) \tag{10.2}$$

where ρ is the density of the sample (nanoparticles system). For small amounts of material, such as NPs, it is customary the use of the expression of specific power dissipated in W/g (see Sect. 10.2.2). The loss power in NPs in the form of heat (Q), will lead to their heating. When the MNPs are found in a biological tissue (after previously being made biocompatible with the enviroment), as a result of applying AMF to obtain a temperature increase of the tissue compared to normal value (ex. ~37 °C), the MNPs suffer hyperthermic effect.

However, it was found that the practical use of MHT as an effect of magnetic hysteresis has some drawbacks (Pankhurst et al. 2003). Mainly for getting an efficient heating of a tissue is needed to increases its temperature in a short time and also to reach the necessary value in a MHT. This requires that the hysteresis loop to be as broad and saturaturated as possible, which implies the existence of high fields for magnetization and magnetic saturation. This is hard to have at frequencies of hundreds of kHz. Also, NPs should be large enough and be structured as magnetic domains in order to have a pronounced hysteresis, and/or to have a large magnetic anisotropy. Using large MNPs raises many aspects of the biological toxicity or biocompatibility and transportation at cellular and intracellular level.

Recent research results have shown that MHT would be more effective if instead of large NPs (average diameter of tens of nm), with magnetic domains structure, its would use NPs that are *smaller*, *single-domain* and *superparamagnetic*, with a (mean) diameters in the ~5–25 nm range. In this case, it was found that MNPs can achieve higher power dissipation than using hysteresis losses (Pankhurst et al. 2003; Hergt et al. 1998), and in addition, their toxicity should be lower, due their smaller size. Moreover, small MNPs should be more effective in intracellular therapy (Fortin et al. 2008).

The main feature of MNPs, FeM or FiM, that have a superparamagnetic (SPM) behavior is the lack of hysteresis (Fig. 10.2a). Their magnetization is done by a function of Langevin type (Jacobs and Bean 1963). At the same time, the magnetic domains structure is missing, NPs having a single-domain (Fig. 10.2b) magnetized spontaneously to saturation. In this case, the spontaneous magnetization $\left(\vec{M}_s\right)$ is not stable, but fluctuates along a direction, called the easy magnetization axis (e.m.a.) under the action of thermal activation, a process known as magnetic relaxation, respectively, in this case *superparamagnetic relaxation* (Néel 1949; Bean and Livingston 1959). The 180° magnetization reversal and the magnetic moment of nanoparticle $\left(\vec{m}\right)$ which has volume V, respectively

$$\vec{m} = V\vec{M}_s, \tag{10.3}$$

are done in the Néel magnetic relaxation time (Néel 1949; Aharoni 1964), which has the formula

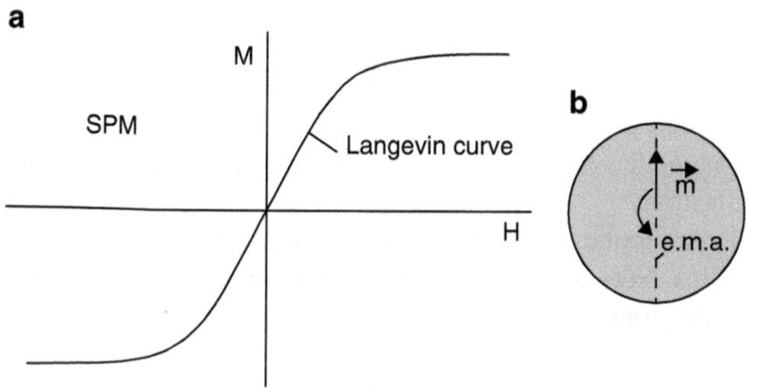

Fig. 10.2 (**a**) The magnetization (M) of nanoparticles that have single domain in the external magnetic field (H); (**b**) magnetic structures in single domain nanoparticles with uniaxial anysotropi

$$\tau_N = \tau_0 \exp\left(\frac{KV}{k_B T}\right), \tag{10.4}$$

where KV is the energy barrier for the magnetic moment of NPs, $k_B T$ is the thermal energy and τ_0 is a time constant (10^{-9} s Back et al. 1998). In Eq. 10.4 K is the magnetocrystalline anisotropy constant, which in the simplest case is considered uniaxial, and k_B is the Boltzmann constant.

This process of superparamagnetic relaxation in an external AMF leads to a dissipation of energy in the form of heat. Thus, they heat up. This magnetic relaxation effect (*iii*) in *superparamagnetic nanoparticles* (SPMNPs) is used in MHT and appears to be more appropriate than the effect given by hysteresis (*ii*) both in terms of increased power dissipation obtained but also of lower toxicity, due to using smaller nanoparticles. According to some researchers in the field, the SPMHT is considered the future cancer therapy method, more efficient and with almost no toxicity compared with chemo- and radiotherapy, which are used today in treatment of cancer and have a high toxicity and in many cases are inefficient.

10.2.2 *Superparamagnetic Hyperthermia*

In an harmonic AMF with amplitude H_0 and frequency f, power (volumetric) dissipated is given by (Rosensweig 2002)

$$P_{MHT} = \pi \mu_0 f \chi'' H^2, \tag{10.5}$$

where χ'' is the component of complex magnetic suspectivity, called the loss component. This component is found (Shliomis 1974) as having the form of the Debye equation

$$\chi'' = \chi_0 \frac{\omega\tau}{1+(\omega\tau)^2}, \tag{10.6}$$

where χ_0 is magnetic susceptibility of echilibrum, τ is magnetic relaxation time and $\omega = 2\pi f$ is pulsation.

When NPs are found in colloidal suspensions, magnetic relaxation can occur by two mechanisms, Néel and Brown, and relaxation time will be given by (Néel 1949; Brown 1963; Aharoni 1964)

$$\tau = \frac{\tau_N \tau_B}{\tau_N + \tau_B}, \tag{10.7}$$

where τ_N is given by Eq. 10.4 and τ_B is (Brown 1963)

$$\tau_B = \frac{3\eta V_h}{k_B T}. \tag{10.8}$$

In Eq. 10.8 η it is the viscosity coefficient of the fluid in which the NPs are dispersed and V_h is the hydrodynamic volume of the NPs (Fig. 10.3), which in a spherical nanoparticles approximation model (Fig. 10.3a have the expression

$$V_h = \frac{\pi}{6}(D + 2\delta)^3. \tag{10.9}$$

In Eq. 10.9 D is the diameter of the NPs (Fig. 10.3a) and δ is the thickness of the biocompatible coating (Fig. 10.3b), such as dextran, liposomes, cyclodextrins, polymers, etc., depending on biomedical application (Veiseh et al. 2010).

Considering Eq. 10.6, the loss power by magnetic relaxation in fluid (Eq. 10.5) will have the formula

$$P_{MHT} = \mu_0 \pi f H^2 \chi_0 \frac{2\pi f\tau}{1+(2\pi f\tau)^2}. \tag{10.10}$$

In many cases, NPs in biological environments are fixed, or thickness of the Bio-L is quite thick, so the condition is achieved $\tau_B \gg \tau_N$, and relaxation time τ is reduced to τ_N ($\tau = \tau_N$). This approximation takes place in most practical cases. In this case, the loss power occurs only through Néel relaxation process, as by rotation of the magnetic moments within the MNPs under the action of the AMF.

In the case of different NPs, to be able to compare results, *specific* loss power is used (Eq. 10.2)

$$(P_{MHT})_S = \pi\mu_0 \chi_0 H^2 f \frac{2\pi f\tau}{\rho\left[1+(2\pi f\tau)^2\right]}, \quad (\text{in W/g}) \tag{10.11}$$

where δ is the density of NPs ($\delta = m/V$, m is the mass).

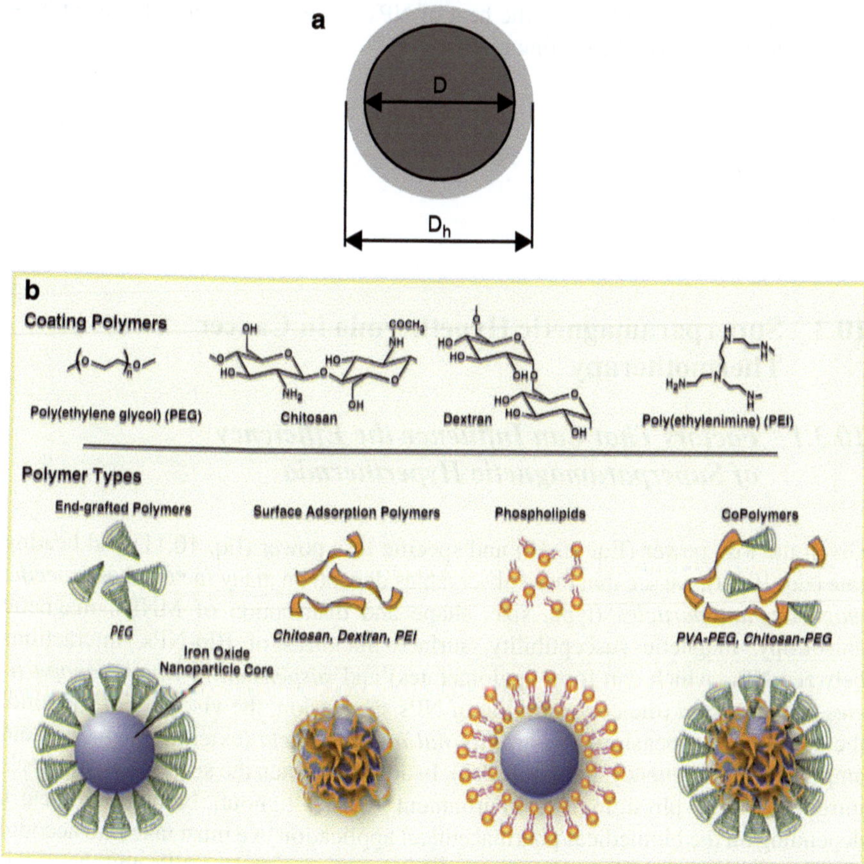

Fig. 10.3 (**a**) The core-shell model of compatible nanoparticles; (**b**) (color online) Illustration depicting the assembly of polymers onto the surface of magnetic nanoparticle cores (Reprinted from Veiseh et al. 2010, Copyright (2010), with permission from Elsevier)

The formula found (10.11) determines the temperature of heating nanoparticles (ΔT), considering the adiabatic system, when entire energy (power) is transformed into heat (Q) in a time period Δt,

$$Q = mc\Delta T = P_{\mathrm{MHT}} V \Delta t \qquad (10.12)$$

and

$$\Delta T = \frac{P_{\mathrm{MHT}} \Delta t}{c\delta}, \qquad (10.13)$$

respectively, where c is the specific heat of NPs. In studies of magnetic hyperthermia often one will use the heating rate (speed)

$$\frac{\Delta T}{\Delta t} = \frac{\left(P_{\mathrm{MHT}}\right)_{\mathrm{S}}}{c} \tag{10.14}$$

which is more suggestive.

10.3 Superparamagnetic Hypethermia in Cancer Thermotherapy

10.3.1 Factors That Can Influence the Efficiency of Superparamagnetic Hyperthermia

Given the loss power (Eq. 10.10) and specific loss power (Eq. 10.11) and heating rate (Eq. 10.14), we see that these observables depend on many factors, both *specific magnetic nanoparticles* (type, size, shape and distribution of MNPs, magnetic anisotropy, magnetic susceptibility, surface thickness of Bio-NPs, interactions between NPs, which can form agglomerates) and *suspension in which nanoparticles are dispersed* (the concentration of NPs suspension, the viscosity of the fluid, the density of suspension) and an *external magnetic field* (external magnetic field amplitude and frequency, harmonic AC). In addition, when the suspension of NPs is introduced in a physiological environment (tissue, tumour, blood, fluid, etc.), depending on the biomedical/pharmaceutical application, we must take into account the characteristics of this environment, the interaction with Bio-NPs and the possibility of movement or no movement (fixing) of NPs. All these factors should be considered when the maximum efficiency of the MHT application is pursued.

However, some of these can be considered *critic* parameters, such as the *size* of NPs. Besides thess, it quite critical is the *magnetic anisotropy* of MNPs that can change the power dissipation and heating rate, in order to get a maximum hyperthermic effect, with direct repercussions on the size of NPs to be used in MHT.

To get a clearer picture of the influence these parameters have on the MHT, Fig. 10.4 shows different curves of heating rates $\Delta T/\Delta t$ (Eq. 10.14) for different O-MNPs with Fe^{2+}, Fe^{3+} and Co^{2+} metal ions (FiM nanostructures of Fe_3O_4 (magnetite), γ-Fe_2O_3 (maghemite), $CoFe_2O_4$ (cobalt ferrite) and Ba ferrite) dispersed in teradecane (carrier liquid) (Rosensweig 2002). The heating rates of these nanosystems presents a maximum for each of the values of diameters (sizes) of the NPs, in range of 3–12 nm, bigger being at Fe_3O_4 and γ-Fe_2O_3. From the diagrams we can see that a small change in the NPs diameter leads to a high decreases of heating rate (ex. curves 3 and 4). This shows that the diameter of NPs is a critical measure in obtaining the maximum heating rate. Also, magnetic anisotropy of NPs changes significantly the maximum positions (the size of NPs where the maximum is obtained), that are obtained at slightly lower diameters when the magnetic anisotropy is high.

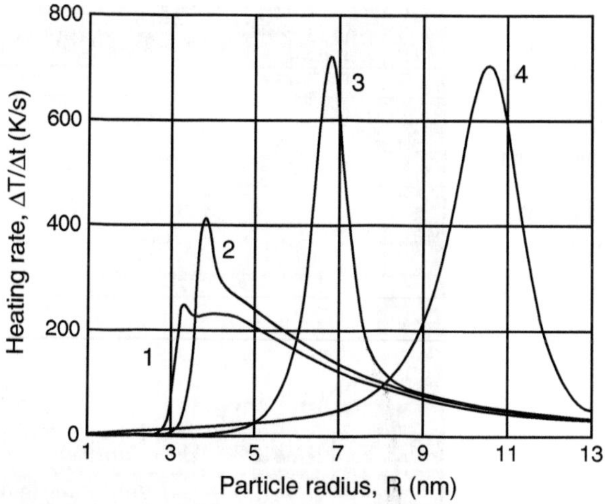

Fig. 10.4 Comparative heating rates for various magnetic solids. *1* Barium ferrite, *2* Cobalt ferrite, *3* Magnetite, *4* Maghemite. $\phi = 0.071$, $\eta = 0.00235$ kg m^{-1} s^{-1}, $f = 300$ kHz, $B = 0.09$ T, $\delta = 2$ nm (Reprinted from Rosensweig 2002, © (2002), with permission from Elsevier)

For example, this can be observed in the case of barium and cobalt ferrite (curves 1 and 2) compared to magnetite and maghemite (curves 2 and 4). The curves were recorded in the magnetic field $H = 71.6$ kA/m ($H = B/\mu_0$, $B = 0.09$ T) and the frequency f of 300 kHz, and the dispersion environment of the NPs had the following characteristics: the volume fraction of the NPs in liquid (ϕ) is 7.1% and the viscosity coefficient of liquid (η) is 0.00235 kgm^{-1} s^{-1}).

In Fig. 10.5 we can properly seen (in a 3D representation) the influence of the external magnetic field (amplitude H and frequency f) and of the mean diameter (<*D*>) of MNPs (taking into account a distribution of size of NPs in a real dispersion) on the specific loss power (SLP) (Eq. 10.11), in the case of Co ferrite NPs and in the absence of relaxation Brown (when the viscosity of the fluid is high or NPs are fixed), and when the magnetic susceptibility of equilibrium (χ_0) is constant (Caizer 2010). For the magnetic field were considered the following values of 10–30 kA/m and the variable parameters, frequency (f) and mean diameter (<*D*>) in the ranges of 100–500 kHz, and 4–7 nm, respectively. The study was done in a biological tissue without cellular damage by magnetic field (amplitude and frequency) (Hergt and Dutz 2007)

$$H \cdot f \leq 5 \times 10^9 \,\text{Am}^{-1}\text{Hz}. \tag{10.15}$$

One such study done with a professional software, *optimizes* MHT (finding the conditions to specify the highest specific loss power for a particular nanosystem) study required both prior to the experiments in vitro and in vivo, and during these tests.

The purpose of my study is the possible future use of Co ferrite NPs encapsulated in LPs, with a SPM behavior in an external magnetic field of up to 150 kA/m

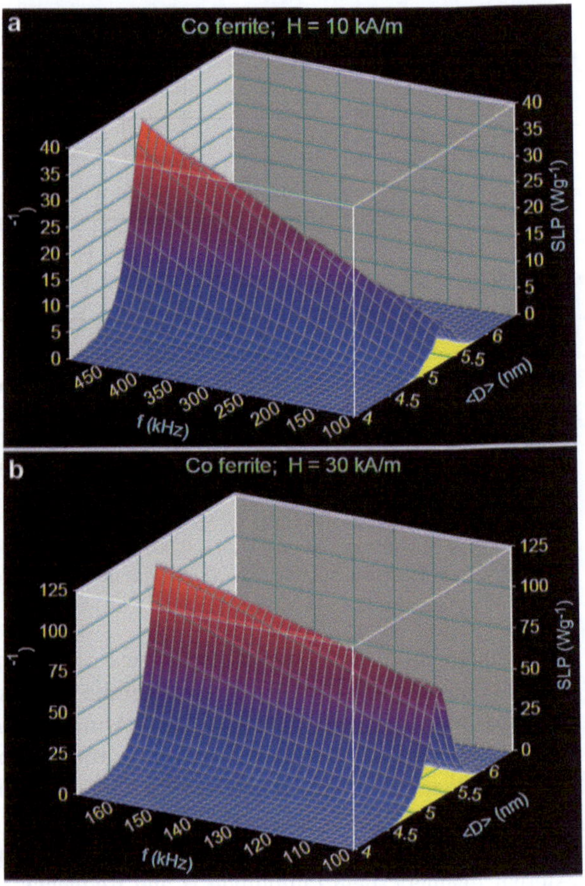

Fig. 10.5 (Color online) 3D SLP for cobalt ferrite nanoparticles as a function of *f* and <*D*>; (**a**) 10 kAm^{-1} and (**b**) 30 kAm^{-1} (Reprinted from Caizer 2010, © (2010), with permission from West University Publising House)

(Fig. 10.6), in the therapy of malignant tumors. The curve was recorded for concentrations of 20 mg/ml of liposomal suspension of nanoparticles (magnetoliposomes in distilled water). Due to the LPs sizes (>100 nm), Brown relaxation (Eq. 10.8) has been neglected, losses taking place during Néel relaxation processes (Eq. 10.4), in the low field of 10–30 kA/m. An important finding of this study is that the high specific loss power is achieved at a lower mean diameter of Co ferrite NPs. This result leads to the idea that these NPs that would be suitable for use in intracellular therapy (Jordan et al. 1999a; Fortin et al. 2008) of hyperthermia may destroy tumour cells more effectively within them.

An excellent review, where the influence of other parameters on the SPMHT can be observed, is found in Gazeau et al. (2008). For example, the viscosity of fluid or the large dispersion of the MNPs that always exists in a real nanosystems, can substantially affect the SPMHT (Rosensweig 2002). In the case of the large dispersion of NPs we must consider the distribution function of the sizes (diameters) of nanoparticles. Experimentally, in most cases, it was found that the nanoparticle size distribution can be approximated by a lognormal function (Bacri et al. 1986; O'Grady and Bradbury 1994; Caizer 2003)

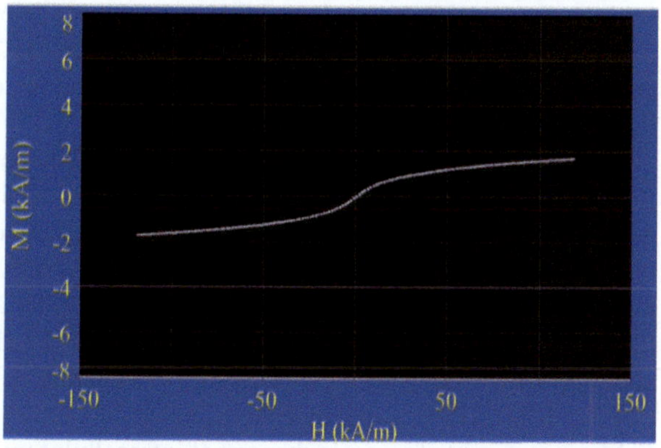

Fig. 10.6 (Color online) magnetisation of CoFe2O4 nanoparticles encapsulated in liposomes (Reprinted from Caizer 2010, © (2010), with permission from West University Publising House)

$$f(D)=\frac{1}{\sqrt{2\pi}\sigma D}\exp\left(-\frac{\ln^2(D/d)}{2\sigma^2}\right) \tag{10.16}$$

where D is the diameter of the NPs (Fig. 10.3) (in spherical nanoparticles approximation, for example) and d and σ are the distribution parameters. In this case, the mean diameter of the NPs will be (Baker et al. 2006)

$$D=d\cdot\exp\left(\frac{\sigma^2}{2}\right) \tag{10.17}$$

Given such ditribution, and the Eqs. 10.10 and 10.11, *mean* specific loss power in this case is calculated by formula

$$P=\int_0^\infty f(D)P_{\mathrm{MHT}}d(D) \tag{10.18}$$

In order to have a small influence of the NPs sizes over the power dissipated, is important to aims at achieving this by various methods used in obtaining nanoparticle systems with more narrow distributions. However, in real terms this condition often is difficult, and the formula (10.18) should be applied for assessing the correct loss power.

10.3.2 *Superparamagnetic Hyperthermia with Metal Magnetic Nanoparticles*

While aiming at increasing the loss power in a short time in a system of MNPs and given the theoretical and experimental results regarding the SPMHT, and other results and recent studies (Baker et al. 2006; Lee et al. 2007; Ondeck et al. 2009; Purushotham and Ramanujan 2010), an alternative route is considered in order to achieve the desired results. In order to reduce the side effects of toxicity on normal cells surrounding the tumour cells targeted through MHT, the use of *metallic* superparamagnetic nanoparticles (Me-SPMNPs) is used instead of those oxides that are limited (Kappiyoor et al. 2010).

Me-MNPs that are FeM would bring additional benefits in SPMHT compared to the oxides, primarily by strong increase in power dissipated due to the high magnetic moment (Eq. 10.3) and the high susceptibility of FeMNPs, which can be tens – hundreds of times higher than the FiM oxides (ferrites). In addition, the magnetic anisotropy of FeM metals such as Fe, Co, brings other advantages due its generally higher values than iron-based FiM oxides, such as Fe_3O_4 or γ-Fe_2O_3, the most suitable and used in experiments of SPMHT. This makes the maximum power dissipated in SPMHT (Eq. 10.10) to be achieved at smaller sizes of NPs, which makes FeM more suitable for use in therapy of intracellular magnetic hyperthermia (Fortin et al. 2008). Me-FeMNPs are smaller than FiM oxides, and thus will have less effect on cell toxicity. Moreover, smaller NPs can easily get inside cells by active transportation mechanisms, making them more efficient in terms of effects of MHT. In addition, even using various biocompatible surfactants the hydrodynamic volume (diameter) (Fig. 10.3 and Eq. 10.9) will be smaller, thus facilitating both relaxation mechanisms in MHT and transportation mechanisms of biocompatible NPs to the target biological tissue.

The result shown in Fig. 10.7 is very suggestive in this regard. It can be observed very clearly that the loss power in Me-FeMNPs of Fe or Co is noticeably higher than in the case of O-FiMNPs of Fe_3O_4 or γ-Fe_2O_3. In addition, the diameter of Me-MNPs where can achieve the maximum power is approx. two times lower than that of the O-MNPs, being about 10 nm, aspect benefit for SPMHT. Another important issue when using Me-FeMNPs in SPMHT is also their usage as metal alloys,

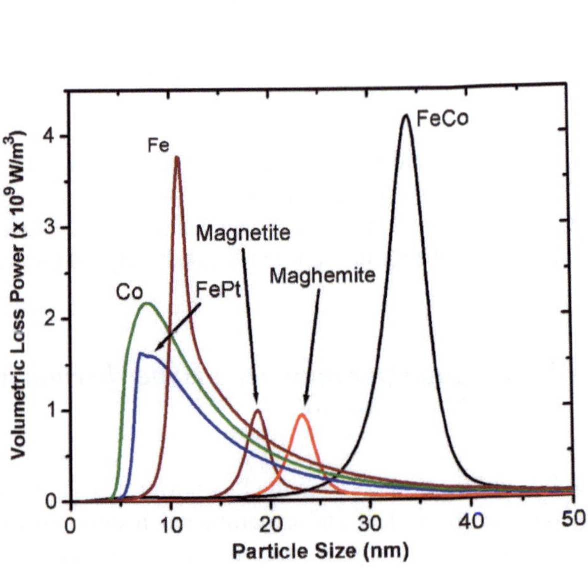

Fig. 10.7 (Color online) volumetric power loss for various magnetic materials at 300 kHz and 50 mT ac field in aqueous dispersion with 10% particle concentration (Reprinted from Habib et al. 2008, with the permission of AIP Publishing)

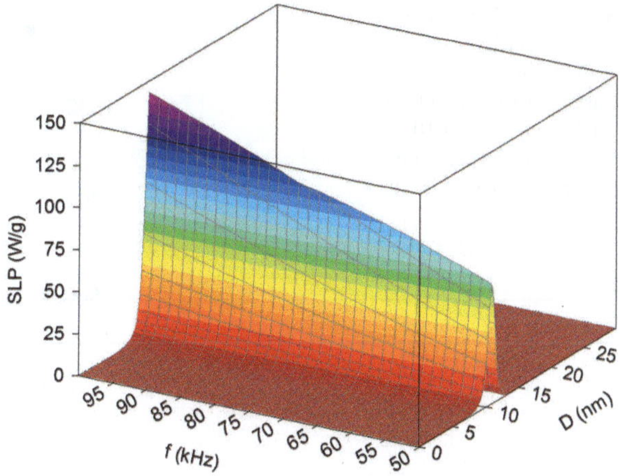

Fig. 10.8 The specific loss power for nanoparticles of Fe, according to the diameter of nanoparticles and frequency of alternating magnetic field

which gives the possibility of obtaining a magnetic anisotropy adjustable depending on the concentration of one or other of the chemicals used (ex. FeCo). This allows a modification of Me-MNPs size in a fairly wide range without the power dissipation having to change too much, which is great for SPMHT.

In addition, the limitation imposed by Eq. 10.15 could be easily met in Me-FeMNPs by the real possibility of using a magnetic field with amplitude or frequency reduced, and that will lead to the same hyperthermic effect as when iron O-MNPs are used.

Given the above advantages, preparation of colloidal suspensions of Me-FeMNPs based on Fe and Co, and their use in SPMHT, has already caught scientist's attention (Zhang et al. 2003; Mendoza-Resendez et al. 2004; Hutten et al. 2005; Hergt et al. 2006; Zeisbergera et al. 2007; Habib et al. 2008; Mehdaoui et al. 2010a, b; Kappiyoor et al. 2010; McNerny et al. 2010; Wu et al. 2011).

In Fig. 10.8 I have a 3D computational result of a study that we did in the case of Me-MNPs of Fe as a possible use of their SPMHT. The study was aimed at finding optimal diameter of the NPs (considered as spherical) that gives the highest specific loss power, starting from simulating real conditions (considering the iron magnetic characteristics, a magnetic packing fraction (volumetric) of 10%, and NPs fixed in the future tissue (neglecting the Brown relaxation processes)). The result is surprising, because one can get a fairly high specific power dissipation even at low H and f (i.e., $H < 40$ kA/m and $f < 100$ kHz), making these MNPs very suitable for SPMHT.

Similar results were obtained by other authors. Mehdaoui et al. (2010a) reported MHT for iron FeMNPs with cubic shape and sizes of 11 nm and 16 nm, where they obtained a specific loss power (SAR – specific absorption rate) of 1,690 W/g in a magnetic field of 66 mT (~53 kA/m). This power exceeds that one reported in other nanosystems. In Fig. 10.9 SAR is shown for the two samples studied.

However, reduced biocompatibility compared to NPs of the iron oxide is a major issue when it comes to their use in SPMHT in tumors. If the problem of

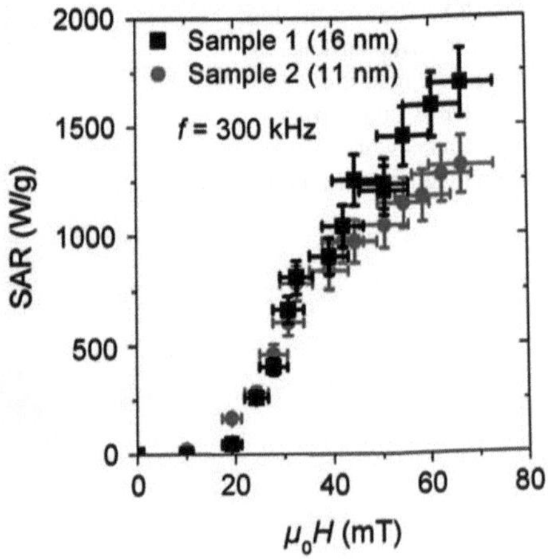

Fig. 10.9 Magnetic field dependence of SAR at 300 kHz for the two samples (Reprinted from Mehdaoui et al. 2010a, Copyright (2010), with permission from Elsevier)

biocompatibility and stability of suspensions of Me-MNPs in aqueous solution will be solved the Me-FeMNPs would be used successfully in SPMHT instead of O-FiMNPs ones, this leading to greater efficiency of the hyperthermic method. The problem of biocompatibility and using oxide/metal NPs in MHT studies in vitro and in vivo will be discussed in the next section.

In this respect, a first step in reducing the toxicity of Me-FeMNPs on healthy tissues could be done by reducing from the start their concentrations in the tissue, respectively the volume fraction of magnetic material in suspension. Because of this there is an additional reserve of loss power that would be obtained in the case of Me-MNPs, compared with the oxide ones. Depending on the application targeted for using MHT, it is beyond doubt that we must find the most suitable organic compounds, in order to obtain a good biocompatibility of Me-MNPs, so they can then be used in SPMHT.

Overall, it is observed that FiMNPs of iron oxide, such as Fe_3O_4 and γ-Fe_2O_3, are regarded today as the most suitable for use in MHT. A special attention should be paid to nanoparticle-based metallic iron, e.g. Fe or Co metal, that increases the power dissipated. In a good biocompatible that can be achieved, they could be used successfully and with high efficiency in SPMHT of tumours.

10.3.3 Biocompatibility of Magnetic Nanoparticles for Cancer Therapy by Magnetic Hyperthermia

One of the most important issues in the application of MHT is avoiding possible toxicity on healthy tissue, which can be caused by the MHT or mostly by the used MNPs. Damage to healthy tissue by a magnetic field can be avoided by using AMF with low amplitudes and frequencies, so that the conditions in Eq. 10.15 are respected. Also, the temperature achieved by the magnetic field must be maintained at around ~43 °C with the effect of inducing apoptosis and necrosis in the tumour tissue, this being controlled by external electronic devices. Only in some cases, the temperature of localized tumors may be slightly higher in the tumor (44–46 °C), healthy tissue being unaffected, warming taking place only on the tumour in which there are MNPs.

It only remains that the NPs that mediate hyperthermia to be biocompatible with the tissues where they are to be introduced. In this direction, since the question of using MNPs in MHT, various methods/techniques were developed for making Bio-MNPs using different agents (polymeric, no-polymeric, biological molecules) coating/covering surface nanoparticles (surfacted, BioC, BioFS, etc.) or encapsulating NPs in different mebranes/biological nanostructures, creating hybrid bionanostructures core-shell (magnetic core and biocompatible biological shell (Molday and MacKenzie 1982; Portet et al. 2001; Safarik and Safarikova 2002; Zhang et al. 2002; Berry et al. 2003; Mart et al. 2009; Guandong et al. 2010 Kobayashi et al. 2014). A synthesis of these and highlighting MNPs used in MHT was made recently by Kobayashi et al. (2014). In Table 10.1 these are summarized (with permission from Kobayashi et al. 2014).

Current trends are in favour of BioC of the surface of MNPs with different specific molecular nanostructures, antibodies or by inducing through MHT an immune

Table 10.1 (Color online) representative magnetic nanoparticles for magnetic nanoparticle-mediated hyperthermia

Name	Core size	Characteristics	Ref.
Magnetite	A few μm	The first demonstration using magnetic particles	Fortin et al. (2008)
Dextran magnetite	6 nm	The first demonstration using magnetite of nanometer sizes	Gazeau et al. (2008)
Aminosilane-coated magnetite	15 nm	Enhaces the uptake by cancer cells and prevents intracellular digestion	Baldi et al. (2007)
Magnetite cationic liposome or cationic protein	10 nm	Enhaces the uptake by cancer cells and stabilizes the colloidal solution	Gordon et al. (1979), Guandong et al. (2010), and Hayashi et al. (2014)
NPrCAP-conjugated magnetic nanoparticle	10 nm	Targets melanoma cells and exerts chemotherapeutic effects	Hergt et al. (2006, 1998) and Hergt and Dutz (2007)
Antibody-conjugated magnetic nanoparticle	20 nm	Targets human breast cancer and is conjugated with radioactive indium	Hilger et al. (2005)
Antibody or aptamer-conjugated magnetic nanoparticle	10 nm	Targets tumor cells and stabilizes the colloidal solution	Hilger et al. (2003), Hutten et al. (2005), and Ito et al. (2005a, b)
Magnetic nanoparticle with encapsulated antitumor drug	20–30 nm	Controlled drug release	Ito et al. (2003a, 2004)

Republished with permission of Future Medicine Ltd., from Kobayashi et al. (2014), permission conveyed through Copyright Clearance Center, Inc

NPrCAP N-propionyl-cysteaminylphenol

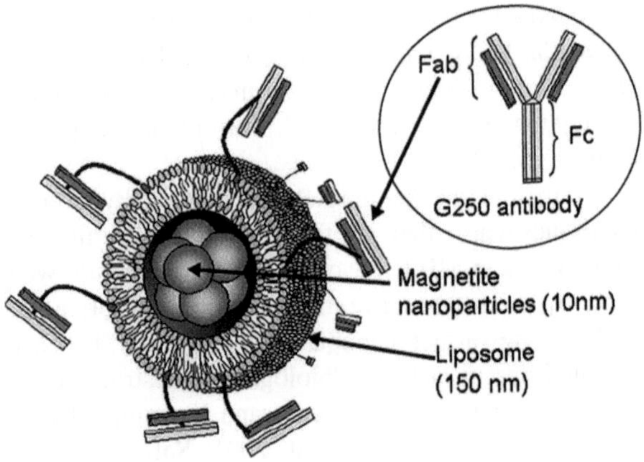

Fig. 10.10 Antibody-conjugated magnetoliposomes (AMLs) for active targeting. Magnetite nanoparticles were wrapped in a neutral liposome, and a G250 antibody was covalently linked to the liposomal surface (Reprinted from Ito et al. 2005a, Copyright (2005) The Society for Biotechnology Japan, with permission from Elsevier)

Table 10.2 Characteristic observables of nanoparticles and alternating magnetic field parameters (Caizer 2014)

Samples	M_s (kA/m)	K (kJ/m^3)	ρ (g/cm^3)	ε	d_{LS} [3] (nm)	d_{CDs} [14] (nm)	η (kgm^{-1} s^{-1})	f (kHz)	H (kA/m)
Fe_3O_4 [13]	477	11	5.24	0.017	35	0.8	7×10^{-4}	100–1,000	10–20

CCL 2014, IOP Publishing. Reproduced by permission of CCL-IOP Publishing

response to specific proteins to stimulate the antitumoral immunity (heat shock protein (HSP) – mediated antitumor Immunity), which would make MHT possible even in tumour metastasis (Schlesinger 1990; Ito et al. 2005a). In Fig. 10.10 the BioC of LPs containing magnetite NPs with antibody G250 for targeting active tumor cells is shown.

Another way, which I have in mind as a possible biocompatibilization of MNPs for cancer therapy through MHT, is through the use of CDs. We can use CDs to decorate the surface of NPs, which could lead to increase in the overall efficiency of the method while minimizing cell toxicity (Fagui et al. 2011; Li et al. 2011; Nigam et al. 2011; Yallapu et al. 2011; Hayashi et al. 2014; Caizer 2014). CDs are fully biocompatible (are natural polysaccharides) being used today in pharmacy and medicine to deliver controlled release drug carriers (Pradhan et al. 2007; Bellia et al. 2009). These form the toroidal cavity guest host for the drugs that can be encapsulated in hydrophobic cavities. CDs are well dispersed in an aqueous solution. Of these, γ-CDs are most stable in aqueous solution and are more suitable for bioconjugation surface of MNPs using chemical agents.

In a recent study, the specific absorption rate in biocompatible Fe_3O_4 NPs monodispersed in an aqueous solution, suitable for MHT, is still being shown (Caizer 2014). The some results are presents below (republished with permission from CCL-IOP Publishing).

"Data used in calculations, according to the formulas in Sect. 10.2, are given in Table 10.2. I considered two cases: (i) Fe_3O_4 encapsulated in LPs and (ii) Fe_3O_4 bioconjugated with CDs, strategy currently used in research on targeted therapy.

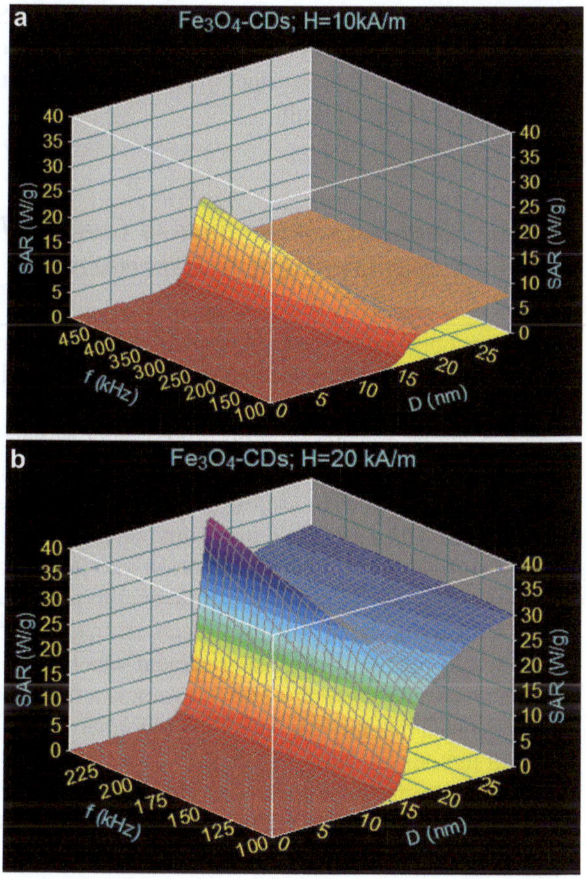

Fig. 10.11 (Color online) 3D diagram of SAR until the biological limit in the case of Fe3O4 nanoparticles bioconjugated with CDs, for magnetic field amplitude of (**a**) 10 kA/m and (**b**) 20 kA/m (Caizer 2014) (CCL 2014, IOP Publishing. Reproduced by permission of CCL-IOP Publishing)

When the NPs are encapsulated in LPs (i), Brown relaxation time becomes much higher than the Néel relaxation time ($\tau_B(167 \times 10^3\ \text{ns}) \gg \tau_N(122\ \text{ns})$), and the relaxation time will be: $\tau = \tau_N/(1 + \tau_N/\tau_B) \cong \tau_N$ (see Sect. 10.2.2). In this case the SAR diagram obtained is similar to that in Fig. 10.5, these being determined only by the Néel dissipation processes.

The results for Fe_3O_4-CDs are shown in Fig. 10.11 for two values of the magnetic field, considering biological accessibility limit, $Hf \leq 5 \times 10^9$ Am^{-1} Hz (Hergt and Dutz 2007). In this case, two magnetic relaxation processes contribute to the SAR, both Néel and Brown, for magnetic diameters greater than 15 nm. Néel relaxation processes become dominant with the increasing frequency to the upper limit at the ~15.5–17 nm. At lower frequencies, such as 150 kHz, contributions to the SAR of the Néel and Brown relaxation processes become comparable and at the frequency of 100 kHz, Brown relaxation prevail, especially in large diameters. When the magnetic field reaches to 20 kA/m (Fig. 10.11b), the highest maximum of SAR increases significantly from ~15 W/g to about ~38 W/g, at the biological limit, for the 17 nm diameter. In this case, the SAR still remains high (~30 W/g) for $D > 22$ nm, due to Brown relaxation.

In conclusion, the 3D study allowed us to optimize SPMHT, in order to apply it in practice: by finding the most suitable bionanoparticles, as material and size, and also as magnetic field parameters (*f*, *H*), in order to obtain the highest SAR within the biological accepted limit."

10.3.4 Main Results Obtained in Cancer Therapy by Using Superparamagnetic Hyperthermia

The results obtained so far, both in vitro and in vivo, have demonstrated to viability of the MHT method (Ito et al. 2003a; Johannsen et al. 2005; Tanaka et al. 2005; Hilger et al. 2005).

For example, Johannsen et al. (2005) demonstrated an in vivo experiment for treatment of prostate cancer done by means of MHT, using a magnetic fluid. In experiment 48 mice were used which had induced prostate cancer. The mice were injected once with a ferrofluid in the tumour, after which they were subjected to two MHT treatments. Ferrofluid was injected into the tumor at a rate of 0.5 ml/cm^3. It was performed with a ferrofluid of magnetite particles in water, which were surfacted with aminosilane, having a mean diameter of 15 nm and a concentration in water of 120 mg/ml. The MHT treatment was made by applying a magnetic field with an amplitude between 0 and 18 kA/m, at a frequency of 100 kHz. The obtained results showed that the magnetic fluid hiperthermia (MFH) treatment leads to inhibition of tumour growth by 44–51%.

Other experiments (Jordan et al. 1999b; Ito et al. 2003a; 2004; Ito et al. 2003b; Hilger et al. 2005; Sunderland et al. 2006; Baldi et al. 2007; Johannsen et al. 2007a) emphasize the viability of the MHT method and the possibility of applying them successfully in the not too distant future for the treatment of various cancers. Ito et al. (2003a) showed the complete regression of mammary carcinoma by *repeating* MHT with magnetite nanoparticles.

A synthesis of tumour regression results obtained by applying MHT and the conditions used are shown in Fig. 10.12 and in the text of Figure (Gazeau et al.

Fig. 10.12 (continued) Although the rectum temperature did not increase by more than 1 °C, tumors injected were heated up to 45 °C (10 °C increase). (*iii*) Prussian blue staining of the tumor after thermotherapy treatment: the injection of the aminosilane-coated particles led to the formation of stable deposit into the tumor: the nanoparticles (in *blue*) were still distributed throughout the tumor after treatment. (*iv*) Effect of treatment temperature on the survival rate after thermotherapy (one round, 30 min). Intratumoral temperature greater than 43 °C significantly increased the lifespan of tumor-bearing rats compared with nontreated controls (survival fraction null at days 8–13, data not shown). The gain of survival was correlated with the intratumoral temperature (Adapted with permission from Jordan et al. 2006). *MCL* magnetic cationic liposomes. (Republished with permission of Future Medicine Ltd., from Gazeau et al. 2008; permission conveyed through Copyright Clearance Center, Inc)

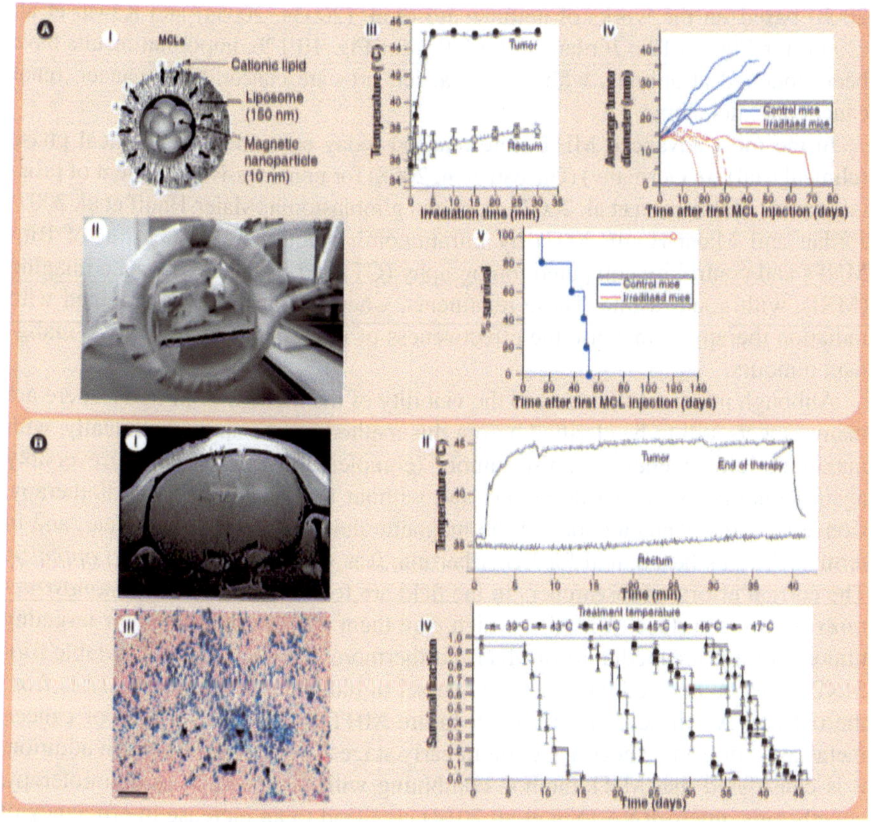

Fig. 10.12 (Color online) examples of tumor regression following magnetic hyperthermia. (**a**) Complete regression of mouse mammary carcinoma by frequent repeated hyperthermia using magnetic cationic liposomes. (*i*) Magnetite nanoparticles were embedded into cationic liposomes (MCLs) and injected directly into 15-mm tumors, which were induced subcutaneously in mice. (*ii*) Alternating magnetic field (AMF) of frequency 118 kHz was generated by a horizontal coil, where the mouse was placed. (*iii*) Temperatures in the rectum and at the surface of the tumor were recorded with an optical fiber probe during AMF exposure. For each irradiation AMF, the tumor was heated to 45 °C for 30 min by controlling the AMF intensity. (*iv*) The therapeutic effect of tumor irradiation. MCLs were injected directly into 15-mm diameter tumors (2 mg magnetite). Tumors were irradiated for 30 min, repetitively (one to six rounds). Compared with control (*blue lines*), the tumor growth was stopped after AMF irradiation (*red lines*). Each line represent a single tumor ($n = 5$ for both control and treated mice). After repeated hyperthermia, complete regression of the tumor was observed. (*v*) Percentage of survival for the same group of mice. Control mice: *full circles*; treated mice: *open circles*. All the treated mice survived for 120 days after treatment, with no regrowth of the tumor (Adapted with permission from Ito et al. 2003a, 2005a; Kawai et al. 2006). (**b**) Effect of magnetic hyperthermia on rat malignant glioma. (*i*) Tumors were induced by implantation of rat glioma RG-2-cells into the brains of rats and were visualized by MRI (T1-weighted MR image). (*ii*) Magnetic nanoparticles with two different coatings and sizes were administered in to the tumors by stereotactic injection (20 µl at 2 mol/l iron concentration). AMF (100 kHz, 0–18 kA/m) was generated using an applicator system (MFH 12-TS®, MagForce Nanotechnology, Germany). Temperature into the tumor was measured using a fiber optic thermometry device.

2008) based on the works of authors: Ito et al. (2003a, 2005a) and Kawai et al. (2006) for Fig. 10.12a; Jordan et al. (2006) for Fig. 10.12b. Important results have been obtained at animals with breast cancer, prostate cancer, brain cancer, renal cancer and melanoma.

In the most advanced MHT state reached today even in the preclinical phase (clinical trial) (in Germany) (Gazeau et al. 2008) for non-invasive treatment of prostate cancer (Johannsen et al. 2007a, b, c) and glioblastoma (Maier-Hauff et al. 2007; Jordan and Maier-Hauff 2007) by intratumoral injection of suspension of Bio-MNPs and control by computed tomography (CT) and medical resonance imaging (MRI), with good results. These treatments, when applicable, are combined with radiation therapy to increase the effectiveness of irreversible destruction of malignant tumours.

Although many results attest to the viability of the method of SPMHT, there are many aspects still to be clarified before this method can be applied clinically, with maximum effectiveness against tumors (complete destruction and irreversibly against tumors) without side effects and without a longer need for radiotherapy. Concerning the characteristics of the magnetic field, things are quite clear, and in terms of MNPs best suited for hyperthermia, it is still necessary here to *optimize*. The current efforts of researchers in the field are focused heavily on *bionanostructures* from the surface of MNPs, which give them *specific* properties for targeting tumor cells and are mediators of MHT. Furthermore, the suspension of suitable Bio-MNPs must be injected into the blood vessel of hidden tumours, inaccessible from the outside. All these could result in future MHT in advanced stages of cancer, metastases, or even cancer detection in early stages (Levy et al. 2010). In addition, it is considered that MHT method combining with others, such as chimioterapy, genetherapy, micro-RNA (Yin et al. 2014) etc., lead to an increase in efficiency of destroying tumorus cells.

10.4 Conclusions

MHT, and in particular, SPMHT based on temperature increase in a tissue at ~43 °C by Néel – Brown relaxation in SPMNPs that are biocompatible and injected inside, it has a great potential for future use in cancer therapy. According to the results obtained so far in vitro and in vivo, and recently even in *clinical trial* in the treatment of prostate cancer and glioblastoma, the chances of using MHT in the not so far future are promising. SPMHT is a advanced stage of the alternative cancer therapy, non-invasive and with very low toxicity compared to chemotherapy and radiotherapy, which are highly toxic and sometimes in advanced stages of cancer are ineffective. SPMHT efficiency depends on: (i) using SPMNPs best suited for this experiment (type, size, size distribution, magnetic anisotropy, etc.), (ii) AMF parameters and how to apply it on tumours (amplitude, frequency, time of exposure, temperature control, etc.), (iii) biological nanostructures coating of SPMNPs to make them biocompatible with the tissue and increase their affinity to targeted tumor cells. A future perspective of increasing the loss power in specific tumours in a short time and reduce the side effects on healthy cells, is also an avantage resulting from using SPMNPs of FeM metals instead of FiM oxides. While the first two issues (i) and (ii) are well developed and addressed in the contemporary research, the third issue (iii) has high development potential.

Future advances in bionanotehnology and finding chemical agents most suitable for coating the MNPs with high specificity and affinity for different target tumors, opens real prospects for the SPMHT method to become, in the not too distant future, an alternative method, non-invasive, the most effective in cancer therapy.

References

Aharoni A. Thermal agitation of single domain particles. Phys Rev A. 1964;135:447–9.

Back CH, Weller D, Heidmann J, Mauri D, Guarisco D, Garwin EL, Siegmann HC. Magnetization reversal in ultrashort magnetic field pulses. Phys Rev Lett. 1998;81:3251.

Bacri JC, Perzinski R, Salin D, Cabuil V, Massart R. Magnetic colloidal properties of ionic ferrofluids. J Magn Magn Mater. 1986;62:36–46.

Baker I, Zeng Q, Li W, Sullivan CR. Heat deposition in iron oxide and iron nanoparticles for localized hyperthermia. J Appl Phys. 2006;99:08H106–08H106-3.

Baldi G, Bonacchi D, Innocenti C, Lorenzi G, Sangregorio C. Cobalt ferrite nanoparticles: the control of the particle size and surface state and their effects on magnetic properties. J Magn Magn Mater. 2007;311:10–16.

Bean CP, Livingston LD. Superparamagnetism. J Appl Phys. 1959;30:S120–9.

Bellia F, La Mendola D, Pedone C, Rizzarelli E, Saviano M, Vecchio G. Selectively functionalized cyclodextrins and their metal complexes. Chem Soc Rev. 2009;38:2756–81.

Berry CC, Wells S, Charles S, Curtis ASG. Dextran and albumin derivatised iron oxide nanoparticles: influence on fibroblasts in vitro. Biomaterials. 2003;24:4551–7.

Brown WF Jr. Thermal fluctuations of a single-domain particle. Phys Rev. 1963;130:1677–86.

Caizer C. T2 law for magnetite-based ferrofluids. J Phys Condens Matter. 2003;15:765–76.

Caizer C. Nano-biomagnetism. West University of Timisoara, Timisoara; 2010

Caizer C. Computational study on superparamagnetic hyperthermia with biocompatible SPIONs to destroy the cancer cells. J Phys Conf Ser. 2014;521:012015–4.

Datta NR, Ordóñez SG, Gaipl US, Paulides MM, Crezee H, Gellermann J, Marder D, Puric E, Bodis S. Local hyperthermia combined with radiotherapy and-/or chemotherapy: recent advances and promises for the future. Cancer Treat Rev. 2015;41:742–53.

Fagui A, Dalmas F, Lorthioir C, Wintgens V, Volet G, Amiel C. Well-defined core-shell nanoparticles containing cyclodextrin in the shell: a comprehensive study. Polymer. 2011;52:3752–61.

Fortin JP, Gazeau F, Wilhelm C. Intracellular heating of living cells through Néel relaxation of magnetic nanoparticles. Eur Biophys J. 2008;37:223–8.

Gazeau F, Lévy M, Wilhelm C. Optimizing magnetic nanoparticle design for nanothermotherapy. Nanomedicine. 2008;3:831–44.

Gordon RT, Hines JR, Gordon D. Intracellular hyperthermia. A biophysical approach to cancer treatment via intracellular temperature and biophysical alterations. Med Hypotheses. 1979;5:83–102.

Guandong Z, Yifeng L, Baker I. Surface engineering of core/shell iron/iron oxide nanoparticles from microemulsions for hyperthermia. Mater Sci Eng C. 2010;30:92–7.

Habib AH, Ondeck CL, Chaudhary P, Bockstaller MR, McHenry ME. Evaluation of iron-cobalt/ferrite core-shell nanoparticles for cancer thermotherapy. J Appl Phys. 2008;103:07A307–07A307-3.

Hayashi K, Nakamura M, Miki H, Ozaki S, Abe M, Matsumoto T, Sakamoto W, Yogo T, Ishimura K. Magnetically responsive smart nanoparticles for cancer treatment with a combination of magnetic hyperthermia and remote-control drug release. Theranostics. 2014;4:834–44.
Hergt R, Dutz S. Magnetic particle hyperthermia – biophysical limitations of a visionary tumour therapy. J Magn Magn Mater. 2007;311:187–92.
Hergt R, Andra W, d'Ambly C, Hilger I, Kaiser W, Richter U, Schmidt H. Physical limits of hyperthermia using magnetite fine particles. IEEE Trans Magn. 1998;34:3745–54.
Hergt R, Dutz S, Muller R, Zeisberger M. Magnetic particle hyperthermia: nanoparticle magnetism and materials development for cancer therapy. J Phys Condens Matter. 2006;18:S2919–34.
Hilger I, Fruhauf S, Linss W, Hiergeist R, Andra W, Hergt R, Kaiser WA. Cytotoxicity of selected magnetic fluids on human adenocarcinoma cells. J Magn Magn Mater. 2003;261:7–12.
Hilger I, Hergt R, Kaiser WA. Towards breast cancer treatment by magnetic heating. J Magn Magn Mater. 2005;293:314–9.
Hutten A, Sudfeld D, Ennen I, Reiss G, Wojczykowski K, Jutzi P. Ferromagnetic FeCo nanoparticles for biotechnology. J Magn Magn Mater. 2005;293:93–101.
Ito A, Tanaka K, Honda H, Abe S, Yamaguchi H, Kobayaschi T. Complete regression of mouse mammary carcinoma with a size greater than 15 mm by frequent repeated hyperthermia using magnetite nanoparticles. J Biosci Bioeng. 2003a;96:364–9.
Ito A, Matsuoka F, Honda H, Kobayashi T. Heat shock protein 70 gene therapy combined with hyperthermia using magnetic nanoparticles. Cancer Gene Ther. 2003b;10:918–25.
Ito A, Kuga Y, Honda H, Kikkawa H, Horiuchi A, Watanabe Y, Kobayashi T. Magnetite nanoparticle-loaded anti-HER2 immunoliposomes for combination of antibody therapy with hyperthermia. Cancer Lett. 2004;212:167–75.
Ito A, Shinkai M, Honda H, Kobayashi T. Medical application of functionalized magnetic nanoparticles. J Biosci Bioeng. 2005a;100:1–11.
Ito A, Ino K, Kobayashi T, Honda H. The effect of RGD peptide-conjugated magnetite cationic liposomes on cell growth and cell sheet harvesting. Biomaterials. 2005b;26:6185–93.
Jacobs IS, Bean CP. Fine particles, thin films and exchange anisotropy. In: Rado GT, Suhl H, editors. Magnetism, vol. III. New York: Academic; 1963. p. 271–350.
Johannsen M, Thiesen B, Jordan A, Taymoorian K, Gneveckow U, Waldofner N. Magnetic fluid hyperthermia (MFH) reduces prostate cancer growth in the orthotopic Dunning R3327 rat model. Prostate. 2005;64:283–92.
Johannsen M, Gneveckow U, Thiesen B, Taymoorian K, Cho CH, Waldofner N, Scholz R, Jordan A, Loening SA, Wust P. Thermotherapy of prostate cancer using magnetic nanoparticles: feasibility, imaging, and three-dimensional temperature distribution. Eur Urol. 2007a;52:1653–62.
Johannsen M, Gneveckow U, Taymoorian K, Thiesen B, Waldöfner N, Scholz R, Jung K, Jordan A, Wust P, Loening SA. Morbidity and quality of life during thermotherapy using magnetic nanoparticles in locally recurrent prostate cancer: results of a prospective phase I trial. Int J Hyperth. 2007b;23:315–23.
Johannsen M, Gneveckow U, Taymoorian K, Cho CH, Thiesen B, Scholz R, Waldöfner N, Loening SA, Wust P, Jordan A. Thermal therapy of prostate cancer using magnetic nanoparticles. Actas Urol Esp. 2007c;31:660–7.
Jordan A, Maier-Hauff K. Magnetic nanoparticles for intracranial thermotherapy. J Nanosci Nanotechnol. 2007;7:4604–6.
Jordan A, Scholz R, Wust P, Schirra H, Schiestel T, Schmidt H, Felix RJ. Endocytosis of dextran and silan-coated magnetite nanoparticles and the effect of intracellular hyperthermia on human mammary carcinoma cells in vitro. J Magn Magn Mater. 1999a;194:185–96.
Jordan A, Scholz R, Wust P, Fähling H, Felix R. Magnetic fluid hyperthermia (MFH): cancer treatment with AC magnetic field induced excitation of biocompatible superparamagnetic nanoparticles. J Magn Magn Mater. 1999b;201:413–9.
Jordan A, Scholz R, Maier-Hauff K, Landeghem FKH, Waldoefner N, Teichgraeber U, Pinkernelle J, Bruhn H, Neumann F, Thiesen B, Deimling A, Felix R. The effect of thermotherapy using magnetic nanoparticles on rat malignant glioma. J Neuro-Oncol. 2006;78:7–14.

Kappiyoor R, Liangruksa M, Ganguly R, Puri IK. The effects of magnetic nanoparticle properties on magnetic fluid hyperthermia. J Appl Phys. 2010;108:094702.

Kawai N, Ito A, Nakahara Y, Honda H, Kobayashi T, Futakuchi M, Shirai T, Tozawa K, Kohri K. Complete regression of experimental prostate cancer in nude mice by repeated hyperthermia using magnetite cationic liposomes and a newly developed solenoid containing a ferrite core. Prostate. 2006;66:718–27.

Kobayashi T, Kakimi K, Nakayama E, Jimbow K. Antitumor immunity by magnetic nanoparticle-mediated hyperthermia. Nanomedicine. 2014;9:1715–26.

Lacava ZGM, Azevedo RB, Martins EV, Lacava LM, Freitas MLL, Garcia VAP, Rebula CA, Lemos APC, Sousa MH, Tourinho FA, Da Silva MF, Morais PC. Biological effects of magnetic fluids: toxicity studies. J Magn Magn Mater. 1999;201:431–4.

Le Renard PE, Jordan O, Faes A, Petri-Fink A, Hofmann H, Rufenacht D, Bosman F, Buchegger F, Doelker E. The in vivo performance of magnetic particle-loaded injectable, in situ gelling, carriers for the delivery of local hyperthermia. Biomaterials. 2010;31:691–705.

Le Renard PE, Lortz R, Senatore C, Rapin JP, Buchegger F, Petri-Fink A, Hofmann H, Doelker E, Jordan O. Magnetic and in vitro heating properties of implants formed in situ from injectable formulations and containing superparamagnetic iron oxide nanoparticles (SPIONs) embedded in silica microparticles for magnetically induced local hyperthermia. J Magn Magn Mater. 2011;323:1054–63.

Lee SW, Bae S, Takemura Y, Shim I-B, Kim TM, Kim J, Lee HJ, Zurn S, Kim CS. Self-heating characteristics of cobalt ferrite nanoparticles for hyperthermia application. J Magn Magn Mater. 2007;310:2868–70.

Levy A, Dayan A, Ben-David M, Gannot I. A new thermography-based approach to early detection of cancer utilizing magnetic nanoparticles theory simulation and in vitro validation. Nanomedicine. 2010;6:786–96.

Li R, Liu S, Zhao J, Otsuka H, Takahara A. Preparation of superparamagnetic β-cyclodextrin functionalized composite nanoparticles with core–shell structures. Polym Bull. 2011;66:1125–36.

Maier-Hauff K, Rothe R, Scholz R, Gneveckow U, Wust P, Thiesen B, Feussner A, Deimling A, Waldoefner N, Felix R, Jordan A. Intracranial thermotherapy using magnetic nanoparticles combined with external beam radiotherapy: results of a feasibility study on patients with glioblastoma multiforme. J Neuro-Oncol. 2007;81:53–60.

Malhi S, Dixit K, Sohi H, Shegokar R. Expedition of liposomes to intracellular targets in solid tumors after intravenous administration. J Pharm Investig. 2013;43:75–87.

Mart RJ, Liem KP, Webb SJ. Creating functional vesicle assemblies from vesicles and nanoparticles. Pharm Res. 2009;26:1701–10.

McNerny KL, Kim Y, Laughlin DE, McHenry ME. Chemical synthesis of monodisperse γγ-Fe–Ni magnetic nanoparticles with tunable curie temperatures for self-regulated hyperthermia. J Appl Phys. 2010;107:09A312–4.

Mehdaoui B, Meffre A, Lacroix LM, Carrey J, Lachaize S, Gougeon M, Respaud M, Chaudret B. Large specific absorption rates in the magnetic hyperthermia properties of metallic iron nanocubes. J Magn Magn Mater. 2010a;322:L49–52.

Mehdaoui B, Meffre A, Lacroix LM, Carrey J, Lachaize S, Respaud M, Gougeon M, Chaudret B. Magnetic anisotropy determination and magnetic hyperthermia properties of small Fe nanoparticles in the superparamagnetic regime. J Appl Phys. 2010b;107:09A324.

Mendoza-Resendez R, Bomati-Miguel O, Morales MP, Bonville P, Serna CJ. Microstructural characterization of ellipsoidal iron metal nanoparticles. Nanotechnology. 2004;15:S254–8.

Molday RS, MacKenzie D. Immunospecific ferromagnetic iron-dextran reagents for the labeling and magnetic separation of cells. J Immunol Methods. 1982;52:353–67.

Naqvi S, Samim M, Abdin MZ, Ahmed FJ, Maitra AN, Prashant CK, Dinda AK. Concentration-dependent toxicity of iron oxide nanoparticles mediated by increased oxidative stress. Int J Nanomedicine. 2010;5:983–9.

Néel L. Théorie du traînage magnétique des ferromagnétiques en grains fins avec application aux terres cuites. Ann Geophys. 1949;5:99–136.

Nigam S, Barick KC, Bahadur D. Development of citrate-stabilized Fe_3O_4 nanoparticles: conjugation and release of doxorubicin for therapeutic application. J Magn Magn Mater. 2011;323:237–43.

O'Grady K, Bradbury A. Particle size analysis in ferrofluids. J Magn Magn Mater. 1994;39:91–4.

Ondeck CL, Habib AH, Ohodnicki P, Miller K, Sawyer CA. Theory of magnetic fluid heating with an alternating magnetic field with temperature dependent materials properties for self-regulated heating. J Appl Phys. 2009;105:07B324–07B324-3.

Pankhurst QA, Connolly J, Jones SK, Dobson J. Applications of magnetic nano-particles in biomedicine. J Phys D Appl Phys. 2003;36:R167–81.

Portet D, Denizot B, Rump E, Lejeune JJ, Jallet P. Nonpolymeric coatings of iron oxide colloids for biological use as magnetic resonance imaging contrast agents. J Colloid Interface Sci. 2001;238:37–42.

Pradhan P, Giri J, Samanta G, Sarma HD, Mishra KP, Bellare J, Banerjee R, Bahadur D. Comparative evaluation of heating ability and biocompatibility of different ferrite-based magnetic fluids for hyperthermia application. J Biomed Mater Res B Appl Biomater. 2007;81B:12–22.

Prasad NK, Rathinasamy K, Panda D, Bahadur D. Mechanism of cell death induced by magnetic hyperthermia with nanoparticles of γ-$Mn_xFe_{2-x}O_3$ synthesized by a single step process. J Mater Chem. 2007;17:5042–51.

Purushotham S, Ramanujan RV. Modeling the performance of magnetic nanoparticles in multimodal cancer therapy. J Appl Phys. 2010;107:114701–114701-9.

Rosensweig RE. Heating magnetic fluid with alternating magnetic field. J Magn Magn Mater. 2002;252:370–4.

Safarik I, Safarikova M. Magnetic nanoparticles and biosciences. Monatchefte Chem. 2002;133:737–59.

Schlesinger MJ. Heat shock proteins. J Biol Chem. 1990;265:12111–4.

Shinkai M, Ito A. Functional magnetic particles for medical application. Adv Biochem Eng Biotechnol. 2004;91:191–220.

Shliomis MI. Magnetic fluids. Sov Phys Usp. 1974;17:153–69.

Sunderland CJ, Steiert M, Talmadge JE, Derfus AM, Barry SE. Targeted nanoparticles for detecting and treating cancer. Drug Dev Res. 2006;67:70–93.

Tanaka K, Ito A, Kobayashi T, Kawamura T, Shimada S, Matsumoto K, Saida T, Honda H. Intratumoral injection of immature dendritic cells enhances antitumor effect of hyperthermia using magnetic nanoparticles. Int J Cancer. 2005;116:624–33.

Veiseh O, Gunn JW, Zhang M. Design and fabrication of magnetic nanoparticles for targeted drug delivery and imaging. Adv Drug Deliv Rev. 2010;62:284–304.

Wu SY, Yang KC, Tseng CL, Chen JC, Lin FH. Silica-modified Fe-doped calcium sulfide nanoparticles for in vitro and in vivo cancer hyperthermia. J Nanopart Res. 2011;13:1139–49.

Yallapu MM, Othman SF, Curtis ET, Gupta BK, Jaggi M, Chauhan SC. Multi-functional magnetic nanoparticles for magnetic resonance imaging and cancer therapy. Biomaterials. 2011;32:1890–905.

Yin PT, Shah BP, Lee K-B. Combined magnetic nanoparticle-based microRNA and hyperthermia therapy to enhance apoptosis in brain cancer cells. Small. 2014;10:4106–12.

Zeisbergera M, Dutza S, Muller R, Hergt R, Matoussevitch N, Bonnemann H. Metallic cobalt nanoparticles for heating applications. J Magn Magn Mater. 2007;311:224–7.

Zhang Y, Kohler N, Zhang M. Surface modification of superparamagnetic magnetite nanoparticles and their intracellular uptake. Biomaterials. 2002;23:1553–61.

Zhang XX, Wen GH, Xiao G, Sun S. Magnetic relaxation of diluted and self-assembled cobalt nanocrystals. J Magn Magn Mater. 2003;261:21–8.

Chapter 11
Biocompatible Magnetic Oxide Nanoparticles with Metal Ions Coated with Organic Shell as Potential Therapeutic Agents in Cancer

Costica Caizer, Alice Sandra Buteica, and Ion Mindrila

Abstract The magnetic nanoparticles (MNPs) of oxide with metal ions (MeIs-O) encapsulated with biological membranes or decorated with different organic shells are widely used in nano-biotechnology and nanomedicine, either as drug carriers in diagnosis and therapeutic biomedical fields or as a result of their specific magnetic effects on biological tissues. The MNP applications in pharma are closely related with their coating, shape, surface charge, core, and hydrodynamic size. The main characteristic of MNPs is the possibility to manipulate them with an external magnetic field (e-MF), which strongly depends on the magnetic structure of nanoparticles (NPs), multi-domain structure, in the case of NPs having sizes greater than approximately 30 nm or single domain in the case of NPs of sizes generally below 30 nm (depending on the type of NPs). A special structure of MNPs is the super paramagnetic (SPM) state when the magnetization of single-domain NPs is not stable and fluctuates along the easy magnetization axis under the action of thermal activation. This effect generally occurs when NPs have sizes smaller than approximately 15 nm. Depending on the targeted application in biotechnology/nanomedicine, only the NPs having the magnetic characteristics suitable for the type of application envisaged should be considered, in order to obtain the expected effect. In this chapter, we present the specific magnetic properties of MNPs depending on their size, the physical principle of handling NPs with an e-MF, magnetization of NPs, biocompatible MNPs (BC-MNPs) with salicylic acid (SA) and other organic

C. Caizer (✉)
Department of Physics, Faculty of Physics, West University of Timisoara, Bv. V. Parvan no. 4, 300223 Timisoara, Romania
e-mail: ccaizer@physics.uvt.ro; costica.caizer@e-uvt.ro

A.S. Buteica
Department of Pharmacy, Faculty of Pharmacy, University of Medicine and Pharmacy of Craiova, P. Rares no. 2, 200349 Craiova, Romania

I. Mindrila
Department of Morphological Sciences, Faculty of Medicine, University of Medicine and Pharmacy of Craiova, P. Rares no. 2, 200349 Craiova, Romania

M. Rai, R. Shegokar (eds.), *Metal Nanoparticles in Pharma*,
DOI 10.1007/978-3-319-63790-7_11

coating, vascular nanoblocking of tumors, in vivo results using the chick embryo chorioallantoic membrane (CAM) model, and cytotoxicity of SA-MNPs.

Keywords Metal ions • Oxides • Magnetic nanoparticles • Biocompatibility • Cytotoxicity • Tumors

Nomenclature

Bio-MNPs	Biocompatible magnetic nanoparticles
CAM	(chick embryo) Chorioallantoic membrane (model)
e-MF	External magnetic field
FeO-MNPs	Iron oxide magnetic nanoparticles
FiM	Ferrimagnetic
FiMNPs	Ferrimagnetic nanoparticles
HR-TEM	High-resolution transmission electron microscopy
LDH	Lactate dehydrogenase
MeIs	Metal ions
MeIs-O	Oxide with metal ions
MF	Magnetic field
MNPs	Magnetic nanoparticles
MONPs	Magnetic oxide nanoparticles
MRI	Magnetic resonance imaging
MTS	3-(4,5-Dimethylthiazol-2-yl)-5-(3-carboxymethoxyphenyl)-2-(4-sulfophenyl)-2*H*-tetrazolium
MTT	3-(4,5-Dimethylthiazol-2-yl)-2,5-diphenyltetrazolium bromide
NPs	Nanoparticles
ROS	Reactive oxygen species
SA	Salicylic acid
SA-Fe_3O_4	Fe_3O_4 (magnetite) nanoparticles coated with salicylic acid
SA-MNPs	Magnetic nanoparticles coated with salicylic acid
SPM	Superparamagnetic
TEM	Transmission electron microscopy

11.1 Introduction

Since the early twentieth century, ultrafine particles have been extensively studied due to their different properties that they present in comparison with the bulk material, and their synthesis became a scientific field of interest. The impact of nanotechnology on the global economy estimated for 2015 was 3.1 trillion dollars (Schmidt 2009). A significant challenge related to the use of the magnetic nanoparticles (MNPs) as therapeutic tools is their in vivo behavior. In vivo behavior of MNPs can now be designed or tailored on demand by various nanofabrication techniques that can control the size, shape, composition, morphology, or surface charge.

Engineering these physicochemical properties of MNPs can directly influence their properties such as responsivity toward external stimuli, including magnetic field (MF) or light, drug release characteristic, blood circulation time, immunogenicity, therapeutic toxicity and their subsequent pharmacokinetics, and biodistribution (Sun et al. 2008; Zhang et al. 2008). After functionalization/envelopment with various coatings, MNPs can be conveniently positioned in vitro and in vivo, by applying an MF. Among the materials commonly used for coating the magnetic core are surfactants, polymers, antibodies, enzymes, specific proteins, and ligands, which make it possible to link other biologically active compounds or receptors on the cell surface. Biocompatibility, biodistribution, and bioavailability evaluation is a mandatory testing stage that MNPs with use in various medical fields have to go through. The lack of standardized methods for rapid assessment of nanoparticles (NPs) impacts on human tissue functions (Hofmann-Amtenbrink et al. 2015) transform in vitro methods and preclinical models in basic tools currently used in testing biological properties of MNPs. In vitro cytotoxicity evaluation and in vivo experimental studies have shown that changes in chemical composition, molecular structures, or surface properties and potential toxicity of NPs influence their work (Jiang et al. 2008). The most used in vivo models to study and evaluate the MNPs cytotoxicity, biodistribution, tissue damage, and biomedical potential were zebra fish model, chorioallantoic membrane model (CAM model), and also murine model.

For MNPs, in order to present a favorable therapeutic potential, they must have (Chomoucka et al. 2010) (i) a magnetic core (to drive the NPs to target's vicinity and also the particle or hyperthermic effects of drug release to be increased at high temperatures), (ii) a recognition layer (to which are attached proper receptor molecule/recognition), and (iii) a therapeutic load (a therapeutic substance inside the pores or inner cavities that envelope the NPs).

Consequently, the biomedical applications of the MNPs in general can be separated into two big categories: *diagnostic* and *therapeutic* with potential promising technologies and potential uses in biosensing, medical imaging, and administration of drugs and in particular, for the treatment of cancer. With its own design, NPs can cover both diagnostic and therapeutic functions, the so-called individualized therapy. Regarding the function of therapy, recent advances in nanotechnology have already been implemented in the pharmaceutical industry.

Investigation on MNPs for their biomedical applications' perspective dates back to more than 40 years and has opened a wide field of applications in medicine (Pouliquen et al. 1989; Cunningham et al. 2005; Ito et al. 2005; Varshney et al. 2005; Jurgons et al. 2006): in magnetic separation techniques, as contrast agents in magnetic resonance imaging (MRI) (Combidex®, Resovist®, Endorem®, Sinerem®), and to produce local hyperthermia and magnetic carriers targeted for multiple transport systems of drugs. Drug delivery and imaging are the most important nanomedical fields in which the MNPs are used successfully (Kim et al. 2005;

Sun et al. 2008). MNPs have attracted scientists' attention because of their bioapplications as heating mediators for cancer therapy (hyperthermia treatment). Magnetic fluid hyperthermia consist of targeting the magnetic nanoparticles to malign tumors and applying an MF oscillating in the tumor region, the NPs disperse the energy in the form of heat energy, and the rise of temperature induces apoptosis of cancer cells. In order to obtain an efficient target of cancerous tissues, NPs must be systemic administered to obtain the following: to extravagate from the bloodstream and pass through the extracellular matrix of the tumor; to bind cells and cross the membrane surface to enter into the targeted cells and linked to intracellular sites (Brannon-Peppas and Blanchette 2004).

In the scientific literature, there are also mention of other innovative MNPs, for example, mutational analysis of the gene for detection of the pathogen immunological agent, separation of nucleic acids and cells, protein purification, and based on the magnetic force, tissue engineering (Pankhurst et al. 2003). MNPs embedded in the cell and routed away from MFs can apply tensile forces which would stretch the neuronal membrane, and it would trigger and lengthen the axons and would lead to a goal without toxic effects on cells (Riggio et al. 2012). The MNPs bind around the nerve, applying an MF, resulting in a mechanical strain that stimulate nerve renewal in the necessary direction by the magnetic force.

Also, the potential application of magnetic techniques in the pharmaceutical industry is growing, especially thanks to the contribution of the MNPs to improving the kinetics and distribution of drugs and in this way opening the prospect of safer and more specific therapies. With the discovery of NPs, as any new biomedical finding, the compromise risk versus benefit should be measured to assess whether risks can be justified.

Magnetic nanoparticles of iron oxide (FeO-MNPs) with metal ions (MeIs), such as Fe_3O_4 (magnetite) and its oxidized form γ-Fe_2O_3 (maghemite), were subjected to a number of research studies because they are nontoxic and are thermally stable (Wu et al. 2008). Besides, the FeO-MNPs may be regarded as an indicator of biosafety because they do not cause oxidative stress and can be used at high doses (Hohnholt et al. 2011).

In this chapter, first, we present some important magnetic aspects which must be considered when using magnetic oxide nanoparticles (MONPs) with MeIs in pharmaceuticals and biomedicine, such as (i) the magnetization of ferrimagnetic nanoparticles (FiMNPs), considering the distribution of MeIs in the tetrahedral and octahedral magnetic lattices, (ii) the magnetic structure and the ferrimagnetic (FiM) and superparamagnetic (SPM) behavior of the NPs depending on their size, (iii) the stability issues of the MNP suspensions in pharmaceutical and biomedical applications, and (iv) the handling and control of the MNPs with an external magnetic field (e-MF), uniform or with a field gradient. We also present the biocompatible MNPs (bio-MNPs) with salicylic acid (SA) and other organic coating, vascular nanoblocking of tumors, in vivo results using the chick embryo chorioallantoic membrane (CAM) model, and cytotoxicity of SA-MNPs.

11.2 Magnetic Properties of Magnetic Oxide Nanoparticles with Metal Ions

MONPs with potential applications in pharma and biomedicine (Safarik and Safarikova 2002; Tartaj et al. 2003; Pankhurst et al. 2003; Shinkai and Ito 2004; Ito et al. 2005) are those containing different MeIs, generally divalent or trivalent, group 3d, such as iron, nickel, cobalt, zinc, manganese, chromium, and copper (Fe, Ni, Co, Zn, Mn, Cr, Cu); ions of the rare earth group 4f as dysprosium, gadolinium, terbium, holmium, erbium, and thulium (Dy, Gd, Tb, Ho, Er, Tm) (Caizer 2004); or ions of other groups, such as magnesium, cadmium (Mg, Cd), or sometimes platinum, gold, silver, palladium (Pt, Au, Ag, Pd), etc. The compounds mostly used are those that have the general chemical formula $X^{2+}Y_2^{3+}O_4^{2-}$ (Smit and Wijin 1961; Caizer 2013), where X^{2+} is a bivalent MeI (Me^{2+}: Ni^{2+}, Zn^{2+}, Cr^{2+} etc.; sometimes it can be also Fe^{2+}) and Y_2^{3+} is a trivalent MeI, usually Fe^{3+}. The most representative compound is the magnetite $Fe^{2+}Fe_2^{3+}O_4^{2-}$. The final positive electric charge on a cation X and two cations Y is 8 units, with the most common valence combination of $2^+(X) - 3^+(Y)$, for example, $Ni^{2+}Fe_2^{3+}O_4^{2-}$, $Co^{2+}Fe_2^{3+}O_4^{2-}$, etc., also called ferrite (ferrite nickel, cobalt ferrite).

The crystal structure of these compounds with MeIs is one of a spinel type (Smit and Wijin 1961) (as in the $MgAl_2O_4$ spinel mineral). The elementary unit cell contains eight molecules of XY_2O_4 and has a cubic symmetry (Fig. 11.1c, d) (Cullity and Graham 2009). The distribution of ions is present below (Cullity and Graham 2009; Caizer 2013).

The structure of the spinel is formed by the O^{-2} ions (anions) which form a face-centered cubic lattice (fcc) with a filling factor of 0.74 (Fig. 11.1d). Cations (MeIs), with size smaller than the ions of oxygen, are disposed in an orderly manner in the interstices, along the spatial diagonals in *tetrahedral* positions (Fig. 11.1a) or in the middle of the cube edges in *octahedral* positions (Fig. 11.1b). Interstices (between the oxygen ions) are therefore of two types: *tetrahedral*, or type A, and *octahedral*, or type B. In the tetrahedral arrangement, the MeI of type A (Me^{2+}) is surrounded by four ions O^{2-} (the closest ones) placed on the peaks of a regular tetrahedron (Fig. 11.1a), while in the octahedral arrangement, the MeI type B (Me^{3+}) is surrounded by six ions O^{2-} (neighbors' first order) placed on the peaks of a regular octahedron (Fig. 11.1b).

The MeIs in tetrahedral positions form a cube having the edge equal to half the edge of the cube formed by the oxygen ions. In the unit cell of the spinel, there are 64 tetrahedral interstices and 32 octahedral interstices, of which only 8 and 16, respectively, are filled with the cations, the rest being free.

Most of the MeIs have their magnetic moments. In Table 11.1, the value of the magnetic moments is given for some metal ions, expressed in Bohr magneton μ_B (Smit and Wijin 1961) ($\mu_B = e\hbar/2m_0$, where e is the electron charge, $\hbar = h/2\pi$, h is Planck's constant ($h = 6,63 \cdot 10^{-34} J \cdot s$), m_0 is the mass of the electron). Bohr magneton has the following value: $\mu_B = 9,27 \times 10^{-24} J \cdot T^{-1}$.

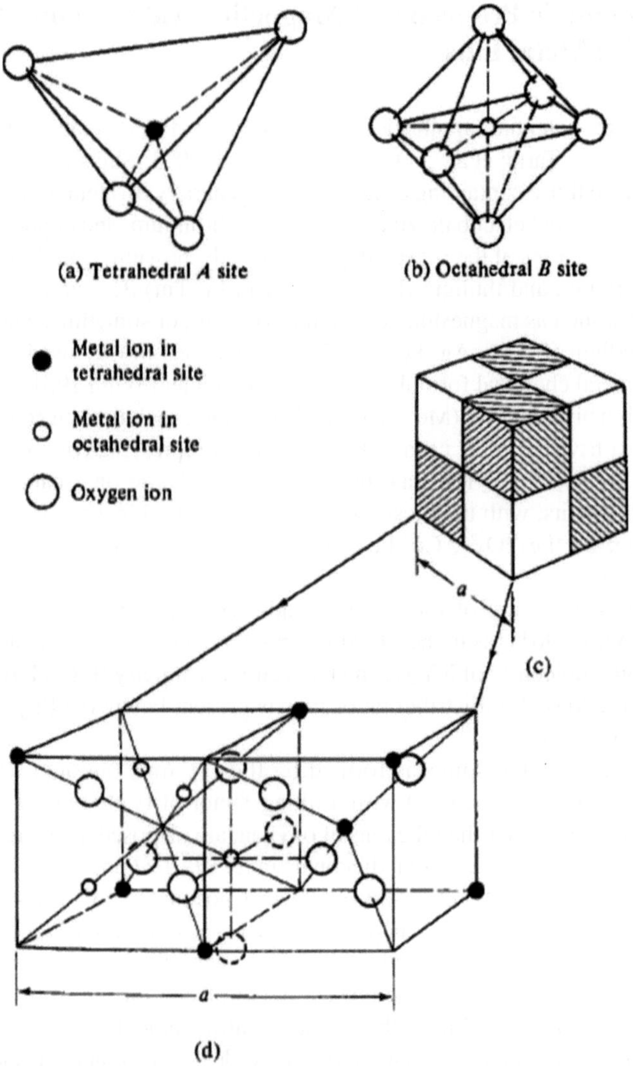

Fig. 11.1 Crystal structure of a cubic ferrite (**a**) metal and oxygen ions in tetrahedral lattice; (**b**) metal and oxygen ions in octahedral lattice; (**c**) cubic structure of unit cell; (**d**) distribution of metal and oxygen ions in two minor cubes along "a" lattice parameter (see fig. (**c**)). (Republished with permission from John Wiley & Sons from Cullity and Graham 2009, © 2009 of the Institute of Electrical and Electronics Engineers, Inc.; permission conveyed through Copyright Clearance Center

Table 11.1 The spin magnetic moments values for different types of 3D ions

Cation	Fe^{3+}	Fe^{2+}	Ni^{2+}	Co^{2+}	Cu^{2+}	Zn^{2+}
Magnetic moment	$5\mu_B$	$4\mu_B$	$2\mu_B$	$3\mu_B$	$1\mu_B$	$0\mu_B$

Reprinted from Caizer 2013, © (2013), with permission from Eurobit

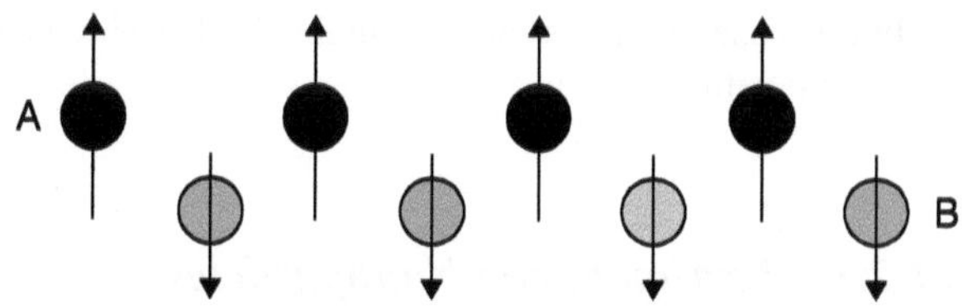

Fig. 11.2 The schematic representation of the ionic magnetic moments orientation within the ferrimagnetic structure

These MeIs with magnetic moments in tetrahedral (A) and octahedral (B) sublattices lead to the existence of two magnetic lattices (Néel 1948) that have magnetic moments aligned parallel to one direction but in opposite (Fig. 11.2), one to the extent of a Weiss domain (Clark 1980) and under superexchange interaction (Anderson 1950; Van Vleck 1951) between the magnetic moments, which takes place in this case through the oxygen ions. Thus, there will be a magnetization (spontaneous) (M) of the magnetic domain (Weiss) given by the difference between the two sublattices' magnetizations, the tetrahedral (M_A) and octahedral (M_B):

$$M_S = M_A - M_B \tag{11.1}$$

If $M_B > M_A$, the difference will be reversed ($M_S = M_B - M_A$). The magnetization is defined as the sum of the magnetic moments $(\vec{\mu}_i)$ per unit volume (V):

$$\vec{M} = \sum_i \vec{\mu}_i / V, \tag{11.2}$$

(in A/m in SI units) and characterizes, from a magnetic standpoint, a substance/material at a macroscopic level. This is the case for the *ferrimagnetic* (FiM) structure, which we will refer further on and which differs significantly from the ferromagnetic one, where the magnetic moments are all aligned in one direction and have the same sense (Rado and Suhl 1973).

Given the $Y^{3+}[X^{2+}Y^{3+}]O_4^{2-}$ structure for a molecule for the reversed spinel, for the purpose of highlighting the A and B magnetic lattices, the bracket contains MeIs in the octahedral positions from the spinel structure, while the rest of Y^{3+} represent the MeIs from the tetrahedral lattice. For example, for the FiM nickel structure, one could use the formula $Fe^{3+}[Ni^{2+}Fe^{3+}]O_4^{2-}$, where the Ni^{2+} MeIs occupy octahedral positions and the Fe^{3+} MeIs occupy both octahedral positions and tetrahedral positions. Considering the magnetic moments of the spin ions 3d Ni^{2+} and Fe^{3+} in Table 11.1, we can theoretically calculate the total magnetic moment per molecule $(2\mu_B + 5\mu_B) - 5\mu_B = 2\mu_B$ (Smit and Wijin 1961), a value which is close to the experimental value of $2.3\mu_B$. The variations which appear between the theoretical and experimental values are due to the fact that, in reality, the spinels are in an intermediate structure, especially at room temperature. Thus, the following formula is more appropriate, where x can have a value between 0 and 1, which reflects more accurately the total magnetic moment per molecule:

$$\left(\mathrm{Me}_x^{2+}\mathrm{Fe}_{1-x}^{3+}\right)\left[\mathrm{Me}_{1-x}^{2+}\mathrm{Fe}_{1+x}^{3+}\right]O_4^{2-} \tag{11. 3}$$

In this case, the parenthesis contains MeIs (which can be fully magnetic) in tetrahedral positions.

11.2.1 Ferrimagnetic Nanoparticles

For large FiM nanostructures in the range of few nm – tens of nm (an area of great interest being ~1 to 100 nm) – we are mainly addressing the nanoparticles/nanocrystallites/nanostructures. However, depending on their size and magnetic anisotropy, those can be *multi-domains* (with a structure of magnetic domains) (Cullity and Graham 2009; Rado and Suhl 1973) (Fig. 11.3a) or *single domain* (without the structure of magnetic domains) (Caizer 2004) (Fig. 11.3b). A certain structure of magnetic domains depends on the crystal symmetry. If the symmetry is *uniaxial*, the characteristic structure is the one in Fig. 11.3a *with magnetic poles free*, and if the symmetry is *cubic*, then the characteristic structure is the one in Fig. 11.3a2 without magnetic poles free or with *magnetic flux closure domains*. These are formed out of magnetic *domains* spontaneously magnetized at saturation, separated by *walls of magnetic domains*.

In general, the FiMNPs are single domain when their diameter is <20 to 30 nm, in the case of *soft* magnetic, and <7–10 nm in the case of *hard* magnetic. At larger sizes than those approximated, the NPs will always have a structure of magnetic domains, depending also on their crystal symmetry. In Fig. 11.4, the transmission electron microscopy (TEM) image of MNPs is shown, where the single crystal structure is revealed in inset by high-resolution transmission electron microscopy (HR-TEM).

Large NP magnetization, with magnetic domain structure, in external magnetic field is with hysteresis, the width of the hysteresis loop (Fig. 11.5) depending on the size of the NPs and also on their magnetic anisotropy, respectively, if they are *soft* (Fig. 11.5a) or *hard* (Fig. 11.5b).

The magnetization (*M*) is a nonlinear magnetic field (*H*) function:

$$M = \chi H \tag{11.4}$$

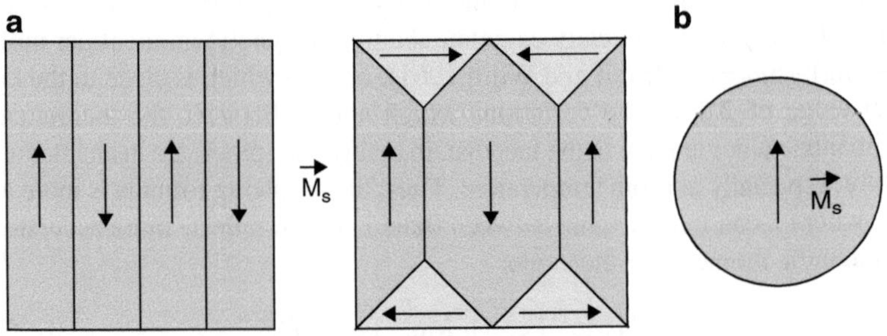

Fig. 11.3 Magnetic structures of nanoparticles: (**a**) multi-domain nanoparticles with **a1** uniaxial and **a2** cubic symmetry, (**b**) single-domain nanoparticle

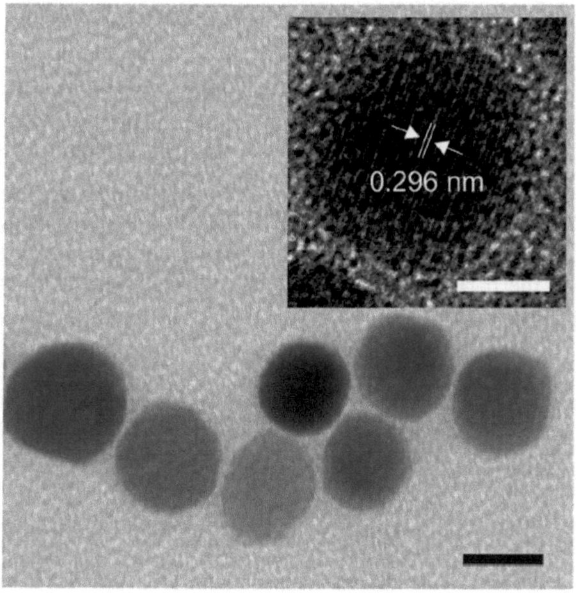

Fig. 11.4 TEM image of the MNPs (scale bar = 20 nm). *Inset*: High-resolution TEM image of the MNPs showing the lattice fringes (scale bar = 10 nm) (Reprinted from Yin et al. 2014, © 2014 Wiley-VCH Verlag GmbH & Co. KGaA (Weinheim), with permission from John Wiley and Sons)

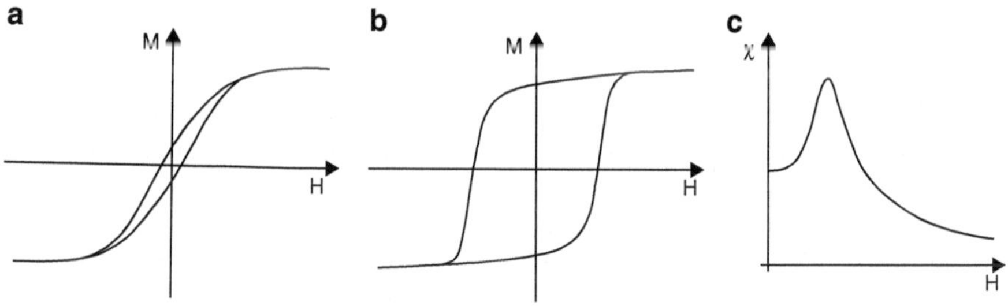

Fig. 11.5 (**a, b**) Hysteresis loops and (**c**) magnetic susceptibility for different magnetic nanoparticles with magnetic domains structure

where χ is the magnetic susceptibility, because χ depends on the field, and $\chi = \chi(H)$, being a complex function of *H* and having a shape similar to that of the variation from Fig. 11.5c.

However, in a high e-MF, toward saturation, also the larger NPs end up consisting of a single magnetic domain, magnetized at saturation. This will cause a magnetic moment (Caizer 2015) for the larger NPs:

$$\vec{m} = V\vec{M}_s \tag{11.5}$$

Since the volume of NPs, now occupied by a single magnetic domain, will be high, this will cause a quick and strong reaction of the NP to the external magnetic field. Such NPs can be used in magnetorheological suspension, with applications in nanotechnology, pharmaceuticals, and biomedicine (Safarik and Safarikova 2002; Pankhurst et al. 2003; Tartaj et al. 2003).

11.2.2 Superparamagnetic Nanoparticles

A special case is that of the behavior of *small* FiMNPs in e-MF (Caizer 2015). For small sizes, generally <10 to 15 nm, in the case of the single-domain NPs, their magnetization will no longer be stable but fluctuant along a direction (the axis of easy magnetization (e.m.a.)) (Fig. 11.6a) under the action of the thermal activation (Néel 1949). Under an e-MF (H), their magnetization (spontaneous magnetization M_s) or the magnetic moment (m) will quasi-instantly point toward the MF, similar to the paramagnetic atoms, following Langevin function (Fig. 11.6b) (Rado and Suhl 1973):

$$M_{\mathrm{SPM}} = M_{\mathrm{sat}}\left(\mathrm{cth}\alpha - \frac{1}{\alpha}\right) \tag{11.6}$$

where

$$M_{\mathrm{sat}} = \mathrm{nm} \tag{11.7}$$

is the saturation magnetization, and

$$\alpha = \frac{\mu_0 m_{\mathrm{p}} H}{k_{\mathrm{B}} T} \tag{11.8}$$

is the Langevin parameter. In Eqs. 11.7 and 11.8 , the observables are n is the NP concentration, μ_0 is the permeability of vacuum, k_{B} is the Boltzmann constant, and T is the temperature (ex. room temperature).

The magnetization of NPs in an e-MF (H) and at the temperature T shows no hysteresis in this case (Fig. 11.6b), and the susceptibility has a variation as the one in Fig. 11.6c, with a constant decrease to zero until reaching a high field. In this case, the behavior of NPs in the e-MF is *superparamagnetic* (SPM) (Bean and Livingston 1959; Rado and Suhl 1973), and the magnetic moment of the nanoparticles follows quasi-instantly the magnetic field.

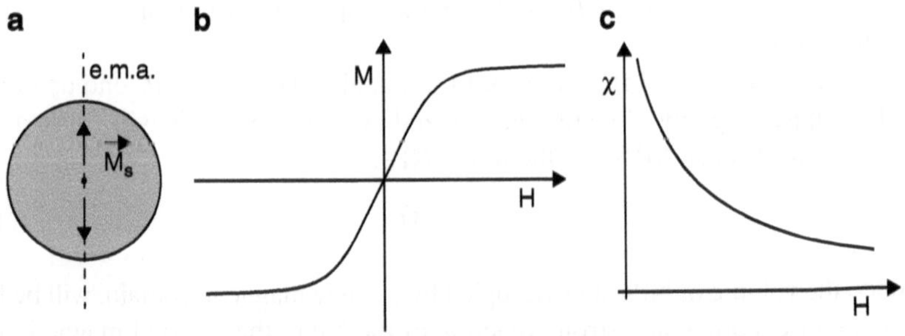

Fig. 11.6 (**a**) Superparamagnetic nanoparticle; (**b**) Magnetization and (**c**) susceptibility as a function of external magnetic field H of superparamagnetic nanoparticles

11.3 The Stability of the Suspension of Nanoparticles Dispersed in a Fluid

For different applications in the pharmaceutical and biomedical, the FiMNPs with MeIs (see Sect. 11.1) must be dispersed in various pharmaceutical liquids, thus forming magnetic suspension (Figuerola et al. 2010; Medeirosa et al. 2011; Bucak et al. 2012; Cherukula et al. 2016; Hajba and Guttman 2016). These can then be handled with various e-MFs (Whitesides et al. 1983; Zborowski 1997; Jeong et al. 2007) for achieving the desired magnetic, mechanical, thermic, and lightning effects.

An essential condition for the suspension of NPs to be applied in the pharmaceutical and biomedical field is its *stability* over time and at room temperature, or at the human body temperature (~37 °C), and *not sedimenting*. The *suspension stability* of the ferrimagnetic relies on *keeping the NPs in suspension* in the pharmaceutical liquid without letting them agglomerate or sedimenting as a result of possible interactions of various internal or external environmental actions. In addition, the NPs also need to be *biocompatible* with the biological environment where they are going to be introduced. For the latter condition, there are different techniques used for making it biocompatible, depending on the intended purpose and nanoparticle size, by coating with various organic molecules, bioencapsulation in different biological membranes, and bioconjugation or by decorating/functionalizing the surface of the NPs with different molecules and organic agents (Safarik and Safarikova 2002; Tartaj et al. 2003; Pankhurst et al. 2003; Shinkai and Ito 2004; Ito et al. 2005; Kobayashi et al. 2014).

In Fig. 11.7a, the image of bioencapsulating NPs in liposomes is shown, which is a natural membrane (cell), perfectly biocompatible, and made up of phospholipids

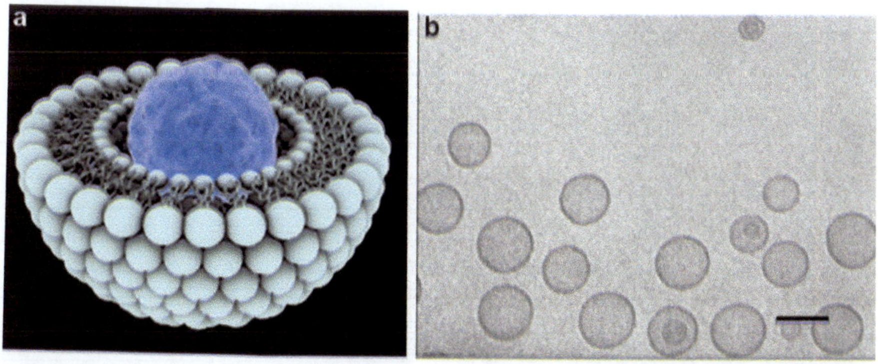

Fig. 11.7 (**a**) Nanoparticle encapsulated/surrounded by the lipid bilayer (Stathopoulos and Boulikas 2012) (CCL: © (2012) CNRS Phototheque/SAGASCIENCE/ Réalisation 3D: François Caillaud/Ligneris Studio. Reproduced by permission of CCL – Hindawi Publishing); (**b**) Cryo-TEM image of EPC liposomes sterically stabilized by DDP $(EO)_{92}$ at copolymer contents of 2.1 mol %. [EPC] = 1 mM, bar = 100 nm (*EPC* egg-phosphatidylcholine, *DDP* (*EO*) 1,3-didodecyloxy-propane-2-ol (ethylene oxide)) (Reproduced with permission from Rangelov et al. 2003, Copyright (2003) American Chemical Society)

arranged in a double layer with the hydrophilic head outward and the hydrophobic tails inward the membrane. The MNPs can be encapsulated in cavities formed by liposomes by different preparation techniques depending on their size. A major disadvantage of using liposomes is their large size (medium diameter) (of the order of 10^2 nm) (Fig. 11.7b) compared with the size of the encapsulated single-domain NPs (of the order of 10–20 nm), which makes the desired magnetic effect to become weak and ineffective.

One of the current methods used in the pharmaceutical industry and medicine, with great potential also for the MNPs, is the bioconjugation of the nanoparticles with cyclodextrins (Fig. 11.8) (Wang et al. 2003). To do this, γ-cyclodextrins are most suitable due to their stability and, in addition, due to the advantage of including in their toroidal cavities the inclusion complexes achieving a double action, magnetic and chemical (Pradhan et al. 2007; Bellia et al. 2009).

More about this issue will be presented in the next section where the effectiveness of our method of efficiently making *biocompatible* Fe_3O_4 (magnetite) MNPs with

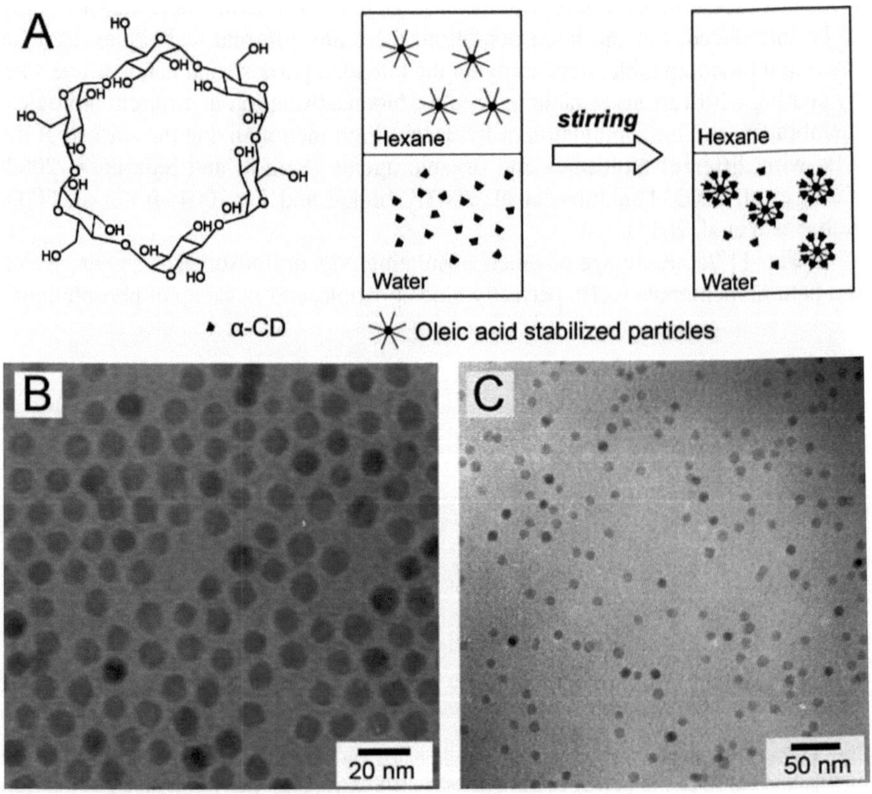

Fig. 11.8 (**a**) Schematic illustration of phase transfer of oleic acid stabilized nanoparticles from an organic to an aqueous medium by surface modification with a-cyclodextrin (*a-CD*). (**b**, **c**) TEM images of iron oxide nanoparticles before and after phase transfer (Reprinted with permission from Wang et al. (2003), Copyright (2003) American Chemical Society)

SA will be discussed also. The magnetite has the formula $Fe^{3+}[Fe^{2+}Fe^{3+}]O_4^2$ containing the MeIs of Fe^{2+} and Fe^{3+} with the magnetic moments $4\mu_B$ and $5\mu_B$ (see Table 11.1), which will give a higher magnetic moment per molecule, ~$4\mu_B$. The magnetite is frequently used in pharmaceutical and biomedical applications due to (*i*) their very good magnetic properties (Smit and Wijin 1961), which makes it suitable for such applications, but also due to the fact that (*ii*) *small* NPs are well tolerated by the living organisms, having a low toxicity (Hilger et al. 2003; Naqvi et al. 2010). However, for the complete elimination of their toxicity, the NPs need to be made biocompatible before actually being used in the biological environment. For the MNPs, making them biocompatible through different techniques and by using different organic substances, they will also determine another desired effect, namely, isolating them from each other and, thus, eliminating the magnetic dipole interaction. As a result, the NPs remain suspended in the pharmaceutical liquid, this being one of the desired effects for practical purposes.

The main *interactions* that may affect the stability of a suspension of FiMNPs in a liquid are (Rosensweig 1985; Odenbach 2002; Caizer 2002, 2004, 2008) the *gravitational interaction*, the *Van der Waals interaction*, the *dipole-dipole interaction*, or *dipole-magnetic field interaction*. The *dipole-external magnetic field interaction*, uniform or irregular, will be discussed in Sect. 19.4.

The *gravitational interaction* can lead to sedimentation of the NPs in suspension under the influence of the gravitational field and the concentration gradients which can occur inside the liquid columns. Considering the particles approximately spherical which are suspended in a liquid, and given the sedimentation rate determined from Stokes formula, and the speed determined by the Brownian motion as a result of the heat activation at a certain temperature *T*, the maximum diameter (*d*) which the particles can have without settling is given by the following:

$$d \le \left(\frac{\eta^2 k_B T}{\rho(\rho - \rho_l)^2 g^2} \right)^{1/7} \tag{11.9}$$

where η is the viscosity coefficient of the fluid, ρ is the particle's density, ρ_l is the liquid density, and g is the gravitational acceleration. For usual liquids, the calculations show that the magnetic particles can have diameters up to hundreds of nm without them sedimenting and the colloidal suspension remaining stable in the gravitational field. At larger diameters, however, one must consider the possible sedimentation and, thus, the instability of the colloidal suspension.

Where the suspension of NPs was found in testing tubes forming columns of liquids with different heights, one must consider also the presence of concentration gradients along the vertical column. Under the barometric law, the height z of the liquid column must meet the following condition:

$$z < \frac{k_B T}{(\rho - \rho_l) g V} \tag{11.10}$$

in order not to have a concentration gradient between the extremities of the column. In the formula, *V* is the volume of the NPs. For MNPs with a diameter <15 to 20 nm, inside the usual liquid and at usual column heights, the sedimentation as a result of the concentration gradients is excluded. However, for larger particles or those embedded in large biocapsules, like the magnetoliposomes, one must consider the effect of the sedimentation inside the liquid column.

The *Van der Waals* interaction may occur between the suspended NPs when they are not surfactants or biocompatible (coated with an organic shell) due to induced electric dipoles at the nanoparticle's surface as a result of the fluctuations in electron orbitals. Given that usually in pharma and biomedicine applications, the MNPs generally have a layer of biocompatible surface, the *Van der Waals* interaction will not show or is very weak when the surfaces of NPs are coated with surfactants that form molecular chains of reduced length (<0.1 nm), allowing the NPs near each other to come closer. Therefore, we will not discuss this type of interaction here.

The *dipole-dipole interaction* can affect the stability of the suspension of MNPs the most, even when they are surfactants, but the thickness of the surfactant layer is not big enough (Caizer 2002, 2008). Therefore, this type of interaction which always exists between NPs, since they are magnetic and have their own magnetic moment (Eq. 11.5), should be considered and, on a case-by-case basis, also evaluated prior to obtaining or using a magnetic suspension.

Below we show (Caizer 2008) a case where assessment of the strength of *dipole-dipole interaction* for the $Zn_{0.15}Ni_{0.85}Fe_2O_4$ FiMNPs, with MeIs of Zn^{2+}, Ni^{2+}, and Fe^{3+}, surrounded with SiO_2 (MNPs dispersed in silica matrix) (Fig. 11.9) is performed (republished (with minor changes) with kind permission from Elsevier © (2007) Elsevier).

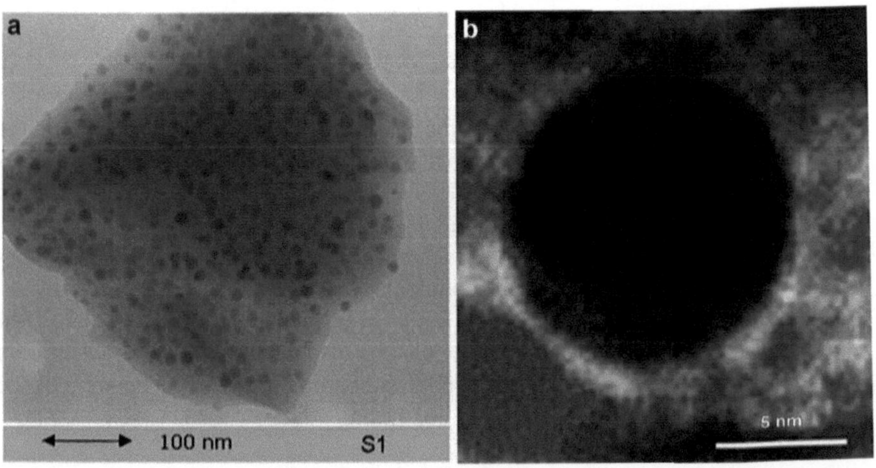

Fig. 11.9 (**a**) TEM images of $(Zn_{0.15}Ni_{0.85}Fe_2O_4)_{0.15}/(SiO2)_{0.85}$ (Reprinted from Caizer et al. 2003, © (2003) Acta Materialia Inc., with permission from Elsevier); (**b**) HR-TEM images of $(Zn_{0.15}Ni_{0.85}Fe_2O_4)_{0.15}/(SiO2)_{0.85}$ (Reprinted from Caizer (2008), © (2007), with permission from Elsevier)

The intensity of the interactions between the NPs is evaluated with the coupling parameter:

$$\lambda = \left(W_{d-d}\right)_{\max} / k_B T \tag{11.11}$$

defined as the ratio between the maximum interaction energy of the magnetic moments $\vec{m}_i$ and $\vec{m}_j$ of two NPs approximated as being spherical and the thermal energy k_BT (Söffge and Schmidbauer 1981). In agreement with Fig. 11.10a, the interaction energy between the magnetic moments $\vec{m}_i$ and $\vec{m}_j$ of the particles i and j, fixed by the position vectors $\vec{r}_i$ and $\vec{r}_j$, is (Jackson 1974; Jonsson et al. 1995)

$$\left(W_{d-d}\right)_{ij} = \left(\mu_0 / 4\pi\right) \cdot \left[\left(\vec{m}_i \cdot \vec{m}_j\right) r_{ij}^2 - 3\left(\vec{m}_i \cdot \vec{r}_{ij}\right)\left(\vec{m}_j \cdot \vec{r}_{ij}\right)\right] / r_{ij}^5 \tag{11.12}$$

In the case of a maximum interaction, the energy is

$$\left(W_{d-d}\right)_{\max} = \left(\mu_0 / 4\pi\right) \cdot m^2 / r_{ij}^3 \tag{11.13}$$

where we considered $m_i = m_j = m$. Taking into account the image HR-TEM from Fig. 11.9b, in the case of NPs from the silica matrix, there is a nonmagnetic interface layer (nanoparticle matrix), because the mean magnetic diameter ($\langle D_m \rangle$ =8.9 nm$\rangle$) (obtained after magnetic measurements Caizer 2003; Caizer et al. 2003) is much smaller than the mean physical diameter ($\langle D \rangle$ =11.6 nm) obtained from TEM (Caizer et al. 2003). This layer has a mean thickness of $\langle \eta \rangle$ =($\langle D \rangle$ – $\langle D_m \rangle$)/2≅1.4nm (Fig. 11.10b). In the presence of this interface layer, the magnetic moment $m = V_m M = \pi D_m^3/6$ (in the approximation of the spherical particles) can be written as

$$m = \pi\left(\langle D \rangle - 2\langle \eta \rangle\right)^3 M / 6. \tag{11.14}$$

Taking into account the average distance between the magnetic moments i and j, $\langle r_{ij} \rangle = \langle D \rangle + \langle d \rangle$, and the magnetic diameter $\langle D_m \rangle = \langle D \rangle - 2\langle \eta \rangle$ (Fig. 11.10b), the maximum energy of dipolar interaction will be (according to Eqs. 11.13 and 11.14)

$$\left(W_{d-d}\right)_{\max} = \left(\mu_0 \pi M^2 / 144\right)\left[\left(\langle D \rangle - 2\langle \eta \rangle\right)^2 / \left(\langle D \rangle + \langle d \rangle\right)\right]^3 \tag{11.15}$$

and in agreement with Eq. 11.11 , the coupling parameter will have the expression

$$\lambda = \left(\mu_0 \pi M^2 / 144 k_B T\right)\left[\left(\langle D_m \rangle\right)^2 / \left(\langle D \rangle + \langle d \rangle\right)\right]^3 \tag{11.16}$$

After replacing in Eq. 11.16 the values for $\langle D \rangle$, $\langle D_m \rangle$, and $\langle d \rangle$ ≅3 nm (values obtained from the processing of the electron micrograph) (Fig. 11.9a), we obtain the value of the coupling parameter $\lambda = 0.12 << 1$. This result shows that the dipolar interactions between the NPs, in the presence of thermal agitation, are very weak and thus can't couple the magnetic moments with each other. Under these conditions, the interactions are negligible.

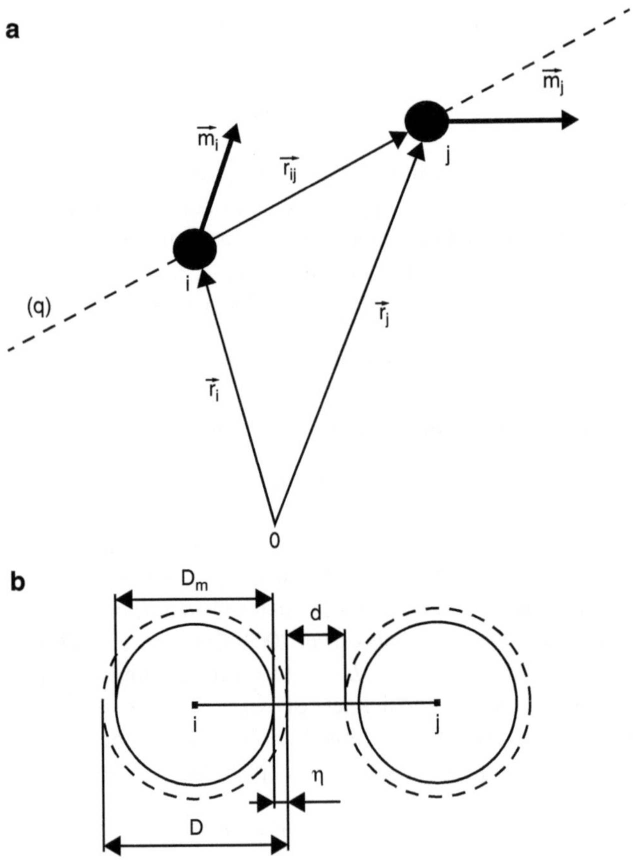

Fig. 11.10 (**a**) Diagram of the interacting magnetic dipole moments (m_i and m_j); (– – –) is the axe (*q*) that passes through the center of the nanoparticles *i* and *j* and (**b**) nanoparticles *i* and *j* from the silica matrix separated by the distance *d* (Reprinted from Caizer 2008; copyright (2007), with permission from Elsevier)

However, the coupling parameter value λ depends strongly on the distance d between the surfacted/biocompatible NPs, this increasing rapidly (with d^3) when the distance between NPs decreases. Moreover, also when the η thickness of the surface layer of the NPs is reduced, the value of the coupling parameter increases rapidly.

Given above, for the Bio-FiMNPs dispersed in various pharmaceutical liquids, there may be situations when $\lambda > 1$ or even $\lambda >> 1$, when the magnetic dipolar interactions can no longer be neglected and could lead to the agglomeration of NPs or even their sedimentation and separation inside the liquid. In preparing the colloidal suspension with biocompatible MNPs and their applications in pharmaceuticals and biomedicine, the abovementioned is a very important aspect which should be considered when the NPs are needed to remain in suspension and form a colloidal solution which is stable over time.

In conclusion, given the possible interactions discussed above, the stability of a suspension where MNPs are dispersed can be made possible, primarily, by increasing the distance between the NPs in the liquid (isolating them from each other), up to a certain value, and maintaining it over time. This is usually done by making an organic shell on the surface of the NPs, by having a certain thickness, by surfactating with various organic compounds (*steric stabilization*) (Molday and Mackenzie 1982; Tiefenauer et al. 1993; Halbreich et al. 1998; Domingo et al. 2001; Pardoe et al. 2001; Sousa et al. 2001), by electrical polarization (+ or −) of the NP surface (*ionic stabilization*) (Massart 1981), or even by using both methods (*stabilization mixed*), leading to a high stability. Besides maintaining the NPs at a certain distance between each other, in order for them to remain suspended, the NPs must be smaller than a certain size (mean diameter smaller than a certain value), taking into account also the thickness of the surface or the size (diameter) of the biocapsules in which the NPs are embedded, such as liposomes or other bio-nanostructures. In addition, sometimes, we must also consider the action of the e-MF upon the MNPs (their magnetic moments), if it is uniform or not and their strength, which can sometimes influence the stability and lead to the agglomeration of NPs or even their separation in liquid, an unexpected effect.

11.4 Handling Magnetic Nanoparticles with an External Magnetic Field

The influence of an e-MF, uniform or nonuniform (field gradient), upon the *stability* of the MNPs inside a colloidal suspension and the *handling of the nanoparticles* with an e-MF, is presented below (Caizer 2004) (republished (translated) with kind permission from Eurobit Publishing © (2004) Eurobit). This issue is very important in practice for assessing the conditions in which a suspension of MNPs can remain *stable* over time, *without the formation of agglomerates* and/or *their sedimentation*, under the action of an e-MF and in the presence of the gravitational field and thermal agitation.

A *nonuniform* e-MF can radically affect the stability of the magnetic suspensions, even those using very small particles (nm).

Comparing the energy of the particles in suspension at the height *h* in the gravitational field,

$$W_{\mathrm{h}} = (\rho - \rho_{\mathrm{l}})\mathrm{ghV}, \tag{11.17}$$

with the energy of the magnetic dipole (particle) in a *uniform* field,

$$W_{\mathrm{m}} = \mu_0 M_{\mathrm{s}}\mathrm{HV}, \tag{11.18}$$

the following ratio is obtained

$$r = \frac{\Delta\rho g h}{\mu_0 M_{\mathrm{s}} H} \tag{11.19}$$

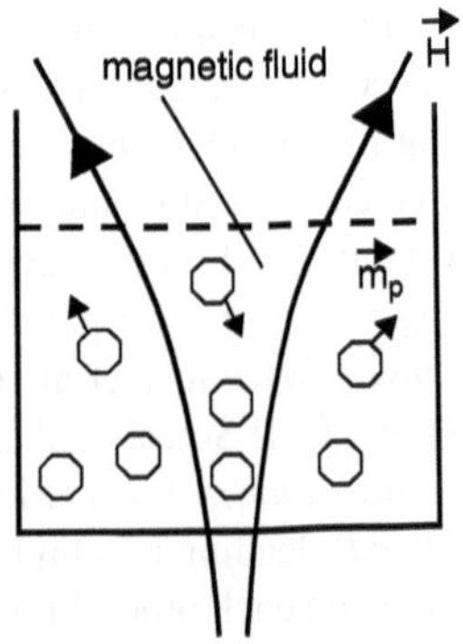

Fig. 11.11 Magnetic field gradient action on magnetic nanoparticles in a fluid (Reprinted from Caizer 2004, © 2004, with permission from Eurobit Publishing)

This relationship allows, in specific cases, to determine which of the two fields (gravitational or magnetic) influence the stability and in what proportion. For example, for r a value of ~0.05 can be obtained in normal conditions of a magnetic fluid, with magnetite particles of ($\Delta\rho = (\rho - \rho_l) = 10\text{g/cm}^2$)in a field of 10 kA/m and h ~ 5 cm. The result shows that for this particular value of the e-MF, the influence of the gravitational field is much weaker than that of the MF, and thus, it can be neglected.

When the e-MF is *nonuniform*, its tendency is to *agglomerate the particles inside the liquid in areas of intense field* (Fig. 11.11) because on a particle with volume V and spontaneous magnetization $\vec{M}_s$, it acts on the following force (Rosensweig 1985):

$$\vec{F} = \mu_0 V\left(\vec{M}_s \cdot \nabla\right)\vec{H}. \tag{11.20}$$

When the magnetization direction is the same with the one of the MF, explaining the vectors and considering the case of a conductive fluid and displacement currents equal to zero $\left(\nabla \times \vec{H} = 0\right)$, Eq. 11.18 becomes

$$\vec{F} = \mu_0 \text{VM}\nabla H, \tag{11.21}$$

where ∇H is the MF gradient. This is a good approximation, since in general the NPs inside the magnetic fluid are free to rotate quasi-instantly, having the magnetization vector along the MF.

The agglomeration trend is opposed by the thermal agitation, whose energy can be expressed by the following:

$$W_T \oplus k_B T \tag{11.22}$$

For the system to be stable, the following condition must be satisfied:

$$\frac{\mu_0 M_s \text{VH}}{k_B T} < 1, \tag{11.23}$$

when the thermal agitation manages to disperse the particles against the MF. Close to the edge, for the spherical particles approximation, the maximum diameter

$$d_{\max} = \left(\frac{6k_{\mathrm{B}}T}{\pi\mu_0 M_{\mathrm{s}} H} \right)^{1/3}, \tag{11.24}$$

up to which the NPs can still be dispersed is obtained.

Thus, for the colloidal system to be stable (not having agglomerations under the e-MF action) the NP diameter must be$d < d_{\max}$. For example, for Fe_3O_4, where $H = 10$ KA/m and $T = 300$ K, the condition of $D < 10$ nm must be met. For NPs with $D \geq 10$ nm, the NPs can agglomerate under the MF action, opposite to the thermal agitation.

According to the formula (11.19), the bio-MNPs found in suspension in a pharmaceutical or biological fluid (blood, blood plasma, etc.) can easily be handled from the outside with an MF gradient. Primarily, this is shown schematically in Fig. 11.12 (Jeong et al. 2007). Small MNPs (SPM) can get inside the tissue (Fig. 11.12) and larger NPs (over 50 nm), which cannot pass through the blood vessel walls, will accumulate in the bloodstream around the MF, forming blockages preventing the blood flow (nanoblocking). The magnetic force acting upon the MNPs depends, besides the strength of the MF gradient (∇H), also on the volume of the nanoparticle (V). Knowing the size of the MNPs, one could determine the MF gradient necessary for an efficient handling of the NPs, either for assuring a certain movement or for maintaining a certain position for a certain amount of time. This problem can be addressed vice versa as well; knowing the MF gradient, produced by a permanent magnet, one could find the proper size (volume) of the MNPs necessary for an efficient handling or fixing/locking their position.

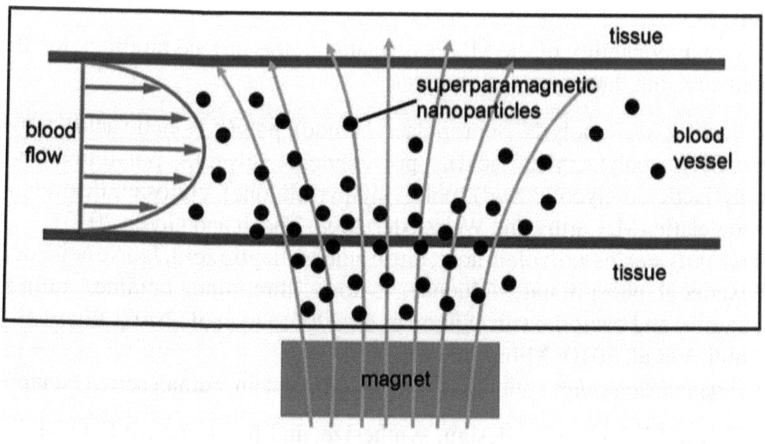

Fig. 11.12 Schematic drawing of a magnetic drug delivery system. A magnet is placed outside or inside the body so that the magnetic gradient can drive the magnetic carriers flowing in the circulatory system into the specific site of a tissue (Reprinted from Jeong et al. 2007, © 2007 Wiley-VCH Verlag GmbH & Co. KgaA (Weinheim), with permission from John Wiley and Sons)

In Sect. 11.6, in order to obtain a vascular nanoblockage, effective and lasting for the therapy of malignant tumors, an e-MF gradient (derived from a permanent magnet) was used for handling and controlling the spatial agglomeration of Fe_3O_4 FiMNPs. The NPs were coated with SA, at a mean diameter between 30 and 90 nm, and injected (from the suspension) directly into the blood vessel that feeds the tumor.

11.5 Biocompatible Magnetic Oxide Nanoparticles with Metal Ions

11.5.1 Magnetic Iron Oxide Nanoparticle Surface Modification

In general, the parameters as size, surface, shape, composition, and coating of MNPs are the most important properties for establishing the cytotoxicity, and the changes of their surface are an important tool to decrease toxicological side effects. A clear example proving a wide range of applicability of MNPs is their versatility and individualized therapy due to their biocompatibility and the absence of toxicity compared to other metal derivatives (Corchero and Villaverde 2009).

Berry et al. (2003) showed that the uncoated FeO-MNPs have a cytotoxicity than those functionalized with specific coating like biomolecules, polymers, different acids, etc. Functionalization of the MNPs minimizes hydrophobic interactions, thereby increasing the colloidal dispersion property, and bioavailability, facilitating the use of NPs in various combinations with drugs and other target molecules in the treatment of various types of cancer and other diseases, and this is being a recent and promising application of NPs (Alexiou et al. 2006; Bucak et al. 2012).

A rising applicability of the MNPs is related to the surface modification that can be obtained using three types of coatings:

(i) *Polymeric shell*: poly(N-isopropylacrylamide), poly(L,L-lactic acid), poly(vinyl alcohol), poly(acrylic acid), polyethylene glycol, poly(ethyleneimine), poly(lactic-co-glycolic acid), poly(vinylpyrrolidone), chitosan, dextran, starch, and gelatin (McCarthy and Weissleder 2008; Thanh and Green 2010)
(ii) *Non-polymeric shell*: oleic acid, citric acid, salicylic acid, lauric acid, dodecyl/hexadecyl phosphonate, glucose, L-dopa, threonine, betaine, amino acid, vitamin, and cyclodextrin (Sahoo et al. 2005; Du et al. 2007; Wu et al. 2008; Buteică et al. 2010; Mihaiescu et al. 2013)
(iii) *Biological molecules*: antibody, biotin, avidin, protein, human serum albumin, etc.

The research's focus is to design, synthesize, and functionalize FeO-MNPs with good dispersability in aqueous medium, in order to obtain magnetic nanomaterials suitable for biomedical applications with higher biological compatibility and a lower toxicity (Bucak et al. 2012). Zhang et al. (2002) showed that the MNP surface modification using polymers generates a high biocompatibility, but induced a low tissular penetration and distribution and also adverse effects. In spite of that, Feridex I.V.® and Resovist are MRI agents for liver diagnosis, as nanoformulations containing FeO-MNPs coated with dextran/carboxydextran, authorized by Food and Drug Administration/European Medicine Agency (Wang 2011).

Du et al. (2007) synthesized and evaluated in vitro and in vivo biocompatible FeO-MNPs coated with betaine (a liver-born S-adenosylmethionine generator) that are useful as drug delivery system for the liver disease treatment. The Fe_3O_4 FiMNPs coated with oleic acid/cephalosporins demonstrated bacteriostatic effect on *E. coli* and *S. aureus* and showed their potency to use as drug delivery nanosystem (Buteică et al. 2010). A therapeutic radionuclide, a pharmaceutical substance or shell, or a magnetic targeting can be bound to a magnetic nanocomposite, introduced into the body, and focused in the target zone through a magnetic field action.

In Table 11.2, the different types of coating agent used for surface modification and functionalization of FeO-MNPs (polymeric shell, non-polymeric shell, biological molecules) are shown.

11.5.2 Magnetic Nanoparticle Uptake and Internalization in Cultured Cells

The complex mechanisms of cell uptake with NPs have been intensively studied lately, and their knowledge could be very useful in the process of NP designing. There are applications regarding some NPs designed in large amounts to be accumulated into the cytosol, such as photothermal and photodynamic therapies (Pekkanen et al. 2014), nonviral gene vectors in gene therapy (Liu and Zhang 2011), and intracellular drug delivery (Paulo et al. 2011). On the other hand, there are also applications which are designed to obtain NPs with long-term intravascular persistence such as vascular nanoblock, and in this case, reduced cell uptake with NPs is mandatory.

Key factors directly affecting NP internalization in cultured cells are the size, coating properties, surface electric charge of NPs, surface hydrophobicity, and the type of cells used. Additionally, cell uptake with NPs could also be influenced by the time of exposure and NP concentration (Li et al. 2012). MNPs penetrate cells by active transport (receptor-mediated endocytosis). Passive transport was observed only on NPs with Au (Taylor et al. 2010). Depending on the size, NPs can be internalized into the cell through multiple pathways: clathrin-mediated endocytosis, caveolae-mediated endocytosis, pinocytosis, macropinocytosis, and phagocytosis (Zhao et al. 2011a; Kou et al. 2013).

NPs with sizes around 500 nm are rapidly opsonized and loaded intracellularly by phagocytosis; pinocytosis is responsible of NP cell uptake smaller than 200 nm, while NPs loaded by clathrin-mediated endocytosis have a diameter of about 100 nm, and those loaded by caveolae-mediated endocytosis have dimensions

Table 11.2 Different types of coating agents used for surface modification and functionalization of iron oxide magnetic nanoparticles

Coating agent		Advantages	References
Polymeric shell	Chitosan	Biocompatible and natural compounds	Lee et al. (2005)
	Dextran, carboxymethylated dextran, carboxydextran		Goodwin et al. (2009), Molday and MacKenzie (1982)
	Alginate		Ma et al. (2007)
	Gelatin		Lee et al. (1996), Zhang et al. (2002), Högemann-Savellano et al. (2003), and Lawaczeck et al. (2004)
	Starch		
	Arabinogalactan		
	Sulfonated styrene-divinylbenzene	Increase the biocompatibility and the stability, low toxicity	
	Polyethylene glycol (PEG)		
	Polyvinyl alcohol (PVA)		
	Polyacrylic acid		
	Poly(D,L-lactide)		
Non-polymeric shell	Folic acid	Label the tumor cells	Jiang et al. (2014)
	Salicylic acid	Improve stabilization in aqueous medium	Laurent et al. (2008) and Mihaiescu et al. (2013)
	Citric acid		
	Oleic acid		
	Lauric acid		
	Gluconic acid		Fauconnier et al. (1996) and Fauconnier et al. (1997)
	Dimercaptosuccinic acid		
	Phosphorylcholine		Portet et al. (2001)
	Dodecylphosphonic acid		
	Hexadecylphosphonic acid	Thermodynamically stable nanoparticle dispersion	Sahoo et al. (2001)
	Dihexadecyl phosphate		
Biological molecules	Human/bovine serum albumin	Improve biocompatibility and are useful in separation techniques	Berry et al. (2003)
	TAT peptide		Lewin et al. (2000)
	Antibody		Smith et al. (2011)
	Biotin		Wu et al. (2008)
	Avidin		

between 60 and 80 nm (Shin et al. 2015). With the exception of caveolae-mediated endocytosis, the other endocytic pathways use lysosome path, where macromolecules could be degraded; in the case of caveolae-mediated endocytosis, NPs can directly reach the Golgi bodies and endoplasmic reticulum avoiding lysosomal degradation. The clathrin-mediated endocytosis is the fastest; NPs which use this path are rapidly internalized by the cell. Cells presenting numerous caveolae (smooth muscle cells, fibroblasts, adipocytes, and endothelial cells) preferentially use this route; these path features of NP cell uptake explain the variability of workload and different intracellular localizations of NPs (Huang et al. 2010). Nonphagocytic cells have higher rates of NP cell uptake with dimension between 20 and 50 nm, with the exception of enterocytes; they load rapidly NPs ranging in size between 100 and 200 nm (Fröhlich 2012).

The amount of NPs which penetrates the cell is directly influenced by the size of NPs. In a study over fibroblastic mouse cells (RAW 264.7), cell uptake with polyvinylpyrrolidone shell FeO-MNPs ranging in size between 30 and 120 nm was strongly influenced by their diameters, yielding the optimum loading for the 37 nm, followed by 65 nm, 8 nm, and 23 nm (Huang et al. 2010). It seems that the size of NPs affects more cell uptake rate rather than the type of loading surface (Yu et al. 2012). The second important factor which influences NP cell uptake is the surface charge. It is well known that NPs with positive surface charge have a greater degree of internalization than those with neutral or negatively charged surface.

This different behavior of NPs has enabled multifunctionalized design of MNPs such as contrast agents for MRI, and intravital fluorescence microscopy. Also the loaded NPs by macrophages is used to improve targeting the hepatocytes, and splenic cells, for tumor imaging (Li et al. 2012). Coating properties can in turn influence the degree of cell uptake by the MNPs, for example, macrophages show a higher degree of carboxylate NP loading, while monocytes load preferentially amino-functionalized NPs (Lunov et al. 2011). Other studies have shown that peptide sequences attached to the surface of NP function as ligands that favor the merging of NPs with cell membrane and work as enhance agents for NP cell uptake (Lee et al. 2014).

11.5.3 Cytotoxicity

The safe use of MNPs in biomedical applications requires detailed studies to investigate the cytotoxicity or to confirm biocompatibility of functionalized MNPs. It is mandatory for NPs designed for diagnostic and therapeutic purposes to go through in vitro cytotoxicity studies. In order to test the effects of the MNPs on cells, generally, several categories of tests are being used to evaluate cell uptakes, cellular viability, reactive oxygen species (ROS) production, cellular stress, or cellular genotoxicity.

Mitochondrial functions are measured using 3-(4,5-dimethylthiazol-2-yl)-5-(3-carboxymethoxyphenyl)-2-(4-sulfophenyl)-2*H*-tetrazolium (MTS) and 3-(4,5-dimethylthiazol-2-yl)-2,5-diphenyltetrazolium bromide (MTT) tests and the integrity of cellular membrane by lactate dehydrogenase (LDH) test. These tests are simple and could be performed rapidly. Both tests are inexpensive and assure reliable and reproducible results without the use of animal models (Patil et al. 2015).

The majority of cytotoxicity studies revealed production of ROS at the site of mitochondria, and it is the principal cause underlying the cytotoxic action of MNPs. Cells have detoxification mechanisms which involve glutathione and several enzymes such as superoxide dismutase and glutathione peroxidase. This mechanism activates when ROS level increases and results in oxidation of glutathione after interaction with ROS; subsequently, glutathione is decreased by one NADPH-dependent reductase; thus, the amount of ROS is correlated with level of oxidated glutathione.

After internalization into the cell, MNPs could be degraded in iron atoms by lysosome enzymes; ferrous iron (2+) could pass in the mitochondria where it reacts with hydrogen peroxide and oxygen to produce hydroxyl radicals and ferric ions (3+), by Fenton reaction. Hydroxyl radicals are linked with the appearance of lipid peroxidation, DNA strand breaks, alteration in gene transcription, alters appearance proteins of protein radicals or affects the immunity system. Cytotoxicity of MNPs could be influenced by size, shape, and surface chemistry, for example, rod-shaped NPs are accumulated mostly in the cytoplasm. On the other hand, spherical-shaped NPs are aggregated in vacuoles, and this different mode of internalization would explain the higher toxicity rod-shaped compared to spherical-shaped nanoparticles (Lee et al. 2014). Additionally, more studies conducted over MNPs synthesized by different techniques have highlighted that cytotoxicity of bare NPs is higher than coated NPs (Mahmoudi et al. 2010; Magdolenova et al. 2015). Another factor which influences the toxicity of MNPs is their chemical composition (stoichiometric ratio of Fe^{2+} and Fe^{3+}) or crystalline nature (magnetite and maghemite). It was noticed that magnetite nanoparticles (Fe_3O_4) have higher cytotoxicity than maghemite (γ-Fe_2O_3), probably because dissolution of maghemite releases only Fe^{3+} ion, while magnetic nanoparticles can release both iron ions (Park et al. 2014).

However, cell uptake is the determining factor for NPs' toxicity. Experiments on normal cells and tumor cells revealed tumor cell uptake of MNPs are higher than normal cells, and this feature would be responsible of MNPs' specificity for tumor cells. FeO-MNPs with sizes below 100 nm have antitumor activity preferentially affecting the mitochondria of tumor cells, at this site decreases cellular production of ATP, which affects membrane potential of the mitochondria, leading to mitochondrial-dependent apoptosis of the tumor cells (He et al. 2016). At the site of tumor cell mitochondria, ferromagnetic NPs affect electron transport chain, which has an important role in the process of oxidative phosphorylation and production of ATP (Birsoy et al. 2015). Intrinsic mitochondrial-dependent apoptosis pathways are activated by the release of apoptotic factors which activates caspases and produces cellular protein alteration (Dix et al. 2008). Other studies have shown that ferromagnetic NPs decrease the antioxidative capacity of tumor cell mitochondria by decreasing synthesis of glutathione and expression of some antioxidant genes (GRLX5 codifies synthesis of glutaredoxin 5 protein with iron homeostasis role; HSPA9 codifies synthesis of heat shock 70 kDa 9 protein with the role of cells protection against oxidative stress), but in the case of an increase in mitochondrial oxidative stress, they contribute on specific cytotoxic affect over tumor cells, a characteristic of ferromagnetic NPs (Camaschella et al. 2007; Ye et al. 2010; Zhao et al. 2011b; Chen et al. 2013; Ryu et al. 2014). In conclusion, ferromagnetic NPs have the ability to destroy specific tumor cells by affecting functions and morphologies of mitochondria (Shiff et al. 2003).

Given these mechanisms of actions, the multitude of coating types, and methods of MNP synthesis, the different results could be explained, more often contradictories obtained from in vitro testing on normal cells or tumor cytotoxicities of MNPs (Naqvi et al. 2010). Coating instability in cellular lining and dissolution by lysosomal enzymes during endocytosis charging are factors which would increase cytotoxicity that should be taken into account in the process of NP designing; antitumor effects of NPs could be enhanced by the sum of cytotoxic effect of NP coat together with the core.

11.5.4 Biocompatible Salicylic Acid-Fe_3O_4 Nanoparticles

Based on observations, we tried to obtain a nanocomposite which would combine antitumor effects of FiMNPs and SA, which finally led to the synthesis of FeO-MNPs functionalized with SA by modified Massart method (Fig. 11.13) (Mihaiescu et al. 2013). From a physicochemical point of view, these FiMNPs functionalized with SA (SA-Fe_3O_4) have a hydrodynamic diameter of 50 nm, a 10–15 nm diameter of iron oxide core, a 35–40 nm coat of SA, very good stability (zeta potential of 40 mV) (Fig. 11.13c), a 0.356 mg/ml concentration of iron, and a handling static increased MF.

A number of studies that have used colon cancer cells have shown that acetylsalicylic acid could prevent tumor growth and induce apoptosis of cancer cells (Thun et al. 2012; Dovizio et al. 2013). Aspirin and its primary metabolite, SA, induce apoptosis of tumor cells (Shiff et al. 2003) and inhibit angiogenesis (Abdelrahim and Safe 2005) by affecting COX-2 expression both at the transcriptional and post-transcriptional level (Xu et al. 1999). By inhibiting cyclooxygenase 2, SA increased the level of arachidonic acid, with favorable effect over the conversion of sphingomyelin in ceramide, which is an important mediator of apoptosis (Chan et al. 1998). On the other hand, acetylation of cyclooxygenase 2 by aspirin has enzymatic activity modification effect resulting in lipoxin generation with antiproliferative and

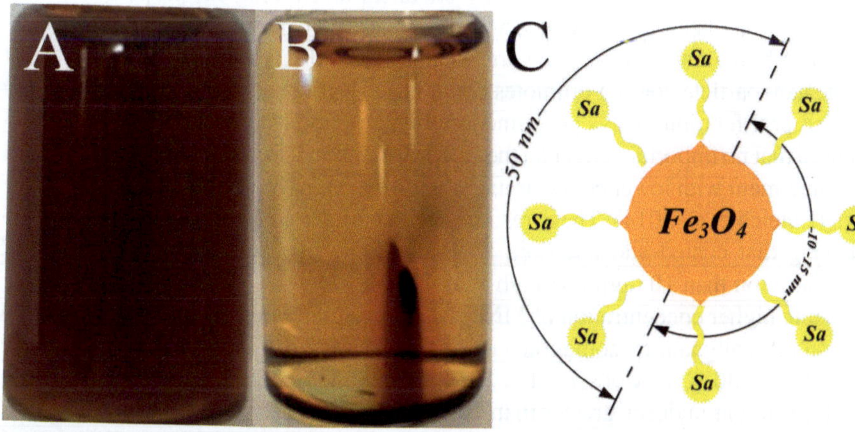

Fig. 11.13 (**a, b**) Macroscopic aspect and (**c**) physicochemical characteristics of iron oxide functionalized with salicylic acid dispersion. Placement of nanoparticles on the wall of the container is being observed (**b**) after exposure to the action of an external static magnetic field; *SA* salicylic acid

antiangiogenetic effect (Ferrandez et al. 2012). In addition to the antitumor action due to the inactivation of cyclooxygenase 2, aspirin and salicylic acid inhibit kinase IkBβ and prevent activation of NF-kB gene, responsible of some proteins involved in anti-inflammatory and angiogenesis response (McCarty and Block 2006). SA also inhibits the penetration of calcium in mitochondria leading to the blockage of cell proliferation (Nunez et al. 2006). Antiproliferative effect of aspirin and SA is also due to the blocking activities of 6-phosphofructo-1-kinase which has a lowering effect on cell glucose consumption (Spitz et al. 2009).

11.6 Salicylic Acid-Fe_3O_4 Nanoparticles as a Therapeutic Carrier in Cancer

11.6.1 In Vitro Cytotoxicity of Salicylic Acid-Fe_3O_4 Nanoparticles

Nowadays, in vitro assays are by far the most used methods to assess the biological properties of the MNPs because they present several advantages such as simplicity, convenience, and cost. In vitro assays can use both normal human cells and a wide variety of tumor cells, which are usually grown as monocultures in a single layer of cells adherent on culture plates (2D cultures). Cells in 2D culture have optimal access to nutrients, oxygen, and tested MNPs. In general, conventional 2D cultures are utilized to evaluate the cytotoxicity, cellular uptake, and intracellular trafficking or antiproliferative effects of the MNPs. Also in vitro 2D culture is the test of choice when trying to assess the effect of MNPs on DNA and RNA.

In vitro cytotoxicity studies over NPs with SA shell showed a high degree in their loading, which could be explained by chemical and biological properties of the coating; it has been shown that SA increases cellular permeability and enhances transmembrane transport of some drugs (Jahn et al. 2012). As well, negative charge to the nanoparticle coating promotes cell uptake (Kajii et al. 1986).

Cell proliferation assays performed on human glioblastoma and fibroblast cells showed that no important effect on the cell proliferation was recorded after SA-Fe_3O_4 NPs treatment with lower concentration than 1 μg/mL of bio-MNPs aqueous dispersion (Buteică et al. 2014). These results are concordant with other published results showing that coated and uncoated FiMNPs don't have cytotoxic effect at lower concentration than 10 ug/mL (Singh et al. 2010).

Using higher concentration of MNPs on human pleural malignant mesothelioma and on rat malignant mesothelioma cell line (CARM-L12 TG3), a decrease in viability below 30 % was observed at 72 h and below 5 % at 120 h after SA-Fe_3O_4 MNP treatment in doses greater than 100 mg/dL.

Prussian blue staining revealed the high degree of transmembrane diffusion of FeO-MNPs coated with SA (Fig. 11.14). Regarding intracellular localization, it was shown that once penetrated into the cell, NPs are accumulated juxtanuclear, and as

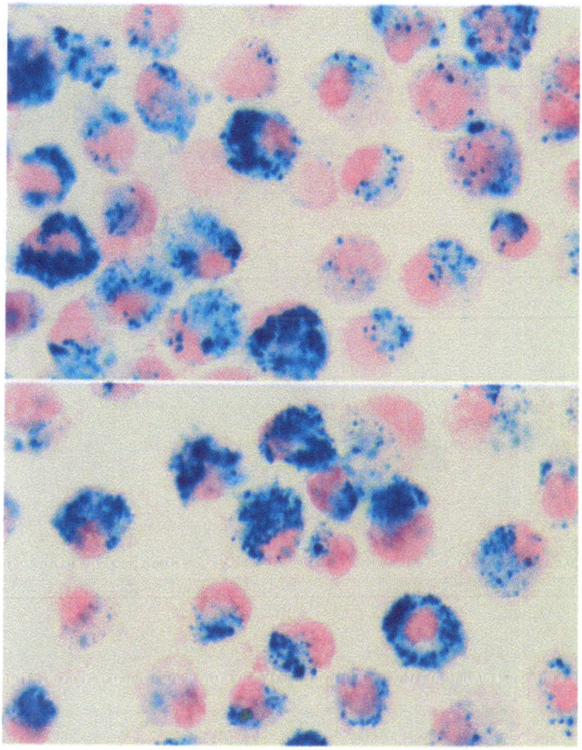

Fig. 11.14 Smear with malignant pleural mesothelioma of murine cells 24 h after treatment with 100 ug/mL iron oxide nanoparticles functionalized with salicylic acid (Prusian blue staining; Ob ×100)

the amount of intracellular NP deposit increases, they are being placed like a perinuclear crown, which eventually completely blocks the core. The blocked core by the intracellular deposits of MNPs is followed by alteration of cell function, which would lead to cellular apoptosis (Fig. 11.15) (unpublished data).

11.6.2 In Vivo Cytotoxicity

The chorioallantoic membrane model (CAM) is a useful preclinical in vivo model for studying and understanding MNPs' biological properties and mechanism of action. It is often used due to their lack of a functional immune system, inexpensive, and easy to manipulate for the development of human tissue xenografts, facile handling, ethically correct, etc. (Mihaiescu et al. 2013). Using CAM model enables evaluation of intravascular embolic risk assessment, distribution to tissues and organs, viability in MF, and toxicities of MNPs coated with SA.

In vitro models based on cell cultures may not provide complete information on NP interactions with biologic systems; on the other hand, animal models are expensive, time-consuming, and laborious, and it has begun to be used less and less by

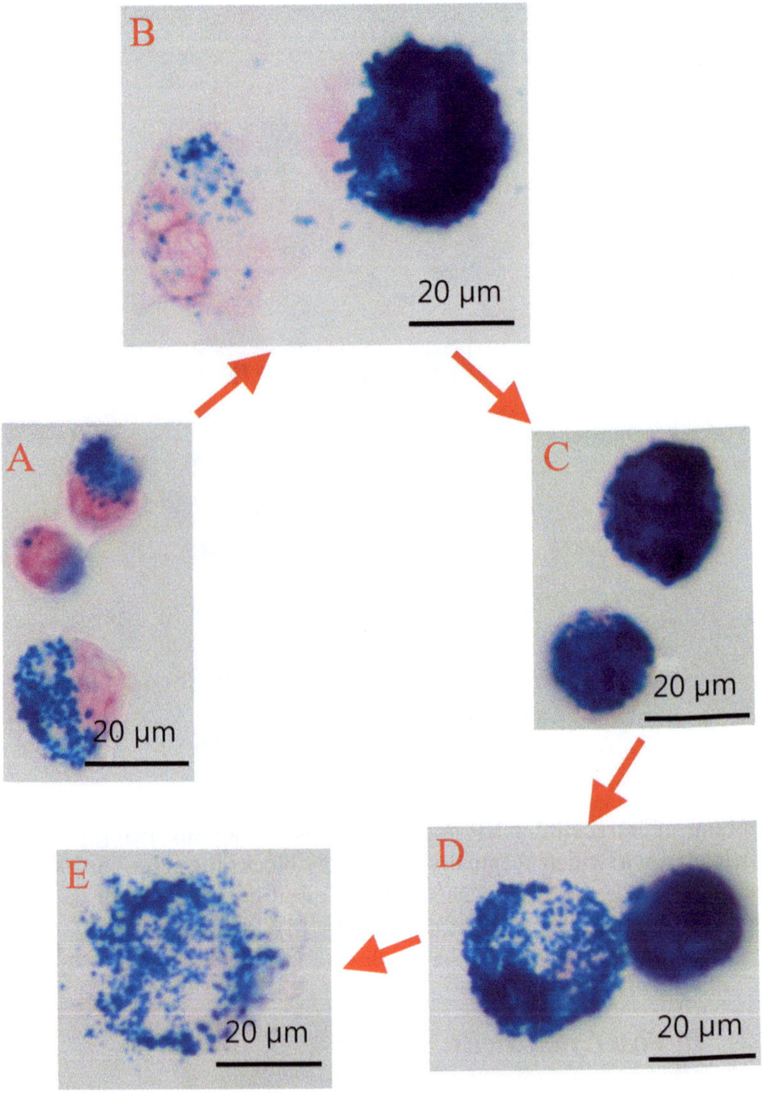

Fig. 11.15 The steps of tumor cells uptake of iron oxide nanoparticles functionalized with salicylic acid deposits. Penetration of nanoparticles into the cell is followed by formation of perinuclear deposits (**a**, **b**), as it accumulates, leads to the blocking of the core (**b–d**) and initiation of cellular apoptosis (**d**, **e**) (Prussian blue staining)

researchers because of the ethical issues it raises. To avoid this inconvenience, in vivo models have been used on nonmammalian vertebrate embryos for the study of NP biologic properties.

Of these, the most used models are models using embryos of zebra fish, frog, and chicken; these models are very reliable for the study of NP adverse effects because embryos are very easily affected if NPs influence intracellular communication or if molecular signaling is needed for the process of development; thus, various developmental disorders may appear that cause morphologic malformations or even death of the embryo (Giannaccini et al. 2014). These development anomalies are easy to pinpoint, given that these models have a short-term development and the embryos allow direct visualization during the development.

Using CAM models allows evaluation of intravascular compartment, evaluation of embolic risk, distribution to the tissue and other organs, handling in e-MF, and toxicity of NPs functionalized with SA, because CAM vessels are easily accessible; it has the advantage that NPs could be administered intravascularly, which makes it an ideal model for directly evaluating NPs into the blood vessel. Another major advantage which CAM presents is the lack of a functional immune system, which makes it an inexpensive bioreactor and easy to manipulate for the development of human tissue xenografts; thus, CAM provides a powerful tool to study the anticancer effects of NPs, because biopsies of tumors implanted into the CAM allow the tumor cells to grow in microenvironment conditions similarly to those of the human body.

Knowing the intravascular behavior is essential when trying to design NPs with long circulating time. Generally, in vitro models could not provide useful information, and other animal models are difficult to work with when trying to analyze NP coating interactions with protein medium of the blood. It is known that proteins attached on the surface of the NPs could form an outer layer (corona protein) which influences NP properties such as cellular charge or tissue distribution (Li et al. 2012).

Using CAM model allows easy evaluation of intravascular compartment of FeO-MNPs because the vascular system allows intravascular administration of NPs, a maneuver which can be done easily starting from day 10 of the embryo's development when the vascular system is almost completely developed and vessels are easily accessible. Additionally, the vessel could be examined in vivo by an optic microscope and could be subjected under directly static e-MF action, which allows follow-up of aggregation of FeO-MNPs in the bloodstream.

The intravascular behavior of FeO-MNPs coated with SA is much influenced by the hydrodynamic diameter of nanoparticles. Administering the same doses of SA-Fe_3O_4 FiMNPs intravascularly, it was noticed that NPs with a diameter greater than 190 nm have superior embolic potential risk when subjected to an MF gradient action. Moreover, SA-Fe_3O_4 NPs with hydrodynamic diameter of 278 nm administered intravenously produce emboli even in the absence of intravascular aggregation under MF action (Fig. 11.16).

SA-Fe_3O_4 bio-MNPs with a hydrodynamic diameter smaller than 50 nm do not have embolic risk, and neither can be aggregated in bloodstream with the help of 0.18 T NdFeB (neodymium-iron-boron) magnet. This behavior may be due to the rapid leaving of the vascular bed by the FeO-MNPs through endothelial fenestration and partly because MNP attraction by the magnet is defeated by the blood viscosity which washes NPs from this level of endothelia subjected to the action of MF.

SA-Fe_3O_4 bio-MNPs with hydrodynamic diameter between 50 and 100 nm could be aggregated easily by an e-MF gradient and do not present embolic risk neither before nor after redispersion of the intravascular NP aggregates (Buteică et al. 2016). Arterioles and venules with the diameter lesser than 100 μ are rapidly blocked by the aggregates of SA-Fe_3O_4 NPs, obtained by the short time (5 15 min) action of

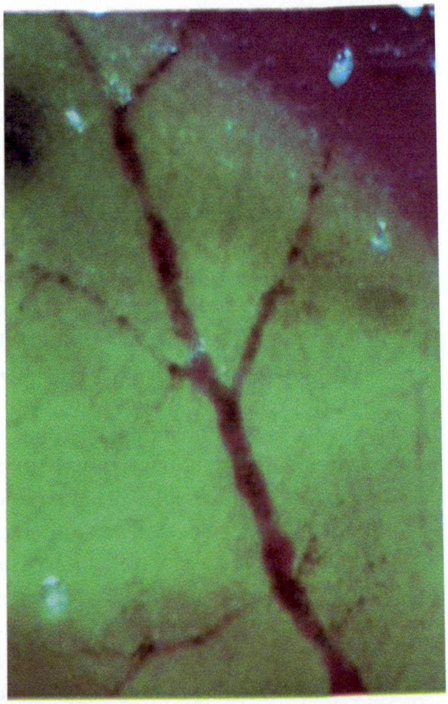

Fig. 11.16 Aspect of an arteriolar chorioallantoic membrane nanoblocked by emboli formed after intravascular injection of 278 nm SA-Fe_3O_4 nanoparticles

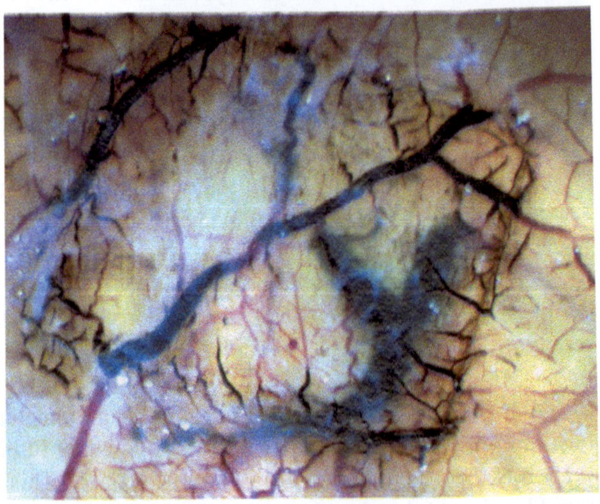

Fig. 11.17 Intravascular aggregates of SA-Fe_3O_4 nanoparticles and nanoblocked CAM vessels, after a magnetic field action

the MF of 0.18 T (Fig. 11.17). If the MF time increases to 30 min, it could block the vessels with the diameter of 1–2 mm.

After removing the MF, veins are rapidly unblocked by redispersion of NP aggregate, but capillaries remain blocked by the aggregates of NPs, thus making the repermeabilization difficult. These observations show that nanoblockage by an MF of nutritive vascularization of a tumor is possible and efficient if the blockage of capillary networks is complete (Mîndrilă et al. 2016).

The cytotoxicity evaluation of SA-Fe_3O_4 bio-MNPs on different types of human tumor xenografts (breast, glioblastoma, malignant mesothelioma xenografts) implanted on the CAM showed a rapid aggregation of the NPs into the xenograft blood vessels under an e-MF action (Fig. 11.18a). The nanoblockage effect on capillaries and arterioles induced a slower tumor xenograft growth due to reduced intake of nutrients and stopped the migration of tumor cells as well local or distance metastasis (Fig. 11.18b–d).

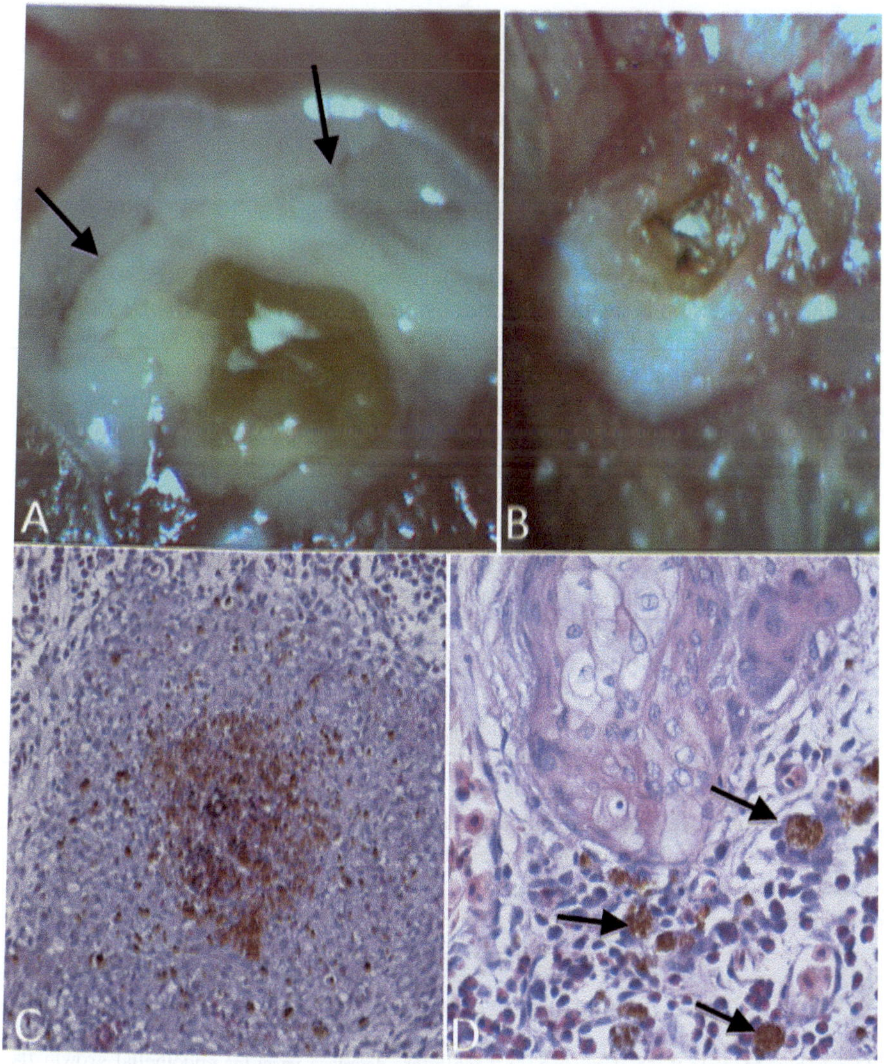

Fig. 11.18 The SA-Fe_3O_4 magnetic nanoparticles effect on human glioblastoma xenografted on the chorioallantoic membrane; (**a**) tumor vessels blocked with SA-Fe_3O_4 magnetic nanoparticles aggregates under magnetic field action (*black arrow*); (**b**) tumor growth inhibition after 4 days from vascular nanoblockage; (**c**) intravascular magnetic nanoparticles aggregates; (**d**) nanoblocked tumor capillaries (*black arrow*)

The possibility to aggregate FeO-MNPs functionalized with SA into the region of the tumor using an e-MF is beneficial because it can be administered at low concentrations, nontoxic at the site of tissues, but which at the tumor site exceeds the toxicity threshold (Buteică et al. 2014).

11.7 Conclusions

The FiMNPs with iron oxides can be successfully used in various pharmaceutical and biomedical applications. However, depending on the application intended, first, one must know very well their magnetic properties in order to efficiently use them. Given the possible vascular nanoblocking presented in this chapter, as a possible alternative method in therapy of malignant tumors, noninvasive and nontoxic, we show some basic magnetic properties in relation to the size of the NPs. However, for their use in a biological environment, without toxicity, the FiMNPs must be biocompatible. Also, in order to be efficiently injected, the MNPs must be dispersed in a fluid in a certain concentration which could determine magnetic suspensions. We also presented and discussed at length the necessary requirement for obtaining a stable suspension for a longer time, without forming agglomerates or particle sedimentation and separation under the action of gravitation, magnetic dipolar, or an e-MF. As a possible effective method for having bio-MNPs, and at lower costs, F_3O_4 (with metallic ions Fe^{2+} and Fe^{3+}) NPs can be enveloped with SA to obtain an SA-Fe_3O_4 nanocarriers with good results when it comes to vascular nanoblocking or other possible future applications. These bio-MNPs, having a magnetic core, can be easily handled and controlled by a MF gradient applied from the outside through various techniques. In addition, the physical principle for determining the field gradient or the nanoparticles' size (volume), depending on the particularity of different types of applications, and how to obtain the field gradient in the area of interest, were studied.

In vitro experiments have demonstrated cytotoxic effects of SA-Fe_3O_4 NPs at higher concentration than 100 μg/mL. Intravascular behavior and in vivo cytotoxicity after intravascular administration of SA-Fe_3O_4 NPs aqueous dispersion showed that they do not produce embolism, neither after aggregation in a selected target under e-MF action. Vascular nanoblockage effects let to slow down tumor xenografts development and reduce metastasis.

References

Abdelrahim M, Safe S. Cyclooxygenase-2 inhibitors decrease vascular endothelial growth factor expression in colon cancer cells by enhanced degradation of Sp1 and Sp4 proteins. Mol Pharmacol. 2005;68:317–29.

Alexiou C, Jurgons R, Seliger C, Iro H. Medical applications of magnetic nanoparticles. J Nanosci Nanotechnol. 2006;6:2762–8.

Anderson PW. Antiferromagnetism. Theory of superexchange interaction. Phys Rev. 1950;79:350.

Bean CP, Livingston LD. Superparamagnetism. J Appl Phys. 1959;30:S120–9.

Bellia F, La Mendola D, Pedone C, Rizzarelli E, Saviano M, Vecchio G. Selectively functionalized cyclodextrins and their metal complexes. Chem Soc Rev. 2009;38:2756–81.

Berry CC, Wells S, Charles S, Curtis ASG. Dextran and albumin derivatised iron oxide nanoparticles: influence on fibroblasts in vitro. Biomaterials. 2003;24:4551–7.

Birsoy K, Wang T, Chen WW, Freinkman E, Abu-Remaileh M, Sabatini DM. Essential role of the mitochondrial electron transport chain in cell proliferation is to enable aspartate synthesis. Cell. 2015;162:540–51.

Brannon-Peppas L, Blanchette JO. Nanoparticle and targeted systems for cancer therapy. Adv Drug Deliv Rev. 2004;64:206–12.

Bucak S, Yavuztürk B Sezer AD. Magnetic nanoparticles: synthesis, surface modifications and application in drug delivery. Chapter in: Recent advances in novel drug carrier systems. Rijeka: InTech; 2012.

Buteică AS, Mihaiescu DE, Grumezescu AM, Vasile BŞ, Popescu A, Mihaiescu OM, Cristescu R. The anti-bacterial activity of magnetic nanofluid: Fe_3O_4/oleic acid/cephalosporins core/shell/adsorption-shell proved on *S. aureus* and *E. coli* and possible applications as drug delivery systems. Dig J Nanomater Biostruct. 2010;5:927–32.

Buteică SA, Mîndrilă I, Mihaiescu DE, Purcaru SO, Dricu A, Nicolicescu C, Neamţu J. In vitro and in vivo effects of Fe_3O_4/salicylic acid magnetic nanoparticles on the human glioblastoma cells. Dig J Nanomater Biostruct. 2014;9:956–65.

Buteică SA, Mihaiescu DE, Rogoveanu I, Mărgăritescu DN, Mîndrilă I. Chick chorioallantoic membrane model as a preclinical tool for nanoparticles biology study. Rom Biotechnol Lett. 2016;21:11684–90.

Caizer C. Magnetic behaviour of $Mn_{0.6}Fe_{0.4}Fe_2O_4$ nanoparticles in ferrofluid at low temperatures. J Magn Magn Mater. 2002;251:304–15.

Caizer C. T2 law for magnetite-based ferrofluids. J Phys Condens Matter. 2003;15:765–76.

Caizer C. Magnetic nanofluids (in Romanian). Timişoara: Eurobit; 2004.

Caizer C. Magnetic properties of the novel nanocomposite $(Zn_{0.15}Ni_{0.85}Fe_2O_4)_{0.15}/(SiO_2)_{0.85}$ at room temperature. J Magn Magn Mater. 2008;320:1056–62.

Caizer C. Bioelectromagnetism (in Romanian). Timişoara: Eurobit; 2013.

Caizer C. Nanoparticle size effect on some magnetic properties. Chapter in: Handbook of nanoparticles. Springer; 2015.

Caizer C, Popovici M, Savii C. Spherical $(Zn_\delta Ni_{1-\delta}Fe_2O_4)_\gamma$ nanoparticles in an amorphous $(SiO_2)_{1-\gamma}$ matrix, prepared with the sol-gel method. Acta Mater. 2003;51:3607–16.

Camaschella C, Campanella A, De Falco L, Boschetto L, Merlini R, Silvestri L, Levi S, Iolascon A. The human counterpart of zebrafish shiraz shows sideroblastic-like microcytic anemia and iron overload. Blood. 2007;110:1353–8.

Chan TA, Morin PJ, Vogelstein B, Kinzler KW. Mechanisms underlying nonsteroidal antiinflammatory drug-mediated apoptosis. Proc Natl Acad Sci U S A. 1998;95:681–6.

Chen WH, Chen JX, Cheng H, Chen CS, Yang J, Xu XD, Wang Y, Zhuo RX, Zhang XZ. A new anti-cancer strategy of damaging mitochondria by pro-apoptotic peptide functionalized gold nanoparticles. Chem Commun. 2013;49:6403–5.

Cherukula K, Manickavasagam LK, Uthaman S, Cho K, Cho CS, Park IK. Multifunctional inorganic nanoparticles: recent progress in thermal therapy and imaging. Nanomaterials. 2016;6:76.

Chomoucka J, Drbohlavova J, Huska D, Adam V, Kizek R, Hubalek J. Magnetic nanoparticles and targeted drug delivering. Pharmacol Res. 2010;62:144–9.

Clark AE. Ferromagnetic materials. Amsterdam: North Holland; 1980.

Corchero JL, Villaverde A. Biomedical applications of distally controlled magnetic nanoparticles. Trends Biotechnol. 2009;27:468–76.

Cullity BD, Graham CD. Introduction to magnetic materials. 2nd ed. Hoboken: Wiley; 2009.

Cunningham CH, Arai T, Yang PC, McConnell MV, Pauly JM, Conolly SM. Positive contrast magnetic resonance imaging of cells labeled with magnetic nanoparticles. Magn Reson Med. 2005;53:999–1005.

Dix MM, Simon GM, Cravatt BF. Global mapping of the topography and magnitude of proteolytic events in apoptosis. Cell. 2008;34:679–91.
Domingo JC, Mercadal M, Petriz J, De Madariaga MA. Preparation of PEG-grafted immunomagnetoliposomes entrapping citrate stabilized magnetite particles and their application in CD34+ cell sorting. J Microencapsul. 2001;18:41–54.
Dovizio M, Bruno A, Tacconelli S Patrignani P. Mode of action of aspirin as a chemopreventive agent. Chapter in: prospects for chemoprevention of colorectal neoplasia. Springer; 2013.
Du L, Chen J, Qi Y, Li D, Yuan C, Lin MC, Yew DT, Kung HF, Yu JC, Lai L. Preparation and biomedical application of a non-polymer coated superparamagnetic nanoparticle. Int J Nanomedicine. 2007;2:805–12.
Fauconnier N, Bee A, Roger J, Pons JN. Adsorption of gluconic and citric acids on maghemite particles in aqueous medium. Trends Colloid Interf Sci X. 1996;100:212–6.
Fauconnier N, Pons JN, Roger J, Bee A. Thiolation of maghemite nanoparticles by dimercaptosuccinic acid. J Colloid Interface Sci. 1997;194:427–33.
Ferrandez A, Piazuelo E, Castells A. Aspirin and the prevention of colorectal cancer. Best Pract Res Clin Gastroenterol. 2012;26:185–95.
Figuerola A, Di Corato R, Manna L, Pellegrino T. From iron oxide nanoparticles towards advanced iron-based inorganic materials designed for biomedical applications. Pharmacol Res. 2010;62:126–43.
Fröhlich E. The role of surface charge in cellular uptake and cytotoxicity of medical nanoparticles. Int J Nanomedicine. 2012;7:5577–91.
Giannaccini M, Cuschieri A, Dente L, Raffa V. Non-mammalian vertebrate embryos as models in nanomedicine. Nanomedicine. 2014;10:703–19.
Goodwin AP, Tabakman SM, Welsher K, Sherlock SP, Prencipe G, Dai H. Phospholipid-Dextran with a Single Coupling Point: a Useful Amphiphile for Functionalization of Nanomaterials. J Am Chem Soc. 2009;131:289–96.
Hajba L, Guttman A. The use of magnetic nanoparticles in cancer theranostics: toward handheld diagnostic devices. Biotechnol Adv. 2016;34:354–61.
Halbreich A, Roger J, Pons JN, Geldwerth D, DaSilva MF, Roudier M, Bacri JC. Biomedical applications of maghemite ferrofluid. Biochimie. 1998;80:379–90.
He C, Jiang S, Jin H, Chen S, Lin G, Yao H, Wang X, Mi P, Ji Z, Lin Y, Lin Z, Liu G. Mitochondrial electron transport chain identified as a novel molecular target of SPIO nanoparticles mediated cancer-specific cytotoxicity. Biomaterials. 2016;83:102–14.
Hilger I, Frühauf S, Linß W, Hiergeist R, Andrä W, Hergt R, Kaiser WA. Cytotoxicity of selected magnetic fluids on human adenocarcinoma cells. J Magn Magn Mater. 2003;261:7–12.
Hofmann-Amtenbrink M, Grainger DW, Hofmann H. Nanoparticles in medicine: current challenges facing inorganic nanoparticle toxicity assessments and standardizations. Nanomedicine. 2015;11:1689–94.
Högemann-Savellano D, Bos E, Blondet C, Sato F, Abe T, Josephson L, Weissleder R, Gaudet J, Sgroi D, Peters PJ, Basilion JP. The transferrin receptor: a potential molecular imaging marker for human cancer. Neoplasia. 2003;5:495–506.
Hohnholt MC, Geppert M, Dringen R. Treatment with iron oxide nanoparticles induces ferritin synthesis but not oxidative stress in oligodendroglial cells. Acta Biomater. 2011;7:3946–54.
Huang J, Bu L, Xie J, Chen K, Cheng Z, Li X, Chen X. Effects of nanoparticle size on cellular uptake and liver MRI with polyvinylpyrrolidone-coated iron oxide nanoparticles. ACS Nano. 2010;4:7151–60.
Ito A, Shinkai M, Honda H, Kobayashi T. Medical application of functionalized magnetic nanoparticles. J Biosci Bioeng. 2005;100:1–11.
Jackson JD. Classical electrodynamics. New York: Wiley; 1974.
Jahn MR, Nawroth T, Fütterer S, Wolfrum U, Kolb U, Langguth P. Iron oxide/hydroxide nanoparticles with negatively charged shells show increased uptake in Caco-2 cells. Mol Pharm. 2012;9:1628–37.
Jeong U, Teng X, Wang Y, Yang H, Xia Y. Superparamagnetic colloids: controlled synthesis and niche applications. Adv Mater. 2007;19:33–60.

Jiang W, Kim BYS, Rutka JT, Chan WCW. Nanoparticle-mediated cellular response is size-dependent. Nat Nanotechnol. 2008;3:145–50.

Jiang QL, Zheng SW, Hong RY, Deng SM, Guo L, Hu RL, Gao B, Huang M, Cheng LF, Liu GH, Wang YQ. Folic acid-conjugated Fe3O4 magnetic nanoparticles for hyperthermia and MRI in vitro and in vivo. Appl Surf Sci. 2014;307:224–33.

Jonsson T, Svedlindh P, Nordblad P. AC susceptibility and magnetic relaxation studies on frozen ferrofluids evidence for magnetic dipole-dipole interactions. J Magn Magn Mater. 1995;140:401–2.

Jurgons R, Seliger C, Hilpert A, Trahms L, Odenbach S, Alexiou C. Drug loaded magnetic nanoparticles for cancer therapy. J Phys Condens Matter. 2006;18:S2893.

Kajii H, Horie T, Hayashi M, Awazu S. Effects of salicylic acid on the permeability of the plasma membrane of the small intestine of the rat: a fluorescence spectroscopic approach to elucidate the mechanism of promoted drug absorption. J Pharm Sci. 1986;75:475–8.

Kim EH, Lee HS, Kwak BK, Kim BK. Synthesis of ferrofluid with magnetic nanoparticles by sonochemical method for MRI contrast agent. J Magn Magn Mater. 2005;289:328–30.

Kobayashi T, Kakimi K, Nakayama E, Jimbow K. Antitumor immunity by magnetic nanoparticle-mediated hyperthermia. Nanomedicine. 2014;9:1715–26.

Kou L, Sun J, Zhai Y, He Z. The endocytosis and intracellular fate of nanomedicines: implication for rational design. Asian J Pharm Sci. 2013;8:1–10.

Laurent S, Forge D, Port M, Roch A, Robic C, Vander Elst L, Muller RN. Magnetic iron oxide nanoparticles: synthesis, stabilization, vectorization, physicochemical characterizations, and biological applications. Chem Rev. 2008;108:2064–110.

Lawaczeck R, Menzel M, Pietsch H. Superparamagnetic iron oxide particles: contrast media for magnetic resonance imaging. Appl Organomet Chem. 2004;18:506–13.

Lee J, Isobe T, Senna M. Preparation of ultrafine Fe_3O_4 particles by precipitation in the presence of PVA at high pH. J Colloid Interface Sci. 1996;177:490–4.

Lee JH, Ju JE, Kim BI, Pak PJ, Choi EK, Lee HS, Chung N. Rod-shaped iron oxide nanoparticles are more toxic than sphere-shaped nanoparticles to murine macrophage cells. Environ Toxicol Chem. 2014;33:2759–66.

Lewin M, Carlesso N, Tung CH, Tang XW, Cory D, Scadden DT, Weissleder R. Tat peptide-derivatized magnetic nanoparticles allow in vivo tracking and recovery of progenitor cells. Nat Biotechnol. 2000;18:410–4.

Li Y, Chen ZW, Gu N. In vitro biological effects of magnetic nanoparticles. Chin Sci Bull. 2012;57:3972–8.

Liu C, Zhang N. Nanoparticles in gene therapy principles, prospects, and challenges. Prog Mol Biol Transl Sci. 2011;104:509–62.

Lunov O, Syrovets T, Loos C, Beil J, Delacher M, Tron K, Nienhaus GU, Musyanovych A, Mailänder V, Landfester K, Simmet T. Differential uptake of functionalized polystyrene nanoparticles by human macrophages and a monocytic cell line. ACS Nano. 2011;5:1657–69.

Ma HL, Qi XR, Maitani Y, Nagai T. Preparation and characterization of superparamagnetic iron oxide nanoparticles stabilized by alginate. Int J Pharm. 2007;333:177–86.

Magdolenova Z, Drlickova M, Henjum K, Rundén-Pran E, Tulinska J, Bilanicova D, Pojana G, Kazimirova A, Barancokova M, Kuricova M, Liskova A, Staruchova M, Ciampor F, Vavra I, Lorenzo Y, Collins A, Rinna A, Fjellsbø L, Volkovova K, Marcomini A, Amiry-Moghaddam M, Dusinska M. Coating-dependent induction of cytotoxicity and genotoxicity of iron oxide nanoparticles. Nanotoxicology. 2015;9:44–56.

Mahmoudi M, Simchi A, Imani M, Shokrgozar MA, Milani AS, Häfeli UO, Stroeve PA. Recent advances in surface engineering of superparamagnetic iron oxide nanoparticles for biomedical applications. J Iran Chem Soc. 2010;7:S1–S27.

Massart R. Preparation of aqueous magnetic liquids in alkaline and acidic media. IEEE Trans Magn. 1981;17:1247–8.

McCarthy JR, Weissleder R. Multifunctional magnetic nanoparticles for targeted imaging and therapy. Adv Drug Deliv Rev. 2008;60:1241–51.

McCarty MF, Block KI. Preadministration of high-dose salicylates, suppressors of NF-kappaB activation, may increase the chemosensitivity of many cancers: an example of proapoptotic signal modulation therapy. Integr Cancer Ther. 2006;5:252–68.
Medeirosa SF, Santosa AM, Fessi H, Elaissari A. Stimuli-responsive magnetic particles for biomedical applications. Int J Pharm. 2011;403:139–61.
Mihaiescu DE, Buteică AS, Neamţu J, Istrati D, Mindrila I. Fe_3O_4/Salicylic acid nanoparticles behavior on chick CAM vasculature. J Nanopart Res. 2013;15:1857.
Mîndrilă I, Buteică AS, Mihaiescu DE, Fudulu D, Badea G, Mărgăritescu DN. Fe_3O_4/salicylic acid nanoparticles versatility in magnetic mediated vascular nanoblockage. J Nanopart Res. 2016;18:10.
Molday RS, Mackenzie D. Immunospecific ferromagnetic iron-dextran reagents for the labeling and magnetic separation of cells. J Immunol Methods. 1982;52:353–67.
Naqvi S, Samim M, Abdin MZ, Ahmed FJ, Maitra AN, Prashant CK, Dinda AK. Concentration-dependent toxicity of iron oxide nanoparticles mediated by increased oxidative stress. Int J Nanomedicine. 2010;5:983–9.
Néel L. Magnetic properties of ferrites: ferrimagnetism and antiferromagnetism. Ann Phys. 1948;3:137–98.
Néel L. Theory of magnetic viscosity of fine grained ferromagnetics with application to baked clays. Ann Geophys. 1949;5:99–136.
Nunez L, Valero RA, Senovilla L, Sanz-Blasco S, Garcia-Sancho J, Villalobos C. Cell proliferation depends on mitochondrial Ca^{2+} uptake: inhibition by salicylate. J Physiol. 2006;571:57–73.
Odenbach S. Magnetoviscous effects in ferrofluids. Berlin: Springer-Verlag; 2002.
Pankhurst QA, Connolly J, Jones SK, Dobson JJ. Applications of magnetic nanoparticles in biomedicine. J Phys D Appl Phys. 2003;36:R167.
Pardoe H, Chua-Anusorn W, Pierre TGS, Dobson J. Structural and magnetic properties of nanoscale iron oxide particles synthesized in the presence of dextran or polyvinyl alcohol. J Magn Magn Mater. 2001;225:41–6.
Park EJ, Umh HN, Choi DH, Cho MH, Choi W, Kim SW, Kim Y, Kim JH. Magnetite-and maghemite-induced different toxicity in murine alveolar macrophage cells. Arch Toxicol. 2014;88:1607–18.
Patil US, Adireddy S, Jaiswal A, Mandava S, Lee BR, Chrisey DB. In vitro/in vivo toxicity evaluation and quantification of iron oxide nanoparticles. Int J Mol Sci. 2015;16:24417–50.
Paulo CSO, Pires-das Neves R, Ferreira LS. Nanoparticles for intracellular-targeted drug delivery. Nanotechnology. 2011;22:494002.
Pekkanen AM, DeWitt MR, Rylander MN. Nanoparticle enhanced optical imaging and phototherapy of cancer. J Biomed Nanotechnol. 2014;10:1677–712.
Portet D, Denizot B, Rump E, Lejeune JJ, Jallet P. Nonpolymeric coatings of iron oxide colloids for biological use as magnetic resonance imaging contrast agents. J Colloid Interface Sci. 2001;238:37–42.
Pouliquen D, Perdrisot R, Ermias A, Akoka S, Jallet P, Le Jeune JJ. Superparamagnetic iron oxide nanoparticles as a liver MRI contrast agent: contribution of microencapsulation to improved biodistribution. Magn Reson Imaging. 1989;7:619–27.
Pradhan P, Giri J, Samanta G, Sarma HD, Mishra KP, Bellare J, Banerjee R, Bahadur D. Comparative evaluation of heating ability and biocompatibility of different ferrite-based magnetic fluids for hyperthermia application. J Biomed Mater Res B: Appl Biomater. 2007;81:12–22.
Rado GT, Suhl H. Magnetism. New York: Academic; 1973.
Rangelov S, Edwards K, Almgren M, Karlsson G. Steric stabilization of egg-phosphatidylcholine liposomes by copolymers bearing short blocks of lipid-mimetic units. Langmuir. 2003;19:172–81.
Riggio C, Calatayud MP, Hoskins C, Pinkernelle J, Sanz B, Torres TE, Ibarra MR, Wang L, Keilhoff G, Goya GF, Raffa V, Cuschieri A. Poly-l-lysine-coated magnetic nanoparticles as intracellular actuators for neural guidance. Int J Nanomedicine. 2012;7:3155–66.
Rosensweig RE. Ferrohydrodynamics. Cambridge, UK: Cambridge University Press; 1985.
Ryu J, Kaul Z, Yoon AR, Liu Y, Yaguchi T, Na Y, Ahn HM, Gao R, Choi IK, Yun CO, Kaul SC, Wadhwa R. Identification and functional characterization of nuclear mortalin in human carcinogenesis. J Biol Chem. 2014;289:24832–44.

Safarik I Safarikova M. Magnetic nanoparticles and biosciences. Chapter in: nanostructured materials. Springer; 2002.

Sahoo Y, Pizem H, Fried T, Golodnitsky D, Burstein L, Sukenik CN, Markovich G. Alkyl phosphonate/phosphate coating on magnetite nanoparticles: a comparison with fatty acids. Langmuir. 2001;17:7907–11.

Sahoo Y, Goodarzi A, Swihart MT, Ohulchanskyy TY, Furlani EP, Prasad PN. Aqueous ferrofluid of magnetite nanoparticles: fluorescence labeling and magnetophoretic control. J Phys Chem B. 2005;109:3879–85.

Schmidt CW. Nanotechnology-related environment, health, and safety research: examining the national strategy. Environ Health Perspect. 2009;117:158–61.

Shiff SJ, Shivaprasad P, Santini DL. Cyclooxygenase inhibitors: drugs for cancer prevention. Curr Opin Pharmacol. 2003;3:352–61.

Shin SW, Song IH, Um SH. Role of physicochemical properties in nanoparticle toxicity. Nanomaterials. 2015;5:1351–65.

Shinkai M, Ito A. Functional magnetic particles for medical application. Recent progress of biochemical and biomedical engineering in Japan II. 2004. p. 761.

Singh N, Jenkins GJS, Asadi R, Doak SH. Potential toxicity of superparamagnetic iron oxide nanoparticles. Nano Rev. 2010;1:5358.

Smit J, Wijin HPJ. Les Ferrites. Bibl. Tech. Philips. 1961.

Smith JE, Sapsford KE, Tan W, Ligler FS. Optimization of antibody-conjugated magnetic nanoparticles for target preconcentration and immunoassays. Anal Biochem. 2011;410:124–32.

Söffge F, Schmidbauer E. AC susceptibility and static magnetic properties of Fe_3O_4 ferrofluid. J Magn Magn Mater. 1981;24:54–66.

Sousa MH, Rubim JC, Sobrinho PG, Tourinho FA. Biocompatible magnetic fluid precursors based on aspartic and glutamic acid modified maghemite nanostructures. J Magn Magn Mater. 2001;225:67–72.

Spitz GA, Furtado CM, Sola-Penna M, Zancan P. Acetylsalicylic acid and salicylic acid decrease tumor cell viability and glucose metabolism modulating 6-phosphofructo-1-kinase structure and activity. Biochem Pharmacol. 2009;77:46–53.

Stathopoulos GP, Boulikas TJ. Lipoplatin formulation. J Drug Deliv. 2012;581363:1–10.

Sun C, Lee JSH, Zhang M. Magnetic nanoparticles in MR imaging and drug delivery. Adv Drug Deliv Rev. 2008;60:1252–65.

Tartaj P, Veintemillas-Verdaguer S, Serna CJ. The preparation of magnetic nanoparticles for applications in biomedicine. J Phys D Appl Phys. 2003;36:R182.

Taylor U, Klein S, Petersen S, Kues W, Barcikowski S, Rath D. Nonendosomal cellular uptake of ligand-free, positively charged gold nanoparticles. Cytometry A. 2010;77:439–46.

Thanh NT, Green LA. Functionalisation of nanoparticles for biomedical applications. Nano Today. 2010;5:213–30.

Thun MJ, Jacobs EJ, Patrono C. The role of aspirin in cancer prevention. Nat Rev Clin Oncol. 2012;9:259–67.

Tiefenauer LX, Kuehne G Andres RY (1993) Antibody-magnetite nanoparticles: in vitro characterization of a potential tumor-specific contrast agent for magnetic resonance imaging. Bioconjug Chem 4:347–352.

Van Vleck JH. The coupling of angular momentum vectors in molecules. Rev Mod Phys. 1951;23:213.

Varshney M, Yang L, Su XL, Li Y. Magnetic nanoparticle-antibody conjugates for the separation of Escherichia coli O157: H7 in ground beef. J Food Prot. 2005;68:1804–11.

Wang YXJ. Superparamagnetic iron oxide based MRI contrast agents: current status of clinical application. Quant Imaging Med Surg. 2011;1:35–40.

Wang Y, Teng X, Wang JS, Lin XZ, Yang H. Solvent-free atom transfer radical polymerization in the synthesis of Fe_2O_3 polystyrene core-shell nanoparticles. Nano Lett. 2003;3:789–93.

Whitesides GM, Kazlauskas RJ, Josephson L. Magnetic separations in biotechnology. Trends Biotechnol. 1983;1:144–8.

Wu W, He Q, Jiang C. Magnetic iron oxide nanoparticles: synthesis and surface functionalization strategies. Nanoscale Res Lett. 2008;3:397–415.

Xu XM, Sansores-Garcia L, Chen XM, Matijevic-Aleksic N, Du M, Wu KK. Suppression of inducible cyclooxygenase 2 gene transcription by aspirin and sodium salicylate. Proc Natl Acad Sci U S A. 1999;96:5292–7.

Ye H, Jeong SY, Ghosh MC, Kovtunovych G, Silvestri L, Ortillo D, Uchida N, Tisdale J, Camaschella C, Rouault TA. Glutaredoxin 5 deficiency causes sideroblastic anemia by specifically impairing heme biosynthesis and depleting cytosolic iron in human erythroblasts. J Clin Investig. 2010;120:1749–61.

Yin PT, Birju PS, Ki-Bum L. Combined magnetic nanoparticle-based microRNA and hyperthermia therapy to enhance apoptosis in brain cancer cells. Small. 2014;10:4106–12.

Yu SS, Lau CM, Thomas SN, Jerome WG, Maron DJ, Dickerson JH, Hubbell JA, Giorgio TD. Size- and charge-dependent non-specific uptake of PEGylated nanoparticles by macrophages. Int J Nanomedicine. 2012;7:799–813.

Zborowski M. Physics of magnetic cell sorting. Chapter in: scientific and clinical applications of magnetic carriers. Springer US; 1997.

Zhang Y, Kohler N, Zhang M. Surface modification of superparamagnetic magnetite nanoparticles and their intracellular uptake. Biomaterials. 2002;23:1553–61.

Zhang L, Gu FX, Chan JM, Wang AZ, Langer RS, Farokhzad OC. Nanoparticles in medicine: therapeutic applications and developments. Clin Pharmacol Ther. 2008;83:761–9.

Zhao F, Zhao Y, Liu Y, Chang X, Chen C, Zhao Y. Cellular uptake, intracellular trafficking, and cytotoxicity of nanomaterials. Small. 2011a;7:1322–37.

Zhao H, Xing D, Chen Q. New insights of mitochondria reactive oxygen species generation and cell apoptosis induced by low dose photodynamic therapy. Eur J Cancer. 2011b;47:2750–61.

Chapter 12
Metal and Metal Oxide Nanoparticles in Photoinactivation of Pathogens

Irena Maliszewska and Katarzyna Popko

Abstract Bacterial infections pose serious health problem that has drawn public attention worldwide. Increased outbreak and infections of pathogenic strains, bacterial antibiotic resistance, emergence of new bacterial mutations, lack of suitable vaccine and nosocomial infections are global health hazard to human. Over the last few years, the increased attention of the researchers was directed to questions related to the biomedical use of different nanoparticles. Nanotechnology is a research hot spot in modern materials science. This technology can provide new applications that range from innovative fabric compounds, food processing and agricultural production to medicinal techniques.

This chapter summarizes the experimental results of the effect of metal (like silver, gold) and metal oxide nanoparticles (like zinc oxide or titanic oxide) and quantum dots on the microorganisms under light exposure. The following sections discuss the properties of gold nanoparticles in photothermal killing of various pathogens, the ability of the conjugates of magnetic and plasmon-resonance nanoparticles with dyes, porphyrins and phthalocyanines to kill microorganisms as well as photocatalytic properties of ZnO and TiO_2 in inactivation of microorganisms.

Keywords Metal nanoparticles • Oxide nanoparticles • Photodynamic inactivation • Microorganisms • Bactericide enhancement • Fungicide activity

12.1 Introduction

Since the discovery, antimicrobial medications have proved remarkably effective for the control of microbial infections. It was, however, found out that some pathogens very soon became resistant to many of the first-generation drugs. Research into

---

I. Maliszewska (✉) • K. Popko
Division of Medicinal Chemistry and Microbiology, Faculty of Chemistry, Wroclaw University of Science and Technology, Wybrzeże Wyspiańskiego 27, 50-370 Wroclaw, Poland
e-mail: irena.helena.maliszewska@pwr.edu.pl; katarzyna.popko@pwr.edu.pl

M. Rai, R. Shegokar (eds.), *Metal Nanoparticles in Pharma*,
DOI 10.1007/978-3-319-63790-7_12

and development of new antimicrobial medicines, vaccines and diagnostic tools call for more innovation and investment.

Photodynamic inactivation (PDI) of microbes is one of the innovative and promising approaches that involves reactive oxygen species (ROS) to kill microorganisms by combining a non-toxic dye termed a photosensitizer (PS) with low-intensity visible light, which, in the presence of oxygen, produces a cytotoxic species (Hamblin and Hasan 2004; Kharkwal et al. 2011). Typically, type I reactions generate radical and radical anion species (e.g. $O_2^{\bullet-}$, $HO^{\bullet}$), while type II reactions produce reactive singlet oxygen ($^1O_2$). For the type II pathway, the impact of PDI is highly dependent on the oxygen content, which is the presence of oxygen. The mechanism of action of PDI is a multi-target damaging process without any specific photosensitizer–receptor interaction on the surface or inside microorganisms. The multiplicity of cellular targets in microorganisms should reduce the risk of selection of photo-mutant resistant strains, and this risk should be further minimized by the lack of mutagenic effects of PDI (Plaetzer et al. 2009). PDI turned out to be a particularly good technique for dental and dermatological applications, involving the light irradiation of a tissue containing microorganisms that were previously exposed to a photosensitizing dye.

It is well known that susceptibility of bacteria to PDI is related to the structures of cell envelopes. Gram-positive bacteria are more sensitive to photodynamic inactivation than Gram-negative cells. The outer membrane of Gram-negative bacteria contains lipopolysaccharides and facilitates non-vesicle-mediated transport through channels. The complex structure of the cell wall makes the Gram-negative bacteria poorly permeable for photosensitizers and generated reactive species.

This therapy has some limitations particularly associated with the delivery of PSs (Ricchelli 1995; Lu et al. 2008). Most of the photosensitizers are hydrophobic and tend to aggregate in aqueous solutions, resulting in a reduced singlet oxygen generation (Kuznetsova et al. 2003). Therefore, in order to overcome this limitation, there is a great interest in developing photosensitizers based on nanoparticles incorporating PSs. Various approaches to utilizing nanoparticles for PDI have been reported, mostly using nanoparticles as a photosensitizer carrier. Nanoscale materials have unique physicochemical properties including extremely small size, high surface-to-mass ratio, high reactivity and special interaction with biological systems. By loading PSs into nanoparticles through physical encapsulation, chemical adsorption or coupling, many aspects of PDI can be improved, such as light dosimetry, photosensitizer bleaching, production of singlet oxygen, pharmacokinetics and drug therapeutic index.

The aim of this chapter is to discuss the use of plasmonic, magnetic, semiconductor and metal oxide nanoparticles in photodynamic inactivation of microbes.

12.2 Gold Nanoparticle-Mediated Photo Killing of Microorganisms

Gold nanoparticles (AuNPs) are considered to be very interesting nanomaterials in the photodynamic inactivation (PDI) field. Features such as oxidation resistance, biocompatibility or physical and chemical stability make AuNPs attractive for various antibacterial technologies. Their optical properties can be evaluated by nonlinear optic measurements (Z-scan) (Olesiak-Bańska et al. 2012). In addition, synthesis or separation of these agents is easy (Guo and Wang 2007; Huang et al. 2009; Gordel et al. 2014).

In 2006, Zharov et al. described a new method for selective laser killing of bacteria targeted with light-absorbing gold nanoparticles conjugated with specific antibodies. AuNP-mediated PDT causes physical damage to the bacterium by using a combination of pulsed laser energy and the nanoparticles attached to the bacterium. When nanoparticles are irradiated, they absorb energy, which is quickly transformed through nonradiative relaxation into heat and other accompanying effects. All these factors eventually lead to irreparable damage to the bacterium. Efficient killing of *S. aureus* (above 90%) was achieved with 40 nm gold nanoparticles conjugated to a secondary IgG that targeted the bacterium by the use of a primary protein A (Zharov et al. 2006).

Norman (2008) performed covalent linking AuNPs to primary antibodies. The selectivity obtained through that linking was essential to destroying pathogenic Gram-negative *Pseudomonas aeruginosa*. There was a 75% decrease in cell viability after exposure of nanorod-coated cells to NIR radiation, which corresponded to a significant increase in the number of dead or compromised cells. It was suggested that cell membrane damage following nanoparticle exposure to NIR radiation could be due to numerous factors, including a nanoparticle explosion, shock waves, bubble formation and thermal disintegration.

Research concerning AuNP-mediated PDI is done with lasers working in the red and/or infrared spectral ranges. Investigation focused on the use of Au nanorods conjugated with photosensitizers or antibiotics to ablate MRSA via simultaneous PACT and NIR photothermal radiation. Au nanorods and nanocages simultaneously served as photodynamic and photothermal agents to deactivate MRSA (Kuo et al. 2009; Meeker et al. 2016).

Huang et al. (2007) proved that vancomycin-bound gold nanoparticles are capable of selective binding onto the cell walls of pathogenic bacteria. A large portion (>99%) of bacteria targeted by the AuNPs was destroyed under illumination by NIR light within 5 min. Next the inhibition of pathogenic bacteria cell growth, including Gram-positive, Gram-negative and antibiotic-resistant bacteria, was investigated (Huang et al. 2009).

Jijie et al. (2016) noticed an additional modality of nanoparticle-based photoinactivation of *E. coli* in their research on gold nanorods. The authors explain that antibacterial activity under pulse laser light is linked to the generation of singlet oxygen with the capacity of reducing bacterial viability by 2.5 $\log_{10}$. As ROS are highly reactive, they are believed to penetrate about 200 nm into solution from the site of the generation.

A photothermal effect of gold nanoparticles was observed during the biogenic AuNP-enhanced, MB-induced phototoxic effect on *S. epidermidis*. Irradiation with He–Ne laser (632 nm) caused significant kills in the presence of the free AuNPs. After 5 and 10 min of laser irradiation, the viable count showed a reduction of 1.84 and 2.04 $\log_{10}$, respectively (Maliszewska et al. 2014).

12.3 Conjugates of Gold Nanoparticles

It is well known that various molecules could be successfully attached to the surface of the gold nanoparticles and achieve their affinity in binding to the target cell. Therefore, improvement of sensitivity or bacteria detection speed in PDI methods by AuNP enhancement can be obtained. The detailed mechanism responsible for such phenomena is yet to be discovered although in many research papers, different hypotheses are advanced. Some of them discuss the AuNP presence that changes the relative distribution of ROS agents or increases their production. It is possible that various mechanisms are apparently involved in killing the bacteria, including the local increase in the concentration of the photosensitizer through targeted delivery of nanoparticles, selective interaction with the cell wall of bacteria and the resonance heating of AuNPs under laser light irradiation (Bucharskaya et al. 2016).

Researchers from Wellman Center for Photomedicine from Massachusetts admit that there are a few main advantages of using PS-containing nanoparticles (Sharma et al. 2012). First of all, the ability of the target cell to pump the drug molecule back out is limited; therefore the possibility of drug resistance is reduced. Passive or active targeting via the charged surface of the nanoparticle improves treatment selectivity, and finally the nanoparticle matrix is non-immunogenic (Konan et al. 2003; Pagonis et al. 2010).

It has been proven that the effectiveness of light-activated antimicrobial agents was amended by AuNPs combined with methylene blue in polysiloxane polymers. A significant antimicrobial activity has been shown against methicillin-resistant *S. aureus* and *E. coli*. A 3.5 $\log_{10}$ reduction in the viable count was obtained, when these bacteria were exposed for 5 min to light from a low-power 660 nm laser (Perni et al. 2009).

Perni et al. (2010) investigated the antimicrobial activity of light-activated silicon containing methylene blue (MB) and AuNPs of different sizes. In their study, they found that only gold nanoparticles of 2 nm diameter enhanced the antibacterial activity of MB, while 20 nm AuNPs reduced the antibacterial activity of the dye.

Another research concerning *S. aureus* was performed by Naik et al. (2011). Polyurethane polymer sheets embedded with MB or toluidine blue (TBO) as photosensitizer and 2 nm AuNPs also exhibited a good antibacterial activity. The incorporation of gold nanoparticles with the dyes enhanced the observed kill from 2.8 to 3.8 $\log_{10}$ (MB) and from 4.3 to 4.8 $\log_{10}$ (TBO).

Maliszewska et al. (2014) presented another promising PDI technique. Monodispersed colloidal AuNPs were synthesized by a reduction in Au^{3+} in the presence of *Trichoderma koningii* cell-free filtrate. Afterwards, these biogenic gold nanoparticles with MB as a photosensitizer were used as successful enhancers of *S. epidermidis* lethal photosensitization. MB showed a significant 1.5 and 1.8 $\log_{10}$ unit reduction in *S. epidermidis* after 5 and 10 min of irradiation by Xe lamplight. The AuNP–MB mixture showed a reduction in CFU of 4.7 $\log_{10}$ after 5 min of irradiation, which is a 99.997% kill compared with a 1.5 $\log_{10}$ or 95.238% kill for MB of the same concentration.

Besides the antibacterial potential described above, the AuNP–PS conjugates (MB or TBO) and 540 nm irradiation inactivated *Candida albicans* fungal biofilm. The AuNP-enhanced photodynamic therapy of MB against recalcitrant pathogenic *C. albicans* and *S. aureus* biofilm was also described (Khan et al. 2012; Darabpour et al. 2017). The treated cells lost their wall integrity and increased gold nanoparticle–methylene blue (AuNP–MB) conjugate permeability. The conjugate penetrated into the cell, causing cell wall disintegration and fragmentation of the nucleus or degradation of nuclear DNA.

It was suggested that the AuNP conjugate based on PDT could be employed effectively for treatment of cutaneous *C. albicans* infections in model animals (Mohd et al. 2015).

In 2007, it was demonstrated that toluidine O–tiopronin–GNP (GNP–gold nanoparticles) agents were much more effective in bacteria killing, compared to the non-enhanced with NPs toluidine O. The conjugate was active under both 632 nm laser and white light that significantly exceeded that of TBO. At a 1 mM concentration and a 30-min white light exposure, the conjugate showed a reduction in CFU of 4.5 $\log_{10}$, while 0.5 $\log_{10}$ value was obtained for TBO at the same concentration. The cited authors connected these results with the increased extinction coefficient of the conjugate compared to free toluidine (Gil-Toma's et al. 2007).

Tuchina et al. (2011) showed a 5% enhancement of photodynamic and photothermal effects after a 30-min laser illumination (λ = 808 nm) of *S. aureus* treated with conjugates of indocyanine green and gold nanocages.

A reduction in size of the *E. coli* has recently been achieved through a combination of gold nanorods with indocyanine green (ICG) photosensitizer (PS) and pulsed laser light (810 nm) (Jijie et al. 2016).

The combination of near-infrared (NIR) photothermolysis and photodynamic therapy against different models of bacteria (*S. aureus*, *S. epidermidis*, both methicillin susceptible and resistant) was shown by Ratto et al. (2011).

The antibacterial effect of conjugates of gold nanoparticles with dyes and porphyrins against *S. aureus* is compared in Table 12.1. These results show the strong bactericidal effect of these conjugates, and it seems that gold nanocages AuNCg2–haematoporphyrins conjugate show the highest antistaphylococcal activity.

One of the latest studies reports on a successful photodynamic inactivation of planktonic and biofilm cells of *C. albicans* using Rose Bengal (RB) in combination with biogenic gold nanoparticles synthesized by the cell-free filtrate of *Penicillium funiculosum* BL1 strain. Spherical gold nanoparticles (24 ± 3 nm) were coated with proteins; a Xe lamp (80 mW) was used as a light source. Rose Bengal showed a significant 1.74 $\log_{10}$ and 2.19 $\log_{10}$ unit reduction in planktonic cells of *C. albicans* after 20 and 30 min of irradiation. The RB + AuNP mixture showed a reduction in CFU of 4.7 $\log_{10}$ and 4.89 $\log_{10}$, after the same amount of time, that is, a 99.91% and 99.99% kill compared with that of 98.21% and 99.37% for RB of the same concentration. The authors presume that the enhanced phototoxic effect of RB by the biogenic gold nanoparticles may involve improvement of RB accumulation in *C. albicans* cells and a change of the main oxidative mechanism of Rose Bengal from type II to type I photosensitization (Maliszewska et al. 2017).

Table 12.1 Antibacterial effect of conjugates of gold nanoparticles against *S. aureus* 209P

Abbr.	Nanoparticle Shape	Photosensitizier (PS)	Average size, nm	Type of radiation	Maximal inhibition of *S.aureus* 209 P after 30 min-light exposure; CFU, % (Reference)
AuNRd1	Nanorods	ICG	30 × 10	808 nm, 50 mW/cm^2	65 (Ratto et al. 2011)
AuNS	Nanoshells	ICG	140	805 nm, 46 mW/cm^2	55 (Tuchina et al. 2011)
AuNCg	Nanocages	ICG	53	808 nm, 60 mW/cm^2	64 (Tuchina et al. 2011)
AuNR2	Nanorods	HP	50 × 10	808 nm, 100 mW/cm^2	90 (Khlebtsov et al. 2013)
AuNCg2	Nanocages	HP	50	625 nm, 100 mW/cm^2	97 (Khlebtsov et al. 2013)
AuNCl	Nanoclusters	PhS	1.8 (25 Au atoms)	660 nm, 50 mW/cm^2	90 (Khlebtsov et al. 2015)

ICG indocyanine green, *HP* hematoporphyrins, *PhS* Photosens™

Intensively studied, a new type of fluorophores–nanoclusters (NCs) is also emerging in PDT. Highly fluorescent bovine serum albumin (BSA)-directed Au–BSA NCs combined with human antistaphylococcal immunoglobulin (antiSAIgG) and Photosens™ used as photosensitizer created an effective complex. Khlebtsov et al. (2015) and Khlebtsov and Dykman (2017) demonstrated that photodynamic treatment of methicillin-sensitive and methicillin-resistant *S. aureus* with Au–BSA–antiSAIgG–PS complexes and 660 nm light irradiation significantly inactivates both types of bacteria.

12.4 Quantum Dots as Photosensitizer

One of the main predictions regarding quantum dots (QDs) as photosensitizer was made by Bakalova et al. in (2004), who suggested the possibility for energy transfer between quantum dot particles and cell molecules (as triplet oxygen, reducing equivalents, pigments). Potentially, this particular feature could induce the generation of reactive oxygen species, which may provoke apoptosis in cells (Bakalova et al. 2004). Quantum dots are known to be highly selective energy donors (Fisher et al. 2004). They are species with well-defined size, shape and composition and can be synthesized by relatively simple and inexpensive methods. QDs have been shown to be non-toxic in the absence of light, but have the potential to be cytotoxic under irradiation. Notwithstanding, it is their tunable optical properties and surface chemistries that are the biggest advantage of quantum dots over molecular photosensitizers (Samia et al. 2009).

In order to improve antibacterial activity, Narband et al. (2008) used CdSe/ZnS quantum dots with toluidine blue O as a photosensitizer. The authors came to the conclusion that mixtures of QD and TBO enhance bacterial kills in solution upon irradiation with white light, but only at low QD concentrations.

According to the latest research by Chong et al. (2016), graphene quantum dots (GQDs) are a promising novelty for PDI application. It was revealed that GQDs can perform both anti- and pro-oxidant activities depending upon light exposure, which will be useful in guiding the safe application and development of GQDs in terms of their antibacterial properties. Upon exposure to blue light, graphene quantum dots accelerate the oxidation of non-enzymic antioxidants and promote lipid peroxidation, contributing to its phototoxicity.

12.5 Conjugates of Magnetic and Plasmon-Resonance Nanoparticles with Porphyrins and Phthalocyanines

One of the most important aspects in photodynamic inactivation of microbes is the synthesis of functional cationic nanomagnet–porphyrin hybrids. Some researchers have investigated the possibility of using the PDT technique not only for clinical but also environmental application, more specifically for the inactivation of pathogenic microorganisms in water and wastewater (Jemli et al. 2002; Carvalho et al. 2007; Kuznetsova et al. 2007; Oliveira et al. 2009). Carvalho et al. (2010) for the first time used magnetic nanoparticles functionalized with neutral and cationic porphyrins as antimicrobial materials. Three cationic hybrids *6*, *8* and *9* (Fig. 12.1) were synthesized, and their photodynamic therapeutic capabilities were investigated for *E. coli*, *Enterococcus faecalis* and T4-like phages. These selected bacteria and the T4-like phages are commonly used as indicators of the presence of pathogenic microorganisms in wastewaters. The obtained results showed that these multi-charged nanomagnet–porphyrin hybrids are very stable in water and highly effective in the photoinactivation of bacteria and phages. Their remarkable antimicrobial activity, associated with their easy recovery, just by applying a magnetic field, makes these materials into novel photosensitizers for water or wastewater disinfection.

More recently, these researchers carried out the synthesis of new nanomagnet-porphyrin hybrids with a $CoFe_2O_4$ core and studied the recycling and reuse capability of this type of hybrids in water contaminated with bacteria. Bacterial inactivation was examined by monitoring the bioluminescence of Gram-negative *Aliivibrio fischeri* during photosensitization. It was shown that the synthesized cationic nanomagnet-porphyrin hybrids were highly efficient in bacterial photoinactivation and sustained several photoinactivation cycles (recycling and reuse) (Alves et al. 2014).

Haematoporphyrin derivative (HpD), a closely related mixture of oligomeric photosensitizers from blood, is the first-generation sensitizer for use in clinical PDT that engages in some activity against both bacteria and viruses. Their use in, for example, the disinfection of open wounds might, therefore, be problematic on the

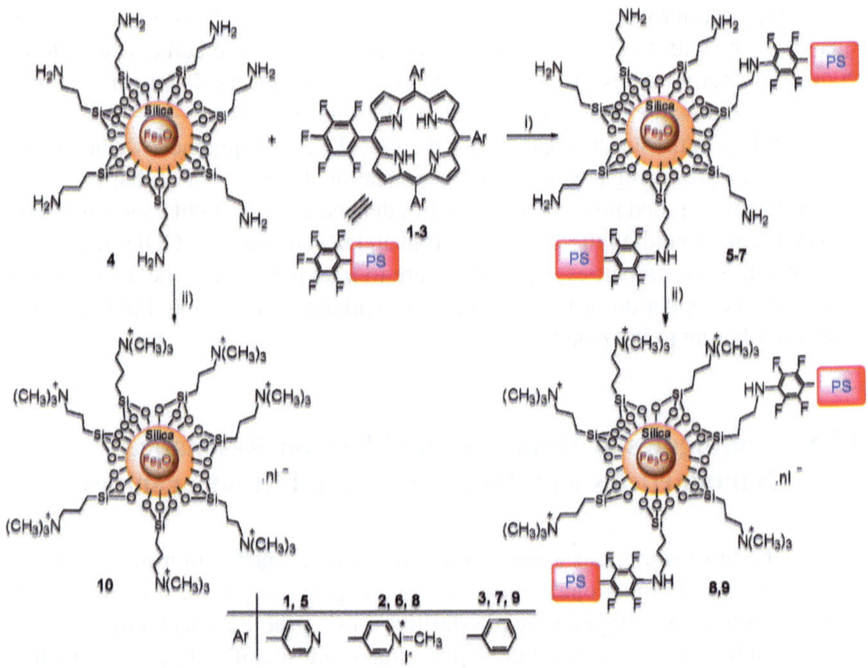

Fig. 12.1 Molecular structure of nanomagnet-porphyrin hybrids (Reproduced from Carvalho et al. 2010; American Chemical Society)

grounds of endogenous light absorption causing a decrease in photosensitizing efficiency, which is why some researchers adopted a new approach to this problem.

For example, Khlebtsov et al. (2013) prepared composite nanoparticles (gold nanorods) consisting of a plasmonic core and used a haematoporphyrin (HP)-doped silica shell. They observed an enhanced inactivation of *S. aureus* 209 P by nanocomposites in comparison with the reference solutions. After a 15-min irradiation with 405 nm light, the PDT efficiency of the composites was four to five times higher than that of free HP, and the percentage of the CFU was less than 3% in the case of incubation with NCs.

A further example of hybrids is presented by Hu et al. (2014), who described a successful synthesis of Ag@mSiO_2@photosensitizer hybrids. Ag@mSiO_2@HPIX (HPIX, haematoporphyrin IX dihydrochloride) was tested against *S. epidermidis* (ATCC 35984). The bacterial culture was mixed with the Ag@mSiO_2@HPIX hybrid and immediately irradiated with white light. The Ag@mSiO_2@HPIX hybrid displayed an enhancement in bacterial killing efficacy of 5 log when the concentration of the adsorbed HPIX was 2 mM. These authors also examined the antibacterial effect of the Ag@mSiO_2@HPIX hybrids against Gram-negative *E. coli* (ATCC 35218) and *A. baumannii* (ATCC 19606). For *E. coli*, the hybrid with an adsorbed HPIX concentration of 1 mM and under a fluence of 400 J cm^{-2} resulted in the complete eradication of the bacterium. In the case of *A. baumannii*, the same hybrid can completely eradicate the bacterium under a fluence of 200 J cm^{-2}. Thus it is clear that the Ag@mSiO_2@photosensitizer hybrids display a synergistic effect in killing both Gram-positive and Gram-negative bacteria. The highly improved PDI efficiency of the Ag@mSiO_2@ photosensitizer hybrids was explained by (1) adsorption of photosensitizers in the mesopores of the silica matrix resulting in a very high local concentration, (2) the surface plasmon-photosensitizer coupling that enhances the singlet oxygen production efficiency and (3) the locally generated singlet oxygen that may reach a higher concentration than when free photosensitizers act individually, causing more damage to the bacteria.

Another study (Ding et al. 2016) showed the synthesis of silver nanoparticles (AgNPS) stabilized by poly(N-isopropylacrylamide-block-styrene; BCP) and used this material to entrap hydrophobic photosensitizing molecules (haematoporphyrin; HP). Synthesized AgNP@BCP@HP demonstrated a very high efficacy in the photoinactivation of *Staphylococcus epidermidis* (ATCC 35984) and *Escherichia coli* (ATCC 35218) under white light illumination. The enhancement of PDI efficacy was defined as $\log_{10}$ (enhancement killing) = $\log_{10}$ (AgNP@BCP@HP killing) − $\log_{10}$ (pristine HP killing) − $\log_{10}$ (AgNPs killing). A significant difference in the PDI efficacy against *S. epidermidis* among AgNP@BCP, pristine HP and AgNP@BCP@HP was thus clearly demonstrated. AgNP@BCP exhibited only a little bacterial killing, while pristine HP had moderate PDI efficiency under white light illumination. In contrast, AgNP@BCP@HP was characterized by higher PDI efficiencies than the sum of AgNP@BCP and HP. A similar effect was also observed in the PDI tests against the Gram-negative rods of *E. coli*, with a PDI efficacy enhancement of up to ~5 orders of magnitude. It should be noticed that AgNP@BCP@HP also photoinactivated bacteria under red/NIR illumination due to its broadened excitation profile. Again, AgNP@BCP@HP displayed a much higher PDI efficacy against both Gram-positive and Gram-negative bacteria than the summary efficacy of AgNP@BCP and HP. Moreover, it was also observed that in these red/NIR illumination experiments, the photoinactivation efficacy of AgNP@BCP@HP against *E. coli* appeared to be higher than that under white light illumination. This phenomenon was attributed to the longer illumination time (20 min) as compared to that under white light illumination (12 min), which may cause photothermal effect on the bacteria. This possibility was indirectly supported by the results of the control experiment involving only AgNP@BCP, where a longer illumination time also led to a higher killing efficacy.

Phthalocyanines have been studied extensively since the first synthesis in 1907 (Braun and Tcherniac 1907). The central cavity of phthalocyanines is known to be capable of accommodating 63 different elemental ions, including hydrogens (metal-free phthalocyanine, H_2-Pc). A phthalocyanine containing one or two metal ions is called a metal phthalocyanine (M-Pc) (Sakamoto and Ohno-Okumura 2009). Phthalocyanines and their derivatives, which have a similar structure to porphyrin, have been applied in important functional materials in many fields. Their useful properties are attributed to their efficient electron transfer abilities. Within the timeframe of the years 1930–1950, the full elucidation of the chemical structure of phthalocyanines was determined, and their X-ray spectra, absorption spectra oxidation and reduction, catalytic properties, magnetic properties, photoconductivity and many more physical properties were investigated. As a result of these studies, it was concluded that phthalocyanines are highly coloured, planar 18 π-electron aromatic ring systems similar to porphyrins (De Diesbach and Von der Weid 1927).

In the last 20 years, phthalocyanine chemistry has been undergoing a renaissance because phthalocyanines and some of their derivatives exhibit singular and unconventional physical properties interesting for applications in materials science (Wöhrle et al. 2012). They are very stable and have strong absorption at short- and long-wavelength ends of the visible spectrum and are thus well suited for optical applications (Bonnett 1995; Kaliya et al. 1999; Loschenov et al. 2000; Makarov et al. 2007; Bohrer et al. 2009; Walter et al. 2010; Wang et al. 2012; Zhang et al. 2014). A great potential application is to use them as photosensitizers for photodynamic inactivation of pathogens (Minnock et al. 1996; Segalla et al. 2002; Caminos et al. 2008; Kussovski et al. 2009; Spesia et al. 2010; Di Palma et al. 2013; Zafar et al. 2016).

Great efforts of some researchers were focused on the efficiency improvement of photoinduced processes in phthalocyanines. In this connection, synthesis of hybrid exciton–plasmon systems based on metal nanostructures is considered to be one of the most promising solutions.

Nombona et al. (2012) described the photoinactivation activity of Zn phthalocyanine–polylysine conjugates in the presence of gold and silver nanoparticles against *S. aureus*. The obtained results showed that the antimicrobial efficacy of photosensitizers can be boosted by the presence of gold and more especially silver nanoparticles.

Masilela et al. (2013) reported on the axial coordination of zinc phthalocyanine and bis-(1,6-hexanedithiol) silicon phthalocyanine to silver and gold nanoparticles. An improvement in the photophysicochemical behaviour and antimicrobial activity was achieved in the presence of metal nanoparticles for both complexes. The bacterial inhibition was found to be best for the bis-(1,6-hexanedithiol) silicon phthalocyanine derivative in the presence of nanoparticles compared to the zinc phthalocyanine counterpart. The highest antimicrobial activity was achieved for both conjugates against *B. subtilis* compared to *S. aureus* both in the dark and under illumination with light.

Mthethwa and Nyokong (2015) reported on the effective complex of aluminium phthalocyanine in combination with gold nanorods (Fig. 12.2) for the photoinactivation of *Candida albicans* and *Escherichia coli*. The efficiency of this complex was evaluated by measuring the log reduction of the studied microorganisms after irradiation with visible light in the presence of photosensitizers. Aluminium phthalocyanine alone showed 1.78 $\log_{10}$ and 2.51 $\log_{10}$ reductions for *C. albicans* and *E. coli*, respectively. The conjugates showed higher photosensitization with 2.53 $\log_{10}$ and 3.71 $\log_{10}$ for *C. albicans* and *E. coli*, respectively.

12.6 Photocatalytic Properties of ZnO and TiO_2 for the Removal of Pathogens

The development of visible light-active materials for the removal of infectious pathogens has become very desirable. In general, wide band gap semiconductors such as ZnO and TiO_2 are recognized as efficient photocatalysts because of their

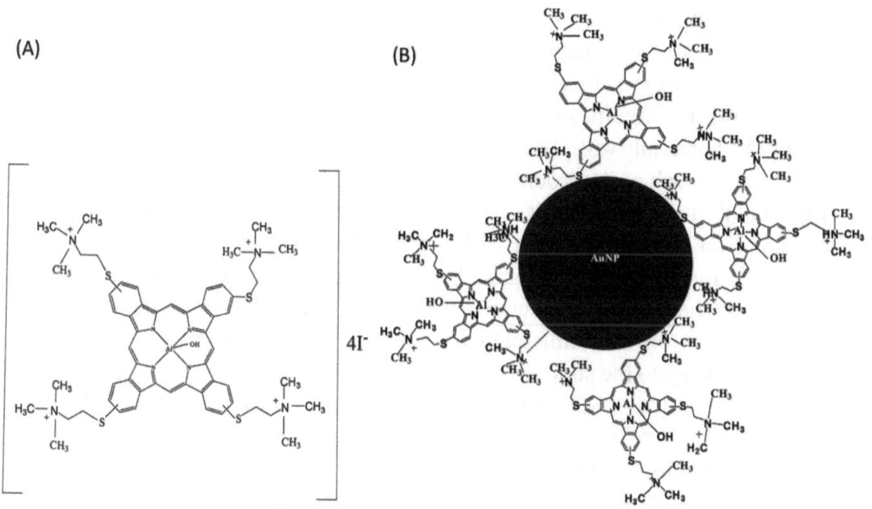

Fig. 12.2 (**a**) Molecular structure of the AlPc derivative (complex *1*). (**b**) Hypothetical structure based on the linking of complex *1* to gold nanoparticles (AuNPs) (Reproduced from Mthethwa and Nyokong 2015; Royal Society of Chemistry)

high redox potential of photocharge carriers. The mechanisms of the photocidal action of nanoparticles activated by light for microbial inactivation are different for various nanoparticles. It was found out that photocatalysis occurs in semiconductors such as ZnO and TiO_2.

12.6.1 Titanium Oxide Nanoparticles

One of the first reports concerning inactivation of bacteria and yeast under the presence of TiO_2 nanoparticles and the UV-A (360–400 nm) is the paper by Matsunaga et al. (1985). The authors reported that *Lactobacillus acidophilus*, *Saccharomyces cerevisiae* and *Escherichia coli* were killed photoelectrochemically with semiconductor powder (platinum-loaded titanium oxide, TiO_2/Pt; halide lamp irradiation for 60–120 min.). It was suggested that coenzyme A was photoelectrochemically oxidized, and the respiration of cells was inhibited, which caused death of the bacterial cells.

Later, it was shown that diffused solar light in the presence of TiO_2 resulted in the inactivation of various bacteria (Saito et al. 1992; Ireland et al. 1993; Wei et al. 1994; Kikuchi et al. 1997; Sunada et al. 1998; Shchukin et al. 2004; Robertson et al. 2005; Rao et al. 2006; Raulio et al. 2006; Kim et al. 2017) and viruses (Watts et al. 1995; Lee et al. 1997; 1998; Gerrity et al. 2008).

A large amount of experimental material related to TiO_2 nanoparticles' bactericidal effect under UV-A light illumination is known. The fact that the increased concentration of photocatalyst TiO_2 and UV-A light intensity results in a more rapid *Bacillus anthracis* killing has been considered non-direct evidence that the photocatalytic mechanism of their action occurs (Prasad et al. 2009; Sun et al. 2016).

An interesting observation was made by Bui et al. (2008). These authors suggested that the effective inactivation of bacteria is due to the contact between TiO_2 nanoparticles and the cellular wall. It was shown that bacterial cell and titanium dioxide adsorption significantly depends on the surface charge of TiO_2 nanoparticles and the bacterial wall. The surface charge of nanoparticles is determined by several factors, such as the isoelectric point, pH and electrolyte content.

Photocatalytic inactivation of microbial cells is schematically shown in Fig. 12.3. The absorption of light quantum with energy exceeding the width of a forbidden band in TiO_2 results in an electron and gap generation in conductivity and valence, respectively. Charges photogenerated in TiO_2 nanoparticles recombine or participate in reactions on the semiconductor–electrolyte division border, and electron reduces air oxygen to the superoxide of anion $O_2^{-\bullet}$ ($HO_2^{\bullet}$). The gap is responsible for the oxidation of hydroxide $H\bar{O}$ anion or the water molecule to the $HO^{\bullet}$ radical (Linsebigler et al. 1995). $HO^{\bullet}$ and $O_2^{-\bullet}$ radicals attack the organic molecules of a cell, which finally results in bacterial killing.

An interesting approach to this topic is modification of TiO_2 photocatalysts via doping by various atoms absorbing the light within the visible range. The effectiveness of bacterial inactivation in some cases increases under visible light and with TiO_2 doped by nitrogen and/or sulphur atoms (Bacsa et al. 2005; Liu et al. 2007; Vacaroiu et al. 2009; Zane et al. 2016) or Fe^{+3} (Rincon and Pulgarin 2007). There was observed an increased photocatalytic activity of TiO_2 under visible light (pure TiO_2 demonstrated rather little absorption under exposure to the visible light). Lately, researchers' interest in the field of the inhibition of pathogenic bacteria by TiO_2 has shifted to the study of TiO_2 nanoparticles containing such metal nanoparticles as Ag, Au, Pt, Pd, etc. The significant enhancement in bacterial killing was observed when the silver nanoparticles were inserted in TiO_2. The complete inactivation of *E. coli* was achieved in 1–2 min under exposure to UV-A light, whereas pure TiO_2 was shown to suppress bacterial viability in a suspension in about 30 min (Sokmen et al. 2001). The authors suggested that the enhanced bactericidal effect after silver application was due to increased effectiveness of organic cellular material oxidation in photocatalytic reactions. Changes in the kinetics of lipid oxidation and the malonic dialdehyde formation were reported while comparing the effects from the pure TiO_2 and Ag/TiO_2.

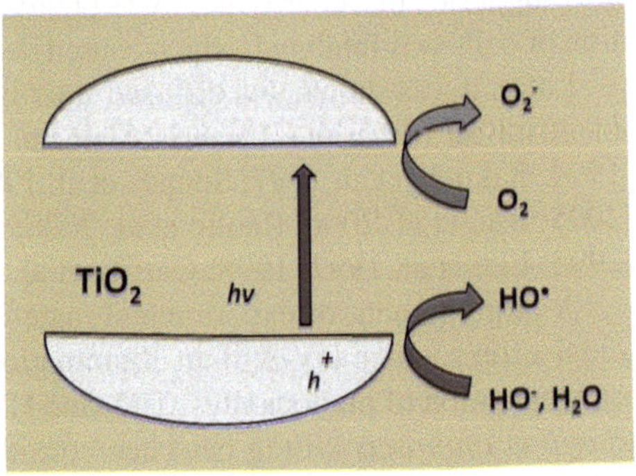

Fig. 12.3 Mechanism of the photocatalytic activity of TiO_2 nanoparticles (Adapted from Nadtochenko et al. 2010)

Some authors (Es-Souni et al. 2008) indicated that Ag/TiO_2 has reproducible bactericidal properties; however, the Ag/TiO_2 system might also possess bactericidal activity under dark conditions (without exposure to UV-A light). The said authors posit that the bactericidal effect of Ag/TiO_2 is due to the possibility of silver nanoparticles forming silver ions that are released into the solution.

The antimicrobial activity of nanostructural TiO_2 and TiO_2:In_2O_3 films and the effect of Ag or bimetallic Ag/Ni nanoparticles against *Pseudomonas fluorescens* and *Lactococcus lactis* were demonstrated by Skorb et al. (2008). The silver-modified TiO_2 film reveals the highest photo biocide efficiency, enhancing the bactericidal activity of UV light ca. 71-fold, which results from a radical improvement of microorganism adsorption and suppression of recombination of photo-produced charge carriers. The inactivation of Gram-negative *P. fluorescens* was higher than that of Gram-positive *L. lactis*. This difference was explained by the various structures of the cell wall in the Gram-negative and Gram-positive bacteria and the resistance of outer membranes to the reactive oxygen species generated by photocatalytic reactions.

Ag–TiO_2 was shown to possess a bactericidal activity against *E. coli* on the surface of hydroxyapatites (Reddy et al. 2007). A hydroxyapatite effectively adsorbs bacterial cells and Ag–TiO_2 inactivated them under the exposure to the visible light. Their combined action results in 100% bacterial death after 2 min. A study of this system using an electron microscopy technique revealed that photocatalyst nanoparticles adsorbed on the bacterial cellular surface obstructed cell nutrition. Moreover, it was shown that Ag–TiO_2 nanoparticles on hydroxyapatite inactivated bacteria in dark conditions.

Copper nanoparticles also demonstrate photocatalytic activity on the surface of TiO_2 nanoparticles (Sunada et al. 2003). Their study focused on the copper-resistant *E. coli* strain. Cu/TiO_2 has no influence on bacterial growth under dark conditions, but it completely inhibited bacterial growth under very weak UV light. It was shown that the decay curve of survival of the bacteria studied on the Cu/TiO_2 film under very weak UV light illumination consisted of two steps, similar to the survival change of normal *E. coli* on TiO_2 films under strong UV illumination. The first step was related to a partial decomposition of the outer membrane caused by photocatalytic oxidation, which further resulted in copper ions penetrating into the cytoplasmic membrane. The second stage was caused by a loss of cytoplasmic membrane integrity due to the copper ions, which brought about the loss of cell integrity. This two-step mechanism explains why the Cu/TiO_2 film system shows an effective bactericidal activity even under very weak UV light illumination.

Palladium nanoparticles were dispersed on two metal oxide substrates TiO_2 and SnO_2 (Erkan et al. 2006). The antimicrobial effect was studied for *E. coli*, *S. aureus*, *Saccharomyces cerevisiae* and *Aspergillus niger*. The antimicrobial efficiencies against different microorganisms and fungal spores were found to decrease in the following order: *E. coli* > *S. aureus* > *S. cerevisiae* > *A. niger* spores for which complexity and strength of the cell walls increased in the same order.

12.6.2 Zinc Oxide Nanoparticles

ZnO is known as a functional, promising and versatile inorganic material with a broad range of applications. ZnO is currently listed as generally recognized as a safe (GRAS) material by the US Food and Drug Administration (21CFR182.8991). Research on ZnO as an antimicrobial agent started in the early 1950s. During the last 50 years, it has been shown that ZnO can be used for many antimicrobial applications (Sirelkhatim et al. 2015).

It has unique optical, chemical and electrical properties (Fan and Lu 2005). It is characterized by a direct wide band gap (~3.3 eV) in the near-UV spectrum, a high excitonic binding energy (60 meV) at room temperature (Wang 2004; Janotti and Van de Walle 2009) and a natural n-type electrical conductivity (Wellings et al. 2008). The wide band gap of ZnO has a significant effect on its properties, such as the electrical conductivity and optical absorption. It was proved that ZnO nanoparticles in aqueous solution under UV radiation produce ROS such as hydrogen peroxide (H_2O_2) and superoxide ions (O^{2-}) (Fiedot et al. 2017; Prado-Prone et al. 2017). These generated active species penetrate into cells and are able to inactivate microorganisms. A detailed reaction mechanism of phototoxic antimicrobial activity of ZnO nanoparticles was proposed by Seven et al. (2004) and Padmavathy and Vijayaraghavan (2011). Photocatalysis was described as a photoinduced oxidation (Baruah et al. 2010).

ZnO is currently being examined as an antibacterial agent in both microscale and nanoscale formulations, and it is well known that this compound exhibits a particularly strong antimicrobial activity when particle size is reduced to a nanometre range. It was discovered that the improved antibacterial activity of ZnO nanoparticles compared to its microparticles was related to the surface area enhancement in the nanoparticles. Padmavathy and Vijayaraghavan (2008) investigated the antibacterial activity of ZnO nanoparticles with various particle sizes (12–2000 nm). These authors demonstrated that the bactericidal efficacy of ZnO nanoparticles increases with decreasing particle size. It was proposed that both the abrasiveness and the surface oxygen species of ZnO nanoparticles promote the biocidal properties of ZnO nanoparticles. More recently, the same authors (Padmavathy and Vijayaraghavan 2011) examined the antibacterial activity of ZnO nanoparticles with various particle sizes. The obtained results demonstrated that the bactericidal efficacy of ZnO nanoparticles also increased by decreasing particle size.

Prasanna and Vijayaraghavan (2015) conducted a systematic and complete antibacterial study on microparticle and nanoparticle of ZnO in both dark and light conditions. It was shown that micro ZnO in the dark revealed no activity against the bacteria studied. In light, ZnO nanoparticles exhibit a bactericidal activity remarkably higher than micro ZnO. Moreover, these studies have conclusively proved that reactive oxygen species (ROS) such as $^{\bullet}OH$, $^{\bullet}O_2^-$ and H_2O_2 are significantly produced from ZnO aqueous suspension even in the dark.

Zhou et al. (2008) showed the results of the antimicrobial activity of nanohydroxyapatite/zinc oxide upon UV exposure toward *E. coli* and S. *aureus*. Their findings also proved that the bactericidal activity can be achieved under UV illumination, ambient light or even in the dark.

Ann et al. (2014) studied the antibacterial activity against *E. coli* and *S. aureus* using ZnO of two forms (ZnO-rod and ZnO plate) which are exposed to UV-A illumination (390 nm). These authors found that UV-A illumination significantly influenced the interaction of both ZnO samples with the tested bacteria compared with unexposed ZnO by 13–21%. The antimicrobial activity depends on the shape of nanoparticles. Release of reactive oxygen species was proposed as an explanation of the mechanism of antimicrobial activity. The reactive oxygen species disrupted the DNA and protein synthesis of the bacterial cell, causing bacteriostatic effects toward *E. coli* and *S. aureus*.

Antimicrobial activity of photoactivated zinc oxide nanoparticles against human pathogens *Escherichia coli* O157:H7, *Listeria monocytogenes* ATCL3C 7644 and plant pathogen *Botrytis cinerea* was investigated by Kairyte et al. (2013). The obtained results suggested that ZnO nanoparticles in the presence of visible light exhibit a strong antibacterial and antifungal activity. Such properties could be used for the development of effective fungicides in agriculture or innovative physical antibacterial agents, so it is important in medicine and food microbial control.

The photocatalytic efficiency of modified ZnO was also evaluated for the inactivation of pathogens. For example, Guo et al. (2015) prepared a novel photocatalyst of Ta-doped ZnO nanoparticles. The antimicrobial study of the impact/influence of these nanoparticles on several bacteria *B. subtilis*, *S. aureus*, *E. coli* and *P. aeruginosa* was performed. The authors showed a particularly strong antimicrobial activity of Ta-doped ZnO nanoparticles under visible light irradiation.

Ta-doped ZnO nanoparticles exhibit an effective bactericidal efficacy due to the synergistic effect of enhanced surface bioactivity and increased electrostatic force (probably Ta^{5+} ions incorporated into ZnO).

ZnO nanoparticles with cobalt doping were synthesized by Oves et al. (2015), and antimicrobial properties were evaluated under sunlight deposition. The most effective bactericidal results were found for *E. coli* and *Vibrio cholerae*.

The F-doping was found to be effective against *S. aureus* (99.99% antibacterial activity) and *E. coli* (99.87% antibacterial activity) when irradiated with visible light. Reactive oxygen species production is one of the major factors that negatively impacts bacterial growth (Podporska-Carroll et al. 2017).

An interesting approach to this topic is the possibility to enhance the pathogen inactivation by combining ZnO and chlorophyllin. Chlorophyllin is a water-soluble food additive (E 140) known for its antimutagenic and anticarcinogenic properties and exhibiting a high antioxidant capacity (Kamat et al. 2000). The results obtained by Aponiene and Luksiene (2015) indicated that inactivation of *E. coli* by ZnO–chlorophyllin-based photosensitization is fairly effective.

In spite of many years of research, the exact mechanism of antimicrobial activity of ZnO particles has not been well understood. A number of mechanisms such as generation of hydrogen peroxide, accumulation of the particles on the bacteria surface, ROS generation on the surface of the particles, zinc ion release, membrane dysfunction, nanoparticle internalization and interruption of transmembrane electron transportation have been proposed as possible explanations.

12.7 Conclusions

The PDI is a truly promising approach to fight with highly pathogenic microorganisms. The variety of opportunities fuels constant development in this area of photochemotherapy. Different kinds of metallic nanoparticles like gold or zinc oxide nanoparticles can be involved in photosensitization process acting as a carrier that is conjugated with the selected dye. For instance, gold nanoparticles are often combined with such photosensitizers like MB, TBO or ICG, whereas plasmon-resonance and magnetic nanoparticles are conjugated with porphyrins and phthalocyanines. Crucial photocatalytic properties of quantum dots, ZnO, TiO_2 and gold nanoparticles are also intensively investigated in order to evaluate their antibacterial activity. Techniques presented here shed light on the potential environmental applications as well as on the improvement of already existing methods of PDI.

Acknowledgements This work was partially financed by a statutory activity subsidy from the Polish Ministry of Science and Higher Education (PMSHE) for the Faculty of Chemistry of Wrocław University of Science and Technology.

References

Alves E, Rodrigues JMM, Faustino MAF, Neves MGPMS, Cavaleiro JAS, Lin Z, Cunha A, Nadais MH, Tome JPC. A new insight on nanomagneteporphyrin hybrids for photodynamic inactivation of microorganisms. Dyes Pigments. 2014;110:80–8.

Ann LC, Mahmud S, Bakhori SKM, Sirelkhatim A, Mohamad D, Hasan H, Seeni A, Rahman RA. Effect of surface modification and UVA photoactivation on antibacterial bioactivity of zinc oxide powder. Appl Surf Sci. 2014;292:405–12.

Aponiene K, Luksiene Z. Effective combination of LED-based visible light, photosensitizer and photocatalyst to combat gram (−) bacteria. J Photochem Photobiol B Biol. 2015;142:257–63.

Bacsa R, Kiwi J, Ohno T, Albers P, Nadtochenko V. Preparation, testing and characterization of doped TiO_2 active in the peroxidation of biomolecules under visible light. J Phys Chem B. 2005;109:5994–6003.

Bakalova R, Ohba H, Zhelev Z, Ishikawa M, Baba Y. Quantum dots as photosensitizers? Nat Biotechnol. 2004;22:1360–1.

Baruah S, Mahmood MA, Myint MTZ, Bora T, Dutta J. Enhanced visible light photocatalysis through fast crystallization of zinc oxide nanorods. Beilstein J Nanotechnol. 2010;1:14–20.

Bohrer FI, Colesniuc CN, Park J, Ruidiaz ME, Schuller IK, Kummel AC, Trogler WC. Comparative gas sensing in cobalt, nickel, copper, zinc, and metal-free phthalocyanine chemiresistors. J Am Chem Soc. 2009;131:478–85.

Bonnett R. Photosensitizers of the porphyrin and phthalocyanine series for photodynamic therapy. Chem Soc Rev. 1995;24:19–33.

Braun A, Tcherniac J. Über die Produkte der Einwirkung von Acetanhydrid auf Phthalamid. Ber Dtsch Chem Ges. 1907;40:2709–14.

Bucharskaya A, Maslyakova G, Terentyuk G, Yakunin A, Avetisyan Y, Bibikova O, Tuchina E, Khlebtsov B, Khlebtsov N, Tuchin V. Towards effective photothermal/photodynamic treatment using plasmonic gold nanoparticles. Int J Mol Sci. 2016;17:1295.

Bui TH, Felix C, Pigeot-Remy S, Herrmann JM, Lejeune P, Guillard C. Photocatalytic inactivation of wild and hyper-adherent *E. coli* strains in presence of suspended or supported TiO_2.

Influence of the isoelectric point of the particle size and of the adsorptive properties of titania. J Adv Oxid Technol. 2008;11:510–8.

Caminos DA, Spesia MB, Pons P, Durantini EN. Mechanisms of *Escherichia coli* photodynamic inactivation by an amphiphilic tricationic porphyrin and 5,10,15,20-tetra(4-N,N,N-trimethylammoniumphenyl) porphyrin. J Photochem Photobiol. 2008;7:1071–8.

Carvalho CMB, Alves E, Costa L, Tome JPC, Faustino MAF, Neves MGPM, Tome A, Cavaleiro JAS, Almeida A, Cunha A, Lin Z, Rocha J. Functional cationic nanomagnet-porphyrin hybrids for the photoinactivation of microorganisms. ACS Nano. 2010;4:7133–40.

Carvalho CMB, Gomes ATPC, Fernandes SCD, Prata ACB, Almeida MA, Cunha MA, Tome JPC, Faustino MAF, Neves MGPM, Tome A. Photoinactivation of bacteria in wastewater by porphyrins: bacterial-galactosidase activity and leucine-uptake as methods to monitor the process. J Photochem Photobiol B Biol. 2007;88:112–8.

Chong Y, Ge C, Fang G, Tian X, Ma X, Wen T, Wamer WG, Chen C, Chai Z, Yin JJ. Crossover between anti- and pro-oxidant activities of graphene quantum dots in the absence or presence of light. ACS Nano. 2016;10:8690–9.

Darabpour E, Kashef N, Amini SM, Kharrazi S, Gholamreza ED. Fast and effective photodynamic inactivation of 4-day-old biofilm of methicillin-resistant *Staphylococcus aureus* using methylene blue-conjugated gold nanoparticles. J Drug Deliv Sci Technol. 2017;37:134–40.

De Diesbach H, Von der Weid E. Quelques Sels Complexes des o-dinitriles avec le cuivre et la pyridine. Helv Chim Acta. 1927;10:886–8.

Ding R, Yu X, Wang P, Zhang J, Zhou Y, Cao X, Tang H, Ayres N, Zhang P. Hybrid photosensitizer based on amphiphilic block copolymer stabilized silver nanoparticles for highly efficient photodynamic inactivation of bacteria. RSC Adv. 2016;6:20392–8.

Di Palma MA, Alvarez MG, Ochoa AL, Milanesio ME, Durantini EN. Optimization of cellular uptake of zinc(II) 2,9,16,23-tetrakis[4-(N-methylpyridyloxy)]phthalocyanine for maximal photoinactivation of *Candida albicans*. Fungal Biol. 2013;117:744–51.

Erkan A, Bakir U, Karakas G. Photocatalytic microbial inactivation over Pd doped SnO_2 and TiO_2 thin films. J Photochem Photobiol A Chem. 2006;184:313–21.

Es-Souni M, Fischer-Brandies H, Es-Souni M. Versatile nanocomposite coatings with tunable cell adhesion and bactericidity. Adv Funct Mater. 2008;18:3179–88.

Fan Z, Lu JG. Zinc oxide nanostructures: synthesis and properties. J Nanosci Nanotechnol. 2005;5:1561–73.

Fiedot M, Maliszewska I, Rac-Rumijowska O, Suchorska-Woźniak P, Lewińska A, Teterycz H. The relationship between the mechanism of zinc oxide crystallization and its antimicrobial properties for the surface modification of surgical meshes. Materials. 2017;10:353.

Fisher BR, Eisler HJ, Scott NE, Bawendi MG. Emission intensity dependence and single-exponential behavior in single colloidal quantum dot fluorescence lifetimes. J Phys Chem B. 2004;108:143–8.

Gerrity D, Ryu H, Crittenden J, Abbaszadegan M. Photocatalytic inactivation of viruses using titanium dioxide nanoparticles and low-pressure UV light. J Environ Sci Health – Part A Toxic/Hazard Subst Environ Eng. 2008;43:1261–70.

Gil-Toma's JG, Tubby S, Parkin IP, Narband N, Dekker L, Nair SP, Wilson M, Street C. Lethal photosensitisation of *Staphylococcus aureus* using a toluidine blue O–tiopronin–gold nanoparticle conjugate. J Mater Chem. 2007;17:3739–46.

Gordel M, Olesiak-Bańska J, Matczyszyn K, Nogues C, Buckle M, Samoć M. Post-synthesis reshaping of gold nanorods using a femtosecond laser. Phys Chem Chem Phys. 2014;16:71–8.

Guo B-L, Han P, Guo L-C, Cao Y-Q, Li A-D, Kong J-Z, Zhai H-F, Wu D. The antibacterial activity of Ta-doped ZnO nanoparticles. Nanoscale Res Lett. 2015;10:336.

Guo SE, Wang E. Synthesis and electrochemical applications of gold nanoparticles. Anal Chim Acta. 2007;598:181–92.

Hamblin MR, Hasan T. Photodynamic therapy: a new antimicrobial approach to infectious disease? Photochem Photobiol Sci. 2004;3:436–50.

Hu B, Cao X, Nahan K, Caruso J, Tang H, Zhang P. Surface plasmon-photosensitizer resonance coupling: an enhanced singlet oxygen production platform for broad-spectrum photodynamic inactivation of bacteria. J Mater Chem B. 2014;2:7073–81.

Huang X, Neretina S, El-Sayed MA. Gold nanorods: from synthesis and properties to biological and biomedical applications. Adv Mater. 2009;21:4880–910.

Huang WC, Tsai PJ, Chen YC. Functional gold nanoparticles as photothermal agents for selective-killing of pathogenic bacteria. Nanomedicine. 2007;2:777–87.

Ireland JC, Klostermann P, Rice EW, Clark RM. Inactivation of *Escherichia coli* by titanium dioxide photocatalytic oxidation. Appl Environ Microbiol. 1993;59:1668–70.

Janotti A, Van de Walle CG. Fundamentals of zinc oxide as a semiconductor. Rep Prog Phys. 2009;72:126501.

Jemli M, Alouini Z, Sabbahi S, Gueddari M. Destruction of fecal bacteria in wastewater by three photosensitizers. J Environ Monit. 2002;4:511–6.

Jijie R, Dumych T, Chengnan L, Bouckaert J, Turcheniuk K, Hage CH, Heliot L, Cudennec B, Dumitrascu N, Boukherroub R, Szunerits S. Particle-based photodynamic therapy based on indocyanine green modified plasmonic nanostructures for inactivation of a Crohn's disease-associated *Escherichia coli* strain. J Mater Chem B. 2016;4:2598–605.

Kairyte K, Kadys A, Luksiene Z. Antibacterial and antifungal activity of photoactivated ZnO nanoparticles in suspension. J Photochem Photobiol B Biol. 2013;128:78–84.

Kaliya OL, Lukyanets EA, Vorozhtsov GN. Catalysis and photocatalysis by phthalocyanines for technology, ecology and medicine. J Porphyrins Phthalocyanines. 1999;3:592–610.

Kamat JP, Boloor KK, Devasagayam TPA. Chlorophyllin as an effective antioxidant against membrane damage in vitro and ex vivo. Biochimica et Biophysica Acta (BBA) – Mol Cell Biol Lipids. 2000;1487:113–27.

Khan S, Alan F, Azam A, Khan AU. Gold nanoparticles enhance methylene blue-induced photodynamic therapy: a novel therapeutic approach to inhibit *Candida albicans* biofilm. Int J Nanomedicine. 2012;7:3245–57.

Kharkwal GB, Sharma SK, Huang YY, Dai T, Hamblin MR. Photodynamic therapy for infections: clinical applications. Lasers Surg Med. 2011;43:755–67.

Khlebtsov BN, Tuchina ES, Khanadeev VA, Panfilova EV, Petrov PO, Tuchin VV, Khlebtsov NG. Enhanced photoinactivation of *Staphylococcus aureus* with nanocomposites containing plasmonic particles and hematoporphyrin. J Biophotonics. 2013;6:338–51.

Khlebtsov BN, Tuchina ES, Tuchin VV, Khlebtsov NG. Multifunctional Au nanoclusters for targeted bioimaging and enhanced photodynamic inactivation of *Staphylococcus aureus*. RSC Adv. 2015;5:61639–49.

Khlebtsov NG, Dykman LA. Immunological properties of gold nanoparticles. Chem Sci. 2017;8:1719–35.

Kikuchi Y, Sunada K, Iyoda T, Hashimoto K, Fujishima A. Photocatalytic bactericidal effect of TiO_2 thin films: dynamic view of the active oxygen species responsible for the effect. J Photochem Photobiol A Chem. 1997;106:51–6.

Kim CH, Lee ES, Kang SM, de Jong EJ, Kim BI. Bactericidal effect of the photocatalytic reaction of titanium dioxide using visible wavelengths on *Streptococcus mutans* biofilm. Photodiagn Photodyn Ther. 2017;18:279–83.

Konan YN, Berton M, Górny R, Allemann E. Enhanced photodynamic activity of meso-tetra(4-hydroxyphenyl). Eur J Pharm Sci. 2003;18:241–9.

Kuo W, Chang CN, Chang YT, Yeh CS. Antimicrobial gold nanorods with dual-modality photodynamic inactivation and hyperthermia. Chem Commun. 2009;32:4853–5.

Kussovski V, Mantareva V, Angelou I, Orozova P, Wöhrle D, Schnurpfeil G, Borisova E, Auramov L. Photodynamic inactivation of *Aeromonas hydrophila* by cationic phthalocyanines with different hydrophobicity. FEMS Microbiol Lett. 2009;294:133–40.

Kuznetsova NA, Gretsova NS, Derkacheva VM, Kaliya OL, Lukyanets EA. Sulfonated phthalocyanines: aggregation and singlet oxygen quantum yield in aqueous solution. J Porphyrins Phthalocyanines. 2003;7:147–54.

Kuznetsova NA, Makarov DA, Kaliya OL, Vorozhtsov GN. Photosensitized oxidation by dioxygen as the base for drinking water disinfection. J Hazard Mater. 2007;146:487–91.

Lee S, Nakamura M, Ohgaki S. Inactivation of phage Q beta by 254 nm UV light and titanium dioxide photocatalyst. J Environ Sci Health – Part A Toxic/Hazard Subst Environ Eng. 1998;33:1643–55.

Lee S, Nishida K, Otaki M, Ohgaki S. Photocatalytic inactivation of phage Qβ by immobilized titanium dioxide mediated photocatalyst. Water Sci Technol. 1997;35:101–6.

Linsebigler AL, Lu G, Yates JT Jr. Photocatalysis on TiO_2 surfaces: principles, mechanisms, and selected results. Chem Rev. 1995;95:735–58.

Liu Y, Li J, Qiu XF, Burda C. Bactericidal activity of nitrogen-doped metal oxide nanocatalysts and the influence of bacterial extracellular polymeric substances (EPS). J Photochem Photobiol A Chem. 2007;190:94–100.

Loschenov VB, Konov VI, Prokhorov AM. Photodynamic therapy and fluorescence diagnostics. Laser Phys. 2000;10:1188–207.

Lu T, Shao P, Mathew I, Sand A, Sun W. Synthesis and photophysics of benzotexaphyrin: a near-infrared emitter and photosensitizer. J Am Chem Soc. 2008;130:15782–3.

Makarov NS, Rebane A, Drobizhev M, Wolleb H, Spahni H. Optimizing two-photon absorption for volumetric optical data storage. J Opt Soc Am B. 2007;24:1874–85.

Maliszewska I, Leśniewska A, Olesiak-Bańska J, Matczyszyn K, Samoć M. Biogenic gold nanoparticles enhance methylene blueinduced phototoxic effect on *Staphylococcus epidermidis*. J Nanopart Res. 2014;16:2457.

Maliszewska I, Lisiak B, Popko K, Matczyszyn K. Enhancement of rose bengal-mediated photodynamic fungicidal efficacy against *Candida albicans* in the presence of biogenic gold nanoparticles. Photochem Photobiol. 2017.

Masilela N, Antunes E, Nyokong T. Axial coordination of zinc and silicon phthalocyanines to silver and gold nanoparticles: an investigation of their photophysicochemical and antimicrobial behaviour. J Porphyrins Phthalocyanines. 2013;17:417–30.

Matsunaga T, Tomoda R, Nakajima T, Wake H. Photoelectrochemical sterilization of microbial cells by semiconductor powders. FEMS Microbiol Lett. 1985;29:211–4.

Meeker DG, Jenkins SV, Miller EK, Beenken KE, Loughran AJ, Powless A, Muldoon TJ, Galanzha EI, Zharov VP, Smeltzer MS, Chen J. Synergistic photothermal and antibiotic killing of biofilm-associated *Staphylococcus aureus* using targeted antibiotic-loaded gold nanoconstructs. ACS Infect Dis. 2016;2:241–50.

Minnock A, Vernon DI, Schofield J, Griffiths J, Parish JH, Brown SB. Photoinactivation of bacteria. Use of a cationic water-soluble zinc phthalocyanine to photoinactivate both gram-positive and gram-negative bacteria. J Photochem Photobiol B Biol. 1996;32:159–64.

Mohd AS, Tufail S, Khan AA, Owais M. Gold nanoparticle-photosensitizer conjugate based photodynamic inactivation of biofilm producing cells: potential for treatment of *C. albicans* infection in BALB/c mice. PLoS One. 2015;10:1–20.

Mthethwa T, Nyokong T. Photoinactivation of *Candida albicans* and *Escherichia coli* using aluminium phthalocyanine on gold nanoparticles. Photochem Photobiol Sci. 2015;14:1346–56.

Nadtochenko VA, Radtsig MA, Khmel IA. Antimicrobial effect of metallic and semiconductor nanoparticles. Nanotechnol Russ. 2010;5:277–88.

Naik AJT, Ismailc S, Kayb C, Wilsonc M, Parkin IP. Antimicrobial activity of polyurethane embedded with methylene blue, toluidine blue and gold nanoparticles against *Staphylococcus aureus*; illuminated with white light. Mater Chem Phys. 2011;129:446–50.

Narband N, Mubarak M, Ready D, Parkin IP, Nair SP, Green MA, Beeby A, Wilson M. Quantum dots as enhancers of the efficacy of bacterial lethal photosensitization. Nanotechnology. 2008;19:445102.

Nombona N, Antunes E, Chidawanyika W, Kleyi P, Tshentu Z, Nyokong T. Synthesis, photophysics and photochemistry of phthalocyanine-Ɛ-polylysine bonjugates in the presence of metal nanoparticles against *Staphylococcus aureus*. J Photochem Photobiol A Chem. 2012;233:24–33.

Norman RS. Targeted photothermal lysis of the pathogenic bacteria, *Pseudomonas aeruginosa*, with gold nanorods. Nano Lett. 2008;8:302–6.

Olesiak-Bańska J, Gordel M, Kołkowski R, Matczyszyn K, Samoć M. Third-order nonlinear optical properties of colloidal gold nanorods. J Phys Chem C. 2012;116:13731–7.

Oliveira A, Almeida A, Carvalho CMB, Tome JPC, Faustino MAF, Neves MGPMS, Tomé AC, Cavaleiro JA, Cunha A. Porphyrin derivatives as photosensitizers for the inactivation of *Bacillus cereus* endospores. J Appl Microbiol. 2009;106:1986–95.

Oves M, Arshad M, Khan MS, Ahmed AS, Azam A, Ismail IMI. Antimicrobial activity of cobalt doped zinc oxide nanoparticles: targeting water borne bacteria. J Saudi Chem Soc. 2015;19:581–8.

Padmavathy N, Vijayaraghavan R. Enhanced bioactivity of ZnO nanoparticles-an antimicrobial study. Sci Technol Adv Mater. 2008;9:1–7.

Padmavathy N, Vijayaraghavan R. Interaction of ZnO nanoparticles with microbes-a physio and biochemical assay. J Biomed Nanotechnol. 2011;7:813–22.

Pagonis TC, Chen J, Fontana CR, Devalapally H, Ruggiero K, Song X, Foschi F, Dunham J, Skobe Z, Yamazaki H, Kent R, Tanner AC, Amiji MM, Soukos NS. Nanoparticle-based endodontic antimicrobial photodynamic therapy. J Endod. 2010;36:322–8.

Perni S, Piccirillo C, Kafizas A, Uppal M, Pratten J, Wilson M, Parkin IP. Antibacterial activity of silicone containing methylene blue and gold nanoparticles of various sizes under laser light irradiation. J Clust Sci. 2010;21:427–38.

Perni S, Piccirillo C, Pratten J, Prokopovich P, Chrzanowski W, Parkin IP, Wilson M. The antimicrobial properties of light-activated polymers containing methylene blue and gold nanoparticles. Biomaterials. 2009;30:89–93.

Plaetzer K, Krammer B, Berlanda J, Berr F, Kiesslich T. Photophysics and photochemistry of photodynamic therapy: fundamental aspects. Lasers Med Sci. 2009;24:259–68.

Podporska-Carroll J, Myles A, Quilty B, McCormack DE, Fagan R, Hinder SJ, Dionysiou DD, Suresh CP. Antibacterial properties of F-doped ZnO visible light photocatalyst. J Hazard Mater. 2017;324:39–47.

Prado-Prone G, Silva-Bermudez P, Garcia-Macedo A, Almaguer-Flores A, Ibarra C, Velasquillo-Martinez C. Photocatalytic antibacterial effect of ZnO nanoparticles into coaxial electrospun PCL fibers to prevent infections from skin injuries. Proceedings of the energy-based treatment of tissue and assessment IX; 22.02.2017. San Francisco: Thomas P. Ryan; 2017.

Prasad GK, Agarwal GS, Singh B, Rai GP, Vijayaraghavan R. Photocatalytic inactivation of *Bacillus anthracis* by titania nanomaterials. J Hazard Mater. 2009;165:506–10.

Prasanna VL, Vijayaraghavan R. Insight into the mechanism of antibacterial activity of ZnO: surface defects mediated reactive oxygen species even in the dark. Langmuir. 2015;31:9155–62.

Rao KVS, Zhuo BX, Cox JM, Chiang K, Brungs M, Amal R. Photoinduced bactericidal properties of nanocrystalline TiO_2 thin films. J Biomed Nanotechnol. 2006;2:71–3.

Ratto F, Tuchina ES, Khlebtsov BN, Centi S, Matteini P, Rossi F, Fusi F, Khlebtsov NG, Pini R, Tuchin VV. Combined near infrared photothermolysis and photodynamic therapy by association of gold nanoparticles and an organic dye. Proceedings of the plasmonics in biology and medicine VIII, San Francisco; 2011.

Raulio M, Pore V, Areva S, Ritala M, Leskela M, Linden M, Rosenholm JB, Lounatmaa K, Salkinoja-Salonen M. Destruction of *Deinococcus geothermalis* biofilm by photocatalytic ALD and sol-gel $TiO_{(2)}$ surfaces. J Ind Microbiol Biotechnol. 2006;33:261–8.

Reddy MP, Venugopal A, Subrahmanyam M. Hydroxyapatite-supported ag–TiO_2 as *Escherichia coli* disinfection photocatalyst. Water Res. 2007;41:379–86.

Ricchelli F. Photophysical properties of porphyrins in biological membranes. J Photochem Photobiol B Biol. 1995;29:109–18.

Rincon AG, Pulgarin C. $Fe^{3+}$and TiO_2 solar-light-assisted inactivation of *E. coli* at field scale: implications in solar disinfection at low temperature of large quantities of water. Catal Today. 2007;122:128–36.

Robertson JMC, Robertson PKJ, Lawton LA. A comparison of the effectiveness of TiO_2 photocatalysis and UVA photolysis for the destruction of three pathogenic microorganisms. J Photochem Photobiol A Chem. 2005;175:51–6.

Saito T, Iwase T, Horie J, Morioka T. Mode of photocatalytic bactericidal action of powdered semiconductor TiO_2 on mutans streptococci. J Photochem Photobiol B Biol. 1992;14:369–79.

Sakamoto K, Ohno-Okumura E. Syntheses and functional properties of phthalocyanines. Materials. 2009;2:1127–79.

Samia AC, Dayal S, Burda C. Quantum dot-based energy transfer: perspectives and potential for applications in photodynamic therapy. Photochem Photobiol. 2009;82:617–25.

Segalla A, Borsarelli CD, Braslavsky SE, Spikes JD, Roncucci GG, Dei D, Chiti G, Jori G, Reddi E. Photophysical, photochemical and antibacterial photosensitizing properties of a novel octacationic Zn(II)-phthalocyanine. Photochem Photobiol Sci. 2002;1:641–8.

Seven O, Dindar B, Aydemir S, Metin D, Ozinel M, Icli S. Solar photocatalytic disinfection of a group of bacteria and fungi aqueous suspensions with TiO_2, ZnO and Sahara Desert dust. J Photochem Photobiol A Chem. 2004;165:103–7.

Sharma SK, Mroz P, Dai T, Huang YY, Denis TGS, Hamblin MR. Photodynamic therapy for cancer and for infections: what is the difference? Israel J Chem. 2012;52:691–705.

Shchukin DG, Ustinovich EA, Kulak AI, Sviridov DV. Heterogeneous photocatalysis in titania-containing liquid foam. Photochem Photobiol Sci. 2004;3:157–9.

Sirelkhatim A, Mahmud S, Seeni A, Kaus NHM, Ann LC, Bakhori SKM, Hasan H, Dasmawati M. Review on zinc oxide nanoparticles: antibacterial activity and toxicity mechanism. Nano-Micro Lett. 2015;7:219–42.

Skorb EV, Antonouskaya LI, Belyasova NA, Shchukin DG, Mohwald H, Sviridov DV. Antibacterial activity of thin-film photocatalysts based on metal-modified TiO_2 and TiO_2:In_2O_3 nanocomposite. Appl Catal B Environ. 2008;84:94–9.

Sokmen M, Candan F, Sumer Z. Disinfection of *E-coli* by the Ag-TiO_2/UV system: lipidperoxidation. J Photochem Photobiol A Chem. 2001;143:241–4.

Spesia MB, Rovera M, Durantini EN. Photodynamic inactivation of *Escherichia coli* and *Streptococcus mitis* by cationic zinc(II) phthalocyanines in media with blood derivatives. Eur J Med Chem. 2010;45:2198–205.

Sun DS, Kau JH, Huang HH, Tseng YH, Wu WS, Chang HH. Antibacterial properties of visible-light-responsive carbon-containing titanium dioxide photocatalytic nanoparticles against anthrax. Nano. 2016;6:237.

Sunada K, Kikuchi Y, Hashimoto K, Fujishima A. Bactericidal and detoxification effects of TiO_2 thin film photocatalysts. Environ Sci Technol. 1998;32:726–8.

Sunada K, Watanabe T, Hashimoto K. Bactericidal activity of copper-deposited TiO_2 thin film under weak UV light illumination. Environ Sci Technol. 2003;37:4785–9.

Tuchina ES, Tuchin VV, Khlebtsov BN, Khlebtsov NG. Phototoxic effect of conjugates of plasmon-resonance nanoparticles with indocyanine green dye on *Staphylococcus aureus* induced by IR laser radiation. Quantum Electron. 2011;41:354–9.

Vacaroiu C, Enache M, Gartner M, Popescu G, Anastasescu M, Brezeanu A, Todorova N, Giannakopoulou T, Trapalis C, Dumitru L. The effect of thermal treatment on antibacterial properties of nanostructured TiO_2 (N) films illuminated with visible light. World J Microbiol Biotechnol. 2009;25:27–31.

Walter MG, Rudine AB, Wamser CC. Porphyrins and phthalocyanines in solar photovoltaic cells. J Porphyrins Phthalocyanines. 2010;14:759–92.

Wang ZL. Zinc oxide nanostructures: growth, properties and applications. J Phys Condens Matter. 2004;16:R829–58.

Wang C, Dong H, Hu W, Liu Y, Zhu D. Semiconducting π-conjugated systems in field-effect transistors: a material odyssey of organic electronics. Chem Rev. 2012;112:2208–67.

Watts RJ, Kong S, Orr MP, Miller GC, Henry BE. Photocatalytic inactivation of coliform bacteria and viruses in secondary wastewater effluent. Water Res. 1995;29:95–100.

Wei W, Lin W-Y, Zainal Z, Williams NE, Zhu K, Kruzic AP, Smith RL, Rajeshwar K. Bactericidal activity of TiO_2 photocatalyst in aqueous media: toward a solar-assisted water disinfection system. Environ Sci Technol. 1994;28:934–8.

Wellings J, Chaure N, Heavens S, Dharmadasa I. Growth and characterisation of electrodeposited ZnO thin films. Thin Solid Films. 2008;516:3893–8.

Wöhrle D, Schnurpfeil G, Makarov SG, Kazarin A, Suvorova ON. Practical applications of phthalocyanines – from dyes and pigments to materials for optical, electronic and photo-electronic devices. Макрогетероциклы / Macroheterocycles. 2012;5:191–202.

Zafar I, Arfan M, Nasir RP, Shaikh AJ. Aluminum phthalocyanine derivatives: potential in antimicrobial PDT and photodiagnosis. Austin Biomol Open Access. 2016;1:1010.

Zane A, Ranfang Z, Villamena FA, Rockenbauer A, Foushee AMD, Flores K, Dutta PK, Nagy A. Biocompatibility and antibacterial activity of nitrogen-doped titanium dioxide nanoparticles for use in dental resin formulations. Int J Nanomedicine. 2016;11:6459–70.

Zhang X, Yu L, Zhuang C, Peng T, Li R, Li X. Highly asymmetric phthalocyanine as a sensitizer of graphitic carbon nitride for extremely efficient photocatalytic H_2 production under near-infrared light. ACS Catal. 2014;4:162–70.

Zharov VP, Mercer KE, Galitovskaya EN, Smeltzery MS. Photothermal nanotherapeutics and nanodiagnostics for selective killing of bacteria targeted with gold nanoparticles. Biophys J. 2006;90:619–27.

Zhou G, Li Y, Xiao W, Zhang L, Zuo Y, Xue J, Jansen JA. Synthesis, characterization, and antibacterial activities of a novel nanohydroxyapatite/zinc oxide complex. J Biomed Mater Res A. 2008;85:929–37.

Chapter 13
Colloidal Bio-nanoparticles in Polymer Fibers: Current Trends and Future Prospects

Zuzana Konvičková, Ondrej Laššák, Gabriela Kratošová, Kateřina Škrlová, and Veronika Holišová

Abstract Biotechnology and bio-nanotechnology are emerging fields that inspire vast scientific and engineering inquiry. Bio-nanotechnology is relatively new and dynamic, applying biological principles to produce new systems and materials at nanoscale level. Eco-friendly nanomaterial development and production by biosynthesis have an interesting niche, where metallic and functionally diverse biosynthesized nanoparticles (bio-NPs) are prepared by exploiting both biological processes in microorganisms and biochemical reactions in plant extracts and other biomass. The major advantage of this approach is one-step chemical reduction and stabilization, with the two principal components providing toxic-free intermediates in the bio-NP genesis. This heralds exciting possibilities for inexpensive NP production and consequent rapid and wide adoption of novel applications, such as incorporation of bio-NPs to augment polymer nanofiber properties.

This chapter presents an overview of critical aspects of the composite materials' design and development. The recognized mechanics of bio-NP formation is followed by idiosyncrasies in choice of the core material and the "host" environment where synthesis occurs and the physical and chemical characterization of resultant bio-NPs. Application potential is then outlined, and highly biocompatible polymers are highlighted in a major review of nanofiber production. Finally, future prospects in bio-NP and nanofiber composition are investigated.

Keywords Metallic nanoparticles • Biosynthesis • Phytosynthesis • Polymer fibers • Electrospinning • Colloids

---

Z. Konvičková (✉) • G. Kratošová • K. Škrlová • V. Holišová
Nanotechnology Centre, VŠB - Technical University of Ostrava,
17. listopadu 15/2172, Ostrava, Poruba 708 33, Czech Republic
e-mail: konvickova.zuzana@gmail.com

O. Laššák
4 Medical Innovations CBTD, Dr. Slabihoudka 6232/11,
Ostrava, Poruba 708 52, Czech Republic

M. Rai, R. Shegokar (eds.), *Metal Nanoparticles in Pharma*,
DOI 10.1007/978-3-319-63790-7_13

Nomenclature

Ag	Silver
Au	Gold
CM	Carboxy-methyl
CuO	Copper oxide
DLS	Dynamic light scattering
Fe_2O_3	Iron (III) oxide
FTIR	Fourier transform infrared spectroscopy
MEMS	Microelectromechanical systems
NP	Nanoparticle
PCL	Poly(ε-caprolacton)
Pd	Palladium
PGA	Poly(glycolic acid)
PLA	Poly(lactic acid)
PLGA	Poly(lactic-co-lactic acid)
Pt	Platinum
PVA	Poly(vinyl alcohol)
PVP	Polyvinylpyrrolidone
SEM	Scanning electron microscopy
TEM	Transmission electron microscopy
XRD	X-ray diffraction
ZnO	Zinc oxide
ZrO_2	Zirconia oxide

13.1 Introduction

Nanotechnology explores strategies for engineering, manipulating matter and energy on atomic and molecular levels. This technology creates nanosystems, assembles them in functional systems, and converts them to devices which facilitate novel applications. It provides access to their unusual properties which are emerging from the quantum realm. As the dimensions shrink to the nanoscale level, the quantum aspect dominates and defines a wide range of properties, including chemical, optical, electrical, mechanical, and thermal capacity (Narayanan and Sakthivel 2010; Krahne et al. 2011; Díez-Pascual et al. 2012; Son et al. 2012).

The two basic approaches in producing nanoscale materials and structures are top-down and bottom-up strategies (Narayanan and Sakthivel 2010). Top-down procedures reduce macro- and mesoscale materials to nanodimensions. Although photolithographic processes play a dominant role in nanofabrication, these are burdened by excessive cost and focus on fixed-bulk substrate nanostructures with very strict requirements for precision and accuracy. Expedient examples include specific conditions such as production of integrated circuits, MEMS, and quantum optics; and this method is unsuitable for mass production of colloidal nanoparticles (NPs).

Further application is ball milling. This is a mechanical method for NP production with long processing time, low level of control over NP parameters, impurities, and common defects in crystal structure (Yadav et al. 2012), but these can be minimized, and ball milling remains a preferred method for some types of materials and applications, such as gas sensing (Pentimalli et al. 2015).

Although the bottom-up approach appears most effective for NP production because it forms nanostructures by assembling single atoms and molecules, combining the bottom-up and top-down procedures offers an optimal strategy for some applications, and their use together also provides the sole solution to some NP production problems (Choi et al. 2008; Pinna et al. 2013; Hu et al. 2014).

An alternative dichotomy discerns between physical and chemical approaches and additive and subtractive methods (Engstrom et al. 2014). An example of a physical approach that also fits the bottom-up and additive categories is atomic force lithography where single atoms are manipulated under high control. However, there is a wide range of chemical methods to prepare nanomaterials in material science and nanotechnology. For example (a) sol-gel is used to fabricate silicon and TiO_2 metal oxides, and a general feature here is the homogenization of metal alkoxides in the solution and their transition to gel, (b) precipitation produces solid material from solution, and (c) synthesis using microwave radiation can be applied to chemical reactions (Nguyen et al. 2010; Junlabhut et al. 2014; Omri et al. 2015).

Biosynthesis fits the combined chemical, additive, and bottom-up approach, and this is further explored throughout the text because this specialized chemical method has emancipated into an individual research field. The application of biological reactions and processes in living organisms contributes to the formation of nanomaterials, especially metal NPs (Castro-Longoria et al. 2011; Schröfel et al. 2011; Zhang et al. 2011). In biosynthesis, an inorganic precursor plays the simple role of metal donor to the biological system. A further example occurs in herbal leachate extract. Although herbs lack living entities, added metal precursors such as tetrachloroauric acid ($HAuCl_4$) are reduced by the active phytochemical leachate in the extract into capped and stable bio-NPs. Although this process in living bio-matter results from an active and complex effort to neutralize and eliminate exogenous elements, it is expected to be similar to that in inanimate species, and while the precise mechanism in both processes remains undetermined, biosynthesis retains a great potential for efficient NP production from different precursors in a scalable and environmentally friendly manner.

13.2 Biosynthesis: A Simple Method for Metallic Nanoparticles Preparation

Processes involved in NP biosynthesis occur in living organisms such as bacteria, yeasts, algae, and/or in their immediate surroundings containing bioactive agents originating from the organisms. Other possible milieu is a biomass completely free of living organisms such as plant extracts. All aforementioned systems are capable of producing nanostructures and nanoparticles via the processes described below. Microorganism/plant extract and metallic precursor are mixed together in a single step, and NP formation starts after mutual contact.

13.2.1 Formation of Nanoparticles

Biomasses contain large amounts of organic compounds composed of positively and negatively charged functional groups. Examples of negatively charged include hydroxyl (−OH), amino ($-NH_2$), and carboxyl (−COOH) groups, and metal ions are reduced to the zero valent (or oxide) form when the biomass and a metal salt precursor are mixed (Vijayaraghavan et al. 2011). NP growth based on thermodynamic models is observed simultaneously with their stabilization by organic compounds. Although NP nuclei are generated homogeneously at the same time, growth and nucleation are discrete processes, and NPs subsequently grow without additional nucleation. Here, the La Mer model with its modifications is the commonly accepted model describing the general mechanism of NP formation, but this describes only the nucleation process followed by growth of the stable nuclei, and it cannot predict or characterize the evolution of NPs' size distribution (Polte 2015). The complex nature of microorganisms and higher plants makes very difficult to determine the organic compounds responsible for NP biosynthesis and stabilization. For example, it was shown that NP biosynthesis can be initiated by enzymes present in live bacteria. Examples include the mechanism of intracellular biosynthesis using nitrate reductase and NP biosynthesis using *Escherichia coli* (Gurunathan et al. 2009) or *Bacillus licheniformis* (Duran et al. 2011). Flavonoids in fruit extracts also play a major role in NP biosynthesis, where *Syzygium cumini* fruit extract, for example, has been successfully utilized in Ag NP biosynthesis (Mittal et al. 2014).

While the stabilizing agent is an organic component from the microorganism or plant extract encasing the NP, the precise stabilization period depends on temperature changes denaturing the protein or acidic/alkaline pH fluctuations (Thatoi et al. 2016; Yuan et al. 2017).

Different biosynthetic processes enable preparation of a wide spectrum of NPs including Ag, Au, Pd, Pt, and the oxidic ZnO, ZrO_2, and Fe_2O_3 NPs. For example, NPs can be synthesized on cell body or plant substrate matrices or directly in colloidal systems, depending on the type of organism involved. NP type and concentration also strongly depend on the following: capping agents, thermodynamic size control, and kinetic and stoichiometric influences. Different NP shapes can be due to organisms' complex nature; while most NPs are spherical, others can be triangular, hexagonal, or rods.

NP synthesis is generally accompanied by color change of the final metal colloid solution (Fig. 13.1a, b) or biomass (Fig. 13.1c, d).For example, Au NPs change the color of the solution to dark red or purple (Fig. 13.1a) and Ag NPs to orange or brown (Fig. 13.1b). NP size also influences sample appearance.

Despite great biosynthetic ability of different organisms and plants and interesting catalytic and antibacterial properties of NPs, not all types of organisms have subsequent nanoparticle application potential. Bio-NPs can be attached to some parts of organism (scales, frustules, scaffolds, husks, etc.) or cellular parts; their separation may be an obstacle for further application (Fig. 13.2). In contrast, metallic NPs embedded on matrices could be easily dried to powder forming so-called bionanocomposite and can be resuspended in different solvents, including water and ethanol (Schröfel et al. 2011; Konvičková et al. 2016; Konvičková et al. 2017; Holišová et al. 2017).

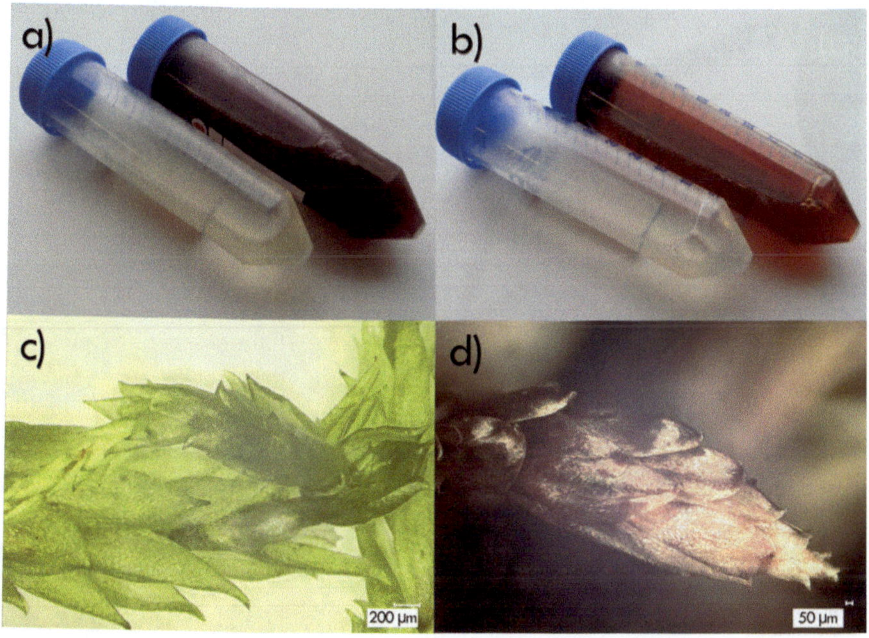

Fig. 13.1 Biosynthesis protocols with different biomass. Colloidal solutions of Au (**a**) and Ag NPs (**b**) are prepared by linden extract. Color change is clearly visible due to reduction of metal precursor and subsequent stabilization by organic compounds in the extract. Another biosynthesis of Au NPs using parts of a moss is presented. Optical microscopy images show a color change between pure biomass (**c**) and the moss with sorbed Au NPs (**d**)

Phytosynthesis is a biosynthetic protocol used by many scientific teams. This involves the preparation of metallic NPs from selected plant extracts and eluates (Velmurugan et al. 2015; Yallappa et al. 2015). It is necessary to consider whether extract preparation or the use of nutritionally significant product fits appropriately with the concept of environmentally and economically friendly biological synthesis. Considerations must also include the nanomaterials' intended use, how often is it to be used, and in what quantities. The use of waste biomass from agriculture and food industries has therefore been intensively studied and tested for bioreduction potential in NP preparation (Yang et al. 2014). The focus should be on wastes which have no further use and are landfilled and on those which are not primarily toxic and used in secondary raw materials.

13.2.2 Phytosynthesized NPs and Their Stabilization

One of the most important aspects of NPs preparation is their stabilization in the dispersal medium. Nanoscale particles are unstable and tend to agglomerate because of short inter-particle distance and their attraction by van der Waals electrostatic

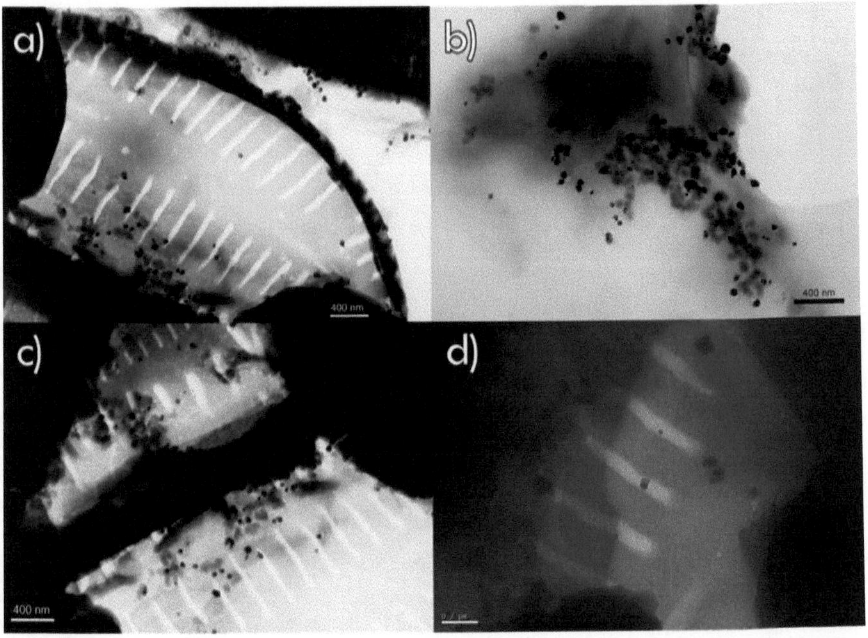

Fig. 13.2 TEM micrographs of Ag NPS on biosilica surface. NPs are synthesized on the diatom silica frustules (**a**). NPs are synthesized on the biosilica surface (**b**) where reduction of metal ions and stabilization of metal NPs occur. Diatoms are unicellular microorganisms with typical ornate frustules with pores on biosilica surface (**a, c, d**)

forces. Aggregation, agglomeration, and coalescence proceed in the absence of counteractive repellent forces. In contrast, electrostatically stabilized NPs have at least one electric double layer due to surface charging, with the resultant Coulomb forces between the particles decaying exponentially with distance between particles. Particle coagulation is prevented when this repulsion is sufficiently strong. The most commonly used stabilizing agents in chemical synthesis are sodium citrate, cetrimonium bromide, and thiols (Polte 2015). In phytosynthesis, the stabilization is mainly provided by phytochemicals released during biosynthesis process.

13.2.3 Preparation of Phytosynthesized NPs and Their Characterization

As outlined above, many NP types can be produced by biosynthesis including Ag NPs, Au NPs, zirconia oxide (ZrO_2 NPs), copper oxide (CuO NPs), and platinum (Pt NPs). The simplest method of extract preparation is a biomass immersion in cold or hot distilled water. For example, this method provides rapid plant extract preparation in up to 30 min, with the final extract obtained by filtering through a strainer. The pure extract is then mixed with the metal precursor, and concentration ranges from 10^{-2} to 10^{-4} mol/dm^{-3} to synthesize NPs.

UV/VIS spectroscopy verifies NP presence in colloidal solutions via absorption of atomic and molecular electron transitions from ground to excited states. Resultant peaks determine NP size, where narrow, sharp peaks estimate the size in tens of nanometers (Baset et al. 2011).

The dynamic light scattering method (DLS) determines the particle size in a colloidal solution via laser beam illuminating the particles and analyzing the scattered light intensity fluctuations. These fluctuations are associated with light interference in unsteady dispersed phase particles. An electrical double layer exists around each particle, and mV potential between this layer and the outer region is the ζ-potential. ζ-potential determination is suitable for colloidal system study because net charge development at the particle surface affects ion distribution at the interface, resulting in increased concentration of "counterions" with opposite charge to that of the particle. The magnitude of the ζ-potential indicates the potential stability of the colloidal system. When the suspension has a highly positive value above +30 mV or highly negative below −30 mV, the system provides no tendency for aggregation. In contrast, when the ζ-potential approaches 0 mV, there is no force to prevent particles conglomerating. However, pH is a determining factor in ζ-potential; addition of alkali to the suspension increases particle negative charge, and this negative charge is neutralized if acid is added, and surfeit of acid causes positive charge. The ζ-potential and pH therefore have a strong relationship; ζ-potential will be positive at low pH and lower or negative at high pH (Bhattacharjee 2016).

Fourier transform infrared spectroscopy (FTIR) is widely applied in biosynthetic protocols. This is based on the absorption of infrared radiation passing through the sample and changing molecule vibrational energy levels according to changes in the molecule's dipole moment. FTIR is thought to determine the specific functional groups such as -OH and -COOH responsible for bioreduction. Other spectroscopic methods useful for bio-NP characterization are also atomic emission (or absorption) spectroscopy, X-ray diffraction, or Raman spectroscopy.

Scanning electron microscopy (SEM) and transmission electron microscopy (TEM) are also integral parts of characterization. Here, the TEM micrographs determine particle size, shape, crystallinity, and chemical composition. The combined analytic capabilities of such methods are most important for identification and further characterization of metal NPs used as antibacterial agents, in catalysis, for enzyme immobilization, drug delivery, sensing, detection, etc. Moreover, metallic bio-NPs can be incorporated in substrates, such as polymer fibers, dependent on their determined properties. For these purposes, biosynthesis and especially *phytosynthesis* are the most simple, rapid, and successful methods of preparing functional metallic NPs.

13.3 Fibers as Matrix for Bio-NPs

New fiber and nanofiber protocols have been developed for the following important applications: material engineering (Chronakis 2005), medicine (Sebe et al. 2013), or biotechnology (Nair and Laurencin 2007). Polymers are large molecules composed of repetitive monomers (McMurry 2004), and initial polymer solutions, copolymers, and melts significantly control the chemical and physical properties of final fibers (Lukáš et al. 2009). Polymer nanofibers can range from tens of nanometers to 1 μm in diameter.

13.3.1 Brief History of Fiber Spinning

Spinning in an electrostatic field has been known since the sixteenth century, and the first paper on fine-fiber spinning was published by William J. Morton at the beginning of twentieth century (Morton 1902). This was then followed by crucial work on modern capillary electrospinning by John Zelený who designed capillary apparatus to study liquid point electrical discharge (Zeleny 1914). Unfortunately, this advance had insufficient industrial application and development terminated, most likely due to lack of optical and analytical instruments to study at the nanometer level, and it was also unnecessary to focus on this topic because of the limited number of applications. Although the first electron microscope prototype was designed in 1931 (Ruska 1980), it was not until the 1980s that application for its use emerged. The unique technique Nanospider™ patented by Czech scientist Oldřich Jirsák at the Technical University in Liberec heralded an important impetus in the field of polymer fiber preparation. Nanospider™ changed the spinning technique from needle-use to the new needleless innovation, producing fibers in the 200–500 nm diameter range (Jirsak et al. 2009).

13.3.2 Fiber Preparation via Spinning Techniques

Electrospinning is a nonmechanical, electrostatic technique using a high-voltage electrostatic field to charge the surface of liquid solutions, especially polymers. The electric field is applied between the polymer solution and a grounded collector, polymer drops are drawn, and the charged jet is ejected when the electric field overcomes liquid surface tension. The path of the charged jet is controlled by the electric field (Fig. 13.3). Electrospinning technology is divided into "capillary" and "needleless" categories, depending on the place where the fibers are extracted. While capillary electrospinning is limited by the amount of injected polymer, the needleless technique has high fiber production from the free surface of the polymer via self-organization (Lukáš et al. 2009).

Electrospinning covers a wide range of physical phenomena in the field of electrohydrodynamics, including electrophoresis, electro-diffusion, and electroosmosis. The liquid in electrospinning acquires surface tension from short intermolecular forces without influence from the external electrostatic field, but this can only occur when the liquid surface thickness lies within the range of the intermolecular forces. The external electrostatic field then acts on the liquid surface and charge is induced (Lukáš et al. 2009).

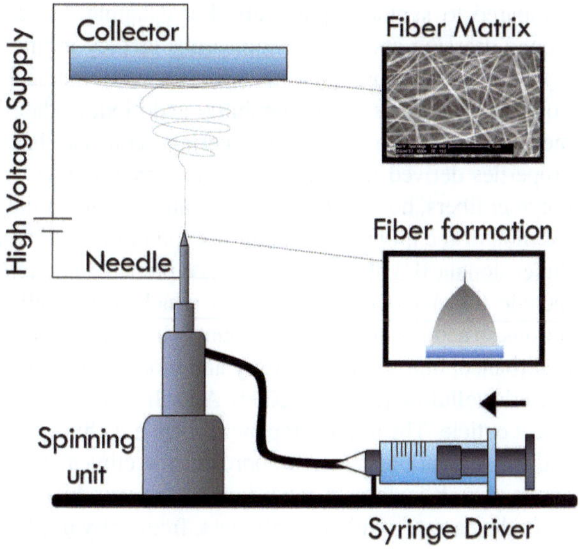

Fig. 13.3 Preparation of polymer fibers by capillary electrospinning. The polymer is pressed from the syringe to the spinning unit and injected during the spinning process. After applied voltage, the polymer jets are taken out and fall on the collector. The jet path is accompanied by solvent evaporation and solid fiber formation

Important technical parameters for polymer fiber preparation include polymer molecular weight and concentration, voltage and conductivity, viscosity, surface tension, flow velocity, distance between the spinneret and collector, and laboratory temperature and humidity (Lukáš et al. 2009). Polymer concentration is important because the fibrous sample becomes a mixture of fibers and defects when the concentration reaches a limiting value. Higher voltages and consequent higher Coulomb forces in the jet induce greater stretching of the polymer solution, leading to fiber diameter decrease and rapid evaporation of solvent from the fibers.

Centrifugal spinning is an alternate technique for fiber preparation, also called rotary-jet or force spinning, because it uses centrifugal force to extrude fibers from the spinning unit. The spinneret rotates around its axis, and a viscous jet is ejected as fibers onto a metal collector under maximized rotation speed (Mellado et al. 2011). Replacement of electrostatics by centrifugal force initiated increase in conductive fiber-spinning materials, and the application of heat near the spinning unit enhanced the melting and spinning of solid materials, thus negating chemical preparation (Sarkar et al. 2010). Centrifugal spinning also enables fiber preparation from solutions with higher concentrations than possible with electrostatic spinning (Lu et al. 2013).

13.3.3 Biocompatible Polymers Suitable for Fiber Formation via Electrospinning

Many materials are suitable for spinning natural polymers fibers with submicron diameters. These are extremely useful in tissue engineering, filtration membranes, and other biomedical fields. The natural collagen, gelatin, chitin, casein, cellulose acetate, and fibrinogen polymers have high biocompatibility and low immunogenicity compared to synthetic polymers. For example, (1) collagen scaffolds are used in wound dressing and tissue engineering, as they mimic the native collagen network, (2) gelatin is a natural polymer widely used in medical and pharmaceutical applications because of its biocompatibility and biodegradability in physiological environments, and (3) in tissue engineering, chitosan has beneficial physicochemical properties derived from its solid-state structure. One drawback of naturally derived polymer fibers, however, is that they can undergo partial denaturation, leading to degradation of the initial material during spinning (Bhardwaj and Kundu 2010). A three-dimensional (3D) fiber network can be prepared naturally by living organisms. These include the *Acetobacter* bacterium which forms cellulose (Yamada 1983). Its 3D fibrous structures are similar to plant cellulose, characterized by high porosity, water absorbance, high chemical purity and biocompatibility, and they are therefore called bacteria cellulose (Hu et al. 2009). A further inspiration from nature is mimicry of the insect cuticle. The main components here are chitosan and fibrin, with biodegradable and biocompatible features approved for clinical products. In addition to medical application, it replaces plastics in consumer products (Fernandez and Ingber 2012).

In the case of synthetic polymers, frequently used polymers are group of polyesters such as poly(ε-caprolacton) PCL, poly(lactic acid) (PLA), or poly(glycolic acid) (PGA) and its copolymers. PCL is used in bone tissue engineering as its fibrous structure and high porosity provide viability, proliferation, and cell adhesion, and its large pores enhance cell growth and bone integration (Erben et al. 2015). The mixture of PCL with gelatin improves the final scaffold for enhanced cell migration to the PCL mesh (Zhang et al. 2005). Poly(lactic-co-lactic acid) (PLGA) is suitable for nonwoven scaffolds because their porosity is over 90% and their high surface area enables cellular attachment and fiber orientation. In addition, their subsequent impregnation with antibiotics reduces post-surgery adhesions (Bhardwaj and Kundu 2010).

Synthetic poly(vinyl alcohol) (PVA) combined with Ag NPs produces a nonwoven membrane which controls antibacterial properties (Jia et al. 2007). It is also advantageous in a wide range of biomedical applications through its water-solubility, biocompatibility, and hydrophilic nature. PVA and carboxy-methyl (CM) chitosan are also promising carriers of Ag NPs for biomedical application, and chitosan and its derivatives' biocompatibility and biodegradability make these the most widely used natural polysaccharides in biomedical applications (Nguyen et al. 2011).

13.4 Prospective Bio-NP Incorporation into the Polymer Fibers

Many types of colloidal NPs such as gold (Deniz et al. 2011), zinc (Virovska et al. 2014), titanium (Fathona and Yabuki 2014), or carbon nanotubes (Salalha et al. 2004) and other particulates (Hansen et al. 2005) can be incorporated into the polymer fibers, polymer mixtures, or copolymers. NPs are usually obtained by chemical and/or thermal synthesis, dispersed in the polymer solution, inserted into the spinning unit, and spun into fibers with solvent evaporation (Reneker and Yarin 2008). However, this approach can result in nonhomogenous NP dispersion in the polymer matrix or aggregation which destroys uniform composite structure (He et al. 2009). Dissolution of the metallic precursor prior to polymer spinning may prevent these problems (Wang et al. 2005).

Examples of conventionally prepared NPs in polymer fibers include:

1. Chemically prepared ZnO is incorporated in a wide range of polymers including PEO, PVP, and PVA, and ZnO nanoparticles are incorporated in PLA fibers by combined electrospinning and electrospraying. This hybrid fibrous material combines the polymer and inorganic filler properties of biodegradability, photocatalytic, and antibacterial activity (Virovska et al. 2014).
2. TiO_2 NPs are also incorporated in polymer fibers, as in cellulose acetate electrospun fibers. While this composite enhances drug delivery systems, material template fabrication, and composite fibril fillers, high NP concentration causes aggregation inside the polymer (Fathona and Yabuki 2014).
3. Examination of chitosan/sericin/PVA fibers with incorporated Ag NPs highlights that chitosan is beneficial in filtration, drug delivery, tissue engineering, and wound dressings. Intermolecular interactions between chitosan and both sericin and PVA are through hydrogen bonding, and these compounds can reduce silver ions to Ag NPs. Impressively, these fibers completely inhibited bacterial growth, with 100% antibacterial activity against *E.coli* (Hadipour-Goudarzi et al. 2014).
4. PVA/Ag NPs and PVP/Ag NPs fibers were tested against *E.coli*, *Staphylococcus aureus*, and *Pseudomonas aeruginosa* in two NP preparation methods: thermal synthesis and chemical reduction of silver nitrate. The NPs are produced by these methods and then incorporated in PVA and PVP. Further antibacterial testing using disk diffusion formed a larger inhibition zone in PVA/Ag NPs than in PVP/Ag NPs (Pencheva et al. 2012).
5. Silver selenide nanoparticle incorporation in PVP fibers highlighted that selenide on the Ag NP surface effects a strong chemical interaction between NPs and the polymer. This hybrid nanocomposite's antibacterial activity emanates mainly from the Ag NPs, and it can also be applied as a filter membrane in systems removing heavy metals and bacteria from water (More et al. 2015).

In addition to chemically prepared particles, metallic bio-NPs, and especially their colloid suspensions, also have spinning potential. Their major advantage is that preparation, stabilization (through organic layer on the NP surface), and reduction can be achieved in one simple step, and since bio-NPs are usually prepared in aqueous medium, the polymer precursor can be dissolved directly in the bio-NP solution (Fig. 13.4).The dissolution can be conducted at ambient temperature, and while continuous heating is also effective, this can have negative effects. Polymers can be hydrophilic and hydrophobic, so prior to commencing the experiment, it is crucial to

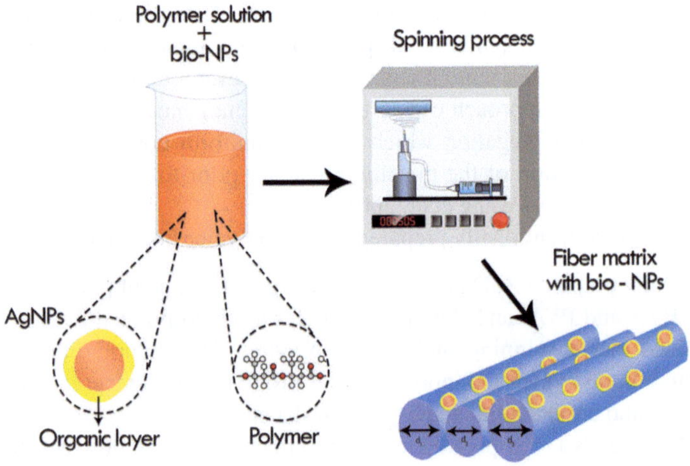

Fig. 13.4 Fiber preparation containing bio-Ag NPs: metallic NPs can be directly dispersed in polymer solution. The mixture is continuously batched to the spinning unit and fibers containing NPs are formed. Calculated diameters highlight that polymer fibers have nonuniform diameter

consider whether a water-soluble or oil-soluble solvent is appropriate for synthesis of chosen bio-NPs and also what type of biosynthesis is the most suitable.

13.5 Conclusions

Biosynthesis, and especially phytosynthesis, is a simple method for effective preparation of metallic nanoparticles suitable for a great variety of applications, most especially in antibacterial, medical, and chemical compound catalysis. This is proven in 4-nitrophenol, COx, and NOx (Schröfel et al. 2014; Holišová et al. 2017). The increasing trend of antibiotic and medicinal overuse has enabled bacteria to adapt in multidrug resistance to bacteriostatic and bactericidal agents, and nanoparticles may be quite effective in countering this problem. The preparation of new composite materials based on natural and synthetic polymers with active antibacterial particles heralds exciting possibilities for medical and pharmaceutical application.

Polymers must comply with exacting demands to satisfy these applications, especially biocompatibility, biodegradability, resistance to temperature changes, mechanical and chemical resistance, and chemical purity. This requires a new generation of active nanocomposites, where phytosynthesis-prepared nanoparticles incorporated in the fiber membranes release nano-sized killing agents against bacteria colonies. Finally, these prepared products can be widely used, especially in medical equipment coatings, antibacterial gels, and air-conditioning systems.

Acknowledgment This work was supported by Ministry of Education of the Czech Republic project No. SP2016/57, SP2017/70.

References

Baset S, Akbari H, Zeynali H, Shafie M. Size measurement of metal and semiconductor nanoparticles vis UV/VIS absorption spectra. Dig J Nanomater Biostruct. 2011;6(2):709–16.

Bhardwaj N, Kundu SC. Electrospinning: a fascinating fiber fabrication technique. Biotechnol Adv. 2010;28(3):325–47.

Bhattacharjee S. DLS and zeta potential – what they are and what they are not? J Control Release. 2016;235:337–51.

Castro-Longoria E, Vilchis-Nestor AR, Avalos-Borja M. Biosynthesis of silver, gold and bimetallic nanoparticles using the filamentous fungus Neurospora crassa. Colloids Surf B: Biointerfaces. 2011;83(1):42–8.

Choi WK, Liew TH, Chew HG, Zheng F, Thompson CV, Wang Y, Hong MH, Wang XD, Li L, Yun J. A combined top-down and bottom-up approach for precise placement of metal nanoparticles on silicon. Small. 2008;4(3):330–3.

Chronakis IS. Novel nanocomposites and nanoceramics based on polymer nanofibers using electrospinning process—a review. J Mater Process Technol. 2005;167(2–3):283–93.

Deniz AE, Vural HA, Ortaç B, Uyar T. Gold nanoparticle/polymer nanofibrous composites by laser ablation and electrospinning. Mater Lett. 2011;65(19–20):2941–3.

Díez-Pascual AM, Naffakh M, Marco C, Ellis G. Mechanical and electrical properties of carbon nanotube/poly(phenylene sulphide) composites incorporating polyetherimide and inorganic fullerene-like nanoparticles. Compos A: Appl Sci Manuf. 2012;43(4):603–12.

Duran N, Marcato PD, Duran M, Yadav A, Gade A, Rai M. Mechanistic aspects in the biogenic synthesis of extracellular metal nanoparticles by peptides, bacteria, fungi, and plants. Appl Microbiol Biotechnol. 2011;90(5):1609–24.

Engstrom DS, Porter B, Pacios M, Bhaskaran H. Additive nanomanufacturing – a review. J Mater Res. 2014;29(17):1792–816.

Erben J, Pilarova K, Sanetrnik F, Chvojka J, Jencova V, Blazkova L, Havlicek J, Novak O, Mikes P, Prosecka E, Lukas D, Kuzelova Kostakova E. The combination of meltblown and electrospinning for bone tissue engineering. Mater Lett. 2015;143(0):172–6.

Fathona IW, Yabuki A. Short electrospun composite nanofibers: effects of nanoparticle concentration and surface charge on fiber length. Curr Appl Phys. 2014;14(5):761–7.

Fernandez JG, Ingber DE. Unexpected strength and toughness in chitosan-fibroin laminates inspired by insect cuticle. Adv Mater. 2012;24(4):480–4.

Gurunathan S, Kalishwaralal K, Vaidyanathan R, Venkataraman D, Pandian SR, Muniyandi J, Hariharan N, Eom SH. Biosynthesis, purification and characterization of silver nanoparticles using Escherichia coli. Colloids Surf B Biointerfaces. 2009;74(1):328–35.

Hadipour-Goudarzi E, Montazer M, Latifi M, Aghaji AAG. Electrospinning of chitosan/sericin/PVA nanofibers incorporated with in situ synthesis of nano silver. Carbohydr Polym. 2014;113(0):231–9.

Hansen LM, Smith DJ, Reneker DH, Kataphinan W. Water absorption and mechanical properties of electrospun structured hydrogels. J Appl Polym Sci. 2005;95(2):427–34.

He D, Hu B, Yao QF, Wang K, Yu SH. Large-scale synthesis of flexible free-standing SERS substrates with high sensitivity: electrospun PVA nanofibers embedded with controlled alignment of silver nanoparticles. ACS Nano. 2009;3(12):3993–4002.

Holišová V, Urban M, Kolenčík M, Němcová Y, Schröfel A, Peikertová P, Slabotinský J, Kratošová G. Biosilica-nanogold composite: Easy-to-prepare catalyst for soman degradation. Arab J Chem. 2017. in press, doi.org/10.1016/j.arabjc.2017.08.003.

Hu W, Chen S, Li X, Shi S, Shen W, Zhang X, Wang H. In situ synthesis of silver chloride nanoparticles into bacterial cellulose membranes. Mater Sci Eng C. 2009;29(4):1216–9.
Hu X, Chen X, Zhang L, Lin X, Zhang Y, Tang X, Wang Y. A combined bottom-up/top-down approach to prepare a sterile injectable nanosuspension. Int J Pharm. 2014;472(1–2):130–9.
Jia J, Duan Y-Y, Wang S-H, Zhang S-F, Wang Z-Y. Preparation and characterization of antibacterial silver-containing nanofibers for wound dressing applications. J US-China Med Sci. 2007;4(2):52–4.
Jirsak O, Sanetrnik F, Lukas D, Kotek V, Martinova L, Chaloupek J. Method of nanofibres production from a polymer solution using electrostatic spinning and a device for carrying out the method, Google Patents 2009;US 7,585,437 B2.
Junlabhut P, Mekprasart W, Noonuruk R, Chongsri K, Pecharapa W. Characterization of ZnO:Sn Nanopowders synthesized by co-precipitation method. Energy Procedia. 2014;56(0):560–5.
Krahne R, Morello G, Figuerola A, George C, Deka S, Manna L. Physical properties of elongated inorganic nanoparticles. Phys Rep. 2011;501(3–5):75–221.
Konvičková Z, Schröfel A, Kolenčík M, Dědková K, Peikertová P, Žídek M, Seidlerová J, Kratošová G. Antimicrobial bionanocomposite–from precursors to the functional material in one simple step. J Nanopart Res. 2016;18(12).
Konvičková Z, Schröfel A, Kolenčík M, Dědková K, Peikertová P, Žídek M, Seidlerová J, Kratošová G. Erratum to: Antimicrobial bionanocomposite–from precursors to the functional material in one simple step (Journal of Nanoparticle Research, (2016), 18, 12, (368), 10.1007/s11051-016-3664-y). J Nanopart Res. 2017;19(7).
Lu Y, Li Y, Zhang S, Xu G, Fu K, Lee H, Zhang X. Parameter study and characterization for polyacrylonitrile nanofibers fabricated via centrifugal spinning process. Eur Polym J. 2013;49(12):3834–45.
Lukáš D, Sarkar A, Martinová L, Vodsed'álková K, Lubasová D, Chaloupek J, Pokorný P, Mikeš P, Chvojka J, Komárek M. Physical principles of electrospinning (electrospinning as a nano-scale technology of the twenty-first century). Text Prog. 2009;41(2):59–140.
McMurry J. Organic chemistry. Belmont: Thomson-Brooks/Cole; 2004.
Mellado P, McIlwee HA, Badrossamay MR, Goss JA, Mahadevan L, Kit Parker K. A simple model for nanofiber formation by rotary jet-spinning. Appl Phys Lett. 2011;99(20): -.
Mittal AK, Bhaumik J, Kumar S, Banerjee UC. Biosynthesis of silver nanoparticles: elucidation of prospective mechanism and therapeutic potential. J Colloid Interface Sci. 2014;415(0):39–47.
More DS, Moloto MJ, Moloto N, Matabola KP. TOPO-capped silver selenide nanoparticles and their incorporation into polymer nanofibers using electrospinning technique. Mater Res Bull. 2015;65(0):14–22.
Morton WJ. Method of dispersing fluids, Google Patents. 1902;US 705,691.
Nair LS, Laurencin CT. Biodegradable polymers as biomaterials. Prog Polym Sci. 2007;32(8–9):762–98.
Narayanan KB, Sakthivel N. Biological synthesis of metal nanoparticles by microbes. Adv Colloid Interf Sci. 2010;156(1–2):1–13.
Nguyen T-H, Lee K-H, Lee B-T. Fabrication of Ag nanoparticles dispersed in PVA nanowire mats by microwave irradiation and electro-spinning. Mater Sci Eng C. 2010;30(7):944–50.
Nguyen TH, Kim YH, Song HY, Lee BT. Nano Ag loaded PVA nano-fibrous mats for skin applications. J Biomed Mater Res B Appl Biomater. 2011;96(2):225–33.
Omri A, Benzina M, Bennour F. Industrial application of photocatalysts prepared by hydrothermal and sol–gel methods. J Ind Eng Chem. 2015;21(0):356–62.
Pencheva D, Bryaskova R, Kantardjiev T. Polyvinyl alcohol/silver nanoparticles (PVA/AgNps) as a model for testing the biological activity of hybrid materials with included silver nanoparticles. Mater Sci Eng C. 2012;32(7):2048–51.
Pentimalli M, Bellusci M, Padella F. High-energy ball milling as a general tool for nanomaterials synthesis and processing. Handbook of mechanical nanostructuring, Wiley-VCH Verlag GmbH & Co. KGaA; 2015. pp. 663–679.

Pinna A, Lasio B, Piccinini M, Marmiroli B, Amenitsch H, Falcaro P, Tokudome Y, Malfatti L, Innocenzi P. Combining top-down and bottom-up routes for fabrication of mesoporous Titania films containing ceria nanoparticles for free radical scavenging. ACS Appl Mater Interfaces. 2013;5(8):3168–75.

Polte J. Fundamental growth principles of colloidal metal nanoparticles – a new perspective. CrystEngComm. 2015;17(36):6809–30.

Reneker DH, Yarin AL. Electrospinning jets and polymer nanofibers. Polymer. 2008;49(10):2387–425.

Ruska E. The early development of electron lenses and electron microscopy. Stuttgart: Hirzel; 1980.

Salalha W, Dror Y, Khalfin RL, Cohen Y, Yarin AL, Zussman E. Single-walled carbon nanotubes embedded in oriented polymeric nanofibers by electrospinning. Langmuir. 2004;20(22):9852–5.

Sarkar K, Gomez C, Zambrano S, Ramirez M, de Hoyos E, Vasquez H, Lozano K. Electrospinning to Forcespinning™. Mater Today. 2010;13(11):12–4.

Schröfel A, Kratošová G, Bohunická M, Dobročka E, Vávra I. Biosynthesis of gold nanoparticles using diatoms—silica-gold and EPS-gold bionanocomposite formation. J Nanopart Res. 2011;13(8):3207–16.

Schröfel A, Kratošová G, Šafařík I, Šafaříková M, Raška I, Shor LM. Applications of biosynthesized metallic nanoparticles – a review. Acta Biomater. 2014;10(10):4023–42.

Sebe I, Szabó B, Nagy ZK, Szabó D, Zsidai L, Kocsis B, Zelkó R. Polymer structure and antimicrobial activity of polyvinylpyrrolidone-based iodine nanofibers prepared with high-speed rotary spinning technique. Int J Pharm. 2013;458(1):99–103.

Son Y, Yeo J, Ha CW, Lee J, Hong S, Nam KH, Yang D-Y, Ko SH. Application of the specific thermal properties of Ag nanoparticles to high-resolution metal patterning. Thermochim Acta. 2012;542:52–6.

Thatoi P, Kerry RG, Gouda S, Das G, Pramanik K, Thatoi H, Patra JK. Photo-mediated green synthesis of silver and zinc oxide nanoparticles using aqueous extracts of two mangrove plant species, Heritiera fomes and Sonneratia apetala and investigation of their biomedical applications. J Photochem Photobiol B Biol. 2016;163:311–8.

Velmurugan P, Cho M, Lim S-S, Seo S-K, Myung H, Bang K-S, Sivakumar S, Cho K-M, Oh B-T. Phytosynthesis of silver nanoparticles by Prunus yedoensis leaf extract and their antimicrobial activity. Mater Lett. 2015;138:272–5.

Vijayaraghavan K, Mahadevan A, Sathishkumar M, Pavagadhi S, Balasubramanian R. Biosynthesis of Au(0) from Au(III) via biosorption and bioreduction using brown marine alga Turbinaria conoides. Chem Eng J. 2011;167(1):5–5.

Virovska D, Paneva D, Manolova N, Rashkov I, Karashanova D. Electrospinning/electrospraying vs. electrospinning: a comparative study on the design of poly(l-lactide)/zinc oxide non-woven textile. Appl Surf Sci. 2014;311(0):842–50.

Wang Y, Yang Q, Shan G, Wang C, Du J, Wang S, Li Y, Chen X, Jing X, Wei Y. Preparation of silver nanoparticles dispersed in polyacrylonitrile nanofiber film spun by electrospinning. Mater Lett. 2005;59(24–25):3046–9.

Yadav TP, Yadav RM, Singh DP. Mechanical milling: a top down approach for the synthesis of nanomaterials and nanocomposites. Nanosci Nanotechnol. 2012;2(3):22–48.

Yallappa S, Manjanna J, Dhananjaya BL. Phytosynthesis of stable Au, Ag and Au–Ag alloy nanoparticles using J. Sambac leaves extract, and their enhanced antimicrobial activity in presence of organic antimicrobials. Spectrochim Acta A Mol Biomol Spectrosc. 2015;137(0):236–43.

Yamada Y. Acetobacter xylinus sp. nov., nom. rev., for the cellulose-forming and cellulose-less, acetate-oxidizing acetic acid bacteria with the Q-10 system. J Gen Appl Microbiol. 1983;29(5):417–20.

Yang N, WeiHong L, Hao L. Biosynthesis of Au nanoparticles using agricultural waste mango peel extract and its in vitro cytotoxic effect on two normal cells. Mater Lett. 2014;134:67–70.

Yuan C-G, Huo C, Yu S, Gui B. Biosynthesis of gold nanoparticles using Capsicum annuum var. grossum pulp extract and its catalytic activity. Physica E. 2017;85:19–26.

Zeleny J. The electrical discharge from liquid points, and a hydrostatic method of measuring the electric intensity at their surfaces. Phys Rev. 1914;3(2):69–91.

Zhang YZ, Venugopal J, Huang ZM, Lim CT, Ramakrishna S. Characterization of the surface biocompatibility of the electrospun PCL-collagen nanofibers using fibroblasts. Biomacromolecules. 2005;6(5):2583–9.

Zhang W, Chen Z, Liu H, Zhang L, Gao P, Li D. Biosynthesis and structural characteristics of selenium nanoparticles by pseudomonas alcaliphila. Colloids Surf B Biointerfaces. 2011;88(1):196–201.

Chapter 14
Decoration of Inorganic Substrates with Metallic Nanoparticles and Their Application as Antimicrobial Agents

Marianna Hundáková, Kateřina Dědková, and Gražyna Simha Martynková

Abstract Effect on antimicrobial activity observed for several types of hybrid materials is described in our chapter. The substrates for functional antimicrobial particles are natural clay minerals and carbon materials for this review limited to graphite/graphene and carbon nanoparticles (nanotubes and fullerenes). Short description of substrate materials and their properties is followed by discussion of the effect of selected most popular antimicrobial metals (silver, copper) and several oxides (zinc, titanium and copper oxides) and it is conferred for Gram positive and Gram negative bacterial strains. The methods for preparation of such particles may vary but the most used are intercalation and decoration methods from solution for the clay minerals. Nanoparticles (NPs) of metals and metal oxides on carbon and nanocarbon materials are prepared using physico-chemical approach. The research confirmed that the shape and size of functional NPs can depend on used substrate, preparation conditions and used method. Interestingly, it was found that Ag-clay sample was as effective as the free Ag+ions. Generally, it was found the size of active surface area, mobility and availability of potential active particles (ions or nanoparticles) and chemical state of them plays an important role in antimicrobial activity.

Keywords Antimicrobial activity • Silver • Copper • Zinc oxide • Titanium oxide • Nanoparticles • Clay minerals • Carbon materials

---

M. Hundáková • K. Dědková (✉)
Nanotechnology Centre, VŠB-Technical University of Ostrava,
17.listopadu 15/2172, 70833 Ostrava, Poruba, Czech Republic

Regional Materials Science and Technology Centre, VŠB-Technical University of Ostrava,
17. listopadu 15/2172, 70833 Ostrava, Poruba, Czech Republic
e-mail: katerina.dedkova@vsb.cz

G.S. Martynková
Nanotechnology Centre, VŠB-Technical University of Ostrava,
17.listopadu 15/2172, 70833 Ostrava, Poruba, Czech Republic

M. Rai, R. Shegokar (eds.), *Metal Nanoparticles in Pharma*,
DOI 10.1007/978-3-319-63790-7_14

14.1 Introduction

The word antimicrobial was derived from three the Greek words anti (against), mikros (little) and bios (life) and refers to all agents that turn against microbial organisms. This is not synonymous with antibiotics, a similar term derived from the Greek word anti (against) and biotikos (concerning life). By strict definition, the word "antibiotic" refers to substances formed by microorganisms that act against another microorganism. Thus, antibiotics do not include antimicrobial substances that are synthetic (sulfonamides and quinolones), or semisynthetic (methicillin and amoxicillin), or those which come from plants (quercetin and alkaloids) or animals (lysozyme). In contrast, the term "antimicrobials" include all agents that turn against all types of microorganisms - bacteria (antibacterial), viruses (antiviral), fungi (antifungal) and protozoa (antiprotozoal).

Antimicrobials are classified in several ways, as: spectrum of activity, effect on bacteria, mode of action or as the basic (cationic), the uncharged (neutral) and the acidic (anionic groups). Antimicrobials often display different minimum inhibition concentration (MIC) values, especially between Gram-negative (G^-) and Gram-positive (G^+) bacteria strains. The G^- strains are less susceptible to antibiotics due to the permeability barrier provided by their outer membrane. Consequently, a full characterization of antibacterial activity should be assessed using both types of organisms. Examples of commonly found G^- bacteria are *Escherichia coli, Salmonella typhimurium* and *Pseudomonas aeruginosa*, and examples of G^+ strains are *Streptococcus aureus* and *Streptococcus pyogenes*.

There is interest to expand of range of antimicrobial agents for other than most popular silver, since extensive and uncontrolled use of silver, is increasing public interest to address and monitor the clinical risk related to silver resistance and in an environmental context to study the sources, fate, transport routes and toxicity of environmentally relevant forms of silver. Over the past few years, various nano-sized antibacterial agents such as metal and metal oxide nanoparticles have been evaluated by researchers. Several types of metal and metal oxide nanoparticles apart from silver (Ag) and silver oxide (Ag_2O), the titanium dioxide (TiO_2), zinc oxide (ZnO), gold (Au), copper oxide (CuO), and magnesium oxide (MgO) have been known to show antimicrobial activity (Jakobsen et al. 2011; Vargas-Reus et al. 2012).

Silver and copper belongs to the metals, which are by tradition known their antibacterial behavior. In last years, silver compounds are studied as antimicrobial agents for many applications, for example, as silver-coated endotracheal tubes to reduce incidence of ventilator-associated pneumonia (Kollef et al. 2008) as a safe preservative for use in cosmetics Kokura et al. 2010, as additive to water based paints (Holtz et al. 2012), in textiles (Üreyen et al. 2012) and other applications (Rai et al. 2009). Copper compounds are studied as antimicrobial agents, for example, for drinking water treatment (Dankovich and Smith 2014) and for other applications (Vincent et al. 2016).

The mechanism of the antimicrobial action of metal nanoparticles (NPs) is not fully understood. Some authors described possible action and interaction of metals with the bacterial cell. The antibacterial tests showed the differences in antibacterial action of materials containing Ag between the G+ and the G- bacteria. This can be explained by differences in structure of bacterial cell. While the cell wall of the G+ bacteria consists of a thick layer of peptidoglycan with lipoteichoic acid or another acidic polymer, the cell wall of G- bacteria is formed by lipopolysaccharides and phospholipids and only a thin layer of peptidoglycan. The thick layer of peptidoglycan can make bacteria more resistant before the metal ions and NPs (Russell 2001; Zhao et al. 2006). It was observed and confirmed that several actions take place during interaction of the Ag+ ions with bacteria: (1) the Ag+ ions penetrate during the cell wall into the bacteria cell, where (2) interact with thiol groups in proteins and caused their inactivation, and moreover, (3) the Ag+ in the bacteria cell caused that DNA molecules become condensed and lose their replication abilities (Feng et al. 2000).

In case of Ag NPs, the size of NPs play important role in antimicrobial action. It was observed that smaller particles exhibited higher antimicrobial activity (Panáček et al. 2006). The small Ag NPs were incorporated into the membrane structure. The treated bacteria cell showed changes and damage to membrane which leads to increase in their permeability. The bacterial cells incapable of properly regulating transport through the plasma membrane; the cytoplasmic content is released to the medium which is causing cell death without affecting proteins or nucleic acids (Sondi and Salopek-Sondi 2004; Rai and Bai 2011). Large specific surface area of NPs can influence the antimicrobial action as well. The antibacterial properties of Ag NPs are related to the total surface area of the NPs. Smaller particles with a larger surface to volume ratio are more efficient agents for antibacterial activity (Baker et al. 2005).

The proteomic analysis of Ag NPs on *E. coli* shows that the Ag NPs destabilized the outer membrane and disrupts the outer membrane barrier components. The mode of action Ag NPs was found to be like that of Ag^+ ions. However, the effective concentrations of Ag NPs and Ag^+ ions were found to be at nanomolar and micromolar levels, respectively (Lok et al. 2006).

The mechanism of the bacteria cells inhibition by Cu^{2+} ions and CuO or Cu_2O NPs work in a similar way as silver agents. The G^- bacteria with an outer membrane covering a thin layer of peptidoglycan which is negatively charged probably binds the Cu^{2+} ions. Both G^+ and G^- strains turned from normal rod-shape into irregular shape (round, ellipse, etc.) after treatment with Cu-montmorillonite. The cell wall was destroyed, bacterial inner vacuoles appeared, there was an efflux of nutrient, the space between the cell wall and cell membrane widened, and the cytoplasm tended to concentrate. The Cu^{2+} can combine with the plasma membrane by electrostatic attraction and penetrate into the cell membrane through opening or closing of the membrane channel. This affects the permeability of cellular membranes and results in leakage of intracellular ions and low molecular-weight metabolites. Moreover, the Cu^{2+} enters into the cell strongly combines with intracellular sulfur-containing amino acids, which leads to denaturation of protein and bacterial death (Tong et al. 2005). The CuO or Cu_2O NPs caused bacterial membrane damage. The NPs due to the appropriate charge and small size can penetrate the membrane and cause bacteria cell death by the production of reactive oxygen species or by the disruption of cell function thereby affecting proteins and DNA (Meghana et al. 2015).

Titanium dioxide (*TiO_2*) is known for its chemical stability, photocatalytic characteristics, durability and antimicrobial activity, which could be attributed to its crystal structure (Chung et al. 2009). Anatase has stronger antimicrobial and photocatalytic activity than rutile (Chung et al. 2007a, b, 2011). Titanium dioxide is known for reactive oxygen species (ROS) production via photoactivation due to UV-light. There are also studies which find ROS production in the absence of photoactivation (Gurr et al. 2005). TiO_2 can promote the decomposition of inorganic and organic compounds, which could be used in potential applications in sanitation and sterilization. Materials coated with TiO_2 are already being used as antibacterial materials (Hashimoto et al. 2005).

Among metal oxide powders, *ZnO* demonstrates significant growth inhibition of broad spectrum of bacteria (Jones et al. 2008). The suggested mechanism for the antibacterial activity of ZnO is based on catalysis of formation of reactive oxygen species (ROS) (Yamamoto et al. 2002). Since the catalysis of radical formation occurs on the particle surface, particles with larger surface area demonstrate stronger antibacterial activity. Therefore, as the size of the ZnO particles decreases their antibacterial activity increases (Jones et al. 2008).

Nevertheless, metal ions and NPs could also be toxic for living organisms mainly at higher concentrations (Braydich-Stolle et al. 2005; Lu et al. 2010; Chang et al. 2012). Therefore, it is desirable to control their gradual release. For this purpose metal ions or metal NPs and their oxides may be prepared anchored on various inorganic substrates. The most commonly used substrates include clay minerals, zeolites or carbon materials.

Clay minerals are widely used as supports for metals or metal oxides preparation. Clay minerals belong to the layered silicates (phyllosilicates) with structure consist from octahedral and tetrahedral sheets. Based on the sheets arrangement to the layers, clay minerals are divided to the two groups, with type of layer 1:1 and 2:1. The 1:1 type consists of the repetition of one tetrahedral sheet and one octahedral sheet. The 2:1 type consists of the repetition of one octahedral sheet sandwiched between two tetrahedral sheets. The composition of the 1:1 phyllosilicates is characterized by a predominance of Al^{3+} as central octahedral cations, although some isomorphous substitution of Mg^{2+}, Fe^{3+}, Ti^{4+} for Al^{3+} can occur. The composition of the 2:1 phyllosilicates is characterized by a predominance of Si^{4+} as central octahedral cations and Al^{3+} and Fe^{3+} as central tetrahedral cations. Central cations are usually substituted by cations with lower valance as Al^{3+}, Fe^{3+}, Fe^{2+}, Mg^{2+}, etc. and a negative charge arising on layers from these substitutions. The charge variability is one of the most important features of 2:1 phyllosilicates, because it induces occupancy of the interlayer space by exchangeable cations which compensate the negative layer charge. Due to the low proportion of substitution in 1:1 phyllosilicates the layer charge is usually close to zero, so the interlayer space is without exchangeable cations (Brigatti et al. 2006). To the 1:1 phyllosilicates using as supports for metal nanoparticles preparation belong, for example, kaolinite (Kao) and halloysite (Hal) and to the 2:1 phyllosilicates belong montmorillonite (Mt), vermiculite (Ver), talc, palygorskite (Pal) and sepiolite (Sep).

Mt is defined as dioctahedral smectite. Smectites are swelling and turbostratically disordered minerals occurring in nature as the main component of bentonites. The term smectite is used for planar dioctahedral and trioctahedral 2:1 clay minerals with a layer charge between −0.2 and −0.6 per formula unit which contain hydrated exchangeable cations. Minerals of the smectite group have high specific surface area and ability of cation exchange capacity. Hydrated exchangeable cations between the layers compensate the negative charge and may be easily exchanged by other metal cations. The cation exchange capacity (CEC) is the measure of the cations, which balance the negative charge sites of the clay (Valášková and Martynkova 2012). Ver is planar dioctahedral and trioctahedral 2:1 clay mineral with a layer charge between −0.6 and −0.9 per formula unit which contain hydrated exchangeable cations. Hal is a 1:1 layered aluminosilicate structure chemically similar to Kao. The size of Hal tubules varies within 0.5–10 microns in length and 15–200 nm in inner diameter, depending on the deposit. Pal structure of 2:1 layers consists of the continuous two-dimensional tetrahedral sheet but lacking continuous octahedral sheet. Pal is characterized by microfibrous morphology, low surface charge and high surface area (Martynková and Valášková 2014).

Other groups of minerals are zeolites. Zeolites (Zeol) belong to the hydrated aluminosilicates with three-dimensional structures with independent component: the alumunosilicate framework, exchangeable cations and zeolitic water. The Zeol framework is composed from tetrahedron, which center is occupied by Si^{4+} or Al^{3+} cations, with four oxygen atoms at the vehicles. Due to the substitution of Si^{4+} by Al^{3+} arise the negative charge on the framework, which is compensated by cations located together with water. The water molecules can be in large cavities and bonded between framework and exchangeable ions. The most common representative of Zeol is clinoptilotite (Cli) (Wang and Peng 2010).

Health and environmental impacts of *graphene-based materials* need to be thoroughly evaluated before their potential applications. Graphene has strong cytotoxicity toward bacteria. To better understand its antimicrobial mechanism, comparison of the antibacterial activity of four types of graphene-based materials (graphite (Gt), graphite oxide (GtO), graphene oxide (GO), and reduced graphene oxide (rGO)) toward a bacterial model – *Escherichia coli* is given. GO dispersion shows the highest antibacterial activity, sequentially followed by rGO, Gt, and GtO. The direct contacts with graphene nanosheets disrupt cell membrane. Conductive rGO and Gt have higher oxidation capacities than insulating GO and GtO. Antimicrobial actions are contributed by both membrane and oxidation stress. Physicochemical properties of graphene-based materials, such as density of functional groups, size, and conductivity, can be precisely tailored to either reducing their health and environmental risks or increasing their application potentials (Liu et al. 2011).

The ability of carbon nanotubes (CNTs) to undergo surface modification allows them to form advanced nanocomposites with different materials such as polymers, metal nanoparticles, biomolecules, and metal oxides. The biocidal nature, protein fouling resistance, and fouling release properties of CNT-NCs render them the perfect material for biofouling prevention (Narayan et al. 2005). Cytotoxicity of CNT can be reduced before applying them as substrates to promote biofilm formation in environmental biotechnology applications.

To understand the inherent antimicrobial nature of pristine CNTs is key issue that may determine the efficiency of biocidal CNT-nanocomposites. Microorganisms lose viability when they come in contact with nanotubes; due to the impingement of nanometer-sized fibers and the needle-like character of CNTs easily penetrate through the cell walls of bacteria (Narayan et al. 2005). CNTs in solution develop nanotube networks on the cell surface, and then destroy the bacterial envelopes with leakage of the intracellular contents (Liu et al. 2010). The antimicrobial effect of CNTs has been demonstrated over a wide range of microorganisms including bacteria provided confirmatory evidence that both, single wall carbon nanotubes (SWCNTs) and multiwall carbon nanotubes (MWCNTs) are able to decrease the metabolic activity of *E. coli*. SWCNTs can produce cytotoxic effects on microbial communities in macro-environmental entities. Efficient contact between the CNTs and bacterial cell surface is crucial the biocidal action of CNTs. However, this effort depends on a variety of factors, such as: (i) physical and structural properties of CNTs (size and length); (ii) physical condition of CNTs (aggregated or dispersed); (iii) type and concentration of impurities associated with CNTs and their availability to bacteria (heavy metal impurities); and (iv) number of layers (single or multi-walled) of CNTs (Martynková and Valášková 2014). Normally, loosely packed, debundled, highly dispersed, and shorter length tubes can easily penetrate through the cell membrane and display higher cell cytotoxicity.

Fullerenes are soccer ball-shaped molecules composed of carbon atoms. Fullerenes showed antimicrobial activity against various bacteria. The antibacterial effect was probably due to inhibition of energy metabolism after internalization of the nanoparticles into the bacteria (Shvedova et al. 2012). It has also been suggested that fullerene derivatives can inhibit bacterial growth by impairing the respiratory chain (Cataldo and Da Ros 2008; Deryabin et al. 2014). In the beginning, a decrease of oxygen uptake (at low fullerene derivative concentration) and then an increase of oxygen uptake (followed by an enhancement of hydrogen peroxide production) are occurred. Another bactericidal mechanism, which has been proposed, was the induction of cell membrane disruption. Hydrophobic surface of the fullerenes can easily interact with membrane lipids and intercalate into them. The discovery of fullerenes ability to interact with biological membranes has encouraged many researchers to evaluate their antimicrobials applications (Tegos et al. 2005; Yang et al. 2014).

The cationic-substituted fullerene derivatives are highly effective in killing a broad spectrum of microbial cells after irradiation with white light. Affecting factor was an increased number of quaternary cationic groups that were widely dispersed around the fullerene cage to minimize aggregation. The quaternized fullerenes could be effectively applied in treatment of superficial infections, e.g. wounds and burns, where light penetration into tissue is not problematic (Mizuno et al. 2011).

Nowadays, many publications are focused on the preparation, characterization and study of the functional properties (especially catalytic) of silver, copper, zinc and titanium oxide nanoparticles on inorganic substrates. In this chapter, we focus only on the research that describes study of antimicrobial properties of these materials.

14.2 Metals on Inorganic Substrates as Antimicrobial Agent

14.2.1 Silver on Clay Minerals

Silver can be synthesized on inorganic substrates via several methods: as Ag^+ ions and Ag or Ag_2CO_3 NPs. Summary of selected preparation techniques of Ag on the various substrates, tested bacterial strains and methods used for studying the antimicrobial activity is shown in Table 14.1.

Montmorillonite is widely used clay mineral for various industrial and medical applications. Antibacterial properties of Ag modified Mt on Gram-negative (G^-) bacterium *Escherichia coli (E. coli)* were studied. Mt was pretreated by calcination at 550 °C or by grinding, followed by enriching with Ag using cation exchange methods. The metallic Ag^0 NPs precipitated on clay surface was confirmed. Nevertheless, the Ag^+ ions were also incorporated in the interlayer of Mt structure. Results of antibacterial activity of Ag-Mt composites, evaluated via the disc susceptibility (DS) test and by determination of the minimum inhibitory concentration (MIC) test, showed good inhibition properties on the growth of *E. coli*. The antibacterial behavior was affected by the availability of the ionic Ag^+ to be in contact with the bacteria (Magaña et al. 2008).

The simple preparation process of Ag NPs on Mt matrix included the stirring of Mt with $AgNO_3$ aqueous solution at room temperature. The mean particle size of Ag NPs on Mt surface was 50 nm and more on the edges of the Mt flakes. Antibacterial activity was tested via the broth dilution method by MIC value determination. The Gram-positive (G^+) bacteria *Staphylococcus aureus* (*S. aureus*) and *Enterococcus faecalis* (*E. faecalis*) and the Gram-negative (G^-) bacteria *Klebsiella pneumoniae* (*K. pneumoniae*) and *Pseudomonas aeruginosa* (*P. aeruginosa*) were used for antibacterial test. Antibacterial action of samples started after 1.5 h against the G^- bacteria and after 3–5 h against the G^+ bacteria and action was persisting for the whole testing period 6 days. Moreover, authors studied release of Ag^+ from Mt matrix into the water environment, which was determined as 0.1–0.3% from the total Ag content in the samples (Valášková et al. 2010). Antibacterial study of the Ag^+- exchanged and Ag^0- covered Mt samples showed good inhibition effect determined by the disk diffusion method against bacteria *P. aeruginosa* (G^-) and *S. aureus* (G^+). The diameter of inhibition zone was in the range 20–21 and 21–22 mm for Ag^+-Mt and Ag^0-Mt, respectively (Özdemir et al. 2010). The antibacterial and antifungal properties of Ag-Mt prepared by cation exchange method on Mt using $AgNO_3$ aqueous solution

Table 14.1 Summary of methods and precursors used for the preparation of silver on the inorganic substrates, used testing methods, and bacterial strains for studying the antimicrobial activity of prepared materials

Method, precursor (conditions)	Inorganic substrate, chemical pretreatment	Tested bacteria	Testing method, test evaluation	Reference
Stirring, $AgNO_3$ (room temperature, 1 week)	Mt (Pellegrini Lake, Argentina) – pretreated by (a) calcination, (b) grinding	*E. coli*	Disk susceptibility test, MIC determination	Magaña et al. (2008)
Shaking, $AgNO_3$ (room temperature, 24 h)	Mt (Ivančice, Czech Republic), Ver (China)	*S. aureus*, *P. aeruginosa*, *K. pneumoniae*, *E. faecalis*	Broth dilution method, MIC determination	Valášková et al. (2010)
Shaking, $AgNO_3$, $Cu(NO_3)_2$, $Zn(NO_3)_2$, CP (room temperature, 24 h)	Mt (Middle Anatolia) – purified	*S. aureus*, *P. aeruginosa*	Disk diffusion method	Özdemir et al. (2010)
Shaking, $AgNO_3$, $ZnSO_4$, $CuSO_4$ (room temperature, 24 h)	Na^+-rich Mt (SWy2) (Crook County, Wyoming)	*E. coli*, *P. cinnabarinus*, *P. ostreatus*	CFU counting, median effective concentration (EC50)	Malachová et al. (2011)
Melting, $AgNO_3$, $NaNO_3$ (435 °C, 4h)	Ben (San Juan, Mexico) – pretreated with acid (HCl or H_2SO_4)	*E. coli*, *S. aureus*	Agar diffusion method, MIC determination	Santos et al. (2011)
(1) Ion exchange, Ag_2O, $NH_3{\cdot}H_2O$ (60 °C, 4 h), (2) UV-photoreduction (PVP, high pressure mercury lamp)	Mt (from Zhejiang Fenghong Clay Chemicals Corp., China)	*E. coli*	Broth microdilution method, MIC determination, sterilizing efficiency	Xu et al. (2011)
Cation exchange, $AgNO_3$ – use in suture	Mt (from Sanding Technology Company in Zhejiang Province, China) – purified	*E. coli*, *S. aureus*	CFU counting, hemolysis test, in vitro cytotoxicity test	Cao et al. (2014)
(1) Shaking, $AgNO_3$ (room temperature, 24 h), (2) reduction, $NaBH_4$ (shaken, several min, room temperature)	Na^+-rich Mt (SWy2) (Crook County, Wyoming)	*E. coli*, *E. faecium*	MIC determination	Malachová et al. (2009)
(1) Absorption, $AgNO_3$, n-hexanol, (2) reduction, $NaBH_4$	Mt (Kunipia G, from Kunimine Co., Ltd.)	*E. coli*	CFU counting	Miyoshi et al. (2010)

(1) Stirring, $AgNO_3$ (60 °C, 8 h), (2) reduction: (a) $NaBH_4$, (b) UV-irradiation, (c) calcination	Mt (from Nanocor, USA) – unmodified (Nanocor PGN) – alkylammonium modified (Nanocor 1.44P)	*E. coli*	CFU counting	Girase et al. (2011)
(1) Stirring, $AgNO_3$ (room temperature, 24 h), (2) reduction, $NaBH_4$	Mt (from Kunipa-F, Japan)	*E. coli*, *S. aureus*, *K. pneumoniae*	Disk diffusion method, inhibition zone determination	Shameli et al. (2011)
Shaking, $AgNO_3$ (room temperature, 24 h)	Mt (Ivančice, Czech Republic) – original and Na form of Mt	*E. faecalis*, *P. aeruginosa*	Broth dilution method, MIC determination	Hundáková et al. (2013a)
(1) Grinding, $AgNO_3$, with or without Na_2CO_3, (2) stirring in EG, → Ag, or Ag_2CO_3 NPs	Commercial Na-MT from Southern Clay Products (Gonzales, TX)	*E. coli*	Disk diffusion method, MIC determination	Sohrabnezdah et al. (2015)
(1) Stirring, $AgNO_3$, formaldehyde, NaOH, citric acid (3 h), (2) microwave-assisted method	Mt K-10 clay (from Sigma Aldrich)	*S. aureus*, *P. aeruginosa*	Disk diffusion and macrodilution methods	Kheiralla et al. (2014)
Adsorption, $AgNO_3$ (8 days)	Natural smectite (Mt) (North Patagonian, Argentina) – modified with silanizing agent APS or surfactant HDTMA	*E. coli*	Inhibition zone determination	Parolo et al. (2011)
Cation exchange, $AgNO_3$, used in package material	Mt – Na^+ form	Mesophilic, psychotrophic, and lactic acid bacteria, coliforms, yeasts, molds	CFU counting	Costa et al. (2011)
Cation exchange, $AgNO_3$, used in package material	Mt – Na^+ form	Mesophilic and psychotrophic bacteria, *Enterobacteriaceae spp.*, *Pseudomonas spp.*, yeasts, molds	CFU counting	Costa et al. (2012)
Shaking, $AgNO_3$ (room temperature, 24 h)	Ver (Santa Luzia, Brazil) – original and acidified by HCl	*P. aeruginosa*, *E. faecalis*	Broth dilution method, MIC determination	Hundáková et al. (2011)

(continued)

Table 14.1 (continued)

Method, precursor (conditions)	Inorganic substrate, chemical pretreatment	Tested bacteria	Testing method, test evaluation	Reference
Shaking, $AgNO_3$, $Cu(NO_3)_2$ (room temperature, 24 h)	Ver (Santa Luzia, Brazil)	*S. aureus*, *E. faecalis*, *K. pneumoniae*, *P. aeruginosa*	Broth dilution method, MIC determination	Hundáková et al. (2013b)
Shaking, $AgNO_3$, $Cu(NO_3)_2$ (room temperature, 24 h)	Ver (Santa Luzia, Brazil) Mt (Ivančice, Czech republic)	*E. coli*	Broth dilution method, MIC determination	Hundáková et al. (2014b)
Shaking (room temperature, 24 h), $AgNO_3$, $Cu(NO_3)_2$, used as filler into PE	Ver (Santa Luzia, Brazil)	*E. faecalis*	Broth dilution method, MIC determination, CFU counting	Hundáková et al. (2014a)
Adsorption, $AgNO_3$, $Cu(NO_3)_2$ (24 h)	Pal from the Mingguang palygorskite mine in Anhui Province of China – calcined and acid activated	*E. coli*, *S. aureus*	CFU counting	Zhao et al. (2006)
(1) stirring, $AgNO_3$ (room temperature, 24 h), (2) reduction, $NaBH_4$, shaking, $AgNO_3$ (room temperature, 24 h)	Talc (from Sigma Aldrich, St. Louis, MO)	*E. coli*, *S. aureus*	Disk diffusion method, inhibition zone determination	Shameli et al. (2011)
Cation exchange, $AgNO_3$, phosphoric acid	Kao, Sep, Cli (Eskisehir region) –Na^+ forms	*E. coli*	Halo test	Karel et al. (2015)
Shaking, $AgNO_3$ (24 h) – used as filler into PP	Kao (Sedlec, CZ) – (1) formamide, (2) NH_4Br, (3) vinylacetate + Bz_2H_2	*P. aeruginosa*, *E. faecalis*	Broth dilution method, MIC determination, CFU counting	Hundáková et al. (2016)
Shaking, $AgNO_3$ (24 h), reduction, $NaBH_4$	Hal (Henan province, China) – modification with silane coupling agent in toluene (refluxing 24 h)	*E. coli*, *S. aureus*	Inhibition zone determination, MIC determination, optical density	Zhang et al. (2013)
Adsorption, $AgNO_3$	Mexican zeolitic mineral clinoptilotite-heulandite (Taxco, Guerrero) – Na^+ form	*E. coli*, *S. faecalis*	CFU counting	Rivera-Garza et al. (2000)

Ag^+ bound electrostatically to Zeol	Zeol (Zeotomic AJ10N) (Sinanen Zeomic Co. Ltd., Nagoya, Japan)	*P. gingivalis, P. intermedia, A. actinomycetemcomitans, S. mutans, S. sanguis, A. viscosus, S. aureus*	MIC determination	Kawahara et al. (2000)
Cation exchange, $AgNO_3$, $ZnNO_3{\cdot}5\ H_2O$, $Cu(NO_3){\cdot}2.5H_2O$ (shaking, 25 °C, 2 days)	Cli-rich Zeol mineral (Gördes, Turkey, Western Anatolia) – Na^+-rich form	*E. coli, P. aeruginosa*	Disk diffusion method	Top and Ülkü (2004)
Reflux, $AgNO_3$ (12 h) – use for disinfection of wastewater	Mexican zeol (Clinoptilotite-Heulandite) from Sonora	*E. coli*	CFU counting	Rosa-Gomez et al. (2008)
Reflux, $AgNO_3$ (12 h)	Mexican Cli (from Oaxaca and Sonora) – Na^+-rich form	*E. coli*	CFU counting	Rosa-Gomez et al. (2010)
Cation exchange, $AgNO_3$ (3× 24 h)	Cli (Marsid, Romania) – Na^+-rich form	*E. coli, S. aureus*	CFU counting	Copcia et al. (2011)
(1) Shaking, $AgNO_3$ (3 h), (2) reducing (550 or 700 °C, H_2 atm., 4 h)	Cli-rich tuff (Etla, Oaxaca, southeast Mexico) – Na^+-rich form	*E. coli, S. typhi*	CFU counting	Guerra et al. (2012)
Shaking, $AgNO_3$, $ZnNO_3{\cdot}5H_2O$, $Cu(NO_3){\cdot}2.5H_2O$, Benzalkonium chloride	Natural Zeol with 70wt.% Cli (the sedimentary deposit Zlatokop, Serbia Zeolite) – Na^+-rich form	*A. baumannii* – European clone I and II	CFU counting, MIC and MBC determination	Hrenovic et al. (2013)
Sol-gel dip coating – used to disinfection of wastewater	Natural Zeol (obtained from Afrand Tooska, Co. (Iran))	*Saprolegnia Sp.*		Johari et al. (2016)
Stirring, CTAB (room temperature, 16 h), $AgNO_3$	NaY Zeol (Zeolyst International, Havennummer, The Netherlands) –Na^+ rich	*E. coli, S. aureus*	MIC determination	Salim and Malek (2016)
Cation exchange, $AgNO_3$, $Cu(NO_3)_2$, $ZnCl_2$	Mt (Ivančice, Czech Republic), Ver (Letovice, Czech Republic)	*P. aeruginosa, E. faecalis, E. coli, T. vaginalis*	Broth dilution method, MIC determination	Pazdziora et al. (2010)

was investigated against *E. coli* (G^-), *Pycnoposus cinnabarinus* (*P. cinnabarinus*) and *Pleurotus ostreatus* (*P. ostreatus*), respectively. It was found that Ag-Mt composite was as effective as the free Ag^+ ions. Also, the inhibition of fungal growth on Ag-Mt composite was similarly active as free Ag^+ ions (Malachová et al. 2011).

Cation exchange method by dry way was used for incorporating of Ag^+ into bentonite (Ben) pre-treatment with HCl or H_2SO_4. Raw and acid activated Ben was subjected to an ion exchange process with melt of $AgNO_3$ and $NaNO_3$ at 435 °C. Authors studied antimicrobial properties of samples using bacteria *E. coli* (G^-) and *S. aureus* (G^+). Only Ben samples with Ag^+ showed antimicrobial effect. Moreover, the microbiological results showed influence of acid pre-treated of Ben on antimicrobial effect. Ben activated by HCl showed better bactericidal properties than Ben activated by H_2SO_4. The MIC value was for *S. aureus* 0.022 g of sample Ag^+-HCl-Ben and 0.038 g of Ag^+-H_2SO_4-Ben and for *E. coli* 0.012 g of sample Ag^+-HCl-Ben and 0.038 g of Ag^+-H_2SO_4-Ben (Santos et al. 2011).

Xu et al. (2011) introduced a novel way of preparation of Ag-Mt material with antibacterial activity exhibiting slow release property. At first, transparent $[Ag(NH_3)_2]$ OH aqueous solution was prepared by adding $NH_3.H_2O$ to Ag_2O powder. The solution was mixed with Mt and stirred at 60 °C in the dark. Mixture was added to poly (N-vinyl-2-pyrrolidone (PVP) and UV-irradiated at room temperature. Results shown, that after UV irradiation the Ag^+ turned into metallic Ag^0 NPs. This indirectly suggested change of Ag-Mt color from milk-white to black after irradiation. Antibacterial activity was tested against *E. coli* (G^-). The MIC value was 100×10^{-6} of Ag-Mt and the sterilizing efficiency (SE) was reaching 100%. Even after five times washes with distilled water, the MIC value of sample was 200×10^{-6} and the SE was more than 99%. Authors confirm slow release property of Ag from Mt substrate which can be explain by strong interactional force between Ag NPs and Mt due to a large Mt surface area and small diameter of Ag NPs (Xu et al. 2011). Antibacterial activity of Ag-Mt prepared using original Mt and Na^+-form of Mt was compared. The monoionic Na^+-form of Mt (Na^+Mt) was prepared using the most common cation exchanged process with NaCl aqueous solution. Original Mt and Na^+Mt were shook with $AgNO_3$ aqueous solution (protected by aluminum foil against the light). Total amount of Ag determined in samples was from 0.64 to 4.47 wt.% depending on initial concentration of $AgNO_3$ solution and used matrix (Mt or Na^+Mt). Samples prepared from monoionic Na^+Mt form contained higher amount of Ag and showed better inhibition effect on both tested bacteria *P. aeruginosa* (G^-) and *E. faecalis* (G^+) (Hundáková et al. 2013a).

The interesting studies were focused to application of Ag-Mt to packaging materials (Costa et al. 2011, 2012). In one study, prepared Ag-Mt samples were used to improve the shelf life of fresh fruit salad. The sensorial and microbiological quantity was determined. The microbiological activity was evaluated by monitoring the principal spoilage microorganisms (mesophilic and psychotropic bacteria, coliforms, lactic acid bacteria, yeasts and molds). Results show that a significant shelf life prolongation of fresh fruit salad can be obtained by a straightforward new packaging system by using Ag-Mt (Costa et al. 2011). In second study, the effect of active coating loaded with Ag-Mt and film barrier properties on shelf life of fresh cut carrots were investigated. The sensorial and microbiological quality was observed. For microbiological quality, the spoilage microorganisms as mesophilic and psychotropic bacteria, *Enterobacteriaceae* spp., *Pseudomonas* spp., yeasts and moulds were used. For sensorial evaluation, color, odor, firmness and product overall quality were evaluated) (Costa et al. 2012). Cao et al. (2014) published very interesting study about sutures modified by Ag^+ loaded Mt (Ag^+ Mt/sutures). The antibacterial properties of prepared materials were tested using bacteria *E. coli* (G^-) and *S. aureus* (G^+). Moreover, hemolysis tests and in vitro cytotoxicity test of Ag^+ Mt/sutures were determined. The Ag^+ Mt/sutures exhibited good blood and tissue biocompatibilities. In addition, the Ag^+ Mt/sutures inhibited growth of *E. coli* by 99% (Cao et al. 2014).

Several authors published study about antibacterial activity of Ag NPs prepared on substrates by the chemical reduction method with borohydride ($NaBH_4$) as reducing agent.

Mt was shaken with $AgNO_3$ aqueous solution at room temperature. To dried sample the $NaBH_4$ aqueous solution was added and reduction under shaking took for several minutes. Antibacterial activity of Ag^+ ions and metallic Ag^0 prepared on Mt substrate was tested on *E. coli* (G^-) and *E. faccium* (G^+). The metallic Ag^0 showed no antibacterial effect (Malachová et al. 2009). The Ag NPson the surface and in the interlayer space of Mt was prepared by $NaBH_4$ reduction in n-hexanol. Antibacterial activity against *E. coli* (G^-) was determined by counting the CFU (colony forming unit) number. Samples were tested in the dark and under room light when a stronger antibacterial effect was observed. Authors also tested and confirmed antibacterial activity of samples after 12 years what confirmed their stability and suitability for application to industry use (Miyoshi et al. 2010). Shameli et al. (2011) synthesized Ag NPs on Mt with the mean diameter of Ag NPs from 4.19 to 8.53 nm. The particles size increased with increased concentration of initial Ag^+ ions. Moreover, intercalated structure of Mt was confirmed in Ag-Mt nanocomposites. Antibacterial activity was tested for $AgNO_3$-Mt suspension and Ag-Mt nanocomposites on the G^- bacteria *E. coli* and *K. pneumoniae* and the G^+ bacteria *S. aureus*. Results showed that measured inhibition zone for $AgNO_3$-Mt and Ag-Mt nanocomposites were similar for both type of bacteria. The antibacterial activity of Ag-Mt nanocomposites decreased with increased of Ag NPs particles size (Shameli et al. 2011). Girase et al. (2011) prepared Ag on Mt substrate, which was used as untreated or organically modified and both ball milling, by three methods (Fig. 14.1). The Ag NPs size on Mt depended on used method. The size distribution of the Ag NPs on Mt followed the sequence: UV > Calcined > $NaBH_4$ reduced. The Ag NPs synthesized in the absence of Mt by all three method (calcination, $NaBH_4$ reduction and UV irradiation) were larger (~60–200 nm) than the particles precipitated on Mt (5–25 nm). Results confirmed that the shape and size of NPs can depend on used substrate, preparation conditions and used method. Difference in size of Ag NPs synthetized on Mt substrate by three various methods is show in Fig. 14.1. Moreover, the Ag NPs precipitated ex-situ in solution prepared to compare. The both ex situ Ag NPs and in situ precipitated Ag NPs on Mt indicated antibacterial activity against *E. coli* (G^-) as a function of time. In situ precipitated

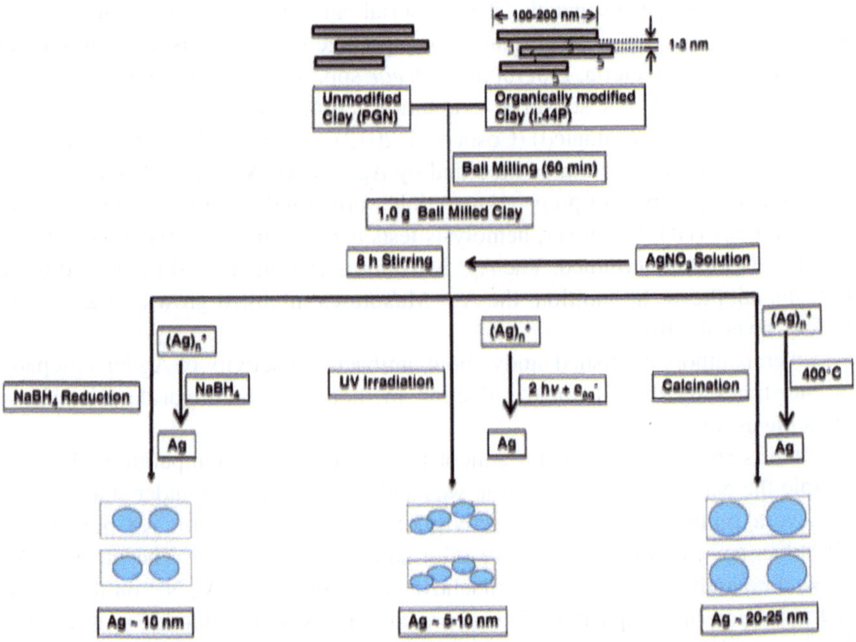

Fig. 14.1 Schematic illustration of the synthesize Ag NPs-Mt nanohybrid material (*in-situ* precipitation of Ag NPs on the Mt surface). Reproduced with permission from (Girase et al. 2011), © 2011, Elsevier B.V

Ag NPs on Mt indicated good antibacterial performance, irrespective of the method used to Ag preparation (reduction, calcination or UV method). In the case of Ag NPs-Mt nanohybrid structure, the CFU number of *E. coli* decreased by four orders of magnitude as compared to samples without Ag. Authors observed that the behavior of antimicrobial activity is related to the behavior of release of Ag ions from Mt (Fig. 14.2) (Girase et al. 2011).

Sohrabnezdah et al. (2015) used Na^+Mt as stabilizer for preparation of Ag_2CO_3 and Ag NPs in aqueous and polyol solvent. Different treatments were used: In first, Mt with $AgNO_3$ and Na_2CO_3 were ground and ethylene glycol (EG) was added under stirring. In second, the same procedure was used without Na_2CO_3. The metallic Ag^0 in these Mt nanocomposites was confirmed. In third, the same method was used with water instead EG. This process led primarily to the formation of Ag_2CO_3 NPs in nanocomposite. Moreover, results also show Ag^0 NPs present and probable the intercalation of both Ag and Ag_2CO_3 NPs into the Mt gallery. Results of antibacterial test on *E. coli* (G^-) showed that the Ag_2CO_3-Mt nanocomposite exhibited an antibacterial activity higher than Ag-Mt. Authors explained higher antibacterial action of the Ag_2CO_3-Mt by higher release of Ag^+ ions from Ag_2CO_3 and interaction with bacterial strain. In case of Ag NPs on Mt surface, less Ag^+ ions are oxidatively released from Ag NPs surface.

Kheiralla et al. (2014) synthetized the Ag NPs on Mt substrate using the microwave assisted method. The Ag NPs-Mt nanocomposites were prepared by mixing of

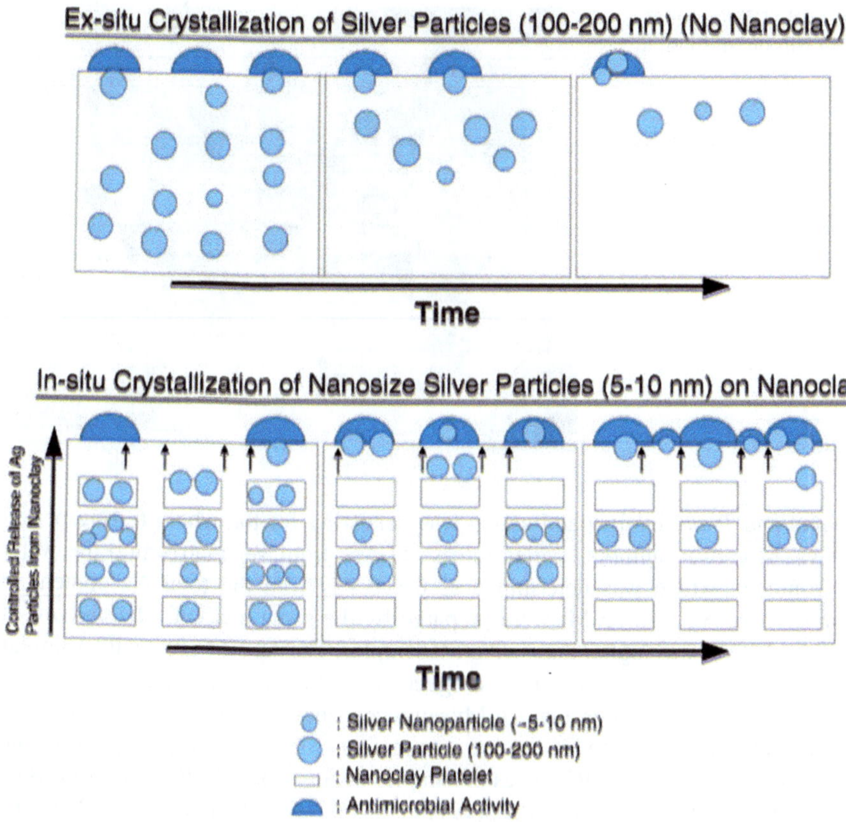

Fig. 14.2 Schematic illustration of diffusion-controlled release of Ag ions from clay platelets (in-situ precipitated Ag NPs) and ex-situ precipitated Ag NPs in solution and their antimicrobial behavior. Reproduced with permission from (Girase et al. 2011), © 2011, Elsevier B.V

aqueous solution $AgNO_3$ with Mt. Formaldehyde and NaOH aqueous solution was added until the pH was 10 and citric acid was added. The solution was placed in a microwave. The samples with the weight ratio 1, 3 and 5% of Ag in Mt were prepared. The diameter of synthetized Ag NPs on Mt was below 15 nm. Antibacterial activity was evaluated against bacteria *S. aureus* (G^+) and *P. aeruginosa* (G^-). The maximum inhibition zone diameters for concentration of Ag 1, 3 and 5% were determined for *S. aureus* as 29, 39 and 45 mm and for *P. aeruginosa* as 10, 25 and 37 mm, respectively. The MIC value of Ag NPs was determined as 0.031 mg for both bacteria, while the MBC value was 0.5 mg for *S. aureus* and 0.25 mg for *P. aeruginosa*. Moreover, the synergistic effect of Ag NPs and antibiotics was observed.

Vermiculite (Ver) is used as substrate for Ag NPs preparation very sporadically. The Ag NPs were prepared on Ver matrix by shaken of Ver with $AgNO_3$ aqueous solution at room temp. The size of Ag NPs on Ver surface was heterogeneous, from smaller than 20–50 nm. Antibacterial action of samples started after 1–1.5 h against G^- bacteria *S. aureus* and *E. faecalis* and after 2–3 h against G^+ bacteria *K. pneu-*

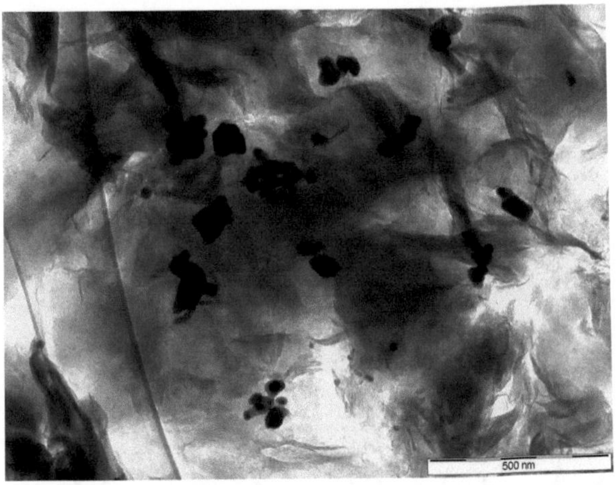

Fig. 14.3 TEM images of Ag NPs on Mt substrate

moniae and *P. aeruginosa* and action was persisting for the whole testing period 6 days. Moreover, authors studied release of Ag^+ from Ver matrix into the water environment, which was determined as 0.1–0.4% from the total Ag content in samples. The Ag NPs on Mt were prepared by the same method. The amount of silver in Ag-Ver was higher than in Ag-Mt samples. The Ag NPs grew with a similar size on Mt and Ag NPs size was heterogeneous on Ver. The antibacterial action of Ag-Ver samples was better than Ag-Mt samples (Valášková et al. 2010). The original and acidified Ver (with HCl) were compared as substrates for precipitation and growth of Ag NPs. The MIC values of samples against bacteria *P. aeruginosa* (G^-) and *E. faecalis* (G^+) were from 10% (w/v) to 0.37% (w/v) (Hundáková et al. 2011). The antibacterial activity of Ag-Ver, Cu-Ver and Ag, Cu-Ver materials was investigated. The inhibition effect on bacterial growth was confirmed in all prepared Ag-, Cu-Ver samples. The synergic effect of Ag and Cu in combined samples was observed (Hundáková et al. 2013b). The antibacterial properties of Ag and Cu prepared on two clay mineral matrices Mt and Ver were compared. The transmission electron microscopy (TEM) images of the Ag NPs reduced on Mt and Ver substrate are show in Figs. 14.3 and 14.4. Results of antibacterial test on *E. coli* (G^-) show good inhibition effect on bacterial growth. Moreover, stability of Ag^+ and Cu^{2+} on both matrices in water environment was studied and the relationship between amount of metal ions released from samples and antibacterial effect was confirmed (Hundáková et al. 2014a). The Ag-Ver, Cu-Ver and Ag-Cu-Ver were used as nanofillers to polyethylene (PE) matrix for a purpose to obtain the antibacterial PE material. Antibacterial properties of powder Ver nanofillers and surfaces of PE/Ver composites was tested on the G^+ bacteria *E. faecalis*. Samples Ag-Ver and AgCu-Ver showed the MIC value 10% (w/v) after 1 h and 3.33% (w/v) after 24 h. The sample Cu-Ver was slightly less effective and showed MIC value 10% (w/v) after 5 h and 3.33% (w/v) after 48 h of inhibition. The surfaces of all tested PE/Ver-Ag,Cu composites showed inhibition effect after 24 h in comparison to pure PE. The CFU number decreased from the countless number to several hundred colonies (Hundáková et al. 2014b).

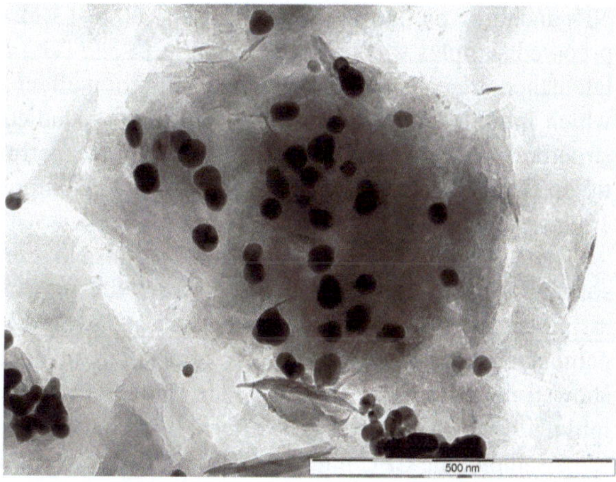

Fig. 14.4 TEM images of Ag NPs on Ver substrate

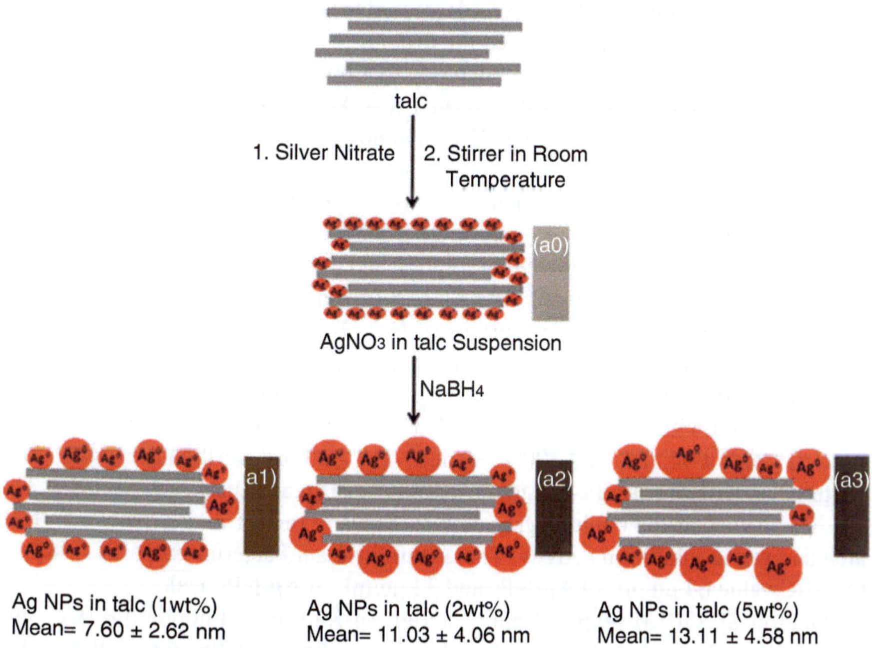

Fig. 14.5 Schematic illustration of the synthesis of Ag NPs-talc composite material by the chemical reduction. The color change of AgNO3/talc suspension during process was from colorless (a0) to light brown (a1), brown (a2) and dark brown (a3) according the Ag NPs content. Reproduced with permission from (Shameli et al. 2013), © Springer Science+Business Media Dordrecht 2013

Clay minerals as talc, palygorskite, kaolinite, halloysite, sepiolite, etc. are very rarely used for metal nanoparticles preparation for a purpose to study their antibacterial activity.

Nevertheless, the Ag NPs were synthetized on the talc surface by the wet chemical reducing method with the $NaBH_4$ solution (Fig. 14.5). The mean diameter of Ag

NPs anchored on talc surface was from 7.60 to 13.11 nm. Antibacterial activity of prepared samples was tested on bacteria *E. coli* (G^-) and *S. aureus* (G^+). Ag NPs-talc nanocomposite did not show any inhibition effect in contrast with $AgNO_3$ -talc which inhibited the bacterial growth. Authors studied also Ag^+ release from talc structure, which was fast at the beginning of experiment and becomes slower in time. The Ag^+ release from Ag NPs-talc nanocomposites can last for more than 16 days (Shameli et al. 2013).

The Ag- or Cu-Pal were prepared for purpose to removing bacteria from aqueous solution. Antibacterial testing was performed using *E. coli* (G^-) and *S. aureus* (G^+) as indicators of fecal contamination of water. The Ag- and Cu-Pal eliminated the pathogenic organisms from water after 12 h of contact time. Antibacterial test showed no activity of untreated Pal. The CFU number of *E. coli* decreased from initial 1.6×10^4 CFU to 0 CFU after 6 h in contact with Ag-Pal (0.6% Ag). For the same sample, the CFU number of *S. aureus* decreased from initial 1.5×10^4 CFU to 0 CFU after 12 h in contact. The CFU number of *E. coli* decreased to 0 CFU after 12 h and the CFU number of *S. aureus* decreased to 0 CFU after 24 h in contact with Ag-Pal (0.57% Cu). Authors showed the differences in the results to the structural differences in bacteria cell walls, concretely to a thicker cellular wall of *S. aureus*. The results of water erosion test indicated that Ag and Cu amount remained on 80.2% and 72.5% in samples after 72 h in contact with water, respectively (Zhao et al. 2006).

Zhang et al. (2013) used Hal nanotubes as a support of Ag NPs. Hal was modified with [3-(2-aminoethyl) aminopropyl] trimethoxysilane in toluene. Modified Hal was mixed with $AgNO_3$ methanol solution and reduction of Ag NPs with $NaBH_4$ aqueous solution was performed. The Ag NPs (average diameter about 5 nm) were uniformly distributed across the surface of Hal. Results of antibacterial test show the diameter of inhibition zone 12 mm for *E. coli* and 13 mm for *S. aureus*. The photographs of samples solutions in agar plates after 16 h exposure to bacteria *E. coli*, compared with the control, shows that Ag NPs and Ag NPs-Hal reaches antibacterial rate of 94.58% and 100%, respectively. Authors ascribed the higher antibacterial activity of Ag-NPs-Hal to the large surface area of the Ag-NPs on Hal. The Ag-NPs in suspension could aggregate what leads to reduce their effective surface area and worse inhibition activity. The quantitative antibacterial properties showed the MIC value 64 μg/mL of Ag-NPs and 32 μg/mL of Ag-NPs-Hal.

Karel et al. (2015) prepared antibacterial Ag-Kao, Ag-Clinoptilotite (Cli) and Ag-Sep. Samples with Ag were prepared by ion exchange of Na^+-forms of Kao, Cli and Sep with $AgNO_3$ aqueous solution during continuously stirring at a dark environment (physical treatments). Prepared materials were subjected to phosphoric acid solution (chemical treatments). The amount of silver incorporated in samples was by chemical treatment: 0.13 wt.% in Kao, 0.65 wt.% in Cli, and 0.75 wt.% in Sep and by physical treatment: 0.84 wt.% in Kao, 0.70 wt.% in Cli, and 0.79 wt.% in Sep. Chemically treated clays contained about 1.5 wt.% of Phosphorous. In samples treatment with phosphoric acid, the amount of Ag in samples was higher. Antibacterial halo test (Fig. 14.6) against *E. coli* (G^-) showed that the physically treated clays exhibited greater circular zone diameter (25–28 mm) in comparison with chemically

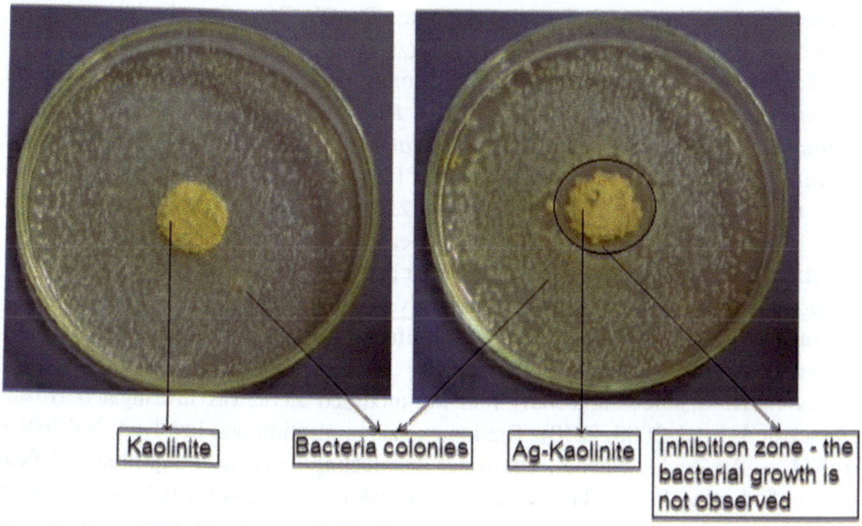

Fig. 14.6 Illustration of inhibition zone determination from antibacterial (modified halo) test. Adapted from (Karel et al. 2015)

treated clays (22–24 nm). Results of Ag release showed that Ag in chemically treated samples is more stable opposite physically treated samples. The amount of residual Ag in samples after 30 days of exposure to water was for chemical treatment samples: 0.06 wt.% in Kao, 0.26 wt.% in Cli, and 0.19 wt.% in Sep and for physical treatment samples: 0.70 wt.% in Kao, 0.54 wt.% in Cli, and 0.36 wt.% in Sep.

Antibacterial Ag-Kao samples were prepared by chemical modification of Kao pre-treatment with formamide, ammonium bromide, benzoylperoxide and vinylacetate monomer (polymerization in-situ) and subsequent treatment with $AgNO_3$ aqueous solution. Prepared materials were used as nanofillers to polypropylene (PP) matrix and antibacterial properties of PP/Ag-Kao composites were studied on the bacteria *E. faecalis* (G^+) and *P. aeruginosa* (G^-). Results show that inhibition effect increased with increasing content of Ag-Kao nanofiller in PP/Ag-Kao composite. Nevertheless, tested *E. faecalis* exhibited higher sensitivity to action of PP/Ag-Kao composites than *P. aeruginosa* and the CFU number of *E. faecalis* decreased about 100% after 24 h of contact time with surface of PP/Ag-Kao composite (Hundáková et al. 2016).

Antibacterial zeolite material as a new antibacterial agent was described already in 1987 (Maeda 1987). Zeol are widely used as adsorbents in wastewater treatment, so their using as water filter media is possible. For example, water adsorption properties of Ag- and Zn-exchange Zeol were studied (Benaliouche et al. 2015).

The antibacterial effect of Zeol exchanged with Ag was tested on *E. coli* (G^-) and *S. aureus* (G^+) by number viable bacterial colonies (NVC) counted. Zeo was pretreated to Na^+Zeo and put into contact with $AgNO_3$ solutions (Ag-Zeo). Antibacterial effect is dependent on Ag content in Ag-Zeo samples. Nevertheless, the NVC of *E. coli* decreased to 0 after 2 h and the NVC of *S. faecalis* decrease to

0 after 6 h in contact with the same samples. The Na^+Zeo mineral without Ag not showed antibacterial behavior (Rivera-Garza et al. 2000). Antibacterial effect of Ag-Zeol on oral bacteria under anaerobic conditions was studied. The oral bacteria *Porphyromonas gingivalis* (*P. gimgivalis*), *Prevotella intermedia* (*P. intermedia*), *Actinobacillus actinomycetencomitans* (*A. actinomycetencomitans*), *Streptococcus mutans* (*S. mutans*), *Streptococcus sanguis* (*S. sanguis*), *A. viscosus* and *S. aureus* were used. The MIC values were between 256 and 2,048 μg/ml of Ag-Zeol which corresponded to 4.8–38.4 μg/ml of Ag^+. Release of Ag^+ into the broth was also measured. Results of antibacterial test show that bacteria *P. gingivalis*, *P. intermedia*, *A. actinomycetencomitans* were more sensitive to the action of Ag-Zeol samples. According the study, Ag-Zeol may be a useful to provide antibacterial function for dentistry materials (Kawahara et al. 2000).

The water disinfecting behavior of Ag-modified Zeol was investigated (Rosa-Gomez de la et al. 2008, 2010). The ion exchange reaction was involved: Na^+Zeol + $AgNO_3$ ↔ Ag-Zeol + $NaNO_3$. The water disinfecting behavior of Ag-modified Zeol was studied using *E. coli* (G^-) bacteria as indicator of microbiological contamination of water in a column system. Glass columns were filled with tested samples and loaded with synthetic wastewater with *E. coli* or municipal wastewater polluted with coliform microorganisms. In both wastewaters the initial microorganism concentration was around 10^7 NVC/100 ml. The wastewater was pumped through the column at a flow rate of 2 ml/min. The Ag concentration and microorganisms were monitored. The average amount of Ag^+ in Ag-Zeol was 4.33 wt.% (0.4009 meq Ag/g). After complete disinfection processes, it was found that 16.6% and 0.6% of *E. coli* survival. The amounts of silver determined in effluents were 220.9 and 305.7 μg corresponding to 5.1% and 3.5% of the initial Ag. When the silver effluent was less than 0.6 μg/ml, the *E. coli* percentage increased and the volume of disinfected water diminished. After complete disinfection processes, 100% of coliform microorganism's survival was recorded. The amount of silver determined in effluents were 263.7 and 222.4 μg corresponding to 6.1% and 2.5% of the initial Ag. The Na-Zeol sample did not show antibacterial activity in both cases. Moreover, authors studied interference of NH^{4+} and Cl^- ions on the disinfection process. They found that the presence of NH^{4+} ions improves the antibacterial activity of Ag-Zeol on *E. coli* and the presence of Cl^- ions notably diminished the antibacterial activity (Rosa-Gomez de la et al. 2008).

Natural Cli-rich tuff was first pre-treated with oxalic acid (followed by treatment with NaCl solution) (P1) and NaOH solution (P2). Samples were cation exchanged with $AgNO_3$ aqueous solution. The amount of Ag in samples was 1.9 meq Ag/g in the P1-Ag^+ and 1.2 meq Ag/g in the P2-Ag^+. The antibacterial effect of P1-Ag^+ sample was observed stronger for the growth of *E. coli* (G^-) (18% CFU/ml) opposite *S. aureus* (G^+) (31.58% CFU/ml). Conversely, the antibacterial effect of P2-Ag^+ sample was observed stronger for the growth of *S. aureus* (62.71% CFU/ml) opposite *E. coli* (74% CFU/ml). The antibacterial action of sample prepared from Cli pre-treated with oxalic acid and NaOH was more effective in comparison with sample pre-treated with NaOH due to the different content of Ag (Copcia et al. 2011).

Guerra et al. (2012) evaluated Ag-Cli as a biocide for the G^- bacteria *E. coli* and *S. typhi*. The Ag-Cli samples were prepared by shaking of Na^+Cli with $AgNO_3$ solution and by reduction at 550 or 700 °C under H_2. The Ag content in samples was determined as 2.1 wt.% and 4.0 wt.%. The X-ray diffraction patterns of samples with higher amount of Ag show well-defined narrow peaks of metallic Ag NPs. The size distribution of Ag NPs (spherical shape) determined was in the range 0.7–9.2 nm depending on the amount of Ag in samples and on reducing temperature. Study of biocidal effect on *E. coli* show that Ag-Cli samples (0.06 g per 18 ml of culture media) with 4 wt.% or 2 wt.% of Ag eliminated all colonies during 30 min and 120 min, respectively. The higher amount of samples was needed to achieve similar inhibition for *S. typhi*. The Ag-Cli samples (0.12 g per 18 ml of culture media) with both 4 wt.% or 2 wt.% of Ag eliminated all colonies during 5 min. It was concluded that the shape and size of the Ag NPs dispersed on the support are not crucial for biocidal effect at high concentration of Ag. Authors also found that Ag did not leach from samples during the experiment with both bacteria, thus the biocidal material can be reused.

Hrenovic et al. (2013) prepared Ag-Zeol, Cu-Zeol and benzalkonium (BC)-Zeol materials and confirmed their antibacterial activity against isolates of *A. baumannii* (Eropean clone I (EUI) and II (EUII). Samples Ag-Zeol, Cu-Zeol and BC-Zeol contained 50.65 mg Ag^+, 20.33 mg Cu^{2+} and 85.0 mg BC per gram, respectively. The *A. baumannii* from EUII was more sensitive to Ag-Zeol and Cu-Zeol with the MBC value 31.2 mg/l and 125 mg/l, respectively opposite EUI with MBC value 250 mg/l for both samples. Authors compared antibacterial activity of Ag, Cu, BC-Zeol samples with Ag, Cu, BC cations. Results show that the modified Zeol did not show the pronounced activity opposite Ag, Cu and BC salts. The main reason is their stability and slow release of cations from modified Zeol.

The starting material NaY-Zeol was pre-treated with cetyltrimethyl ammonium bromide (CTAB). The CTAB-modified NaY Zeol was regenerated to NaY Zeol using thermal treatment calcination at 550 °C. The regenerated NaY Zeol was cation exchange to Na-form because Na^+ in Zeol could be easily exchange with Ag^+. The AgY Zeol samples with different Ag amount were prepared by mixing of regenerated NaY Zeol with $AgNO_3$ aqueous solution. The amount of silver in AgY samples was 9 mg/g, 63 mg/g and 90 mg/g with increasing concentration of initial $AgNO_3$ solution. Results of antibacterial testing show that AgY Zeol sample had better inhibition effect on *E. coli* in comparison with *S. aureus* in distilled water, while both bacteria showed low sensitivity to samples in saline solution. The MIC value decreased with increasing content of Ag in samples (Salim and Malek 2016). The next interesting study was published about studied of Ag NPs-coated Zeol as water filter media for fungal disinfection of rainbow trout eggs (Johari et al. 2016).

14.2.2 Copper on Clay Minerals

Copper can be prepared on inorganic substrates by several methods as Cu^{2+} cations or Cu, CuO and Cu_2O nanoparticles. The most frequently used method of Cu^{2+} is preparation by cation exchange on clay substrates, while the most widely used is montmorillonite. Summary of some preparation methods of copper on the various substrates, bacterial strains and testing methods used for studying the antimicrobial activity is shown in Table 14.2.

The antibacterial ability of Cu^{2+}-Mt was studied on *E. coli* (G^-) and *S. aureus* (G^+) bacteria. The MIC value of Cu^{2+}-Mt on these bacteria was 10 mg/l (equal to amount of Cu^{2+} ~ 100 ppm) after 24 h and 50 mg/l (equal to Cu^{2+} ~ 500 ppm) after 6 h and 4 h, respectively. The MIC value of Cu^{2+}- Mt for both bacteria was 200 mg/l after 2 h of action. The Cu^{2+} concentration released from Mt was determined after 2 h of incubation as 0.07 mg/l from 10 mg/l Cu^{2+}-Mt and 1.61 mg/l from 200 mg/l Cu^{2+}-Mt (Zhou et al. 2004). The antibacterial activity of Cu^{2+}-Mt was tested against the G^- bacteria *Aeromonas hydrophila* (*A. hydrophila*). The MIC and minimum bactericidal concentration (MBC) of Cu^{2+}-Mt were found to be 150 and 600 mg/l, respectively (Hu et al. 2005). The other authors published study about antibacterial effect of Cu^{2+}-Mt on two G^- bacterial strains *E. coli* and *Salmonella choleraesius* (*S. choleraesius*) and the MIC value of Cu^{2+}-Mt was found as 1,024 μg/ml and 2,048 μg/ml, respectively. Authors divided the antibacterial process to two stages: adsorption of bacteria from solution and immobilization on the Cu^{2+}-Mt surface and action related to accumulation of Cu^{2+} on the Mt surface. The release concentration of Cu^{2+} ions from Mt into the broth increased with their increasing amount in Cu^{2+}-Mt. Nevertheless, the released amount of Cu^{2+} from Cu^{2+}-Mt into the broth was very low. The effect of Cu^{2+}-Mt on bacterial cell walls, on enzyme activity of bacteria and on the respiratory metabolism of bacteria was confirmed (Tong et al. 2005). The original CaMt, monoionic form NaMt and acid activated (AA)Mt were cation exchange with $CuSO_4$ to Cu^{2+}-CaMt, Cu^{2+}-NaMt and Cu^{2+}-AAMt. The antibacterial activity was tested on *E. coli* (G^-). The initial CaMt, NaMt and AAMt showed reducing of bacterial plate counts by 14.2%, 13.4% and 37.4%, respectively. Nevertheless, the Cu^{2+}-CaMt, Cu^{2+}-NaMt and Cu^{2+}-AAMt showed reducing of bacterial plate counts by 95.6%, 97.5% and 98.6%, respectively (Hu and Xia 2006).

In order to use in veterinary medicine, the Cu^{2+}-Mt was also studied in relating to the effect on the growth performance and intestinal microflora of weanling pigs. The weanling pigs (total 128 pigs with initial average weight 7.5 kg) were divided into four groups according the dietary treatments in single doses adjusted follows: (1) basal diet, (2) basal diet +1.5 g/kg Mt, (3) basal diet +36.75 mg/kg $CuSO_4$ (with the Cu^{2+} equivalent to that in Cu^{2+}-Mt) and (4) basal diet +1.5 g/kg Cu^{2+}-Mt. Experiment was carried out for 45 days. Effect of diet was studied on intestinal microflora (total aerobes, total anaerobes, *Bifidobacterium*, *Lactobacillus*, *Clostridium*, *E. coli*) in both small intestine and proximal colon of pigs. The diet supplemented with Mt and $CuSO_4$ had no effect on growth performance, intestinal microflora and enzyme activities. Pigs fed with Cu^{2+}-Mt had reduced total viable counts of *Clostridium* and *E. coli* in the small intestine and proximal colon opposite control. The improvement of growth performance of weanling pigs and reduced of bacterial enzyme activities were found (Xia et al. 2005). The effect of two Cu^{2+} exchanged Mt, concrete Cu^{2+}-CaMT and Cu^{2+}-NaMt were studied as alternative agents to chlortetracycline. The growth performance, diarrhea, intesti-

Table 14.2 Summary of methods and precursors used for the preparation of copper on the inorganic substrates, used testing methods, and bacterial strains for studying the antimicrobial activity of prepared materials

Method (conditions), precursor	Inorganic substrate, chemical pretreatment	Tested bacteria	Testing method/test evaluation	Reference
Cation exchange (stirring, 60 °C, 6 h), $CuSO_4 \cdot 5H_2O$	Mt (Chifeng, the Inner Mongolia Autonomous Region, China)	*E. coli*, *S. faecalis*	MIC determination	Zhou et al. (2004)
Cation exchange (stirring, 60 °C, 6 h, pH 5), $CuSO_4 \cdot 5H_2O$	Mt (Inner Mongolia Autonomous Region, China) – pretreated to Na^+ form	*A. hydrophila*	MIC and MBC determination – broth dilution method	Hu et al. (2005)
Cation exchange (agitating 24 h), $CuSO_4 \cdot 5H_2O$	Mt (Inner Mongolia Autonomous Region, China)	*E. coli*, *S. choleraesuis*	MIC determination	Tong et al. (2005)
Cation exchange (stirring, 60 °C, 6 h, pH 5), $CuSO_4 \cdot 5H_2O$	Ca-Mt (Inner Mongolia Autonomous Region, China) – Na^+ form and acid activated Mt	*E. coli*	CFU counting	Hu and Xia (2006)
Cation exchange (stirring, 60 °C, 6 h, pH 5), $CuSO_4 \cdot 5H_2O$	Mt (Inner Mongolia Autonomous Region, China)	Intestinal microflora of pigs	CFU counting	Xia et al. (2005)
Cation exchange (stirring, 60 °C, 6 h, pH 5), $CuSO_4 \cdot 5H_2O$	Mt (Inner Mongolia Autonomous Region, China), Mt pretreated to Na^+ form	Intestinal microflora of pigs		Song et al. (2013)
Cation exchange (stirring, 60 °C, 6 h, pH 5), $CuSO_4 \cdot 5H_2O$	Mt (Inner Mongolia Autonomous Region, China), Mt pretreated to Na^+ form	Intestinal microflora of Nile tilapia	CFU counting	Hu et al. (2007)
Shaking (24 h, room temperature), $AgNO_3$, $Cu(NO_3)_2$, $Zn(NO_3)_2$, CP	Mt (Middle Anatolia) – purified	*S. aureus*, *P. aeruginosa*	Disk diffusion method	Özdemir et al. (2010)
Shaking (24 h, room temp.), $AgNO_3$, $ZnSO_4$, $CuSO_4$	Na^+-rich Mt (SWy2) (Crook County, Wyoming)	*E. coli*, *P. cinnabarinus*, *P. ostreatus*	CFU counting, median effective concentration (EC50)	Malachová et al. (2011)

(continued)

Table 14.2 (continued)

Method (conditions), precursor	Inorganic substrate, chemical pretreatment	Tested bacteria	Testing method/test evaluation	Reference
In situ reduction of a copper ammonium complex, → Cu NPs $CuCl_2{\cdot}2H_2O$, hydrazine	Mt (Nanocor Inc., SA) – pretreated to Na^+ form	*E. coli*, *S. aureus*, *P. aeruginosa*, *E. faecalis*	CFU counting, MBC calculation, cytotoxicity test	Bagchi et al. (2013)
Thermal decomposition method, → CuO NPs (600°C) $CuSO_4{\cdot}5H_2O$, Na_2CO_3	Commercial Na-MT purcheased. from Southern Clay Products (Gonzales, TX)	*E. coli*	MIC determination, disk diffusion method	Sohrabnezdah et al. (2014)
Cation exchange (stirring, 80 °C, 3 h), $CuCl_2{\cdot}2H_2O$ (Cu-Ver heated at 200 or 400 °C, 3 h)	Ver (Yli County in Xinjiang, China)	*E. coli*	Halo method	Li et al. (2002)
Cation exchange (80 °C, 3 × 2 h) $CuCl_2$, $AgNO_3$, $ZnCl_2$	Mt (Ivančice, Czech Republic), Ver (Letovice, Czech Republic)	*E. faecalis*, *E. coli*, *P. aeruginosa*, *T. vaginalis*	MIC determination	Pazdziora et al. (2010)
Cation exchange (stirring, 70–80 °C, 4–12h), $CuSO_4{\cdot}5H_2O$ H_2 reduction (Ar flow to 400–600°C, H_2 or air flow 2–3h) → Cu NPs	Two commercial exfoliated Ver from Virginia Vermiculite LLC – Grade No. 5 and Milled No. 7	*S. aureus*	Disk diffusion test	Drelich et al. (2011)
Shaking (24 h, room temperature), $AgNO_3$, $Cu(NO_3)_2$	Ver (Santa Luzia, Brazil)	*S. aureus*, *E. faecalis*, *K. pneumoniae*, *P. aeruginosa*	Broth dilution method, MIC determination	Hundáková et al. (2013b)
Shaking (24 h, room temperature), $AgNO_3$, $Cu(NO_3)_2$	Ver (Santa Luzia, Brazil) Mt (Ivančice, Czech Republic)	*E. coli*	Broth dilution method, MIC determination	Hundáková et al. (2014b)
Shaking (24 h, room temperature), $AgNO_3$, $Cu(NO_3)_2$, PE	Ver (Santa Luzia, Brazil)	*E. faecalis*	Broth dilution method, MIC determination, CFU counting	Hundáková et al. (2014a)

Cation exchange (stirring, 70–80 °C, 4–12h), $CuSO_4 \cdot 5H_2O$, reduction (Ar/H_2 atm. 500°C, 2 h) → Cu NPs	Sep – purified with HNO_3	*E. coli, S. aureus*	CFU counting	Esteban-Cubillo et al. (2006)
Cation exchange, $CuCl_2$, $ZnCl_2$, $NiCl_2$ → Cu_2O, ZnO, NiO NPs	Natural Zeol with 70wt.% Cli (the sedimentary deposit Zlatokop, Serbia) – Na^+-rich form	*E. faecalis, E. coli, P. caudatum, E. affinis*	CFU counting	Hrenovic et al. (2012)
Cation exchange (stirring 24 h, room temperature) $ZnCl_2$, $CuSO_4 \cdot 5H_2O$	Commercial Zeol X (Loyang Jianlong Chemical Industrial Co.)	*E. coli, P. aeruginosa, S. aureus, C. albicans, A. niger*	Disk diffusion method	Tekin and Bac (2016)
Ion exchange by melting, $CuSO_4 \cdot 5H_2O$, (550–560 °C, 10–90 min)	Ben (Na-Mt) (Rasht, Iran)	*E. coli, S. aureus*	CFU counting	Pourabolghasem et al. (2016)

nal permeability and proinflammatory cytokine in weanling pigs were investigated. Similar as in previous study, the weanling pigs (total 96 pigs with initial average weight 5.6 kg) were divided into four groups according the dietary treatments adjusted follows: (1) basal diet (control), (2) basal diet +1.5 g/kg Cu^{2+}-CaMt, (3) basal diet +1.5 g/kg Cu^{2+}-CaMt and (4) basal diet +75 mg/kg chlortetracycline. Experiment was carried out for 14 days. Results showed that diet with Cu^{2+}-Mt was as effective as chlortetracycline against diarrhea and inflammation, also improving intestinal microflora and mucosal barrier integrity of weanling pigs (Song et al. 2013).

The interesting study focused on potential use of Cu^{2+}-Mt in fish farming industry was also published. The effect of Cu^{2+}-Mt was studied on growth performance, microbial ecology and intestinal morphology of Nile tilapia (*Oreochromis niloticus*). The Nile tilapia (total 360 fingerlings with initial average weight 3.9 g) were divided into four groups according the dietary treatments adjusted follows: (1) basal diet, (2) basal diet +1.5 g/kg Mt, (3) basal diet +30 mg/kg $CuSO_4$ (with the Cu^{2+} equivalent to that in Cu^{2+}-Mt) and (4) basal diet +1.5 g/kg Cu^{2+}-Mt. Experiment was carried out for 56 days. The Cu^{2+}-Mt in diet improved growth performance, reduced the total intestinal aerobic bacterial counts and affected the composition of intestinal microflora in comparison with control and diet with Mt or $CuSO_4$. The composition of microflora was evaluated according the counts of *Aeromonas*, *Flavobacterium*, *Enterobacteriaceae*, *Vibrio*, *Pseudomonas*, *Acinetobacter*, *Alcaligence*, *Corynebactrium* and *Micrococcus* (Hu et al. 2007).

Antibacterial effect of Cu^{2+}-, Zn^{2+}-, Ag^{+}- and Ag^{0}- and cetylpyridinium (CP – exchanged Mt against *P. aeruginosa* (G^-) and *S. aureus* (G^+) was determined. Both bacteria are highly resistant to antibiotics and cause infections in hospitalized patients. The Ag^+-Mt, Cu^{2+}-Mt and Ag^0-Mt samples showed good antibacterial activity against both bacteria. The CP-Mt did not show antibacterial activity (Özdemir et al. 2010). The Ag-Mt, Cu-Mt and Zn-Mt were prepared by cation exchange of Mt. The antibacterial activity against *E. coli* (G^-), and antifungal activity against *P. cinnabarinus* and *P. ostreatus* were investigated. It was found that inhibition effect on *E. coli* decreased as Ag-Mt > Cu-Mt ≈ Zn-Mt. The inhibition effect on *P. cinnabarinus* decreased as Zn-Mt ≥ Cu-Mt > Ag-Mt. The inhibition effect on *P. ostreatus* decreased as Cu-Mt > Zn-Mt > Ag-Mt. The samples were as effective as the free Ag^+ ions. The free Ag^+, Cu^{2+}, Zn^{2+} cations inhibited the bacterial and fungal growth in similar manner (Malachová et al. 2011).

The Cu NPs were synthetized on Mt substrate by in situ reduction of copper ammonium complex ion. As first, the Mt-$[Cu(NH_3)_4(H_2O)_2]^{2+}$ powder was prepared followed: $CuCl_2$ aqueous solution was made ammoniacal and copper ammonia complex ion $[Cu(NH_3)_4(H_2O)_2]^{2+}$ was formatted (indicating according a dark blue color) and Mt was added. The dried powder was suspended in ammoniacal water, the hydrazine hydrate wad added and stirred to a bluish purple color, washed and dried. The results showed that the Cu NPs were both intercalated and adsorbed by the Mt. Antibacterial activity of prepared material was studied on the G^- bacteria *E. coli* and *P. aeruginosa* and the G^+ bacteria *S. aureus* and *E. faecalis*. Results showed mortality over 80% after 12 h of incubation. The MBC value of Cu NPs-Mt (Cu

NPs) determined for individual bacteria was 5.7 mg/mL (285 μg/ml) for *S. aureus*, 5.2 mg/mL (260 μg/ml) for *E. coli*, 5.1 mg/mL (255 μg/ml) for *E. faecalis*, 7.2 mg/mL (360 μg/ml) for *P. aeruginosa*. The antibacterial action of Cu NPs-Mt is based on Cu NPs/Cu ions mediated cell disruption which is aided by the Mt substrate which serves as a stable carrier for the Cu NPs and also increases the contact frequency with the bacterial cell. The cytotoxicity study performed on two human cells showed a decrease in viability in both cell lines. Nevertheless, the stabilizing effect was observed with increase of Cu NPs-Mt concentration (Bagchi et al. 2013).

The CuO NPs were synthetized on Mt substrate by thermal decomposition method. The CuO-Mt was prepared by adding of Na_2CO_3 and $CuSO_4$ to deionized water and stirred at 60 °C. Mt was added and suspension was stirred to forming green precipitate $Cu_4(SO_4)(OH)_6$, separated by filtration and kept to a muffle furnace at 600 °C. The mean diameter of small spherical CuO NPs agglomerated on Mt was determined as ~3–5 nm. The antibacterial test on *E. coli* (G^-) shows that the MIC values for samples Mt, CuO and CuO-Mt were 100, 10 and 0.1 ng, respectively. The results of disc susceptibility test show no antibacterial activity of Mt. Contrary, CuO and CuO-Mt show antibacterial behavior, and the inhibition zone was higher for CuO-Mt (Sohrabnezhad et al. 2014).

There are not many studies published on antibacterial activity of Cu^{2+} ions and Cu or CuO NPs on vermiculite as substrate.

The Cu-Ver was prepared by cation exchange and antibacterial activity was tested on *E. coli* (G^-). Moreover, in order to determine the influence of heating on the antibacterial activity, the Cu-Ver sample was heated at 200 or 400 °C. Results of halo test show good antibacterial activity of Cu-Ver and no significant decrease in activity of samples after heating (Li et al. 2002). The antibacterial and antiprotozoal effects of Ag^+, Cu^{2+} and Zn^{2+} cation exchanged in Mt and Ver were compared. Antibacterial activity was tested on the G^- bacterial strains *E. coli* and *P. aeruginosa* and the G^+ bacterial strain *E. faecalis*. Antiprotozoal effect was tested on *Trichomonas vaginalis* (*T. vaginalis*). Results show the high antibacterial and antiprotozoal effect. Moreover, it was found that *T. vaginalis* was the most sensitive on tested samples (Pazdziora et al. 2010).

Ver substrate was decorated with Cu NPs and antibacterial activity was studied against bacteria *S. aureus* (G^+). The Cu NPs were prepared by heat treatment and H_2 reduction. First, the Cu^{2+}-Ver samples were prepared by cation exchanged with $CuSO_4$ aqueous solution. Second, the Cu^{2+}-Ver were heat treated in a Lindberg hydrogen furnace filled with Ar. After achieved the reduction temperature 400 or 600 °C, the H_2 or air flow was introduced to the furnace. The Cu content in samples was from 1.7 to 3.5 wt.%. The size of Cu NPs distributed primarily on Ver surface was in the broad range from ~1 to 400 nm. The good antibacterial activity of prepared Cu NPs-Ver samples was confirmed (Drelich et al. 2011). The antibacterial activity of Ag-Ver, Cu-Ver and Ag,Cu-Ver materials was compared. The inhibition effect on bacterial growth was observed in all prepared Ag-, Cu-Ver samples. The synergic effect of Ag and Cu in combined samples was observed. Nevertheless, the antibacterial activity of samples contained Ag was better compared to samples contained Cu (Hundáková et al. 2013b, 2014a, b).

The antibacterial activity of Cu NPs prepared on *Zeol* mineral sepiolite was also studied. The Cu NPs on Sep were synthetized using the process of cation exchange with $CuSO_4$ aqueous solution and adjusting the pH with NaOH to precipitate the metallic cations. Then, the prepared powder was subjected to the reduction in a 90% Ar/10% H_2 atmosphere at 500 °C to obtain Cu NPs. The size of Cu NPs distributed on Sep was 2–5 nm. The antibacterial effect of Cu NPs-Sep was tested on *E. coli* (G^-) and *S. aureus* (G^+). The concentration of both bacteria was reduced about 99.99% after 24 h of action (Esteban-Cubillo et al. 2006). Hrenovic et al. (2012) compared antimicrobial activity of Cu_2O, ZnO and NiO NPs supported on natural Cli. The numbers of viable bacterial cells of *E. coli* and *S. aureus* were reduced for four to six orders of magnitude after 24 h contact with Cu_2O and ZnO NPs. Moreover, the Cu2O and ZnO NPs showed 100% of antiprotozoal activity against *Paramecium caudatum* (*P. caudatum*) and *Euplotes affinis* (*E. affinis*) after 1 h of contact. The antibacterial and antiprotozoal activity of NiO NPs was less efficient. The Cu^{2+}-Zeol and Zn^{2+}-Zeol materials were prepared from commercial Zeol by cation exchanged with $CuSO_4$ or $ZnCl_2$ aqueous solution. Moreover, the encapsulation of a fragrance molecule - triplal, was studied. The antimicrobial activity of materials before and after encapsulation was studied. The Cu^{2+}-Zeol and Zn^{2+}-Zeol inhibited the growth of *S. aureus* (G^+) more than the growth of *E. coli* (G^-) and *P. aeruginosa* (G^-). The Cu^{2+}-Zeol samples showed larger inhibition zone compared Zn^{2+}-Zeol against all bacteria. On the contrary, in antifungal activities test, the Zn^{2+}-Zeol samples showed larger inhibition zone compared Cu^{2+}-Zeol against yeast *Candida albicans* (*C. albicans*) and fungus *Aspergillus niger* (*A. niger*) (Tekin and Bac 2016).

14.2.3 Silver and Copper NPs on Carbon Materials

Nanostructured Graphite Oxide (GO) and low-cost GO coated with silver or sand nanoparticles were developed and characterized. Antibacterial efficacy of these nanoparticles was investigated using waterborne pathogenic *E. coli* strain. Inhibition of pathogenic *E. coli* with sand and GO NPs. Highest bacterial removal efficiency (100%) by the Coated Graphite Oxide and the lowest by sand filters with 17.9–88.9% reduction rate from 0 to 24 h, respectively. The filter system with GO composite can be used as an effective filter for water disinfection and production of potable and pathogens free drinking water (Jakobsen et al. 2011). GO based nanocomposites have raised significant interests in many different areas silver nanoparticle (AgNPs) anchored GO (GO-Ag) has shown promising antimicrobial potential. Factors affecting its antibacterial activity as well as the underlying mechanism remain unclear. GO-Ag nanocomposites with different Ag NPs to GO ratios examined for their antibacterial activities against both the G^- bacteria *E. coli* and the G^+ bacteria *S. aureus*. GO-Ag nanocomposite with an optimal ratio of AgNPs to GO is much more effective and shows synergistically enhanced, strong antibacterial activities at rather low dose (2.5 μg/ml) compare to pure AgNPs. The GO-Ag nanocomposite is more toxic to *E. coli* than that to *S. aureus*. The antibacterial effects of GO-Ag nanocomposite are further investigated, revealing distinct, species-specific mechanisms. The results demonstrate that GO-Ag nanocomposite functions as a bactericide against the *E. coli* (G^-) through disrupting bacterial cell wall integrity, whereas it exhibits bacteriostatic effect on the *S. aureus* (G^+) by dramatically inhibiting cell division (Tang et al. 2013). The effect of bactericide dosage and pH on antibacterial activity of GO-Ag was examined. GO-Ag was much more destructive to cell membrane of *E. coli* than that of *S. aureus*. Experiments were carried out using catalase, superoxide dismutase and sodium thioglycollate to investigate the formation of reactive oxygen species and free silver ions in the bactericidal process. The activity of intracellular antioxidant enzymes was measured to investigate the potential role of oxidative stress. According to the consequence, synergetic mechanism including destruction of cell membranes and oxidative stress accounted for the antibacterial activity of GO-Ag nanocomposites. All the results suggested that GO-Ag nanocomposites displayed a good potential for application in water disinfection (Jones and Hoek 2010) (Fig. 14.7).

The combination of silver with copper gave enhanced result. Ag, Cu monometallic and Ag/Cu bimetallic NPs were in situ grown on the surface of graphene, which was produced by chemical vapor deposition using ferrocene as precursor and further functionalized to introduce oxygen-containing surface groups. The antibacterial performance of the resulting hybrids was evaluated against *E. coli* cells and compared experiments of varying metal type and concentration. It was found that both Ag- and Cu-based monometallic graphene composites significantly suppress bacterial growth, yet the Ag-based ones exhibit higher activity compared to that of their Cu-based counterparts. Compared with well-dispersed colloidal Ag NPs of the same metal concentration, Ag- and Cu-based graphene hybrids display weaker antibacterial activity. However, the bimetallic Ag/Cu NPs-graphene hybrids exhibit superior performance compared to that of all other materials tested, i.e., both the monometallic graphene structures as well as the colloidal NPs, achieving complete bacterial growth inhibition at all metal concentrations tested. A systematic analysis of antibacterial activity of GO nanosheets, Ag and Cu NPs, and combinations of Cu-Ag NPs, and GO-Cu-Ag nanocomposites against *E. coli*, *P. aeruginosa*, *S. aureus*, *K. pneumoniae* and Methicillin-resistant *S. aureus* (MRSA) was performed. MRSA showed highest resistance in all cases (Perdikaki et al. 2016; Jankauskaitė et al. 2016).

Transition metal NPs such as Ag and Cu have been grafted onto carbon nanotube surface through wet chemical approach leading to the development of densely packed NP decorated carbon nanotubes. Chemically active surface and high-temperature stability are the basic attributes to use carbon nanotubes as the template for the growth of NPs. The antimicrobial properties of acid-treated MWCNT (MWCNT-COOH), Ag-MWCNT, and Cu-MWCNT are investigated against *E. coli* (G^-) bacteria. Ag-MWCNT and Cu-MWCNT (97% kill vs. 75% kill), whereas MWCNT-COOH only killed 20% of bacteria. Possible mechanisms are proposed to explain the higher antimicrobial activity by NP-coated MWCNT. These findings suggest that

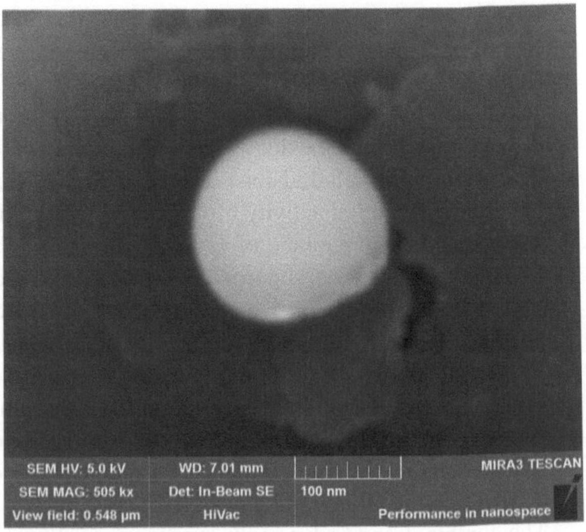

Fig. 14.7 Silver nanoparticle on graphite flake (SEM image acquired using secondary electron detector)

Ag-MWCNT and Cu-MWCNT may be used as effective antimicrobial materials that find applications in biomedical devices and antibacterial controlling system.

Ag NPs attached to CNTs have also shown enhanced activity against *E. coli*. Although, this strategy gives enhanced antibacterial effect but parameters like structural defects, agglomeration and impurities could also lead to partial loss in the ultimate expected results. Thus, precise control is required to further achieve the improved results. In the case of Ag-NPs attached to CNTs, it is believed that the antibacterial activity is due to Ag NPs. Several reports are available on the antibacterial activity of Ag NPs (Jones and Hoek 2010). When Ag-NPs are exposed to air, oxidation at the surface occurs (Eq. 14.1) and when in acidic environment, Ag^+ ions are released (Eq. 14.2) (Xiu et al. 2012). The released Ag^+ ions bind to the thiol group (SH) in enzymes and proteins on cellular surface and produce holes to enter the cell.

$$4Ag(0) + O_2 \rightarrow 2Ag_2O \tag{14.1}$$

$$Ag_2O + 4H^+ \rightarrow 4Ag^+ + 2H_2O \tag{14.2}$$

Another important aspect is the formation of Ag-NPs-CNT interface, which is believed to be quite stable without adversely affecting the antibacterial activity of Ag-NPs. In fact, presence of substrate (CNTs) serves two purposes in the experiment: (i) act as a host to increase the long term stability of NPs. Li et al. (2011) reported the long term stability for the about 1 month of MWCNTs and Ag-NPs composites. (ii) CNTs as possible substrates or a platform for drug delivery application. Yuan et al. (2008) showed the improvement of antibacterial efficiency against *S. aureus* (G^+) bacteria after the deposition of Ag-NPs on MWCNTs grafted with hyperbranched poly amidoamin (dMWCNTs), both d-MWNTs and d-MWNTs/Ag were found equally effective against *E. coli* (G^-) and *P. aeruginosa* (G^-) bacteria. Jung et al. (2011) reported the antibacterial efficiency with pure MWCNTs, pure Ag-NPs and deposited Ag/CNTs against *E. coli* (G^-) and *S. epidermidis* (G^+), while higher inactivation of *E. coli* was observed relative to *S. epidermidis*. The present work dealt with the CVD growth of CNTs and synthesis of Ag-NPs and formation of composite by mixing CNTs in a mixture of $AgNO_3$ resorcinol and ethanol in one reaction.

14.3 Metal Oxides on Inorganic Substrates as Antimicrobial Agents

14.3.1 ZnO on Clay Minerals

Several studies dealing with preparation a characterization of nanocomposites where ZnO NPs are attached or bounded at clay matrix were published. Preparation of ZnO/Pal composites was developed by Huo and Yang (2010). Fibers of Pal were uniformly coated by ZnO NPs with average size equal to 15 nm. This composite exhibited antibacterial activity against *E. coli* (G^-) stronger than pure ZnO NPs. Authors proposed that clay matrix positively affected production of H_2O_2, which is harmful to living cells, and led to the cell membrane destruction and growth inhibition of bacteria. Zinc-enhanced montmorillonites were prepared and their influence to intestinal microbiota and barrier function in weaned pigs was evaluated (Jiao et al. 2015). Study revealed that providing 150 mg/kg Zn, supplementation prepared Zn-Mt improved postweaning diarrhea and enhanced growth performance in the weaned pigs. Explanation of results was attributed to the adsorption of the bacteria and immobilization on the surface of the composite. Alternatively, Zn was released from structure of the composite and directly exerted its antimicrobial effect on the bacteria. Therefore, the antibacterial effect of Zn-Mt may be explained by interactions of Mt with Zn. Also ZnO-Mt hybrid composite was investigated on performance, diarrhea, intestinal permeability and morphology (Hu et al. 2012). It was discovered that supplementing weaned pigs diets with 500 mg/kg of Zn from ZnO-Mt was as efficacious as 2,000 mg/kg of Zn from ZnO in promoting growth performance, alleviating diarrhea, improving intestinal microflora and barrier function. Pigs fed with 500 mg/kg of Zn from ZnO-Mt had higher performance and intestinal barrier function than those fed with Mt or 500 mg/kg of Zn from ZnO. ZnO-Mt composite was also tested as a material with potency to inhibit the cyanobacterial bloom (Gu et al. 2015). *Microcystis aeruginosa* (*M. aeruginosa*) is cyanobacteria which has a negative impact on water quality and therefore to human health. It was proved, that ZnO-Mt has strong flocculation effect on the tested cyanobacteria in comparison with pure Mt or ZnO under visible light and better photocatalytic degradation under UV irradiation. Synergistic effect of flocculation and photocatalysis of ZnO-Mt promoted removal of *M. aeruginosa*. Antibacterial activity of nanocomposite ZnO/

kaoline (ZinKa) was evaluated (Dědková et al. 2015a). ZinKa with 50 wt.% exhibited antibacterial activity under artificial day light against four common human pathogens (*S. aureus* (G^+), *E. faecalis* (G^+), *E. coli* (G^-), *P. aeruginosa* (G^-)). The highest antibacterial activity was observed against *S. aureus*, where the lowest value of MIC was determined equal to 0.41 mg/ml. As a comparison to this study, nanocomposite with the same chemical composition but prepared from different precursor and at slightly different conditions denoted Kao/ZnO (KAZN) was tested against the same four human pathogens at the same antibacterial assay (Dědková et al. 2016).

It was found that even chemical composition o ZinKa and KAZN is the same resulting antibacterial activity is different and therefore can be assumed that the way of preparation nanocomposites with clay matrix have an influence on resulting activity, nevertheless authors did not provide any detail explanation of this result because it needs to be investigated more. However, authors proposed possible mechanism of antibacterial activity, which could be based on photocatalytic reaction and production of ROS. Antimicrobial activity of ZnO-Ben composite was introduced by Pouraboulghasem et al. (2016). *E. coli* (G^-) served as a testing bacteria and pure Ben did not exhibit any antibacterial activity. After alkaline ion exchange treatment, antibacterial activity was observed. The most promising composite was ZnO/Ben after alkaline ion exchange for 60 and 90 min. Leaching test, which was also done, showed that ZnO/Ben did not present any risk to drinking water treatment due to the amount of leached zinc below to 4 mg/L, which is in the acceptable range according to World Health Organization regulations. Motshekga et al. (2013) published study dealing with composites where ZnO, Ag and combination of Ag- ZnO NPs supported on Ben clay were prepared. Disc diffusion method was used as a method for antibacterial assay where *E. coli* (G^-) with *E. faecalis* (G^+) were selected as tested bacteria. ZnO-clay composite and Ag-clay composite exhibited antibacterial activity however Ag/ZnO-clay exhibited the highest antibacterial activity from tested composites and according to authors it could be used as a material for drinking water treatment to target the G^+ also the G^- bacteria.

14.3.2 TiO_2 on Clay Minerals

There are many published papers dealing with photocatalytic activity of TiO_2/clay based nanocomposites, however, only a few are dealing with the antibacterial activity or describing possible applications in medicine. Kao/TiO_2 (KATI) nanocomposite was prepared and antibacterial activity in relation to irradiation time was investigated (Dědková et al. 2014). A standard microdilution test was used to determine the antibacterial activity using four human pathogenic bacterial strains (*S. aureus* (G^+), *E. coli* (G^-), *E. faecalis* (G^+), *P. aeruginosa* (G^-)). Artificial day light was applied to induce photocatalytic reaction. It was observed that UV light is not essential for antibacterial activity of KATI composite which is important in terms of potential applications for antibacterial modification of various surfaces.

Authors proposed that these nanocomposites would be used in future e.g. for surface treatment of external fixators in the treatment of complicated fractures of humans or animals for reduction of potential infection. TiO_2 nanotubes loaded with ZnO and Ag NPs were investigated by Roguska et al. (2015) with respect functional coatings to enhance bio-compatibility and antibacterial activity of implant materials. Authors observed considerable antibacterial activity of pure TiO_2 nanotubes to *S. epidermidis* (G^+), however, loading of nanotubes with nanoparticles significantly increased its effect on tested bacteria, where cell viability and adhesion extenuated after 1.5 h contact with modified surface. This functional coating seems to be promising delivery system to reduce bone implant related infections after operations; nevertheless, the amount of loaded nanoparticles has to be balanced to avoid overdose and potential risk to patients.

14.3.3 ZnO on Carbon Materials

Different forms of carbon may serve as a matrix for anchoring of nanoparticles. Simple hydrothermal method was used for decoration of graphene sheets with ZnO NPs (Bykkam et al. 2015). *S. typhi* (G^-) and *E. coli* (G^-) were used during well diffusion test to evaluate the antibacterial activity of prepared composite material. Provided results shown that few layered graphene sheets decorated by ZnO NPs exhibited good antibacterial activity. ZnO:Cu:Graphene nanopowder and its photocatalytic and antibacterial activity was studied (Ravichandran et al. 2016). It was demonstrated that this material has better photocatalytic and antibacterial properties in comparison with pure ZnO and ZnO:Cu nanopowders. Authors proposed that graphene layers could efficiently separate the photoinduced charge carriers and delay their recombination. This material is potential candidate not only in the field of decontamination of organic pollutants but also in medicine as an effective antibacterial agent. Plasmonic sulfonated graphene oxide-ZnO-Ag (SGO-ZnO-Ag) composites were developed by Gao et al. (2013) via nanocrystal-seed-directed hydrothermal method. Surface plasmon resonance of Ag shifted the light absorption ability of this material to visible region of spectra. Moreover, hierarchical structure of SGO-ZnO-Ag improved the incident light scattering and reflection. Finally, sulfonated graphite sheets enabled charge transfer and reduce the recombination of electron-hole pairs. These synergistic effects led to much faster rate of photodegradation of Rhodamine B and disinfection of *E. coli* than in the case of pure ZnO. SGO-ZnO-Ag composite is promising candidate in the field of disinfection and photodegradation. Dědková et al. (2015c) introduced nanostructured composite material ZnO/graphite (ZnGt). Micromilled and high purity natural graphite (Gt) were used as a matrix to anchoring of NPs, where 50 wt.% of ZnO NPs were given. Antibacterial assay using four human pathogens showed that ZnGt composited exhibited antibacterial activity under artificial daylight irradiation. Those materials could find the potential application in the field of surface modification of materials used in medicine especially when UV light is not essential for induction of its antibacterial activity.

Carbon fibers may also serve as a material for anchoring of NPs, however, it is still not frequently studied. Carbon nanofibers decorated with TiO_2/ZnO NPs were prepared via electrospinning method by Pant et al. (2016). The small amount of ZnO was introduced to fibers through electrospinning and that provided nucleation sites for the crystal growth of ZnO during hydrothermal synthesis and enabled holding of TiO_2/ZnO particles on the fiber surface. This process of preparation provided composite material with better stability. Antibacterial activity of the prepared material was tested against *E. coli* under UV irradiation. It was demonstrated that carbon-TiO_2/ZnO has higher antibacterial activity than TiO_2/ZnO. This material is promising for air and water purification purposes (Pant et al. 2013).

Carbon nanotubes are being studying intensively in terms of potential applications in several fields of human lives. Antimicrobial properties of carbon nanotubes were already published (Liu et al. 2009) and currently surface modifications and functionalization of carbon nanotubes are investigated. Metal and metal oxides based nanoparticles may serve as one of many possible modificants. ZnO coated multi-walled carbon nanotubes (ZnO/MWCNTs) prepared by Sui et al. (2013) were tested in terms of potential antimicrobial material. *E. coli* was selected as a target organism. Raw and purified MWCNTs were used as a reference material. Obtained results indicate that raw and purified MWCNTs only adsorbed cells of *E. coli*, whereas ZnO/MWCNTs showed bactericidal effect. In other study functionalized multi-walled carbon nanotubes in ZnO thin films (MWCNT-ZnO) were synthesized for photoinactivation of bacteria (Akhavan et al. 2011). MWCNT–ZnO nanocomposite thin films with various MWCNT contents were prepared. Photoinactivation of *E. coli* under UV-visible light irradiation on the surface of unfuctionalized film as well as on the surface of MWCNT–ZnO functionalized film was observed, where functionalized composites with various MWCNT content showed significantly higher photoiactivation of bacteria. Result showed that 10% of MWCNT-ZnO is the optimum content in the film. Proposed mechanism of the antibacterial activity is assigned to charge transfer through Zn-O-C bonds, which are originated, between atoms of zinc of ZnO film and oxygen atoms of carboxylic functional groups of the MWCNTs.

14.3.4 TiO_2 on Carbon Materials

Direct redox reaction served a method for preparation of composites of ultrafine TiO_2 nanoparticles and graphene sheets (TiO_2/GSs) (Cao et al. 2013). The composite possessed extended light absorption, which means that the composite could be excited by visible light. Antibacterial potency induced via visible-light irradiation was investigated against *E. coli*. Results shown, that the antibacterial activity of TiO_2/GSs was significantly higher than to pure TiO_2 nanoparticles. Authors proposed possible application of this material in the field of indoor air disinfection. Titanium dioxide-reduced graphene oxide (TiO_2-RGO) were synthesized by the photocatalytic reduction of exfoliated graphene oxide (GO) by TiO_2 under UV irradiation in the presence of methanol as a hole acceptor (Fernández-Ibáñez et al. 2015). Water contaminated with *E. coli* and *Fusarium solani* (*F. solani*) was treated by TiO_2-RGO and P25 (TiO_2) under real sunlight. Fast disinfection of the contaminated water was observed. Inactivation of *F. solani* was similar for both tested materials; nevertheless, the higher inactivation of *E. coli* was caused by TiO_2-RGO composite. Proposed explanation is given to the production of singlet oxygen via visible light excitation of the composite. Another study was focused on the magnetic graphene oxide–TiO_2 (MGO–TiO_2) (Chang et al. 2015) composites and its antibacterial activity against *E. coli* under solar irradiation. It was observed that MGO–TiO_2 caused complete inaction of the *E. coli* within 30 min under solar irradiation.

Graphene oxide-TiO_2-Ag (GO-TiO_2-Ag) composites were introduced as highly efficient water disinfectants by Liu et al. (2013). Antibacterial activity of GO-TiO_2 and GO-Ag against *E. coli* was observed, however, the highest activity showed GO-TiO_2-Ag composite. The explanation is given to enhanced photocatalytic reaction, which led to bacterial inactivation. It is expected that Ag nanoparticles significantly suppress recombination of photoinduced electrons and holes, which enhance the photocatalytic performance of GO-TiO_2-Ag nanocomposites. Simple one step hydrothermal reaction of TiO_2 nanoparticles (Degusa P25) with $AgNO_3$ and reduced graphene oxide led to creation of multifunctional Ag-TiO_2/rGO composite (Pant et al. 2016). Enhanced optical response of the nanocomposite led to the absorption of light in the visible spectrum via localized surface plasmon resonance effects. Antibacterial activity was performed on *E. coli* as a tested organism. The nanocomposite showed synergistic effect of photocatalytic disinfection of E. coli than pure TiO_2 or TiO_2-Ag and therefore its antibacterial activity was significantly higher. It seems that Ag-TiO_2/rGO is a promising candidate for water treatment due to its improved photocatalytic and antibacterial properties. One study introduced nanostructured composite material graphite/TiO_2 (GrafTi) (Dědková et al. 2015c) and its antibacterial activity against four common human pathogens (*S. aureus* (G^+), *E. coli* (G^-), *E. faecalis* (G^+), *P. aeruginosa* (G^-)) under artificial day light. Antibacterial activity of this material is attributed to photocatalytic reaction with subsequent interaction of ROS with bacterial cells. Difference between antibacterial activity against selected bacteria and difference in the onset of activity were observed. The developed nanocomposite exhibited a potential for future applications as antibacterial agents for modification of surfaces to control bacterial growth e.g. in biomedical fields.

Nanocomposites based on TiO_2 nanoparticles and various amounts of functionalized MWCNTs (TiO_2-MWCNTs) were synthesized (Koli et al. 2016) and its antibacterial activity against *E. coli* and *S. aureus* under artificial visible light irradiation has been performed. Experiments revealed high antibacterial activity of TiO_2-MWCNTs, while bare TiO_2 nanoparticles did not show any inhibitory effect. Authors expect that it is caused by smaller particle size and visible light activation; however, detail explanation is not provided. This material is efficient against wide range of bacteria and may be used to control the persistence and spreading of bacterial infections.

14.4 Conclusion

Various types of antimicrobial materials are researched and developed based on both inorganic and organic matter. One of them are purely inorganic based on inorganic substrate and enhanced by metals or metal oxides. Clay minerals are suitable substrates for various metals as silver, copper their oxides or oxides of zinc or titanium. Improved and more stable antimicrobial action is proven for wide range of bacterial strains, and anchored interaction in hybrids provides lower mobility of nanoparticles therefore lowered negative effect on environment. Similar interaction is observed in case of carbon materials (nanocarbons), however, the forces attracting the active particles are non-bonding interactions. Utilization of such hybrid antimicrobial materials are very wide. There are applications in packaging using those particles as nanofillers for polymeric nanocomposites or medicinal application, where active particles anchored on clay or carbon substrate can be promising delivery system to reduce bone implant related infections after operations, for surface treatment of external fixators in the treatment of complicated fractures of humans or animals for reduction of potential infection or dental applications. They can be used for waste water treatment and sanitary purposes as well. However, the application of antimicrobial materials should be carefully tailored and the toxicity effect considered.

Acknowledgement This chapter was created with support of the Project No. LO1203 "Regional Materials Science and Technology Centre – Feasibility Program" funded by Ministry of Education, Youth and Sports of the Czech Republic and by the MSMT (SP2016/75 and SP2017/86).

References

Akhavan O, Azimirad R, Safa S. Functionalized carbon nanotubes in ZnO thin films for photoinactivation of bacteria. Mater Chem Phys. 2011;130(1–2):598–602.

Bagchi B, Kar S, Dey SK, Bhandary S, Roy D, Mukhopadhyay TK, et al. In situ synthesis and antibacterial activity of copper nanoparticle loaded natural montmorillonite clay based on contact inhibition and ion release. Colloid Surf B-Biointerfaces. 2013;108:358–65.

Baker C, Pradhan A, Pakstis L, Pochan DJ, Shah SI. Synthesis and antibacterial properties of silver nanoparticles. J Nanosci Nanotechnol. 2005;5:244–9.

Benaliouche F, Hidous N, Guerza M, Zouad Y, Boucheffa Y. Characterization and water adsorption properties of Ag and Zn-exchanged A zeolites. Microporous Mesoporous Mater. 2015;209:184–8.

Braydich-Stolle L, Hussain S, Schlager J, Hofmann MC. In vitro cytotoxicity of nanoparticles in mammalian germ line stem cells. Toxicol Sci. 2005;88:412–9.

Brigatti MF, Galan E, Theng BKG. Structures and mineralogy of clay minerals. In: Bergaya F, Theng BKG, Lagaly G, editors. Handbook of clay science. Oxford: Elsevier; 2006. p. 19–86.

Bykkam S, Narsingam S, Ahmadipour M, Dayakar T, et al. Few layered graphene sheet decorated by ZnO nanoparticles for anti-bacterial application. Superlattice Microst. 2015;83:776–84.

Cao B, Cao S, Dong P, Gao J, Wang J. High antibacterial activity of ultrafine TiO_2/graphene sheets nanocomposites under visible light irradiation. Mater Lett. 2013;93:349–52.

Cao GF, Sun Y, Chen JG, Song LP, Jiang JQ, Liu ZT, Liu ZW. Sutures modified by silver-loaded montmorillonite with antibacterial properties. Appl Clay Sci. 2014;93-94:102–6.

Cataldo F, Da Ros T. Medicinal chemistry and pharmacological potential of fullerenes and carbon nanotubes. Trieste: Springer; 2008.

Chang YN, Zhang M, Xia L, Zhang J, Xing G. The toxic effects and mechanisms of CuO and ZnO nanoparticles. Materials. 2012;5:2850–71.

Chang Y, Ou X, Zeng G, Gong J, Deng C. Synthesis of magnetic graphene oxide-TiO_2 and their antibacterial properties under solar irradiation. Appl Surf Sci. 2015;343:1–10.

Chung CJ, Su CW, He JL. Antimicrobial efficacy of photocatalytic TiO2 coatings prepared by arc ion plating. Surf Coat Technol. 2007a;202(4–7):1302–7.

Chung CJ, Su CW, He JL. Microstructural effect on the antimicrobial efficacy of arc ion plated TiO_2. J Mater Res. 2007b;22:3137–43.

Chung CJ, Lin HI, Hsieh PY, Chen KC, He JL, Leyland A, et al. Growth behavior and microstructure of arc ion plated titanium dioxide. Surf Coat Technol. 2009;204:915–22.

Chung CJ, Tsou HK, Chen HL, Hsieh PY, He JL. Low temperature preparation of phase-tunable and antimicrobial titanium dioxide coating on biomedical polymer implants for reducing implant-related infections. Surf Coat Technol. 2011;205:5035–9.

Copcia VE, Luchian C, Dunca S, Bilba N, Hristodor CM. Antibacterial activity of silver-modified natural clinoptilolite. J Mater Sci. 2011;46:7121–8.

Costa C, Conte A, Buonocore GG, Del Nobile MA. Antimicrobial silver-montmorillonite nanoparticles to prolong the shelf life of fresh fruit salad. Int J Food Microbiol. 2011;148:164–7.

Costa C, Conte A, Buonocore GG, Lavorgna M, Del Nobile MA. Calcium-alginate coating loaded with silver-montmorillonite nanoparticles to prolong the shelf-life of fresh-cut carrots. Food Res Int. 2012;48:164–9.

Dankovich TA, Smith JA. Incorporation of copper nanoparticles into paper for point-of-use water purification. Water Res. 2014;63:245–51.

Dědková K, Matějová K, Lang J, Peikertová P, Mamulová Kutláková K, Neuwirthová L, et al. Antibacterial activity of kaolinite/nanoTiO_2 composites in relation to irradiation time. J Photochem Photobiol B Biol. 2014;135:17–22.

Dědková K, Janíková B, Matějová K, Peikertová P, Neuwirthová L, Holešinský J, Kukutschová J. Preparation, characterization and antibacterial properties of ZnO/kaoline nanocomposites. J Photochem Photobiol B Biol. 2015a;148:113–7.

Dědková K, Lang J, Matějová K, Peikertová P, Holešinsky J, Vodárek V, et al. Nanostructured composite material graphite/TiO_2 and its antibacterial activity under visible light irradiation. J Photochem Photobiol B-Biol. 2015b;149:265–71.

Dědková K, Janíková B, Matějová K, Čabanová K, Váňa R, Kalup A, et al. ZnO/graphite composites and its antibacterial activity at different conditions. J Photochem Photobiol B Biol. 2015c;151:256–63.

Dědková K, Mamulová Kutlákova K, Matějová K, Kukutschová J. The study of the antibacterial activity of kaolinite/ZnO composites. Adv Sci Lett. 2016;22(3):695–8.

Deryabin DG, Davydova OK, Yankina ZZ, Vasilchenko AS, Miroshnikov SA, Kornev AB, et al. The activity of [60] fullerene derivatives bearing amine and carboxylic solubilizing groups against *Escherichia coli*: a comparative study. J Nanomater. 2014;2014:1–9.

Drelich J, Li B, Bowen P, Hwang JY, Mills O, Hoffman D. Vermiculite decorated with copper nanoparticles: novel antibacterial hybrid material. Appl Surf Sci. 2011;257:9435–43.

Esteban-Cubillo A, Pecharromán C, Aguilar E, Santarén J, Moya JS. Antibacterial activity of copper monodispersed nanoparticles into sepiolite. J Mater Sci. 2006;41:5208–12.

Feng QL, Wu J, Chen GQ, Cui FZ, Kim TN, Kim JO. A mechanistic study of an antibacterial effect of silver ions on *Escherichia coli* and *Staphylococcus aureus*. J Biomed Mater Res. 2000;662–8.

Fernández-Ibáñez P, Polo-López MI, Malato S, Wadhwa S, Hamilton JWJ, Dunlop PSM, et al. Solar photocatalytic disinfection of water using titanium dioxide graphene composites. Chem Eng J. 2015;261:36–44.

Gao P, Ng K, Sun DD. Sulfonated graphene oxide-ZnO-Ag photocatalyst for fast photodegradation and disinfection under visible light. J Hazard Mater. 2013;262:826–35.
Girase B, Depan D, Shah JS, Xu W, Misra RDK. Silver–clay nanohybrid structure for effective and diffusion-controlled antimicrobial activity. Mater Sci Eng C. 2011;31:1759–66.
Gu N, Gao J, Wang K, Yang X, Dong W. ZnO-montmorillonite as photocatalyst and flocculant for inhibition of cyanobacterial bloom. Water Air Soil Pollut. 2015;226(5):136–47.
Guerra R, Lima E, Viniegra M, Guzman A, Lara V. Growth of *Escherichia coli* and *Salmonella typhi* inhibited by fractal silver nanoparticles supported on zeolites. Microporous Mesoporous Mater. 2012;147:267–73.
Gurr JR, Wang ASS, Chen CH, Jan KY. Ultrafine titanium dioxide particles in the absence of photoactivation can induce oxidative damage to human bronchial epithelial cells. Toxicology. 2005;213(1–2):66–73.
Hashimoto K, Irie H, Fujishima A. TiO_2 photocatalysis: a historical overview and future prospects. Jpn J Appl Phys. 2005;44(12):8269–85.
Holtz RD, Lima BA, Souza Filho AG, Brocchi M, Alves OL. Nanostructured silver vanadate as a promising antibacterial additive to water-based paints. Nanomed-Nanotechnol Biol Med. 2012;8:935–40.
Hrenovic J, Milenkovic J, Daneu N, Matonickin Kepcija R, Rajic N. Antimicrobial activity of metal oxide nanoparticles supported onto natural clinoptilolite. Chemosphere. 2012;88:1103–7.
Hrenovic J, Milenkovic J, Goic-Barisic I, Rajic N. Antibacterial activity of modified natural clinoptilolite against clinical isolates of *Acinetobacter baumannii*. Microporous Mesoporous Mater. 2013;169:148–52.
Hu CH, Xia MS. Adsorption and antibacterial effect of copper-exchanged montmorillonite on *Escherichia coli* K88. Appl Clay Sci. 2006;31:180–4.
Hu CH, Xu ZR, Xia MS. Antibacterial effect of Cu^{2+}-exchanged montmorillonite on *Aeromonas hydrophila* and discussion on its mechanism. Vet Microbiol. 2005;109:83–8.
Hu CH, Xu Y, Xia MS, Xiong L, Xu ZR. Effects of Cu^{2+}-exchanged montmorillonite on growth performance, microbial ecology and intestinal morphology of Nile tilapia (*Oreochromis niloticus*). Aquaculture. 2007;270:200–6.
Hu CH, Gu LY, Luan ZS, Song J, Zhu K. Effects of montmorillonite-zinc oxide hybrid on performance, diarrhea, intestinal permeability and morphology of weanling pigs. Anim Feed Sci Technol. 2012;177:108–15.
Hundáková M, Valášková M, Pazdziora E, Matějová K, Študentová S. Structural and antibacterial properties of original vermiculite and acidified vermiculite with silver. In: Proceedings of the 3rd International Conference NANOCON 2011; 2011 Sep 21–23; Brno, CZ; Ostrava: Tanger Ltd, 2011. p. 617–22.
Hundáková M, Valášková M, Seidlerová J, Pazdziora E, Matějová K. Preparation and evaluation of different antibacterial Ag-montmorillonites. In: Proceedings of the 4th International Conference NANOCON 2012; 2012 Oct 23–25; Brno, CZ. Ostrava: Tanger Ltd; 2013a. p. 591–5.
Hundáková M, Valášková M, Tomášek V, Pazdziora E, Matějová K. Silver and/or copper vermiculites and their antibacterial effect. Acta Geodyn Geomater. 2013b;10(1):97–104.
Hundáková M, Valášková M, Samlíková M, Pazdziora E. Vermiculite with Ag and Cu used as an antibacterial nanofiller in polyethylene. GeoSci Eng. 2014a;LX(3):28–39.
Hundáková M, Valášková M, Seidlerová J. Stability of silver and copper on clay minerals in water and their antibacterial activity. In: Proceedings of the 5th International Conference NANOCON 2013; 2013 Oct 16–18; Brno, CZ; Ostrava: Tanger Ltd; 2014b. p. 632–7.
Hundáková M, Simha Martynková G, Valášková M, Měřínská D, Pazdziora E. Antibacterial polypropylene/Ag-kaolinite, preparation and characterization. Adv Sci Lett. 2016;22(3):656–60.
Huo C, Yang H. Synthesis and characterization of ZnO/palygorskite. Appl Clay Sci. 2010;50(3):362–6.
Jakobsen L, Andersen AS, Friis-Møller A, Jørgensen B, Krogfelt KA, Frimodt-Møller N. Silver resistance: an alarming public health concern? Int J Antimicrob Agents. 2011;38(5):454–5.
Jankauskaitė V, Vitkauskienė A, Lazauskas A, Baltrusaitis J, Prosyčevas I, Andrulevičius M. Bactericidal effect of graphene oxide/Cu/Ag nanoderivatives against *Escherichia coli*,

Pseudomonas aeruginosa, Klebsiella pneumoniae, Staphylococcus aureus and methicillin-resistant *Staphylococcus aureus*. Int J Pharm. 2016;511(1):90–7.

Jiao LF, Ke YL, Xiao K, Song ZH, Lu JJ, Hu CH. Effects of zinc-exchanged montmorillonite with different zinc loading capacities on growth performance, intestinal microbiota, morphology and permeability in weaned piglets. Appl Clay Sci. 2015;113:40–3.

Johari SA, Kalbassi MR, Soltani M, Yu IJ. Application of nanosilver-coated zeolite as water filter media for fungal disinfection of rainbow trout (*Oncorhynchus mykiss*) eggs. Aquacult Int. 2016;24:23–38.

Jones CM, Hoek EMV. A review of the antibacterial effects of silver nanomaterials and potential implications for human health and the environment. J Nanopart Res. 2010;12:1531–51.

Jones N, Ray B, Ranjit KT, Manna AC. Antibacterial activity of ZnO nanoparticle suspensions on a broad spectrum of microorganisms. FEMS Microbiol Lett. 2008;279(1):71–6.

Jung JH, Hwang GB, Lee JE, Bae GN. Preparation of airborne Ag/CNT hybrid nanoparticles using an aerosol process and their application to antimicrobial air filtration. Langmuir. 2011;27:10256–64.

Karel FB, Koparal AS, Kaynak E. Development of silver ion doped antibacterial clays and investigation of their antibacterial activity. Adv Mater Sci Eng. 2015;2015:1–6. ID: 409078

Kawahara K, Tsuruda K, Morishita M, Uchida M. Antibacterial effect of silver-zeolite on oral bacteria under anaerobic conditions. Dent Mater. 2000;16:452–5.

Kheiralla ZMH, Rushdy AA, Betiha MA, Yakob NAN. High-performance antibacterial of montmorillonite decorated with silver nanoparticles using microwave assisted method. J Nanopart Res. 2014;16:2560.

Kokura S, Handa O, Takagi T, Ishikawa T, Naito Y, Yoshikawa T. Silver nanoparticles as a safe preservative for use in cosmetics. Nanomed-Nanotechnol Biol Med. 2010;6:570–4.

Koli VB, Dhodamani AG, Raut AV, Thorat ND, Pawar SH, Delekar SD. Visible light photo-induced antibacterial activity of TiO_2-MWCNTs nanocomposites with varying the contents of MWCNTs. J Photochem Photobiol A-Chem. 2016;328:50–8.

Kollef MH, Afessa B, Anzueto A, Veremakis C, Kerr KM, Margolis BD, et al. Silver coated endotracheal tubes and incidence of ventilator-associated pneumonia. JAMA. 2008;300(7):805–13.

Li B, Yu S, Hwang JY, Shi S. Antibacterial vermiculite nano-material. J Minerals Mater Charact Eng. 2002;1:61–8.

Li Z, Fan L, Zhang T, Li K. Facile synthesis of Ag nanoparticles supported on MWCNTs with favorable stability and their bactericidal properties. J Hazard Mater. 2011;187:466–72.

Liu S, Wei L, Hao L, Fang N, Chang MW, Xu R, et al. Sharper and faster 'nano darts' kill more bacteria: a study of antibacterial activity of individually dispersed pristine single-walled carbon nanotube. ACS Nano. 2009;3(12):3891–902.

Liu S, Keong A, Xu R, Wei J, Tan CM, Yanga Y, Chen Y. Antibacterial action of dispersed single-walled carbon nanotubes on *Escherichia coli* and *Bacillus subtilis* investigated by atomic force microscopy. Nanoscale. 2010;2:2744.

Liu S, Zeng TH, Hofmann M, Burcombe E, Wei J, Jiang R, et al. Antibacterial activity of graphite, graphite oxide, graphene oxide, and reduced graphene oxide: membrane and oxidative stress. ACS Nano. 2011;5(9):6971–80.

Liu L, Bai H, Liu J, Sun DD. Multifunctional graphene oxide-TiO_2-Ag nanocomposites for high performance water disinfection and decontamination under solar irradiation. J Hazard Mater. 2013;261:214–23.

Lok CN, Ho CM, Chen R, He QY, Yu WY, Sun H, et al. Proteomic analysis of the mode of antibacterial action of silver nanoparticles. J Proteome Res. 2006;5:916–24.

Lu W, Senapati D, Wang S, Tovmachenko O, Kumar Singh A, Hongtao Y, et al. Effect of surface coating on the toxicity of silver nanomaterials on human skin keratinocytes. Chem Phys Lett. 2010;487:92–6.

Maeda K. Basic studies on possible clinical application of antibacterial zeolite. Nesshou. 1987;13:316–21.

Magaña SM, Quintana P, Aguilar DH, Toledo JA, Ángeles-Chávez C, Cortés MA, et al. Antibacterial activity of montmorillonites modified with silver. J Mol Catal A-Chem. 2008;281:192–9.

Malachová K, Praus P, Pavlíčková Z, Turicová M. Activity of antibacterial compounds immobilised on montmorillonite. Appl Clay Sci. 2009;43:364–8.

Malachová K, Praus P, Rybková Z, Kozák O. Antibacterial and antifungal activities of silver, copper and zinc montmorillonites. Appl Clay Sci. 2011;53(4):642–5.

Martynková GS, Valášková M. Antimicrobial nanocomposites based on natural modified materials: a review of carbons and clays. J Nanosci Nanotechnol. 2014;14(1):673–93.

Meghana S, Kabra P, Chakraborty S, Padmavathy N. Understanding the pathway of antibacterial activity of copper oxide nanoparticles. RSC Adv. 2015;5:12293–9.

Miyoshi H, Ohno H, Sakai K, Okamura N, Kourai H. Characterization and photochemical and antibacterial properties of highly stable silver nanoparticles prepared on montmorillonite clay in n-hexanol. J Colloid Interface Sci. 2010;345:433–41.

Mizuno K, Zhiyentayev T, Huang L, Khalil S, Nasim F, Tegos GP, et al. Antimicrobial photodynamic therapy with functionalized fullerenes: quantitative structure-activity relationships. J Nanomed Nanotechnol. 2011;2(2):1–9.

Motshekga SC, Ray SS, Onyango MS, Momba MNB. Microwave-assisted synthesis, characterization and antibacterial activity of Ag/ZnO nanoparticles supported bentonite clay. J Hazard Mater. 2013;262:439–46.

Narayan RJ, Abernathy H, Riester L, Berry CJ, Brigmon R. Antimicrobial properties of diamond-like carbon-silver-platinum nanocomposite thin films. J Mater Eng Perform. 2005;14:435–40.

Özdemir G, Limoncu MH, Yapar S. The antibacterial effect of heavy metal and cetylpridinium-exchanged montmorillonites. Appl Clay Sci. 2010;48:319–23.

Panáček A, Kvítek L, Prucek R, Kolář M, Večeřová R, Pizúrová N, et al. Silver colloid nanoparticles: synthesis, characterization, and their antibacterial activity. J Phys Chem. 2006;110:16248–53.

Pant B, Raj H, Barakat NAM, Park M, Jeon K. Carbon nanofibers decorated with binary semiconductor (TiO_2/ZnO) nanocomposites for the effective removal of organic pollutants and the enhancement of antibacterial activities. Ceram Int. 2013;39(6):7029–35.

Pant B, Singh P, Park M, Park S, Kim H. General one-pot strategy to prepare Ag- TiO_2 decorated reduced graphene oxide nanocomposites for chemical and biological disinfectant. J Alloys Compd. 2016;671:51–9.

Parolo ME, Fernández LG, Zajonkovsky I, Sánchez MP, Baschini M. Antibacterial activity of materials synthesized from clay minerals. In: Méndez-Vilas A, editor. Science against microbial pathogens: communicating current research and technological advances, vol. 1: FORMATEX; 2011. p. 144–51. ISBN-13: 978-84-939843-1-1.

Pazdziora E, Matějová K, Valášková M, Holešová S, Hundáková M. Comparison of antibacterial and antiprotozoal effects of nanoparticles Zn^{2+}, Cu^{2+} a Ag^{+} intercalated on clay minerals. In: Proceedings of the 2nd International Conference NANOCON 2010; 2010 Oct 12–14; Olomouc, CZ; Ostrava: Tanger Ltd; 2010. p. 460–4.

Perdikaki A, Galeou A, Pilatos G, Karatasios I, Kanellopoulos KH, Prombona A, Karanikolos GN. Ag and Cu monometallic and Ag/Cu bimetallic nanoparticle–graphene composites with enhanced antibacterial performance. ACS Appl Mater Interfaces. 2016;8(41):27498–510.

Pourabolghasem H, Ghorbanpour M, Shayegh R. Antibacterial activity of copper-doped montmorillonite nanocomposites prepared by alkaline ion exchange method. J Phys Sci. 2016;27(2):1–12.

Pouraboulghasem H, Ghorbanpour M, Shayegh R, Lotfiman S. Synthesis, characterization and antimicrobial activity of alkaline ion-exchanged ZnO/bentonite nanocomposites. J Cent South Univ. 2016;23(4):787–92.

Rai VR, Bai AJ. Nanoparticles and their potential application as antimicrobials. In: Méndez-Vilas A, editor. Science against microbial pathogens: communicating current research and technological advances. Badajoz: FORMATEX; 2011. p. 197–209.

Rai M, Yadav A, Gade A. Silver nanoparticles as a new generation of antimicrobials. Biotechnol Adv. 2009;27:76–83.

Ravichandran K, Chidhambaram N, Gobalakrishnan S. Copper and graphene activated ZnO nanopowders for enhanced photocatalytic and antibacterial activities. J Phys Chem Solids. 2016;93:82–90.

Rivera-Garza M, Olguín MT, García-Sosa I, Alcántara D, Rodríguez-Fuentes G. Silver supported on natural Mexican zeolite as an antibacterial material. Microporous Mesoporous Mater. 2000;39:431–44.

Roguska A, Belcarz A, Pisarek M, Ginalska G, Lewandowska M. TiO_2 nanotube composite layers as delivery system for ZnO and Ag nanoparticles – an unexpected overdose effect decreasing their antibacterial efficacy. Mater Sci Eng C Mater Biol Appl. 2015;51:158–66.

Rosa-Gomez de la I, Olguina MT, Alcantara D. Antibacterial behavior of silver-modified clinoptilolite-heulandite rich tuff on coliform microorganisms from wastewater in a column system. J Environ Manag. 2008;88:853–63.

Rosa-Gomez de la I, Olguina MT, Alcantara D. Silver-modified mexican clinoptilolite-rich tuffs with various particle sizes as antimicrobial agents against *Escherichia coli*. J Mex Chem Soc. 2010;54(3):139–42.

Russell AD. Principles of antimicrobial activity and resistance. In: Block SS, editor. Disinfection, sterilization and preservatives. Philadelphia: Lippincott Williams & Wilkins; 2001. p. 31–56.

Salim MM, Malek NANN. Characterization and antibacterial activity of silver exchanged regenerated NaY zeolite from surfactant-modified NaY zeolite. Mater Sci Eng C. 2016;59:70–7.

Santos MF, Oliveira CM, Tachinski CT, Fernandes MP, Pich CT, Angioletto E, et al. Bactericidal properties of bentonite treated with Ag^+ and acid. Int J Miner Process. 2011;100:51–3.

Shameli K, Ahmad MB, Zargar M, Wan Yunus WMZ, Rustaiyan A, Ibrahim NA. Synthesis of silver nanoparticles in montmorillonite and their antibacterial behavior. Int J Nanomedicine. 2011;6:581–90.

Shameli K, Ahmad MB, Al-Mulla EAJ, Shabanzadeh P, Bagheri S. Antibacterial effect of silver nanoparticles on talc composites. Res Chem Intermed. 2013;41(1):251–63.

Shvedova AA, Pietroiusti A, Fadeel B, Kagan VE. Mechanisms of carbon nanotube-induced toxicity: focus on oxidative stress. Toxicol Appl Pharmacol. 2012;261(2):121–33.

Sohrabnezhad S, Mehdipour Moghaddam MJ, Salavatiyan T. Synthesis and characterization of CuO-montmorillonite nanocomposite by thermal decomposition method and antibacterial activity of nanocomposite. Spectrochim Acta Pt A-Mol Biomol Spectr. 2014;125:73–8.

Sohrabnezhad S, Pourahmad A, Mehdipour Moghaddam MJ, Sadeghi A. Study of antibacterial activity of Ag and Ag_2CO_3 nanoparticles stabilized over montmorillonite. Spectroc Acta Pt A-Mol Biomol Spectr. 2015;136:1728–33.

Sondi I, Salopek-Sondi B. Silver nanoparticles as antimicrobial agent: a case study on *E. coli* as a model for Gram-negative bacteria. J Colloid Interface Sci. 2004;275:177–82.

Song J, Li Y, Hu CH. Effects of copper-exchanged montmorillonite, as alternative to antibiotic, on diarrhea, intestinal permeability and proinflammatory cytokine of weanling pigs. Appl Clay Sci. 2013;77-78:52–5.

Sui M, Zhang L, Sheng L, Huang S, She L. Synthesis of ZnO coated multi-walled carbon nanotubes and their antibacterial activities. Sci Total Environ. 2013;452-453:148–54.

Tang J, Chen Q, Xu L, Zhang S, Feng L, Cheng L, et al. Graphene oxide–silver nanocomposite as a highly effective antibacterial agent with species-specific mechanisms. ACS Appl Mater Interfaces. 2013;5(9):3867–74.

Tegos GP, Demidova TN, Arcila-Lopez D, Lee H, Wharton T, Gali H, et al. Cationic fullerenes are effective and selective antimicrobial photosensitizers. Chem Biol. 2005;12(10):1127–35.

Tekin R, Bac N. Antimicrobial behavior of ion-exchanged zeolite X containing fragrance. Microporous Mesoporous Mater. 2016;234:55–60.

Tong G, Yulong M, Peng G, Zirong X. Antibacterial effects of the cu(II)-exchanged montmorillonite on *Escherichia coli* K88 and *Salmonella choleraesius*. Vet Microbiol. 2005;105:113–22.

Top A, Ülkü S. Silver, zinc, and copper exchange in a Na-clinoptilolite and resulting effect on antibacterial activity. Appl Clay Sci. 2004;27:13–9.

Üreyen ME, Doğan A, Koparal AS. Antibacterial functionalization of cotton and polyester fabrics with a finishing agent based on silver-doped calcium phosphate powders. Text Res J. 2012;82(17):1731–42.

Valášková M, Martynkova GS. Vermiculite: structural properties and examples of the use, clay minerals in nature-their characterization, modification and application. InTech.; 2012. p. 326. ISBN 978-953-51-0738-5.

Valášková M, Hundáková M, Mamulová Kutláková K, Seidlerová J, Čapková P, Pazdziora E, et al. Preparation and characterization of antibacterial silver/vermiculites and silver/montmorillonites. Geochim Cosmochim Acta. 2010;74:6287–300.

Vargas-Reus MA, Memarzadeha K, Huang J, Renc GG, Allakera RP. Antimicrobial activity of nanoparticulate metal oxides against peri-implantitis pathogens. Int J Antimicrob Agents. 2012;40:135–9.

Vincent M, Hartemann P, Engels-Deutsch M. Antimicrobial applications of copper. Int J Hyg Environ Health. 2016;219:585–91.

Wang S, Peng Y. Natural zeolites as effective adsorbents in water and wastewater treatment. Chem Eng J. 2010;156:11–24.

Xia MS, Hu CH, Xu ZR. Effects of copper bearing montmorillonite on the growth performance, intestinal microflora and morphology of weanling pigs. Anim Feed Sci Technol. 2005;118:307–17.

Xiu ZM, Zhang QB, Puppala HL, Colvin VL, Alvarez PJJ. Negligible particle-specific antibacterial activity of silver nanoparticles. Nano Lett. 2012;12:4271–5.

Xu G, Qiao X, Qiu X, Chen J. Preparation and characterization of nano-silver loaded montmorillonite with strong antibacterial activity and slow release property. J Mater Sci Technol. 2011;27(8):685–90.

Yamamoto O, Sawai J, Sasamoto T. Activated carbon sphere with antibacterial characteristics. Mater Trans. 2002;43(5):1069–73.

Yang X, Ebrahimi A, Li J, Cui Q. Fullerene-biomolecule conjugates and their biomedicinal applications. Int J Nanomedicine. 2014;9:77–92.

Yuan W, Jiang G, Che J, Qi X, Xu R, Chang MW, et al. Deposition of silver nanoparticles on multiwalled carbon nanotubes grafted with hyperbranched poly(amidoamine) and their antimicrobial effects. J Phys Chem C. 2008;112:18754–9.

Zhang Y, Chen Y, Zhang H, Zhang B, Liu J. Potent antibacterial activity of a novel silver nanoparticle-halloysite nanotube nanocomposite powder. J Inorg Biochem. 2013;118:59–64.

Zhao D, Zhou J, Liu N. Preparation and characterization of Mingguang palygorskite supported with silver and copper for antibacterial behavior. Appl Clay Sci. 2006;33:161–70.

Zhou Y, Xia M, Ye Y, Hu C. Antimicrobial ability of Cu^{2+}-montmorillonite. Appl Clay Sci. 2004;27:215–8.

Chapter 15
Antimicrobial Activities of Metal Nanoparticles

Adriano Brandelli, Ana Carolina Ritter, and Flávio Fonseca Veras

Abstract Metals have been used since ancient times to combat infectious diseases. With the introduction of nanotechnology, metal nanoparticles have gaining increased attention as antimicrobial agents due to their broad inhibitory spectrum against bacteria, fungi, and viruses. Although silver nanoparticles have been mostly investigated due to their recognized antimicrobial properties, while other metal nanoparticles have received increasing interest as antimicrobials. These include gold, zinc oxide, titanium oxide, copper oxide, and magnesium oxide nanoparticles, since their antibacterial effects have been described. Metal nanoparticles can exert their effect on microbial cells by generating membrane damage, oxidative stress, and injury to proteins and DNA. In addition, metal nanoparticles can be associated with other nanostructures and used as carriers to antimicrobial drugs, improving the array of potential applications.

Keywords Antibacterial activity • Antifungal activity • Antiviral activity • Nanotechnology • Silver nanoparticles

15.1 Introduction

Nanotechnology encompasses the development, management, and application of structures in the nanometer size range and is a newly advanced discipline with great impact on diverse scientific fields (Farokhzad and Langer 2006). It provides new tools for the molecular treatment and rapid detection of diseases, showing a great potential to transform pharmacy, biology, and medicine as well. Advanced materials with nanometric dimensions provide several means for innovative design of nanosized drug delivery systems to overcome biological barriers in order to direct the

A. Brandelli (✉) • A.C. Ritter • F.F. Veras
Laboratório de Bioquímica e Microbiologia Aplicada, Instituto de Ciência e Tecnologia de Alimentos, Universidade Federal do Rio Grande do Sul, Porto Alegre, Brazil
e-mail: abrand@ufrgs.br

M. Rai, R. Shegokar (eds.), *Metal Nanoparticles in Pharma*,
DOI 10.1007/978-3-319-63790-7_15

drug to specific targets. Diverse nanostructures such as nanotubes, nanoparticles, and nanofibers have been used as effective systems for drug delivery, becoming a part of nanomedicine (Faraji and Wipf 2009; Brandelli et al. 2017).

Nanoparticles have attracted much interest because of their unique physical and chemical properties, which originate from the high area to volume ratio and elevated quantity of surface atoms. In fact, as the diameter decreases, the available surface area of the particle itself dramatically increases, and, consequently, there is an increase over the original properties of the corresponding bulk material. This feature makes nanoparticles superior and exceptional candidates for biomedical applications as a variety of biological processes occur at nanometer level (Sharma et al. 2009; Prabhu and Poulose 2012). Thus, nanoparticles hold an incredible potential in various biomedical uses including effective drug delivery systems.

Since ancient times, heavy metals such as copper, silver, and gold have been widely used for the control and treatment of infectious diseases. Among these, silver is the principal metal that has been frequently used because of its recognized antimicrobial properties, and, being an antimicrobial agent, it is preferred in medical applications. Ag also shows potential activity against antibiotic-resistant organisms, which is one of the major concerns in public healthcare (Prabhu and Poulose 2012). Currently, metallic Ag has been replaced by Ag nanoparticles (AgNPs). Metal nanoparticles are used in industry, agriculture, and healthcare. Moreover, several studies performed in the past few years have proved the potential of AgNPs as a significant antimicrobial agent, which can combat various infectious diseases and be used as novel nanomedicine (Rai et al. 2009; Murphy et al. 2015). In addition, other metallic nanoparticles, including noble metal nanoparticles and metal oxide nanoparticles, have been also investigated for their antimicrobial potential or as effective antimicrobial drug carriers (Brandelli 2012; Rai et al. 2015).

In this chapter, the antimicrobial activity of metal nanoparticles against bacteria, fungi, and viruses is discussed. Specific examples for applications of silver, gold, and metal oxide nanoparticles are presented, and the possible mechanisms of action on microbial cells and viruses are described.

15.2 Metal Nanoparticles as Antibacterial Agents

In the last years, bacterial resistance to bactericides and antibiotics has increased. Many organic antimicrobial agents are toxic to humans and other animals, furthermore, can be the cause of different allergic reactions (Rajawat and Qureshi 2012; Hossain et al. 2015). In order to solve this problem, inorganic antibacterial agents have attracted interest for bacteria control, due to their good safety, sustainability, heat resistance, and improved stability under harsh processing conditions.

Currently, nanotechnology provides a sound platform for adjusting the physicochemical properties of numerous materials to generate effective antimicrobials (Beyth et al. 2015). Silver (Ag), gold (Au), titanium oxide (TiO_2), copper oxide (CuO), zinc oxide (ZnO), and magnesium oxide (MgO) are the principal metal nanoparticles (NPs) used as antibacterial agents once their potent antibacterial effects are well known (Beyth et al. 2015; Zhang 2015).

Recently, many studies have demonstrated that different metal oxide nanoparticles exhibit biocidal action against Gram-positive and Gram-negative bacteria. The antimicrobial activity of the nanoparticles is known to be a function of the surface area in contact with the microorganisms (Ravishankar and Jamuna 2011; Franci et al. 2015; Chiriac et al. 2016). The bactericidal effect of metal nanoparticles has been attributed to their small size and high surface to volume ratio, which allows them to interact closely with microbial membranes and is not merely due to the release of metal ions in solution (Ruparelia et al. 2008).

Specific metal ions (essential metals) are indispensable for the biochemistry of life in all organisms, and their privation can cause damages in the structure of cell membranes and DNA. However, the excess of these ions or the presence of other non-essential ions (such as Ag) can be lethal to cells, often caused by oxidative stress, protein dysfunction, or membrane damage (Fig. 15.1) (Lemire et al. 2013; Beyth et al. 2015).

It is clear that some metal nanoparticles are effective antimicrobial agents against several pathogenic microorganisms. This antimicrobial activity is responsible by water treatment, synthetic textiles, biomedical and surgical devices, and food processing and packaging (Ravishankar and Jamuna 2011).

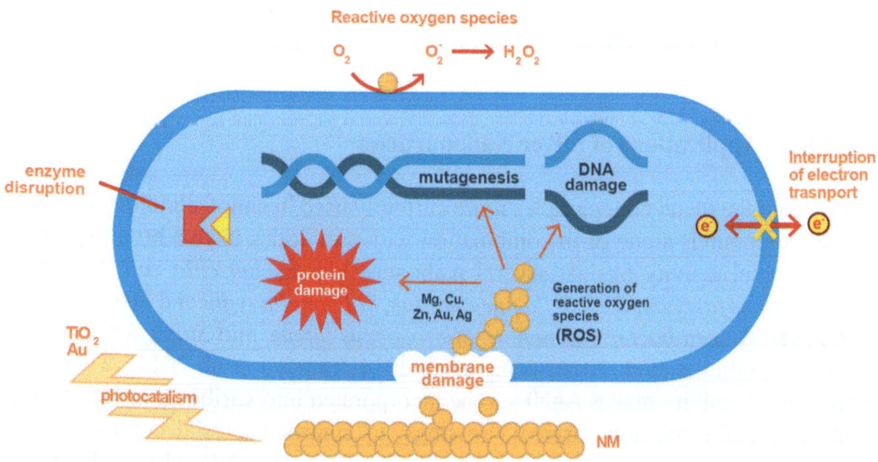

Fig. 15.1 Antibacterial effects of metal nanoparticles. The bactericidal effect of metal nanoparticles is attributed to their small size and high surface to volume ratio, which allows a close interaction with bacterial membranes. After interaction with the target cells, metal nanoparticles can induce loss of membrane integrity, oxidative stress, and injury to proteins and DNA

15.2.1 Silver Nanoparticles

AgNPs have been used against Gram-positive and Gram-negative bacteria due to its relatively low cytotoxicity. The highly antibacterial effect of AgNPs has been described by many studies, but their mechanism of action is not fully understood (Ravishankar and Jamuna 2011; Beyth et al. 2015).

In Fig. 15.1, a summary of mechanisms of action against Gram-positive and Gram-negative bacteria is presented, but it is assumed that the high affinity of silver toward sulfur and phosphorus is the key element for the antimicrobial effect (Ravishankar and Jamuna 2011; Maiti et al. 2014; Franci et al. 2015). The Ag ion, which has an affinity for sulfur and nitrogen, can inhibit and disrupt protein structure by binding to thiol and amino groups, which in turn affects bacterial cell viability (Beyth et al. 2015). In addition, silver ions released from AgNPs can interact with phosphorus moieties in DNA, resulting in inactivation of DNA replication, or they can react with sulfur-containing proteins, leading to the inhibition of enzyme functions (León-Silva et al. 2016).

Another mechanism frequently reported is the induction of reactive oxygen species (ROS) by AgNPs, forming free radicals with a bactericidal action (Franci et al. 2015). ROS cause damage to the intracellular structures, in particular mitochondria and DNA, besides ribosome denaturation, generating inhibition of protein synthesis (León-Silva et al. 2016).

In addition, some studies have shown that the bactericidal activity of AgNPs is also dependent on their size and zeta potential. Smaller dimensions (<30 nm) have a superior ability to penetrate into bacteria, and the electrostatic force developed when nanoparticles with a positive zeta potential encounter the negative surface charge of bacteria promotes a closer attraction and interaction between the two entities and possibly the penetration across the bacterial membranes (Franci et al. 2015).

15.2.1.1 Applications of Silver Nanoparticles

AgNPs are the metal nanoparticles more currently used in nanotechnology research. Silver compounds alone or in combination with antibiotics have a broad-spectrum antimicrobial activity against several pathogens like *Escherichia coli*, *S. aureus*, *Enterococcus faecium*, *Klebsiella pneumoniae*, *Salmonella typhi* and *Listeria monocytogenes*, *Acinetobacter baumannii*, *Micrococcus luteus*, and *Staphylococcus epidermidis* (Salunke et al. 2014; Wang et al. 2014; Tamayo et al. 2014). This broad spectrum of activity makes AgNPs to be incorporated into various matrices such as bandages, catheters, and surgical dressings, to control bacterial colonization on diverse materials to prevent infections. Applications of AgNPs also include burn treatment, elimination of microorganisms on textile fabrics, and water treatment (Maiti et al. 2014; Franci et al. 2015).

Currently, AgNPs are being added to many common household products such as bedding, washers, water purification systems, toothpaste, shampoo, fabrics, deodorants, filters, paints, kitchen utensils, toys, and humidifiers to impart antimicrobial properties (Ravishankar and Jamuna 2011). Additionally, Samsung has created and marketed a material called Silver Nano, which includes silver nanoparticles on the surfaces of household appliances (Mody et al. 2010).

The use of silver in food is restricted, but there may be residues on fruit and vegetables that have been disinfected by washing with suspensions of AgNPs. In the food packaging, the AgNPs are attractive due to their capability to extend the shelf life, and even though their use in food packaging is regulated by the EU and the USA, the food safety authorities recommend that they be used in a prudent way, due to the inability to make conclusive statements on their toxicity (Carbone et al. 2016). Youssef and Abdel-Aziz (2013) observed that polystyrene with AgNPs exhibits low volatility and stability at high temperatures and in addition to AgNPs can be hosted in different matrices such as polymers and stabilizing agents (citrates and long-chain alcohols) (Toker et al. 2013).

The antibacterial and antibiofilm capabilities of AgNPs decorated on graphene nanosheets were also demonstrated. The nanocomposite showed antibacterial activity against *Pseudomonas aeruginosa* with minimum inhibitory concentration ranging from 2.5 to 5.0 μg/ml and a 100% inhibition rate toward bacterial cells adhered on stainless surface (Faria et al. 2015). In addition, AgNPs can be incorporated into nanofiber scaffolds, generating an interesting material for application as antimicrobial wound dressing (Dubey et al. 2015).

15.2.2 Gold Nanoparticles

Gold nanoparticles (AuNPs) are considered nontoxic to human cells and presents higher stability when in contact with biological fluids (Zhang 2015). Alone, AuNPs are considered biologically inert and lack antimicrobial properties but can be modified to possess chemical or photothermal functionality (Zhang 2015).

AuNPs when combined with photosensitizers can kill many multidrug-resistant pathogens and cancer cells (León-Silva et al. 2016). If gold nanoparticles are tuned to absorb near-infrared (NIR) radiation, they can transfer this energy into the surrounding environment as heat, and, once attached to bacterial cells, the localized heating that occurs during NIR irradiation should cause irreparable cellular damage (Norman et al. 2008; Beyth et al. 2015).

Many studies also demonstrated the high antibacterial activity of AuNPs connected to antibiotics such as ampicillin, vancomycin, cefaclor, and the antibacterial enzyme lysozyme against Gram-positive and Gram-negative bacteria (Zhang 2015; Payne et al. 2016). In case of cefaclor, the mode of action which consists in the inhibition of the synthesis of the peptidoglycan layer, and when this antibiotic is bound to gold nanoparticles (that generate holes in the cell wall), results in the leakage of cell contents and cell death (Ravishankar and Jamuna 2011).

Another possibility of increasing the effect of AgNPs is combined with platinum (Pt). This blend proved strong bactericidal effect, and interestingly in contrast to other metal nanoparticles, this effect was ROS-independent, because the cell death resulting from membrane damage and a severe elevation of ATP (Beyth et al. 2015).

15.2.2.1 Applications of Gold Nanoparticles

Gold was discovered around 2500 BC, and its therapeutic use has been widely reported. In the middle ages, Paracelso used "potable gold" to treat some mental disorders and syphilis. After some years, Giovanni Andrea applied *aurum potabile* to the treatment of lepra, ulcer, epilepsy, and diarrhea. In the 1920s, Robert Koch found out that gold cyanide has bacteriostatic effect against the tubercle bacillus and still in this decade, the use of colloidal gold was proposed to ease the suffering of inoperable cancer patients (Ravishankar and Jamuna 2011; Dykmana and Khlebtsov 2012).

Today, AuNPs have been proposed for use in medical, prophylaxis, and hygiene applications. Their major use is in diagnostic assays, drug and gene delivery to target tissues or tumors, and as enhancers/sensitizers of radiotherapy; targeted delivery of drugs, peptides, DNA, and antigens; detection and photothermolysis of microorganisms and cancer cells; biosensorics; immunoassays; clinical chemistry; water purification; and treatment of rheumatic diseases including psoriasis, juvenile arthritis, palindromic rheumatism, and discoid lupus erythematosus (Dykmana and Khlebtsov 2012; Zhang 2015).

AuNPs are generally considered to be biologically inert but can be engineered to possess chemical or photothermal functionality. On near-infrared (NIR) irradiation, the Au-based nanomaterials with characteristic NIR absorption can destroy cancer cells and bacteria via photothermal heating (Ravishankar and Jamuna 2011). Li et al. (2014) reported the use of cationic and hydrophobic functionalized AuNPs against 11 clinical MDR bacterial isolates and concluded that these nanoparticles suppressed the growth of all bacteria. Recently, Payne et al. (2016) reported the activity of kanamycin-capped AuNPs (Kan-AuNPs) against different bacteria (*S. bovis*, *S. epidermidis*, *E. aerogenes*, *P. aeruginosa*, *Y. pestis*) and demonstrated that AuNP conjugation dramatically improved the efficacy of the antibiotic against these bacteria (both Kan-sensitive and Kan-resistant bacteria). Roshmi et al. (2015) demonstrated the highly stable AuNPs fabricated by a *Bacillus* sp. and functionalized with ciprofloxacin, gentamycin, rifampicin, and vancomycin are effective against selected coagulase-negative staphylococci (CoNS), like *Staphylococcus epidermidis* and *S. haemolyticus*.

Lima et al. (2013) concluded that gold nanoparticles dispersed on zeolites are excellent biocide to eliminate food pathogens such as *Escherichia coli* and *Salmonella typhi* at short times. Thirumurugan et al. (2013) reported the combined effect of bacteriocin-like peptides with AuNPs against food spoiling organism and concluded that the antibacterial activities were increased on combination of bacteriocin with AuNPs and the bacteriocin nisin with nanoparticles.

15.2.3 Copper Oxide Nanoparticles

Copper oxide (CuO) is cheaper, easily mixes with polymers, is relatively stable in terms of chemical and physical properties, and is highly ionic (Gawande et al. 2016). The exact mechanism behind bactericidal effect of copper nanoparticles is not clear. It is known that CuO nanoparticle activity depends on bacterial species, and, like other metallic nanoparticles, its antibacterial effect takes place by membrane disruption and ROS production (Beyth et al. 2015). Lemire et al. (2013) reported that loss of bacterial cell viability has been correlated with the uptake of Cu ions and increased production of ROS.

Other authors observed that Cu surfaces do not induce significant oxidative DNA damage in vivo, and multiple lines of evidence suggest that DNA degradation is not the primary cause of Cu-mediated surface killing (Warnes et al. 2012). However, genetic material released from dead cells is gradually degraded at a rate that increases with respect to the Cu content of the surface alloy.

Beyth et al. (2015) reported that copper nanoparticles have a high antimicrobial activity against *B. subtilis* and *B. anthracis* due to greater abundance of amines and carboxyl groups on their cell surfaces and greater affinity of copper toward these groups.

15.2.3.1 Applications of Copper Nanoparticles

Copper nanoparticles (CuNPs) are important semiconductors and exhibit a range of potentially useful physical properties such as superconductivity, electron correlation effects, and spin dynamics and further have a low production cost (Ananth et al. 2015). Because of their catalytic, optical, electrical, and antimicrobial applications, CuNPs have attracted more attention in recent years (Letfullin et al. 2011; Ingle et al. 2014; Umer et al. 2014; Alzahrani and Ahmed 2016). Despite the large range of applications, the use of CuNPs is restricted by inherent copper's instability under atmospheric conditions, which makes it prone to oxidation (Gawande et al. 2016.) Scientists are using different inert media such as argon or nitrogen to overcome this oxidation problem and also using reducing, capping, or protecting agents for the reduction of copper salt used (Umer et al. 2014).

Before the development of antibiotics and other chemotherapeutics, silver and copper were used to treat microbial infections. Unfortunately, copper compounds may be toxic to some organisms and may cause environmental hazards. However, copper in the form of copper nanoparticles can be the substitute to avoid these consequences (Ingle et al. 2014). CuNPs have shown to be effective against various bacterial pathogens, although their antibacterial efficacy is inferior to that of Ag or ZnO. Nevertheless, many reports demonstrated the efficacy of CuNPs against bacteria such as *E. coli*, *S. aureus* (Cubillo et al. 2006), *B. subtilis* (Ruparelia et al. 2008), *Micrococcus luteus*, *Klebsiella pneumoniae*, and *Bacillus megaterium* (Esteban-Tejeda et al. 2009; Raffi et al. 2010; Theivasanthi and Alagar, 2011). On the other hand, Ramyadevi et al. (2012) observed that *P. aeruginosa* was resistant to CuNPs; it is suggested that CuO nanoparticle activity varies greatly depending on the challenged bacterial species.

Ananth et al. (2015) emphasized on the importance of CuNPs as a sterilization agent in water, arsenic removal in water treatment, and antitumor therapy. Ingle et al. (2014) also observed that the nanoparticles have a specific capacity for drug loading and efficient photoluminescence ability, and, therefore, they are important materials in the targeted delivery of imaging agents and anticancer drugs.

In food, CuNPs could be useful in food packaging. Longano et al. (2012) developed a composed CuNPs embedded in polylactic acid, combining the antibacterial properties of CuNPs with the biodegradability of the polymer matrix. The authors concluded that the nanocomposites proposed are extremely attractive nanomaterials since they possess good antibacterial activity against *Pseudomonas* spp. Conte et al. (2013) proposed films of PLA embedded with copper nanoparticles (used to package fior di latte cheese), and the results showed that in vivo and in vitro tests presented an excellent potential for PLA-copper films for food contact applications as well as in active packaging technologies.

15.2.4 Zinc Oxide Nanoparticles

ZnO nanoparticles have been proved to be highly toxic against Gram-positive and Gram-negative bacteria as well as the spores, which are resistant to high temperature and high pressure (Ma et al. 2013; Chiriac et al. 2016). On the other hand, the ZnO exhibits minimal effect on human cells, which makes it safe for use as antimicrobial agent (Espitia et al. 2012; Zhang et al. 2013).

Studies have shown that the antimicrobial activity of zinc oxide nanoparticles depends on the chosen concentration and particle size (Ma et al. 2013; Zhang et al. 2013). Among the mechanisms responsible by the antibacterial activity of ZnO nanoparticles, the generation of hydrogen peroxide (H_2O_2) from the surface of ZnO is considered more effective means for the inhibition of bacterial growth. The inhibitory effect of H_2O_2 is due to the fact that most living cells contain the enzyme catalase, which converts H_2O_2 into water (H_2O) and oxygen (O_2). Oxygen atoms attract electrons from the cell walls of bacteria (microorganisms carry a negative charge), thus destroying the basic molecular structure of the cell proteins and the layer of lipids on the cell surface. The oxidized cell walls become damaged or even completely break apart (Ravishankar and Jamuna 2011; Chiriac et al. 2016).

15.2.4.1 Applications of Zinc Oxide Nanoparticles

ZnO nanoparticles constitute an attractive material. The unique semiconducting, optical piezoelectric properties make this material highly useful for a wide variety of applications. Nano-electronic/nano-optical devices, energy storage, cosmetic products, nanosensors, gas sensors, ceramics, and additive in food products and paints are some examples (Bogutska et al. 2013; Zhang et al. 2013).

Many studies have demonstrated the antimicrobial activities of ZnO nanoparticles against different bacteria such as *Escherichia coli*, *Pseudomonas fluorescens*, *Listeria monocytogenes*, *Salmonella enteritidis*, *E. coli* O157:H7, and *S. aureus* (Zhang et al. 2007; Jin et al. 2009). Due to this fact, ZnO nanoparticles are used to prevent biofilm formation, making them suitable for coating materials designated for medical and other devices (wallpapers in hospitals, active ingredient for dermatological applications in creams, etc.) (Bogutska et al. 2013).

In the last years, nanozinc compounds have been acquiring great importance in biomedicine. For example, due to their excellent UV absorption and reflective properties, ZnO nanoparticles are common constituents of UV blocking (sunscreens, cosmetic creams, ointments) (Bogutska et al. 2013; Ma et al. 2013). ZnO has been approved for cosmetic uses by the FDA (Zhang et al. 2013).

Due to its good chemical stability and rapid electrochemical response, ZnO nanoparticles can be used as biosensors for detecting glucose, H_2O_2, cholesterol, and urea (Zhang et al. 2013). Still in the biomedical area, Bogutska et al. (2013) reported that ZnO nanoparticles can selectively kill some cancerous cells, thus helping the problem with most drugs used in chemotherapy (inability of anticancer drugs to distinguish between healthy and cancerous cells).

Lastly, ZnO is one of the five zinc compounds that are listed as a generally recognized as safe (GRAS) material by the US Food and Drug Administration (Espitia et al. 2012) and is available as a food additive. Some foods may be fortified with the trace mineral to help boost nutrition in the food such as breakfast cereals, nutrition drinks, and nutrition bars. Many studies have used ZnO in food packaging. Due to the fact ZnO nanoparticles have antimicrobial properties against food-borne pathogens and present positive changes in the structure and properties of packaging materials such as increased mechanical and thermal resistance, their incorporation in different materials including glass, low-density polyethylene (LDPE), polypropylene (PP), polyurethane (PU), paper and chitosan has been successfully demonstrated (Prasad et al. 2010; Eskandari et al. 2011; Espitia et al. 2012).

15.2.5 Titanium Dioxide Nanoparticles

Titanium oxide (TiO_2) nanoparticles have attracted significant interest because they exhibit exclusive and improved properties compared to their bulk material counterparts. These nanoparticles show quantum size effects in which the physical and chemical properties of materials are strongly dependent on particle size (Othaman et al. 2014). Titanium dioxide nanoparticles (TiO_2 NPs) are photocatalytic. TiO_2 NPs decompose organic compounds by the formation and constant release of hydroxyl radicals and superoxide ions when exposed to nonlethal ultraviolet light (UV) wavelength of less than 385 nm (Beyth et al. 2015). The use of nanometer-sized TiO_2 particles has the potential to further enhance the antimicrobial activity of TiO_2 (Othaman et al. 2014).

15.2.5.1 Applications of Titanium Dioxide Nanoparticles

TiO_2 has been extensively used as a pigment because of its brightness, high refractive index, and resistance to discoloration. TiO_2 is mostly used as a pigment in paints, but it is also employed as a pigment in plastics, paper, pharmaceuticals, cosmetics, toothpaste, and foods (Weir et al. 2012; Yin et al. 2013). TiO_2 has been approved by the American Food and Drug Administration (FDA) for use in human food, drugs, cosmetics, and food contact materials because it is considered nontoxic (Othaman et al. 2014).

With regard to their microbiological effects, TiO_2 nanoparticles exhibit antibacterial activity against multi-antibiotic-resistant strains of *E. coli* (Mamonova et al. 2013), *P. aeruginosa*, *Staphylococcus aureus*, and *Pseudomonas putida* and spores of *Bacillus* (Yao and Yeung 2011; Bonetta et al. 2013). For this reason, TiO_2 has been used in combination with silver as an antimicrobial agent and as a self-cleaning and self-disinfecting material for surface coatings in many applications (Yoshimura et al. 1995). Moreover, due to the photocatalytic activity, TiO_2 has been used in air and wastewater purification (Weir et al. 2012).

In food industries, besides the disinfection of equipments, TiO_2 can be added into food packaging. Recently, Othman et al. (2014) developed TiO_2 nanoparticle-coated films and observed that the antimicrobial activity of the films exposed to both fluorescent and UV light increased with an increase in the TiO_2 nanoparticle concentration. The UV light was found to be more effective when compared to fluorescent light, due to the suitable bandgap energy of UV light and the higher hydroxyl radical concentration on the surface of the coated films. Some studies also revealed that the TiO_2 nanoparticles are able to control organic compounds such as ethylene (Luo et al. 2013), thus demonstrating another reason to use TiO_2 in food industries and packaging. Silicon dioxide and titanium dioxide are the two most commonly used nanoparticles in food packaging (Pradhan et al. 2015).

Among all the potential uses of TiO_2, Yin et al. (2013) name the importance of medical applications. For those authors, the biomedical applications of TiO_2 can be classified as follows: photodynamic therapy for cancer, drug delivery systems, cell imaging, biosensors for biological assay, and genetic engineering.

15.2.6 Magnesium Oxide Nanoparticles

Magnesium is the fourth most common mineral and the second intracellular cation after potassium in the human body. Compared with silver, copper, and TiO_2, magnesium oxide (MgO) nanoparticles have the advantage that it can be prepared from available and economical precursors. The mechanism of action of this compound in bacteria is by inducing ROS and inhibiting essential microbial enzymes. The alkaline effect has been considered as another primary factor in the antibacterial action of MgO nanoparticles (Tang and Lv 2014). According to Hossain et al. (2014), MgO can inactivate the pathogens by forming ROS; however, adsorption process and direct cell membrane penetration by the particles may be other possible ways for pathogen inactivation, and these possibilities need to be investigated in more detail. In contrast, Leung et al. (2014) investigated the toxicity of three different MgO nanoparticle samples in *Escherichia coli* bacterial cells and demonstrated the absence of ROS production for two MgO nanoparticle samples. The authors suggest that proteomics data clearly demonstrate the absence of oxidative stress and indicate that the primary mechanism of cell death is related to the cell membrane damage, which does not appear to be associated with lipid peroxidation.

15.2.6.1 Applications of Magnesium Oxide Nanoparticles

Compared to other metal nanoparticles, MgO nanoparticles have the advantage that it can be prepared from available and economical precursors. Furthermore, high thermal stability, being readily available, biocompatible, and strong antibacterial activity are attracting the interest of researchers for these nanoparticles (Leung et al. 2014).

MgO nanoparticles have been used in pharmaceuticals (used for the relief of dyspepsia, for sore stomach, and for bone regeneration), toxic waste remediation, toxic gas removal, paint and semiconductors, catalyst supports, refractory materials, and adsorbents, among others (Tang and Lv 2014; Hossain et al. 2015). MgO is one of the six magnesium compounds that are currently recognized as safe by the US Food and Drug Administration (21CFR184.1431). In view of that, Di et al. (2012) and Krishnamoorthy et al. (2012) recently showed the promise for application of MgO nanoparticles in cancer therapy.

MgO prepared by aerogel procedure (AP-MgO) has many properties that are desirable for a potent disinfectant. Their ability to adsorb and retain for a long time significant amounts of elemental chlorine and bromine, high surface area, and enhanced surface reactivity are some of these properties (Ravishankar and Jamuna 2011). Stoimenov et al. (2002) demonstrated the antimicrobial efficiency of AP-MgO against different bacteria. As a result, MgO showed excellent activity against *E. coli* and *Bacillus megaterium* and a good activity against spores of *Bacillus subtilis*. According to Hossain et al. (2014), the pathogen inactivation might depend on factors like contact of MgO particles with bacteria, alkaline condition of the particle surface, particle-specific surface area, contact time, and availability of active oxygen on surface.

Jin and He (2011) observed the antibacterial activities of MgO nanoparticles alone or in combination with other antimicrobials (nisin and ZnO nanoparticles) against *Escherichia coli* O157:H7 and *Salmonella Stanley*. They concluded that the antibacterial activity of MgO nanoparticles increased as the concentrations of MgO increased and that MgO nanoparticles alone or in combination with nisin could be used as an effective antibacterial agent in food formulations directly or through the slow release from packaging materials to kill or inhibit pathogens in foods. Mirhosseini and Afzali (2016) also investigated the effect of MgO nanoparticles on food pathogens (*Escherichia coli* and *Staphylococcus aureus*) in milk, and the results showed that the nanoparticles have strong bactericidal activity against the pathogens. The synergistic effect of MgO in combination with nisin and heat damaged the cell membrane, resulting in a leakage of intracellular contents and eventually the death of bacterial cells.

15.3 Metal Nanoparticles as Antifungal Agents

The presence of fungi in the environment significantly contributes to the rise of problems in health and agriculture (Hsu et al. 2012; Gauthier and Keller 2013; Baxi et al. 2016); therefore, the search for alternative strategies to avoid or minimize fungal development must also be extensively investigated. Considering the same aspects to the use of antibacterial agents involving safety and resistance (Sequeira et al. 2012; Morace et al. 2014), metallic nanoparticles are also promising nanomaterials to combat the growth of pathogenic and spoilage-causing fungi. In this case, the effectiveness of nanoparticles, including stability and long-term activity, their lower toxicity, and little evidence on development of fungal resistance are important features that make them favorable as disinfection method when compared with antifungal chemicals (Hanus and Harry 2013; Alghuthaymi et al. 2015). However, the amount of research investigating the antifungal activity of metallic nanoparticles is still considerably lower than those who evaluated the effects of nanoparticles against bacteria (Tran et al. 2013).

15.3.1 Antifungal Activity of Metal Nanoparticles

Among the most common metal/metal oxide nanoparticles, those composed by silver, copper, zinc, and gold are currently attracting attention as broad-spectrum antifungal agents owing to their antimicrobial properties (Murphy et al. 2015; Zazo et al. 2016) besides the possibility of using biological methods for their synthesis (Kharissova et al. 2013; Elumalai and Velmurugan 2015; Ravichandran et al. 2016). Other metallic nanoparticles such as aluminum, iron, magnesium, nickel, palladium, platinum, and titanium have been also investigated against some fungi (Zazo et al. 2016). Antifungal activity of these nanoparticles is summarized in Table 15.1.

There are few specific reports about the antifungal mechanism of metal nanoparticles because most studies have focused on bacterial activity (Ravishankar and Jamuna 2011; Warnes et al. 2012; Beyth et al. 2015; Chiriac et al. 2016). Therefore, the role played by these nanoparticles as antifungal agents is still controversial. While some studies consider that such mechanisms are similar to those that occur in bacteria, others disprove this basis.

Some authors stated that AgNPs are able to penetrate cell wall and fungal membrane causing cell death due to interaction with biological functional groups of membrane proteins. They also mentioned that these nanoparticles can generate silver ions, which produce ROS leading to protein denaturation, DNA damage, and

Table 15.1 Studies performed with metallic nanoparticles against different types of fungi

Metal	Used condition	Size (nm)	Fungal species	Reference
Silver	A	55[a]	*Penicillium brevicompactum*, *Aspergillus fumigatus*, *Cladosporium cladosporioides*, *Chaetomium globosum*, and *Stachybotrys chartarum*	Ogar et al. (2015)
	A	30–90	*Colletotrichum coccodes*, *Monilinia* sp., and *Pyricularia* sp.	Lee et al. (2013)
	A	20–50	*Aspergillus versicolor*	Ravichandran et al. (2016)[b]
	A	10–50	*Phytophthora infestans* and *Phytophthora capsici*	Velmurugan et al. (2015)[b]
	A	10–60	*Candida albicans*	Cakić et al. (2016)
	B	10–80	*Aspergillus niger*, *A. fumigatus*, *Candida* sp., and *Aspergillus flavus*	Rajeshkumar et al. (2016)[b]
	A	9–130	*C. albicans*, *Cryptococcus neoformans*, and *Fusarium oxysporum*	Ashajyothi et al. (2016)[b]
	B	8–15	*C. albicans* and *Candida tropicalis*	Ahmad et al. (2016)[b]
	B	6–16	*A. niger*	Pinto et al. (2013)
	A	14–28	*A. niger*, *A. flavus*, *Penicillium notatum*, *Saccharomyces cerevisiae*, and *C. albicans*	Rao et al. (2016)[b]
Zinc oxide	B	2–28	*Fusarium* sp.	Sharma et al. (2010)
	B	47–63	*C. albicans*	Hameed et al. (2015)
	A	18[a]	*C. albicans* and *C. tropicalis*	Elumalai and Velmurugan (2015)[b]
	A	16–96	*C. albicans*, *C. neoformans*, and *F. oxysporum*	Ashajyothi et al. (2016)[b]
	A	70[a]	*Botrytis cinerea and Penicillium expansum*	He et al. (2011)
	A	25–40	*A. flavus*, *Aspergillus nidulans*, *Trichoderma harzianum*, and *Rhizopus stolonifer*	Gunalan et al. (2012)[b]
	A	57[a]	*Trichophyton mentagrophytes*, *Microsporum canis*, *C. albicans*, and *A. fumigatus*	El-Diasty et al. (2013)

(continued)

Table 15.1 (continued)

Metal	Used condition	Size (nm)	Fungal species	Reference
Copper	A	NM	*A. flavus*, *A. niger*, and *C. albicans*	Ramyadevi et al. (2012)
	A	3–10	*Phoma destructiva*, *Curvularia lunata*, *Alternaria alternata*, and *F. oxysporum*	Kanhed et al. (2014)
	A	20–90	*C. albicans* and *C. neoformans*	Ashajyothi et al. (2016)[b]
	B	20–30	*Trichophyton rubrum*, *C. albicans*, and *Chrysosporium indicum*	Yallappa et al. (2016)[b]
Gold	A	3–20	*C. albicans*, *C. tropicalis*, and *Candida glabrata*	Ahmad et al. (2013)
	A	20–70	*A. niger*	Ashajyothi et al. (2016)[b]
	A	10–55	*Candida* species	Wani and Ahmad (2013)
Titanium dioxide	A	73–106	*A. niger* and *Trichoderma reesei*	Durairaj et al. (2015)[b]
	B	3–33	*A. niger*	Yu et al. (2013)
Silver oxide	B	10–32	*A. niger*	Suresh et al. (2016a)
Copper oxide	A	20–60	*A. niger* and *C. albicans*	Katwal et al. (2015)
Platinum	A	10–50	*Colletotrichum acutatum* and *Cladosporium fulvum*	Velmurugan et al. (2016)[b]
Palladium	A	4[a]	*A. niger*	Mallikarjuna et al. (2013)
Nickel oxide	B	12–30	*A. niger*	Suresh et al. (2016b)
Zirconium dioxide	A	50–100	*C. albicans* and *A. niger*	Gowri et al. (2014)
Iron oxide	B	32–39[a]	*C. albicans*	Hussein-Al-Ali et al. (2014)

NM not mentioned, *A* only nanoparticles, *B* nanoparticle systems: nanoparticles coupled with another nanostructure or a chemical compound

[a]Average values

[b]Authors obtained nanoparticles by biosynthesis process (green synthesis)

leakage of cell material (Kim et al. 2009, 2012; Kumar et al. 2014; Abkhoo and Panjehkeh 2016). On the other hand, a recent study demonstrated that neither oxidative stress nor membrane permeabilization is the principal toxicity mechanism of Ag ions and AgNPs against *Saccharomyces cerevisiae* (Käosaar et al. 2016).

Lipovsky et al. (2011) reported that ROS production by ZnO nanoparticles induce significantly growth inhibition of *Candida albicans* (similar to that described with bacteria). However, the process of oxidative stress by ROS generation and any interaction of the fungal cell wall with ZnO nanoparticles must be better investigated in order to full elucidation (Hameed et al. 2015).

Evaluating the antimicrobial activity of CuO nanoparticles against *E. coli*, *S. aureus*, *A. niger*, and *C. albicans*, Katwal et al. (2015) suggested the same mode of action for both the bacteria and fungi. A considerable reduction effect on H^+ efflux of the *Candida* species by AuNPs was earlier confirmed (Ahmad et al. 2013). The authors attribute this effect to inhibition of H^+-ATPase-mediated proton pumping that was caused by interaction between AuNPs and the ATP-driven enzyme (H^+-ATPase).

Other mechanisms of action are still speculated, such as interruption of electron transport, interference in absorption of nutrients, DNA destruction, or even the possibility of simultaneous occurrence of various mechanisms (Alghuthaymi et al. 2015). Some authors assume that the particles may affect the bacterial growth by multiple targets of action (Knetsch and Koole 2011; Alghuthaymi et al. 2015; He et al. 2016), and this behavior could also be suggested to the effects on fungi. Therefore, there are still open questions about the real mechanisms of antifungal activity of metal nanoparticles, and further investigation on this subject is required.

It has also been shown that metal nanoparticles cause different effects on the fungal growth, such as disruption of cell membrane and wall, inhibition of sporulation and germination of spores, color changes of colonies on filamentous fungi, or inhibition of budding in yeast (Ogar et al. 2015; Hameed et al. 2015). Hameed et al. (2015) verified that the accumulation of Mg-doped ZnO nanoparticles in membranes of *C. albicans* causes cell damage. Xia et al. (2016) described various effects of AgNPs on the emerging fungal pathogen *Trichosporon asahii* including deformation and shrinkage of hyphae, cellular damage causing leakage of cytoplasmic contents, and degeneration of organelles (Fig. 15.2).

Several factors can interfere with the action mode and effectiveness of nanoparticles, for example, metal type, size, shape, and concentration of particles and fungal species, among others (Kim et al. 2012; Hanus and Harris 2013; Lee et al. 2013; Ogar et al. 2015). Kim et al. (2012) verified that inhibitory activity of AgNPs against phytopathogenic fungi was influenced by concentration and type of particles applied. In another study, Ahmad et al. (2013) noted a change in antifungal activity according to size of AuNPs. In this case, the greatest effect was caused by the particles of smaller size due to increase in their surface area. The use of ZnO nanoparticles combined with visible light illumination increased their deleterious effect on *C. albicans* viability (Lipovsky et al. 2011).

It is also important to mention that the fungal toxicity is generally lower than in bacteria (Rathnayakea et al. 2012; Hanus and Harris 2013). Ravichandran et al. (2016) evaluated the antimicrobial activity of biosynthesized AgNPs and found that tested bacteria were more susceptible to the nanoparticles than the fungus *Aspergillus versicolor*. Some biological mechanisms may contribute to fungal tolerance such as production of pigments and other substances, including metal chelation and secretion as glutathione and oxalate (Pócsi 2011). Researchers observed formation of droplets in some fungi after contact with AgNPs suggesting that this event could serve as a water and metabolite reservoir necessary for fungal growth (Ogar et al. 2015). Those authors also proposed that the ability of fungi to produce low-molecular-weight organic acids could lead metal-complex formation as a protective mechanism.

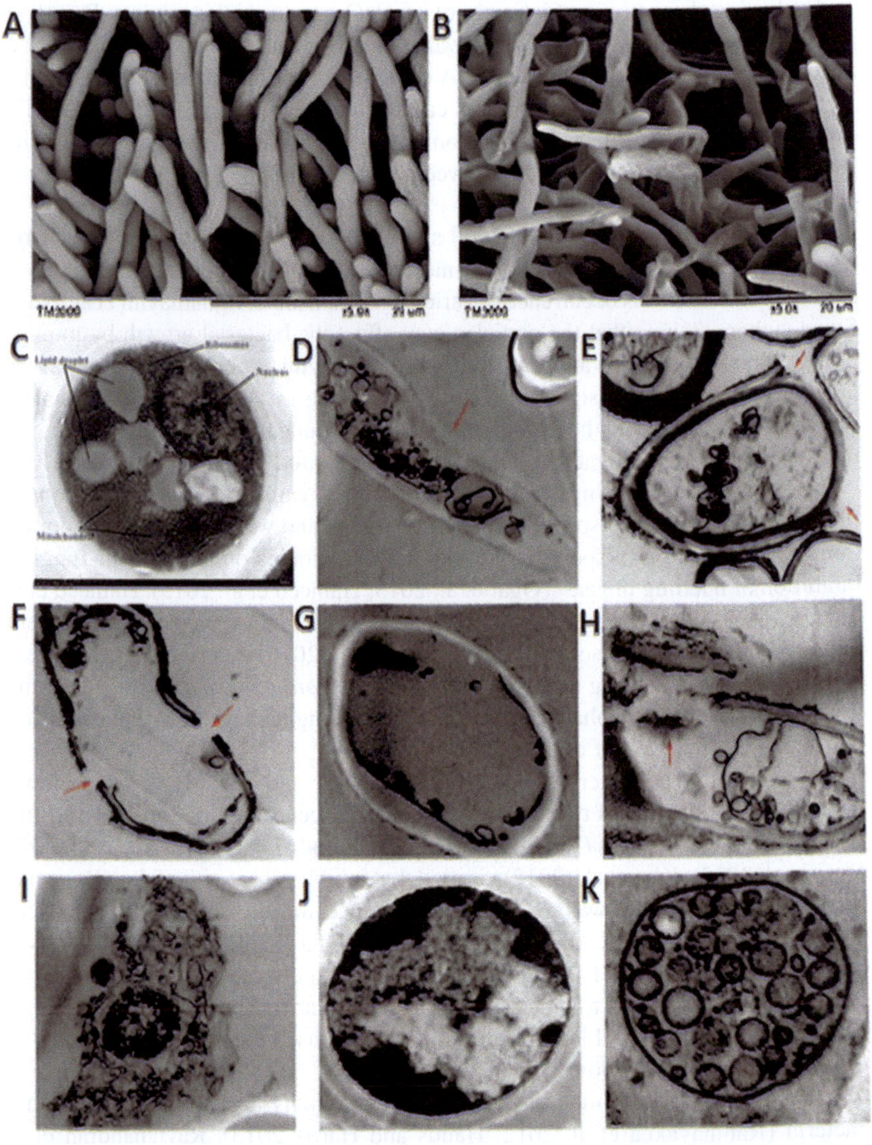

Fig. 15.2 Electron microscopy observations showing the effect of silver nanoparticles (AgNPs) on *Trichosporon asahii*. On scanning electron microscopy images, it is possible to see a normal micelial growth on untreated fungus (**a**) and an irregular growth after treatment with AgNPs solution. Transmission electron microscopy images (**c–k**) allow viewing the cell structure intact of *T. asahii* without the nanoparticles (**c**) and the presence of the following damage when the fungus was treated with AgNPs: undefined, fallen, and ruptured cell wall (**d, e** and **f**, respectively), cell emptying due to shrinkage of cell membrane and degradation of organelles (**g**), rupture of cell membrane (**h**), absence of cell wall and cell membrane (**i**), organelle deterioration (**j**), and occurrence of multivesicular bodies. *Arrows* indicate the effects (Images reproduced from Xia et al. (2016), doi 10.1016/j.jmii.2014.04.013, under the terms of the Creative Commons Attribution-Non-Commercial-NoDerivatives License (CC BY NC ND))

Considering this, some strategies can be employed by their functionalization with different compounds or adding to other structures. Suresh et al. (2016b) reported better antifungal activity of nickel oxide nanoparticles after their functionalization with 5-amino-2-mercaptobenzimidazole, which provided an increase on dispersibility of particles enhancing their interaction with membrane and intracellular proteins of the fungus *A. flavus*. Nanofibers with antifungal functionality are another option to improve activity of metal nanoparticles against fungi (Ifuku et al. 2015; Quirós et al. 2016; Tan et al. 2016). Ifuku et al. (2015) have successfully incorporated AgNPs on chitin nanofiber surfaces, resulting in a marked reduction in spore germination of several phytopathogenic fungi. Graphene is a single layer of graphite with carbon atoms hexagonally arranged and has certain features that make it attractive as a drug delivery (Liu et al. 2013). Recent research offers the possibility of synthesis of graphene oxide containing nanoparticles of metals in order to improve their antifungal properties (Li et al. 2013; Pusty et al. 2016).

15.3.2 Antifungal Applications of Metal Nanoparticles

Antifungal properties of metallic NPs favor their use in various applications and may be considered promising for prevention and treatment of fungal infections in different situations. In order to solve problems about fungal resistance to commercial antimicrobial agents, studies with metallic nanoparticles have been performed against clinical pathogens. The antifungal activity of AgNPs was evaluated toward *Candida* sp., *Aspergillus niger*, *A. fumigatus*, and *A. flavus* (Rajeshkumar et al. 2016). In addition to this activity, the authors found a synergistic effect of these particles with the antifungal ketoconazole. Another study reported that plastic catheters coated with AgNPs effectively inhibited the growth of *C. albicans* (Roe et al. 2008).

Some studies have demonstrated that metal nanoparticles could be used for controlling plant diseases (Alghuthaymi et al. 2015). The antifungal ability of AgNPs was confirmed to control the following plant pathogenic fungi: *Colletotrichum coccodes*, *Monilinia* sp., *Pyricularia* sp. (Lee et al. 2013), *Gloeophyllum abietinum*, *G. trabeum*, *Chaetomium globosum*, *Phanerochaete sordida* (Narayanan and Park 2014), *Colletotrichum* spp. (Lamsal et al. 2011), *Alternaria alternata*, *Sclerotinia sclerotiorum*, *Macrophomina phaseolina*, *Rhizoctonia solani*, *Botrytis cinerea*, and *Curvularia lunata* (Krishnaraj et al. 2012). Kanhed et al. (2014) proved the potential antifungal activity of CuNPs against the phytopathogens *A. alternata*, *C. lunata*, *Fusarium oxysporum*, and *Phoma destructiva*.

Ogar et al. (2015) found that nanosized Ag exhibited strong antimicrobial activity against common indoor fungal species such as *A. fumigatus*, *Cladosporium cladosporioides*, *Chaetomium globosum*, *Penicillium brevicompactum*, and *Stachybotrys chartarum*. Besides, gypsum drywall sprayed with these nanoparticles reduced the growth of *C. globosum* and *S. chartarum*. Similarly, Hochmannova and Vytrasova (2010) showed that an interior paint containing ZnO nanoparticles was effective in control of *A. niger* and *Penicillium chrysogenum*.

15.3.3 Metal Nanoparticles as Antiviral Agents

Viruses are a major cause of disease and death in the world. Viruses infect all forms of cellular life including both eukaryotes and prokaryotes and can be found in all different environmental places. Although some viral diseases that caused massive death, such as smallpox, have been eradicated with the help of the vaccines, no current prevention or treatment is available for diverse pressing and emerging viral pathogens. In this regard, metallic nanoparticles have been proposed as a novel therapeutic approach against lethal viruses (Galdiero et al. 2011; Khandelwal et al. 2014).

Ag and Au nanoparticles have been essentially tested for antiviral activity, and several viruses like HIV-1, hepatitis B virus, influenza virus, and Tacaribe virus, among others, can be inhibited by AgNPs ranging 1–100 nm (Khandelwal et al. 2014). The effectiveness of AgNPs smaller than 100 nm has been investigated against hemorrhagic fever viruses, with a dose-dependent reduction of $TCID_{50}$ for Tacaribe virus reaching 100% neutralization with 50 mg/mL of 10 nm nanoparticles (Speshock and Hussain 2009). They also showed that AgNPs have an inhibitory activity on cathepsin B, which have an essential role on Ebola virus replication within the host cells.

The size of the nanoparticles may have a substantial effect on their effectiveness, since the direct interaction with the viral surface proteins and the genome of the virus appear as important factors for the antiviral activity (Galdiero et al. 2011). As the size of the nanoparticle is smaller, more interaction with viral molecules and more inhibition occur. Smaller nanoparticles can also enter into the host cell and then enter the viral genome where they block the cellular factors and/or viral factors, which help in the viral replication (Fig. 15.3).

15.3.3.1 Antiviral Applications of Metal Nanoparticles

Lara et al. (2010) investigated the specific mode of action of AgNPs against HIV-1. These nanoparticles may exert anti-HIV activity at an early stage of viral replication, most likely as a virucidal agent or as an inhibitor of viral entry. The antiviral activity of AgNPs results from their inhibition of the interaction between gp120 and the target cell membrane receptors. AgNPs bind to gp120 in a manner that they prevent CD4-dependent virion binding, fusion, and infectivity, acting as an effective virucidal agent against cell-free virus (laboratory strains, clinical isolates, T and M tropic strains, and resistant strains) and cell-associated virus. Moreover, AgNPs prevent postentry stages of the HIV-1 life cycle. These properties indicate that AgNPs have a broad-spectrum antiviral effect not predisposed to inducing resistance that could be used preventively against a wide variety of circulating HIV-1 strains.

The antiviral activity of AgNPs and AuNPs capped with mercaptoethane sulfonate was tested against wild-type herpes simplex virus (HSV) type 1. The binding of HSV glycoprotein C to heparan sulfate is the key interaction during attachment of the virus to the cell surface. The viral inhibition suggests that AgNPs and AuNPs

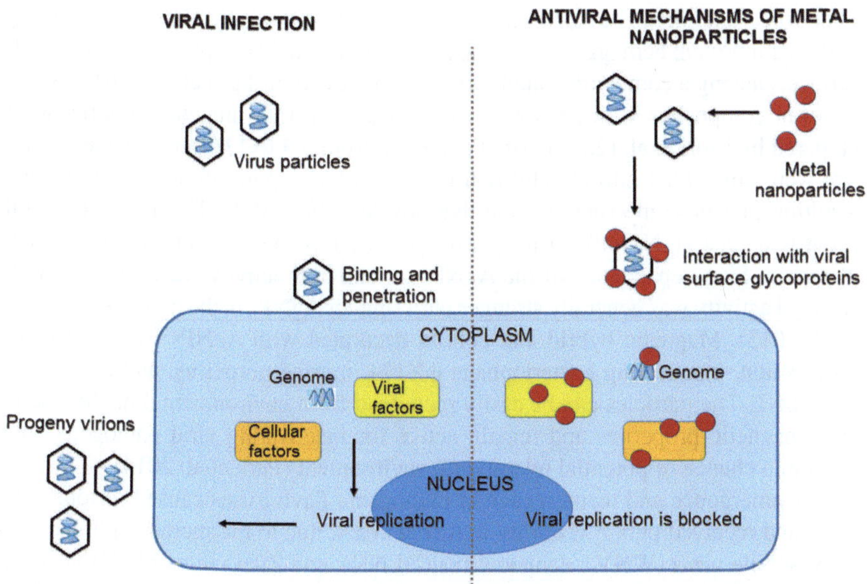

Fig. 15.3 Antiviral effects of metal nanoparticles. The antiviral activity of metal nanoparticles has been attributed to their interaction of surface glycoproteins of the viral envelope. Smaller nanoparticles can enter the host cell and bind to viral and cellular factors blocking the mechanism of viral replication

capped with mercaptoethane sulfonate block the attachment of HSV-1 to the host cell based on their ability to mimic cell-surface-receptor heparan sulfate. Further mechanistic studies revealed that capped Au-NPs interfere with viral attachment, entry, and cell-to-cell spread, thus preventing subsequent viral infection in a multimodal manner (Baram-Pinto et al. 2010).

AgNPs have a size-dependent interaction with HSV types 1 and 2 and with human parainfluenza virus type 3 (Gaikwad et al. 2013). Five fungal cell filtrates (*Alternaria* sp., *F. oxysporum*, *Curvularia* sp., *C. indicum*, and *Phoma* sp.) supported the synthesis of AgNPs when treated with aqueous $AgNO_3$, and their antiviral activity was dependent on the production system used. A consistent decrease in replication efficiency for HSV-1 and HPIV-3 and a minor effect on the replication of HSV-2 were observed. AgNPs produced by *F. oxysporum*, having a smaller size than the other AgNPs, showed the strongest inhibitory activity, especially toward HSV-1 and HPIV-3. Silver nanoparticles are capable of reducing viral infectivity, probably by blocking interaction of the virus with the cell, which might depend on the size and zeta potential of the AgNPs. Smaller-sized nanoparticles were able to inhibit the infectivity of the viruses analyzed.

AuNPs of 2 and 14 nm were functionalized with sialic acid and tested against influenza virus. The activity was dependent on the particle size of the interacting ligand/receptor molecules, since only the 14 nm AuNPs strongly bind to the protein hemagglutinin on the virus surface at concentrations in the nanomolar range.

Therefore, functionalized AuNPs are capable to reduce viral infection, as the binding of the viral protein hemagglutinin is mediated by sialic acid receptors on the host cell surface, causing a competitive inhibition of viral infection (Papp et al. 2010). AgNP/chitosan composites with antiviral activity against H1N1 influenza A virus were prepared by Mori et al. (2013). AgNPs ranging from 3.5 to 12.9 nm average diameters were embedded into the chitosan matrix without aggregation or size change, resulting in nanocomposites as floc-like powders. For all AgNPs tested, antiviral activity against H1N1 influenza A virus increased as the concentration of AgNPs increased. Size dependence of the AgNPs on antiviral activity was also observed; antiviral activity was generally stronger with smaller AgNPs in the composites (Mori et al. 2013). Magnetic hybrid and colloid decorated with AgNPs were tested for disinfection, inactivating bacteriophage ϕX174, murine norovirus, and adenovirus serotype 2. The particles can be easily recovered from environmental media due to their magnetic properties and remain active for inactivating viral pathogens with minimum chance of potential release into environment (Park et al. 2014).

The emergence and reemergence of pathogenic flaviviruses caused serious concerns and received public visibility in recent years, due to the increasing incidence of West Nile virus (WNV), dengue virus (DENV), and Zika virus (ZIKV). Despite vaccines are under development, the prevention and control essentially depend on effective vector control measures. A recent study showed that AgNPs act as inhibitors of the production of dengue viral envelope (E) protein in Vero cell cultures and downregulated the expression of dengue viral E gene (Murugan et al. 2015). This finding confirms the broad antiviral spectrum of AgNPs and opens a new perspective to combat flaviviruses.

15.4 Conclusion and Perspectives

Several metal nanoparticles display antimicrobial properties against a broad range of microorganisms, including bacteria, fungi, and viruses. Several important microbial pathogens have been effectively inhibited by metal nanoparticles, often in a size-dependent manner, as smaller particles facilitate the entry to target cells. AgNPs have been mostly studied, and their broad antimicrobial activity has been described for major pathogenic species of bacteria and against several viruses and fungi. However, there is a growing interest on the antimicrobial activities of other metal nanoparticles. The amount of scientific evidence on their biological activities and their unique properties make metal nanoparticles useful tools as broad range antimicrobial agents.

Despite metal nanoparticles are considered as valuable tools for controlling various pathogens, there are concerns about the release of metal nanoparticles to the environment, since they can generate threats to human health and ecological effects. In this regard, innovative strategies for application of metal nanoparticles have been recently proposed, including the use of magnetic and decorated nanoparticles that are effective antimicrobial materials. The use of metal nanoparticles supported on different materials may allow a controlled release, maintaining their antimicrobial effect and reducing the risks of environmental contamination.

Nanotechnology is an exciting emerging field, and the use of metal nanoparticles as an alternative to combat microbial pathogens has provided interesting strategies to conventional therapy and may help to reduce the impacts of infectious diseases.

References

Abkhoo J, Panjehkeh N. Evaluation of antifungal activity of silver nanoparticles on *Fusarium oxysporum*. Int J Inf Secur. 2016; doi:10.17795/iji-41126.

Ahmad T, Wania IA, Lone IH, Ganguly A, Manzoorb N, Ahmad A, Ahmedc J, Al-Shihri AS. Antifungal activity of gold nanoparticles prepared by solvothermal method. Mater Res Bull. 2013;48:12–20.

Ahmad A, Wei Y, Syed F, Tahir K, Taj R, Khan AU, Hameed MU, Yuan Q. Amphotericin B-conjugated biogenic silver nanoparticles as an innovative strategy for fungal infections. Microb Pathog. 2016;99:271–81.

Alghuthaymi MA, Almoammar H, Rai M, Said-Galiev E, Abd-Elsalam KA. Myconanoparticles: synthesis and their role in phytopathogens management. Biotechnol Biotechnol Equip. 2015;29:221–36.

Alzahrani E, Ahmed RA. Synthesis of copper nanoparticles with various sizes and shapes: application as a superior non-enzymatic sensor and antibacterial agent. Int J Electrochem Sci. 2016;11:4712–23.

Anantha A, Dharaneedharanb S, Heob MS, Mok YS. Copper oxide nanomaterials: synthesis, characterization and structure-specific antibacterial performance. Chem Eng J. 2015;262:179–88.

Ashajyothi C, Prabhurajeshwar C, Handral HK, Kelmani CR. Investigation of antifungal and anti-mycelium activities using biogenic nanoparticles: an eco-friendly approach. Environ Nanotechnol Monit Manag. 2016;5:81–7.

Baram-Pinto D, Shukla N, Gedanken A, Sarid R. Inhibition of HSV-1 attachment, entry, and cell-to-cell spread by functionalized multivalent gold nanoparticles. Small. 2010;6:1044–50.

Baxi SN, Portnoy JM, Larenas-Linnemann D, Phipatanakul W. Exposure and health effects of fungi on humans. J Allerg Clin Immunol: In Pract. 2016;4:396–404.

Beyth N, Houri-Haddad Y, Domb A, Khan W, Hazan R. Alternative antimicrobial approach: Nano-antimicrobial materials. Evid-Based Complement Altern Med. 2015;2015:246012.

Bogutska KI, Sklyarov YP, Prylutskyy YI. Zinc and zinc nanoparticles: biological role and application in biomedicine. Ukr Bioorg Acta. 2013;1:9–16.

Bonetta S, Motta F, Strini A, Carraro E. Photocatalytic bacterial inactivation by TiO_2-coated surfaces. AMB Express. 2013;3:59.

Brandelli A. Nanostructures as promising tools for delivery of antimicrobial peptides. Mini Rev Med Chem. 2012;12:731–41.

Brandelli A, Lopes NA, Boelter JF. Food applications of nanostructured antimicrobials. In: Grumezescu AM, editor. Food preservation. London: Elsevier; 2017. p. 35–74.

Cakić M, Glišić S, Nikolić G, Nikolić GM, Cakić K, Cvetinov M. Synthesis, characterization and antimicrobial activity of dextran sulphate stabilized silver nanoparticles. J Mol Struct. 2016;1110:156–61.

Carbone M, Donia DT, Sabbatella G, Silver RA. Nanoparticles in polymeric matrices for fresh food packaging. J King Saud Univ Sci. 2016;28:273–9.

Chiriac V, Stratulat DN, Calin G, Nichitus S, Burlui V, Stadoleanu C, Popa M. Antimicrobial property of zinc based nanoparticles. IOP Conf Ser: Mater Sci Eng. 2016;133:012055.

Conte A, Longano D, Costa C, Ditaranto N, Ancona A, Cioffi N, Scrocco C, Sabbatini L, Contòd F, Del Nobile MA. A novel preservation technique applied to fiordilatte cheese. Innov Food Sci Emerg Technol. 2013;19:158–65.

Cubillo AE, Pecharromán C, Aguilar E, Santarén J, Moya JS. Antibacterial activity of copper monodispersed nanoparticles into sepiolite. J Mater Sci. 2006;41:5208–12.

Di DR, He ZZ, Sun ZQ, Liu J. A new nano-cryosurgical modality for tumor treatment using biodegradable MgO nanoparticles. Nanomedicine. 2012;8:1233–41.

Dubey P, Brushan B, Sachdev A, Matai I, Kumar SU, Gopinath P. Silver-nanoparticle-incorporated composite nanofibers for potential wound-dressing applications. J Appl Polym Sci. 2015;132:42473.

Durairaj B, Muthu S, Xavier T. Antimicrobial activity of *Aspergillus niger* synthesized titanium dioxide nanoparticles. Adv Appl Sci Res. 2015;6:45–8.

Dykmana L, Khlebtsov N. Gold nanoparticles in biomedical applications: recent advances and perspectives. Chem Soc Rev. 2012;41:2256–82.

El-Diasty EM, Ahmed MA, Okasha N, Mansour SF, El-Dek SI, El-Khalek HMA, Youssif MH. Antifungal activity of zinc oxide nanoparticles against dermatophytic lesions of cattle. Rom J Biophys. 2013;23:191–202.

Elumalai K, Velmurugan S. Green synthesis, characterization and antimicrobial activities of zinc oxide nanoparticles from the leaf extract of *Azadirachta indica* (L.). Appl Surf Sci. 2015;345:329–36.

Eskandari M, Haghighi N, Ahmadi V, Haghighi F, Mohammadi SR. Growth and investigation of antifungal properties of ZnO nanorod arrays on the glass. Phys B Condens Matter. 2011;406:112–4.

Espitia PJP, Soares NFF, Coimbra JSR, de Andrade NJ, Cruz RS. Zinc oxide nanoparticles: synthesis, antimicrobial activity and food packaging applications. Food Bioprocess Technol. 2012;5:1447–64.

Esteban-Tejeda L, Malpartida F, Esteban-Cubillo A, Pecharromán C, Moya JS. The antibacterial and antifungal activity of a soda-lime glass containing silver nanoparticles. Nanotechnology. 2009;20:85103.

Faraji AM, Wipf P. Nanoparticles in cellular drug delivery. Bioorg Med Chem. 2009;17:2950–62.

Faria AF, Martinez DST, Meira SMM, Moraes ACM, Brandelli A, Filho AGS, Alves OL. Antiadhesion and antibacterial activity silver nanoparticles supported on graphene oxide sheets. Colloid Surf B: Biointerf. 2015;113:115–24.

Farokhzad OC, Langer R. Nanomedicine: developing smarter therapeutic and diagnostic modalities. Adv Drug Deliv Rev. 2006;58:1456–9.

Franci G, Falanga A, Galdiero S, Palomba L, Rai M, Morelli G, Galdiero M. Silver nanoparticles as potential antibacterial agents. Molecules. 2015;20:8856–74.

Gaikwad S, Ingle A, Gade A, Rai M, Falanga A, Incoronato N, Russo L, Galdiero S, Galdiero M. Antiviral activity of mycosynthesized silver nanoparticles against herpes simplex vírus and human parainfluenza virus type 3. Int J Nanomedicine. 2013;8:4303–14.

Galdiero S, Falanga A, Vitiello M, Cantisani M, Marra V, Galdiero M. Silver nanoparticles as potential antiviral agents. Molecules. 2011;16:8894–918.

Gauthier GM, Keller NP. Crossover fungal pathogens: the biology and pathogenesis of fungi capable of crossing kingdoms to infect plants and humans. Fungal Genet Biol. 2013;61:146–57.

Gawande MB, Goswami A, Felpin FX, Asefa T, Huang T, Silva R, Zou X, Zboril R, Varma RS. Cu and Cu based nanoparticles: synthesis and applications in catalysis. Chem Rev. 2016;116:3722–811.

Gowri SR, Gandhi R, Sundrarajan M. Structural, optical, antibacterial and antifungal properties of zirconia nanoparticles by biobased protocol. J Mater Sci Technol. 2014;30:782–90.

Gunalan S, Sivaraj R, Rajendran V. Green synthesized ZnO nanoparticles against bacterial and fungal pathogens. Progr Nat Sci: Mater Int. 2012;22:693–700.

Hameed ASH, Karthikeyan C, Kumar VS, Kumaresan S, Sasikumar S. Effect of Mg^{2+}, Ca^{2+}, Sr^{2+} and Ba^{2+} metal ions on the antifungal activity of ZnO nanoparticles tested against *Candida albicans*. Mater Sci Eng C. 2015;52:171–7.

Hanus MJ, Harris AT. Nanotechnology innovations for the construction industry. Prog Mater Sci. 2013;58:1056–102.

He L, Liu Y, Mustapha A, Lin M. Antifungal activity of zinc oxide nanoparticles against *Botrytis cinerea* and *Penicillium expansum*. Microbiol Res. 2011;166:207–15.

He Y, Ingudam S, Reed S, Gehring A, Strobaugh TP Jr, Irwin P. Study on the mechanism of antibacterial action of magnesium oxide nanoparticles against foodborne pathogens. J Nanobiotechnol. 2016;14:1–9.

Hochmannova L, Vytrasova J. Photocatalytic and antimicrobial effects of interior paints. Progr Org Coat. 2010;67:1–5.

Hossain F, Perales-Perez OJ, Hwang S, Román F. Antimicrobial nanomaterials as water disinfectant: applications, limitations and future perspectives. Sci Total Environ. 2014;466(467):1047–59.

Hossain KMZ, Patel U, Ahmed I. Development of microspheres for biomedical applications: a review. Prog Biomater. 2015;4:1.

Hsu LY, Wijaya L, Ng EST, Gotuzzo E. Tropical fungal infections. Infect Dis Clin N Am. 2012;26:497–512.

Hussein-Al-Ali SH, El Zowalaty ME, Hussein MZ, Geilich BM, Webster TJ. Synthesis, characterization, and antimicrobial activity of an ampicillin-conjugated magnetic nanoantibiotic for medical applications. Int J Nanomedicine. 2014;9:3801–14.

Ifuku S, Tsukiyama Y, Yukawa T, Egusa M, Kaminaka H, Izawa H, Morimotoc M, Saimoto H. Facile preparation of silver nanoparticles immobilized on chitin nanofiber surfaces to endow antifungal activities. Carbohydr Polym. 2015;117:813–7.

Ingle AP, Duran N, Rai M. Bioactivity, mechanism of action, and cytotoxicity of copper-based nanoparticles: a review. Appl Microbiol Biotechnol. 2014;98:1001–9.

Jin T, He YP. Antibacterial activities of magnesium oxide (MgO) nanoparticles against foodborne pathogens. J Nanopart Res. 2011;13:6877–85.

Jin T, Sun D, Su Y, Zhang H, Sue HJ. Antimicrobial efficacy of zinc oxide quantum dots against *Listeria monocytogenes*, *Salmonella enteritidis* and *Escherichia coli* O157:H7. J Food Sci. 2009;74:46–52.

Kanhed P, Birla S, Gaikwad S, Gade A, Seabra AB, Rubilar O, Duran N, Rai M. In vitro antifungal efficacy of copper nanoparticles against selected crop pathogenic fungi. Mater Lett. 2014;115:13–7.

Käosaar S, Kahru A, Mantecca P, Kasemets K. Profiling of the toxicity mechanisms of coated and uncoated silver nanoparticles to yeast *Saccharomyces cerevisiae* BY4741 using a set of its 9 single-gene deletion mutants defective in oxidative stress response, cell wall or membrane integrity and endocytosis. Toxicol In Vitro. 2016;35:149–62.

Katwal R, Kaur H, Sharma G, Naushad M, Pathania D. Electrochemical synthesized copper oxide nanoparticles for enhanced photocatalytic and antimicrobial activity. J Ind Eng Chem. 2015;31:173–84.

Khandelwal N, Kaur G, Kumar N, Tiwari A. Application of silver nanoparticles in viral inhibition: a new hope for antivirals. Dig J Nanomater Biostruct. 2014;9:175–96.

Kharissova OV, Dias HVR, Kharisov BI, Pérez BO, Pérez VMJ. The greener synthesis of nanoparticles. Trends Biotechnol. 2013;31:240–8.

Kim KJ, Sung WS, Suh BK, Moon SK, Choi JS, Kim JG, Lee DG. Antifungal activity and mode of action of silver nano-particles on *Candida albicans*. Biometals. 2009;22:235–42.

Kim SW, Jung JH, Lamsal K, Kim YS, Min JS, Lee YS. Antifungal effects of silver nanoparticles (AgNPs) against various plant pathogenic fungi. Mycobiology. 2012;40:53–8.

Knetsch ML, Koole LH. New strategies in the development of antimicrobial coatings: the example of increasing usage of silver and silver nanoparticles. Polymers. 2011;3:340–66.

Krishnamoorthy K, Manivannan G, Kim SJ, Jeyasubramanian K, Premanathan M. Antibacterial activity of MgO nanoparticles based on lipid peroxidation by oxygen vacancy. J Nanopart Res. 2012;14:1063.

Krishnaraj C, Ramachandran R, Mohan K, Kalaichelvan PT. Optimization for rapid synthesis of silver nanoparticles and its effect on phytopathogenic fungi. Spectrochim Acta A. 2012;93:95–9.

Kumar N, Palmer GR, Shah V, Walker VK. The effect of silver nano-particles on seasonal change in arctic tundra bacterial and fungal assemblages. PLoS One. 2014;9:e99953. doi:10.1371/journal.pone.0099953.

Lamsal K, Kim SW, Jung JH, Kim YS, Kim KS, Lee YS. Application of silver nanoparticles for the control of *Colletotrichum* species *in vitro* and pepper anthracnose disease in field. Mycobiology. 2011;39:194–9.

Lara HH, Ayala-Nuñez N, Ixtepan-Turrent L, Rodriguez-Padilla C. Mode of antiviral action of silver nanoparticles against HIV-1. J Nanobiotechnol. 2010;8:1.

Lee KJ, Park SH, Govarthanan M, Hwang PH, Seo YS, Cho M, Lee WH, Lee JY, Kamala-Kannan S, Oh BT. Synthesis of silver nanoparticles using cow milk and their antifungal activity against phytopathogens. Mater Lett. 2013;105:128–31.

Lemire JA, Harrison JJ, Turner RJ. Antimicrobial activity of metals: mechanisms, molecular targets and applications. Nat Rev Microbiol. 2013;11:371–84.

León-Silva S, Fernández-Luqueño F, López-Valdez F. Silver nanoparticles (AgNP) in the environment: a review of potential risks on human and environmental health. Water Air Soil Pollut. 2016;227:306.

Letfullin RR, Iversen CB, George TF. Modeling nanophotothermal therapy: kinetics of thermal ablation of healthy and cancerous cell organelles and gold nanoparticles. Nanomed: Nanotechnol Biol Med. 2011;7:137–45.

Leung YH, Ng AM, Xu X, Shen Z, Gethings LA, Wong MT, Chan CM, Guo MY, Ng YH, Djurišić AB, Lee PK, Chan WK, Yu LH, Phillips DL, Ma AP, Leung FC. Mechanisms of antibacterial activity of MgO: non-ROS mediated toxicity of MgO nanoparticles towards *Escherichia coli*. Small. 2014;10:1171–83.

Li C, Wang X, Chen F, Zhang C, Zhi X, Wang K, Cui D. The antifungal activity of graphene oxide-silver nanocomposites. Biomaterials. 2013;34:3882–90.

Li X, Robinson SM, Gupta A, Saha K, Jiang Z, Moyano DF, Sahar A, Riley MA, Rotello VM. Functional gold nanoparticles as potent antimicrobial agents against multi-drug-resistant bacteria. ACS Nano. 2014;8:10682–6.

Lima E, Guerra R, Lara V, Guzmán A. Gold nanoparticles as efficient antimicrobial agents for *Escherichia coli* and *Salmonella typhi*. Chem Cent J. 2013;7:11.

Lipovsky A, Nitzan Y, Gedanken A, Lubart R. Antifungal activity of ZnO nanoparticles-the role of ROS mediated cell injury. Nanotechnology. 2011;22:105101.

Liu J, Cui L, Losic D. Graphene and graphene oxide as new nanocarriers for drug delivery applications. Acta Biomater. 2013;9:9243–57.

Longano D, Ditaranto N, Cioffi N, Di Niso F, Sibillano T, Ancona A, Conte A, Del Nobile MA, Sabbatini L, Torsi L. Analytical characterization of laser-generated copper nanoparticles for antibacterial composite food packaging. Anal Bioanal Chem. 2012;403:1179–86.

Luo ZS, Ye QY, Li D. Influence of nano-TiO_2 modified LDPE film packaging on quality of strawberry. Mod Food Sci Technol. 2013;29:2340–4.

Ma H, Williamsb PL, Diamond SA. Ecotoxicity of manufactured ZnO nanoparticles - a review. Environ Pollut. 2013;172:76–85.

Maiti S, Krishnan D, Barman G, Ghosh SK, Laha JK. Antimicrobial activities of silver nanoparticles synthesized from *Lycopersicon esculentum* extract. J Anal Sci Technol. 2014;5:40.

Mallikarjuna K, Sushma NJ, Reddy BVS, Narasimha G, Raju BDP. Palladium nanoparticles: single-step plant-mediated green chemical procedure using *Piper betle* leaves broth and their anti-fungal studies. Int J Chem Anal Sci. 2013;4:14–8.

Mamonova IA, Matasov MD, Babushkina V, Losev OE, Ye G, Chebotareva EV, Gladkova Y, Borodulina V. Study of physical properties and biological activity of copper nanoparticles. Nanotechnol Russia. 2013;8:303–8.

Mirhosseini M, Afzali M. Investigation into the antibacterial behavior of suspensions of magnesium oxide nanoparticles in combination with nisin and heat against *Escherichia coli* and *Staphylococcus aureus* in milk. Food Control 2016;15:208–15.

Mody VV, Siwale R, Singh A, Mody HR. Introduction to metallic nanoparticles. J Pharm Bioallied Sci. 2010;2:282–9.

Morace G, Perdoni F, Borghi E. Antifungal drug resistance in *Candida* species. J Glob Antimicrob Resist. 2014;2:254–9.

Mori Y, Ono T, Miyahira Y, Nguyen VQ, Matsui T, Ishihara M. Antiviral activity of silver nanoparticle/chitosan composites against H1N1 influenza A virus. Nanoscale Res Lett. 2013;8:93.

Murphy M, Ting K, Zhang X, Soo C, Zheng Z. Current development of silver nanoparticle preparation, investigation, and application in the field of medicine. J Nanomater. 2015;2015:696918.

Murugan K, Dinesh D, Paulpandi M, Althbyani AD, Subramaniam J, Madhiyazhagan P. Nanoparticles in the fight against mosquito-borne diseases: bioactivity of *Bruguiera* cylindrical-synthesized nanoparticles against dengue virus DEN-2 (in vitro) and its mosquito vector *Aedes aegypti* (Diptera: Culicidae). Parasitol Res. 2015;114:4349–61.

Narayanan KB, Park HH. Antifungal activity of silver nanoparticles synthesized using turnip leaf extract (*Brassica rapa* L.) against wood rotting pathogens. Eur J Plant Pathol. 2014;140:185–92.

Norman RS, Stone JW, Gole A, Murphy CJ, Sabo-Attwood TL. Targeted photothermal lysis of the pathogenic bacteria, *Pseudomonas aeruginosa*, with gold nanorods. Nano Lett. 2008;8:302–6.

Ogar A, Tylko G, Turnau K. Antifungal properties of silver nanoparticles against indoor mould growth. Sci Total Environ. 2015;521(522):305–14.

Othman SH, Salam NRA, Zainal N, Basha RK, Talib RA. Antimicrobial activity of TiO_2 nanoparticle-coated film for potential food packaging applications. Int J Photoenergy. 2014;2014:945930.

Papp SC, Ludwig K, Roskamp M, Bottcher C, Schlecht S, Herrmann A, Haag R. Inhibition of influenza virus infection by multivalent sialic-acid-functionalized gold nanoparticles. Small. 2010;6:2900–6.

Park SJ, Park HH, Kim SY, Kim SJ, Woo K, Ko GP. Antiviral properties of silver nanoparticles on a magnetic hybrid colloid. Appl Environ Microbiol. 2014;80:2343–50.

Payne JN, Waghwani HK, Connor MG, Hamilton W, Tockstein S, Moolani H, Chavda F, Badwaik V, Lawrenz MB, Dakshinamurthy R. Novel synthesis of kanamycin conjugated gold nanoparticles with potent antibacterial activity. Front Microbiol. 2016.

Pinto RJB, Almeida A, Fernandes SCM, Freire CSR, Silvestre AJD, Pascoal Neto C, Trindade T. Antifungal activity of transparent nanocomposite thin films of pullulan and silver against *Aspergillus niger*. Colloid Surf B: Biointerf. 2013;103:143–8.

Pócsi I. Toxic metal/metalloid tolerance in fungi - a biotechnology-oriented approach. In: Bánfalvi G, editor. Cellular Effects of Heavy Metals. Dordrecht: Springer Science+Business Media B.V.; 2011. p.31–58.

Prabhu S, Poulose EK. Silver nanoparticles: mechanism of antimicrobial action, synthesis, medical applications, and toxicity effects. Int Nano Lett. 2012;2:32.

Pradhan N, Singh S, Ojha N. Facets of nanotechnology as seen in food processing, packaging, and preservation industry. BioMed Research International. 2015; Article ID 365672, doi:10.1155/2015/365672.

Prasad V, Shaikh AJ, Kathe AA, Bisoyi DK, Verma AK, Vigneshwaran N. Functional behaviour of paper coated with zinc oxide-soluble starch nanocomposites. J Mater Process Technol. 2010;210:1962–7.

Pusty M, Rana AK, Kumar Y, Sathe V, Sen S, Shirage P. Synthesis of partially reduced graphene oxide/silver nanocomposite and its inhibitive action on pathogenic fungi grown under ambient conditions. Chem Select. 2016;1:4235–45.

Quirós J, Gonzalo S, Jalvo B, Boltes K, Perdigón-Melón JA, Rosa R. Electrospun cellulose acetate composites containing supported metal nanoparticles for antifungal membranes. Sci Total Environ. 2016;563(564):912–20.

Raffi M, Mehrwan S, Bhatti MT, Javed IA, Abdul H, Wasim Y, Hasan. Investigations into the antibacterial behavior of copper nanoparticles against *Escherichia coli*. Ann Microbiol. 2010;60:75–80.

Rai M, Yadav A, Gade A. Silver nanoparticles as a new generation of antimicrobials. Biotechnol Adv. 2009;27:76–83.

Rai M, Ingle AP, Gupta I, Brandelli A. Bioactivity of noble metal nanoparticles decorated with biopolymers and their application in drug delivery. Int J Pharm. 2015;496:159–72.

Rajawat S, Qureshi MS. Comparative study on bactericidal effect of silver nanoparticles, synthesized using green technology, in combination with antibiotics on *Salmonella typhi*. J Biomater Nanobiotechnol. 2012;3:480–5.

Rajeshkumar S, Malarkodi C, Vanaja M, Annadurai G. Anticancer and enhanced antimicrobial activity of biosynthesized silver nanoparticles against clinical pathogens. J Mol Struct. 2016;1116:165–73.

Ramyadevi J, Jeyasubramanian K, Marikani A, Rajakumar G, Rahuman AA. Synthesis and antimicrobial activity of copper nanoparticles. Mater Lett. 2012;71:114–6.

Rao NH, Lakshmidevi N, Pammi SVN, Kollu P, Ganapaty S, Lakshmi P. Green synthesis of silver nanoparticles using methanolic root extracts of *Diospyros paniculata* and their antimicrobial activities. Mater Sci Eng C. 2016;62:553–7.

Rathnayakea WGIU, Ismail H, Baharin A, Darsanasiri AGND, Rajapakse S. Synthesis and characterization of nano silver based natural rubber latex foam for imparting antibacterial and antifungal properties. Polym Test. 2012;31:586–92.

Ravichandran V, Vasanthi S, Shalini S, Shah SAA, Harish R. Green synthesis of silver nanoparticles using *Artocarpus altilis* leaf extract and the study of their antimicrobial and antioxidant activity. Mater Lett. 2016;180:264–7.

Ravishankar RV, Jamuna BA. Nanoparticles and their potential application as antimicrobials. In: Méndez V, editor. Science against microbial pathogens, communicating current research and technological advances. Badajoz: Formatex; 2011. p. 197–209.

Roe D, Karandikar B, Bonn-Savage N, Gibbins B, Roullet JB. Antimicrobial surface functionalization of plastic catheters by silver nanoparticles. J Antimicrob Chemother. 2008;61:869–76.

Roshmi T, Soumya KR, Jyothis M, Radhakrishnan EK. Effect of biofabricated gold nanoparticle-based antibiotic conjugates on minimum inhibitory concentration of bacterial isolates of clinical origin. Gold Bull. 2015;48:63–71.

Ruparelia JP, Chatterjee AK, Duttagupta SP, Mukherji S. Strain specificity in antimicrobial activity of silver and copper nanoparticles. Acta Biomater. 2008;4:707–16.

Salunke GR, Ghosh S, Santosh Kumar RJ, Khade S, Vashisth P, Kale T, Chopade S, Pruthi V, Kundu G, Bellare JR. Rapid efficient synthesis and characterization of silver, gold, and bimetallic nanoparticles from the medicinal plant *Plumbago zeylanica* and their application in biofilm control. Int J Nanomedicine. 2014;9:2635–53.

Sequeira S, Cabrita EJ, Macedo MF. Antifungals on paper conservation: an overview. Int Biodeterior Biodegrad. 2012;74:67–86.

Sharma VK, Yngard RA, Lin Y. Silver nanoparticles: green synthesis, and their antimicrobial activities. Adv Colloid Interf Sci. 2009;145:83–96.

Speshock J, Hussain S. Novel nanotechnology-based antiviral agents: silver nanoparticle neutralization of hemorrhagic fever viruses. Air Force Research Laboratory, Unclassified Document 88AWB-2009-4491. 2009.

Stoimenov PK, Klinger RL, Marchin GL, Klabunde KJ. Metal oxide nanoparticles as bactericidal agents. Langmuir. 2002;18:6679–86.

Suresh S, Karthikeyan S, Saravanan P, Jayamoorthy K, Dhanalekshmi KI. Comparison of antibacterial and antifungal activity of 5-amino-2-mercapto benzimidazole and functionalized Ag_3O_4 nanoparticles. Karbala Int J Mod Sci. 2016a;2:129–37.

Suresh S, Karthikeyan S, Saravanan P, Jayamoorthy K. Comparison of antibacterial and antifungal activities of 5 amino 2 mercaptobenzimidazole and functionalized NiO nanoparticles. Karbala Int J Mod Sci. 2016b;2:188–95.

Tamayo LA, Zapata PA, Vejar ND, Azocar MI, Gulppi MA, Zhou X, Thompson GE, Rabagliati FM, Paez MA. Release of silver and copper nanoparticles from polyethylene nanocomposites and their penetration into *Listeria monocytogenes*. Mater Sci Eng C. 2014;40:24–31.

Tan G, Sağlam S, Emül E, Erdönmez D, Sağlam N. Synthesis and characterization of silver nanoparticles integrated in polyvinyl alcohol nanofibers for bionanotechnological applications. Turk J Biol. 2016;40:643–51.

Tang ZX, Lv BF. MgO nanoparticles as antibacterial agent: preparation and activity. Braz J Chem Eng. 2014;31:591–601.

Theivasanthi T, Alagar M. Studies of copper nanoparticles effects on micro-organisms. Int J Phys Sci. 2011;6:3662–71.

Thirumurugan A, Ramachandran S, Gowri AS. Combined effect of bacteriocin with gold nanoparticles against food spoiling bacteria - an approach for food packaging material preparation. Int Food Res J. 2013;20:1909–12.

Toker RD, Kayaman-Apohan N, Kahraman MV. UVcurable nano-silver containing polyurethane based organic–inorganic hybrid coatings. Progr Org Coat. 2013;76:1243–50.

Tran QH, Nguyen VQ, Le AT. Silver nanoparticles: synthesis, properties, toxicology, applications and perspectives. Adv Nat Sci Nanosci Nanotechnol. 2013;4:033001.

Umer A, Naveed S, Ramzan N, Rafique MS, Imran M. A green method for the synthesis of copper nanoparticles using L-ascorbic acid. Rev Matér. 2014;19:197–203.

Velmurugan P, Sivakumar S, Young-Chae S, Seong-Ho J, Pyoung-In Y, Jeong-Min S, Sung-Chul H. Synthesis and characterization comparison of peanut shell extract silver nanoparticles with commercial silver nanoparticles and their antifungal activity. J Ind Eng Chem. 2015;31:51–4.

Velmurugan P, Shim J, Kim K, Oh BT. *Prunus* x *yedoensis* tree gum mediated synthesis of platinum nanoparticles with antifungal activity against phytopathogens. Mater Lett. 2016;174:61–5.

Wang C, Huang X, Deng W, Chang C, Hang R, Tang B. A nano-silver composite based on the ion-exchange response for the intelligent antibacterial applications. Mater Sci Eng C. 2014;41:134–41.

Wani IA, Ahmad T. Size and shape dependant antifungal activity of gold nanoparticles: a case study of *Candida*. Colloid Surf B: Biointerf. 2013;101:162–70.

Warnes SL, Caves V, Keevil CW. Mechanism of copper surface toxicity in *Escherichia coli* O157:H7 and *Salmonella* involves immediate membrane depolarization followed by slower rate of DNA destruction which differs from that observed for Gram-positive bacteria. Environ Microbiol. 2012;14:1730–43.

Weir A, Westerhoff P, Fabricius L, Hristovski K, Goetz N. Titanium dioxide nanoparticles in food and personal care products. Environ Sci Technol. 2012;46:2242–50.

Xia ZK, Ma QH, Li SY, Zhang DQ, Cong L, Tian YL, Yang RY. The antifungal effect of silver nanoparticles on *Trichosporon asahii*. J Microbiol Immunol Infect. 2016;49:182–8.

Yallappa S, Manjanna J, Dhananjaya BL, Vishwanatha U, Ravishankar B, Gururaj H, Niranjana P, Hungund BS. Phytochemically functionalized cu and ag nanoparticles embedded in MWCNTs for enhanced antimicrobial and anticancer properties. Nano-Micro Lett. 2016;8:120–30.

Yao N, Yeung L. Investigation of the performance of TiO_2 photocatalytic coatings. Chem Eng J. 2011;167:13–21.

Yin ZF, Wu L, Yang HG, Su YH. Recent progress in biomedical applications of titanium dioxide. Phys Chem Chem Phys. 2013;15:4844–58.

Yoshimura M, Namura S, Akamaysu H, Horio T. Antimicrobial effects of phototherapy and photochemotherapy in vivo and in vitro. Br J Dermatol. 1995;135:528–32.

Youssef AM, Abdel-Aziz MS. Preparation of polystyrene nanocomposites based on silver nanoparticles using marine bacterium for packaging. J Polym Plast Technol Eng. 2013;52:607–13.

Yu KP, Huang YT, Yang SC. The antifungal efficacy of nano-metals supported TiO_2 and ozone on the resistant *Aspergillus niger* spore. J Hazard Mater. 2013;261:155–62.

Zazo H, Colino CI, Lanao JM. Current applications of nanoparticles in infectious diseases. J Control Release. 2016;224:86–102.

Zhang X. Gold nanoparticles: recent advances in the biomedical applications. Cell Biochem Biophys. 2015;72:771–5.

Zhang LL, Jiang YH, Ding YL, Povey M, York D. Investigation into the antibacterial behaviour of suspensions of ZnO nanoparticles (zno nanofluids). J Nanopart Res. 2007;9:479–89.

Zhang Y, Nayak TR, HongH CW. Biomedical applications of zinc oxide nanomaterials. Curr Mol Med. 2013;13:1633–45.

Chapter 16
Gold Nanoparticles in Molecular Diagnostics and Molecular Therapeutics

Ana S. Matias, Fábio F. Carlos, P. Pedrosa, Alexandra R. Fernandes, and Pedro V. Baptista

Abstract Gold nanoparticles, due to their unique physicochemical properties, are among the most widely used nanoscale-based platforms for molecular diagnostics. The intrinsic chemical stability and apparent lack of toxicity have also prompted for application in therapeutics, e.g., for imaging modalities and as vectorization strategies for molecular modulators, i.e., gene silencing, specific targeting of cellular pathways, etc. Because of their common molecular ground, these approaches have been synergistically coupled together into molecular theranostic systems that allow for radical new in vivo diagnostics modalities with simultaneous tackling of molecular disequilibria leading to disease. Despite this tremendous potential, gold nanoparticle-based systems still have to make their effective translation to the clinics. This chapter focuses on the use of gold nanoparticles for molecular diagnostics and molecular therapeutics and their application in theranostics. Attention is paid to those systems that have moved toward the clinics.

Keywords Nanomedicine • Nanodiagnostics • Molecular diagnostics • Targeted delivery • Nanotheranostics • Molecular therapeutics

Nomenclature

AIE	Aggregation-induced emission
AuNRs	Gold nanorods
CT	Computed tomography
DLS	Dynamic light scattering
DOX	Doxorubicin
EPR	Enhanced permeability and retention effect
LFBs	Lateral flow biosensor

A.S. Matias • F.F. Carlos • P. Pedrosa • A.R. Fernandes • P.V. Baptista (✉)
UCIBIO, Departamento de Ciências da Vida, Faculdade de Ciências e Tecnologia, Universidade Nova de Lisboa, Campus da Caparica, 2829-516 Caparica, Portugal
e-mail: pmvb@fct.unl.pt

M. Rai, R. Shegokar (eds.), *Metal Nanoparticles in Pharma*,
DOI 10.1007/978-3-319-63790-7_16

LSPR	Localized surface plasmon resonance
MTBC	Mycobacterium tuberculosis complex
NASBA	Nucleic acid sequence-based amplification
PA	Photoacoustic imaging
PNB	Plasmonic nanobubbles
QCM	Quartz-crystal microbalance
SERS	Surface-enhanced Raman scattering
siRNA	Small interfering RNA
SNP	Single nucleotide polymorphism
SPR	Surface plasmon resonance
TPL	Two-photon luminescence
US	Ultrasounds

16.1 Introduction

Colloidal gold, defined as a stable solution of sub-micrometer gold particles eluted usually in water, has been used as therapeutic agent since ancient times (Higby 1982). Modern scientific assessment of proprieties of colloidal gold began in 1850 when Michael Faraday produced a ruby red solution while rising pieces of gold leaf onto microscope slides. Faraday was the first to note the light-scattering properties of suspended gold microparticles, which is nowadays called Faraday-Tyndall effect (Faraday 1847). A few decades later, Robert Koch used gold cyanide against the tubercle bacilli (*Mycobacterium tuberculosis*) in culture (Benedek 2004), whose efficacy was later dismissed by subsequent works (Gradmann 2001; Mackenzie and Cantab 1913). In 1959, Richard Feynman's talk at the American Physical Society meeting, "There's plenty of room at the bottom," highlighted the possibility of manipulating matter at the nanoscale – nanotechnology (Feynman 1960). This multidisciplinary area which encompasses chemistry, physics, biology, and technology at nanometer-scale provided for new materials, structures, and devices in several areas, such as medicine (Edwards and Thomas 2007; Roco 2003). One of such nanomaterials is gold nanoparticles (AuNPs). AuNPs exhibit unique nanoscale characteristics (e.g., high surface-to-volume ratio, easiness of bioconjugation, etc.) and physicochemical proprieties (e.g., optical, electrochemical, etc.) that have allowed the development of several strategies for sensing (i.e., DNA, RNA, peptides, ions, etc.) (Baptista et al. 2008; Doria et al. 2012; Tsai et al. 2013; Franco et al. 2015; Nunes Pauli et al. 2015; Zheng et al. 2015), therapy (i.e., cancer) (Cai et al. 2008; Jain et al. 2012), and imaging (Ahn et al. 2013) (e.g., contrast agent). The nanoscale properties and manipulation have put forward new strategies for in vivo diagnostic coupled to therapy modalities – nanotheranostics.

In this chapter, we highlighted three distinct settings: (i) molecular diagnostics, (ii) molecular therapeutics, and (iii) nanotheranostics. For each one of these topics, the most relevant AuNP properties (e.g., optical, electrochemical, vectorization, etc.) are emphasized, and examples of translation toward the clinical setting (e.g., validate against clinical samples, clinical trials, etc.) are discussed.

16.2 AuNPs for Molecular Diagnostics

Molecular diagnostics can be a useful tool to provide qualitative and quantitative information in the genome and proteome (Tsongalis and Silverman 2006). Since the completion of the human genome sequence in 2003, molecular diagnostic has undergone remarkable changes. Millions of genetic variants in the human genome were revealed leading to a better comprehension of the genome architecture and the biological routes of complex diseases (Hofker et al. 2014). The technological advances made in this field have made it possible for diagnostic laboratories to conduct high-accuracy molecular diagnostic procedures, where prevention, early detection, and specific therapeutics are the driving force for a personalized medicine (Alyass et al. 2015). However, application of conventional molecular diagnostic and therapy at point-of-need is still far to be a reality. Integration of nanotechnology within standard methods could in principle circumvent time-consuming and expensive procedures that require specialized laboratory personnel (Debnath et al. 2010). Additionally, this combination will allow miniaturization and translation into portable platforms, allowing their use in nontraditional environments (Debnath et al. 2010).

The last decade of nanotechnology research has introduced numerous approaches for molecular diagnostics. Specifically, nanoparticle-based detection systems have been developed, improved, and integrated with standard technologies and are today new benchmarks for diagnostics (Zhou et al. 2015). The promise of nanoparticle technology lies in the idea that the composition and size of the material grant properties that are unique from that observed in the bulk material, which include high surface-to-volume ratio, high surface energy, and unique mechanical, thermal, electrical, magnetic, and optical behavior (Chen et al. 2016; Yamada et al. 2015). Particularly, colloidal AuNPs are used as labels in diagnostics due to a unique combination of chemical and physical properties that allow new sensing concepts (Wilson 2008). AuNPs exhibit a localized surface plasmon resonance (LSPR), which can be described as the coherent oscillation of conduction electrons present on the surface of the AuNPs when interacting with an incident electromagnetic wave. This resonance effect is dependent on the size, shape, composition, and dielectric of the medium. Due to this surface plasmon enhancement, the molecular extinction coefficient of metal nanoparticles (10–100 nm) are 5 or more orders of magnitude greater than that of dyes, allowing higher sensitivity in surface interaction-based detection methods (Jain et al. 2006). AuNPs' easiness of synthesis and surface modification are suitable features for their use in biological systems (Baptista et al. 2008; Franco et al. 2015). There are several molecular diagnostic approaches making use of the exceptional characteristics of AuNPs, where the most frequently reported are (i) colorimetric sensing; (ii) plasmonic sensing; (iii) electrochemical, electrical sensing; and (iv) optical imaging. In this section, we will highlight AuNP application for detection of relevant disease biomarkers.

16.2.1 Ex Vivo Sensing

16.2.1.1 Colorimetric

AuNP-based colorimetric approaches for biomolecules sensing (e.g., nucleic acid and proteins) are frequently used due to their simplicity, high levels of sensitivity, specificity, and portability, making them ideal for point-of-care (POC) diagnostic applications (Doria et al. 2012). These methods rely on the colorimetric change of AuNPs' solution upon aggregation, which can be either mediated by changes in the solution's dielectric or recognition (cross-linking) of a specific target. These systems rely on the ability of complementary targets to modulate and control the interparticle interactions, which define the dispersion state of AuNPs (Franco et al. 2015). There are numerous techniques that are based on color change to identify molecular targets (i.e., DNA/RNA); among the most common approaches are the cross- and non-cross-linking methodology. The cross-linking method led to the first application of AuNPs in nucleic acid sensing using modified AuNPs with thiolated oligonucleotides by Mirkin et al. (1996). Briefly, this method relies on a two-probe functionalized with thiolated oligonucleotides to detect ssDNA as target. Each set of AuNPs functionalized is complementary to one end of the DNA target, and the two probes align in a tail-to-tail design onto the target. Hybridization of the probes to the target forces a small interparticle distance and a consequent change in the solution color from red to blue, where a noncomplementary target does not efficiently hybridize to the nanoprobes, and they remain dispersed, and the solution remains red (Fig. 16.1a) (Storhoff et al. 2004). Based on this method, several groups could develop similar strategies to sense other relevant clinical targets (Table 16.1) (He et al. 2008) without the need of enzymatic DNA amplification. Nevertheless, enzymatic amplification techniques (i.e., PCR, nested PCR, isothermal NASBA) have also been employed with the AuNP cross-linking detection method for sensing human relevant pathogens (e.g., *Mycobacterium tuberculosis*). Gill et al. (2008) were able to detect RNA from *M. tuberculosis* as low as 10 CFU.ml^{-1} using an isothermal NASBA amplification and Au-nanoprobes with the cross-linking approach, and Soo et al. (2009) were able to identify *M. tuberculosis* complex in clinical samples (detection of only 0.5 pmol) through the integration of a nested PCR and a specific set of Au-nanoprobes (Soo et al. 2009).

Baptista et al. (2005) developed a different approach – the non-cross-linking method – where hybridization between Au-nanoprobes and complementary targets resulted in heteroduplex formation at the Au-nanoprobe surface, stabilizing it against salt-induced aggregation. The presence of a noncomplementary or mismatched target would not be sufficient to stabilize the Au-nanoprobes, and aggregation occurs

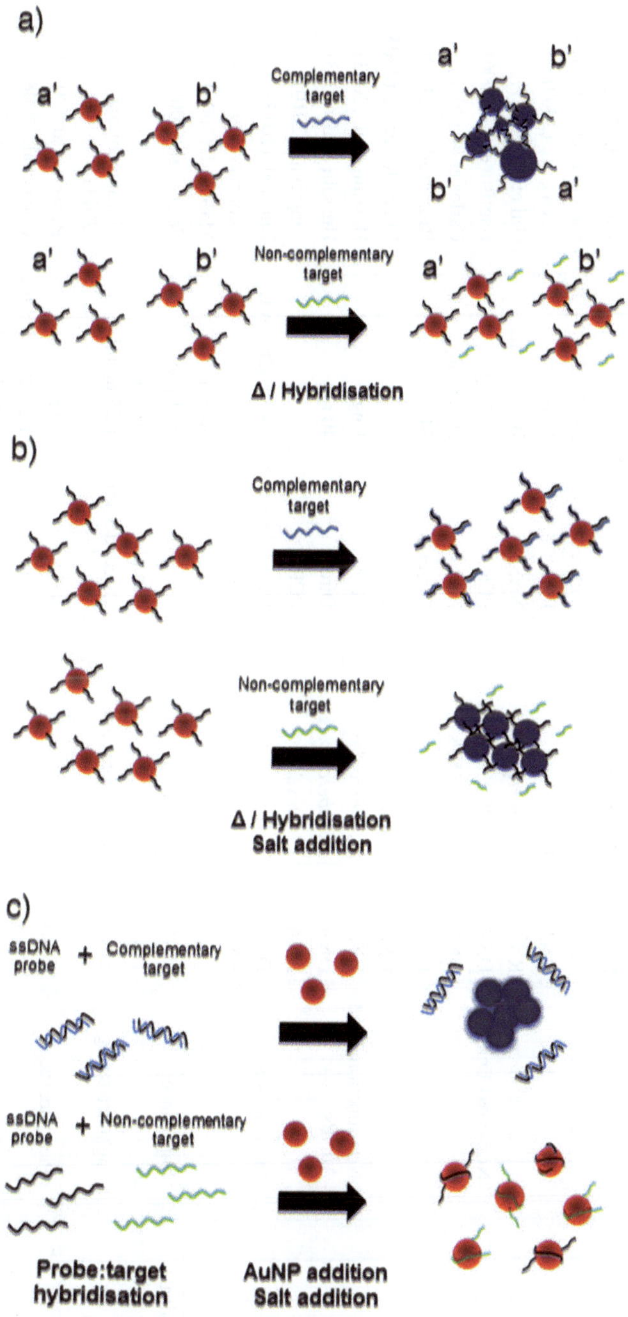

Fig. 16.1 Gold nanoparticle-based colorimetric assays. (**a**) Cross-linking method; (**b**) non-cross-linking method; (**c**) citrate-capped gold nanoparticles (Reproduced and adapted from Larguinho and Baptista (2012) with permission from Elsevier)

Table 16.1 Most relevant AuNP-based molecular diagnostics strategies

Detection technique	Approach	Target	Biological	Type of sample	Detection limit	References
Colorimetric	Cross-linking	DNA	Human	RCA	70 fM	Li et al. (2010)
		DNA	MTBC	PCR	0.5 pmol	Soo et al. (2009)
		DNA	*P. aeruginosa*	PCR	–	Wang et al. (2012)
		DNA	*Salmonella*	Direct detection	37 fM	Kalidasan et al. (2013)
		DNA	HPV	PCR	14 pM	Chen et al. (2009)
		RNA	*M. tuberculosis*	NASBA	10 CPU/mL	Gill et al. (2008)
	Non-cross-linking	DNA	MTBC	PCR	0.75 ug	Baptista et al. (2006)
		DNA	Human SNPs	PCR	20 μg/mL	Carlos et al. (2013)
		DNA	*E. coli*	Direct detection	54 ng	Padmavathy et al. (2012)
		RNA	Human	Direct	–	Li and Rothberg (2005)
	Non-functionalized AuNPs	DNA	*M. tuberculosis*	PCR	1 ng	Hussain et al. (2013)
		DNA	*M. tuberculosis*	Direct	40 ng	Hussain et al. (2013)
		DNA	*Chlamydia*	PCR	–	Jung et al. (2010)
	Immunoassays	DNA	*E. coli*	Direct detection	1×10^7 CFU/mL	Liu et al. (2013)
		RNA	HIV-1	NASBA		Rohrman et al. (2012)
		DNA	Human	PCR	50 pM	Mao et al. (2009)
Electrochemical	Impedance	DNA	–	–	0.31 pM	Feng et al. (2008)
		DNA	Human	Sample	10 aM	Fayazfar et al. (2014)
	Light scattering	DNA	*Mycobacterium*	Direct detection	1.25 ng/mL	Thiruppathiraja et al. (2011)
Electrical	Piezoelectric	DNA	*E. coli*	RCA	10 ng	Russell et al. (2014)
		DNA	*E. coli*	PCR	1.2×10^2 CFU/mL	Chen et al. (2008)
	QCM	DNA	*E. coli*	PCR	2.0×10^3 CFU/mL	Wang et al. (2008)

Abbreviations: RCA rolling circle amplification, *HPV* human papilloma virus, *HIV* human immunodeficiency virus associated with a color change from red to blue (Fig. 16.1b). The first clinical application of this strategy was applied for rapid and sensitive detection of *M. tuberculosis* in clinical samples (Baptista et al. 2006). Subsequently, the same strategy was applied for rapid detection of *M. tuberculosis* complex (MTBC) members and characterization of mutations linked to antibiotic resistance (Veigas et al. 2010), single nucleotide polymorphisms (SNPs) related to human diseases (β-thalassemia and obesity) in biological samples (Carlos et al. 2014), and direct detection of the e14a2 *BCR-ABL* fusion transcript in myeloid leukemia patient samples without the need for retro-transcription (Vinhas et al. 2016), among others (Guirgis et al. 2012; Padmavathy et al. 2012; Carlos et al. 2013). The non-cross-linking method proved its translation into microfluidics for POC (Silva et al. 2011) and to a paper-based platform for *M. tuberculosis* detection (Veigas et al. 2012). Non-functionalized AuNPs have also been used for the detection of DNA/RNA biomarkers. AuNPs are negatively charged and highly sensitive to changes in solution; therefore, the addition of biomolecules to the solution of non-functionalized AuNPs causes an unspecific adsorption to their surface (Baptista et al. 2011) (Fig. 16.1c).

Colorimetric detection has always been an integral part of conventional enzyme-linked immunosorbent immunoassay (ELISA). The developed AuNPs' format is based on the linkage of the catalytic cycle of enzyme label to the growth of AuNPs. The absence of analyte leads to the reduction of Au ions by hydrogen peroxide, resulting in the formation of non-aggregated AuNPs and the red solution. On the other hand, the presence of the analyte leads to the consumption of the hydrogen peroxide by catalase, which results in aggregation AuNPs associated with blue color (Liu et al. 2008). Other AuNPs' approaches for antibody-antigen binding activity and enhancement of the specific binding signal have been also described (Karakus 2015).

The integration of these approaches on lateral flow biosensors (LFBs) had a major impact in different fields with the first application of AuNPs on LFBs as label for the detection of ricin. Recently, LFB AuNP-based detection systems were improved via the integration of a wider set of biomarker types (e.g., aptamers and DNA probes). Xu et al. (2009) designed a LFBs system with an AuNP-aptamer conjugate as label for the detection of thrombin, with high sensitivity (comparable or even superior to the systems which use antibodies). The experiment shows that the color intensity of AuNPs can be detected in diluted plasma samples with lower detection limit compared to antibody-based assays (Xu et al. 2009). The integration of AuNPs this LFBs was also applied for the detection of different targets such as toxins and metallic cations (Quesada-González and Merkoçi 2015).

16.2.1.2 Optical

The use of AuNP plasmonics can also be a powerful tool when integrated with light-scattering techniques (Stewart 2008; Tong et al. 2014). One of these techniques is the detection of light scattering, where nanoparticles of different sizes and composition will allow a different scattering of light due to their distinct surface plasmon resonances. The highly efficient light-scattering properties of gold nanoparticles have been exploited to detect large targets with multiple epitopes, such as virus (Chen et al. 2016). Driskell et al. developed a method to detect and quantify the presence of influenza virus via influenza-specific antibodies conjugated to AuNPs, and aggregation of AuNPs probes is induced by the target virus (Driskell et al. 2011).

Other typology of detection is based on surface-enhanced Raman scattering (SERS), where the metallic nanosurface serves as scaffold to adsorb molecules. Due to the relatively low fabrication cost and high enhancement factor, SERS has been used in a variety of biological sensing applications (Ngo et al. 2016, 2013). Several distinct Raman reporter molecules can be used simultaneously allowing the development of a SERS nanotag. This approach was successfully used for the detection of cancer biomarkers. Dinish et al. could detect three intrinsic biomarkers – EGFR, CD44, and TGFßRII – in a breast cancer model using biocompatible SERS nanotags (Dinish et al. 2014; Viswambari Devi et al. 2015). SERS has also been applied for the detection of other tumor markers in several types of cancer (Kah et al. 2007; Qian et al. 2007; Lin et al. 2011; Wang et al. 2011).

Integration of AuNPs into electrochemical-based detection systems has created a new class of biosensors. The use of enzyme- or nanoparticle-modified electrodes improved significantly the specificity, stability, and sensitivity of this detection approach. Among these strategies, the most common consist on the direct deposition of nanoparticles onto the surface of the electrode (Pingarrón et al. 2008). An electronic method of detection of DNA with AuNPs was developed using microgravimetric quartz crystal microbalance (QCM) measurements. This method depends on the use of a frequency analyzer and quartz crystals sandwiched between two electrodes. The detection is based on the fact that hybridization of the target causes a change to the mass associated with the crystal, which results in a detectable frequency change (Weizmann et al. 2001). Table 16.1 provides several examples of the application of AuNPs for molecular diagnostics.

16.2.2 *In Vivo Sensing*

Imaging plays a critical role not only in diagnostics but also in the evaluation of treatment efficiency and disease management. Standard clinical images modalities include computed tomography (CT), magnetic resonance imaging (MRI), X-rays, and ultrasounds that can be categorized as structural imaging modalities. These imaging modalities can identify anatomical patterns and provide basic information regarding location, size, and disease spread based on the endogenous contrast. However, for smaller features, such as tumors and metastases smaller than 0.5 cm, these imaging modalities lack enough resolution and barely distinguish between benign and cancerous tumors (Popovtzer et al. 2008). Molecular imaging integrates molecular biology with in vivo imaging with the objective of gathering information of biological processes and identifying diseases based on molecular markers. Currently, positron emission tomography (PET) and single photon emission tomography (SPECT) are the main molecular imaging techniques in clinical use; however, they only provide information regarding molecular processes and metabolites, which is direct and nonspecific to distinct cells or diseases. Recently, AuNPs are being used as imaging probes (Popovtzer et al. 2008) in three diagnostic approaches: blood pool, passive targeting, and active targeting. Blood pool contrast agents are tailored to remain in the bloodstream for a prolonged amount of time, by limiting diffusion through the vascular endothelium, to enable a longer imaging window. Passive targeting relies on the nonspecific accumulation of AuNPs within a site of interest through an enhanced permeability and retention (EPR) effect leading to accumulation in tumor tissue of sized molecules or nanoparticles. Active targeting is the ability to deliver and retain a contrast agent at a specific site of interest through surface functionalization with molecules, such as antibodies and/or peptides, which exhibit a specific affinity for that site (Cole et al. 2015). AuNPs have been used to target cancer cells and tumors by exploiting specific receptors on cancer cells. Additional requirements are necessary in AuNPs contrast agent construct for clinical application. The requirements are based in the deliver "go where we want," nontoxic "do not harm along the way," targeting "stay where we want," and contrast enhancement "show me what we want" (Cole et al. 2015).

16.2.2.1 X-Rays

Several groups have proposed the use of gold for imaging, which presents a higher atomic number (Au, 79 vs I, 53) and a higher absorption coefficient (at 100 keV: gold, 5.16 cm^2/g; iodine, 1.94 cm^2/g; soft tissue, 0.169 cm^2/g; and bone, 0.186 cm^2/g) and provides about 2.7 times great contrast per unit weight than iodine (Hainfeld et al. 2006). Also, AuNPs exhibit a longer vascular retention time compared with traditional molecules, which potentially increases the available imaging window (Ahn et al. 2013). Hainfield et al. used 1.9 nm AuNPs to image kidneys and tumor with unusual clarity and high resolution. Blood vessels less than 100 nm in diameter were delineated, thus enabling in vivo vascular casting and regions of increased vascularization and angiogenesis could be distinguished (Hainfeld et al. 2006). Chien et al. tested different contrast agents based on AuNPs for the detection of cancer-related angiogenesis with high-resolution X-ray. One of the most successful approaches uses bare AuNPs (15 nm) injected with heparin to allow in vivo detection of small capillary species. The vessel density was three to seven times higher than with other nanoparticles. Additionally, bare AuNPs with heparin allow to detect symptoms of local extravascular nanoparticle diffusion in tumor areas where capillary leakage appeared (Chien et al. 2012).

16.2.2.2 Computed Tomography

Computed tomography (CT) is among the most interesting imaging/diagnostic tools in hospitals today in terms of availability, efficiency, and cost. Undisputedly, this is one of the leading technologies applied in overall cancer management; however, the sensitivity of CT is limited in the detection of some lesions and its specificity is relatively low, resulting in ~15% false positives [noncancerous findings that are interpreted as malignant tumors]. Therefore, improving the current CT capabilities providing it with functional and molecular-based imaging capacities is critical for cancer detection.

Recently, several different nanoprobes have been developed as blood pool CT contrast agents, such as gold nanoprobes and nanotags (Popovtzer et al. 2008). AuNPs have gained recent attention as CT contrast agents due to the fact that they display a high X-ray attenuation in comparison with other contrasting agents, especially at the energy level used for clinical CT scans (Ahn et al. 2013). Hainfeld et al. (2011) recently showed that AuNPs can enhance the visibility of millimeter-sized human breast tumors (1.5 mm) in mice and that active tumor targeting (with anti-Her2 antibodies) is more efficient than passive targeting. Cai et al. (2007) reported that 14 nm AuNPs functionalized with PEG can be used in CT scans, reaching a longer imaging window when compared with iodine compounds and allowing for more imaging detail when doing a CT scan. Reuveni et al. (2011) demonstrate the efficacy of active molecular targeting in imaging of gold nanoparticles conjugated with the anti-epidermal growth factor receptor (anti-EGFR) (30 nm).

16.2.2.3 Photoacoustic

Photoacoustic (PA) is a noninvasive hybrid imaging modality that combines optical contrast with high-resolution ultrasound imaging. PA imaging capitalizes on the PA effect, which is simply the generation of an acoustic wave resulting from the absorption of optical energy (Larguinho et al. 2015) (Fig. 16.2). Gold nanoshells and gold nanorods conjugated with epidermal growth factor receptor (EGFR) targeting molecules of human epidermal growth factor receptor 2 (HER2) were successfully used for performing multiplex imaging (Li and Chen 2015). Wang et al. (2004) demonstrated the feasibility of using nanoshells in vivo as a new contrast-enhancing agent for photoacoustic tomography. Agarwal et al. (2007) reported a gold nanorod

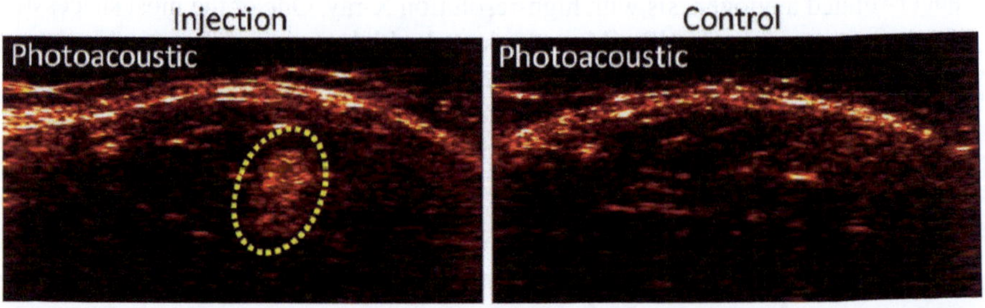

Fig. 16.2 In vivo photoacoustic imaging of stem cells via labeling with AuNPs. In the *left* – stem cells gel encapsulated with AuNP-labeled injected into a rat hind limb (*yellow circle* highlighted) were observed with photoacoustic images. On the *right* – gel-encapsulated stem cells not labeled with AuNPs not observed by photoacoustic images (Reproduced and adapted from Mieszawska et al. (2013) with permission from American Chemical Society)

conjugated with an antibody (Ab-17) for high-contrast photoacoustic imaging; with this conjugation, an effective cell targeting and sensitive photoacoustic detection of a single layer of cells were demonstrated.

16.3 AuNPs for Molecular Therapeutics

The therapeutic value of AuNPs is based on their ability to interact with tumors and damage cancer cells (Kodiha et al. 2015; Mieszawska et al. 2013). This knowledge is critical for two key aspects of nanomedicine: it will define the AuNP-induced events at the subcellular and molecular level, thereby possibly identifying new targets and strategies to improve AuNP-dependent treatment, e.g., photothermal agents, contrast agents, and radiosensitizers (Kodiha et al. 2015). The already described EPR effect constitutes an important mechanism for the passive targeting and selective accumulation of AuNP infiltration (Yu et al. 2012). AuNPs can be engineered to trick the immune system, avoiding removal from circulation or to cross a biological barrier in order to increase therapeutic efficacy and allow systemic tracking (Yu et al. 2012). Passive targeting approaches suffer from several limitations, targeting cancer cells using the EPR effect is not feasible in all tumors, and it depends on the degree of vascularization and porosity of tumor vessels. In addition, AuNPs can be constructed to display a reduced number of opsonin interactions that lead to cellular internalization. PEG functionalization can reduce interactions between nanoparticles and cell surfaces. Another approach to overcome these limitations is to attach targeting moieties (e.g., antibodies, peptides, nucleic acids or small molecules) to the nanoparticle surface (Yhee et al. 2014).

16.3.1 Drug Delivery

An ideal targeted drug delivery system should satisfy several requirements, one of the most important is a selective cellular uptake of the drug carrier to the targeted cells (Chen et al. 2016). AuNPs are suitable for drug delivery because their noble metal core is inert, contributing to low toxicity and biocompatibility, which is a requirement for biological applications. The flexibility of AuNP sizes and shapes facilitates the selection of optimal dimensions for loading therapeutics. Enhanced therapeutic outcomes were observed for cancer cells treated with AuNP drug conjugates when compared to free drugs and to assist overcoming drug resistance (Chen et al. 2016). In addition to the passage of vasculature of tumor, drug carriers must be internalized by cancer cells and achieve cytotoxic effects via endocytosis or other mechanisms (Ajnai et al. 2014). Heo et al. (2012) reported a complex AuNP platform that included biotin as a targeting ligand (due to biotin receptor overexpression on certain cancer cells), paclitaxel as an anticancer drug, rhodamine B to facilitate fluorescent detection, and PEG for enhancement of biocompatibility. Cyclodextrin, a drug host molecule, was attached to the AuNP surface for noncovalent paclitaxel inclusion. This combination allowed for reducing by half the viability rate when compared to the control cells. The multiplexed AuNPs showed higher internalization in three different cancer cell lines (Heo et al. 2012). Kumar et al. (2012) combined therapy with active targeting by functionalizing AuNPs with therapeutic and targeting agents (a peptide inhibitor termed PMI). The PMI peptide interferes with the p53 pathway and can induce cancer cell apoptosis. The peptide binds to the neuripilin-1 receptors, which are overexpressed on cancer cells. Incubation with these AuNPs led to a stronger in vitro toxicity for cancer cells (MDA-MB-321) than MCF-7 (Kumar et al. 2012).

16.3.2 Gene Delivery

The versatility and multifunctionality of AuNPs have facilitated several different approaches for encapsulation and release of nucleic acids since they prevent enzymatic degradation of DNA delivered, efficient cellular uptake, release of DNA from endosomes/lysosomes into the cytoplasm, crossing the cytoplasm and overcoming the nuclear barrier, and successful gene transfection to produce efficient gene expression (Mieszawska et al. 2013; Chen et al. 2016). Plasmid and minivector DNA can be used to repair defective genes, and small interfering RNA (siRNA) can be used to regulate the therapeutic process (Ding et al. 2014; Chen et al. 2016). Conde et al. (2013) have developed an AuNP platform for efficient siRNA delivery to silence the *c-Myc* proto-oncogene. The formulation contained PEG for increased stability, arginyl-glycyl-aspartic acid (RGD) targeting peptides, cell penetrating TAT peptides, and siRNA either covalently or ionically attached to AuNPs. The optical properties of AuNPs may also be used to trigger release of nucleic acids due to strong absorption of light and resultant heating of the nanoparticle, causing the bonds between nucleic acid and nanoparticle to break (Mieszawska et al. 2013). Cui et al. (2011) investigated dendrimer-coated gold nanorods as a delivery vehicle for BRCA1-shRNA into MCF7 cancer cells. Dendrimers are often used as delivery systems to increase the biocompatibility and cellular uptake. Near-infrared laser irradiation triggered RNA release from gold nanorods encapsulated in dendrimers, which led to successful silencing of *BRCA1* gene in MCF7 cells (Cui et al. 2011).

16.3.3 Photothermal Therapy

In hyperthermia, the heat generated externally by using instruments that produce electromagnetic fields (microwaves, radio waves, or ultrasounds) allows to heat to ~ 40–45 °C a region of the body containing the tumor (Abadeer and Murphy 2016). Photothermal therapy, particularly used in cancer applications, is based in the properties of photon absorption and conversion into thermal energy. Several fundamental prerequisites are required for effective photothermal therapy: (i) template nanomaterials should have high photothermal conversion efficiency; (ii) the photothermal effects should occur in response to near-infrared (NIR) light to ensure deep tissue penetration; and (iii) the size and surface of the nanomaterials should be modified with ease for facile and efficient photothermal therapy. AuNPs are especially suited to thermal destruction of cancer due to their ease of surface functionalization and photothermal heating capability (Kim et al. 2016). El-Sayed et al. (2006) functionalized gold nanospheres with anti-EGFR antibodies facilitating the binding and uptake of AuNPs by cancer cells that are known to overexpress EGFR. Irradiation of cells incubated with the nanorods demonstrated a higher rate of death in carcinoma cells; this could be attributed to higher nanorod light absorbance and enhanced light penetration in the NIR (El-Sayed et al. 2006). Generally NIR light is most commonly used due to its low absorption by tissues, resulting in increased depth penetration; however, some studies also report the efficient use of visible light for hyperthermia (Shao et al. 2013).

16.4 AuNPs for Theranostics

Theranostics can be defined as the use of the same platform for diagnostic and therapeutic purposes, promising precise and effective treatment, by tailoring on demand approaches especially on heterogeneous diseases such as cancer (Pedrosa et al. 2015). Theranostic allows customized patient treatment in response to its needs, adjusting dosages and reducing side effects by three approaches:

(i) Therapeutic is followed by diagnostic, e.g., a drug that is efficient, but not for all patients, enabling the diagnosis of individuals for whom it will have effect.
(ii) Diagnostic is followed by the therapeutic, e.g., a diagnostic that distinguishes patients and/or pathology and allows selection of adequate therapy.
(iii) Both diagnostic and therapeutic are performed simultaneously, e.g., when the same platform that performs diagnostic responds with adequate therapy (Conde et al. 2015).

16.4.1 Plasmonic Bubbles

One of the most promising nanotheranostic techniques is plasmonic nanobubbles (PNBs). It was first described by Lapotko group as transient vapor bubbles induced by NIR laser irradiation of plasmonic nanoparticles (Lukianova-Hleb et al. 2010a, b; Lukianova-Hleb et al. 2011). When the pulse laser interval is short (picoseconds), the nanoparticles overheat evaporating the surrounding medium, like a bubble, that expands and then collapses with its lifespan longer than the duration of the incident laser pulse. PNBs can be used as probes that can be triggered on demand with a single cell precision and are observable by UV-Vis due to its increased scattering or by acoustic sound due to shock wave formation. By using antibody-conjugated nanoparticles specific for cancer cells, it is possible to promote self-assembly of nanoparticles inside the cancer cells through endocytosis while normal cells will have an unspecific uptake. Due to laser pulse threshold dependency, it is possible to detect only the accumulated clusters inside cancer cells. Also by finely setting the duration of the laser pulse, it is possible to disrupt only the endocytic vesicles and release a cargo, while longer laser pulses disrupt the whole cell without affecting neighbor cells. This allows the use of PNB as vehicles for controlled gene delivery (Lukianova-Hleb et al. 2016b) and chemotherapy (Lukianova-Hleb et al. 2012a, b) (Fig. 16.3).

The most remarkable work using this technique is against residual microtumors (Lukianova-Hleb et al. 2016a). Using head and neck squamous cell carcinoma as model, authors describe in this preclinical study the conjugation of 60 nm gold spheres with panitumumab antibody, for clinical surgery of mice. PNBs allowed residual microtumor identification, something hardly observable by common imaging tools. Also, in response to single laser pulses, cancer cells are selectively destroyed, without affecting normal cells. The "nanosurgery" showed a significant improvement in the surgical outcome with diagnostic indices after plasmonic nanobubble-guided resections almost coinciding with those for microscopic residual disease-negative tissues, thus showing that this technique allows real-time observation and elimination of microtumors (Lukianova-Hleb et al. 2016a).

16.4.2 Combined Strategies

Several approaches, such as imaging, gene, chemo-, and photothermal therapy, may be combined in a single nanoconstruct due to the plasticity of AuNPs. By using different cellular targets, the probability of drug resistance and dosage for each therapy is reduced. Yin et al. (2015) describe a combination of light-driven, chemo-, and gene therapy to pancreatic cancer cells achieving superior anticancer efficacy. In their work polyelectrolyte polymer-coated gold nanorods were capable of co-delivering the anticancer drug doxorubicin (DOX) and siRNA mutant *K-RAS* gene after irradiation with a 665 nm laser. The use of the laser source increased DOX and siRNA release and the silencing effect. The triple approach showed a synergistic effect and was able to inhibit the tumor growth in vivo by 90% (Yin et al. 2015). Also, Conde et al. (2016) described a triple combination of gene, drug, and phototherapy mediated by gold nanoparticles and hydrogel patch leads. Spherical gold nanoparticles were firstly administered to deliver siRNAs against *K-RAS* gene, followed by gold nanorods. After exposing the particles to NIR laser, the rods convert light into heat, promoting hyperthermia, and release a chemotherapeutic (Avastin)

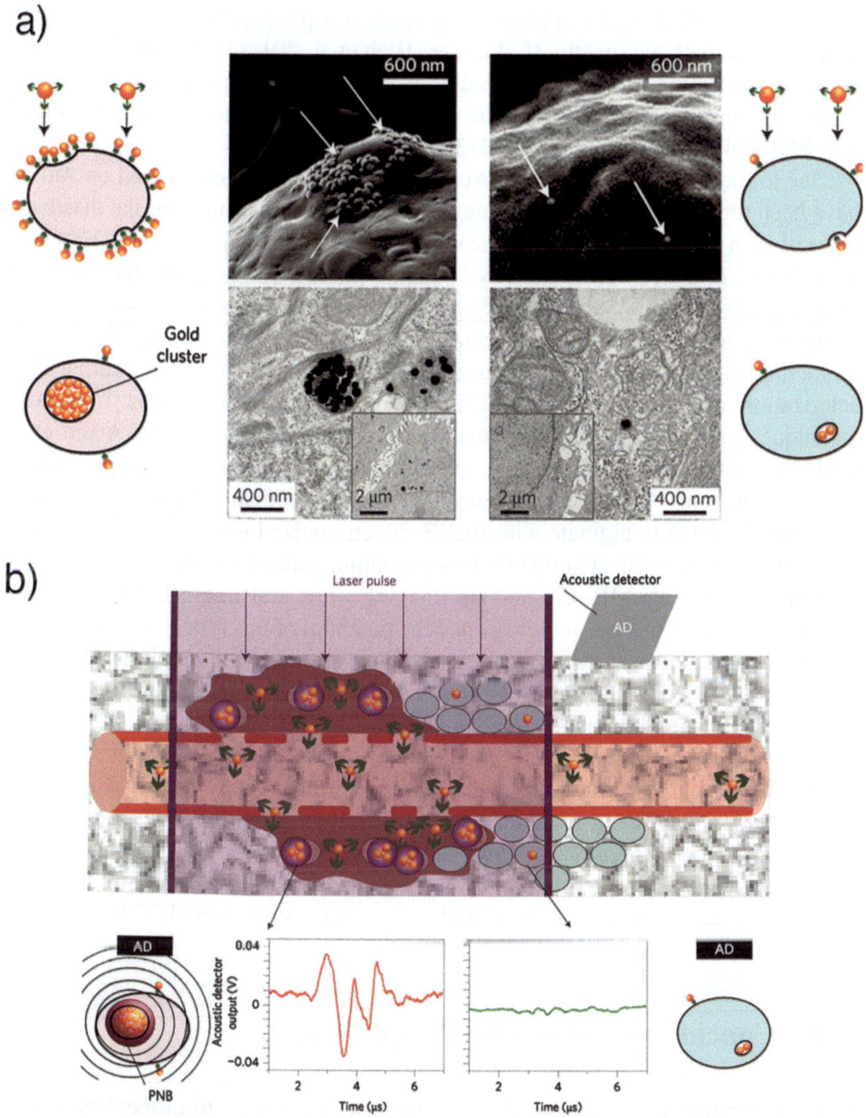

Fig. 16.3 Illustration of the PNB diagnostic mechanism in cancer cells in vivo. (**a**) Clustering of gold conjugates via receptor-mediated endocytosis in cancer cells (*Left*) and normal cells (*right*). (**b**) Schematic representation of gold cluster, when exposed to a single laser pulse. PNBs are selectively generated only in cancer cells, while normal cells do not generate PNBs (*upper*). The acoustic signal of a PNB reports even in single cancer cell in solid tissue (*bottom left*), but not in normal cells (*bottom right*) (Reproduced and adapted from Lukianova-Hleb and Lapotko (2015) with permission from Macmillan Publishers Limited)

(Conde et al. 2016). A few approaches describe a triple combination of phototherapy, X-ray, and chemotherapy (Lukianova-Hleb et al. 2014; Park et al. 2015; Baek et al. 2016) taking advantage of radiosensitization effect promoted by AuNPs, which concentrates X-rays on specific sites and allow computed tomography imaging (Mesbahi 2010; Jeremic et al. 2013; Deng et al. 2014).

Due to the imaging plasticity of AuNPs, theranostic platforms based on AuNPs have been reported in multimodal imaging approaches, that supplant the disadvantages of each imaging technique alone while performing therapeutic ends (Hembury et al. 2015). In a study conducted by Liu et al. (2015), 30 nm gold nanostars were used as probes for SERS detection, X-ray CT, and two-photon luminescence (TPL) imaging while performing photothermal therapy (PTT). The advantage of using different modalities is that CT can be used as a preoperative method for accurate characterization of the tumors while SERS and TPL as intraoperative biosensing techniques guiding localized photothermal therapy (Liu et al. 2015). Arifin et al. (2011) use the multimodal imagining properties of AuNPs for cell therapy tracking of type I diabetes. In this work, transplant cells of human cadaveric islets were microencapsulated in alginate with AuNPs functionalized with dithiolated diethylenetriamine pentaacetic acid (DTDTPA)/gadolinium chelates (GG) and were tested as contrast agents for MR, CT, and US imaging while assisting cell therapy. These in vitro and in vivo mouse studies opened the potential of this type of image-assisted therapy for treatment of type I diabetes without the need for immunosuppressive therapy thus providing a versatile tool for noninvasively monitoring engrafted microcapsules in real time (Arifin et al. 2011).

Theranostic approaches using combination therapy and imaging may significantly save resources, lowering treatment failure rate, fatality ratios, side effects, and development of drug resistance (Webster 2016). Such example is the use of gemcitabine chemotherapy vs PEGylated liposomal doxorubicin nanotherapy, where chemotherapy pretreatment costs were cheaper by €1,285 but €2,670 more expensive in administration and hospitalization costs (Bosetti et al. 2014).

16.5 Conclusion and Future Perspectives

Translating nanotechnology-based platforms from the bench to clinical practice is still far from accomplished (Heath 2015). Conventional molecular diagnostic techniques are still the gold standard for most of the diagnostics procedures. Nanomaterials have been providing for novel approaches in molecular diagnostics, especially for POC applications. Among these, several platforms relying on the use of AuNPs have been included into diagnostic platforms or devices ready to operate in the market. For example, lateral flow immunoassays incorporating AuNPs were capable to increase the sensitivity of detection tests for infectious agents, such as HIV, malaria, Ebola, influenza, etc., or for high-throughput molecular screening of human biomarkers, such as coagulation factors, myoglobin, troponin, etc. (Lefferts et al. 2009; Delaney et al. 2011; Gils et al. 2015). These approaches have been explored by several diagnostics companies that have been paving the way for nanodiagnostics at point-of-care (Vashist et al. 2015). AuNPs have also been used for improving efficacy for highly specific and selective targeted therapies. However, full assessment of toxicity (mainly chronic accumulation) still lacks to fully evaluate their potential. What is more, there are no enforced guidelines to assist nanomaterial characterization to allow comparison of efficiency, since the overall properties greatly depend from method of synthesis, size, shape, functionalization coverage, concentration, and even administration route affect toxicity. There are also great expectations from coupling the power of diagnostics provided by AuNPs to the selective and efficient therapeutic effect, mainly toward the development of nanotheranostic solutions. As data from the first set of clinical trials using AuNPs are stepping into the open, the effective translation of AuNPs to the clinics is just around the corner.

Acknowledgments The authors acknowledge Fundação para a Ciência e Tecnologia (FCT/MEC) for funding, UCIBIO (UID/Multi/04378/2013), co-financed by the ERDF under the PT2020 Partnership Agreement (POCI-01-0145-FEDER-007728), WaterJPI/0003/2013, and PD/BD/105734/2014 for PP.

References

Abadeer NS, Murphy CJ. Recent progress in cancer thermal therapy using gold nanoparticles. J Phys Chem C. 2016;120:4691–716.

Agarwal A, Huang S W, O'Donnell M, Day K C, Day M, Kotov N Ashkenazi S. Targeted gold nanorod contrast agent for prostate cancer detection by photoacoustic imaging. J Appl Phys. 2007;102: 64701-1-64701–4.

Ahn S, Jung SY, Lee SJ. Gold nanoparticle contrast agents in advanced X-ray imaging technologies. Mol. 2013;18:5858–90.

Ajnai G, Chiu A, Kan T, Cheng CC, Tsai TH, Chang J. Trends of gold nanoparticle-based drug delivery system in cancer therapy. J Exp Clin Med. 2014;6:172–8.

Alyass A, Turcotte M, Meyre D. From big data analysis to personalized medicine for all: challenges and opportunities. BMC Med Genet. 2015;8:1–12.

Arifin DR, Long CM, Gilad AA, Alric C, Roux S, Tillement O, Link TW, Arepally A, Bulte JWM. Trimodal gadolinium-gold pancreatic islet cells restore normoglycemia in diabetic mice and can be tracked by using US, CT, and positive contrast MR imaging. Radiology. 2011;260:790–8.

Baek S, Singh RK, Kim TH, Seo JW, Shin US, Chrzanowski W, Kim HW. Triple hit with drug carriers: PH- and temperature-responsive Theranostics for multimodal chemo- and photothermal therapy and diagnostic applications. ACS Appl Mater Interfaces. 2016;8:8967–79.

Baptista P, Doria G, Henriques D, Pereira E, Franco R. Colorimetric detection of eukaryotic gene expression with DNA-derivatized gold nanoparticles. J Biotechnol. 2005;119:111–7.

Baptista PV, Koziol-Montewka M, Paluch-Oles J, Doria G, Franco R. Gold-nanoparticle-probe-based assay for rapid and direct detection of Mycobacterium tuberculosis DNA in clinical samples. Clin Chem. 2006;52:1433–4.

Baptista P, Pereira E, Eaton P, Doria G, Miranda A, Gomes I, Quaresma P, Franco R. Gold nanoparticles for the development of clinical diagnosis methods. Anal Bioanal Chem. 2008;391:943–50.

Baptista PV, Doria G, Quaresma P, Cavadas M, Neves CS, Gomes I, Eaton P, Pereira E, Franco R. Nanoparticles in molecular diagnostics. In: Progress in molecular biology and translational science. 2011. p. 427–88.

Benedek TG. The history of gold therapy for tuberculosis. J Hist Med Allied Sci. 2004;59:50–89.

Bosetti R, Ferrandina G, Marneffe W, Scambia G, Vereeck L. Cost–effectiveness of gemcitabine versus PEGylated liposomal doxorubicin for recurrent or progressive ovarian cancer: comparing chemotherapy with nanotherapy. Nanomedicine. 2014;9:2175–86.

Cai Q, Kim SH, Choi KS, Kim SY, Byun SJ, Kim KW, Park SH, Juhng SK, Yoon K-H. Colloidal gold nanoparticles as a blood-pool contrast agent for X-ray computed tomography in mice. Investig Radiol. 2007;42:797–806.

Cai W, Gao T, Hong H, Sun J. Applications of gold nanoparticles in cancer nanotechnology. Nanotechnol Sci Appl. 2008;1:17–32.

Carlos FF, Silva Nunes J, Flores O, Brito M, Doria G, Veiga L, Baptista P. Association of FTO and PPARG polymorphisms with obesity in Portuguese women. Diabetes Metab Syndr Obes: Targets Ther. 2013;6:241–5.

Carlos FF, Flores O, Doria G, Baptista PV. Characterization of genomic single nucleotide polymorphism via colorimetric detection using a single gold nanoprobe. Anal Biochem. 2014;465:1–5.

Chen SH, Wu VCH, Chuang YC, Lin CS. Using oligonucleotide-functionalized Au nanoparticles to rapidly detect foodborne pathogens on a piezoelectric biosensor. J Microbiol Methods. 2008;73:7–17.

Chen SH, Lin KI, Tang CY, Peng SL, Chuang YC, Lin YR, Wang JP, Lin CS. Optical detection of human papillomavirus type 16 and type 18 by sequence sandwich hybridization with oligonucleotide-functionalized au nanoparticles. IEEE Trans Nanobioscience. 2009;8:120–31.

Chen G, Roy I, Yang C, Prasad PN. Nanochemistry and nanomedicine for nanoparticle-based diagnostics and therapy. Chem Rev. 2016;116:2826–85.

Chien CC, Chen HH, Lai SF, Wu KC, Cai X, Hwu Y, Petibois C, Chu Y, Margaritondo G. Gold nanoparticles as high-resolution X-ray imaging contrast agents for the analysis of tumor-related micro-vasculature. J Nanobiotechnol. 2012;10:1–12.

Cole LE, Ross RD, Tilley JM, Vargo-Gogola T, Roeder RK. Gold nanoparticles as contrast agents in x-ray imaging and computed tomography. Nanomedicine. 2015;10:321–41.

Conde J, Tian F, Hernández Y, Bao C, Cui D, Janssen KP, Ibarra MR, Baptista PV, Stoeger T, de la Fuente JM. In vivo tumor targeting via nanoparticle-mediated therapeutic siRNA coupled to inflammatory response in lung cancer mouse models. Biomaterials. 2013;34:7744–53.

Conde J, Oliva N, Artzi N. Implantable hydrogel embedded dark-gold nanoswitch as a theranostic probe to sense and overcome cancer multidrug resistance. Proc Natl Acad Sci U S A. 2015;112:1–10.

Conde J, Oliva N, Zhang Y, Artzi N. Local triple-combination therapy results in tumor regression and prevents recurrence in a colon cancer model. Nat Mater. 2016;15:1128–38.

Cui D, Huang P, Zhang C, Ozkan CS, Pan B, Xu P. Dendrimer-modified gold nanorods as efficient controlled gene delivery system under near-infrared light irradiation. J Control Release. 2011;152:e137–9.

Debnath M, Prasad G, Bisen P. Molecular diagnostic: promises and possibilities., Chap. 17. London: Springer; 2010.

Delaney KP, Branson BM, Uniyal A, Phillips S, Candal D, Owen SM, Kerndt PR. Evaluation of the performance characteristics of 6 rapid HIV antibody tests. Clin Infect Dis. 2011;52:257–63.

Deng H, Zhong Y, Du M, Liu Q, Fan Z, Dai F, Zhang X. Theranostic self-assembly structure of gold nanoparticles for NIR photothermal therapy and X-ray computed tomography imaging. Theranostics. 2014;4:904–18.

Ding Y, Jiang Z, Saha K, Kim CS, Kim ST, Landis RF, Rotello VM. Gold nanoparticles for nucleic acid delivery. J Am Soc Gene Cell Therapy. 2014;22:1075–83.

Dinish US, Balasundaram G, Chang YT, Olivo M. Actively targeted in vivo multiplex detection of intrinsic cancer biomarkers using biocompatible SERS nanotags. Sci Rep. 2014;4:1–7.

Doria G, Conde J, Veigas B, Giestas L, Almeida C, Assunção M, Rosa J, Baptista PV. Noble metal nanoparticles for biosensing applications. Sensors. 2012;12:1657–87.

Driskell JD, Jones CA, Tompkins SM, Tripp RA. One-step assay for detecting influenza virus using dynamic light scattering and gold nanoparticles. Analyst. 2011;136:3083–90.

Edwards PP, Thomas JM. Gold in a metallic divided state—from faraday to present-day nanoscience. Angew Chem Int Ed. 2007;46:5480–6.

El-Sayed IH, Huang X, El-Sayed MA. Selective laser photo-thermal therapy of epithelial carcinoma using anti-EGFR antibody conjugated gold nanoparticles. Cancer Lett. 2006;239:129–35.

Faraday M. Experimental relations of gold and other metals to light. Philos Trans R Soc Lond. 1847;147:145–81.

Fayazfar H, Afshar A, Dolati M, Dolati A. DNA impedance biosensor for detection of cancer, TP53 gene mutation, based on gold nanoparticles/aligned carbon nanotubes modified electrode. Anal Chim Acta. 2014;836:34–44.

Feng Y, Yang T, Zhang W, Jiang C, Jiao K. Enhanced sensitivity for deoxyribonucleic acid electrochemical impedance sensor: gold nanoparticle/polyaniline nanotube membranes. Anal Chim Acta. 2008;616:144–51.

Feynman RP. There's plenty of room at the bottom. Eng Sci. 1960;23:22–36.

Franco R, Pedrosa P, Carlos F, Veigas B. Gold nanoparticles for DNA/RNA-based diagnostics. Handbook of nanoparticles. 2015. p. 1–25.

Gill P, Ghalami M, Ghaemi A, Mosavari N, Abdul-Tehrani H, Sadeghizadeh M. Nanodiagnostic method for colorimetric detection of mycobacterium tuberculosis 16S rRNA. NanoBiotechnology. 2008;4:28–35.

Gils C, Ramanathan R, Breindahl T, Brokner M, Christiansen AL, Eng Ø, Hammer IJ, Herrera CB, Jansen A, Langsjøen EC, Løkkebo ES, Osestad T, Schrøder AD, Walther L. NT-proBNP on Cobas h 232 in point-of-care testing: performance in the primary health care versus in the hospital laboratory. Scand J Clin Lab Invest. 2015;75:602–9.

Gradmann C. Robert Koch and the pressures of scientific research: tuberculosis and tuberculin. Med Hist. 2001;45:1–32.

Guirgis BSS, Sá E, Cunha C, Gomes I, Cavadas M, Silva I, Doria G, Blatch GL, Baptista PV, Pereira E, Azzazy HME, Mota MM, Prudêncio M, Franco R. Gold nanoparticle-based fluorescence immunoassay for malaria antigen detection. Anal Bioanal Chem. 2012;402:1019–27.

Hainfeld JF, Slatkin DN, Focella TM, Smilowitz HM. Gold nanoparticles: a new X-ray contrast agent. Br J Radiol. 2006;79:248–53.

Hainfeld JF, O'Connor MJ, Dilmanian FA, Slatkin DN, Adams DJ, Smilowitz HM. Micro-CT enables microlocalisation and quantification of Her2-targeted gold nanoparticles within tumor regions. Br J Radiol. 2011;84:526–33.

He W, Cheng ZH, Yuan FL, Jian PX, Rong GY, Pei FZ, Wang J. One-step label-free optical genosensing system for sequence-specific DNA related to the human immunodeficiency virus based on the measurements of light scattering signals of gold nanorods. Anal Chem. 2008;80:8424–30.

Heath JR. Nanotechnologies for biomedical science and translational medicine. Proc Natl Acad Sci. 2015;112:14436–43.

Hembury M, Chiappini C, Bertazzo S, Kalber TL, Drisko GL, Ogunlade O, Walker-Samuel S, Krishna KS, Jumeaux C, Beard P, Kumar CSSR, Porter AE, Lythgoe MF, Boissière C, Sanchez C, Stevens MM. Gold–silica quantum rattles for multimodal imaging and therapy. Proc Natl Acad Sci. 2015;112:1959–64.

Heo DN, Yang DH, Moon HJ, Lee JB, Bae MS, Lee SC, Lee WJ, Sun IC, Kwon IK. Gold nanoparticles surface-functionalized with paclitaxel drug and biotin receptor as theranostic agents for cancer therapy. Biomaterials. 2012;33:856–66.

Higby GJ. Gold in medicine – a review of its use in the west before 1900. Gold Bull. 1982;15:130–40.

Hofker MH, Fu J, Wijmenga C. The genome revolution and its role in understanding complex diseases. Biochim Biophys Acta (BBA) – Mol Basis Dis. 2014;1842:1889–95.

Hussain MM, Samir TM, Azzazy HME. Unmodified gold nanoparticles for direct and rapid detection of mycobacterium tuberculosis complex. Clin Biochem. 2013;46:633–7.

Jain PK, Lee KS, El-Sayed IH, El-Sayed M. Calculated absorption and scattering properties of gold nanoparticles of different size, shape, and composition: applications in biological imaging and biomedicine. J Phys Chem B. 2006;110:7238–48.

Jain S, Hirst DG, O'Sullivan JM. Gold nanoparticles as novel agents for cancer therapy. Br J Radiol. 2012;85:101–13.

Jeremic B, Aguerri AR, Filipovic N. Radiosensitization by gold nanoparticles. Clin Transl Oncol. 2013;15:593–601.

Jung YL, Jung C, Parab H, Li T, Park HG. Direct colorimetric diagnosis of pathogen infections by utilizing thiol-labeled PCR primers and unmodified gold nanoparticles. Biosens Bioelectron. 2010;25:1941–6.

Kah JCY, Kho KW, Lee CGL, James C, Sheppard R, Shen ZX, Soo KC, Olivo MC. Early diagnosis of oral cancer based on the surface plasmon resonance of gold nanoparticles. Int J Nanomedicine. 2007;2:785–98.

Kalidasan K, Neo JL, Uttamchandani M. Direct visual detection of salmonella genomic DNA using gold nanoparticles. Mol BioSyst. 2013;9:618–21.

Karakus C. Development of a lateral flow immunoassay strip for rapid detection of CagA antigen of helicobacter pylori. J Immunoass Immunochem. 2015;36:324–33.

Kim J, Kim J, Jeong C, Kim WJ. Synergistic nanomedicine by combined gene and photothermal therapy. Adv Drug Deliv Rev. 2016;98:99–112.

Kodiha M, Wang YM, Hutter E, Maysinger D, Stochaj U. Off to the organelles – killing cancer cells with targeted gold nanoparticles. Theranostics. 2015;5:357–70.

Kumar A, Ma H, Zhang X, Huang K, Jin S, Liu J, Wei T, Cao W, Zou G, Liang XJ. Gold nanoparticles functionalized with therapeutic and targeted peptides for cancer treatment. Biomaterials. 2012;33:1180–9.

Larguinho M, Baptista PV. Gold and silver nanoparticles for clinical diagnostics - from genomics to proteomics. J Proteomics 2012;75:2811–23.

Larguinho M, Figueiredo S, Cordeiro A, Carlos FF, Cordeiro M, Pedrosa P Baptista PV. Frontiers in nanomedicine, Chap. Nanoparticles for diagnostics and imaging Bentham. 2015.

Lefferts JA, Jannetto P, Tsongalis GJ. Evaluation of the nanosphere verigene system and the verigene F5/F2/MTHFR nucleic acid tests. Exp Mol Pathol. 2009;87:105–8.

Li W, Chen X. Gold nanoparticles for photoacoustic imaging. Nanomedicine. 2015;10:299–320.

Li H, Rothberg L. Detection of specific sequences in RNA using differential adsorption of single-stranded oligonucleotides on gold nanoparticles. Anal Chem. 2005;77:6229–33.

Li J, Deng T, Chu X, Yang R, Jiang J, Shen G, Yu R. Rolling circle amplification combined with gold nanoparticle aggregates for highly sensitive identification of single-nucleotide polymorphisms. Anal Chem. 2010;82:2811–6.

Lin D, Feng S, Pan J, Chen Y, Lin J, Chen G, Xie S, Zeng H, Chen R. Colorectal cancer detection by gold nanoparticle based surface-enhanced Raman spectroscopy of blood serum and statistical analysis. Opt Express. 2011;19:13565–77.

Liu BH, Tsao ZJ, Wang JJ, Yu FY. Development of a monoclonal antibody against ochratoxin A and its application in enzyme-linked immunosorbent assay and gold nanoparticle immunochromatographic strip. Anal Chem. 2008;80:7029–35.

Liu CC, Yeung CY, Chen PH, Yeh MK, Hou SY. Salmonella detection using 16S ribosomal DNA/RNA probe-gold nanoparticles and lateral flow immunoassay. Food Chem. 2013;141:2526–32.

Liu Y, Ashton JR, Moding EJ, Yuan H, Register JK, Fales AM, Choi J, Whitley MJ, Zhao X, Qi Y, Ma Y, Vaidyanathan G, Zalutsky MR, Kirsch DG, Badea CT, Vo-Dinh T. A plasmonic gold nanostar theranostic probe for in vivo tumor imaging and photothermal therapy. Theranostics. 2015;5:946–60.

Lukianova-Hleb EY, Hu Y, Latterini L, Tarpani L, Lee S, Drezek RA, Hafner JH, Lapotko DO. Plasmonic nanobubbles as transient vapor nanobubbles generated around plasmonic nanoparticles. ACS Nano. 2010a;4:2109–23.

Lukianova-Hleb EY, Hanna EY, Hafner JH, Lapotko DO. Tunable plasmonic nanobubbles for cell theranostics. Nanotechnol. 2010b;21:1–19.

Lukianova-Hleb EY, Oginsky AO, Samaniego AP, Shenefelt DL, Wagner DS, Hafner JH, Farach-Carson MC, Lapotko DO. Tunable plasmonic nanoprobes for theranostics of prostate cancer. Theranostics. 2011;1:3–17.

Lukianova-Hleb EY, Belyanin A, Kashinath S, Wu X, Lapotko DO. Plasmonic nanobubble-enhanced endosomal escape processes for selective and guided intracellular delivery of chemotherapy to drug-resistant cancer cells. Biomaterials. 2012a;33:1821–6.

Lukianova-Hleb EY, Ren X, Zasadzinski JA, Wu X, Lapotko DO. Plasmonic nanobubbles enhance efficacy and selectivity of chemotherapy against drug-resistant cancer cells. Adv Mater. 2012b;24:3831–7.

Lukianova-Hleb EY, Ren X, Sawant RR, Wu X, Torchilin VP, Lapotko DO. On-demand intracellular amplification of chemoradiation with cancer-specific plasmonic nanobubbles. Nat Med. 2014;20:778–84.

Lukianova-Hleb EY, Lapotko DO. Rapid Detection and Destruction of Squamous Cell Carcinoma of the Head and Neck by Nano-Quadrapeutics. Head Neck 2015;37:1547–55.

Lukianova-Hleb EY, Kim Y-S, Belatsarkouski I, Gillenwater AM, O'Neill BE, Lapotko DO. Intraoperative diagnostics and elimination of residual microtumors with plasmonic nanobubbles. Nat Nanotechnol. 2016a:25–8.

Lukianova-Hleb EY, Yvon ES, Shpall EJ, Lapotko DO. All-in-one processing of heterogeneous human cell grafts for gene and cell therapy. Mol Ther – Methods Clin Dev. 2016b;3:1–8.

Mackenzie BYH, Cantab MD. Tuberculin treatment: general survey. Lancet. 1913;4695:521–4.

Mao X, Ma Y, Zhang A, Zhang L, Zeng L, Liu G. Disposable nucleic acid biosensors based on gold nanoparticle probes and lateral flow strip. Anal Chem. 2009;81:1660–8.

Mesbahi A. A review on gold nanoparticles radiosensitization effect in radiation therapy of cancer. Rep Pract Oncol Radiother. 2010;15:176–80.

Mieszawska AJ, Mulder WJM, Fayad ZA, Cormode DP. Multifunctional gold nanoparticles for diagnosis and therapy of disease. Mol Pharm. 2013;10:831–47.

Mirkin CA, Letsinger RL, Mucic RC, Storhoff JJ. A DNA-based method for rationally assembling nanoparticles into macroscopic materials. Nature. 1996;382:607–9.

Ngo HT, Wang H-N, Fales AM, Vo-Dinh T. Label-free DNA biosensor based on SERS molecular sentinel on nanowave chip. Anal Chem. 2013;85:6378–83.

Ngo HT, Wang HN, Fales AM, Vo-Dinh T. Plasmonic SERS biosensing nanochips for DNA detection. Anal Bioanal Chem. 2016;408:1773–81.

Nunes Pauli GE, de la Escosura-Muñiz A, Parolo C, Helmuth Bechtold I, Merkoçi A. Lab-in-a-syringe using gold nanoparticles for rapid immunosensing of protein biomarkers. Lab Chip. 2015;15:399–405.

Padmavathy B, Vinoth Kumar R, Jaffar Ali BM. A direct detection of *Escherichia coli* genomic DNA using gold nanoprobes. J Nanobiotechnol. 2012;10:1–10.

Park J, Park J, Ju EJ, Park SS, Choi J, Lee JH, Lee KJ, Shin SH, Ko EJ, Park I, Kim C, Hwang JJ, Lee JS, Song SY, Jeong SY, Choi EK. Multifunctional hollow gold nanoparticles designed for triple combination therapy and CT imaging. J Control Release. 2015;207:77–85.

Pedrosa P, Vinhas R, Fernandes A, Baptista P. Gold nanotheranostics: proof-of-concept or clinical tool? Nanomaterials. 2015;5:1853–79.

Pingarrón JM, Yáñez-Sedeño P, González-Cortés A. Gold nanoparticle-based electrochemical biosensors. Electrochim Acta. 2008;53:5848–66.

Popovtzer R, Agrawal A, Kotov NA, Popovtzer A, Balter J, Carey TE, Kopelman R. Targeted gold nanoparticles enable molecular CT imaging of cancer. Nano Lett. 2008;8:4593–6.

Qian X, Peng X, Ansari DO, Yin-Goen Q, Chen GZ, Shin DM, Yang L, Young AN, Wang MD, Nie S. In vivo tumor targeting and spectroscopic detection with surface-enhanced Raman nanoparticle tags. Nat Biotechnol. 2007;26:83–90.

Quesada-González D, Merkoçi A. Nanoparticle-based lateral flow biosensors. Biosens Bioelectron. 2015;73:47–63.

Reuveni T, Motiei M, Romman Z, Popovtzer A, Popovtzer R. Targeted gold nanoparticles enable molecular CT imaging of cancer: an in vivo study. Int J Nanomedicine. 2011;6:2859–64.

Roco MC. Nanotechnology: convergence with modern biology and medicine. Curr Opin Biotechnol. 2003;14:337–46.

Rohrman BA, Leautaud V, Molyneux E, Richards-Kortum RR. A lateral flow assay for quantitative detection of amplified HIV-1 RNA. PLoS One. 2012;7:1–8.

Russell C, Welch K, Jarvius J, Cai Y, Brucas R, Nikolajeff F, Al RET. Gold nanowire based electrical DNA detection using rolling circle amplification. ACS Nano. 2014;8:1147–53.

Shao J, Griffin RJ, Galanzha EI, Kim JW, Koonce N, Webber J, Mustafa T, Biris AS, Nedosekin DA, Zharov VP. Photothermal nanodrugs: potential of TNF-gold nanospheres for cancer theranostics. Sci Rep. 2013;3:1–9.

Silva LB, Veigas B, Doria G, Costa P, Inácio J, Martins R, Fortunato E, Baptista PV. Portable optoelectronic biosensing platform for identification of mycobacteria from the mycobacterium tuberculosis complex. Biosens Bioelectron. 2011;26:2012–7.
Soo PC, Horng YT, Chang KC, Wang JY, Hsueh PR, Chuang CY, Lu CC, Lai HC. A simple gold nanoparticle probes assay for identification of mycobacterium tuberculosis and mycobacterium tuberculosis complex from clinical specimens. Mol Cell Probes. 2009;23:240–6.
Stewart M. Nanostructured plasmonic sensors. Chem Rev. 2008;108:494–521.
Storhoff JJ, Lucas AD, Garimella V, Bao PY, Müller UR. Homogeneous detection of unamplified genomic DNA sequences based on colorimetric scatter of gold nanoparticle probes. Nat Biotechnol. 2004;22:883–7.
Thiruppathiraja C, Kamatchiammal S, Adaikkappan P, Santhosh DJ, Alagar M. Specific detection of mycobacterium sp. genomic DNA using dual labeled gold nanoparticle based electrochemical biosensor. Anal Biochem. 2011;417:73–9.
Tong L, Wei H, Zhang S, Xu H. Recent advances in plasmonic sensors. Sensors. 2014;14:7959–73.
Tsai TT, Shen SW, Cheng CM, Chen CF. Paper-based tuberculosis diagnostic devices with colorimetric gold nanoparticles. Sci Technol Adv Mater. 2013;14:1–7.
Tsongalis GJ, Silverman LM. Molecular diagnostics: a historical perspective. Clin Chim Acta. 2006;369:188–92.
Vashist SK, Luppa PB, Yeo LY, Ozcan A, Luong JHT. Emerging technologies for next-generation point-of-care testing. Trends Biotechnol. 2015;33:692–705.
Veigas B, Machado D, Perdigão J, Portugal I, Couto I, Viveiros M, Baptista PV. Au-nanoprobes for detection of SNPs associated with antibiotic resistance in mycobacterium tuberculosis. Nanotechnology. 2010;21:1–7.
Veigas B, Jacob JM, Costa MN, Santos DS, Viveiros M, Inácio J, Martins R, Barquinha P, Fortunato E, Baptista PV. Gold on paper–paper platform for Au-nanoprobe TB detection. Lab Chip. 2012;12:4802–8.
Vinhas R, Correia C, Ribeiro P, Lourenço A, Botelho de Sousa A, Fernandes AR, Baptista PV. Colorimetric assessment of BCR-ABL1 transcripts in clinical samples via gold nanoprobes. Anal Bioanal Chem. 2016;408:5277–84.
Viswambari Devi R, Doble M, Verma RS. Nanomaterials for early detection of cancer biomarker with special emphasis on gold nanoparticles in immunoassays/sensors. Biosens Bioelectron. 2015;68:688–98.
Wang Y, Xie X, Wang X, Ku G, Gill KL, O'Neal DP, Stoica G, Wang LV. Photoacoustic tomography of a nanoshell contrast agent in the in vivo rat brain. Nano Lett. 2004;4:1689–92.
Wang L, Wei Q, Wu C, Hu Z, Ji J, Wang P. The *Escherichia coli* O157:H7 DNA detection on a gold nanoparticle-enhanced piezoelectric biosensor. Chin Sci Bull. 2008;53:1175–84.
Wang X, Qian X, Beitler JJ, Chen ZG, Khuri FR, Lewis MM, Shin HJC, Nie S, Shin DM. Detection of circulating tumor cells in human peripheral blood using surface-enhanced Raman scattering nanoparticles. Cancer Res. 2011;71:1526–32.
Wang X, Li Y, Wang J, Wang Q, Xu L, Du J, Yan S, Zhou Y, Fu Q, Wang Y, Zhan L. A broad-range method to detect genomic DNA of multiple pathogenic bacteria based on the aggregation strategy of gold nanorods. Analyst. 2012;137:4267–73.
Webster RM. Combination therapies in oncology. Nat Rev Drug Discov. 2016;15:81–2.
Weizmann Y, Patolsky F, Willner I. Amplified detection of DNA and analysis of single-base mismatches by the catalyzed deposition of gold on Au nanoparticles. Analyst. 2001;126:1502–4.
Wilson R. The use of gold nanoparticles in diagnostics and detection. Chem Soc Rev. 2008;37:2028–45.
Xu H, Mao X, Zeng Q, Wang S, Kawde AN, Liu G. Aptamer-functionalized gold nanoparticles as probes in a dry-reagent strip biosensor for protein analysis. Anal Chem. 2009;81:669–75.
Yamada M, Foote M, Prow TW. Therapeutic gold, silver, and platinum nanoparticles. Wiley Interdiscip Rev Nanomed Nanobiotechnol. 2015;7:428–45.
Yhee JY, Lee S, Kim K. Advances in targeting strategies for nanoparticles in cancer imaging and therapy. Nanoscale. 2014;6:13383–90.

Yin F, Yang C, Wang Q, Zeng S, Hu R, Lin G, Tian J, Hu S, Lan RF, Yoon HS, Lu F, Wang K, Yong KT. A light-driven therapy of pancreatic adenocarcinoma using gold nanorods-based nanocarriers for co-delivery of doxorubicin and siRNA. Theranostics. 2015;5:818–33.

Yu MK, Park J, Jon S. Targeting strategies for multifunctional nanoparticles in cancer imaging and therapy. Theranostics. 2012;2:3–44.

Zheng P, Li M, Jurevic R, Cushing SK, Liu Y, Wu N. A gold nanohole array based surface-enhanced Raman scattering biosensor for detection of silver(I) and mercury(II) in human saliva. Nanoscale. 2015;7:11005–12.

Zhou W, Gao X, Liu D, Chen X. Gold nanoparticles for in vitro diagnostics. Chem Rev. 2015;115:10575–636.

Chapter 17
Nanometals in Bhasma: Ayurvedic Medicine

Dilipkumar Pal and Vinod Kumar Gurjar

Abstract Bhasmas are unique Ayurvedic metallic preparations with herbal juices or fruits, known in the Indian subcontinent since the seventh century BC and widely recommended for treatment of a variety of chronic ailments. More than 18 Bhasmas based on calcium, iron, zinc, mercury, silver, potassium, arsenic, copper, tin, and gemstones were analyzed for up to 18 elements by instrumental neutron activation analysis method, including their C, H, N, and S contents. In addition to the major constituent elements found at percent level, several other essential elements such as Na, K, Ca, Mg, V, Mn, Fe, Cu, and Zn have also been found in μg/g amounts and ultratrace (ng/g) amounts of Au and Co. The Bhasmas are biologically produced nanoparticles and are taken along with milk, butter, honey, or ghee thus; this makes these elements easily assimilable, eliminating their harmful effects and enhancing their biocompatibility. Particle size (1–2 μm) is reduced significantly, which may assist absorption and assimilation of the drug into the body system. Standardization of *Bhasma* is utmost necessary to confirm its identity and to determine its quality, purity safety, effectiveness, and suitability of the product. But the most important challenges faced by these formulations are the lack of complete standardization by physiochemical parameters.

Keywords Ayurveda • *Bhasma* • Nanoparticle • Nanotechnology • *Lauha Bhasma* • *Shankha Bhasma* • Nanomedicine

Abbreviations

AFM	Atomic Force Microscopy
Ag	Silver
Au	Gold
CaO	Calcium Oxide

D. Pal (✉) • V.K. Gurjar
Institute of Pharmaceutical Sciences, Guru Ghasidas Vishwavidyalaya
(A Central University), Koni, Bilaspur 495 009, Chhattisgarh, India
e-mail: drdilip71@gmail.com

M. Rai, R. Shegokar (eds.), *Metal Nanoparticles in Pharma*,
DOI 10.1007/978-3-319-63790-7_17

DLS	Dynamic Light Scattering
EDAX	Elemental Analysis with Energy Dispersive X-Ray Analysis
GIT	Gastro-Intestinal Tract
ICP	Inductively Coupled Plasma
NIH	National Institute of Health
NPs	Nanoparticles
PbS	Lead Suilphide
PXRD	Powder X-Ray Diffraction
TEM	Transmission Electron Microscopy
XPS	X-ray Photoelectron Spectroscopy

17.1 Introduction

Nanotechnology is a recent technology and is being used in almost all industry for the development of cost-effective and eco-friendly products. The term "nanotechnology" was first defined by Tokyo Science University, Norio Taniguchi, in a 1974 paper as follows: "Nanotechnology" mainly consists of the processing of, separation of, consolidation of, and deformation of materials by one atom or one molecule. It is anticipated to open some new aspects to fight and prevent diseases using atomic scale tailoring of materials (Geetha et al. 2016). The ability to uncover the structure and function of biosystems at the nanoscale stimulates research primary to improve biology, biotechnology, medicine, and healthcare. The size of nanomaterials is similar to that of most biological molecules and structures; therefore, nanomaterials can be helpful for both in vivo and in vitro biomedical research and applications. Nanotechnology is currently employed as a tool to discover the darkest avenues of medical sciences in several ways like imaging, sensing, targeted drug delivery and gene delivery systems, and artificial implants. The new-age drugs are nanoparticles of polymers, metals, or ceramics, which can combat conditions like cancer and fight with human pathogens like bacteria and other microorganisms (Pandey and Pandey 2013; Othayoth et al. 2014). This emerging technology has multiple possible applications and thus affects various technological domains including advanced materials, biotechnology and pharmacy, electronics, scientific tools, and industrial manufacturing processes. The emergence of nanotech has been enabled by the development of specialist instruments, which in turn facilitated the observation and management of nanostructures at the atomic or molecular scale, as well as the discoveries of new nanomaterials. Nanotech offers also new opportunities in rapid development of neatness techniques (so-called "top-down" approach, involving decomposition into the smallest manageable entities) and building macrostructures (so-called "bottom-up" approach, allowing reengineered materials at nanolevel and using them in developing new and improved products) (Miyazaki and Nazrul Islam 2007).

Nanotechnology, being so promising and lurking with possibilities, doesn't escape this bitter truth. Owing to their various applications in the field of medical science, the sad fact is that these potentially dwarf particles can expose negative effects to our body's system. Nanoparticles (NPs) can potentially cause adverse effects on organ, tissue, cellular, subcellular, and protein levels due to their abnormal physicochemical properties (e.g., small size, high surface area-to-volume ratio, chemical composition, electronic properties, surface structure reactivity and functional groups, inorganic or organic coatings, solubility, shape, and aggregation behavior). Metal NPs, in particular, have received increasing attention due to their widespread medical, consumer, industrial, and military applications. But these metal-based NPs have been confirmed to show tremendous toxicity, though the same material is relatively inert in its bulk form (e.g., Ag, Au, and Cu). Metals at the nanoscale level owing to their difference in bulk form properties can pose potential harmful biological interactions. Scientific validation and the documentation of Ayurvedic drugs are very crucial for its quality evaluation and worldwide acceptance. Metal nanoparticles have a high specific surface area and a high fraction of surface atoms and have been studied extensively because of their unique physicochemical characteristics including catalytic activity, optical properties, electronic properties, antimicrobial activity, and magnetic properties (Shahverdi et al. 2007). Nanotechnology promises important improvements of advanced materials and manufacturing techniques, which are significant for the future competitiveness of national industries (Miyazaki and Nazrul Islam 2007).

Therapeutic usefulness of Ayurvedic herbs may be improved with high quality, which can be achieved by uniqueness, purity, safety, drug content, and physical and biological properties. Ayurvedic medicines need to be explored with the modern scientific approaches for its validation. Therefore, an effort has been made in the present chapter to highlight the essential aspects that need to be considered for the promotion and development of Ayurvedic medicine. Size reduction is one of the basic unit operation having important applications in pharmacy. It helps in improving solubility and bioavailability, reducing toxicity, enhancing release, and providing better formulation opportunities for drugs. Drugs in the nanometer size range enhance performance in a variety of dosage forms. The recent status of nanotechnology in pharmaceutical field includes development of nanomedicine, tissue engineering, nanorobots, biosensors, biomarkers, etc. Pharmaceutical nanotechnology provides opportunities to improve materials and medical devices and help to build up new technology where existing and more conventional technologies may be reaching their limits (Varshney and Shailender 2012).

17.2 Distinctive Features of Nanoscience and Nanotechnology

Nanoscience and nanotechnology are extensively seen as having enormous potential to many areas of scientific research such as physics, chemistry, material sciences, biology, and engineering and technological applications (such as healthcare and life sciences, energy and environment, electronics, communications and computing, manufacturing, and materials) because of its nanoscale where the materials' properties are significantly different from those of the same materials in bulk or macroscopic form. Nanotechnology encompasses the work of nanoscale science and enlarges understandings of interactions in the atomic or molecular scale and the capacity to characterize and control materials using nano-tools (Miyazaki and Nazrul Islam 2007).

Nanotechnology has a multidisciplinary character, affecting multiple traditional technologies, scientific disciplines, and industries. Additionally, through the nanotech revolution, boundaries between previously distinctive disciplines such as mechanics and chemistry begin to blur, stimulating knowledge transfer and cross-fertilization (Nicolau 2004). Many scientists believe that nanomaterials will induce a new generation of consumer products, based on miniaturized computer chips, nanoscale sensors, and devices for sorting DNA molecules and integrating microsystems and biotechnology (Ikezawa 2001). Nanotechnology innovation can be characterized as evolutionary from micro to nano. An important feature of nanotech is that it is not restricted to the realm of advanced materials, extending also to manufacturing processes, biotechnology and pharmacy, electronics and information technology, as well as other technologies (Miyazaki and Nazrul Islam 2007).

17.3 Nanotechnology in Medicine

Nanotechnology has been setting benchmarks for the last two to three decades, but the origins of this technology achieve back to ancient history. Today, nanoparticles of both metallic and nonmetallic origin are under research and development for applications in different fields of biology/therapeutics (Sengupta et al. 2014). Applying nanotechnology for treatment, diagnosis, monitoring, and control of diseases has been referred to as "nanomedicine." While the application of nanotechnology to medicine appears to be a relatively recent development, the basic nanotechnology approaches for medical application date back several decades (Singh et al. 2008).

17.4 Nanotechnology in Ayurveda

Nanotechnology has been the focus of significant attention in medicine due to the facility with which nanostructures interact with the body at the molecular scale. Pharmacokinetics and biodistribution of active ingredients can be improved remarkably with nano-drug delivery systems by targeting them to the specified site; thereby efficiency and bioavailability can be improved, and drug toxicity reduces. In eighth century AD, the Indian alchemist *Nagarjuna* first introduced the use of metals and minerals like *Swarna* (gold), *Rajat* (silver), *Tamra* (copper), *Abhrak* (mica), and *Makshika* (pyrites), *Rasa* (mercury) as medicinal agents (Conde et al. 2014). The branch of Ayurveda dealing with herbo-metallic preparation is known as *Rasa*

Shastra. One major gain of nanotechnology is its flexibility, which enables the nanomedicines to take different shapes such as liposomes, dendrimers, nanoparticles, nanocrystals, etc., so as to the meet the needs of desired or required biomedical applications (Pal 2015).

17.5 Types of Metal Nanoparticles

Concept of reduction in particle size of metals is prevailing since *Charaka Santhita* (1500 BC). For a metallic preparation of *Lauhadi Rasayana*, the tip of iron is heated to red hot and quenched in some liquid media immediately until flakes of iron become fine powder form. Nanotechnology has ability to work at these levels to produce larger structures with new molecular organization (Sarkar and Chaudhary 2010). The invention of nanoparticles is not less than a miracle due to the distinctive properties that offer innovative and life-changing products and technologies in the fields of medicine. Nanotechnology is however also questioned in terms of safety for humans and also animals, plants, and ecosystem at large. There has been an exponential increase in this field with a wide variety of products, which are questioned for safety by many international organizations (Palkhiwala and Bakshi 2014). Particles of "nano" size have been shown to exhibit improved or novel properties including reactivity, greater sensing capability, and increased mechanical strength. The nanotechniques offer simple, clean, fast, efficient, and economic process for the synthesis of a variety of organic molecules and have provided the thrust for many chemists to switch from traditional method (Arivalagan et al. 2011).

Nanoparticles are the new approach to the scientists in the field of biomedical and commercial application. Nanotechnology is the field of advanced technology for medicine and imaging of various critical diseases. It is built as nanodevices and particles on the scale of 10^{-9} m, whereas size of the cell is 10 μm, and cell organelles have size in nanometer range. Nano-sized particles can easily enter into the cell and take part in the cell metabolism. Metal nanoparticles now are being applied in the field of drug delivery and imaging. Scientists are trying to develop noble nanoparticles, which can release the drug at exact site of targeted tissue and to be able to escape from the degradation system of the body. Metal nanoparticles have characteristics, small size and unique chemical properties, which are the important features for future development for various therapeutics and imaging (Yadav et al. 2016).

Ayurvedic nanomaterials, especially the gold *Bhasmas*, have unique physicochemical properties such as biocompatibility and ease of surface fictionalization. The most chronic illnesses, including cancer, diabetes, and cardiovascular and pulmonary diseases, are mediated through constant inflammation and it has the potential to delay the suppressing chronic inflammation and prevent and treat various chronic diseases, including cancer. The Ayurvedic *Bhasmas*-coated nano-tablets have been studied for their anticancer activity (Amin et al. 2009). The formulations in the "Bhasma" have nanoparticles. It is surprising for many researchers that 5,000-year-old Indian medical system have the knowledge of nanoscience and technology. *Charaksamhitha* is the oldest classical way of Ayurveda with the thought of reduction in particle size of metals. The *Bhasmas* are used for treatments of various diseases in Ayurveda for the past several centuries in the form of nanotechnology; some of the common properties in the Ayurvedic *Bhasmas* are "Rasayana" (immune-modulation and antiaging quality) and "Yogavahi" (ability of drug carry and targeting drug delivery). They were prescribed in minute dosage (15–250 mg/day) (Othayoth et al. 2014).

17.6 Bhasma: An Ayurvedic Medicine

The traditional medicinal system practicing in India for several centuries is well known as Ayurveda. According to this medicinal system, metal-based drugs known as "Bhasma" involve the modification of a metal into its mixed oxides. During these transformations, the zerovalent metal state gets converted into a form with higher oxidation state, and the most important aspect of this synthesis (traditionally known as "Bhasmikarana") is that the toxic nature (i.e., systemic toxicity causing nausea, vomiting, stomach pain, etc.) of the resulting metal oxide is completely destroyed while inducing the medicinal properties into it. The important step implicated in the procedure for making "Bhasma" is repeated treatment (Wadekar et al. 2005). *Bhasmas* are such kind of dosage forms which gained their position as effective formulations for any disease compromising the aspects of nanotechnology and overcoming the limitations of usual dosage forms (Rasheed et al. 2014).

Bhasmas are particulate matters that are thought to be readily assimilated in the body's system. They are highly inert in nature because of insolubility. *Bhasma* is important in maintaining optimum alkalinity for good health, neutralizing harmful acids that lead to illness, because *Bhasma* does not get metabolized so they don't produce any harmful metabolite, rather it breaks down heavy metals in the body (Hareshwar et al. 2017). Most of the *Bhasmas* are mixed with cardamom, cinnamon, ghee, and honey and are taken orally. In clinical practice, *Bhasma* is not reported to have any serious untoward effects. Although honey is one of the most frequently suggested vehicles in Ayurvedic texts, royal jelly, another honeybee product, is apparently not mentioned. Honey is considered as highly nutritious with nine elements (lithium, sodium, potassium, rubidium, magnesium, iron, manganese, copper, and zinc) reported in a recent study in which metal profiles have been used for its classification. Natural food users claim that royal jelly very quickly lowers blood sugar when taken orally by diabetic patients. In fact, it contains a polypeptide that is similar to bovine insulin. The crude royal jelly and a fraction that co-migrates chromatographically metabolize glucose in vitro incubation with rat adipose fat tissues. The molecule is about 5,000–6,000 Da, contains disulfide bonds, and has an amino acid composition similar to that of bovine insulin (Kumar et al. 2006). Traditionally used *Bhasmas* and their ingredients are summarized in Tables 17.1 and 17.2.

Table 17.1 Bhasma and their ingredients

Bhasma	Ingredients
Abhrak Bhasma	Mica
Halthiana Bhasma	Charcoal of elephant tusk
Jasada Bhasma	Zinc oxide
Lauha Bhasma	Iron oxide
Mandura Bhasma	Iron oxide
Mayrapicha Bhasma	Ash of peacock feather
Mukta Bhasma	Oxide of pearl
Naga Bhasma	Lead
Parade Bhasma	Mercury compound
Pravala Bhasma	Oxide of coral
Rajat Bhasma	Silver oxide
Shankha Bhasma	Oxide of conch *Bhasma*
Muktashukti Bhasma	Oxide of pearl, oyster shell
Talaka Bhasma	Arsenic sulfide
Tamra Bhasma	Cupric oxide
Vanga Bhasma	Tin compound

Source: Pal et al. 2014

Table 17.2 Ayurvedic Bhasma ingredients, dosage, and uses

Name	Ingredients	Dosage	Uses
Navratan kalpamrit ras	Calcined ash of expensive gems, minerals like ruby, sapphire, emerald, cat's eye stone, pearl, coral, silver, gold, iron, zinc	62.5 mg twice daily	Cancers of all types, anemia, complications of diabetes
Heerak Bhasma	Diamond	12.5–25 mg twice daily	Useful in cancers, immunity disorders, crippling rheumatoid arthritis, bone marrow depression
Trailokya chintamani ras	Diamond, gold, silver, iron	62.5 mg twice daily	Severe respiratory tract infections, bone marrow depression, ovarian cysts, uterine fibroids
Swarna basant malti ras	Gold, *Piper nigrum*, white pearl powder	62.5 mg twice daily	Tonsillitis, fevers, cough, bronchitis, decreased immunity, cancers, autoimmune disorders
Kamdudha ras	Ochre, *Tinospora cordifolia*, mica (calcined)	250–500 mg twice daily	Hyperacidity, headache, fever, blood pressure
Vasant kusumakar ras	Gold, silver, coral	62.5–125 mg twice daily	Complications of diabetes, neuropathy, general weakness

(continued)

Table 17.2 (continued)

Name	Ingredients	Dosage	Uses
Kumar kalyan ras	Gold, iron, mica, copper pyrite, red sulfide of mercury	62.5–125 mg twice daily	General debility in children, fever, respiratory tract infections
Tamra Bhasma	Copper, mercury, sulfur	62.5–250 mg twice daily	Anemia, jaundice, digestive disturbance, abdominal disorders
Lauha Bhasma	Iron, cinnabar	125–250 mg twice daily	Enlargement of liver, anemia, jaundice
Vaikrant Bhasma	Manganese, sulfur (tourmaline)	62.5–125 mg twice daily	Diabetes, can be used in place of diamond ash in case of poor patients
Loknath ras	Mercury, sulfur, conch shell	62.5–125 mg twice daily	Diarrhea, respiratory disorders, immunity disorders, cancers, ovarian cysts
Abhrak Bhasma	Calcined purified mica ash	125–250 mg twice daily	Respiratory disorders, diabetes, anemia, general weakness
Swarna Bhasma	Ash of gold (calcined gold)	12.5–62.5 mg twice daily	Improves body immunity, general weakness, anemia, energetic
Rajat Bhasma	Silver ash (calcined silver)	62.5–125 mg twice daily	Irritable bowel syndrome, acidity, pitta disorders
Ras raj ras	Red sulfide of mercury, mica, gold, iron, silver, *Withania somnifera*, *Syzygium aromaticum*	62.5–125 mg twice daily	Paralysis, hemiplegia, rheumatism, insomnia, stroke
Shwas kuthar ras	Black sulfide of mercury, *Aconitum ferox*, sodium bicarbonate, *Piper nigrum*, "Trikatu"	125–250 mg twice daily	Cough, pneumonia, bronchitis
Swarn makshik Bhasma	Copper pyrite (calcined), mercury, sulfur	125–250 mg twice daily	Anemia, jaundice, stomatitis, chronic fever
Kaharva pishti	Amber of succinite (Trinkantmani), *Rosa centifolia* (rose)	125–250 mg twice daily	Bleeding
Yogendra rasa	Red sulfide of mercury, gold (calcined), magnetic iron, mica, *Myristica fragrans*	62.5–125 mg twice daily	Polio, paralysis, muscular weakness, insomnia, headache
Bolbadh ras	Black sulfide of mercury, *Tinospora cordifolia*, *Commiphora mukul*	125–250 mg twice daily	Bleeding

(continued)

Table 17.2 (continued)

Name	Ingredients	Dosage	Uses
Praval pishti	Purified powder of corals	125–250 mg twice daily	Calcium deficiency, blood pressure, insomnia, agitation
Praval panchamrit	Powder of corals, pearls, conch shells	125–250 mg twice daily	Richest source of natural calcium, agitation, acidity, burning sensation
Jahar mohra pishti	Powder of serpentine orephite	125–250 mg twice daily	Natural source of calcium, useful in burning sensation, acidity, heart burn,
Sarvatobhadra Vati	Mercury, sulfur (purified and calcined), with gold	62.5–125 mg twice daily	Renal failure, nephrotic syndrome, dialysis, high urea and creatinine
Punarnava mandoor	Iron ore ash, *Boerhavia diffusa*, *Picrorhiza kurroa*, *Embelia ribes*	125–250 mg twice daily	Diuretic, anemia, swelling around joints, blood pressure, liver cirrhosis, ascites
Akik pishti	Agate stone calcined	125–250 mg twice daily	Heat/pitta diseases, blood pressure, acidity, ulcers
Mukta pishti	Pearl powder (motipishti)	62.5–125 mg twice daily	Calcium, cooling and soothing, blood pressure, acne, headaches, acidity, ulcers, heat disorders
Vriht vat chintamani ras	Herbs and minerals for vitiated vata-calcined mercury, sulfur (purified), and other metals and minerals	62.5–125 mg twice daily	Stroke, paralysis, parkinsonism, epilepsy, tetany, muscle stiffness, joint pains

Source: Pal et al. 2014

17.7 Importance of Bhasma

1. Maintain optimum alkalinity for optimum health
2. Provide easily absorbed and usable calcium
3. Cleanse the kidneys, intestines, and liver
4. Maintain stronger bones and healthier teeth
5. Alleviate insomnia and depression
6. Regulates rhythmic heart beating
7. Maintain arrhythmias and mineral balance
8. Help metabolize iron in body
9. Aid nervous system
10. Break down heavy metals and drug residues in the body
11. Neutralize harmful acids that lead to illness
12. Achieve a healthy alkaline level by neutralizing acid
13. Protect body from free radical damage (Pal et al. 2014)

17.7.1 Therapeutic Applications of Bhasmas

Although *Rasa Shastra* is a very important branch of *Ayurveda* since the eighth century report of large-scale randomized clinical trials involving *Bhasmas* is less. One of the reasons for this may be the fact that *Rasa Shastra* is a well-tested science; therefore, there was no need of fresh proof. However, a few properly conducted clinical studies indicated that nutritional anemia in nonpregnant teenager girls can be improved by a daily dose of *Sootshekhar Rasa* (250 mg) plus *Sitopaladi Churna* (400 mg). Another clinical study of *Kukkutandatwak Bhasma* reveals statistically significant improvement in *Swetapradara*, an important gynecological disorder. Likewise, "Swarna Bhasma" has shown some responses in the treatment of solid tumor, and certain herbo-mineral preparations were found to be effective in leukemia. *Swarna Bhasma* also has antioxidant/restorative effects against global and focal models of ischemia (stroke). *Naga Bhasma* (lead calx) is a potent metallic formulation mainly indicated in the treatment of *Prameha* (diabetes) (Pal 2015). Traditionally used Ayurvedic *Bhasma*, their ingredients, dosage, and uses have been summarized in Tables 17.2 and 17.3.

Table 17.3 Bhasmas, description and utility

Bhasma	Description	Utility
Calcium	Pearls and ghee (milk preparation)	Cough impotency, eye disorders, tuberculosis, spree, nervine sedative, used in hyperacidity, asthma, cough and nervous excitement in growing children and pregnant women
1. Mukta moti	Pearls and rose water pearls	
2. Muktashukti	Conch shell	
3. Praval pishti		
4. Shankh		Respiration, cough, heart diseases, stomach, liver, intestine
		Antacid, used in cough, phthisis, scrofulous, affections, spermatorrhea
		Pulmonary hemorrhage and calcium deficiency
		Antiperiodic, carminative, and analgesic, used in colic flatulence and tympanites
Iron	Magnetic iron (purified)	Spree, stomach disorders, anemia, diabetes, blood disorders, restorative
5. Vanaspati yog Lauh	Magnetic iron (purified), ash of incinerated magnetic iron	Hematinic, astringent, jaundice, disorders of liver and spleen
6. Kant Lauh	Ash of incinerated purified ferric oxide	Antirheumatic, hematinic, and used in anemia
7. Mandoor	Ferrum (purified), incinerated and potentiated, rubbed with trifala decoction	Alterative, hematinic, diuretic, used in anemia, edema, chlorosis, rickets, and jaundice
8. Trifala Yog Lauh		Strengthening the body, deficiency of iron, anemia, indigestion

(continued)

Table 17.3 (continued)

Bhasma	Description	Utility
Zinc	Zinc/ShudhYashad	Dysentery, sweating, phthisis, tuberculosis, diabetes, hypoglycemic, astringent used in urinary disorders
9. Yashad	Zinc carbonate/ash of incinerated purified zinc carbonate	
(a) Baidyanath		
(b) Deshrakshak		
10. Kharpar		Antacid, bone strengthening,
Mercury	Mercury	Physical disorders, strengthening the body, fever, malaria, asthma
11. Siddha Makardhwaj		
12. Parad	Mercury	Syphilis, genital disorders, rejuvenation
Silver	Silver	Wasting, nerve disorders, brain functions, eye disorders, tuberculosis
13.Rajat		
Potassium	Potassium nitrate, alum, ammonium nitrate (crystal powder)	Acidity, calculi, urinary tract infection, enlargement of prostate
14. Swet Parpati		
Arsenic	Arsenic	Nervine tonic, asthma, leukoderma, paralysis, and impotency
15. Kushta khas		
Copper	Ash of incinerated purified copper	Acidity, ascites, jaundice, piles, leprosy, leukoderma, asthma, tuberculosis, cough, skin diseases, obesity, chronic bloating, spleen and liver enlargement, cirrhosis
16. Tamra		
Tin	Tin	Asthma, cough, sweating, blood disorders, diabetes, diuretic and urinary antiseptic, semen disorder, syphilis, and gonorrhea
17. Vanga		
Stone	Ash of incinerated purified	Heart-related disorders, blood pressure, vomiting, burning sensation (pitta) cholera, antidote to poison, provides strength, potency, and vigor
18. Jahar Mohra Khatai Pishti	Serpentine orephite	Anemia, ascites, asthma, tuberculosis, diabetes and cancer, substitute of diamond Bhasma
19. Vaikrant	Ash of incinerated purified	
	Tourmaline	

Source: Kumar et al. 2006

17.7.2 Bhasmas as Nanoparticles

A scientific analysis of *Swarna Bhasma* by TEM and AFM has demonstrated that the principle ingredient of *Swarna Bhasma* is globular gold nanoparticles of 56–57 nm. Atomic absorption spectroscopy and IR spectroscopy studies reveal that *Swarna Bhasma* is devoid of any other heavy metal or organic material. Likewise, *Ras Sindoor* (sublimed mercury compound) contains mercury sulfide (crystalline; size, 25–50 nm). This is an organic macromolecule derived from plant extract. Several macro-/trace elements may be present in different amounts, which are bio-available and accountable for adding to medicinal value of *Ras Sindoor*. Study reveals in physicochemical characterization of *Jasada Bhasma* by XPS, ICP, EDAX,

DLS, and TEM that the particles are in oxygen-deficient state and many of them are in nanometer size range. This reports size range of *Jasada Bhasma* might impart its therapeutic property (Pal 2015).

17.7.2.1 Swarna Bhasma

Nobel metals and their compounds as a therapeutic agent, mainly of gold, have a long and distinguished history in medicine. *Swarna Bhasma* has a unique place in the Ayurvedic system of medicine. It is an integral part of Ayurveda, which describes its usage for the treatment of patients with various chronic disorders (Thakur et al. 2017). Containing gold particles, it has shown its anticancer activity. Since ancient times, *Swarna Bhasma*, prepared from gold, is used in Ayurvedic treatment of TB, infertility, asthma, tissue wasting, poisoning, etc. The last few years have witnessed extremely rapid development of nanotechnology, which seamlessly integrates many disciplines including biotechnology, medicine, chemistry, engineering, materials science, and physics. As the size of matter decreases from micrometric to nanometric dimensions, it exhibits novel physical and chemical properties because of increase in surface to volume ratio. Gold nanoparticles have found wide range of applications for diagnosis and targeted drug delivery in nanomedicine because of their chemical stability, surface chemistry, and unique optical properties. Some reports suggest that *Bhasmas* that are metallomedicines in powder form contain nanoparticles. It is suggested that because of nanometric dimensions of these particles, they may provide physiological basis of their action (Das et al. 2012; Rathore et al. 2013).

Swarna makshika is a mineral having different therapeutic uses and has been used since long in Ayurveda. *Swarna makshika* is used for the treatment of anemia, insomnia, convulsions, and skin diseases. It is also used as a single constituent formulation or in multi-ingredient formulation. Swarna makshika contains iron. *Bhasma* contains Fe_2O_3, FeS_2, CuS, and SiO_2 (Rathore et al. 2013).

Swarna Bhasma (incinerated gold) acts as *Vrishya* (aphrodisiac), *Hridya* (cardiac stimulant), and *Rasayana* (immunomodulator). It increases *Valya* (potentiality), *Kantikara* (complexion), *Ayushkara* (longevity), and *Medha Smriti Mati Pradam* (intellect, memory, and attentiveness). It diminishes disorders caused by all the three vitiated *doshas* and is used in the management of poisoning (*Visha Mukti)*. *Swarna Bhasma* is indicated in *Yakshma* (tuberculosis), *Unmada* (schizophrenia), *Jwara* (fever), *Shoka* (grief), *Pandu* (anemia), *Shwasa* (dyspnea), *Kasa* (cough), *Krimi* (worm infestation), *Aruchi* (anorexia), *Chakshuroga* (ophthalmic disorders), and *Visha* (poisoning) (Prajapati et al. 2006; Sarkar et al. 2010).

17.7.2.2 Rajat Bhasma

It has been shown that *Rajat Bhasma* based on Ag acts on the brain and nervous system through a nutritive mechanism. In lower doses, it acts as an anxiolytic, but in higher doses (10–20 mg/kg), it reduces behavioral despair. It is also recommended for eye disorders and tuberculosis. *Rajat Bhasma* contained 23.4 % Ag in addition to As (14.2 %), P (5.14 %), and Na (1.28 %), with Mn (183 μg/g) and Au (140 ng/g) in trace amounts. It also showed 19.9 % S, suggesting the possibility of silver sulfide (Ag_2S) or other sulfides (possibly As_2S_5) in addition to C (0.63 %) and H (0.25 %) resulting from some minor organic constituents such as polycyclic aromatic hydrocarbons (Kumar et al. 2006).

Rajat Bhasma (incinerated silver) possesses *Vrishya* (aphrodisiac), *Vayasthapana* (antiaging), *Lekhana* (scraping), and *Rasayana* (immunomodulator) properties. It increases potentiality (*Vaya*) and intellect *(Medha)*. It eradicates diseases caused by all the three vitiated *doshas. Rajat Bhasma* is used in *Prameha* (diabetes), *Switra* (vitiligo), *Yakshma* (tuberculosis), *Pandu* (anemia), *Shwasa* (dyspnea), *Kasa* (cough), *Nayanaroga* (ophthalmic disorders), *Arsha* (piles), *Trishna* (thirst), *Shosha* (emaciation), and *Visha* (poisoning) (Sarkar et al. 2010).

17.7.2.3 Parada (Mercury) Bhasma

Mercury is used in therapeutics in a compound *(Murchtai)* form. These mercurial compounds are called *Murchita Parada* and possess *Vrishya* (aphrodisiac), *Vardhakya Harana* (antiaging), and *Rasayana* (immunomodulatory) properties. These increase potentiality (*Valakara*), intellect, memory, attentiveness, complexion *(Buddh, Smirlti, Prabha, Kanti Pradam*), and tissue elements (*Dhatu*). These eliminate diseases caused by all the three vitiated *doshas (humoral principles)* even restricting death *(MrityuNasaka)*. Mercurial preparations are used in *Pandu* (anemia), *Shwasa* (dyspnea), *Kasa* (cough), *Kamala* (jaundice), *Jwara* (fever), *Shula* (spasmodic pain), *Mutrakriccha* (nephritis), *Vamana* (vomiting), *Udara Pida* (acute abdomen), *Krimi Dosa* (worm infestation), *Atisara* (diarrhea), etc. (Sarkar et al. 2010).

17.7.2.4 Tamra Bhasma

Copper is one such metal in the Ayurvedic system of medicine, which is used for preparation of *Tamra Bhasma* and is recommended in the dose of 10–30 mg for an adult (70 kg body weight; 0.2 mg/kg) to manage liver disorders, gastrointestinal tract (GIT) disorders, old age diseases, leukoderma, cardiac problems, and various other free radical-mediated disorders, besides alone or as herbo-mineral compositions. Deficiency of copper in the body causes weight loss, bone disorders, microcytic hypochromic anemia, hypopigmentation, graying of hair, demyelination of nerves, etc. (Pattanaik et al. 2003).

Ayurvedic *Tamra Bhasma* is derived from metallic copper that is recommended for different ailments of the liver and spleen, abdominal pains, colitis, heart problems, anemia, tumors, loss of appetite, dropsy, eye troubles, and tuberculosis. Recently, pharmacological investigations have been reported on the use of *Tamra Bhasma* for treatment of gastric ulcers and secretion and management of lipid peroxidation in the liver of albino rats and free radical-scavenging properties. Patil et al. (1987) examine the effect of *Tamra Bhasma* on lipases and lypolytic activities in CCl_4-induced hepatic injury in rats. In case of *Tamra Bhasma*, crystallite size of CuO is found to be higher than that of a standard copper oxide, causing the reduction in its surface area. The preparation process of *Tamra Bhasma* involves repeated calcination cycles, thus facilitating agglomeration and hence bigger crystallites (Wadekar et al. 2005; Waghmare et al. 2016).

It's used for its rejuvenating and antioxidant property; it is also having beneficial effect as aphrodisiac agent. It is known to have properties of maintaining body circulation and tonicity. *Tamra* is included in the group of *Lauha/Dhatu* (metals). It is classified in *Sar Sadharana Lauh* group. Apart from availability of *Tamra* in native form, its different mineral and animal sources (earthworms and feathers of peacock) are also mentioned in the classics (Rai et al. 2008).

17.7.2.5 Abhrak Bhasma

Abhrak Bhasma, a herbo-mineral product of Ayurveda, acts as a tremendous antimicrobial agent. It acts as a synergistic agent, restoring the libido of men. Being a life-promoting drug, it helps in proliferation and synthesis of the sperms. It has a property of oleation. *Abhrak Bhasma* is called as a wonder drug due to its curative property in various ailments. It is the *Bhasma* of the mineral, mica. It contains Fe as a major element and Ca, K, and Si in low concentrations. Its synthesis involves repeated calcinations which transforms the metallic state into corresponding oxide form. It is widely used for the treatment of hepatic dysfunction, leukemia, sex debility, azoospermia, cystic fibrosis, postencephalic dysfunction, and cervical dysplasia (Buwa et al. 2001; Yadav et al. 2016).

17.7.2.6 Lauha Bhasma

Lauha Bhasma is an iron-based Bhasma, which is prepared from iron ore by a process known as *Bhasmikaran*. The procedure of preparation involves several steps, which include *Shodan* (purification), *Maran* (powdering), *Chalan* (stirring), *Dhavan* (washing), *Galan* (filtering), *Putan* (heating), and *Mardan* (triturating). The *Shodan* or the purification process removes the unwanted materials from the raw material and makes it suitable for the next step. During this process, the raw material is heated to red-hot condition and dipped into various agents such as oil, buttermilk, cow's urine, and rice gruel and horse-gram decoction (Rajendran et al. 2012). *Lauha Bhasma* is prescribed for the treatment of anemia due to iron deficiency. Consisting of Fe_2O_3 and Fe_3O_4, the preparation of *Lauha Bhasma* involves a rigorous procedure meant to convert the metal in to a fine, nontoxic and bioavailable form (Yadav et al. 2016). According to *Rasaratna Samuchchaya*, *kanta lauha* (magnetite Fe ore) is measured as best raw material variety of Fe for *Lauha Bhasma*. However in the absence of *kanta lauha*, *Teekshna lauha* (Fe turning) is used for the preparation of *Lauha Bhasma* (Singh et al. 2016).

17.7.2.7 Mandura Bhasma

Mandura Bhasma, an iron-containing preparation, has been used in therapeutics of anemia, jaundice, poor digestion, edema, skin diseases, etc. Generally, *Mandura Bhasma* is prepared in two steps: purification, procedure performs by heating to red-hot state and quenching in liquid media like cow's urine or *Triphala* decoction, and calcination, by *puta* system of heating in *Gaja puta* (specific amount of heat given through fixed quantity of fuel). During calcination, purified *Mandura* is levitated with *Triphala* (fruits *of Terminalia chebula Retz, Terminalia belerica Rox, Emblica officinalis Gaertn*) decoction and *Aloe vera* (*Aloe barbadensis* Mill) juice, etc. Mandura contain iron and silicate and *Mandura Bhasma* uses sensitive tools and techniques. *Mandura Bhasma* contains Fe_2O_3 and SiO_2. *Mandura Bhasma* are uniformly arranged in agglomerates of sizes 200–300 nm as compared to the raw *Mandura* which show a scattered arrangement of grains of sizes 10–2 μm (Mulik and Jha 2011).

17.7.2.8 Hiraka Bhasma

A gemstone *Hiraka* (diamond) is a popular one. The *Bhasma* of *Hiraka* is a well-known organomineral preparation, used for *Rasayana*, *Ayushya*, *Vrishya*, *Tridoshagna*, *Prameha*, *Arbuda*, etc. *Hiraka Bhasma* is not generally available in the market due to lack of standards and high cost, but at the same time it is being manufactured by some physicians and industries in the day-to-day practice of life. Analytically *Hiraka Bhasma* contains Fe_2O_3 as a major compound. Color of the *Hiraka Bhasma* is *malina rakta* which is due to presence of Fe_2O_3 (Zala et al. 2016). *Hiraka Bhasma* prepared from natural diamond is an important drug. It is an excellent remedy for heart troubles, heart pains, contraction of veins, and blood clotting. It is also a powerful tonic and antitumor agent. It's a carbon-based drug. It contains C, O_2, Na, Mg, Al, Si, P, S, K, Ca, Cr, and Fe (Acharya et al. 2014).

17.7.2.9 Jasada Bhasma

Jasada Bhasma is a unique preparation of zinc belonging to this class of *Bhasma*. The *Jasada Bhasma* is zinc oxide with nano- to micron-sized particles. The particles are polycrystalline. Its in-process intermediate on the other hand is mixture of zinc sulfide and oxide, with predominance of sulfide phase. Intermediate too has nano- to micron-sized particles (Chavare et al. 2017). This particular preparation has been successfully used by traditional practitioners for the treatment of diabetes and age-related eye diseases (Bhowmick et al. 2009; Chandran et al. 2016). *Jasada Bhasma* is cited for use in several other conditions including anemia, neuromuscular diseases, and eye diseases and as a wound healing, antimicrobial, and antiaging agent. An early anecdotal study shows antidiabetic activity of *Jasada Bhasma* in diabetic patients. The traditional use of *Jasada Bhasma* is suggestive of antioxidant and

Fig. 17.1 Naga Bhasma (Source: Nagarajan et al. 2014)

immunomodulatory effects. There are sporadic reports on reduction of fasted glucose levels, improved glucose tolerance, and antidiabetic activity in rats treated with *Jasada Bhasma.* In fact, a modern version of *Jasada Bhasma*, viz., zinc oxide nanoparticles, has been systematically investigated for their antidiabetic effect (Umrani et al. 2013).

17.7.2.10 Naga Bhasma

Naga Bhasma (Fig. 17.1) has been used to treat a variety of ailments, and different Ayurvedic formulations containing *Naga Bhasma* are available (Nagarajan et al. 2014a, b); it has its history of medicinal applications dating several centuries back. *Naga Bhasma*, with its principal chemical species being PbS, administered at 6 mg/kg body weight is found to be nontoxic in animal model. Naga Bhasma has specific regenerative potential on germinal epithelium of testes in $CdCl_2$-administered albino rats. In addition to treating diabetes mellitus, it has been prescribed for certain disorders related to the liver, spleen, and skin. Few clinical trials have also shown that *Naga Bhasma* significantly reduces blood glucose level in diabetic patients (Nagarajan et al. 2014a). *Naga (lead)* has been administered in various diseases since *Vedic* period. It has been used in treating diabetes, diarrhea, and spleen and skin disorders and has shown testis regenerative potential on partially degenerated testis (Rajput et al. 2013; Shweta and Thakare 2013).

17.7.2.11 Vanga Bhasma

Vanga Bhasma is a tin-based herbo-metallic preparation given for the treatment of urinary diseases, loss of appetite, and inflammatory disorders, among others (Hiremath et al. 2010). It contains Sn (43.8%) and significant amounts of Ca (7.35 %), Fe (0.3 %), and K (0.88 %) in addition to μg/g amounts of P (720), Mn (257), Zn (67), and In (17.1), but the main content is SnO_2 which is the main constituent, and indium is a rare element frequently present in stannite and other complex sulfides of tin because of similarities in properties. It also contains C (4.2 %), H (0.64 %), and S (0.15 %), suggesting the presence of some sulfides of tin or organosulfur compounds that might remain chelated with various metals. In Ayurvedic literature, it has been recommended for diabetes, semen disorder, impotency, skin disease, syphilis, and gonorrhea. It is also prescribed for asthma, cough, and blood disorders (Umrani et al. 2013).

17.7.2.12 Praval Bhasma

Praval which is well known as coral in English is used in the form of *Bhasma* and *pisti* in order to cure a variety of ailments since ancient times in Ayurvedic system of medicine. Praval Bhasma is used for treatment of inflammation, cough due to phthisis, unnecessary sweating, cardiac fibrillation, osteoporosis, dysuria, and ligourea. *Praval Bhasma* is also an important ingredient of many formulations such as *Sutika-bharana-rasa*, *Vasantakusumakara rasa*, *Muktapanchamrta rasa*, *Brihat vata chintamani rasa*, *Mahatarunarka rasa*, and *Brahmi vati*. It contains CaO as main ingredient (Mishra et al. 2014). *Praval Bhasma* is well known to increase the intestinal absorption of calcium; it remains to be seen to what extent it would be useful in correcting the metabolic conditions that are beneficial and conducive to bone remineralization (Reddy et al. 2003).

17.7.2.13 Trivanga Bhasma

Trivanga Bhasma is a calcinated metal and mineral based on a trimetallic compound used to treat diabetes and as diuretic and *Napunasakta*, *Prameha*, *Ikshumeha*, *Vandhyatva*, *Swetapradara*, *Vata-Pitta dosha*, and *Shaktivardhaka*. It contains elements like lead, zinc, and tin (Rasheed et al. 2014; Sharma and Singh 1987).

17.7.2.14 Shankha Bhasma

Shankha Bhasma derived from conch shell (Gastropoda, class: Mollusca) is used in the treatment of ulcers, dysentery, dyspepsia, indigestion, and jaundice. The constituent of *Shankha Bhasma* is mainly silicate of magnesia. It induces dose-dependent protection against experimental gastric ulcers. It is known to have antacid property (Pandit et al. 2000).

17.7.2.15 Gemstone Bhasma

Bhasmas based on gemstones, mica, and diamonds find much importance in Ayurveda, but their incineration process is very important. For example, mica (*Abhraka*) is believed to be an excellent rejuvenator for the lungs and for *Rasa*

Dhatu. *Vaikrant* alleviates excess *Vata-Pitta-Kapha*, increases vitality, and can be used as a substitute of diamond Bhasma. *Jahar Mohara* is a mineral stone, also called magnesium silicate and green in color. It is commonly used to neutralize poisonous effects of snakebite, causing vomiting, and is mostly recommended by Unani physicians. Some physicians have used it for heart palpitation, nervousness, depression, and irregular heartbeats (Kumar et al. 2006).

17.7.2.16 Muktashukti Bhasma

Muktashukti Bhasma is a compound consisting of pearl (*moti*), *Aloe vera* Linn (*Guar Patha*), and vinegar (*kanji*). The compound is prepared from the outer covering of the shell (pearl oyster), ground and triturated with *Aloe vera* and vinegar in sufficient amount to make a homogenous paste. Recommended proportions of pearl oyster and *Aloe vera* are in the ratio of 1:4. Medicinal properties have been attributed to this preparation in ancient Ayurveda and Unani systems of medicine. *Muktashukti Bhasma* is used in treatment of tuberculosis, cough, chronic fever, conjunctivitis, abdominal discomfort, biliary disturbances, asthma, heart disease, vomiting, acidity, dyspepsia, dysmenorrhea, general weakness, arthritis, rheumatism, and musculo-skeletal disorders. It is recommended in a dose of 125–375 mg twice or thrice daily in treatment of abovementioned disorders (Chouhan et al. 2010; Parmar et al. 2012). Recent studies have shown that adding heated oyster shells to the diet of elderly patient increased the bone mineral density of the lumbar spine. It is one third to one half as potent as anti-inflammatory as the amino salicylic acid. Further, even as *Muktashukti Bhasma* is widely used for its antipyretic activity, there are no scientific reports on antipyretic activity of *Muktashukti Bhasma* (Dubey et al. 2009). Chemically *Muktashukti Bhasma* consists of calcium carbonate, calcium phosphate, aluminum oxide, magnesium oxide, and organic matter. Its ash or paste is used in ancient Ayurvedic medicine to manage various gastric disorders (Chouhan et al. 2010).

17.7.2.17 Varatika Bhasma

Varatika Bhasma is a herbo-mineral formulation coming under the holistic concept of Ayurveda. It is identified as the outer shell of sea animal *Cypraea moneta* Linn or commonly known as money cowry (Rasheed and Shivashankar 2017). It occurs in the coastal areas of the sea (Figs. 17.2 and 17.3). *Cypraea moneta* is commonly known as the money cowry (Fig. 17.4). Chemically it is carbonate of calcium (Pal et al. 2014).

17.7.2.18 Kasisa Bhasma

Kasisa Bhasma is described very briefly in some of the most valuable *Rasa Gransthas* including *Rasa Tarangini*, *Rasaratna Samuchchaya*, and *Ayurveda Prakasha* (Rajput and Tekale 2011).

Fig. 17.2 *Varatika* (before purification) (Source: Pal et al. 2014)

Fig. 17.3 *Varatika* (after purification) (Source: Pal et al. 2014)

Fig. 17.4 Varatika Bhasma (Source: Pal et al. 2014)

Ayurvedic Bhasmas: Toxicity Issues

Ayurvedic medicines are used widely in India; in spite of that, their long-term safety is till date a question due to presence of toxic metals in them. The American medical research community has sounded a heavy metal warning against Ayurvedic cures. Manufacturers of Ayurvedic medicines are now facing different problems such as inferior quality of raw material, lack of authentication of raw material, nonavailability of standards, deficiency in proper standardization method for single drugs and formulations, and no quality control parameters. The use of inferior grade of raw material, adulteration, and deviations in standard manufacturing practice either intentionally or unintentionally leads to the production of inferior quality products, which not only rise the concern over the efficacy but also the safety. Because of widespread use of Ayurvedic medicines, it has become necessary to lay down stringent parameters to ensure batch to batch consistency and reproducibility (Pal 2015).

The use of metals in medicine is often associated with the question of toxicity. Many studies have so far clearly shown that these are nontoxic but exhibit free radical-scavenging activity because of their antioxidant property. The *Bhasmas* are associated with organic compounds and show significant increased superoxide dismutase and catalase activity, two enzymes that reduce the free radical concentration in the body. The preparation and purification of *Bhasmas* undergo elaborate traditional purification procedures and are well mixed with extracts of herbs, fruits, juices, and so forth. The presence of irrelevant elements present at minor or trace levels is the result of the medium in which they are prepared and, thus, might help in enhancing their potentiation. Metallic preparations offer some advantages over plant drugs by virtue of their stability over a long period, lower doses, easy storability, and sustained availability. These have been in use since ancient times and are still considered useful. However, strict quality control by using contamination-free raw materials is necessary (Kumar et al. 2006). Nanoparticles can have many adverse effects at the cellular level. Adverse outcomes may include organelle or DNA damage, oxidative stress, apoptosis, mutagenesis, and protein up-/downregulation. Interestingly, according to some reports, gold nanoparticles have been claimed to be "nontoxic" in nature. In Ayurveda structural and chemical transformation of metal into metal compounds (*Bhasma*) which are bioabsorbable is the main objective of *marana*. Ayurvedic practices aim to avoid toxicity and adverse effects of these products (Rathore et al. 2013).

17.8 Toxicity of the Metal Nanoparticles

Heavy metals in Ayurvedic formulations have been used for centuries with claimed effectiveness and safety. However, concerns are often raised about the toxicity due to heavy metals used in Ayurvedic formulations. On the basis of preference, 18.7 %

of the population uses Ayurveda for normal ailments, 7.1 % in case of sickness, and 5 % in case of serious ailments. A report by the World Health Organization (WHO) indicates that many people in developing countries still rely on herbal medicine. Majority of people believe that herbal medicines are safe and nontoxic, unlike modern chemotherapeutic agents. Individuals usually use herbal medicine for prolonged periods to achieve a desirable effect (Sengupta et al. 2014).

In *Rasa Shastra*, the metals and the minerals are also termed as "Dhatus" and "Updhatus" because of their definite role in biological systems, i.e., they can sustain body tissues by supplementing some of the essential elements to the tissues, whose deficiency causes many unwanted problems or disease in the body. The available Ayurvedic literature emphasizes the need of metals in maintaining the metabolic equilibrium of the human body. These metals are mercury, gold, silver, copper, iron, tin, lead, zinc, etc. Any deficiency, excess, or imbalance in the composition of these metals leads to certain metabolic and anabolic disorders. Equilibrium state of metals in the human body provides the basis for strong immunity. Therefore, any imbalance in the composition of these metals can cause diseases, and equilibrium of these metals is seen as a preconditioning for a normal immune defense and general health. Each time before burning, the metallic powders are processed with fresh herb juices to neutralize their toxicity. One of the numerous tests the *Bhasma* has to pass through is called "Varitar" which means the *Bhasma*, once ready for internal use, floats on water indicating nonexistence of heavy metal in it. The "Bhasma" is then transformed to compound formulas by mixing herbal powders. Special herbal juices are used for processing the compound formula for no more toxic metals and for nontoxic herbo-metallic compounds (Kumar and Gupta 2012).

In a study it has been shown that metal oxide nanoparticles of with particle sizes ranging from 30 to 45 nm have a significant toxicity effect on mammalian cells. Study data indicate toward a more serious concern and issue of teratogenic toxicity exhibited by these particles, which needs to be addressed further carefully before these particles become a part of our day-to-day medicine and other products. Human exposure to nanoparticles can cause severe risks in terms of health, and therefore toxicity issues should be avoided by chemical approaches such as by surface treatment, functionalization, and composite formation. Immune response of nanoparticle exposure is another perspective of nanoparticle interaction that needs to be studied in detail since nanoparticles have immunomodulatory potential and also to stimulate or to suppress the immune system. However, both conditions are undesirable, and the successful nanoparticle-based therapeutics should avoid immunostimulatory or immunosuppressive reactions to the nanomaterials once administered into the body. Nanoparticles can penetrate deeply into the body, having larger surface area because of small size, and possess charge on their surface; therefore, they can cause a greater inflammatory response. Mercury toxicity leads to Alzheimer's disease, Parkinson's disease, gastrointestinal symptoms, renal dysfunction, and neuropsychiatric abnormalities. According to Ayurvedic literature, the toxic effects of Hg are neutralized in the presence of sulfur. Adaptogenic effects (growth promoting, rejuvenating, and facilitating the learning process) on the central nervous system (CNS) in small doses of 15 mg/kg have been reported. The most renowned of all Hg preparations is *Makaradhwaja*, which acts as rejuvenator (Kumar et al. 2006).

17.9 Physiologically Important Cofactor and Metal Nanoparticles

On a biochemical point of view, cofactors are defined as the ion or molecule which binds to the catalytic site of an enzyme, thereby rendering it active, and are capable of catalyzing cytotoxic reactions. Iron-containing enzymes are mononuclear in nature utilizing molecular oxygen which transfers one or both oxygen atoms to substrates, catalyzing many processes including the biosynthesis of hormones, the metabolism of drugs, DNA and RNA base repair, and the biosynthesis of antibiotics. Magnesium is an important consistent controller of glycolysis and Krebs cycle. Copper exhibits different spectroscopic and chemical properties due to its different ligand environments and coordination number, thereby contributing in various biological processes. Zinc is an essential component of many enzymes involved in many metabolic reactions, thus playing an important role in biological activities (Sengupta et al. 2014).

17.10 Metal Nanoparticles and Physiological Implications

Nanoparticles are just not part of a molecular chess game but have potentially evolved as prodigious tools to interact with biological systems. Nanomedicines and nanotherapeutics are presently hot topics in the researcher's domain owing to their versatile application opportunities and molecular signatures. Uptake of nanoparticles into a broad variety of cells seems to be specific for materials in the range of 50–200 nm. Cellular uptake of nanoparticles has been seen to depend on size, surface properties, cell type, and endocytic pathways, enabling optimization of labeling and selection of cells and nanoparticles for applications in vitro and in vivo. Sunscreens containing metal oxide nanoparticles, mainly zinc oxide, appear transparent on the skin, provide excellent protection against sunburn caused by UV radiation, and are readily absorbed by the skin (Sengupta et al. 2014; Mukkavalli et al. 2017). A study is conducted on the elemental and structural characteristics of the traditional Indian medicine, *Rajat Bhasma* with an average particle size of 350 nm. The PXRD pattern in combination with the elemental analysis indicates a complex, heterogeneous, and nonsymmetric arrangement of atoms within the *Bhasma*. Interaction of these nanoparticles with the different types of biological systems suggested that the *Bhasma* is nontoxic to cells compared to pure silver nanoparticles and facilitates transportation of small and large molecules across epithelial cell monolayer, probably through loosening cell-cell tight junctions.

17.11 Metal Nanoparticles and Biological Applications

Using materials at the nanoscale level provides a range of modification flexibility based on requirements in the biomedical field. Nanoparticles as drug delivery agents not only targeted drug release but also monitor release beside with two or more drugs to give combined effects, lesser side effects, extended half-life of the drug in the systemic solution, and better solubility. Several physiologically important metals for the human body like iron, iron oxides, copper, cobalt, and zinc have found an application in biomedical science. Zinc and zinc oxide nanoparticles are extensively used in sunscreens, biosensors, and cancer therapy and show no adverse effects. The most abundantly used nanoparticle is gold, a recent craze among the researchers, owing to its unique optical signature, flexibility in changes of its properties, easy synthesis, easy surface modification, surface plasmon resonance, fluorescence, and chemical stability, which makes it a potential candidate in the field of biomedical science. It is obvious that in recent years iron and iron oxide nanoparticles are the most abundantly used nanoparticles in the biomedical field. On the basis of their excellent properties for qualifying them as potential biomedical tools at the nanoscale level, iron is also physiologically important to the human body. The other elements such as zinc, cobalt, manganese, selenium, and magnesium are also synthesized and are also of physiological importance. Therefore, these metal nanoparticles can be healthier candidates for biological tools and medicines owing to their human body compatibility and acceptability. Even combinations of physiologically important metals and their use in biomedical science, diagnostics, and medicine may counter the possibilities and limitations of metal toxicity and body rejection (Sengupta et al. 2014).

17.12 Conclusion and Future Perspectives

Nanotechnology is beginning to the change the scale and methods of vascular imaging and drug delivery. Indeed, the NIH roadmap's "Nanomedicine initiatives" envisage that nanoscale technologies will begin to produce more medical benefits within the next 10 years. This includes the progress of nanoscale laboratory-based diagnostic and drug delivery platform devices such as nanoscale cantilevers for chemical force microscopes, microchip devices, nanopore sequencing, etc. The National Cancer Institute has related programs too, with the goal of producing nanometer scale multifunctional entities that can diagnose and deliver therapeutic agents and monitor cancer treatment growth. These include design and engineering of targeted contrast agents that improve the resolution of cancer cells to the single cell level and nanodevices competent of addressing the biological and evolutionary diversity of the multiple cancer cells that make up a tumor within an animal (Sujatha et al. 2016).

At present there are three types of *Bhasmas*: metal based, mineral based, and herbal based. In the future this composition of *Bhasma* can be modified, or new composition of *Bhasma* can be introduced. The method of manufacturing of *Bhasma* can be modified or improved for better quality and nano-property. Future application of this nanomedicine "Ayurvedic Bhasma" is immense in the field of healthcare and treatment. But official guidelines have to be set regarding standardization, toxicity and safety studies, mass production issues, labeling rules, clinical studies, and others.

Nanotechnology of tomorrow, to become a next-door technology in our day-to-day life, needs concern and justification about its physiological acceptability and compatibility. A few recommendations and improvements based on biological recognition of these particles can heighten the therapeutic potential and use of them, making them a part of our regular life:

- While designing nano-based products used as a part of human health, more stress should be given to physiologically important metals, which can simply be accepted by the body system, avoiding unnecessary toxicity and being more compatible.
- Studies on biodistribution, retention, and clearance of these particles should be one of the important concerns.
- Need to appraise the role of physiologically important metals, which form an important part of cofactors, and vitamins required for normal physiological function.
- The physiologically important metals which, when used in nanoforms, can be administered along with dietary supplements such as vitamins.
- Initiatives of designing nanoparticles with two or more metals can be probably better by having multi-domain action.
- Even while designing particles with elements more suitable to the body, it should be kept in mind that the total therapeutic system and intake should not exceed the value of recommended dietary intake, which may elevate unnecessary questions of toxicity.
- Initiatives of minimizing toxicity should be taken while synthesizing metal nanoparticles. This can either be done by modifying or functionalizing the nanoparticle to make it more attuned and less toxic.
- Conventional physical and chemical methods of nanoparticle synthesis can be replaced by synthesizing nanoparticles from biological synthesis, "green synthesis," which would be devoid of toxicity and have more body tolerability.
- It would be better if a comparative study on the bulk and nanomaterial of the same dose, route, and other standard physiologic conditions is studied, which will help in developing better nano-based drugs.
- There is a need to reestablish the RDA requirement of humans based on the metal nanoparticle needs of the body.
- A technology and its development can only be successful if both the good and dark side are countered simultaneously, along with keeping in mind the biological aspect, because eventually the technology-based products have to be finally implemented in living systems. A multidisciplinary field like nanotechnology therefore needs special attention in every perspective, especially on the physiological point of view for its success.

References

Acharya P, Ranjan R, Kirar P, Srivastva S, Singh P. Properties of traditional ayurvedic herbo–mineral formulation bhasma. Indian J Adv Plant Res. 2014;1:18–20.

Amin AR, Kucuk O, Khuri FR, Shin DM. Perspectives for cancer prevention with natural compounds. J Clin Oncol. 2009;27:2712–25.

Arivalagan K, Ravichandran S, Rangasamy K, Karthikeyan E. Nanomaterials and its potential applications. Int J Chem Tech Res. 2011;3:534–8.

Bhowmick TK, Suresh AK, Kane SG, Joshi AC, Bellare JR. Physicochemical characterization of an Indian traditional medicine, *Jasada Bhasma*: detection of nanoparticles containing non–stoichiometric zinc oxide. J Nanopart Res. 2009;11:655–64.

Buwa S, Patil S, Kulkarni PH, Kanase A. Hepatoprotective action of *Abhrak Bhasma*, an Ayurvedic drug in albino rats against hepatitis induced by CCl_4. Indian J Exp Biol. 2001;39:1022–7.

Chandran S, Patgiri B, Galib R, Prasanth D. A review through therapeutic attributes of *Yashada Bhasma*. IJPBA. 2016;7:6–11.

Chavare A, Chowdari P, Ghosh S, et al. Safety and bioactivity studies of *Jasad Bhasma* and its in-process intermediate in Swiss mice. J Ethnopharmacol. 2017;197:73–86.

Chouhan O, Gehlot A, Rathore R, Choudhary R. Anti–peptic ulcer activity of *Muktashukti Bhasma*. Pak J Physiol. 2010;6:29–31.

Conde J, Dias JT, Grazú V, Moros M, Baptista PV, de la Fuente JM. Revisiting 30 years of biofunctionalization and surface chemistry of inorganic nanoparticles for nanomedicine. Front Chem. 2014;2:48.

Das S, Das MC, Paul R. *Swarna Bhasma* in cancer: a prospective clinical study. AYU. 2012;33:365–7.

Dubey N, Dubey N, Mehta RS, Saluja AK, Jain DK. Physicochemical and pharmacological assessment of a traditional biomedicine: *Muktashukti Bhasma*. Songklanakarin J Sci Technol. 2009;31:501–10.

Geetha M, Murthy KNC, Basavraj BV, Ahalya N. Pharmaceutical nanotechnology: present past and future. Int J Pharm Sci Nanotech. 2016;9:3061–72.

Hareshwar S, Mayuri D, Raman B. Preparation of *Abhrak Bhasma* and its evaluation on modern parameters. Int J Ayur Pharm Res. 2017;5:30–6.

Hiremath R, Jha CB, Narang KK. *Vanga Bhasma* and its XRD analysis. Anc Sci Life. 2010;29:24–8.

Ikezawa N. Nanotechnology: encounters of atoms, bits and genomes. NRI papers 37. 2001.

Kumar A, Nair AGC, Reddy AVR, Garg AN. Bhasmas unique Ayurvedic metallic–herbal preparations, chemical characterization. Biol Trace Elem Res. 2006;109:231–54.

Kumar G, Gupta YK. Evidence for safety of Ayurvedic herbal, herbo-metallic and *Bhasma* preparations on neuro behavioral activity and oxidative stress in rats. AYU. 2012;33:569–74.

Mishra A, Mishra AK, Tiwari OP, Jha S. In–house preparation and characterization of an Ayurvedic Bhasma: *Praval bhasma*. J Integr Med. 2014;12:52–8.

Miyazaki K, Nazrul Islam N. Nanotechnology systems of innovation—an analysis of industry and academia research activities. Technovation. 2007;27:661–75.

Mukkavalli S, Chalivendra V, Singh BR. Physico-chemical analysis of herbally prepared silver nanoparticles and its potential as a drug bioenhancer. Open Nano. 2017;2:19–27.

Mulik SB, Jha CB. Physicochemical characterization of an iron based Indian traditional medicine: *Mandura Bhasma*. Anc Sci Life. 2011;31:52–7.

Nagarajan S, Sivaji K, Krishnaswamy S, Pemiah B, Rajan KS, Krishnan UM, Sethuraman S. Safety and toxicity issues associated with lead–based traditional herbo–metallic preparations. J Ethnopharmacol. 2014a;151:1–11.

Nagarajan S, Sivaji K, Krishnaswamy S, Pemiah B, et al. Scientific insights in the preparation and characterization of a lead–based *Naga Bhasma*. Indian J Pharm Sci. 2014b;76:38–45.

Nicolau D. Challenges and opportunities for nanotechnology policies: an Australian perspective. Nanotechnol Law Bus J. 2004;1(4):12.

Othayoth R, Kalivarapu S, Botlagunta M. Nanophytomedicine and drug formulations. Int J Nanotechnol Appl. 2014;4:1–8.

Pal SK. The *Ayurvedic Bhasma*: the ancient science of nanomedicine. Recent Patents Nanomedicine. 2015;5:12–8.

Pal D, Sahu CK, Haldar A. *Bhasma:* the ancient Indian nanomedicine. J Adv Pharm Technol Res. 2014;5:4–12.

Palkhiwala S, Bakshi SR. Engineered nanoparticles: revisiting safety concerns in light of ethno medicine. AYU. 2014;35:237–42.

Pandey A, Pandey G. Usefulness of nanotechnology for herbal medicines. Plant Arch. 2013;13:617–21.

Pandit S, Sur TK, Jana U, Bhattacharyya D, Debnath PK. Anti–ulcer effect of *Shankhabhasma* in rats: a preliminary study. Indian J Pharm. 2000;32:378–80.

Parmar KKG, Galib R, Patgiri BJ. Pharmaceutical standardization of *Jalashukti Bhasma* and *Muktashukti Bhasma*. AYU. 2012;33:136–42.

Patil S, Kanase AK, Varute AT. Effect of hepatoprotective Ayurvedic drugs on lipases following CCl_4 induced hepatic injury in rats. Indian J Exp Biol. 1987;27:955–8.

Pattanaik N, Singh AV, Pandey RS, Singh BK, Kumar M, Dixit SK, Tripathi YB. Toxicology and free radicals scavenging property of *Tamra Bhasma*. Indian J Clin Biochem. 2003;18:181–9.

Prajapati PK, Sarkar PK, Nayak SV, Joshi RD, Ravishankar B. Safety and toxicity profile of some metallic preparations of Ayurveda. Anc Sci Life. 2006;25:57–63.

Rai RK, Jha CB, Chansuriya JPN, Kohli KR. Comparative assessment of antihyperlipidaemic action of *Tamra Bhasma*. Indian J Trad Knowl. 2008;7:335–40.

Rajendran N, Pemiah B, Sekar RK, et al. Role of gallic acid in the preparation of an iron–based Indian traditional medicine – *Lauha Bhasma*. Int J Pharm Pharm Sci. 2012;4:45–8.

Rajput DS, Tekale GS. Study on *Bhasma kalpana* with special reference to the preparation of *Kasisa bhasma*. AYU. 2011;32:554–9.

Rajput D, Patgiri BJ, Galib R, Prajapati PK. Anti-diabetic formulations of *Naga Bhasma* (lead calx): a brief review. Anc Sci Life. 2013;33:52–9.

Rasheed SP, Shivashankar M. Preparation and characterization of metal oxide as nanoparticles – *Varatika Bhasma*. Mech Mat Sci Eng. 2017;9:1–6.

Rasheed A, Naik M, Pillanayil K, et al. Formulation, characterization and comparative evaluation of *Trivanga Bhasma*: a herbo–mineral Indian traditional medicine. Pak J Pharm Sci. 2014;27:793–800.

Rathore M, Joshi DS, Kadam SN, Bapat RD. *Swarna Bhasmas* do contain nanoparticles? Int J Pharm Bio Sci. 2013;4:243–9.

Reddy PN, Lakshmana M, Udupa UV. Effect of *Praval Bhasma* (Coral calx), a natural source of rich calcium on bone mineralization in rats. Pharmacol Res. 2003;48:593–9.

Sarkar PK, Chaudhary AK. Ayurvedic *Bhasma*: the most ancient application of nanomedicine. J Sci Ind Res. 2010;69:901–5.

Sarkar PK, Das S, Prajapati PK. Ancient concept of metal pharmacology based on Ayurvedic literature. Anc Sci Life. 2010;29:1–6.

Sengupta J, Ghosh S, Datta P, Gomes A, Gomes A. Physiologically important metal nanoparticles and their toxicity. J Nanosci Nanotechnol. 2014;14:990–1006.

Shahverdi AR, Fakhimi A, Shahverdi HR, Minaian S. Synthesis and effect of silver nanoparticles on the antibacterial activity of different antibiotics against Staphylococcus aureus and Escherichia coli. Nanomedicine. 2007;3(2):168–71.

Sharma PV, Singh VP. Standardization of an Ayurvedic drug: *Trivanga Bhasma*. Anc Sci Life. 1987;6:148–9.

Shweta Z, Thakare GD. Pharmaceutico analytical study of *Naga* (Lead) *Bhasma*. IAMJ. 2013;1:1–4.

Singh M, Singh S, Prasad S, Gambhir IS. Nanotechnology in medicine and antibacterial effect of silver nanoparticles. Digest J Nanomaters Biostruct. 2008;3:115–22.

Singh TR, Gupta LN, Kumar N. Standard manufacturing procedure of *Teekshna Lauha Bhasma*. J Ayurveda Integ Med. 2016;7:100–8.

Sujatha PL, Kumarasamy P, Preetha SP, Balachandran C, Karthick J. Nanomedicine: fate and fortune in future to tailor a device at a billionth of a meter, about half the width of a DNA. Int J Gen Med Pharm. 2016;5:19–24.

Thakur K, Gudi R, Mahesh Vahalia M, et al. Preparation and characterization of *Suvarna Bhasma* parada marit. J Pharm. 2017;20:36–44.

Umrani RD, Agrawal DS, Paknikar KM. Antidiabetic activity and safety assessment of Ayurvedic medicine, *Jasad Bhasma* (zinc ash) in rats. Indian J Exp Biol. 2013;51:811–22.

Varshney HM, Shailender M. "Nanotechnology" current status in pharmaceutical science: a review. Int J Therap Appl. 2012;6:14–24.

Wadekar MP, Rode CV, Bendale YN, et al. Preparation and characterization of a copper based Indian traditional drug: *Tamra Bhasma*. J Pharm Biomed Anal. 2005;39:951–5.

Waghmare AH, Dharkar NS, Waykole YC, et al. Pharmaceutical and physico–chemical study of *Tamra Bhasma* incinerated copper. World J Pharm Pharm Sci. 2016;5:1432–41.

Yadav D, Upadhyay SK, Anwar MF, Unnithan JS. A review on the patents of various metal nanoparticles: preparations and formulations. World J Pharm Pharm Sci. 2016;5:1309–17.

Zala UU, Roa SS, Prajapati PK, Vaishnav PU. A comparative pharmaceutical study of *Hiraka* bhasma. Pharm Sci Monitor. 2016;7:25–7.

Part II
Toxicity Issues

Chapter 18
Gold Nanoparticle Biodistribution and Toxicity: Role of Biological Corona in Relation with Nanoparticle Characteristics

Catherine Carnovale, Gary Bryant, Ravi Shukla, and Vipul Bansal

Abstract It has long been acknowledged that parameters such as nanoparticle size, shape, and surface charge play distinct roles in the way nanomaterials interact with their surrounding biological environment. However, the importance of a nanoparticle's biological corona and the magnitude of its effect have become subject of recent attention. For the purpose of this chapter, we intend to consider the biological corona as the layer of organic molecules, typically proteins, derived from biological systems that bind to nanoparticle surfaces when nanoparticles are exposed to a biological environment. The protein corona has profound implications on cellular uptake and toxicity of gold nanoparticles (AuNPs), though a deeper understanding of the properties that govern corona formation would provide a greater opportunity for scientists to create tailored AuNPs for specific applications. Specifically, the information provided in this chapter predominantly deals with AuNPs; however these principles are likely to be extended to other nanomaterials.

Keywords Gold nanoparticle (AuNPs) • Size • Shape • Protein corona • Cellular uptake • Toxicity

C. Carnovale
Ian Potter NanoBioSensing Facility, NanoBiotechnology Research Laboratory, School of Science, RMIT University, GPO Box 2476, Melbourne, VIC 3001, Australia

Centre for Molecular and Nanoscale Physics, School of Science, RMIT University, GPO Box 2476, Melbourne, VIC 3001, Australia

G. Bryant
Centre for Molecular and Nanoscale Physics, School of Science, RMIT University, GPO Box 2476, Melbourne, VIC 3001, Australia

R. Shukla • V. Bansal (✉)
Ian Potter NanoBioSensing Facility, NanoBiotechnology Research Laboratory, School of Science, RMIT University, GPO Box 2476, Melbourne, VIC 3001, Australia
e-mail: vipul.bansal@rmit.edu.au

M. Rai, R. Shegokar (eds.), *Metal Nanoparticles in Pharma*,
DOI 10.1007/978-3-319-63790-7_18

18.1 Introduction

It is thought that the protein corona confers a nanoparticle its biological identity, whereby cellular interactions take place based on the abundance and conformation of proteins bound on the particle's surface, rather than the bare surface itself (Lundqvist et al. 2004; Alkilany et al. 2009; Carnovale et al. 2016). However the properties of the underlying AuNP – that is, its size, shape, surface charge, and subsequent functionalization – are thought to have a bearing on the eventual corona that forms (Fig. 18.1) (Casals et al. 2011; Rahman et al. 2013). Using this knowledge, it may be possible to design AuNP-based therapies which target specific cell types and cellular locations or, indeed, even evade cellular uptake, by linking these material properties with coronal formation behavior. This chapter therefore attempts to correlate the effect of modifications in AuNP size, shape, and surface characteristics with protein corona formation.

The so-called protein corona is the term given to the protein-rich coating that is formed at the nanoparticle surface, during a process which commences almost immediately upon the introduction of inorganic materials such as AuNPs to biological fluids (Lynch et al. 2007; Casals et al. 2010; Mahmoudi et al. 2011; Wolfram et al. 2014). The immediate phase, which is often invoked within seconds, involves the

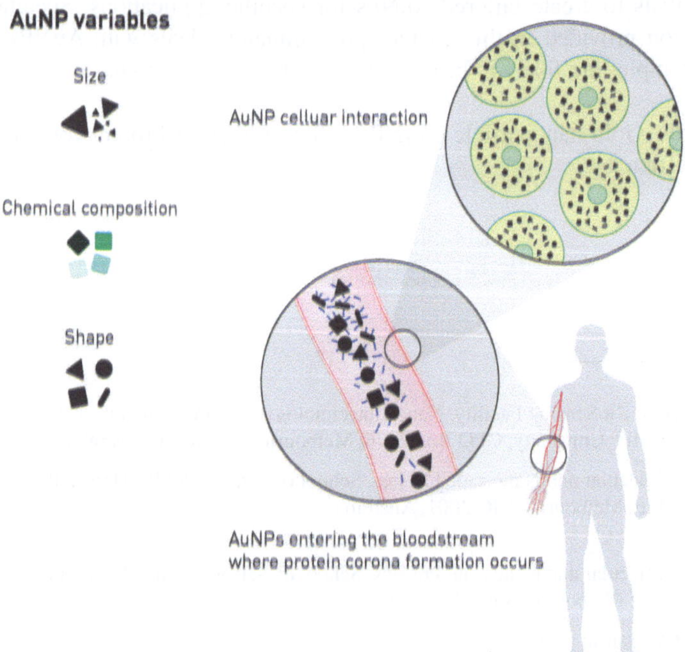

Fig. 18.1 Different physico-chemical properties of gold nanoparticles (AuNPs) that may influence their interaction with proteins. Reprinted with permission from (Carnovale et al. 2016) Progress in Materials Science

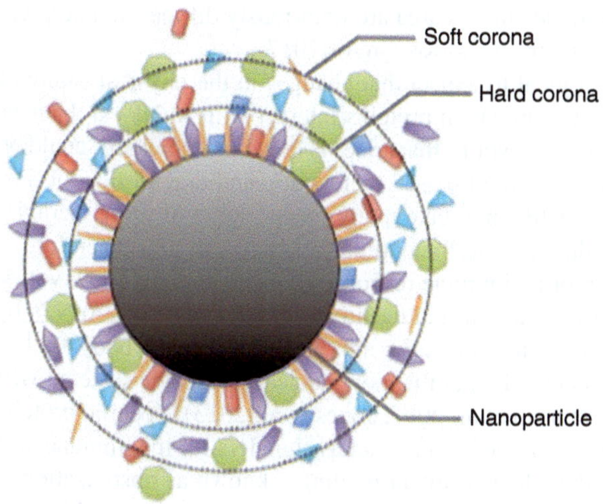

Fig. 18.2 Schematic representation of the progressive formation of the soft and hard coronas around a nanoparticle surface. Reprinted with permission from (Wolfram et al. 2014) Colloids and Surfaces B: Biointerfaces

creation of a soft corona (Fig. 18.2). During this initial period, highly mobile proteins within the surrounding environment decorate the surface of the particle. Over a longer time scale, via means of Vroman's effect, these initial proteins are displaced by others (Ehrenberg et al. 2009). In what is deemed the hard corona, proteins which possess greater affinity for the surface of the particle replace the loosely bound species to form a more stable complex.

This kinetic process ensures that the nature of the corona will transition over time, with respect to the species bound and their relative abundance on the surface. While initially several hundred protein species may compete for a place on the particle's surface as part of the soft corona, it is thought that the hard corona has a more simplified composition, comprising approximately 10–50 different proteins (Mahmoudi et al. 2011). Most commonly, these proteins include albumin, immunoglobulin G, fibrinogen, and apolipoproteins due to their abundance in the blood. However, it should be noted that their concentration within the blood does not necessarily reflect their relative concentration within the corona (Rahman et al. 2013).

To observe the interplay between such proteins and nanomaterials in vitro, models employing bovine serum albumin (BSA) or human serum albumin (HSA) are often used (Lacerda et al. 2010; Sen et al. 2011; Boulos et al. 2013; Khan et al. 2015). Both systems allow observation of the interactions which occur between nanomaterials and serum albumin, which is normally produced by the liver and makes up approximately half of the total serum proteins found in vivo. This more simplified system allows researchers to create predictions for the behavior of AuNPs at the biological interface. Due to the complexities of these interactions, the vast majority of studies carried out to date have been on simplified in vitro systems, while in vivo studies in this area are notoriously difficult to interpret and relatively scarce (Hadjidemetriou and Kostarelos 2017).

While it is logical to expect that changes in the physical nature of the particles will affect the way in which proteins interact with AuNPs at the nanoscale, one of the most pertinent questions that arises is how this knowledge could be exploited for the development of AuNP-based therapeutic and diagnostic tools. While therapies incorporating AuNPs may take many forms, some of these therapies may require extended circulation time for optimal efficacy and must therefore be designed to avoid initiation of an immune response (Aggarwal et al. 2009; von Maltzahn et al. 2009). Such a response would eventuate in the capture of AuNPs by specific systems which operate to remove foreign material from the body, such as the reticuloendothelial system (RES) or the mononuclear phagocyte system (MPS) (Mosqueira et al. 1999; Pastorino et al. 2007; Jones et al. 2013). Specific proteins called opsonins circulate within the blood to act as signalers for the immune system, and upon locating an AuNP, their resultant binding – known as opsonization – may result in clearance of the particles from the body (Gref et al. 2000; Moghimi et al. 2001; Soppimath et al. 2001; Jones et al. 2013). Known opsonins include but are not limited to immunoglobulins (IgG and IgM) as well as specific complement proteins, such as C3, thrombospondin, and fibrinogen. Naturally the human body is capable of an equal and opposing process, known as disopsonization, whereby specific proteins are capable of having the opposite effect. An example of such a protein is albumin; however one of the most commonly used artificial disopsonins is polyethylene glycol (PEG). Employed by researchers for decades to improve biocompatibility, coating an object in PEG macromolecules or PEGylation makes it possible to decrease the extent of protein adsorption, thereby diminishing the risk of opsonization (Romberg et al. 2008; Jokerst et al. 2011; Jones et al. 2013).

Through a combination of well-known practices and newly emerging research related to the effect of various AuNP traits on protein corona formation, researchers will be able to optimally craft AuNP-based therapies which carry the characteristics necessary for their success. Said relationships are invariably complicated, and as such a critical review of the current state of research is necessary to establish the current level of understanding in this area. Therefore, the overall aim of this chapter is to correlate the effect of modifications in AuNP size, shape, and surface characteristics with the formation of protein corona on the nanoparticle surface.

18.2 Establishing a Relationship Between Gold Nanoparticle Size and Protein Corona Formation

In general, an observable difference has been noted in the protein-binding behavior of AuNPs of different sizes. More specifically, however, changes in the degree of curvature present on the surface of spherical AuNPs are thought to be responsible for the differences observed (Lundqvist et al. 2004; Chithrani and Chan 2007; Klein 2007; Casals et al. 2010; Benetti et al. 2013). When considering spherically shaped particles, the concept of size can be thought of not only as an increase in the particle diameter as the particle gets larger but also a decrease in surface curvature. This effect can be examined by comparing the binding which occurs on the surface of spherical AuNPs with the binding of proteins onto flat gold surfaces. At this point, it may again be noted that such size-dependent effects appear to be relevant to all nanoparticles, regardless of their elemental composition.

As an evidence of this point, Chah et al. (2005) studied the binding of yeast iso-1-cytochrome c (Cyt c) onto the surface of 19 nm AuNPs as well as flat Au films to monitor the conformational changes that proteins undergo upon binding to various Au surfaces. While the group observed Cyt c undergoing reversible conformational changes after binding to the flat Au supports, the same was not observed after binding to the surface of the spherical AuNPs. While the exact reason behind this difference was not established, it can be understood that differences in the geometry of the Au surfaces may either permit or prohibit certain proteins from binding, depending on their own three-dimensional constraints (Karajanagi et al. 2004; Lundqvist et al. 2004; Roach et al. 2006). Furthermore, upon binding, the amount of surface curvature may dictate the degree to which the protein is able to contact and effectively bind to the particle surface (Saptarshi et al. 2013). This determines not only the stability of the protein in its bound position but also has implications on the protein functionality (Fig. 18.3). In the case of Cyt c, it appears that the binding of the protein to surfaces of a particular curvature may diminish protein functionality by inducing conformational or structural changes.

In an effort to understand the implication of AuNP size on the species of proteins which bind, Dobrovolskaia et al. incubated citrate-stabilized 30 and 50 nm AuNPs with plasma proteins (Dobrovolskaia et al. 2009). While this study could not discern

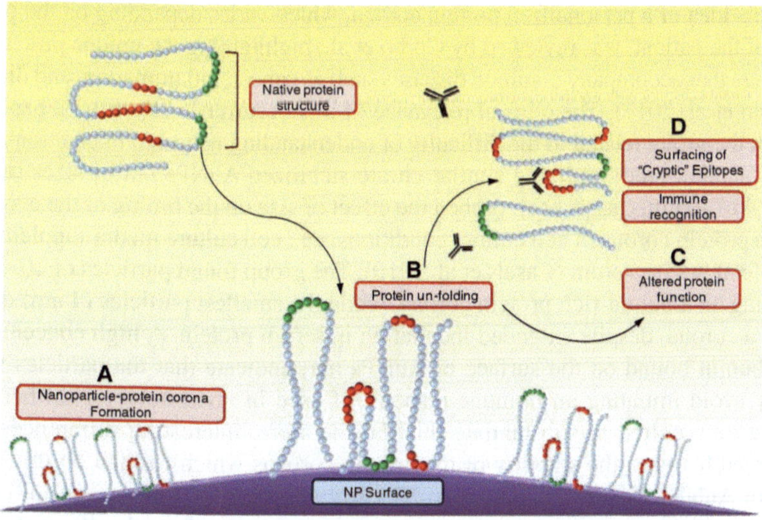

Fig. 18.3 Schematic representation of the likely outcomes that affect protein structure and function after their binding to an AuNP surface. Reprinted with permission from (Saptarshi et al. 2013). Licensed under CC BY 2.0

a size-dependant effect on the amount of bound protein (even after normalizing the dose of the particles for surface area), the findings from this study indicate that 30 nm AuNPs bound a larger range of protein species. This study identified 48 individual proteins on the surface of the 30 nm AuNPs as compared to 21 proteins on the 50 nm particles. Furthermore, despite fibrinogen being found to be the most abundant protein on both AuNPs, there were only 14 proteins common to both particles. These findings indicate that certain proteins may preferentially bind to particles due to their size or surface curvature which will, in turn, have implications on the kinetics, distribution, and clearance of particles within and from the bloodstream in vivo.

Due to their antimicrobial activity, the use of silver nanoparticles as a coating material for implantable medical devices has been a subject of much research (Sondi and Salopek-Sondi 2004; Davoudi et al. 2014). Despite this, Durán et al. have noted a significant gap in the literature, whereby the interaction of silver nanoparticles with plasma proteins has gone relatively unreported (Durán et al. 2015). The review highlights the need for further investigation in this area by discussing an interesting study which showed key differences in the interaction of normal and cancerous human plasma with silver nanoparticles. In the study performed by Feng et al. using surface-enhanced Raman spectroscopy (SERS), the group was able to distinguish the behavior of cervical cancer plasma proteins by measuring the intensity of the SERS peak after incubation with silver nanoparticles (Feng et al. 2013). The Raman feature generated from the amide I band of human serum albumin (HSA) was found to be more intense due to increased expression in the relative amount of proteins in the α-helix conformation, which highlights that specific protein corona variations can be detected in a cancerous environment. For further discussion on the relationship between silver nanoparticles and the evolution of the protein corona, we refer the readers to Durán et al. (2015).

This idea of a personalized protein corona, which varies depending on the pathology of the patient, was reviewed by Corbo et al., highlighting the unique protein biomarkers that accompany common diseases such as cancer, inflammation, and diabetes (Corbo et al. 2017). The clinical relevance of this occurrence is yet to be exploited, which the group related to the difficulty of understanding nano-bio interactions.

In another study utilizing similar citrate-stabilized AuNPs but of sizes ranging from 4 to 40 nm, Casals et al. probed the effect of size on the timing of the evolution of the protein corona in cell culture conditions, i.e., cell culture media supplemented with fetal bovine serum (Casals et al. 2010). The group found particles of 10–40 nm forming an albumin-rich protein corona, while the smallest particles (4 nm) did not form a corona, despite extended incubation in serum protein. A high concentration of albumin bound on the surface of AuNPs may indicate that the particles would likely avoid initiating an immune response if used in vivo; however further work would be required to substantiate this. Furthermore, interesting differences were observed between the stability of the protein coronas which formed on the 10 and 40 nm AuNPs. While the 10 nm AuNPs were capable of forming an initial soft corona, which transitioned to permanent hard corona over time, the 40 nm particles possessed a hard corona that was by comparison both less dense and less persistent. This study inferred that the 10 nm AuNPs were capable of generating the optimal

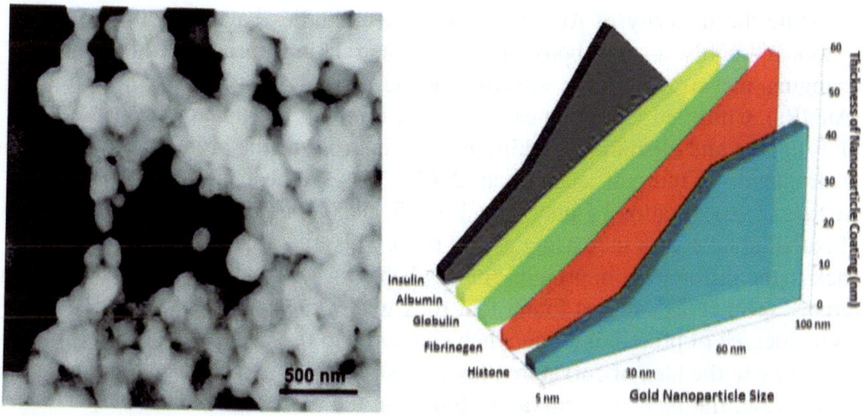

Fig. 18.4 Graphical representation of the increasing protein binding trends with an increase in AuNP size. Adapted with permission from (Lacerda et al. 2010). Copyright 2009 American Chemical Society

protein coverage for evasion of the immune system and might be the ideal size for AuNP-based therapies which require this characteristic.

Encompassing a wider size range of citrate-stabilized AuNPs ranging from 5 to 100 nm, Lacerda and coworkers investigated similar AuNP-protein interactions (Lacerda et al. 2010). While the group observed binding of blood plasma proteins to all sizes examined, the trends indicated stronger binding and a thicker coating of proteins with increasing AuNP size (Fig. 18.4). The trend was shown to be size dependent up to approximately 50 nm, after which saturation was encountered. The group reasoned that the larger AuNPs were able to more efficiently organize and "pack" proteins bound to the surface. Conversely the cooperativity of binding, i.e., the affinity of a particular protein for a nanoparticle after the binding of successive proteins, was shown to decrease with AuNP size in most cases.

Seeking to develop AuNPs with optimal properties for in vivo applications, Walkey et al. explored the relationship between AuNP size and PEG grafting density (Walkey et al. 2012). The group synthesized citrate-stabilized AuNPs of 15–90 nm size and subsequently grafted 5 kDa PEG molecules with grafting densities ranging from 0 (ungrafted) to 1.25 PEG/nm^2. While the ability of PEG to assist in evasion of the MPS is well documented, the specific parameters that control this effect are not fully understood. Macrophage uptake could not be completely avoided; however the grafting density of PEG, and AuNP size, were found to impact on the degree of AuNP uptake by macrophages. As expected, high-density PEG grafting was more effective in blocking macrophage uptake due to blockage of serum protein absorption onto the AuNP surface. Additionally, decreasing AuNP size and therefore increasing curvature on the particle surface led to an increase in the amount of serum proteins adsorbed. It was speculated by the authors that through controlling these two parameters, the degree of macrophage uptake may be potentially manipulated for optimal therapeutic gain.

While the majority of AuNP size-dependent studies have been performed on spherical AuNPs, a small body of work exists relating to rod-shaped AuNPs. By changing the size, or more specifically the aspect ratio of the gold nanorods (AuNRs), Chithrani and Chan examined the effect of multiple parameters on cellular uptake and clearance – taking note of the role of the protein corona in this environment (Chithrani and Chan 2007). By synthesizing transferrin-coated AuNRs of aspect ratios 1 (spherical), 1.5, 3.5, and 6, the group noted a reduced rate of AuNP uptake with increasing aspect ratio. This trend can potentially be related back to the differences in protein-binding behavior which occur on AuNRs of different aspect ratios. It was found that rods with a lower aspect ratio displayed a higher degree of protein binding when compared to rods with higher aspect ratios. Returning to the idea of curvature playing a strong role in protein-binding activity, in this case, the increased curvature found on the extremities of low-aspect ratio rods was hypothesized to be responsible for the trend. Furthermore, this study noted that a higher degree of curvature in this case leads to more strongly bound surface proteins.

While there has been a great volume of research activity around AuNPs, we are currently unable to say if results obtained using a specific cell line are predictive of the behavior of the same AuNP with another cell type. Recently, Cheng and coworkers explored the difference in behavior of phagocytic and non-phagocytic cells interacting with differently sized PEG-coated spherical AuNPs (5–50 nm) (Cheng et al. 2015). This study noted that particles of 50 nm are preferred candidates for cellular uptake. By performing experiments in the absence and presence of serum proteins, a commonly documented trend showing decreased levels of particle uptake in the presence of serum was observed. However, this trend was found to be both size and cell type dependent, being more pronounced for larger particles and in phagocytic cells (Fig. 18.5).

These results illustrate one of the greatest challenges faced by researchers working on in vitro testing of AuNPs and indeed of nanomaterials in general. The difficult nature of grouping results to extrapolate the knowledge we currently possess and apply it in varied conditions dictates that each material be tested individually and in a variety of settings. It is easy to see that this approach is both labor and time intensive. In the attempt to relate AuNP size and protein corona formation, general trends have been observed, most evidently the ability of AuNPs with increased curvature to accommodate proteins with greater ease on their surface.

18.3 Establishing a Relationship Between Gold Nanoparticle Shape and Protein Corona Formation

While arguably holding equal importance to the effect of size, the effect of shape on the evolution of a protein corona around AuNPs is less understood. Spherical nanoparticles are often seen by researchers as a logical starting point for biological studies, and thus, there are markedly fewer studies performed on AuNPs of differing

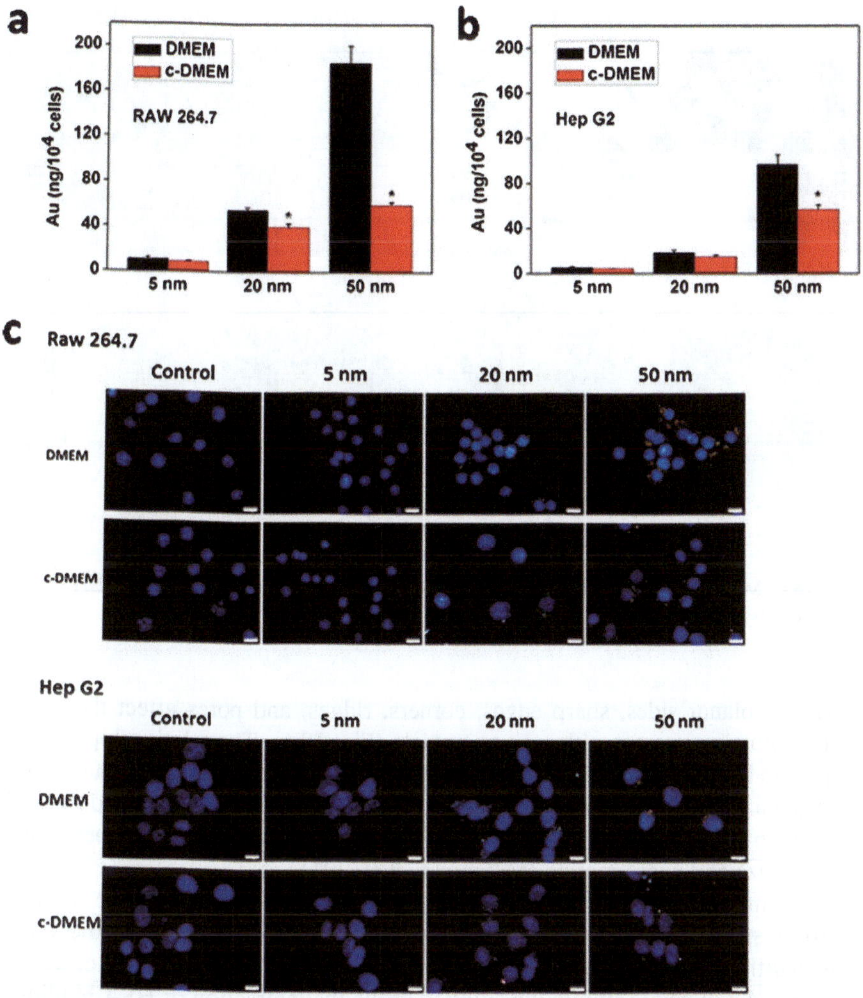

Fig. 18.5 Comparison of size-dependent nanoparticle uptake in phagocytic and non-phagocytic cells in the absence or presence of serum supplemented media (c-DMEM). Adapted with permission from (Cheng et al. 2015). Copyright 2015 American Chemical Society

shape. Furthermore, synthesizing AuNPs of varying shape brings an additional layer of complexity to the synthesis process, particularly when doing comparative studies. For the observation of true shape effects, parameters such as chemical structure, surface charge, and, to an extent, surface area must be kept constant. In this way, only systematic studies which keep these factors constant while solely varying the particle shape can be treated as true indicators of the effect of AuNP shape on protein corona formation. From the small pool of studies conducted, it has been observed that the geometric properties that determine shape such as curved

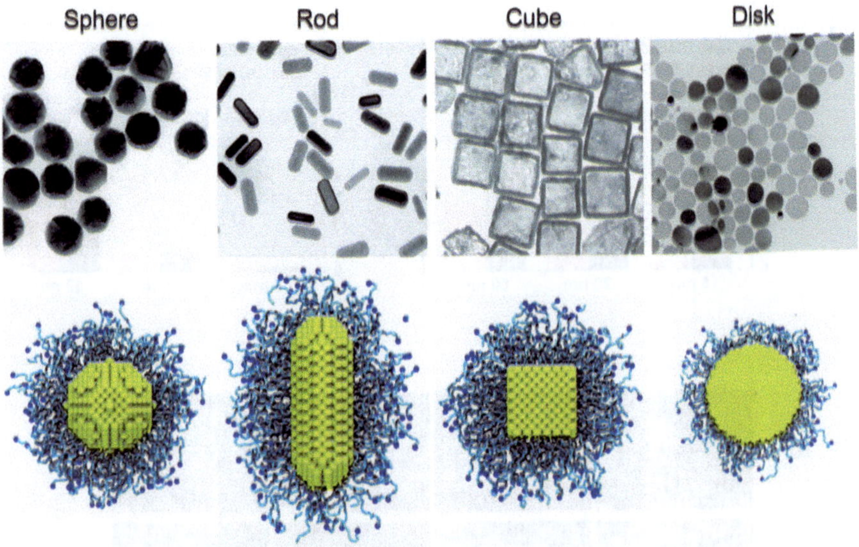

Fig. 18.6 Molecular simulation studies of protein binding on different shapes of AuNPs. Reprinted with permission from (Li et al. 2015). Copyright Royal Society of Chemistry

surfaces, planar sides, sharp edges, corners, ridges, and pores affect the way in which proteins interact with a given particle (Fig. 18.6). The relationship does not appear to be governed by general rules but rather by the specific nature of each protein and nanoparticle. In this way, geometric features have been described as exhibiting both blocking and assistive effects for the binding of proteins depending on protein size and conformation (Mahmoudi et al. 2011).

Due, in part, to the difficulty of performing highly precise systematic studies in vitro, in silico simulations are utilized as a predictive tool to observe the way in which different shaped AuNPs and proteins may interact. Ramezani and coworkers performed a dynamic simulation study to probe the interaction of HSA binding to cubic and spherical AuNPs (Ramezani et al. 2014). The study maintained the common idea that curved surfaces such as those found on spherical AuNPs better accommodate the binding of HSA molecules. Furthermore, this study calculated the distance which would theoretically separate the protein molecule from the AuNPs should they interact. These calculations determined that HSA molecules are indeed able to position closer to the surface of a spherical particle when compared to the cubic particle. However despite the increased distance, the planar surfaces which confine the cubic particle cause the albumin to unfold, generating structural changes much more strongly than spherical AuNPs. These results highlight the possibility that shape can elicit effects on secondary protein structures and may therefore be capable of altering the activity of proteins as they attempt to bind to the particle surface.

Carefully controlled in vitro experiments, such as those by Gagner and coworkers, examined the effect of shape on binding with two specific enzymes – lysozyme and α-chymotrypsin (Gagner et al. 2011). The study attempted to facilitate comparability between the synthesized particles (rods and spheres) by performing surface functionalization to provide similar surfaces to both shapes, despite variations which occurred during the synthesis. Lysozyme showed a minor loss of secondary structure upon binding to both particles, which subsequently caused particle aggregation and led to a major loss of activity. While α-chymotrypsin was able to preserve its structure after interaction with both particles at low concentrations, at higher surface densities where a monolayer of protein was formed on AuNRs, this resulted in significant secondary structure disruption and loss of activity. However, if conditions permitted the development of multiple layers of coverage over the AuNR surface, α-chymotrypsin recovered its original activity level. Such results strongly indicate the effects that different shapes may exert on protein structure and function; however further work is required to understand the rules that govern these interactions. Overall it was noted that both proteins achieved higher binding densities on the surface of AuNRs as compared to AuNSs, which the authors concluded was due to the long cylindrical surface, which is comparatively flatter than that of the AuNS and accommodated a higher number of proteins with greater ease.

Adding an additional layer of complexity to this story, in publications first from Caswell et al. (Caswell et al. 2003) and then also Chang et al. (2005), a contrasting effect was reported. In both cases, the groups utilized AuNRs synthesized using CTAB for which it can be assumed that a residual bilayer of CTAB remains on the surface of the AuNRs, preventing agglomeration. Given Gagner's hypothesis, we would expect preferential binding of proteins to the flat cylindrical surface of the rods; however in the case of AuNRs synthesized with CTAB, the CTAB molecules appear to tightly pack along these surfaces, preventing subsequent molecules from binding to this surface. Thus, both groups found that proteins preferentially bound to the ends of the rods, where it is thought that CTAB is unable to bind at a high density.

While there is undoubtedly more weight to support the notion of AuNP shape exerting some effect on protein adsorption, it is possible that the effect is less pronounced or may not exist at all in some instances. This is evident from the research performed by Boulos et al. that examined the way BSA binds to AuNPs of different shapes (Boulos et al. 2013). The team synthesized AuNRs and AuNSs, surface functionalizing both shapes such that they were either positively charged, or negatively charged, or neutral. While many surface charge-specific effects were observed, this study noted no difference which could be related to the shape of the particles. Similarly, Mirsadeghi et al. observed no shape-dependent effects in protein corona formation around CTAB-stabilized AuNRs and AuNSs (Mirsadeghi et al. 2015). Thus it is understood that without the knowledge of the underlying mechanisms that determine these interactions, case-by-case investigations must be carried out. It is hoped that in this way, the factors that dominate these effects will become more apparent as the pool of research grows.

18.4 Establishing a Relationship Between Gold Nanoparticle Surface Characteristics and Protein Corona Formation

While all AuNPs possess commonality in the context of sharing an inner gold core, the degree to which they may differ on the surface remains significant (Fig. 18.7). Surface coating and functionalization steps are commonly employed post-synthesis to influence biodistribution, alter circulation time, or increase biocompatibility (Gref et al. 2000; Rahman et al. 2013). When considering that it is the outermost surface of an AuNP that encounters the biological interface initially, it is unsurprising that these characteristics are thought to impact on the formation and composition of the protein corona.

It is in this area of AuNP design that researchers may have great influence on the fate of an AuNP after it enters the body. Research has long shown that without steps being taken to modify the surface of nanoparticle prior to its administration in the blood stream, the immune system will rapidly launch a response to clear and eliminate the introduced particles (Gref et al. 1994). With this knowledge, a number of studies have been performed, not only to create surface coatings highly specific for the particle's intended use but also to gain knowledge on the general principles that govern AuNP surface-protein corona interactions.

Casals et al. studied the effect of citrate-stabilized AuNPs which were modified to be either positively or negatively charged to determine the effect of surface charge on protein corona formation (Casals et al. 2010). Most interestingly, the particles which were modified with mercaptoundecanoic acid to provide a negatively charged surface were unable to form a hard or persistent protein corona despite prolonged incubation time. This effect was attributed to the repulsion of negatively charged proteins from within the cell culture medium. In contrast, aminoundecanethiol-

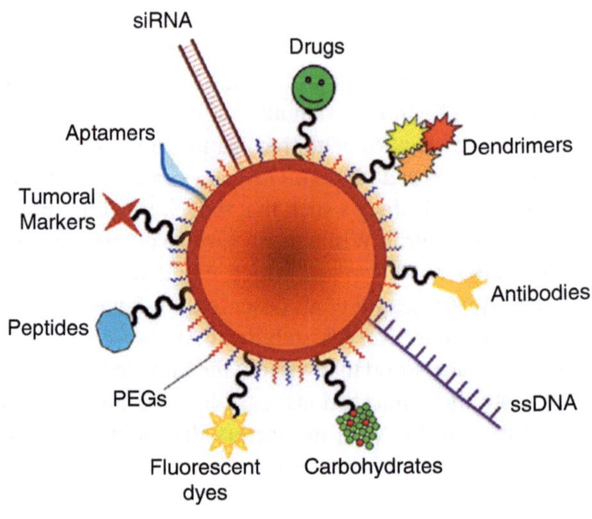

Fig. 18.7 A number of potential surface functionalisation strategies available for altering the surface of AuNPs. Reprinted with permission from (Conde et al. 2014). Licensed under CC BY 2014

modified, positively charged AuNPs were capable of forming a soft corona which was persistent and highly stable.

The realm of AuNP surface modification is boundless, with modifications of increasing complexity being reported particularly for the development of AuNP-based therapies. While many applications of AuNPs will require evasion of the immune system, it can also be understood that certain applications will require increased cellular uptake or prompt clearance from the blood stream. One such study by Chinen and colleagues utilized spherical nucleic acid (SNA) nanoparticles possessing an AuNP core surrounded by a 3′ thiol-modified guanine-rich sequence (Chinen et al. 2015). The study determined that the modified structure of the oligonucleotide affected protein corona dramatically in its composition. This altered protein coating was shown to significantly increase the uptake of AuNPs by macrophages.

The impact of protein adsorption on cellular uptake was also probed by Zhu and coworkers who synthesized four AuNPs, each possessing the same inner core, but bearing different functional groups to impart a wide spectrum of hydrophobicity levels (Zhu et al. 2012). When tested in the absence of serum, AuNPs had similar uptake levels; however when tested in serum-supplemented media, increased hydrophobicity led to decreased cellular uptake (Fig. 18.8). This supports many other studies which show decreased cellular uptake in the presence of protein. The authors reason that BSA binds to the particle surface with increased strength as the hydrophobicity of the particles increase. This presumably creates a repulsion between the negative charges present on BSA and the cellular membrane which leads to decreased levels of cellular uptake; however the authors also noted that other proteins within the medium may also contribute to this effect.

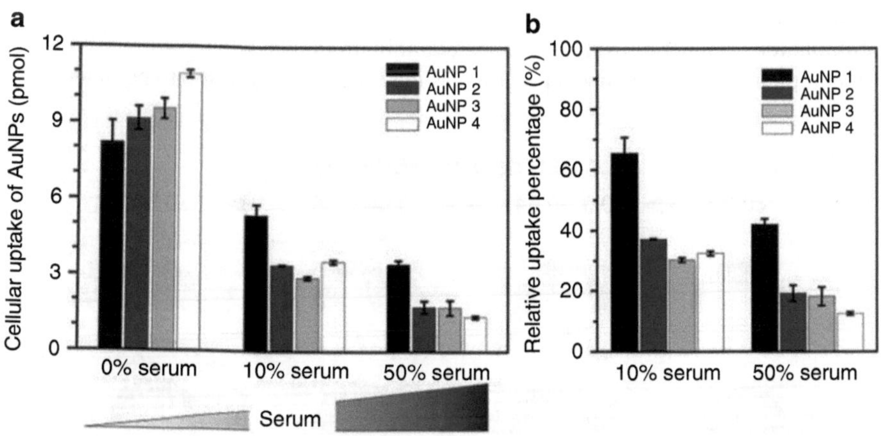

Fig. 18.8 The correlation between increase in AuNP hydrophobicity and decrease in cellular uptake in the presence of serum. The properties of AuNP 1-4 are consistent but for increasing alkane chain lengths to confer varying degrees of hydrophobicity. Reprinted with permission from (Zhu, et al. 2012). Copyright Wiley-VCH

The impact of the composition of the protein corona on biological behavior of nanoparticles has been investigated by several groups; however it must be stated that although changes in composition are reported, we are yet to know how to maximally exploit this effect to create particles that carry a specific composition of proteins. Additionally, the effect on the biomolecule structure and activity after its incorporation into the protein corona is largely unknown. While Radauer-Preiml and coworkers were able to detect modulated immune reactions after conjugation of common allergens to 50 nm AuNPs as compared with exposure to the allergens alone, they were unable to report predictable or consistent results in this area of research (Radauer-Preiml et al. 2016). While the authors were able to conclude that allergen inclusion in the protein corona can increase protease activity and therefore change the process of allergen sensitization, numerous unknowns, including a high degree of allergic donor variability, made it challenging for the results to be clearly interpreted.

While clarity in this area may be seen in the future, current studies have uncovered a consistent message in the case of PEG-modified AuNPs with numerous studies finding a correlation between the molecular weight of the PEG molecule used (or its chain length) and the amount of protein that is bound to the nanoparticle. In this case it is typically noted that using PEG with a high molecular weight leads to a particle which interacts less with proteins and forms a lower-density protein coating.

With this knowledge, Walkey et al. examined the difference in the compositions of the protein coronas that developed on the surface of AuNPs with citrate and PEG surface coatings (Walkey et al. 2012). As expected, the citrate-stabilized AuNPs demonstrated a higher degree of interaction with plasma proteins (binding a greater number of proteins) as compared with the PEG-coated particles (Fig. 18.9). However, the compositional study yielded interesting results with respect to the levels of complement protein C3 on the surface of the particles. Complement protein C3 is an opsonin, known to increase macrophage response and the subsequent elimination of material from the bloodstream. In this study, when the bound fraction of complement component C3 protein was measured, it was found to be made up of

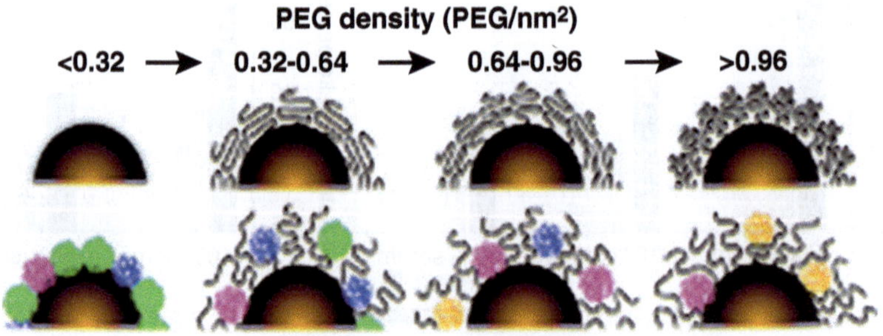

Fig. 18.9 Schematic representation of the effect of PEG density on AuNPs on their ability to adsorb serum proteins. AuNPs with increasing PEG density show reduced protein binding. Adapted with permission from (Walkey et al. 2012). Copyright 2012 American Chemical Society

>30% w/w of total protein adsorbed onto the citrate AuNPs in contrast to <5% w/w on PEG-modified particles. These results add complexity to the effect of PEGylation, whereby it is unclear if reduced uptake of nanoparticles is due to decreased protein interactions or the formation of a specific protein coating which results in macrophage evasion. It is possible that the effect observed may be a combination of both actions; however further research is needed to clarify this argument.

The use of PEG as a nanoparticle surfactant is due to its ability to provide extended circulation times within the blood; however attachment of such molecules to the nanoparticle surface must be performed covalently, which is notably laborious and difficult to control. Recently Muller et al. devised a novel alternative to non-covalently coat the surface of nanoparticles with biodegradable polymeric surfactants derived from poly(phosphoester). The surface coating, which can be applied to a variety of nanoparticles, carries the ability to be easily tuned in terms of length and hydrophobicity to create nanoparticles with tailored surfaces to suit their specific purpose (Müller et al. 2017).

While the vast majority of protein corona studies are designed to investigate the behavior of AuNPs at the nano-blood interface, Mahmoud and coworkers investigated protein corona effects at the nano-skin interface (Mahmoud et al. 2016). By using excised human skin samples, the group was able to investigate the behavior of AuNRs with particular respect to stability by using particles of different surface charges (positive, negative, and neutral). Findings suggested that positively charge AuNRs interacted with proteins released from the inner layers (dermis) to form a protein corona which resulted in rapid particle agglomeration. The lack of interaction between negatively charged proteins and biomolecules allowed the neutral and negatively charged AuNRs to maintain their stability when in contact with the skin for up to 24 h. This outcome has large implications for the development of dermal-based nanotherapeutics, particularly for therapies that require the preservation of characteristic AuNR absorption spectrum, which was lost in the case of the cationic particles after agglomeration.

18.5 Conclusion and Future Perspective

It must be considered that research relating to the evolution and influence of the protein corona on AuNP uptake and toxicity is a relatively new field, and as such, the area is continually advancing. While we are yet to know exactly which parameters dominate in governing AuNP-protein interactions, certain trends are beginning to emerge from the research with an increasing clarity. Over time, it is hoped that these trends will solidify to become a guide for researchers developing novel AuNP-based therapies. In this way, it will be possible to synthesize AuNPs possessing highly specific characteristics which predetermine their toxic effects, uptake constraints, and biodistribution in vivo. Overall, an increase in the commercialization of AuNP-based therapies is forecast to be the likely outcome.

References

Aggarwal P, Hall JB, McLeland CB, Dobrovolskaia MA, McNeil SE. Nanoparticle interaction with plasma proteins as it relates to particle biodistribution, biocompatibility and therapeutic efficacy. Adv Drug Deliv Rev. 2009;61:428–37.

Alkilany AM, Nagaria PK, Hexel CR, Shaw TJ, Murphy CJ, Wyatt MD. Cellular uptake and cytotoxicity of gold nanorods: molecular origin of cytotoxicity and surface effects. Small. 2009;5:701–8.

Benetti F, Fedel M, Minati L, Speranza G, Migliaresi C. Gold nanoparticles: role of size and surface chemistry on blood protein adsorption. J Nanopart Res. 2013;15:1–9.

Boulos SP, Davis TA, Yang JA, Lohse SE, Alkilany AM, Holland LA, Murphy CJ. Nanoparticle–protein interactions: a thermodynamic and kinetic study of the adsorption of bovine serum albumin to gold nanoparticle surfaces. Langmuir. 2013;29:14984–96.

Carnovale C, Bryant G, Shukla R, Bansal V. Size, shape and surface chemistry of nano-gold dictate its cellular interactions, uptake and toxicity. Prog Mater Sci. 2016;83:152–90.

Casals E, Pfaller T, Duschl A, Oostingh GJ, Puntes V. Time evolution of the nanoparticle protein corona. ACS Nano. 2010;4:3623–32.

Casals E, Pfaller T, Duschl A, Oostingh GJ, Puntes VF. Hardening of the nanoparticle-protein corona in metal (au, ag) and oxide (Fe_3O_4, CoO, and CeO_2) nanoparticles. Small. 2011;7:3479–86.

Caswell K, Wilson JN, Bunz UH, Murphy CJ. Preferential end-to-end assembly of gold nanorods by biotin-streptavidin connectors. J Am Chem Soc. 2003;125:13914–5.

Chah S, Hammond MR, Zare RN. Gold nanoparticles as a colorimetric sensor for protein conformational changes. Chem Biol. 2005;12:323–8.

Chang J-Y, Wu H, Chen H, Ling Y-C, Tan W. Oriented assembly of Au nanorods using biorecognition system. Chem. Commun. 2005:1092–4.

Cheng X, Tian X, Wu A, Li J, Tian J, Chong Y, Chai Z, Zhao Y, Chen C, Ge C. Protein corona influences cellular uptake of gold nanoparticles by phagocytic and nonphagocytic cells in a size-dependent manner. ACS Appl Mater Interfaces. 2015;7:20568–75.

Chinen AB, Guan CM, Mirkin CA. Spherical nucleic acid nanoparticle conjugates enhance G-quadruplex formation and increase serum protein interactions. Angew Chem Int Ed. 2015;54:527–31.

Chithrani BD, Chan WCW. Elucidating the mechanism of cellular uptake and removal of protein-coated gold nanoparticles of different sizes and shapes. Nano Lett. 2007;7:1542–50.

Conde J, Dias JT, Grazú V, Moros M, Baptista PV, de la Fuente JM. Revisiting 30 years of biofunctionalization and surface chemistry of inorganic nanoparticles for nanomedicine. Front Chem. 2014;2:48.

Corbo C, Molinaro R, Tabatabaei M, Farokhzad OC, Mahmoudi M. Personalized protein corona on nanoparticles and its clinical implications. Biomater Sci. 2017;5:378–87.

Davoudi ZM, Kandjani AE, Bhatt AI, Kyratzis IL, O'Mullane AP, Bansal V. Hybrid antibacterial fabrics with extremely high aspect ratio Ag/AgTCNQ nanowires. Adv Funct Mater. 2014;24:1047–53.

Dobrovolskaia MA, Patri AK, Zheng J, Clogston JD, Ayub N, Aggarwal P, Neun BW, Hall JB, McNeil SE. Interaction of colloidal gold nanoparticles with human blood: effects on particle size and analysis of plasma protein binding profiles. Nanomed Nanotechnol Biol Med. 2009;5:106–17.

Durán N, Silveira CP, Durán M, Martinez DST. Silver nanoparticle protein corona and toxicity: a mini-review. J Nanobiotechnol. 2015;13:55.

Ehrenberg MS, Friedman AE, Finkelstein JN, Oberdorster G, McGrath JL. The influence of protein adsorption on nanoparticle association with cultured endothelial cells. Biomaterials. 2009;30:603–10.

Feng S, Lin D, Lin J, Li B, Huang Z, Chen G, Zhang W, Wang L, Pan J, Chen R, Zeng H. Blood plasma surface-enhanced Raman spectroscopy for non-invasive optical detection of cervical cancer. Analyst. 2013;138:3967–74.

Gagner JE, Lopez MD, Dordick JS, Siegel RW. Effect of gold nanoparticle morphology on adsorbed protein structure, function. Biomaterials. 2011;32:7241–52.

Gref R, Lück M, Quellec P, Marchand M, Dellacherie E, Harnisch S, Blunk T, Müller RH. 'Stealth' corona-core nanoparticles surface modified by polyethylene glycol (PEG): influences of the corona (PEG chain length and surface density) and of the core composition on phagocytic uptake and plasma protein adsorption. Colloids Surf B: Biointerfaces. 2000;18:301–13.

Gref R, Minamitake Y, Peracchia MT, Trubetskoy V, Torchilin V, Langer R. Biodegradable long-circulating polymeric nanospheres. Science. 1994;263:1600–3.

Hadjidemetriou M, Kostarelos K. Nanomedicine: evolution of the nanoparticle corona. Nat Nanotechnol. 2017;12:288–90.

Jokerst JV, Lobovkina T, Zare RN, Gambhir SS. Nanoparticle PEGylation for imaging and therapy. Nanomedicine. 2011;6:715–28.

Jones SW, Roberts RA, Robbins GR, Perry JL, Kai MP, Chen K, Bo T, Napier ME, Ting JPY, DeSimone JM, Bear JE. Nanoparticle clearance is governed by Th1/Th2 immunity and strain background. J Clin Invest. 2013;123:3061–73.

Karajanagi SS, Vertegel AA, Kane RS, Dordick JS. Structure and function of enzymes adsorbed onto single-walled carbon nanotubes. Langmuir. 2004;20:11594–9.

Khan S, Gupta A, Verma N, Nandi C. Kinetics of protein adsorption on gold nanoparticle with variable protein structure and nanoparticle size. J Chem Phys. 2015;143:164709.

Klein J. Probing the interactions of proteins and nanoparticles. Proc Natl Acad Sci. 2007;104:2029–30.

Lacerda SH, Park JJ, Meuse C, Pristinski D, Becker ML, Karim A, Douglas JF. Interaction of gold nanoparticles with common human blood proteins. ACS Nano. 2010;4:365–79.

Li Y, Kröger M, Liu WK. Shape effect in cellular uptake of pegylated nanoparticles: comparison between sphere, rod, cube and disk. Nanoscale. 2015;7:16631–46.

Lundqvist M, Sethson I, Jonsson B-H. Protein adsorption onto silica nanoparticles: conformational changes depend on the particles' curvature and the protein stability. Langmuir. 2004;20:10639–47.

Lynch I, Cedervall T, Lundqvist M, Cabaleiro-Lago C, Linse S, Dawson KA. The nanoparticle – protein complex as a biological entity; a complex fluids and surface science challenge for the 21st century. Adv Colloid Interf Sci. 2007;134-35:167–74.

Mahmoud NN, Al-Qaoud KM, Al-Bakri AG, Alkilany AM, Khalil EA. Colloidal stability of gold nanorod solution upon exposure to excised human skin: effect of surface chemistry and protein adsorption. Int J Biochem Cell Biol. 2016;75:223–31.

Mahmoudi M, Lynch I, Ejtehadi MR, Monopoli MP, Bombelli FB, Laurent S. Protein-nanoparticle interactions: opportunities and challenges. Chem Rev. 2011;111:5610–37.

Mirsadeghi S, Dinarvand R, Ghahremani MH, Hormozi-Nezhad MR, Mahmoudi Z, Hajipour MJ, Atyabi F, Ghavami M, Mahmoudi M. Protein corona composition of gold nanoparticles/nanorods affects amyloid beta fibrillation process. Nanoscale. 2015;7:5004–13.

Moghimi SM, Hunter AC, Murray JC. Long-circulating and target-specific nanoparticles: theory to practice. Pharmacol Rev. 2001;53:283–318.

Mosqueira VC, Legrand P, Gref R, Heurtault B, Appel M, Barratt G. Interactions between a macrophage cell line (J774A1) and surface-modified poly (D,L-lactide) nanocapsules bearing poly(ethylene glycol). J Drug Target. 1999;7:65–78.

Müller J, Bauer KN, Prozeller D, Simon J, Mailänder V, Wurm FR, Winzen S, Landfester K. Coating nanoparticles with tunable surfactants facilitates control over the protein corona. Biomaterials. 2017;115:1–8.

Pastorino F, Marimpietri D, Brignole C, Di Paolo D, Pagnan G, Daga A, Piccardi F, Cilli M, Allen TM, Ponzoni M. Ligand-targeted liposomal therapies of neuroblastoma. Curr Med Chem. 2007;14:3070–8.

Radauer-Preiml I, Andosch A, Hawranek T, Luetz-Meindl U, Wiederstein M, Horejs-Hoeck J, Himly M, Boyles M, Duschl A. Nanoparticle-allergen interactions mediate human allergic responses: protein corona characterization and cellular responses. Part Fibre Toxicol. 2016;13:3.

Rahman M, Laurent S, Tawil N, Yahia L H, Mahmoudi M. Protein-nanoparticle interactions: the bio-nano interface. Springer-Verlag, Berlin Heidelberg, 2013.

Ramezani F, Amanlou M, Rafii-Tabar H. Gold nanoparticle shape effects on human serum albumin corona interface: a molecular dynamic study. J Nanopart Res. 2014;16:1–9.
Roach P, Farrar D, Perry CC. Surface tailoring for controlled protein adsorption: effect of topography at the nanometer scale and chemistry. J Am Chem Soc. 2006;128:3939–45.
Romberg B, Hennink WE, Storm G. Sheddable coatings for long-circulating nanoparticles. Pharm Res. 2008;25:55–71.
Saptarshi SR, Duschl A, Lopata AL. Interaction of nanoparticles with proteins: relation to bio-reactivity of the nanoparticle. J Nanobiotechnol. 2013;11:1–12.
Sen T, Mandal S, Haldar S, Chattopadhyay K, Patra A. Interaction of gold nanoparticle with human serum albumin (HSA) protein using surface energy transfer. J Phys Chem C. 2011;115:24037–44.
Sondi I, Salopek-Sondi B. Silver nanoparticles as antimicrobial agent: a case study on E. coli as a model for gram-negative bacteria. J Colloid Interface Sci. 2004;275:177–82.
Soppimath KS, Aminabhavi TM, Kulkarni AR, Rudzinski WE. Biodegradable polymeric nanoparticles as drug delivery devices. J Control Release. 2001;70:1–20.
von Maltzahn G, Park J-H, Agrawal A, Bandaru NK, Das SK, Sailor MJ, Bhatia SN. Computationally guided photothermal tumor therapy using long-circulating gold nanorod antennas. Cancer Res. 2009;69:3892–900.
Walkey CD, Olsen JB, Guo H, Emili A, Chan WC. Nanoparticle size and surface chemistry determine serum protein adsorption and macrophage uptake. J Am Chem Soc. 2012;134:2139–47.
Wolfram J, Yang Y, Shen J, Moten A, Chen C, Shen H, Ferrari M, Zhao Y. The nano-plasma interface: implications of the protein corona. Colloids Surf B: Biointerfaces. 2014;124:17–24.
Zhu ZJ, Posati T, Moyano DF, Tang R, Yan B, Vachet RW, Rotello VM. The interplay of monolayer structure and serum protein interactions on the cellular uptake of gold nanoparticles. Small. 2012;8:2659–63.

Chapter 19
The Local and Systemic Effects of Cobalt-Chromium Nanoparticles on the Human Body: The Implications for Metal-on-Metal Hip Arthroplasty

James Drummond, Phong Tran, and Camdon Fary

Abstract Prostheses made from metal alloys have been successfully utilised within medicine for hundreds of years. Among these, cobalt-chromium (Co-Cr) alloys have seen extensive use in orthopaedic applications, including hip and knee joint replacements. Despite all the research and development that has gone into optimising these implants, however, a small proportion of them ultimately fail and require revision after a number of years. While the reasons for this are diverse, the reaction of metal nanoparticles to human tissues is a recognised complication of implanting these alloys within the body. This chapter explores the orthopaedic use of metal alloys within the human body as well as the local and systemic effects of these metal nanoparticles, with emphasis on large-diameter metal-on-metal hip replacements.

Keywords Metal-on-metal • Hip arthroplasty • Hip replacement • Metallosis • Pseudotumour • Aseptic lymphocyte-dominated vasculitis-associated lesion (ALVAL)

J. Drummond (✉)
Department of Orthopaedics, Western Health, Melbourne, VIC, Australia
e-mail: jdrummondmd@gmail.com

P. Tran
Department of Orthopaedics, Western Health, Melbourne, VIC, Australia

Australian Institute for Musculoskeletal Science, Melbourne, VIC, Australia

C. Fary
Department of Orthopaedics, Western Health, Melbourne, VIC, Australia

Australian Institute for Musculoskeletal Science, Melbourne, VIC, Australia

Epworth Musculoskeletal Institute, Melbourne, VIC, Australia

M. Rai, R. Shegokar (eds.), *Metal Nanoparticles in Pharma*,
DOI 10.1007/978-3-319-63790-7_19

Abbreviations

ALVAL	Aseptic lymphocyte-dominated vasculitis-associated lesion
ARMD	Adverse reaction(s) to metal debris
Co-Cr	Cobalt-chromium
EDTA	Ethylenediaminetetraacetate
MARS	Metal artefact reduction sequence
MoM	Metal-on-metal
MoP	Metal-on-polyethylene
MRI	Magnetic resonance imaging
SCENIHR	Scientific committee on emerging and newly identified health risks
THA	Total hip arthroplasty
THR	Total hip replacement(s)

19.1 Introduction

Total hip arthroplasty (THA) is an extremely successful orthopaedic surgical intervention performed most commonly to alleviate hip pain caused by osteoarthritis. Thought to originate in Germany in 1891, the operation involves replacing the native hip joint with a biomechanically similar prosthesis (Knight et al. 2011). Throughout the years, different materials have been utilised for the components of the prosthesis, including ivory, glass, polyethylene, ceramic and metal. Each material carries its own unique properties, with advantages and disadvantages. Currently, the most commonly implanted prosthesis consists of a metal femoral head atop a metal femoral stem that articulates with a polyethylene liner locked inside of a metal acetabular cup. The metal components of this prosthesis are usually made from cobalt-chromium-molybdenum alloys. Furthermore, the dimensions of the head, neck and cup can be manipulated to give the prosthesis different properties, such as reduced dislocation risk with larger diameter heads.

Large diameter metal-on-metal (MoM) total hip replacements (THR) have recently come under the spotlight for higher than expected revision rates (Australian Orthopaedic Association National Joint Replacement Registry 2016a). This issue of prosthesis failure is thought to relate to local and systemic reactions to the metal nanoparticles released from the prosthesis due to wear and/or corrosion (Fehring and Fehring 2015). The severity of these reactions is extremely varied and ranges from asymptomatic changes such as small cysts, all the way through to complete failure of a prosthesis with surrounding bony destruction (Scientific Committee on Emerging Newly Identified Health Risks 2014). Systemic reactions involving neurological damage, hypothyroidism and/or cardiomyopathy have also been documented as a result of high circulating cobalt concentrations (Bradberry et al. 2014). Because of the often progressive and destructive nature of such complications, definitive management usually involves revision of the prosthesis to one without cobalt-chromium (Co-Cr) components.

This chapter explores the biomechanics behind large-diameter MoM hip replacements in order to explain the currently understood theory for these adverse reactions. It details the pathophysiology behind local and systemic reactions and briefly touches on their diagnosis, including Co-Cr ion level testing. Finally, an overview of different options for management of such complications, including when to intervene surgically, is described.

19.2 Metal-on-Metal Hip Replacements

The term metal-on-metal describes the articulation of the prosthesis, which consists of a metal femoral head atop a metal femoral stem, articulating with either a metal acetabular cup or liner within a cup. This is opposed to the more commonly utilised metal-on-polyethylene (MoP) hip replacements, which consist of a metal femoral head atop a metal femoral stem articulating with a highly cross-linked polyethylene liner that is locked within a metal acetabular cup, as previously described. MoM hip prostheses most commonly consist of Co-Cr alloys, often with trace amounts of molybdenum. These metals are selected because of their inert nature, tending not to react to biological tissues as commonly as other elements.

Metal articulating surfaces are utilised because of their low coefficients of friction, creating an extremely smooth surface, significantly reducing wear on the components. Volumetric wear of MoM implants has been shown to be up to 60 times less than that of conventional metal-on-polyethylene articulating surfaces (Cuckler 2005). MoM implants also allow for larger diameter femoral head components, which are thought to reduce both wear and dislocation risk (Cuckler 2005). Finally, osteolysis due to polyethylene particle reactions with surrounding bone is entirely avoided.

19.3 Issues with MoM Articulating Surfaces

All arthroplasties are subject to both wear and corrosion. Mechanical wear in THA generates free metal particles through the contact of metal surfaces that are both intended and unintended to act as bearing surfaces. Furthermore, corrosion, caused by electrochemical oxidation of a metal in the presence of an oxidant such as oxygen, also leads to release of metal ions from components. Corrosion and wear occur simultaneously and often synergistically. This combination of electrochemical corrosion and mechanical (tribological) wear is termed tribocorrosion (Cheung et al. 2016).

Despite the reduced volumetric wear of MoM articulating surfaces, the pattern of wear is actually that of increased release of smaller particles, when compared with MoP articulating surfaces (Mahendra et al. 2009). The Co-Cr particles released from MoM articulations are in the order of nanometres as opposed to micrometre-sized polyethylene particles (Doorn et al. 1998; Brown et al. 2006). This means a higher total surface area of the metal particles and hence increased biological reactivity within periprosthetic tissues, increasing the likelihood of adverse soft tissue reactions. The smaller size also allows the particles to diffuse more easily into surrounding tissues where they can cause reactions (Watters et al. 2010).

It is understood that metal particles are released not only at the articulating bearing surface but also at the head-neck junction and the neck-body junction of the stem in modular stem designs (Cooper et al. 2013). Modular designs are used for increased prosthesis flexibility in an attempt to better restore normal joint biomechanics. Trunnionosis is the term used to describe wear at the femoral head-neck junction and is now acknowledged as an important cause of prosthesis failure in both MoM and MoP articulations utilising modular designs (Mistry et al. 2016). The release of metal ions from this process is thought to relate to a combination of fretting and crevice corrosion (Gilbert et al. 1993; Goldberg et al. 2002). Furthermore, larger diameter femoral heads, such as those commonly used in MoM THR, are thought to increase the effective horizontal lever arm, causing greater torsional forces at the head-neck junction and increasing wear (Langton et al. 2012). Despite the inert nature of Co-Cr, release of large numbers of ions from any of these processes can lead to destructive tissue reactions.

Edge loading is another factor that has been highlighted as a significant cause of increased wear and potential failure of large-diameter MoM total hip replacements (De Haan et al. 2008; Underwood et al. 2012). Edge loading occurs when the contact patch between the acetabular cup and femoral head extends over the cup rim, causing increased wear of the cup edge. This is commonly seen in replacements with large-diameter heads and shallow acetabular cups, such as those utilised in the DePuy ASR hip resurfacing device, which was recalled in 2009 due to higher-than-expected revision rates. This phenomenon can also be caused by surgical technique. If the acetabular cup is implanted at a steep angle of inclination, the contact patch created with the femoral head tends to extend over the rim of the cup, creating huge increases in local contact pressures, leading to increased metal wear and particle release (Underwood et al. 2012). Edge loading is thought to also expel the synovial fluid from the bearing surface, reducing lubrication and further increasing wear (Underwood et al. 2012). Increased wear of MoM bearings leads to further release of metal ions, increasing the likelihood of adverse local tissue reactions.

Inflammatory reactions within local soft tissues are grouped under the term "adverse reactions to metal debris" or ARMD. These reactions are varied in histological and morphological appearance and include disease processes such as pseudotumour, metallosis and aseptic lymphocyte-dominated vasculitis-associated lesion (ALVAL). Due to the differing processes seen histologically, along with the diverse wear patterns, classification of these processes based solely on wear patterns is extremely difficult. When histologic features are compared between THRs revised for high wear and those revised for suspected metal sensitivity, there is extreme variability in the amount of metal debris, inflammatory cell pattern and degree of necrosis (Campbell et al. 2010). As such, there is no consensus regarding whether these adverse reactions are caused by the toxic effect of the metal particles, the immune response to the particles, or a combination of both (Langton et al. 2011).

ALVAL is a histological diagnosis used to describe non-infectious chronic inflammatory changes within periprosthetic tissue. These changes are thought to be caused by metal ions acting as haptens, binding with native proteins and causing a type IV hypersensitivity response in local tissues (Watters et al. 2010). For example, cobalt does not exist in its elemental state within the body, but instead as either a bivalent or trivalent cation that can complex with intra- or extracellular molecules (Cheung et al. 2016). The defining feature of ALVAL is the presence of a dense perivascular inflammatory infiltrate (Willert et al. 2005). This infiltrate consists of diffuse lymphocytes, often accompanied by plasmacytes and/or eosinophils, with a lack of polymorphonuclear leucocytes, indicating its chronicity (Cooper et al. 2013).

The term metallosis describes a collection of macrophages/histiocytes in periprosthetic tissues that have phagocytosed metal particles. It is often associated with aseptic fibrosis and necrosis (Langton et al. 2011). Metallosis can be considered the starting point of the ARMD spectrum, which, if left long enough, can lead to pseudotumour development.

Pseudotumour is the term given to a cystic or solid aseptic mass in periprosthetic tissue, associated with clinical, radiological and histopathological findings of chronic inflammation related to ALVAL or metallosis. While pseudotumours may be asymptomatic, they often cause pain and can lead to osteolysis and aseptic loosening of prostheses. They are usually characterised by a superficial layer of necrotic surface exudate with underlying layers comprising fibrous tissue, granulomas, lymphocytic aggregates and varying degrees of necrosis (Watters et al. 2010; Campbell et al. 2010).

19.4 How Common Are These Adverse Reactions?

The best estimate of the prevalence of these reactions has been attained from national joint registry data (Australian Orthopaedic Association National Joint Replacement Registry 2016b; National Joint Registry for England Wales and Northern Ireland 2016). MoM hip replacements can be classified by head diameter, with small head replacements being ≤32 mm diameter and large head replacements being >32 mm diameter. According to the Australian Orthopaedic Association National Joint Replacement Registry, the use of MoM bearings for total hip arthroplasty is dwindling, with only 18 procedures performed in 2015, all of which utilised small head replacements (Australian Orthopaedic Association National Joint Replacement Registry 2016b). MoM bearing total hip replacements have been shown to have a 15-year cumulative percent revision of 21.8%, with larger head sizes (particularly >40 mm) having even higher revision rates (Australian Orthopaedic Association National Joint Replacement Registry 2016b). This is in comparison to conventional MoP bearings, which carry a 5.8% 15-year cumulative percent revision (Australian Orthopaedic Association National Joint Replacement Registry 2016b). The most common reason for revision of MoM THRs is metal-related pathology (42.1%), which includes ALVAL, metallosis and pseudotumour (Australian Orthopaedic Association National Joint Replacement Registry 2016b). Furthermore, when examining head sizes >32 mm, females and those under the age of 65 had significantly higher revision rates (Australian Orthopaedic Association National Joint Replacement Registry 2016b). This data is backed up by the UK National Joint Registry, reporting that cemented and uncemented MoM THR have a 12-year cumulative percentage probability of revision of 18.96% and 22.14%, respectively (National Joint Registry for England Wales and Northern Ireland 2016). This is in comparison to cemented and uncemented MoP replacements, which have a 3.84% and 5.46% revision probability, respectively (National Joint Registry for England Wales and Northern Ireland 2016).

Because not all patients with adverse local tissue reactions are symptomatic, the true number of these reactions may be significantly higher than what is reported. Asymptomatic pseudotumours have been identified on ultrasound and magnetic resonance imaging (MRI) in 27–32% of patients with MoM THR and resurfacings (Williams et al. 2011; van der Weegen et al. 2013). It is unclear whether, if left long enough, these pseudotumours would grow to a point where revision would be necessary (van der Weegen et al. 2013).

19.5 Systemic Reactions to Metal Ions

It has been postulated that high concentrations of circulating blood metal ions are associated with greater articular junction wear in MoM bearing implants (Langton et al. 2010). Diffusion of soluble metal ions, as well as lymphovascular transport and subsequent deposition of these ions, is thought to lead to systemic adverse reactions (Cheung et al. 2016). As circulating blood concentrations of cobalt and chromium rise, patients are more likely to develop systemic features of toxicity. From patterns identified on blood ion testing in those with MoM hip replacements, it has been shown that cobalt levels can rise up to 663 times the body's normal upper limit, compared with chromium, at 24 times the upper limit (Bradberry et al. 2014). As such, it is believed that cobalt causes the majority of systemic symptoms. Furthermore, the effects of cobalt on the heart, thyroid and peripheral and central nervous system have been extensively documented in research examining oral cobalt chloride treatment of anaemia, vocational exposure to cobalt powder and ingestion of cobalt-based beer additives (Smith and Carson 1981; Cheung et al. 2016). Local reactions, on the other hand, have only been shown to occur when human lymphocytes react in the presence of synergistic action between cobalt and chromium nanoparticles and ions released from tribocorrosion (Posada et al. 2015).

Cobalt blood metal ion levels can be tested to assist with diagnosis of systemic reactions. Due to variability in cobalt blood ion levels relative to symptoms, however, systemic cobalt toxicity is largely a clinical diagnosis. Symptoms tend to occur with increased frequency at serum cobalt levels above 100 μg/L (Zywiel et al. 2016). Despite the trend for higher levels of circulating ions causing symptoms, systemic toxicity has also been shown to occur in patients with blood ion concentrations less than 20 μg/L (Gessner et al. 2015). Certain risk factors are also thought to predispose to cobalt toxicity. Systemic ion accumulation is mediated by renal excretion of metal ions. As such, renal failure will inevitably result in higher serum ion concentrations, leading to increased propensity for toxicity (Zywiel et al. 2016). In patients with hip replacements, younger age and male gender are two factors that have been closely linked to increased implant wear, leading to higher metal ion release (Schmalzried et al. 2000). Finally, hypoalbuminaemia, often secondary to malnutrition, reduces protein-bound cobalt in the blood, leading to increased toxicity from higher levels of ionised cobalt (Paustenbach et al. 2013).

Since 2001, there have been at least 16 documented cases of systemic cobaltism due to a primary MoM THA (Gessner et al. 2015). Systemic cobaltism from potentially malfunctioning MoM THA has been linked to cardiomyopathy, central and peripheral neuropathy, hypothyroidism and retinopathy (Ikeda et al. 2010; Apel et al. 2013; Zywiel et al. 2016; Cheung et al. 2016). Cobalt particles have been shown to impair mitochondrial metabolism, resulting in neuro- and cardiotoxicity due to cellular dysfunction and/or death (Rona and Chappel 1973; Catalani et al. 2012). Cobalt is also thought to preferentially deposit in the myocardium and pericardial fluid, hence its ability to cause isolated cardiac symptoms (Gessner et al. 2015). This can be so severe that hospitalisation for heart failure and even cardiac transplantation may be necessary (Gillam et al. 2017; Moniz et al. 2017). The most common neurological symptoms are that of paraesthesia, tinnitus and visual loss (Cheung et al. 2016). These symptoms are thought to occur due to depletion of essential neurotransmitters, as well as formation of reactive oxygen species, which increase oxidative stress and lead to demyelination and axonal loss (Catalani et al. 2012). Hypothyroidism secondary to cobalt toxicity is due to inhibition of both iodine uptake and thyroperoxidase activity, reducing the body's ability to synthesise thyroxine (Cheung et al. 2016).

A number of large cohort studies have examined the potential link between MoM THA and cancer. As the kidneys excrete cobalt and chromium, there is concern regarding bladder cancer secondary to the particles these ions create. A large observational study examining nearly 300,000 patients with a THR found that patients with MoM articulations had a similar prevalence of cancer when compared to patients with different bearing surfaces (Smith et al. 2012). This is reinforced by statistics showing no increase in the incidence of cancer in those with MoM THA compared with the general population (Visuri et al. 1996). This is despite cobalt being labelled a "probable carcinogen" within the lung by the International Agency for Research on Cancer (Rousseau et al. 2005). While these studies are reassuring, due to the limited amount of available research into the carcinogenic effects of cobalt, definitive conclusions regarding the risk of cancer from systemic cobalt have not yet been reached.

19.6 Diagnosis of Adverse Reactions to Metal Debris

A thorough history and examination, when paired with targeted investigations, will often allow a medical practitioner to diagnose a local or systemic ARMD. Patients with local reactions most commonly present with groin pain that can radiate to the greater trochanter or down the thigh. This pain is thought to be the strongest predictor for pseudotumour presence (Bosker et al. 2015). It often causes a patient to adopt an antalgic gait and may progress to sensations of instability or clicking/clunking when ambulating. Reduced range of movement and gluteal muscle weakness are also common findings.

When an adverse local tissue reaction is suspected in a patient with a MoM THA, urgent referral to an orthopaedic specialist is necessary. Blood tests including inflammatory markers as well as plain X-ray imaging can help to rule out other causes of pain such as infection, periprosthetic fracture, loosening or dislocation. Blood metal ion levels may also assist in determining the likelihood of a local or systemic ARMD. Patients with pseudotumours following implantation of the DePuy ASR hip resurfacing device, for example, have been shown to have higher cobalt and chromium blood ion levels than those without pseudotumours (Hailer et al. 2014). While ion levels in poorly functioning MoM THA are understood to be generally higher than those in well-functioning prostheses, they have also been shown to vary widely (Gunther et al. 2013). As such, blood ion levels should not be used in isolation when assessing a patient with a potential ARMD. Internationally recognised guidelines published by the Scientific Committee on Emerging and Newly Identified Health Risks (SCENIHR) advise that the threshold cobalt and chromium blood metal ion concentrations for clinical concern are expected to lie between 2 and 7 μg/L (Scientific Committee on Emerging Newly Identified Health Risks 2014). Despite much research into the area, the substantial variability in blood ion levels at presentation means that a cut-off for clinical concern is difficult to determine. For a concise summary regarding up to date diagnostic and management guidelines for patients with MoM THA, refer to Table 19.1 (Drummond et al. 2015).

Imaging plays a major role in diagnosis and monitoring of ARMD. The gold standard imaging technique for diagnosis of ARMD is metal artefact reduction sequence (MARS) MRI. It can not only be used as a screening modality for asymptomatic pseudotumours, but its ability to assist with preoperative planning and longitudinal comparison means that it is also recommended over ultrasound as the first-line modality for assessment of periprosthetic tissues in patients with MoM replacements (Siddiqui et al. 2014).

As previously discussed, systemic cobaltism is largely a clinical diagnosis. New-onset shortness of breath, fatigue and peripheral oedema after MoM THA should alert a practitioner to potential cobalt-related cardiomyopathy. Paraesthesia, tinnitus or visual changes are the most common neurological symptoms reported by patients with systemic toxicity. Finally, MoM THA patients presenting with generalised lethargy, weight gain, weakness and/or a clinical picture of goitre and diminished reflexes should be investigated for hypothyroidism secondary to systemic cobaltism.

Table 19.1 Guidelines for diagnosis and management of patients with metal-on-metal total hip arthroplasty

	Small head MoM THA (femoral head <36 mm)		Large head MoM THA (femoral head ≥36 mm)		Hip resurfacing arthroplasty	
	Asymptomatic	Symptomatic	Asymptomatic	Symptomatic	Asymptomatic	Symptomatic
Imaging required	MARS MRI or USS if concern or clinical/blood metal ion abnormality)	Plain X-ray and MARS MRI or USS	MARS MRI or USS if concern clinical/ blood metal ion abnormality)	Plain X-ray and MARS MRI or USS	MARS MRI or USS if concern or clinical/ blood metal ion abnormality)	Plain X-ray and MARS MRI or USS
Blood metal ion testing (cobalt and chromium)	Performed at each follow-up	Performed at and between each follow-up	Performed at each follow-up	Performed at and between each follow-up	Performed at each follow-up	Performed at and between each follow-up
Abnormal blood metal ions	Repeat test after 3 months if concentration >7 µg/L (or >2 µg/L with concerns)	Repeat test after 3 months if concentration >7 µg/L (or >2 µg/L with concerns)	Repeat test after 3 months if concentration >7 µg/L (or >2 µg/L with concerns)	Repeat test after 3 months if concentration >7 µg/L (or >2 µg/L with concerns)	Repeat test after 3 months if concentration >7 µg/L (or >2 µg/L with concerns)	Repeat test after 3 months if concentration >7 µg/L (or >2 µg/L with concerns)
Follow-up timeframe	As per local protocol for conventional THA	No less than annually for life of MoM implant	Annually for life of MoM implant	No less than annually for life of MoM implant	Annually for first 5 years and then as per local protocol for conventional THA (annual for life of implant if ≤50 mm diameter, female or low coverage arc)	No less than annually for life of MoM implant
When to consider revision	Cross-sectional imaging abnormalities and/ or where blood Co/ Cr levels progressively rising >7 µg/L	Persistent symptoms, cross-sectional imaging abnormalities, and/or where blood Co/Cr levels progressively rising >7 µg/L	Cross-sectional imaging abnormalities and/or where blood Co/Cr levels progressively rising >7 µg/L	Persistent symptoms, cross-sectional imaging abnormalities, and/or where blood Co/Cr levels progressively rising >7 µg/L	Cross-sectional imaging abnormalities and/or where blood Co/Cr levels progressively rising >7 µg/L	Persistent symptoms, cross-sectional imaging abnormalities, and/ or where blood Co/ Cr levels progressively rising

19.7 Treatment of Adverse Reactions to Metal Debris

Treatment of ARMD can be divided into medical and surgical management. Medical management involves the use of chelating agents that are thought to alleviate symptoms by reducing circulating blood metal ions. Despite agents such as ethylenediaminetetraacetate (EDTA) and N-acetylcysteine having been shown to reduce blood cobalt ion concentrations, current evidence has revealed a lack of improvement in clinical symptoms with their use (Devlin et al. 2013; Giampreti et al. 2014).

Definitive management of both systemic and local ARMD is achieved through removal of the offending articulation and local soft tissue debridement. This most commonly involves revision of the MoM articulation to ceramic-on-ceramic or ceramic-on-polyethylene components. This is aimed at reducing the local metal ion release from both the articulating surface and trunnion, hence lowering the chance of ARMD recurrence. While MoP articulations can be utilised in revision surgery, due to the potential for further corrosion secondary to trunnionosis, ceramic femoral heads are preferred. In the majority of cases, the primary stem does not need to be revised. When revision is impossible due to extensive local tissue damage or patient factors, excision arthroplasty may be considered. Revision surgery should be considered when symptoms become persistent, unmanageable and progressive. Furthermore, patients with large or expanding pseudotumours, progressive osteolysis, femoral neck thinning or excessively high blood metal ion concentrations should also be considered for revision surgery (Scientific Committee on Emerging Newly Identified Health Risks 2014).

When performing revision surgery, the extent of soft tissue debridement is an important consideration. Excision of pseudotumours has been compared to oncologic resections, where a surgeon should aim for clear margins surrounding the resected pathology in order to reduce the risk of recurrence (Matharu et al. 2014). Unfortunately, the outcomes for revision surgery are often poor. Up to 50% of MoM resurfacing patients and 68% of MoM THR patients revised for inflammatory pseudotumours encounter major complications after revision surgery (Matharu et al. 2014). The most common complications include dislocation, ARMD recurrence and acetabular loosening. The significance of these complications is highlighted in re-revision rates, with up to 38% of revised MoM resurfacings and 21% of revised MoM THR requiring re-revision (Matharu et al. 2014).

19.8 Conclusion

Orthopaedic surgeons have been successfully utilising metal composites in total hip arthroplasty for over 100 years. It wasn't until the popularity spike in large-diameter MoM THR in the early to mid-2000s, however, that extensive data has become available on their outcomes and survivorship. Since then, focus has shifted to addressing local and systemic adverse reactions due to metal debris released from

MoM THR. Patients with MoM THR who present with hip or groin pain or systemic features should be thoroughly investigated for these potentially debilitating reactions. Systemic cobaltism is a rare phenomenon seen in some patients with often malfunctioning MoM THR and can present with cardiomyopathy, central and peripheral neuropathy and/or hypothyroidism. When local or systemic reactions become progressive and severe, revision of the MoM bearing to a ceramic-based bearing is often the only way to alleviate symptoms. Unfortunately, revision outcomes for local reactions are poor, and many patients experience significant complications, with up to a third requiring re-revision. Because of this, the number of MoM THR implanted over the last 10 years has significantly decreased, indicating the prosthesis design will inevitably be phased out. As such, unless medical engineering companies can provide a MoM articulating surface that is not associated with significant risk of ARMD, focus needs to be shifted towards further refinement of lower risk articulating surfaces such as ceramic and polyethylene. Finally, additional focus on prevention and non-operative management of osteoarthritis may help to delay and even avert a THR, avoiding the problem of ARMD altogether.

References

Apel W, Stark D, Stark A, O'hagan S, Ling J. Cobalt-chromium toxic retinopathy case study. Doc Ophthalmol. 2013;126:69–78.

Australian Orthopaedic Association National Joint Replacement Registry. Hip, knee and shoulder arthroplasty annual report 2016. 2016a. https://aoanjrr.sahmri.com/documents/10180/275066/Hip%2C%20Knee%20%26%20Shoulder%20Arthroplasty.

Australian Orthopaedic Association National Joint Replacement Registry. Metal on metal bearing surface conventional hip arthroplasty supplement report 2016. 2016b. https://aoanjrr.sahmri.com/documents/10180/275107/Metal%20on%20Metal%20Bearing%20Surface%20Conventional%20Hip%20Arthroplasty.

Bosker BH, Ettema HB, Van Rossum M, Boomsma MF, Kollen BJ, Maas M, Verheyen CC. Pseudotumor formation and serum ions after large head metal-on-metal stemmed total hip replacement. Risk factors, time course and revisions in 706 hips. Arch Orthop Trauma Surg. 2015;135:417–25.

Bradberry SM, Wilkinson JM, Ferner RE. Systemic toxicity related to metal hip prostheses. Clin Toxicol (Phila). 2014;52:837–47.

Brown C, Fisher J, Ingham E. Biological effects of clinically relevant wear particles from metal-on-metal hip prostheses. Proc Inst Mech Eng H. 2006;220:355–69.

Campbell P, Ebramzadeh E, Nelson S, Takamura K, De Smet K, Amstutz HC. Histological features of pseudotumor-like tissues from metal-on-metal hips. Clin Orthop Relat Res. 2010;468:2321–7.

Catalani S, Rizzetti MC, Padovani A, Apostoli P. Neurotoxicity of cobalt. Hum Exp Toxicol. 2012;31:421–37.

Cheung AC, Banerjee S, Cherian JJ, Wong F, Butany J, Gilbert C, Overgaard C, Syed K, Zywiel MG, Jacobs JJ, Mont MA. Systemic cobalt toxicity from total hip arthroplasties: review of a rare condition part 1 – history, mechanism, measurements, and pathophysiology. Bone Joint J. 2016;98-B:6–13.

Cooper HJ, Urban RM, Wixson RL, Meneghini RM, Jacobs JJ. Adverse local tissue reaction arising from corrosion at the femoral neck-body junction in a dual-taper stem with a cobalt-chromium modular neck. J Bone Joint Surg Am. 2013;95:865–72.

Cuckler JM. The rationale for metal-on-metal total hip arthroplasty. Clin Orthop Relat Res. 2005;441:132–6.

De Haan R, Campbell PA, Su EP, De Smet KA. Revision of metal-on-metal resurfacing arthroplasty of the hip: the influence of malpositioning of the components. J Bone Joint Surg (Br). 2008;90:1158–63.

Devlin JJ, Pomerleau AC, Brent J, Morgan BW, Deitchman S, Schwartz M. Clinical features, testing, and management of patients with suspected prosthetic hip-associated cobalt toxicity: a systematic review of cases. J Med Toxicol. 2013;9:405–15.

Doorn PF, Campbell PA, Worrall J, Benya PD, Mckellop HA, Amstutz HC. Metal wear particle characterization from metal on metal total hip replacements: transmission electron microscopy study of periprosthetic tissues and isolated particles. J Biomed Mater Res. 1998;42:103–11.

Drummond J, Tran P, Fary C. Metal-on-metal hip arthroplasty: a review of adverse reactions and patient management. J Funct Biomater. 2015;6:486–99.

Fehring KA, Fehring TK. Modes of failure in metal-on-metal total hip arthroplasty. Orthop Clin North Am. 2015;46:185–92.

Gessner BD, Steck T, Woelber E, Tower SSA. systematic review of systemic cobaltism after wear or corrosion of chrome-cobalt hip implants. J Patient Saf. 2015;00(0):1–8. http://journals.lww.com/journalpatientsafety/Abstract/publishahead/A_Systematic_Review_of_Systemic_Cobaltism_After.99652.aspx.

Giampreti A, Lonati D, Locatelli CA. Chelation in suspected prosthetic hip-associated cobalt toxicity. Can J Cardiol. 2014;30(465):e13.

Gilbert JL, Buckley CA, Jacobs JJ. In vivo corrosion of modular hip prosthesis components in mixed and similar metal combinations. The effect of crevice, stress, motion, and alloy coupling. J Biomed Mater Res. 1993;27:1533–44.

Gillam MH, Pratt NL, Inacio MC, Roughead EE, Shakib S, Nicholls SJ, Graves SE. Heart failure after conventional metal-on-metal hip replacements. Acta Orthop. 2017;88:2–9.

Goldberg JR, Gilbert JL, Jacobs JJ, Bauer TW, Paprosky W, Leurgans S. A multicenter retrieval study of the taper interfaces of modular hip prostheses. Clin Orthop Relat Res. 2002:149–61.

Gunther KP, Lutzner J, Hannemann F, Schmitt J, Kirschner S, Goronzy J, Stiehler M, Lohmann C, Hartmann A. Update on metal-on-metal hip joints. Orthopade. 2013;42:373–87. quiz 388–9

Hailer NP, Bengtsson M, Lundberg C, Milbrink J. High metal ion levels after use of the ASR device correlate with development of pseudotumors and T cell activation. Clin Orthop Relat Res. 2014;472:953–61.

Ikeda T, Takahashi K, Kabata T, Sakagoshi D, Tomita K, Yamada M. Polyneuropathy caused by cobalt-chromium metallosis after total hip replacement. Muscle Nerve. 2010;42:140–3.

Knight SR, Aujla R, Biswas SP. Total hip arthroplasty – over 100 years of operative history. Orthop Rev (Pavia). 2011;3:e16.

Langton DJ, Jameson SS, Joyce TJ, Hallab NJ, Natu S, Nargol AV. Early failure of metal-on-metal bearings in hip resurfacing and large-diameter total hip replacement: a consequence of excess wear. J Bone Joint Surg (Br). 2010;92:38–46.

Langton DJ, Joyce TJ, Jameson SS, Lord J, Van Orsouw M, Holland JP, Nargol AV, De Smet KA. Adverse reaction to metal debris following hip resurfacing: the influence of component type, orientation and volumetric wear. J Bone Joint Surg (Br). 2011;93:164–71.

Langton DJ, Sidaginamale R, Lord JK, Nargol AV, Joyce TJ. Taper junction failure in large-diameter metal-on-metal bearings. Bone Joint Res. 2012;1:56–63.

Mahendra G, Pandit H, Kliskey K, Murray D, Gill HS, Athanasou N. Necrotic and inflammatory changes in metal-on-metal resurfacing hip arthroplasties. Acta Orthop. 2009;80:653–9.

Matharu GS, Pynsent PB, Dunlop DJ. Revision of metal-on-metal hip replacements and resurfacings for adverse reaction to metal debris: a systematic review of outcomes. Hip Int. 2014;24:311–20.

Mistry JB, Chughtai M, Elmallah RK, Diedrich A, Le S, Thomas M, Mont MA. Trunnionosis in total hip arthroplasty: a review. J Orthop Traumatol. 2016;17:1–6.

Moniz S, Hodgkinson S, Yates P. Cardiac transplant due to metal toxicity associated with hip arthroplasty. Arthroplasty Today. 2017. (Article in press).

National Joint Registry for England Wales and Northern Ireland. 13th annual report. (2016). http://www.njrcentre.org.uk/njrcentre/Portals/0/Documents/England/Reports/13th. Annual Report/07950 NJR Annual Report 2016 ONLINE REPORT.pdf.

Paustenbach DJ, Tvermoes BE, Unice KM, Finley BL, Kerger BD. A review of the health hazards posed by cobalt. Crit Rev Toxicol. 2013;43:316–62.

Posada OM, Tate RJ, Grant MH. Toxicity of cobalt-chromium nanoparticles released from a resurfacing hip implant and cobalt ions on primary human lymphocytes in vitro. J Appl Toxicol. 2015;35:614–22.

Rona G, Chappel CI. Pathogenesis and pathology of cobalt cardiomyopathy. Recent Adv Stud Cardiac Struct Metab. 1973;2:407–22.

Rousseau MC, Straif K, Siemiatycki J. IARC carcinogen update. Environ Health Perspect. 2005;113:A580–1.

Schmalzried TP, Shepherd EF, Dorey FJ, Jackson WO, Dela Rosa M, Fa'vae F, Mckellop HA, Mcclung CD, Martell J, Moreland JR, Amstutz HC. The John Charnley Award. Wear is a function of use, not time. Clin Orthop Relat Res. 2000;381:36–46.

Scientific Committee on Emerging Newly Identified Health Risks. Opinion on: the safety of metal-on-metal joint replacements with a particular focus on hip implants. 2014. http://ec.europa.eu/health/scientific_committees/emerging/docs/scenihr_o_042.pdf.

Siddiqui IA, Sabah SA, Satchithananda K, Lim AK, Cro S, Henckel J, Skinner JA, Hart AJ. A comparison of the diagnostic accuracy of MARS MRI and ultrasound of the painful metal-on-metal hip arthroplasty. Acta Orthop. 2014;85:375–82.

Smith IC, Carson BL. Trace metals in the environment. Volume 6: cobalt an appraisal of environmental exposure. Ann Arbor: Ann Arbor Science Publishers; 1981.

Smith AJ, Dieppe P, Porter M, Blom AW, National Joint Registry Of E and Wales. Risk of cancer in first seven years after metal-on-metal hip replacement compared with other bearings and general population: linkage study between the National Joint Registry of England and Wales and hospital episode statistics. BMJ. 2012;344:e2383.

Underwood R J, Zografos A, Sayles R S, Hart A, Cann P. Edge loading in metal-on-metal hips: low clearance is a new risk factor. Proc Inst Mech Eng H. 2012;226:217–26.

Van Der Weegen W, Smolders JM, Sijbesma T, Hoekstra HJ, Brakel K, Van Susante JL. High incidence of pseudotumours after hip resurfacing even in low risk patients; results from an intensified MRI screening protocol. Hip Int. 2013;23:243–9.

Visuri T, Pukkala E, Paavolainen P, Pulkkinen P, Riska EB. Cancer risk after metal on metal and polyethylene on metal total hip arthroplasty. Clin Orthop Relat Res. 1996:S280–9.

Watters TS, Cardona DM, Menon KS, Vinson EN, Bolognesi MP, Dodd LG. Aseptic lymphocyte-dominated vasculitis-associated lesion: a clinicopathologic review of an underrecognized cause of prosthetic failure. Am J Clin Pathol. 2010;134:886–93.

Willert HG, Buchhorn GH, Fayyazi A, Flury R, Windler M, Koster G, Lohmann CH. Metal-on-metal bearings and hypersensitivity in patients with artificial hip joints. A clinical and histomorphological study. J Bone Joint Surg Am. 2005;87:28–36.

Williams DH, Greidanus NV, Masri BA, Duncan CP, Garbuz DS. Prevalence of pseudotumor in asymptomatic patients after metal-on-metal hip arthroplasty. J Bone Joint Surg Am. 2011;93:2164–71.

Zywiel MG, Cherian JJ, Banerjee S, Cheung AC, Wong F, Butany J, Gilbert C, Overgaard C, Syed K, Jacobs JJ, Mont MA. Systemic cobalt toxicity from total hip arthroplasties: review of a rare condition part 2. Measurement, risk factors, and step-wise approach to treatment. Bone Joint J. 2016;98-B:14–20.

Chapter 20
Biological Synthesis, Pharmacokinetics, and Toxicity of Different Metal Nanoparticles

Raúl A. Trbojevich and Adriana M. Torres

Abstract Nanotechnology is an emerging technology to produce new materials with unique chemical and physical properties. Properties and applications of nanoparticles are strongly dependent on their size, shape, chemical composition, surface charge, and solubility. The synthesis and characterization are very important factors to biological applications. Metal nanoparticles such as silver (AgNPs), titanium dioxide (TiO_2NPs), and mercury (HgNPs) are especially interesting nanomaterials to be studied because of their broad applications in different fields. In this chapter, we approach topics such as some important biological synthesis of these three metal nanoparticles; pharmacokinetics (absorption, distribution, and elimination) and toxic effects of these metals will be described as well as their biological and pharma applications.

Keywords Silver • Mercury • Titanium dioxide • Nanoparticles • Toxicology • Biological applications • Synthesis of metal nanoparticles

20.1 Introduction

Nanotechnology is an emerging technology to produce new materials with unique chemical and physical properties. This new technology allows scientists to work on the scale of 1–100 nm. At this size scale, the matter has a greater volume ratio per gram compared with the bulk, showing new chemical and physical capabilities.

This book chapter is not an official US Food and Drug Administration (FDA) guidance or policy statement. No official support or endorsement by the US FDA is intended or should be inferred.

R.A. Trbojevich (✉)
Division of Biochemical Toxicology. National Center for Toxicological Research, U.S. Food and Drug Administration, Jefferson AR 72079, AR, USA
e-mail: raul.trbojevich@fda.hhs.gov

A.M. Torres
Pharmacology, Faculty of Biochemical and Pharmaceutical Sciences, Rosario National University- National Council of Scientific and Technical Research (CONICET), Rosario, Argentina

M. Rai, R. Shegokar (eds.), *Metal Nanoparticles in Pharma*,
DOI 10.1007/978-3-319-63790-7_20

Metals at these scales have shown an important impact on every aspect of science and technology (Schmidt 2004; Gupta and Kompella 2006; Rao et al. 2007; Rocco et al. 2010). Properties and application of nanoparticles are strongly dependent on their size, shape, chemical composition, surface charge, and solubility (Schmidt 2004; Gupta and Kompella 2006; Rao et al. 2007; Rocco et al. 2010). Developments of new synthesis methods are leading to new applications and have increased year to years (Vance et al. 2015).

Even though there is rapid increase of nanotechnology and applications in many areas and products, there are, just only, a few studies that reported about the human risks and the environment effects of engineered nanomaterials (Wiegand et al. 2009; Kaegi et al. 2013; Hunt et al. 2013; Li et al. 2016; Boudreau et al. 2016). In this regard, cytotoxicity, oxidative stress, and inflammatory effects are some issues that researchers and regulatory stakeholders need to decrease uncertainties on the exposure and risks caused by engineered nanomaterials.

Metal nanoparticles such as silver (AgNPs), titanium dioxide (TiO_2NPs), and mercury (HgNPs) are especially interesting nanomaterials to be studied because of their broad applications in different fields. AgNPs has a history that goes back almost two centuries of being used as a biocidal agent and other numerous applications (Nowack et al. 2010). TiO_2NPs have been found to have biocidal activity but required UV light (Kim et al. 2003). Additional effects of TiO_2NPs are protecting from effects of UV irradiation, making these nanoparticles efficient short-wavelength light absorbers with high photostability which have a broad use in sunscreens (Barnard 2010) and other applications (Zhang et al. 2009; Saha et al. 2011). Mercury metal and HgNPs are extremely toxic and could be present in the environment from too many sources (e.g., combustion of coal) making human exposure rate quite high.

In this chapter, we discuss pharmacokinetics from silver, titanium, and mercury nanoparticles (absorption, distribution, and elimination). Also, some important synthesis and biosynthesis methods for AgNPs, TiO_2NPs, and HgNPs focusing in their biological applications are presented. In addition for: TiO_2, a general methodology for characterization of nanomaterial is discussed. Finally, notes on toxicological point of view of these metal nanocomposites are added.

20.2 Silver (Ag) and AgNPs

20.2.1 Applications

For many years, ionic silver has been used for antimicrobial applications (Alexander 2009). AgNPs have been produced with the aim of slow release of silver ions (Ag^+) in order to extend antibacterial effect (Wijnhoven et al. 2009). Increasing evidence of improved antimicrobial effects of AgNPs and possible immunomodulatory activities are stimulating. This fact is relevant in an age when multiple antibiotic-resistant bacteria are becoming widespread. Silver has a broad antibacterial activity on a range of Gram-negative and Gram-positive bacteria and antibiotic-resistant bacteria strains. AgNPs have been described as effective agent against a wide spectrum of common fungi. Antiviral silver activity has been reported against HIV-1, hepatitis B virus, respiratory syncytial virus, and herpes simplex virus (Ge et al. 2014). The antimicrobial silver effect is dependent upon bioavailability of Ag^+. Ag^+ has a great spectrum of activity and interacts with the cell membrane resulting in uncoupling of the respiratory electron transport system from oxidative phosphorylation. It also interferes with membrane permeability and the proton motive force; inhibits respiratory chain enzymes and intracellular enzymes reacting with electron donor groups, mainly sulfhydryl groups; and intercalates with DNA (Teng et al. 2000). Other authors (Wright et al. 2002) have reported that nanocrystalline Ag^+ inhibits matrix metalloproteases promoting wound healing and binds metallothioneins contributing to tissue repair. It has also been described that Ag^+ modulates the immune system by decreasing the expression of proinflammatory cytokines and gelatinase activity (Nathworthy et al. 2008). Moreover, AgNPs are postulated for anticancer therapy because of its effects on cell survival and signaling (Lee et al. 2016). They have been shown to cause the death of breast cancer cells and to have antileukemia activities (dos Santos et al. 2014).

Silver ion and its nanoparticle form are recently being applied to consumer products, e.g., shampoo and rinse, reusable bottles, nipples, toothpaste, toothbrushes, detergents, deodorants, cosmetics, water purification devices, kitchen utensils, washing machines, refrigerators, humidifiers, food packing materials, and toys (Edwards-Jones 2009). Moreover, AgNPs are being employed in medical uses due to its strong antimicrobial activities. AgNPs have been incorporated into wound dressing, bone prostheses, dental alloys, surgical instruments, cardiovascular implants, urinary tract catheters, endotracheal tubes, and contraceptive devices (Chen and Schluesener 2008). They are also used in drug delivery as cancer and retinal therapies and for bio-diagnosis (Kalishwaralal et al. 2010). Additionally, AgNPs have been included into textiles for the fabrication of clothes and in the food industry to restrict bacterial growth (Vigneshwaran et al. 2007).

20.2.2 *Synthesis*

20.2.2.1 Chemical Synthesis

In medical and biological applications, the most important in AgNP synthesis is to control the size and shape, which could be used as biological labels (surface plasmon resonant band), but also the stabilizer agent selected is important as drug carrier and delivery.

The conventional synthesis of AgNPs, both, for materials and biological applications is the Turkevich method (Turkevich et al. 1951). However, there are some drawbacks in this method (varied in the size range 60–200 nm), compared to other methods. In this method, the citrate is used as a reducing and simultaneously stabilizing agent.

The borohydride method is another way to synthetize AgNPs (Evanoff and Chumanov 2005). It is explained by the relatively high reactivity of borohydride compared with the citrate that needs temperature in the reaction procedure.

An interesting chemical and green method for AgNPs embedded antimicrobial paints based on vegetable oil was developed by Kumar et al. (2008). Shortly, the method produces AgNPs in a single step from any kind of house paint. The synthesis uses a natural oxidative process involving the reaction of free radicals (generated during the natural drying process of oils-alkyd paints) with a reducing silver salt. The process not requires any reducing or stabilizing agents. In addition, the authors have shown that surfaces coated with these AgNPs (e.g., wood, glass) showed excellent biocidal activities for both Gram-positive and Gram-negative bacteria, mainly AgNPs with a 15 nm of size or less.

Microwave-assisted synthesis is a very good method for AgNPs. The method has shorter reaction time and produces nanoparticles with smaller sizes, narrower size distributions, and high degree of crystallization (Nadagouda et al. 2011) which is an important factor for biocidal activities.

Starch (as reducing and stabilizing agent) was used to synthetized AgNPs (Sreeram et al. 2008); the method produces AgNPs with an average size of 12 nm preventing aggregation.

There are many other chemical methods to synthetize AgNPs depending on their use and application (Krutyakov et al. 2008; Iravani et al. 2014).

20.2.2.2 Physical Synthesis

One interesting physical method to the synthesis of AgNPs is a green method that uses amino acid (L-lysine or L-arginine) as reducing agent and soluble starch as stabilizer (Hu et al. 2008; Trbojevich et al. 2016). The method is microwave assisted. It is rapid and produces AgNPs in large quantities. The resulted AgNPs are very uniform in size and distribution but also water soluble which provides excellent biological applications. Other authors have presented an interesting synthesis method for AgNPs for medical and biological applications (Kemp et al. 2009). In this method, the authors used heparin and hyaluronan as both reducing and protecting agents, respectively; the AgNPs show good stability under physiological conditions with narrow size distribution. The AgNPs synthetized have shown medical applications (Kemp et al. 2009). There is a good review that has shown detailed physical method of the AgNP synthesis (Krutyakov et al. 2008). In this review, the authors have shown several physical synthesis methods, but also optical and antibacterial properties of AgNPs are discussed with details.

20.2.2.3 Biological Synthesis

Table 20.1 shows biosynthesis of AgNPs and other metal nanoparticles using fungi, bacteria, and other microorganisms.

Table 20.1 Biosynthesis of AgNPs and other metals using living organisms

Nanoparticles	Living organism	Condition temperature (°C)	Particle size (nm) and shape	Reaction place	References
Silver-gold	*Brevibacterium casei*	37	10–50, spherical	Intracellular	Kalishwaralal et al. (2010)
Silver	*Bacillus cereus*	37	4–5, spherical	Intracellular	Babu and Gunasekaran (2009)
Silver	*Escherichia coli*	37	50, not available	Extracellular	Gurunathan et al. (2009)
Silver	*Aspergillus fumigatus*	25	8.92, spherical	Extracellular	Bhainsa et al. (2006)
Silver	*Corynebacterium glutamicum*	30	5–50, irregular	Extracellular	Sneha et al. (2010)
Silver	*Fusarium oxysporum*	25	5–50, spherical	Extracellular	Senapati et al. (2004)
Silver-gold	*Yeast*	30	9–25, Irregular	Extracellular	Zheng et al. (2010)
Silver	*Verticillium* sp.	25	25, spherical	Extracellular	Senapati et al. (2004)
Silver-gold	*Neurospora crassa*	28	20–50, spherical	Intracellular-extracellular	Castro-Longoria et al. (2011)
Platinum	*Shewanella algae*	25	5, not available	Intracellular	Konishi et al. (2007)
Mercury	*Enterobacter* sp.	30	2–5, spherical	Intracellular	Sinha and Khare (2011)
Titanium oxide	*Lactobacillus* sp.	25	8–35, spherical	Extracellular	Jha et al. (2009)
Titanium oxide	*Fusarium oxysporum*	300	6–13, spherical	Extracellular	Bansal et al. (2005)

The biological synthesis of AgNPs using microorganisms is an option for medical use (Kouvaris et al. 2012). The authors synthesized AgNPs simply from an aqueous silver nitrate solution and leaf broth of *Arbutus unedo* which was acting as a reductant and stabilizer agent. The particles produced were very stable with potential application in biotechnology. Some microorganisms are capable of biosynthesis to produce inorganic and biodegradable materials. In this regard, *Fusarium oxysporum* (a fungus) can make an extracellular reduction of aqueous silver nitrate solution to form AgNPs with 20–50 nm in size (Krutyakov et al. 2008). The biosynthesis of Ag^{+} reduction by different strains of *Fusarium oxysporum* was studied (Duran et al. 2005). AgNPs (water soluble) were in size of 20–50 nm when silver ion solution was exposed to several *Fusarium oxysporum* strains reducing silver ion (Duran et al. 2005). An excellent review that has shown biological synthesis of silver and other metal nanoparticles (Li et al. 2011) has shown the production of several metal nanoparticles in a friendly way technology using bacteria and different microorganisms. The size, shape, stability, the potential application to drug delivery, gene therapy , DNA analysis and cancer treatment are well explained (Li et al. 2011).

The biosynthesis of AgNPs using a fungus (*Aspergillus fumigatus*) was presented (Bhainsa and D'Souza 2006). The synthesis produces AgNPs in a quite fast process with a size of 5–25 nm. The biological process showed an easy route to produce nanoparticles with potential applications to bio-market.

20.2.3 Pharmacokinetics: Absorption, Distribution, and Elimination

Silver oral absorption ranges from 0.4% to 18% in mammals, depending on the species. AgNPs are less bioavailable than ionic silver after oral administration (van der Zande et al. 2012; Hadrupand and Lam 2014; Boudreau et al. 2016).

After oral absorption, ionic and nanoparticulated silver are deposited in all tissues. The majority of the silver was found in the stomach and intestines followed in descending order by the liver, testes, kidneys, brain, lungs, blood, bladder, and heart. In the kidneys, silver was localized in the glomerular basement membranes (Creasey and Moffat 1973) and in the brush border membranes of proximal tubular cells (Loeschner et al. 2011). It is well known that silver can be deposited in the skin causing the blue-gray hyperpigmentation of human skin described during argyria (Chang et al. 2006). Moreover, it was observed that silver was deposited only in certain brain regions, maybe as the result of local differences in the permeability of blood-brain barrier (Rungby and Danscher 1983). Other authors have also described that silver can traverse the placental barrier (Lee et al. 2012).

Silver fecal excretion ranges from 49% to 99.6% depending on the species and on the different forms of silver (Furchner et al. 1968). It has been described that when the bile duct is ligated, the fecal excretion of oral absorbed silver is reduced and silver accumulation in the liver is observed (Scott and Hamilton 1950). Moreover, after parenteral silver administration, 50% of the silver is recovered in rat bile (Dijkstra et al. 1996). The urinary silver excretion after oral administration was described to be in the range of 0.0002–0.0004% in humans and 0.005–0.06% in rats (East et al. 1980).

20.2.4 Toxic Effects

Some studies (Lee et al. 2007) have demonstrated that effects on cell and bacterial are due to a low level of Ag^+ release from AgNPs. The Ag^+ release is a function of AgNP's size and pH, which is higher for smaller nanoparticles and also with the temperature increase, presence of oxygen, sulfur, and light. In addition, it has been shown that AgNPs are less toxic than the equivalent mass from Ag^+ (Kim et al.

2010). Ag^+ has cytotoxic effects to different types of cells. It has been suggested that ionic silver activates ionic channels and changes cell membrane permeability to sodium and potassium, interacts with mitochondria, and induces apoptosis by means of reactive oxygen species production, conducing to cell death. Lee et al. (2007) demonstrated that AgNPs less than 12 nm in size altered early development of fish embryos, produced chromosomal aberrations and DNA damage, and induced proliferation of zebra fish cell lines.

In oral toxicity experiments, silver deposition was observed in the gastrointestinal tract. Abnormal pigmentation of the ileum, damage to the epithelial cell microvilli and intestinal glands, and bile duct hyperplasia with eosinophil infiltration of the hepatic lobules and portal tract have been reported after subchronic and chronic oral administration of AgNPs. Increased activity of alkaline phosphatase and alanine aminotransferase in plasma was also observed (Kim et al. 2010; Hadrup et al. 2012). Heart hypertrophy, increased hemoglobin levels, and hematocrit values were observed in rats after chronic AgNP treatment (Espinosa-Cristobal et al. 2013). It has been described that AgNPs are accumulated in all regions of the kidney cortex and medulla. Silver was mainly deposited in the cortical glomeruli within the cytoplasm of mesangial phagocytes, in thc basement membrane, and in the renal tubules, which may lead to renal function alterations (Kim et al. 2008). Moreover, other authors (Sardari et al. 2012) have reported necrosis of glomerular cells, Bowman's capsule, and proximal tubular cells in the kidneys.

20.3 Titanium Dioxide (TiO_2) and TiO_2NPs

20.3.1 Applications

TiO_2 exists in three crystallographic forms: rutile, brookite, and anatase. The anatase forms have wide bandgap where excitation with photons greater than 300 nm generates long-lived electron-hole pairs (semiconductor materials). Both electron and electron-hole pair can reach the crystal's surface and participate in redox reactions with absorbed substrates. This is a birefringent material with a high refractive index (4.0 for rutile and 3.6 for anatase) in UV and visible range. The photocatalytic property of TiO_2 (anatase) can be used for biocidal proposal.

TiO_2 and TiO_2NPs are inert and poorly soluble materials. It is mainly employed as a pigment because of its brightness, high refractive index, and resistance to discoloration. It is a common additive in personal care, food, and other consumer products. It is used as a pigment in paints and also as a pigment in plastics, paper, ceramics, fibers, enamels, foods (candies, sweets, taste enhancers, chewing gums), and personal care products (cosmetics, shampoos, lip balms, toothpastes, deodorants, sunscreen). All these products after use can enter the sewage system and, subsequently, access to the environment as treated effluent discharged to surface waters or biosolids applied to agricultural land, incinerated wastes, or landfill soils (Smijs and Pavel 2011; Trbojevich et al. 2016).

TiO_2 or TiO_2NPs are also used for antimicrobial applications, catalysts for air and water purification, medical devices, and energy storage (Niinomi 2008; Weir et al. 2012). Titanium alloys are very important in orthopedic surgery (such as hip and knee implants) due to their high resistance to the attack of body fluids, strength, flexibility, biocompatibility, or lack of allergenicity (Niinomi 2008; Smijs and Pavel 2011; Weir et al. 2012; Trbojevich et al. 2016).

20.3.2 Synthesis

There are many routes to synthesize TiO_2 or TiO_2NPs (Mahshid et al. 2006; Smijs and Pavel 2011; Trbojevich et al. 2016). Sol-gel technique is one of the most important to prepare TiO_2 and its nano-form (Mahshid et al. 2006). In this sol-gel technique, TiO_2 was synthetized in a hydrolysis and condensation reaction of Ti-alkoxide precursor in acidic aqueous solution with the following reactions (Mahshid et al. 2006):

$$\mathrm{Ti(OR)_4 + 4\,H_2O \rightarrow 2\,Ti(OH)_4 + 4\,ROH(hydrolysis)}$$
$$\mathrm{Ti(OH)_4 \rightarrow TiO_2 \times H_2O + (2 - x)H_2O(condensation)}$$

R is ethyl, isopropyl, n-butyl, or any organic radical. The nanoparticle size, stability, and morphology could be controlled using the ratio r = $[H_2O]/[Ti]$ (Mahshid et al. 2006; Smijs and Pavel 2011; Trbojevich et al. 2016).

TiO_2 could be selectively synthetized with a morphological control, including anatase, rutile, and brookite in a solvothermal reaction using water-soluble titanium complexes as initial precursors (Kobayashi et al. 2013). The authors have shown the production of TiO_2 with polymorphs, as well as diverse morphological control which can control the biological properties (Kobayashi et al. 2013).

20.3.3 Pharmacokinetics: Absorption, Distribution, and Elimination

As this metal was considered almost completely biologically inert until recently, there is little information regarding the behavior of titanium in the body.

Titanium is widely distributed in the environment and can enter the organism through different routes: oral, dermal, respiratory, or from surgical implants made of titanium alloys. The implants suffer corrosion under unfavorable conditions in the internal medium of the body. Many factors such as constant and high temperature, the presence of dissolved oxygen, ions, proteins, and acidic environment, occurring at the site of the implanted medical devise, provoke degradation of the protective layer on the surface of implants. Consequently, titanium ions or TiO_2NPs are released from the biomaterial into the neighboring tissues (Bhola et al. 2011).

It has been described that TiO_2NPs have very limited bioavailability after oral exposure. Nevertheless there are evidences that gastrointestinal tract absorption is possible since titanium levels are detected in the livers and mesenteric lymph nodes after oral administration (Geraets et al. 2014).

After systemic exposure, TiO_2NPs are distributed to the kidney, liver, spleen, and lung. It has been shown that over a period of 90 days after exposure, there is redistribution from the liver to the spleen. Nuevo-Ordoñez et al. (2011) have demonstrated that titanium has the capacity of binding to serum transferrin. As transferrin is the main transporter of iron, titanium competes with iron for transferrin-binding site.

After a single i.v. dose of titanium (IV) citrate (6 mg Ti/kg b.w.) in rats, the elimination half-life times were 3.3 h, 2.1 h, and 1.9 h in the kidney, spleen, and liver, respectively (Golasik et al. 2016). Other authors (Elgrabli et al. 2015) have reported that after five repeated i.v. doses of TiO_2NPs in rats (total dose between 42.3 and 71.9 mg/kg b.w.), the half-lives for the kidneys, liver, and spleen were in the range of 49–531 days, 54–248 days, and 650 days, respectively. Xie et al. (2011) have demonstrated that TiO_2NPs could be excreted through urine and feces, being the urine excretory rate higher than that of the feces.

20.3.4 Toxic Effects

Exposure of the skin to TiO_2NPs causes barrier dysfunction and can initiate and/or promote skin disease (Yanagisawa et al. 2009).

Inhalation constitutes the main route of exposure of the human body to TiO_2NPs especially at workplaces during handling processes. Chronic lung inhalation studies in experimental animals have reported increased incidence of pneumonia and squamous metaplasia, enhanced proliferation of pulmonary cells, and defects in macrophage function, among others. Inhaled TiO_2NPs after nasal absorption may deposit in the brain altering monoaminergic neurotransmission (Lee et al. 1985). It has been reported titanium may cause liver and kidney toxicity after oral exposure to TiO_2NPs. Hepatocyte apoptosis in the liver, followed by increased accumulation of reactive oxygen species and a decrease in stress-related gene expression levels, histopathological changes, and damage in the function of mice liver, has been described (Li et al. 2010). Wang et al. (2007) and Gui et al. (2013) have described increase in blood urea nitrogen and creatinine as well as histopathological changes in renal tubules and glomerulus, increased production of kidney reactive oxygen species and peroxidation of lipids, proteins, and DNA. Spleen pathological changes leading to the reduction of mice immunity have also been described (Shakeel et al. 2015).

Maternal exposure to titanium might alter the development of the central dopaminergic system in offspring, and the induced stress could be implicated in depressive-like behaviors in adulthood (Takahashi et al. 2010). Moreover, TiO_2NPs reduce fertility and damage mouse ovaries (Shakeel et al. 2015). Other authors (Savi et al. 2014) showed that titanium also alters cardiac excitability and increases the chances of arrhythmic events.

20.4 Mercury (Hg) and HgNPs

20.4.1 *Applications*

The most important natural sources of Hg are gases released during volcanic eruptions. In addition, the burning of fossil fuel (mainly coal and municipal waste incineration) is the main anthropogenic source of Hg to the atmosphere. All living organisms are exposed to Hg but also to HgNPs as a consequence of its ubiquitous presence in the environment. Organic matter and sediment in water could help microorganisms to form HgNPs. The exposure to Hg or HgNPs leading to health hazards is generally due to a multitude of specific human activities: unintentional occupational exposure and ingestion of Hg as an ingredient in food, folk remedies, amalgam fillings, and a preservative in vaccines. This metal is still employed in blood pressure cuffs and thermometers, batteries, switches, and fluorescent light bulbs (Clarkson 1997; Clarkson et al. 2003).

The four main chemical forms of Hg are elemental Hg, Hg salts, organic Hg, and HgNPs. Elemental Hg is the most volatile of Hg inorganic forms, and the exposition to its vapor is primarily occupational; nevertheless, it can also be released from other sources (Clarkson 1997; Clarkson et al. 2003).

Hg salts are present as monovalent mercurous or as divalent mercuric. Mercurous chloride (calomel) has been employed in some skin creams as an antiseptic and was also used as a cathartic and diuretic. Hg salts are still used in industry, which discharged into rivers contaminating the environment (Clarkson 1997; Clarkson et al. 2003).

Inorganic Hg is bio-methylated by microorganisms in sediments of fresh- and ocean water and bioaccumulated as methylmercury in fishes. Moreover, methylmercury has been used as fungicide. Therefore, methylmercury is the most common exposure to organic Hg in humans (Arvind and Sunil 2011).

Ethylmercury is another variety of organic Hg which is the active ingredient of thimerosal employed in vaccines as preservative (Clarkson 1997; Clarkson et al. 2003; Arvind and Sunil 2011).

HgNPs are naturally ubiquitous in the environment and are formed by processes that have taken place for millions of years (Xu and Carraway 2012). They can also be synthetized in the laboratory or produced in mercury-resistant bacteria (Xu 2009; Arvind and Sunil 2011; Sinha and Khare 2011; Sathyavathi et al. 2013).

20.4.2 *Synthesis*

There are several methods to prepare mercury oxides and salt HgNPs, but none have produced pure metal mercury HgNPs. In 2002, synthesis of mercury selenide using sonication was reported (Wang et al. 2002). The method used mercury acetate and sodium selenosulfate in aqueous solution at room temperature. The authors used ethylenediamine (EN), ammonia, and triethanolamine (TEA) to control HgNP size and size distribution. The authors have explained the probable mechanism as follows (Wang et al. 2002):

$$H_2O \rightarrow H^{\bullet} + OH^{\bullet}$$
$$2H^{\bullet} + SeSO_3{}^{2-} \rightarrow Se^{2+} + H_2SO_3$$
$$Hg(\text{TEA or } NH_3 \text{or EN})_n^{2+} + Se^{2-} \rightarrow HgSe + n(\text{TEA or } NH_3 \text{or EN})$$
$$n\,HgSe \rightarrow (HgSe)_n\ (\text{nanocomposite})$$

Xu (Ph.D. thesis 2009) has discussed the synthesis of mercury sulfide-HgNPs. The author used $HgCl_2$ and sulfur powder as precursors, adjusting the pH with NaOH. The reaction was carried out in an ultrasound sonication bath to yield smaller and uniform HgNPs with the sonication and nitrogen flow as control factors. The mercury sulfide-HgNPs with a size of 10–15 nm, but also some aggregates (100 or more nm), were obtained in the final reaction (Xu 2009). Arvind and Sunil (2011) reported simultaneous HgNP synthesis and mercury bioaccumulation by bacteria. The mercury-resistant strain of *Enterobacter* sp. exhibited a mercury bioaccumulation and simultaneous synthesis of HgNPs. HgNPs are synthetized in culture conditions and promote spherical, uniform, and monodispersed intracellular HgNPs with sizes of 2–5 nm. The HgNPs could be recoverable (Arvind and Sunil 2011).

A simple synthesis method of HgO-HgNPs (nanocomposites) was carried out by the sublimation reaction of the mercury (II) acetate powder at 150 °C for 2 h (Mohadesi et al. 2014). The final product was pure orthorhombic HgO-HgNPs with an average size of 100 nm.

Mercury telluride-HgNPs (nanocomposites) have been synthesized using hydrothermal reduction (Sangsefidi et al. 2014). The reaction was carried out with N′-bis(salicylaldehydo)ethylenediamine)mercury(II) (as a new mercury precursor) or $HgCl_2$ and $TeCl_4$ in the presence of $N_2H_4.H_2O$ or KBH_4 as reductant agents. The mercury telluride-HgNPs synthetized, with a size of 15–25 nm, have shown strong dependence of the Hg^{2+} source, temperature, and reaction time. Also, mercury telluride-HgNP morphology depends on temperature and with the presence of KBH_4 as a reductant (Sangsefidi et al. 2014).

20.4.3 Pharmacokinetics: Absorption, Distribution, and Elimination

Elemental mercury shows a low absorption from the gastrointestinal tract. Inhaled mercury vapor is highly absorbed by the lungs and then is oxidized to the divalent mercuric cation by the action of erythrocyte catalase. Mercury vapor arrives to the brain before its oxidation and produces central nervous system toxicity (Clarkson 1997; Zalups 2000; Clarkson et al. 2003; Torres 2013).

After oral administration, soluble inorganic mercuric salts (Hg^{2+}) are absorbed in the gastrointestinal tract. The highest concentration of Hg^{2+} is observed in the kidneys. They are eliminated in the urine and feces. Inorganic mercurials do not cross the placenta and the blood-brain barrier. Inorganic mercury half-life is approximately about 60 days (Zalups 2000; Clarkson 1997; Clarkson et al. 2003; Torres 2013).

Organic mercurials, which are lipid soluble, are highly absorbed from the gastrointestinal tract. Approximately 90% of methylmercury is absorbed from the gastrointestinal tract, crosses the placenta and blood-brain barrier, and causes neurological and teratogenic effects. The carbon-mercury bond of organic mercurials is cleaved after absorption, and inorganic mercuric is generated. Methylmercury is mainly excreted in the feces conjugated with glutathione, and around 10% of its doses are excreted in urine. Methylmercury has a half-life between 40 and 105 days. Mercury forms covalent bonds with sulfur. It binds to endogenous molecules such as glutathione, cysteine, homocysteine, N-acetylcysteine, metallothionein, and albumin which contain thiols. This characteristic accounts for most of the biological properties of the metal. Mercurials (even in low concentrations) are capable of inactivating sulfhydryl enzymes and thus interfering with cellular metabolism and function. The affinity of mercury for thiols gives the support for treatment of mercury poisoning with agents as dimercaprol and penicillamine (Clarkson 1997; Zalups 2000; Clarkson et al. 2003; Torres 2013).

Pharmacokinetics of HgNPs is an interest topic that is currently assess in our laboratories, since there are no reports from other authors, at least to our knowledge.

20.4.4 *Toxic Effects*

The toxic effects of mercury are mainly produced from the exposure of people to methylmercury in fish (Zalups 2000). The major toxic effects of methylmercury are on the central nervous system. The first symptom to appear at the lower dose is paresthesia, which may progress to cerebellar ataxia, dysarthria, constriction of the visual fields, and loss of hearing. It has also been reported that severe brain damage takes place from high prenatal exposure (Clarkson et al. 2003).

Following exposure to elemental or inorganic forms of mercury, the kidneys are the primary sites of mercury accumulation. Organic forms of mercury also accumulate in the kidneys (Torres 2013).

Mercury accumulation causes acute renal failure. This is characterized by profound renal vasoconstriction, reduction in glomerular filtration rate, histological damage, activation of the renin-angiotensin system, and modulation in the expression of membrane transporters (Zalups 2000; Stacchiotti et al. 2004; Di Giusto et al. 2009a; Torres 2013). Torres et al. (2011) have demonstrated using organic anion transporter 1 (Oat1) knockout mice that mercuric chloride-induced acute kidney injury is mainly mediated by the Oat1. Moreover, the lower expression of this transporter in females justified the lower renal damage induced by mercury in females (Hazelhoff et al. 2012). Additionally, the upregulation of the renal expression of Oat1 and multidrug resistance-associated protein 2 (Mrp2) by repeated administration of furosemide improves the proximal damage induced by mercuric chloride (Hazelhoff et al. 2015). Prof. Torres's group has also been pioneer in detecting the organic anion transporter 5 (Oat5) in urine (Di Giusto et al. 2009b). They postulated that urinary excretion of Oat5 could be an early biomarker of proximal tubular damage induced by mercury (Di Giusto and Torres 2010).

There are no reports, at least to our knowledge, regarding the toxic effects of HgNPs. Current research is performed in the laboratory of Prof. Torres in collaboration with Dr. Trbojevich to evaluate the potential toxic effects of HgNPs administration to rats.

Mercury toxic effects can also be prevented by reducing Hg^{2+} to nontoxic form. Bacteria resistant to toxic metals and capable of converting them into nontoxic forms have relevant uses in the bioremediation of contaminated places. In this connection, a mercury-resistant strain of *Enterobacter* sp. isolated from soil, which has the novel characteristic of mercury accumulation with simultaneous synthesis of HgNPs, has been reported (Sinha and Khare 2011). Moreover, another mercury-resistant bacterium (*Bacillus cereus MRS-1*) was isolated from electroplating industrial effluent. This strain converts mercury into extracellular sulfide-HgNPs (Sathyavathi et al. 2013). The remediated mercury trapped in the form of nanoparticles is unable to vaporize back into the environment. These strains can be potentially exploited for the production of HgNPs as well as for detoxification of mercury in the environment without producing secondary pollution of mercury methylation or Hg volatilization.

20.5 Conclusion and Future Perspectives

There are many publications about the synthesis of metal nanoparticles, and all of them have some advantages and disadvantage, but the main thing for the synthesis is the control of size, shape, and stability. In this regard, metal nanoparticles in pharmacological and medical applications required a strict control of those parameters to assure the absence of the collateral effect and the target of their function.

There are many studies related to heavy metal (such as Ag, Hg, and Ti) toxicity in biological systems and in the environment, but few of them are regarding of these metals in their nanoparticle forms. In our opinion, to elucidate the main toxic mechanisms of those metal nanoparticles, it is necessary to combine a good characterization of each nanocomposite (surface chemistry, ROS generation, solubility, and morphology) with in vitro and in vivo data which allow evaluation of the mechanism at the cellular level. Information about translocation, accumulation, and biodistribution of these nanoparticles in vivo as well as their environment effects are necessary. In addition, development and validation of new analytical methods for toxicity evaluation of metal nanoparticles in biological system should be addressed.

Acknowledgments The authors would like to acknowledge the support of this book chapter through project # E0736801 from the National Center for Toxicological Research-US Food and Drug Administration.

References

Alexander JW. History of the medical use of silver. Infect (Larchmt). 2009;10:289–92.

Arvind S, Sunil KK. Mercury bioaccumulation and simultaneous nanoparticles synthesis by Enterobacter sp. Cells Bioresource Technol. 2011;102:4281–4.

Babu MMG, Gunasekaran P. Production and structural characterization of crystalline silver nanoparticles from Bacillus cereus isolate. Colloids Surf B. 2009;74(1):191–5.

Bansal V, Rautaray D, Bharde A, Ahire K, Sanyal A, Ahmad A, Sastry M. Fungus-mediated biosynthesis of silica and titania particles. J Mater 2005;15(26):2583–9.

Barnard AS. One to one comparison of sunscreen efficacy, aesthetics and potential nanotoxicity. Nat Nanotechnol. 2010;5:271–4.

Bhainsa KC, D'Souza SF, Kottaisamy M, BarathManiKanth S, Kartikeyan B, Gurunathan S. Extracellular biosynthesis of silver nanoparticles using the fungus Aspergillus fumigatus. Colloids Surf B: Biointerfaces. 2006;47(2):160–4.

Bhola R, Bohla SM, Mishra B, Olson DL. Corrosion in titanium dental implants/prostheses-a review. Trends Biomater Artif Organs. 2011;25:34–46.

Boudreau MD, Imam MS, Paredes AM, Bryant MS, Cunningham CK, Felton RP, Jones MY, Davis KJ, Olson GR. Differential effects of silver nanoparticles and silver ions on tissue accumulation, distribution, and toxicity in the Sprague Dawley rat following daily oral gavage administration for 13 weeks. Toxicol Sci. 2016;150(1):131–60.

Castro-Longoria E, Vilchis-Nestor AR, Avalos-Borja M. Biosynthesis of silver, gold and bimetallic nanoparticles using the filamentous fungus Neurospora crassa. Colloids Surf B. 2011;83(1):42–8.

Chang AL, Khosravi V, Egbert B. A case of argyria after colloidal silver ingestion. J Cutan Pathol. 2006;33:809–11.

Chen X, Schluesener HJ. Nanosilver: a nanoproduct in medical application. Toxicol Lett. 2008;176:1–12.

Clarkson TW. The toxicology of mercury. Crit Rev Clin Lab Sci. 1997;34:369–403.

Clarkson TW, Magos L, Myers GJ. The toxicology of mercury – current exposures and clinical manifestations. N Engl J Med. 2003;349:1731–7.

Creasey M, Moffat DB. Deposition of ingested silver in the rat kidney at different ages. Experientia. 1973;29:326–7.

Di Giusto G, AM Torres AM (2010). Organic anion transporter 5 renal expression and urinary excretion in rats exposed to mercuric chloride: a potential biomarker of mercury-induced nephropathy. Arch Toxicol 84: 741–749.

Di Giusto G, Anzai N, Ruiz ML, Endou H, Torres AM. Expression and function of Oat1 and Oat3 in rat kidney exposed to mercuric chloride. Arch Toxicol. 2009a;83:887–97.

Di Giusto G, Anzai N, Endou H, Torres AM. Oat5 and NaDC1 protein abundances in kidney and urine following renal ischemic reperfusion injury. J Histochem Cytochem. 2009b;57:17–27.

Dijkstra M, Havinga R, Vonk RJ, Kuipers F. Bile secretion of cadmium, silver, zinc and copper in the rat. Involvement of various transport systems. Life Sci. 1996;59:1237–46.

Duran N, Marcato PD, Alves OL, De Souza GIH, Esposito E. Mechanistic aspects of biosynthesis of silver nanoparticles by several fusarium oxysporum strains. J Nanobiotechnol. 2005;3:8.

East BW, Boddy K, Williams ED, Macintyre D, McLay AL. Silver retention, total body silver and tissue silver concentrations in argyria associated with exposure to an anti-smoking remedy containing silver acetate. Clin Exp Dermatol. 1980;5:305–11.

Edwards-Jones V. The benefits of silver in hygiene, personal care and healthcare. Lett Appl Microbiol. 2009;49:147–52.

Elgrabli D, Beaudouin R, Jbilou N, Floriani M, Pery A, Rogerleux F. Biodistribution and clearance of TiO2 nanoparticles in rats after intravenous injection. PLoS One. 2015;10:e0124490.

Espinosa-Cristobal LF, Martinez Castañon GA, Loyola-Rodriguez JP, Patiño-Marin R-MJF, Vargas-Morales JM. Toxicity, distribution and accumulation of silver nanoparticles in Wistar rats. J Nanopart Res. 2013;15:1702.

Evanoff DD Jr, Chumanov G. Synthesis and optical properties of silver nanoparticles and arrays. ChemPhysChem. 2005;6:1221–31.

Furchner JE, Richmond CR, Draje GA. Comparative metabolism of radionuclides in mammals-IV. Retention of silver-110 m in the mouse, rat, monkey, and dog. Health Phys. 1968;15:505–14.

Ge L, Li Q, Wang M, Ouyang J, Li X, Xing MMQ. Nanosilver particles in medical applications: synthesis, performance, and toxicity. Int J Nanomedicine. 2014;9:2399–407.

Geraets L, Oomen AG, Krystek P, Jacobsen NR, Wallin H, Laurentie M. Tissue distribution and elimination after oral and intravenous administration of different titanium dioxide nanoparticles in rats. Part Fibre Toxicol. 2014;11:30.

Golasik M, Herman M, Olbert M, Librowski T, Szklarzewicz J, Piekoszewski W. Toxicokinetics and tissue distribution of titanium in ionic form after intravenous and oral administration. Toxicol Lett. 2016;247:56–61.

Gui S, Sang X, Zheng L, Ze Y, Zhao X, Sheng L. Intragastric exposure to titanium dioxide nanoparticles induced nephrotoxicity in mice, assessed by physiological and gene expression modifications. Part Fibre Toxicol. 2013;10:4.

Gupta RB, Kompella UB. Nanoparticles Technology for Drug Delivery. Drugs and Pharmaceutical Sciences. New York: Taylor & Francis Group, LLC; 2006.

Gurunathan S, Kalishwaralal K, Vaidyanathan R, Venkataraman D, Pandian SRK, Muniyandi J, Hariharan N, Eom SH. Biosynthesis, purification and characterization of silver nanoparticles using Escherichia coli. Colloids Surf B. 2009;74(1):328–35.

Hadrup N, Loeschner K, Bergström A, Wilcks A, Gao X, Vogel U. Subacute oral toxicity investigation of nanoparticulate and ionic silver in rats. Arch Toxicol. 2012;86:543–51.

Hadrupand N, Lam HR. Oral toxicity of silver ions, silver nanoparticles and colloidal silver-a review. Regul Toxicol Pharmacol. 2014;68:1–7.

Hazelhoff MH, Bulacio RP, Torres AM. Gender related differences in kidney injury induced by mercury. Int J Mol Sci. 2012;13:10523–36.

Hazelhoff MH, Trebucobich MS, Stoyanoff T, Chevalier A, Torres AM. Amelioration of mercury nephrotoxicity after pharmacological manipulation of organic anion transporter 1 (Oat1) and multidrug associated resistance protein 2 (Mrp2) with furosemide. Toxicol Res. 2015;4:1324–32.

Hu B, Wang S-B, Wang K, Zhang M, Yu S-H. Microwave-assisted rapid facile green synthesis of uniform silver nanoparticles: self-assembly into multilayered films and their optical properties. J Phys Chem C. 2008;112:1169–74.

Hunt G, Lynch I, Cassee F, Handy RD, Fernandez TF, Berges M. Towards a consensus view on understanding nanomaterials hazards and managing exposure: knowledge gaps and recommendations. Materials. 2013;6:1090–117.

Iravani S, Korbekandi H, Mirmohammadi SV, Zolfaghari B. Synthesis of silver nanoparticles: chemical, physical and biological methods. Res Pharm Sci. 2014;9(6):385–406.

Jha AK, Prasad K, Kulkarni AR. Synthesis of TiO2 nanoparticles using microorganisms. Colloids Surf B. 2009;71(2):226–9.

Kaegi R, Voegelin A, Ort C, Sinnet B, Thalmann B, Krismer J, Hagendorfer H, Elumelu M, Mueller E. Fate and transformation of silver nanoparticles in urban wastewater systems. Water Res. 2013;47:3866–77.

Kalishwaralal K, Barathmanikanth S, Pandian SR, Deepak V, Gurunathan S. Silver nano- a trove for retinal therapies. J Control Release. 2010;145:76–90.

Kalishwaralal K, Deepak V, Ram Pandian Kumar S. Biosynthesis of silver and gold nanoparticles using Brevibacterium casei. Colloids Surf B. 2010;77(2):257–62.

Kemp MM, Kumar A, Mousa S, Park T-J, Ajayan P, Kubotera N, Mousa S, Linhardt RJ. Synthesis of gold and silver nanoparticles stabilized with glycosaminoglycans having distinctive biological activities. Biomacromolecules. 2009;10:589–95.

Kim B, Kim D, Cho D, Cho S. Bactericidal effect of TiO2 photocatalyst on selected food-borne pathogenic bacteria. Chemosphere. 2003;52:277–81.

Kim YS, Kim JS, Cho HS, Rha DS, Kim JM. Twenty-eight-day oral toxicity, genotoxicity, and gender-related tissue distribution of silver nanoparticles in Sprague-Dawley rats. Inhal Toxicol. 2008;20:575–83.

Kim YS, Song MY, Park JD, Song KS, Ryu HR, Chung YH. Subchronic oral toxicity of silver nanoparticles. Part Fibre Toxicol. 2010;7:20.

Kobayashi M, Kato H, Kakihana M. Synthesis of titanium dioxide nanocrystals with controlled crystal- and micro-structures from titanium complexes. Nanomater. Nanotechno. 2013;Vol. 3, art. 23.

Konishi Y, Ohno K, Saitoh N, Nomura T, Nagamine S, Hishida H, Takahashi Y, Uruga, T (2007) et al., Bioreductive deposition of platinum nanoparticles on the bacterium Shewanella algae, J Biotechnol, 128 (3): 648–653.

Kouvaris P, Delimitis A, Zaspalis V, Papadopoulis D, Tsipas SA, Michailidis N. Green synthesis and characterization of silver nanoparticles produced using Arbutus Unedo leaf extract. Mater Lett. 2012;76:18–20.

Krutyakov YA, Kudrinskiy AA, Olenin AY, Lisichkin GV. Synthesis and properties of silver nanoparticles: advances and prospects. Russsian Chem Rev. 2008;77(3):233–57.

Kumar A, Kumar Vemula P, Ajayan PM, George J. Silver-nanoparticle-embedded antimicrobial paints based on vegetable oil. Nat Mater. 2008;7:236–41.

Lee KP, Trochimowicz HJ, Reinhardt CF. Pulmonary response of rats exposed to titanium dioxide by inhalation for two years. Toxicol Appl Pharmacol. 1985;79:179–92.

Lee KJ, Nallathamby PD, Browning LM, Osgood CJ, Xu XH. In vivo imaging of transport and biocompatibility of single silver nanoparticles in early development of zebrafish embryos. ACS Nano. 2007;1:133–43.

Lee Y, Choi J, Kim P, Choi K, Kim S, Shon W. A transfer of silver nanoparticles from pregnant rat to offspring. Toxicol Res. 2012;28:139–41.

Lee MJ, Lee SJ, Uyn SJ, Jang JY, Kang H, Kim K, In HC, Park S. Silver nanoparticles affect glucose metabolism in hepatoma cells through production of reactive oxygen species. Int J Nanomedicine. 2016;11:11–68.

Li N, Duan Y, Hong M, Zheng L, Fei M, Zhao X. Spleen injury and apoptotic pathway in mice caused by titanium dioxide nanoparticles. Toxicol Lett. 2010;195:161–8.

Li X, Xu H, Chen ZS, Chen G. Biosynthesis of nanoparticles by microorganisms and their applications. J Nanomater. 2011;2011:1–17.

Li Y, Qin T, Ingle T, Yan J, He W, Yin JJ, Chen T. Differential genotoxicity mechanisms of silver nanoparticles and silver ions. Genotoxicity and Carcinogenicity. Arch Toxicol. 2016; doi:10.1007/s00204-016-1730-y.

Loeschner K, Hadrup N, Qvortrup K, Larsen A, Gao X, Vogel U. Distribution of silver in rats following 28 days of repeated oral exposure to silver nanoparticles or silver acetate. Part Fibre Toxicol. 2011;8:18.

Mahshid S, Ghamsari MS, Afshar N, Lahuti S. Synthesis of TiO2 nanoparticles by hydrolysis and peptization of titanium isopropoxide solution. Semicond Phys Quant Electron Optoelectron. 2006;N2(9):65–8.

Mohadesi A, Ranjbar M, Hosseinpour-Mashkani SM. Solvent-free synthesis of mercury oxide nanoparticles by simple thermal decomposition method. Superlattices Microst. 2014;66:48–53.

Nadagouda MN, Speth TF, Varma R. Microwave-assisted green synthesis of silver nanostructures. Acc Chem Res. 2011;44:469–78.

Nathworthy PL, Wang J, Tredget EE, Burrell RE. Anti-inflammatory activity of nanocrystalline silver in a porcine contact dermatitis model. Nanomedicine. 2008;4:241–51.

Niinomi M. Metallic biomaterials. J Artif Organs. 2008;11:105–10.

Nowack B, Krug HF, Height M. 120 years of nanosilver history: implications for policy makers. ACS. 2010; dx.doi.org/10.1021/es103316q/environ. Sci. Technology. Hippocrates, on ulcers, F Adams (trans.), ca. 400 B.C.E. http://www.classics.mit.edu/hippocrates/ulcers.html.

Nuevo-Ordoñez Y, Montes-Bayon M, Blanco-González E, Sanz-Medel A. Titanium preferential binding sites in human serum transferrin at physiological concentrations. Metallomics. 2011;3:1297–303.

Rao CNR, Vivekchand SRC, Biswas K, Govindaraj A. Synthesis of inorganic nanomaterials. Dalton Trans. 2007:3728–49.

Rocco MC, Mirkin CA, Hersam MC. Nanotechnology research directions for societal needs in 2020: retrospective and outlook World Technology Evaluation Center (WTEC) and the National Science Foundation. Springer; 2010. http://www.wtec.org/nano2/Nanotechnology_Research_Directions_to_2020/.

Rungby J, Danscher G. Localization of exogenous silver in brain and spinal cord of silver exposed rats. Acta Neuropathol. 1983;60:92–8.

Saha S, Kocaefe D, Sarkar DK, Boluk Y, Pichette AJ. Coat Technol Res. 2011;8:183–90.

Sangsefidi FS, Salavati-Niasari M, Esmaeili-Zare M. Synthesis and characterization of mercury telluride nanoparticles using a new precursor. J Ind Eng Chem. 2014;20:3415–20.

dos Santos CA, Seckler MM, Ingle AP, Gupta I, Galdiero S, Galdiero M, Gade A, Rai M. Silver nanoparticles: therapeutical uses, toxicity, and safety issues. J Pharm Sci. 2014;103:1931–44.

Sardari RR, Zarchi SR, Talebi A, Nasri S, Imani S, Khoradmehr A. Toxicological effects of silver nanoparticles in rats. Afr J Microb Res. 2012;6(27):5587–93.

Sathyavathi S, Manjula A, Rajendhran J, Gunzsekaran P. Biosynthesis and characterization of mercury sulphide nanoparticles produced by Bacillus cereus MRS-1. Indian J Exp Biol. 2013;51(11):973–8.

Savi M, Rossi S, Bocchi L, Gennaccaro L, Cacciani F, Perotti A. Titanium dioxide nanoparticles promote arrhythmias via a direct interaction with rat cardiac tissue. Part Fibre Toxicol. 2014;11:63.

Schmidt G. Nanoparticles from theory to application. Weinheim: Wiley-VCH; 2004.

Scott KD, Hamilton JG. Metabolism of silver in the rat with radiosilver as an indicator. Univ Calif Publ Pharmacol. 1950;2:241–62.

Senapati S, Mandal D, Ahmad A. Fungus mediated synthesis of silver nanoparticles: a novel biological approach. Indian J Phys A. 2004;78A(1):101–5.

Shakeel M, Jabeen F, Shabbir S, Asghar MS, Khan MS, Chaudhry AS. Toxicity of nano-titanium dioxide (TiO2-NP) through various routes of exposure: a review. Biol Trace Elem Res. 2015;172:1–36.

Sinha A, Khare SK. Mercury bioaccumulation and simultaneous nanoparticle synthesis by Enterobacter sp. cells. Bioresour Technol. 2011;102:4281–4.

Smijs TG, Pavel S. Titanium dioxide and zinc oxide nanoparticles in sunscreens: focus on their safety and effectiveness. Nanotechnol Sci Appl. 2011;4:95–112.

Sneha K, Sathishkumar M, Mao J, Kwak IS, Yun YS. Corynebacterium glutamicum-mediated crystallization of silver ions through sorption and reduction processes. Chem Eng J. 2010;162(3):989–96.

Sreeram KJ, Nidhin M, Nair BU. Microwave assisted template synthesis of silver nanoparticles. Bull Mater Sci. 2008;31:937–42.

Stacchiotti A, Lavazza A, Rezzani R, Borsani E, Rodella L, Bianchi R. Mercuric chloride-induced alterations in stress protein distribution in rat kidney. Histol Histopathol. 2004;19:1209–18.

Takahashi Y, Mizuo K, Shinkai Y, Oshio S, Takeda K. Prenatal exposure to titanium dioxide nanoparticles increases dopamine levels in the prefrontal cortex and neostriatum of mice. J Toxicol Sci. 2010;35:749–56.

Teng QL, Wu J, Chan GQ, Gui FZ, Kim TM, Kim JO. Mechanistic study of the antibacterial effect of silver ions on Escherichia coli and Staphylococcus aureus. J Biomed Mater Res. 2000;52:662–8.

Torres AM. Effects of acute mercury exposition on expression and function of organic anion transporters in kidney. In: Kim K-H, Brown R, editors. Chapter 5 Mercury: sources, applications and health impacts. Hauppauge: Editorial Nova Science Publishers; 2013. p. 99–108.

Torres AM, Dnyanmote AV, Bush KT, Wu W, Nigam SK. Deletion of multispecific organic anion transporter Oat1/Slc22a6 protects against mercury-induced kidney injury. J Biol Chem. 2011;286:26391–5.

Trbojevich R, Fernandez A, Watanable F, Mustafa T, Bryant MS. Comparative study of silver nanoparticles permeation using side-by-side and Franz diffusion cells. J Nanopart Res. 2016;18(3):1–12.

Turkevich J, Stevenson PC, Hiller J. A study of the nucleation and growth processes in the synthesis of colloidal gold. Discuss Faraday Soc. 1951;11:55–75.

Vance ME, Kuiken T, Vejerano EP, McGinnis SP, Hochella MF Jr, Rejeski D, Hull MS. Nanotechnology in the real world: redeveloping the nanomaterial consumer products inventory. Beilstein J Nanotechnol. 2015;6:1769–80.

Vigneshwaran N, Kathe AA, Varadarajan P, Nachane RP, Balasubramanya RH. Functional finishing of cotton fabrics using silver nanoparticles. J Nanosci Nanotechnol. 2007;7:1893–7.

Wang H, Xu S, Zhao X-N, Zhu J-J, Xin X-Q. Sonochemical synthesis of size-controlled mercury selenide nanoparticles. Mater Sci Eng B. 2002;96:60–4.

Wang J, Zhou G, Chen C, Yu H, Wang T, Ma Y. Acute toxicity and biodistribution of different size titanium dioxide particles in mice after oral administration. Toxicol Lett. 2007;168:176–85.

Weir A, Westerhoff P, Fabricius L, Hristovski K, van Goetz N. Titanium dioxide nanoparticles in food and personal care products. Environ Sci Technol. 2012;46:2242–50.

Wiegand HJ, Krüger N, Norppa H, Carmichael N, Greim H, Vrijhof H. Toxicology of engineered nanomaterials. Toxicol Lett. 2009;186:147.

Wijnhoven SWP, Peijnenburg WJGM, Herberts CA, Hagens WI, Oormen AG, Heugens BR, Bisschop J, Gosens L, Van De Meent D, Dekkers S, De Jong WH, van Zijverden M, Spis AJAM, Geertsma RE. Nano-silver, a review of available data and knowledge gaps in humans. Nanotoxicology. 2009;3:109–38.

Wright JB, Lam K, Buret AG, Olson ME, Burrell RE. Early healing events in a porcine model of contaminated wounds: effects of nanocrystalline silver on matrix metalloproteinases, cell apoptosis, and healing. Wound Repair Regen. 2002;10:141–51.

Xie G, Wang C, Sun J, Zhong G. Tissue distribution and excretion of intravenously administered titanium dioxide nanoparticles. Toxicol Lett. 2011;205:55–61.

Xu X. Metal oxide and mercury sulfide nanoparticles synthesis and characterization. Thesis presented to the graduate school of Clemson University, Clemson University, TigerPrints; 2009. December.

Xu X, Carraway ER. Sonication-assisted synthesis of mercuric sulphide nanoparticles. Nanomater Nanotechnol. 2012;2 art. 17:2012.

Yanagisawa R, Takanto H, Inoue K, Koike E, Kamachi T, Sadakane K, Ichinose T. Titanium dioxide nanoparticles aggravate atopic dermatitis-like skin lesions in NC/Nga mice. Exp Biol Med. 2009;234:314–22.

Zalups RK. Molecular interactions with mercury in the kidney. Pharmacol Rev. 2000;52:113–43.

van der Zande M, Vandebriel RJ, Van Doren E, Kramer E, Herrera Rivera Z, Serrano-Rojero CS. Distribution, elimination, and toxicity of silver nanoparticles and silver ions in rats after 28-day oral exposure. ACS Nano. 2012;6:7427–42.

Zhang H, Millington KR, Wang X. The photostability of wool doped with photocatalytic titanium dioxide nanoparticles. Polym Degrad Stab. 2009;94:278–83.

Zheng D, Hu C, Gan T, Dang X, Hu S. Preparation and application of a novel vanillin sensor based on biosynthesis of Au-Ag alloy nanoparticles. Sensors Actuators B: Chem. 2010;148(1):247–52.

Chapter 21
Bio-distribution and Toxicity of Noble Metal Nanoparticles in Humans

Indarchand Gupta, Avinash Ingle, Priti Paralikar, Raksha Pandit, Silvio Silvério da Silva, and Mahendra Rai

Abstract Nanotechnology is attracting the interest of scientists since the last few decades and is the technology of twenty-first century. It deals with the materials having at least one dimension in the range of nanometre scale. There is developing enthusiasm towards the application of nanotechnology in diverse fields. Carbon nanotubes, dendrimers, quantum dots and metal nanoparticles are mostly preferred for applications in various fields. Among them metal nanoparticles, especially the noble metallic nanoparticles, are of great importance due to their ease of synthesis, characterization, surface functionalization and size- and shape-dependent unique optoelectronic properties. They include silver, gold, platinum, palladium and rhodium nanoparticles. But with their increased use and likelihood of potential exposure, the concerns over their potential risk are also rising. Many studies have highlighted the toxicological considerations of those nanoparticles.

The present chapter showcases some of the key studies on the in vitro toxicity of noble metal nanoparticles by using human cell lines. It is general observation that toxicity of noble nanoparticles occurs due to exposure mainly via three routes, i.e. through injection or damaged skin, ingestion with food so as to reach gastrointestinal tract and inhalation causing accumulation in the respiratory tract. Hence, in vitro systems representing each of these exposure routes are generally exploited for finding the risk. Here, we discuss several critical investigations exploring the nanoparticle-wise toxicity. The goal is to explore the current knowledge about the potential

I. Gupta
Department of Biotechnology, Government Institute of Science, Nipatniranjan Nagar, Caves Road, Aurangabad 431004, Maharashtra, India

Nanobiotechnology Laboratory, Department of Biotechnology, SGB Amravati University, Amravati 444602, Maharashtra, India

A. Ingle • P. Paralikar • R. Pandit • M.K. Rai (✉)
Nanobiotechnology Laboratory, Department of Biotechnology, SGB Amravati University, Amravati 444602, Maharashtra, India
e-mail: pmkrai@hotmail.com

S.S. da Silva
Escola de Engenharia de Lorena, University of Sao Paulo, Sao Paulo, Brazil

M. Rai, R. Shegokar (eds.), *Metal Nanoparticles in Pharma*,
DOI 10.1007/978-3-319-63790-7_21

risk of noble nanoparticles while drawing the general cytosolic and molecular mechanism lying behind the induction of harmful effects on various cell models.

Keywords Nanotoxicity • Noble nanoparticle • Nanoparticle exposure • ROS • DNA damage

21.1 Introduction

Nanotechnology refers to the study of materials having dimensions approximating 1–100 nm (Rai et al. 2009). With the decrease in the size from micrometre to nanometre scale, the material exhibits increased mechanical strength, chemical reactivity and biological properties. This is due to the fact that as particle becomes smaller, the number of atoms on the surface of particles increases as compared to its volume. Owing to these properties, they have wide applications in areas including agriculture, fighting pathogens, cosmetics, food, healthcare products, drug delivery, electronics and allied industries. Therefore, it is expected that till the year 2020, approximately 3400 nanotechnology-enabled products will be available in the commercial market (PEN 2009). Till date, many types of nanomaterials have been used for various purposes. Among all of those nanomaterials, the metal nanoparticles are of utmost important as they have extensive applications in medical sector. Noble metal nanoparticles are specifically more popular for those applications and, therefore, largely being produced. Due to a large production and use, there is huge possibility of its release in aquatic, terrestrial and atmospheric environments. The fate of nanoparticles in this environment is still largely unknown, and therefore, the concerns are rising regarding their harmful effects due to size-, dose- (Boudreau et al. 2016) and time-dependent exposure (Efeoglu et al. 2016). The rising concerns of nanoparticles lead to the evolution of branch of nanotechnology called nanotoxicology which deals with the study of the potential toxic effects of nanoparticles on human and environmental systems.

Various studies on nanoparticles provide evidence that workers dealing with their production in nanomaterial-based industry and consumers utilizing those products are vulnerable to the nanomaterials. During an average day, people are getting exposed to nanoparticles through commercially available products such as cosmetics, sunscreens, etc. During the last decade, nanotoxicity studies have been increased with rapid pace addressing the public and regulatory issues related to their risk. Hence, nanotoxicology is designed for characterizing the correlation of physicochemical properties of nanoparticles with toxicity to cell. The present chapter, therefore, aims to highlight the routes of nanoparticle exposure, followed by the discussion on the concern associated with the several noble nanoparticles. Lastly the known mechanism of their toxicity at cellular and molecular level has also been discussed.

21.2 Sources of Nanoparticle Exposure in Humans

There are various possible exposure routes by which nanomaterials can enter into the human body. Mainly it includes inhalation, oral administration, intravenous injection and dermal exposure (Gupta et al. 2012; Wang et al. 2012). The human skin, lungs and gastrointestinal tract are in constant contact with nanoparticles (Patel and Patel 2014; Aragao-Santiago et al. 2016). They are most likely to be the port of entry for nanoparticles in the human body.

Due to the application of metal and metal oxide nanoparticles as food additives, they can easily enter into the human digestive system. As mentioned previously, nanoparticles have small size and high reactivity, and hence they can pass through cell membrane and tissues and can be very easily taken up by the organs (Gupta et al. 2015; Suliman et al. 2015). Hence, researchers have paid attention towards the study of cytotoxicity of noble metal nanoparticles. They can be translocated from the intestinal lumen to intestinal lymphatic tissues, i.e. Peyer's patches having M-cells (Powell et al. 2010). The translocation of nanoparticles in the intestine depends on diffusion and accessibility through mucus, cellular trafficking and post-translocation events (Hoet et al. 2004). Some investigators have reported that nanoparticles with smaller diameter (14 nm) could invade the mucus rapidly (within 2 min) so as to reach the enterocytes, while bigger particles (415 nm) took more time (30 min) to translocate the intestinal barrier (Szentkuti et al. 1997). Therefore, it is general observation that shape, size and surface chemistry of nanomaterials decide its toxic nature (Wang et al. 2015). The smaller size of nanoparticles facilitates easy penetration of nanomaterials inside the cell across the membrane and other biological barriers (Nel et al. 2006). Moreover, increase in size of nanomaterials decreases the amount of their cellular uptake (He et al. 2010; Sakai et al. 2011).

Nanoparticles released in air can be inhaled by human during respiration causing their accumulation in the lungs. Several studies reported that the nanoparticles enter through aforesaid organs leading to increased morbidity and mortality (Powell and Kanarek 2006; Jacobsen et al. 2015). Inhalation is the most important route of airborne nanoparticles (Yang et al. 2008).Their toxicity depends upon the dose, dimensions and exposure time. The dose of nanoparticles plays a significant role in deciding the level of their deposition in the lung. In fact, the deposition also relies on the size and duration of exposure (Larsen et al. 2009; Albanese et al. 2012). The rate of deposition in the lungs increases with the decrease in nanoparticle size (Braakhuis et al. 2014). However, most of the accumulation of nanoparticles will be at fragile epithelial structures of the terminal airways and gas exchange region (Borm and Kreyling 2004). If non-soluble, nondegradable nanoparticles such as noble metal nanoparticles enter into the respiratory tract, then they can be accumulated for longer time upon exposure.

The skin is an anatomical barrier for protection against the pathogens, invaders and harmful effects from environment. It consists of three layers, viz. dermis, epidermis and subcutaneous layer. Many nanoformulations, used for skin-related ailments, have been found to get penetrated in the skin. Moreover, if the skin is damaged, then there are high chances of higher harmful effects to the skin tissue. Therefore, the damaged skin is more vulnerable to the nanoparticle toxicity.

21.3 Cytotoxicity of Noble Metal Nanoparticles: Bio-distribution and Its Effects on Human Cell Lines

It is essential to study the impact of nanomaterials on public health and environment. Cell-based assays are very useful for determining the effect of nanoparticles on human cells. Primary cell lines are ideal for conducting in vitro toxicity studies, as these are the representative of tissues. No animal model is required while performing cell cytotoxicity experiments. In vitro studies are easy to perform and economically viable and can be controlled in the laboratory (Gupta et al. 2012).

21.3.1 Silver Nanoparticles (AgNPs)

Among all metal nanoparticles, AgNPs have attracted the attention of scientists, because of their versatile role in different areas of research. But the toxicological studies of nanomaterials are utmost important before its applications (dos Santos et al. 2014). Suliman et al. (2015) made an attempt to study the toxic effect of AgNPs (>100 nm) on human lung epithelial cells A549. The study demonstrated that AgNPs induce dose- and time-dependent cytotoxicity in human epithelial cells. AgNPs-induced cytotoxic and immunotoxic effect on A549 cells was mainly due to the generation of reactive oxygen species (ROS) and oxidative stress. Soares et al. (2016) evaluated the effect of polyvinylpyrrolidone (PVP)-coated AgNPs with 10 nm and 50 nm in human neutrophils. The study found PVP-coated AgNPs with size of 10 nm to be more toxic to human neutrophils as compared to AgNPs whose size was 50 nm, indicating the size-dependent toxicity. Smaller nanoparticles (10 nm) damage membrane, impaired lysosomal activity and thereby result in neutrophil's oxidative burst. The study further found that AgNPs possess time- and concentration-dependent cytotoxicity.

The cytotoxic effect of chemically synthesized AgNPs was examined in MCF-7 cells. IC_{50} value of 80 nm AgNPs was found to be 40 μg/ml. After exceeding AgNPs' concentration beyond 40 μg/ml, it showed potential cytotoxic effect on human cell lines (Çiftçi et al. 2013). The cytotoxicity of AgNPs with three different sizes (5 nm, 20 nm and 50 nm) was investigated on four different human cell lines such as A549, SGC-7901, HepG2 and MCF-7. Study showed that all the size of AgNPs demonstrated apparent toxicity against all the four cell lines. AgNPs with smaller size showed higher toxicity as compared to nanoparticles with bigger size (Liu et al. 2010). Gaiser et al. (2013) studied the effect of AgNPs on human hepatocyte cell line C3A. AgNPs were found to be highly cytotoxic after 24 h of incubation. The results provide the evidence that AgNPs were highly toxic and inflammogenic to human liver cells (Gaiser et al. 2013). The starch-coated AgNPs were prepared and its in vitro cytotoxicity and genotoxicity were assessed on normal human cells such as lung fibroblast cells (IMR-90) and human glioblastoma cells (U251). While studying toxic effect on normal human cell lines, it was found that several changes occurred in cell morphology, oxidative stress and metabolic activity (AshaRani et al. 2009).

Sambale et al. (2015) investigated the toxic effect of AgNPs on different human cell lines such as fibroblasts (NIH-3T3), human lung epithelial cell line (A-549) and HEP-G2-cells, a human hepatocellular cell line. Different concentrations of AgNPs were taken, and cell viability assay was performed by using MTT assay. Electric cell-substrate impedance sensing (ECIS) is an automated method to monitor the behaviour of cell; from ECIS result it was clear that only the addition of AgNPs on the human cell line represents the toxicity of AgNPs. Milića et al. (2015) investigated the effect of citrate-capped AgNPs on porcine kidney (Pk15) cells. AgNPs were taken up by human kidney cells via endocytosis which can induce toxicity depending on the concentration of nanoparticles. Toxic effect of silver ions was also tested for comparative analysis. The ionic form of AgNPs was found to be more toxic as compared to AgNPs. The comet assay suggested that concentration of AgNPs above 25 mg/l can induce genotoxicity in Pk15 cells.

After the exposure of AgNPs to A549 cell line, they were found to be accumulated in the membrane-bound cytoplasmic vacuoles and in enlarged lysosomes. Moreover, in exposed cells, AgNPs found to induce formation of many multivascular and membrane-rich autophagosomes, indicating the AgNPs-mediated induction of autophagosome formation. The oxidative stress induced by AgNPs was suggested to be the reason behind the accumulation of autophagosomes and enlarged lysosomes (Han et al. 2014).

21.3.2 Gold Nanoparticles (AuNPs)

AuNPs if exposed to environment or inhaled by humans can interact with biomolecule, cells or tissues (Maurer-Jones et al. 2009). Small AuNPs can pass through cell membrane which can interact with DNA (Tsoli et al. 2005). A study conducted by Connor et al. (2005) evaluated the effect of 18 mm diameter (AuNPs) on human leukaemia cells. Results indicated that AuNPs exposed to human leukaemia cell for 3 days did not show any toxic effect.

Pernodet et al. (2006) investigated the effect of citrate-capped AuNPs at different concentration and interval of time on human dermal fibroblast (CF-31) cell. Similar to AgNPs, AuNPs can also pass through the cell membrane, accumulate in vacuole and can affect the extracellular matrix of fibroblast cell. Thakor et al. (2011) investigated the effect of AuNPs in human HeLa and HepG2 cell lines. The study revealed that AuNPs were non-toxic to both cell lines, but as the time interval and dose of nanoparticles were increased, it exhibited toxic effect on cell line.

Paino et al. (2012) investigated the effect of citrate-capped AuNPs on human hepatocellular cells (HepG2) and peripheral blood mononuclear cells (PBMC). Both the cells were kept in the presence of different concentration of AuNPs coated with sodium citrate or polyamidoamine dendrimers. It was found that AuNPs interact with both the cells, but it exhibits cytotoxicity and genotoxicity at very low concentration. Jo et al. (2015) studied the effect of AuNPs on human intestinal cell line (INT-407) as well as on rats. AuNPs did not show cytotoxicity till 24 h in terms of inhibition of cell proliferation, membrane damage and ROS generation. When cells were exposed to AuNPs for 7 days, they found to be toxic to the cells. Sultana et al. (2015) compared the cytotoxicity of spherical-shaped AuNPs and flower-shaped AuNPs on human endothelial cells. Two different sizes of AuNPs (15 nm and 50 nm diameter) with two different types of surface chemistries, uncoated and PEG coated, were studied. Flower-shaped AuNPs were claimed to be more toxic than spherical AuNPs. Surface chemistry did not show any effect on the cytotoxicity of nanoparticles against human cell lines. It was hypothesised that flower-shaped AuNPs exhibited rough surface which showed more cytotoxic effect compared with spherical AuNPs.

21.3.3 Platinum Nanoparticles (PtNPs)

Apart from killing cancer cells, nanoparticle-mediated cancer therapy usually causes damage to healthy cells and thus produces toxic effect. One of the major limitations for using PtNPs is its nonspecific untargeted toxicity. Recently, Shim et al. (2017) reported cytotoxic effect of dendritic PtNPs (DPNs) on normal human cells (human embryonic kidney cells HEK-293). The study showed that after exposing DPNs, the cells showed different viability depending on the incubation time and concentration of the PtNPs. The cell viability decreased, with increase in concentration of the nanoparticles, suggesting a dose-dependent cytotoxicity of PtNPs.

Moreover, the number of viable cells decreased, when the cells incubated for an extended period of time. It clearly indicates that PtNPs showed dose- and exposure time-dependent proliferation and cytotoxicity of cells. Further, authors also observed that 15 $\mu g\,mL^{-1}$ and 30 $\mu g\,mL^{-1}$ concentration of PtNPs showed continuous cell growth along with aggregation of cells. When the DPN concentration increased to 45 $\mu g\,mL^{-1}$, it results in dense aggregation of cells, thus difficult to distinguish single cell. Such agglomeration of cell indicates cell death (Shim et al. 2017). Similar effect of PtNPs was studied by Konieczny et al. (2013) on normal human epidermal keratinocytes (NHEKs). Smaller size PtNPs triggered a strong decrease in metabolic activity of treated cells with stronger DNA damage than bigger ones. It also triggered a toxic effect on primary keratinocytes. The cytotoxicity of PtNPs towards human cells depends on the cell type to which it is exposed (Bendale et al. 2017).

PtNPs in size range of 5–8 nm are capable to induce cytotoxicity in dose-dependent manner (Asharani et al. 2010). It was noted that PtNPs exert considerable genotoxicity and cytotoxicity probably due to increased ROS generation in cell.

Smaller (5–8 nm) nanoparticles play a crucial role by facilitating increased diffusion into cells as compared to larger nanoparticles (11–35 nm) that are commonly taken up by the cells through endocytosis (Asharani et al. 2010). Folate-capped PtNPs were tested on mammary breast cells (both cancer and normal cells) by Mironava et al. (2013). The results showed that IC50 value was significantly higher for normal mammary breast cells as compared to cancerous cells, indicating that these nanoparticles preferably target the cancer cells.

There has been ongoing debate on cytotoxicity of PtNPs whether the nanoparticles are toxic to normal cells or cancer cells. A survey of literature showed that PtNPs were considered to be non-toxic for human health, but massive use can put at risk.

21.3.4 Palladium Nanoparticles (PdNPs)

As compared to other noble nanoparticles, which were exploited for diverse biomedical applications, PdNPs remain far behind. There are a few reports available on cytotoxic effect of PdNPs; however, Shanthi et al. (2015) demonstrated the dose-dependent cytotoxicity of PdNPs. The study found that PdNPs are able to reduce cell viability of HeLa cancerous cell line in a dose-dependent manner. However, no cytotoxic effect was observed against normal cervical cell line (H8) at lower concentration, while cytotoxicity increases with increase in concentration. Another cytotoxicity study of porous PdNPs was carried out by Xiao et al. (2014) on HeLa cell line and A549 cell line. The study revealed that PdNPs showed good biocompatibility for A549 cells. The dose-dependent cytotoxicity was also reported by Gurunathan et al. (2015). However, phytochemical stabilized PdNPs were reported as non-toxic to normal cells (Siddiqi and Husen 2016). Most of the literature focused on toxicity of PdNPs on cancer cells (Fang et al. 2012; Balbin et al. 2014). They induce concentration-dependent cytotoxicity against human cancer cell lines (Bascolo et al. 2010). However, more focus on this research area is needed in order to explore the possible harmful effects of PdNPs.

21.3.5 Rhodium Nanoparticles (RhNPs)

Harmful effect of RhNPs has not been investigated yet. To the best of our knowledge, there is no such report on cytotoxicity of RhNPs on human cells. But, it is well reported that rhodium metal complexes like rhodium (II) citrate, rhodium carboxylate, etc. show antitumour and cytotoxic activity, and it has been widely used as promising agent in chemotherapy (Carneiro et al. 2011). Rhodium (II) citrate was reported cytotoxic to normal cell (Carneiro et al. 2011). Likewise clinical use of rhodium carboxylate was limited due to its toxic effect on normal cells (Katsaros and Anagnostopoulou 2002; de Souza et al. 2006).

One of the major limitations of using noble metal nanoparticles is their nonspecific untargeted toxicity. It is necessary to understand various cytotoxic effects of these nanoparticles on human normal cell. Hence, for development of any therapeutic product, determination of its toxicity to human cell is highly important.

21.4 Cellular and Molecular Mechanism of Toxicity

The widespread uses of different nanomaterials including nanoparticles in various biomedical applications increase their interference, which leads to toxicity in human cells. Various studies have been performed on the evaluation of toxicity of various nanomaterials in normal human cells and proposed different possible cellular and molecular mechanisms involved in it. Many studies claimed that the higher surface area-to-volume ratio of nanoparticles resulted in higher chemical reactivity leading to generation of ROS, which directly or indirectly plays a vital role in cytotoxicity and genotoxicity (Fard et al. 2015).

Generally, generation of ROS induces oxidative stress. It is one of the several mechanisms leading to nanotoxicity, including lipid peroxidation, damage to plasma membrane, mitochondrial damage, protein denaturation, DNA damage, etc. (Xia et al. 2008; Fu et al. 2014). It is a well-known fact that plasma membrane encloses the entire cell cytoplasm, thereby keeping the structure intact, and also plays an important role in the selective or non-selective transport of different types of molecules. The interaction of nanoparticles with cell plasma membrane causes its disruption followed by leakage of the cellular components in the surrounding medium leading to the loss of structure and function of cell (Kim et al. 2009; Napierska et al. 2009).

Actually, plasma membrane is a lipid bilayer structure containing a large amount of phosphate molecules, which makes the cell membrane negatively charged. In such situation, different positively charged nanoparticles like gold nanoparticles or any other cationic polymers can be easily attracted towards negatively charged cell membrane electrostatically and form nanoparticles-micelles, resulting in formation of nanosized pits and causing damage to the cell membrane (Ginzburg and Balijepalli 2007). Moreover, the adsorption of such positively charged nanoparticles having cationic side chains on the surface of cell membrane increases the cell membrane permeability and significantly reduces cell viability (Lin et al. 2010).

Apart from the cell membrane, mitochondrion is another important target site for nanomaterial toxicity. It mainly includes interaction of nanomaterials with mitochondrial DNA. The increased permeability of outer membrane of mitochondria may cause apoptosis. Apoptosis is one of the best described mechanisms by which nanomaterials like nanoparticles can exert their toxic effects through an intrinsic pathway, mediated by mitochondria, or by an extrinsic pathway, mediated by death receptors. As mentioned earlier, TiO_2NPs were reported to induce apoptosis BEAS-2B through heightened ROS and pro-inflammatory response (Shah et al. 2017). Herein, nanoparticles increase the permeability of mitochondrial membrane, causing the release of caspase-3 which consequently induces caspase-9 and effector caspase-3 for apoptosis.

Moreover, TiO_2NPs upregulates the p53 gene, which promotes Bax (BCL-2-associated X protein) gene expression via suppressing Bcl-2 (β-cell lymphoma 2) family regulatory proteins. It leads to the opening of mitochondrial channels releasing the cytochrome C. The similar mechanism was proposed by Piao et al. (2011) while working on exposure of silver nanoparticles (AgNPs) to human liver cells.

The generation of ROS in excess found to cause severe damage to cellular macromolecules such as proteins, lipids and DNA, resulting in detrimental effects on cells (Khanna et al. 2015). An important toxicity mechanism for AgNPs involved the interaction of both the AgNPs and Ag ions release from it with sulphur-containing proteins because of strong affinity of silver towards sulphur, thereby leading to denaturation of proteins (Choi et al. 2009; Banerjee and Das 2013; Hou et al. 2013; Khanna et al. 2015). Apart from these, Saptarshi et al. (2013) reviewed that AgNPs can cause the formation of protein corona, protein unfolding and altered protein function in cell. Moreover, AgNPs also interact with various other proteins such as human serum albumin (Chen et al. 2012), human blood protein haemoglobin (Mahato et al. 2011) and cytoskeletal proteins (Wen et al. 2013) which cause conformational changes in the protein structure and ultimately damage the protein.

In addition to the damage of mitochondrial DNA, nanoparticles can also interact with nuclear DNA after penetration within the cell. The interaction of nanoparticles with DNA can damage it by various means such as gene chromosomal damage and DNA strand break (Wang et al. 2007; Kang et al. 2008). Gurr et al. (2005) demonstrated that the direct interaction of TiO_2NPs with DNA leads to the DNA damage; alternatively, oxidative stress induced by ROS and inflammatory responses may also cause DNA damage (Federici et al. 2007; Li et al. 2008). Figure 21.1 shows that nanoparticles induce ROS, which lead to oxidative stress and cause cytotoxicity and genotoxicity by various mechanisms such as lipid peroxidation, mitochondrial damage, protein denaturation and DNA damage.

21.5 Findings-Based Recommendation for Future Studies

The nanotechnology industry is developing rapidly, and therefore, evaluation of implications associated with various nanomaterials is now a challenge for scientists and regulatory organizations. Based on the research discussed herein, it is observed that there is pronounced lack of data regarding the toxicity of noble nanoparticles. Although majority of studies indicated the toxicity of noble nanoparticles at low concentrations, more extensive research for finding potential hazard of nanoparticles is needed. We also recommend that more advanced level of investigations should be performed to understand the long-term low and high exposure of noble nanoparticles. There is also need to perform analysis, which can correlate the in vitro and in vivo studies. Additionally, there is requirement to set up an international panel of researchers, scientists and representatives from various countries and agencies to review timely status of the obtained data, which could be helpful for the peaceful applications of noble nanoparticles.

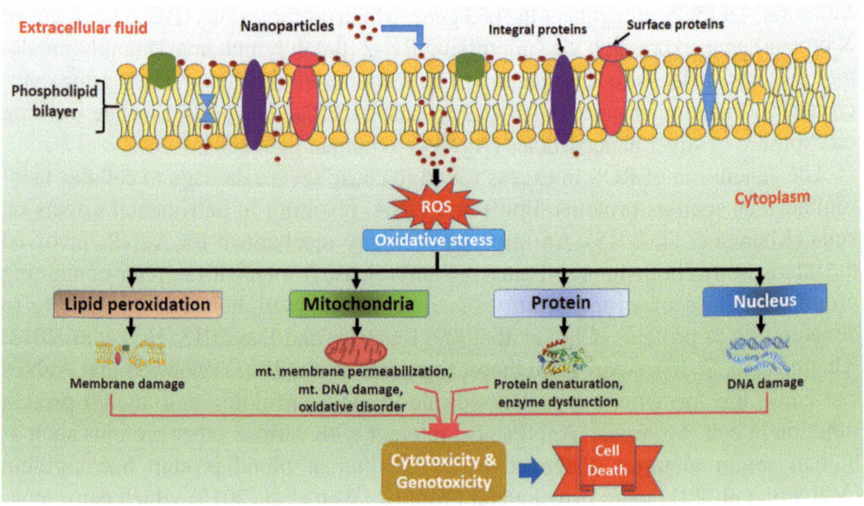

Fig. 21.1 Schematic representation of possible mechanisms involved in cytotoxicity by nanoparticles

21.6 Conclusions

Since the last decade, many revolutionary advances have been made in the field of nanotechnology. Different nanomaterials including nanoparticles are extensively being used in the field of nanomedicine due to their unique and potential physicochemical properties. However, the widespread use of nanomaterials in various consumer products without following safety guidelines and regulations leads to nanotoxicity, which has become a growing concern globally. Various possible routes such as inhalation, oral administration, intravenous injection and dermal exposure have been confirmed by which nanomaterials can enter into the human body. The smaller size of nanoparticles facilitates easy penetration of nanomaterials into the cells, which may be further distributed to different body organs and cause cytotoxicity.

Although the shape, size and surface chemistry of nanomaterials play a vital role in toxicity of nanomaterials, extensive in vitro and in vivo studies are required to elucidate the mechanisms involved in the cellular and genotoxicity of nanomaterials. Most of the studies reported that various mechanisms are involved in the toxicity of nanomaterials, which includes interaction of nanoparticles with phospholipids of the membrane resulting into the formation of pits. However, in other mechanisms generation of oxidative stress due to induction of ROS can cause damage to mitochondria, cellular proteins/enzymes and nuclear DNA, which ultimately reduces cell viability and causes cell death. Therefore, there is a need to further explore the toxicological aspects of noble nanoparticles, which could ensure their safe applications for the benefit of human population.

Acknowledgement Raksha Pandit acknowledges the Department of Science and Technology, New Delhi, India, for DST INSPIRE fellowship, Grant No. IF150452 for pursuing her research work. We also thankfully acknowledge the financial assistance by UGC, New Delhi, under the Special Assistance Programme (SAP) DRS I. Mahendra Rai and Silvio Silvério da Silva thankfully acknowledge the financial help rendered by Conselho Nacional de Desenvolvimento Científico e Tecnológico (CNPq-Brazil, process number 449609/2014-6)

References

Albanese A, Tang PS, Chan WCW. The effect of nanoparticle size, shape, and surface chemistry on biological systems. Annu Rev Biomed Eng. 2012;14:1–16.

Aragao-Santiago L, Hillaireau H, Grabowski N, Mura S, Nascimento TL, Dufort S, Coll JL, Tsapis N, Fattal E. Compared in vivo toxicity in mice of lung delivered biodegradable and non-biodegradable nanoparticles. Nanotoxicology. 2016;10(3):292–302.

AshaRani PV, Mun GLK, Hande MP, Valiyaveettil S. Cytotoxicity and genotoxicity of silver nanoparticles in human cells. ACS Nano. 2009;3(2):279–90.

Asharani PV, Xinyi N, Hande MP, Valiyaveettil S. DNA damage and p53-mediated growth arrest in human cells treated with platinum nanoparticles. Nanomedicine (Lond). 2010;5:51–64.

Balbin A, Gaballo F, Ceballos-Torres J, Prashar S, Fajardo M, Kaluderović GN, Gómez-Ruiz S. Dual application of Pd nanoparticles supported on mesoporous silica SBA-15 and MSU-2: supported catalysts for C-C coupling reactions and cytotoxic agents against human cancer cell lines. RSC Adv. 2014;4:54775–87.

Bendale Y, Bendale V, Paul S. Evaluation of cytotoxic activity of platinum nanoparticles against normal and cancer cells and its anticancer potential through induction of apoptosis. Integr Med Res. 2017. https://doi.org/10.1016/j.imr.2017.01.006.

Banerjee V, Das KP. Interaction of silver nanoparticles with proteins: a characteristic protein concentration dependent profile of spr signal. Colloids Surf B: Biointerfaces. 2013;111:71–9.

Borm PJ, Kreyling W. Toxicological hazards of inhaled nanoparticles – potential implications for drug delivery. J Nanosci Nanotechnol. 2004;4(5):521–31.

Boscolo P, Bellante V, Leopold K, Maier M, di Giampaolo L, Antonucci A, Iavicoli I, Tobia L, Paoletti A, Montalti M, Petrarca C, Qiao N, Sabbioni E, Di Gioacchino M. Effects of palladium nanoparticles on the cytokine release from peripheral blood mononuclear cells of non-atopic women. J Biol Regul Homeost Agents. 2010;24:207–14.

Boudreau MD, Imam MS, Paredes AM, Bryant MS, Cunningham CK, Felton RP, Jones MY, Davis KJ, Olson GR. Differential effects of silver nanoparticles and silver ions on tissue accumulation, distribution, and toxicity in the sprague dawley rat following daily oral gavage administration for 13 weeks. Toxicol Sci. 2016;150(1):131–60.

Braakhuis HM, Gosens I, Krystek P, Boere JA, Cassee FR, Fokkens PH, Post JA, van Loveren H, Park MV. Particle size dependent deposition and pulmonary inflammation after short-term inhalation of silver nanoparticles. Part Fibre Toxicol. 2014;11:49.

Carneiro ML, Nunes ES, Peixoto RC, Oliveira RG, Lourenço LH, da Silva IC, Simioni AR, Tedesco AC, de Souza AR, Lacava ZG, Báo SN. Free rhodium (II) citrate and rhodium (II) citrate magnetic carriers as potential strategies for breast cancer therapy. J Nanobiotechnol. 2011;9:11.

Chen R, Choudhary P, Schurr RN, Bhattacharya P, Brown JM, Ke PC. Interaction of lipid vesicle with silver nanoparticle-serum albumin protein corona. Appl Phys Lett. 2012;100(1):013703–34.

Choi O, Deng KK, Kim N-J, Ross L Jr, Surampalli RY, Hu Z. The inhibitory effects of silver nanoparticles, silver ions, and silver chloride colloids on microbial growth. Water Res. 2009;42(12):3066–74.

Çiftçi H, Türk M, Tamer U, Karahan S, Menemen Y. Silver nanoparticles: cytotoxic, apoptotic, and necrotic effects on MCF-7 cells. Turk J Biol. 2013;37:573–81.

Connor EE, Mwamuka J, Gole A, Murphy CJ, Wyatt MD. Gold nanoparticles. Small. 2005;1(3): 325–7.

de Souza AR, Coelho EP, Zyngier SB. Comparison of the anti-neoplastic effects of dirhodium(II) tetrapropionate and its adducts with nicotinate and isonicotinate anions in mice bearing Ehrlich tumors. Eur J Med Chem. 2006;41:1214–6.

dos Santos CA, Seckler MM, Ingle AP, Gupta I, Galdiero S, Galdiero M, Gade A, Rai M. Silver nanoparticles: therapeutical uses, toxicity, and safety issues. J Pharm Sci. 2014;103:1931–44.

Efeoglu E, Casey A, Byrne HJ. In vitro monitoring of time and dose dependent cytotoxicity of aminated nanoparticles using Raman spectroscopy. Analyst. 2016;141:5417–31.

Fang W, Tang S, Liu P, Fang X, Gong J, Zheng N. Pd nanosheet-covered hollow mesoporous silica nanoparticles as a platform for the chemo-photothermal treatment of cancer cells. Small. 2012;8:3816–22.

Fard JK, Jafari S, Eghbal MA. Review of molecular mechanisms involved in toxicity of nanoparticles. Adv Pharm Bull. 2015;5(4):447–54.

Federici G, Shaw BJ, Handy RD. Toxicity of titanium dioxide nanoparticles to rainbow trout (*Oncorhynchus mykiss*): gill injury, oxidative stress, and other physiological effects. Aquat Toxicol. 2007;84:415–30.

Fu PP, Xia Q, Hwang HM, Ray PC, Yu H. Mechanisms of nanotoxicity: generation of reactive oxygen species. J Food Drug Anal. 2014;22:64–75.

Gaiser BK, Hirn S, Kermanizadeh A, Kanase N, Fytianos K, Wenk A, Haberl N, Brunelli A, Kreyling WG, Stone V. Effects of silver nanoparticles on the liver and hepatocytes in vitro. Toxicol Sci. 2013;131(2):537–47.

Ginzburg VV, Balijepalli S. Modeling the thermodynamics of the interaction of nanoparticles with cell membranes. Nano Lett. 2007;7:3716–22.

Gupta I, Duran N, Rai M. Nano-silver toxicity: emerging concerns and consequences in human health. In: Cioffi, Rai M, editors. Nano-antimicrobials progress and prospects. Heidelberg/New York: Springer; 2012. p. 525–55.

Gupta I, Gaikwad S, Ingle A, Kon K, Duran N, Rai M. Nanotoxicity a mechanistic approach. In: Prokopovich P, editor. Biological and pharmaceutical applications of nanomaterials. Boca Raton, FL, USA: CRC Press; 2015. p 393–410.

Gurr JR, Wang AS, Chen CH, Jan KY. Ultrafine titanium dioxide particles in the absence of photoactivation can induce oxidative damage to human bronchial epithelial cells. Toxicology. 2005;213:66–73.

Gurunathan S, Kim E, Han JW, Park JH, Kim JH. Green chemistry approach for synthesis of effective anticancer palladium nanoparticles. Molecules. 2015;20:22476–98.

Han JW, Gurunathan S, Jeong J-K, Choi Y-J, Kwon D-N, Park J-K, Kim J-H. Oxidative stress mediated cytotoxicity of biologically synthesized silver nanoparticles in human lung epithelial adenocarcinoma cell line. Nanoscale Res Lett. 2014;9:459.

He C, Hu Y, Yin L, Tang C, Yin C. Effects of particle size and surface charge on cellular uptake and biodistribution of polymeric nanoparticles. Biomaterials. 2010;31:3657–66.

Hoet PHM, Brüske-Hohlfeld I, Salata OV. Nanoparticles – known and unknown health risks. J Nanobiotechnol. 2004;2:12.

Hou W-C, Stuart B, Howes R, Zepp RG. Sunlight-driven reduction of silver ions by natural organic matter: formation and transformation of silver nanoparticles. Environ Sci Technol. 2013;47(14): 7713–21.

Jacobsen NR, Stoeger T, van den Brule S, Saber AT, Beyerle A, Vietti G, Mortensen A, Szarek J, Budtz HC, Kermanizadeh A, Banerjee A, Ercal N, Vogel U, Wallin HA, Møller P. Acute and subacute pulmonary toxicity and mortality in mice after intratracheal instillation of ZnO nanoparticles in three laboratories. Food Chem Toxicol. 2015;85:84–95.

Jo M, Bae SH, Go MR, Kim HJ, Hwang YG, Choi SJ. Toxicity and biokinetics of colloidal gold nanoparticles. Nano. 2015;5:835–50.

Kang SJ, Kim BM, Lee YJ, Chung HW. Titanium dioxide nanoparticles trigger p53-mediated damage response in peripheral blood lymphocytes. Environ Mol Mutagen. 2008;49:399–405.

Katsaros N, Anagnostopoulou A. Rhodium and its compounds as potential agents in cancer treatment. Crit Rev Oncol Hematol. 2002;42:297–308.

Khanna P, Ong C, Bay BH, Baeg GH. Nanotoxicity: an interplay of oxidative stress, inflammation and cell death. Nanomaterials. 2015;5(3):1163–80.

Kim S, Choi JE, Choi J, Chung KH, Park K, Yi J, Ryu DY. Oxidative stress-dependent toxicity of silver nanoparticles in human hepatoma cells. Toxicol In Vitro. 2009;23:1076–84.

Konieczny P, Goralczyk AG, Szmyd R, Skalniak L, Koziel J, Filon FL, Crosera M, Cierniak A, Zuba-Surma EK, Borowczyk J, Laczna E, Drukala J, Pyza E, Semik D, Woznicka O, Klein A, Jura J. Effects triggered by platinum nanoparticles on primary keratinocytes. Int J Nanomedicine. 2013;8:3963–75.

Larsen EK, Nielsen T, Wittenborn T, Birkedal H, Vorup-Jensen T, Jakobsen MH, Ostergaard L, Horsman MR, Besenbacher F, Howard KA, Kjems J. Size-dependent accumulation of PEGylated silane-coated magnetic iron oxide nanoparticles in murine tumors. ACS Nano. 2009;3(7):1947–51.

Li S, Zhu H, Zhu R, Sun X, Yao S, Wang S. Impact and mechanism of TiO2 nanoparticles on DNA synthesis in vitro. Sci China Ser B: Chem. 2008;51:367–72.

Lin JQ, Zhang HW, Chen Z, Zheng YG. Penetration of lipid membranes by gold nanoparticles: insights into cellular uptake, cytotoxicity, and their relationship. ACS Nano. 2010;4:5421–9.

Liu W, Wu Y, Wang C, Li HC, Wang T, Liao CY, Cui L, Zhou QF, Yan B, Jiang GB. Impact of silver nanoparticles on human cells: effect of particle size. Nanotoxicology. 2010;4:319–30.

Mahato M, Pal P, Tah B, Ghosh M, Talapatra GB. Study of silver nanoparticle–hemoglobin interaction and composite formation. Colloids Surf B: Biointerfaces. 2011;88:141–9.

Maurer-Jones MA, Bantz KC, Love SA, Marquis BJ, Haynes CL. Toxicity of therapeutic nanoparticles. Nanomedicine. 2009;4(2):219–41.

Milića M, Leitingerb G, Pavičića I, Avdičevićd MZ, Dobrovićd S, Goesslere W, Vinković Vrčeka I. Cellular uptake and toxicity effects of silver nanoparticles in mammalian kidney cells. J Appl Toxicity. 2015;35(6):581–92.

Mironava T, Simon M, Rafailovich MH, Rigas B. Platinum folate nanoparticles toxicity: cancer vs. normal cells. Toxicol In Vitro. 2013;27:882–9.

Napierska D, Thomassen LC, Rabolli V, Lison D, Gonzalez L, Kirsch-Volders M, Martens JA, Hoet PH. Size-dependent cytotoxicity of monodisperse silica nanoparticles in human endothelial cells. Small. 2009;5:846–53.

Nel A, Xia T, Madler L, Li N. Toxic potential of materials at the nanolevel. Science. 2006;311:622–7.

Painoa IMM, Marangonia VS, de Cássia Silva, de Oliveirab R, Maria L, Antunesb G, Zucolotto V. Cyto and genotoxicity of gold nanoparticles in human hepatocellular carcinoma and peripheral blood mononuclear cells. Toxicol Lett. 2012;215:119–25.

Patel J, Patel A. Toxicity of nanomaterials on liver, kidney, and spleen. In: Pathak, Sutariya, editors. Bio-interaction of nanomaterial. Boca Raton: Taylor and Francis; 2014. p. 286–306.

PEN. Project of the emerging nanotechnologies (PEN). 543. 2009. Available at http://www.nanotechproject.org.

Pernodet N, Fang X, Sun Y, Bakhtina A, Ramakrishnan A, Sokolov J, Ulman A, Rafailovich M. Adverse effects of citrate/gold nanoparticles on human dermal fibroblasts. Small. 2006;2(6):766–73.

Piao MJ, Kang KA, Lee IK, Kim HS, Kim S, Choi JY, Choi J, Hyun JW. Silver nanoparticles induce oxidative cell damage in human liver cells through inhibition of reduced glutathione and induction of mitochondria-involved apoptosis. Toxicol Lett. 2011;201:92–100.

Powell MC, Kanarek MS. Nanomaterial health effects-part 1: background and current knowledge. Wis Med J. 2006;105:16–20.

Powell JJ, Faria N, Thomas-McKay E, Pele LC. Origin and fate of dietary nanoparticles and microparticles in the gastrointestinal tract. J Autoimmun. 2010;34:226–33.

Rai M, Yadav A, Gade A. Silver nanoparticles as a new generation of antimicrobials. Biotechnol Adv. 2009;27:76–83.

Sakai N, Matsui Y, Nakayama A, Tsuda A, M Yoneda M. Functional dependent and size-dependent uptake of nanoparticles in pc12. J Phys Conf Ser. 2011;304:012049.

Sambale F, Wagner S, Stahl F, Khaydarov RR, Scheper T, Bahnemann D. Investigations of the toxic effect of silver nanoparticles on mammalian cell lines. J Nanomater. 2015, Article ID 136765, 9 pages, http://dx.doi.org/10.1155/2015/136765.

Saptarshi SR, Duschl A, Lopata AL. Interaction of nanoparticles with proteins: relation to bio-reactivity of the nanoparticle. J Nanobiotechnol. 2013;11:26.

Shah SNA, Shah Z, Hussain M, Khan M. Hazardous effects of titanium dioxide nanoparticles in ecosystem. Bioinorg Chem Appl. 2017: Article ID 4101735, 12 pages, https://doi.org/10.1155/2017/4101735.

Shanthi K, Sreevani V, Vimala K, Kannan S. Cytotoxic effect of palladium nanoparticles synthesized from *syzygium aromaticum* aqueous extracts and induction of apoptosis in cervical carcinoma. Proc Nat Acad Sci India Sect B: Biol Sci. 2015; doi:10.1007/s40011-015-0678-7.

Shim K, Kim J, Heo YU, Jiang B, Li C, Shahabuddin M, Wu KC, Hossain MS, Yamauchi Y, Kim JH. Synthesis and cytotoxicity of dendritic platinum nanoparticles with HEK-293 cells. Chem Asian J. 2017;12:21–6.

Siddiqi KS, Husen A. Green synthesis, characterization and uses of palladium/platinum nanoparticles. Nanoscale Res Lett. 2016;11:482.

Soares T, Ribeiro D, Proença C, Chisté RC, Fernandes E, Freitas M. Size-dependent cytotoxicity of silver nanoparticles in human neutrophils assessed by multiple analytical approaches. Life Sci. 2016;145:247–54.

Suliman YAO, Ali D, Alarifi S, Harrath AH, Mansour L, Alwasel SH. Evaluation of cytotoxic, oxidative stress, proinflammatory and genotoxic effect of silver nanoparticles in human lung epithelial cells. Environ Toxicol. 2015;30(2):149–60.

Sultana S, Djaker N, Boca-Farcau S, Salerno M, Charnaux N, Astilean S, Hlawaty H, de la Chapelle ML. Comparative toxicity evaluation of flowershaped and spherical gold nanoparticles on human endothelial cells. Nanotechnology. 2015 26 (2015) 055101 (12pp) doi:10.1088/0957-4484/26/5/055101.

Szentkuti L. Light microscopical observations on luminally administered dyes, dextrans, nanospheres and microspheres in the pre-epithelial mucus gel layer of the rat distal colon. J Control Release. 1997;46:233–42.

Thakor AS, Paulmurugan R, Kempen P, Zavaleta C, Sinclair R, Massoud TF. Oxidative stress mediates the effects of Raman-active gold nanoparticles in human cells. Small. 2011;7(1):126–36.

Tsoli M, Kuhn H, Brandau W, Esche H, Schmid G. Cellular uptake and toxicity of AU(55) clusters. Small. 2005;1(9):841–4.

Wang JJ, Sanderson BJ, Wang H. Cyto- and genotoxicity of ultrafine TiO2 particles in cultured human lymphoblastoid cells. Mutat Res. 2007;628:99–106.

Wang B, He X, Zhang ZY, Zhao YL, Feng WY. Metabolism of nanomaterials in vivo: blood circulation and organ clearance. Acc Chem Res. 2012;46:761.

Wang P, Wang X, Wang L, Hou X, Liu W, Chen C. Interaction of gold nanoparticles with proteins and cells. Sci Technol Adv Mater. 2015;16:034610. (15pp). doi:10.1088/1468-6996/16/3/034610.

Wen Y, Geitner NK, Chen R, Ding F, Chen P, Andorfer RE, Govindan PN, Ke PC. Binding of cytoskeletal proteins with silver nanoparticles. RSC Adv. 2013;3:22002–7.

Xia T, Kovochich M, Liong M, Mädler L, Gilbert B, Shi H, Yeh JI, Zink JI, Nel AE. Comparison of the mechanism of toxicity of zinc oxide and cerium oxide nanoparticles based on dissolution and oxidative stress properties. ACS Nano. 2008;2:2121–34.

Xiao JW, Fan SX, Wang F, Sun LD, Zheng XY, Yan CH. Porous Pd nanoparticles with high photothermal conversion efficiency for efficient ablation of cancer cells. Nanoscale. 2014;6(8):4345–51.

Yang W, Peters JI, Williams RO III. Inhaled nanoparticles—a current review. Int J Pharm. 2008;356(1–2):239–47.

Index

M. Rai, R. Shegokar (eds.), *Metal Nanoparticles in Pharma*,
DOI 10.1007/978-3-319-63790-7

H

I

T

MIX
Papier aus verantwortungsvollen Quellen
Paper from responsible sources
FSC® C105338

If you have any concerns about our products,
you can contact us on
ProductSafety@springernature.com

In case Publisher is established outside the EU,
the EU authorized representative is:
Springer Nature Customer Service Center GmbH
Europaplatz 3, 69115 Heidelberg, Germany

Printed by Libri Plureos GmbH
in Hamburg, Germany